EARTH

EARTH

An Introduction to Physical Geology

Sixth Edition

Edward J. Tarbuck
Frederick K. Lutgens

Illinois Central College

Illustrated by

Dennis Tasa

PRENTICE HALL
Upper Saddle River, New Jersey 07458

Library of Congress Cataloging-in-Publication Data

Tarbuck , Edward J.
        Earth : an introduction to physical geology / Edward J. Tarbuck,
Frederick K. Lutgens : illustrations by Dennis Tasa. — 6th ed.
             p.   cm.
      Includes index.
      ISBN 0-13-974122-4
      1. Physical geology. I. Lutgens , Frederick K. II. Title.
QE28.2.T37 1999
550—dc21   98-20384
                                 CIP

Senior Editor: *Daniel Kaveney*
Editor in Chief: *Paul F. Corey*
Editorial Director: *Tim Bozik*
Director of Production and Manufacturing: *David W. Riccardi*
Executive Managing Editor: *Kathleen Schiaparelli*
Assistant Managing Editor: *Lisa Kinne*
Production Editor: *Edward Thomas*
Associate Editor in Chief, Development: *Carol Trueheart*
Editor in Chief of Development: *Ray Mullaney*
Executive Marketing Manager: *Leslie Cavaliere*
Creative Director: *Paula Maylahn*
Art Director: *Joseph Sengotta*
Art Manager: *Gus Vibal*
Photo Editors: *Lorinda Morris-Nantz and Melinda Reo*
Photo Researcher: *Clare Maxwell*
Copy Editor: *James Tully*
Assistant Editor: *Wendy Rivers*
Editorial Assistant: *Margaret Ziegler*
Marketing Assistant: *Cheryl Adam*
Cover/Interior Designer: *Judith A. Matz-Coniglio*
Cover Design: *Amy Rosen*
Manufacturing Manager: *Trudy Pisciotti*
Manufacturing Buyer: *Benjamin Smith*
Text Composition: *Molly W. Pike/Lido Graphics*
Cover Photo: *The Chamonix Needles Lac Blanc/French Alps, France. Photo by Art Wolfe.*
Title Page Photo: *Banner Peak in Ansel Adams Wilderness, California. Photo by Carr Clifton.*

10   9   8   7   6   5   4   3

ISBN 0-13-974122-4

Prentice-Hall International (UK) Limited, *London*
Prentice-Hall of Australia Pty. Limited, *Sydney*
Prentice-Hall Canada Inc., *Toronto*
Prentice-Hall Hispanoamericana, S.A., *Mexico*
Prentice-Hall of India Private Limited, *New Delhi*
Prentice-Hall of Japan, Inc., *Tokyo*
Pearson Education Asia Pte. Ltd., *Singapore*
Editora Prentice-Hall do Brasil, Ltda., *Rio de Janeiro*

Brief Contents

Contents

This icon for the GEODe II CD-ROM appears whenever a text discussion has a corresponding GEODe II activity.

Chapter 6

Sedimentary Rocks   143

Chapter 7

Metamorphics Rocks   169

Chapter 22

Planetary Geology   567

Preface

Earth is a *very* small part of a vast universe, but it is our home. It provides the resources that support our modern society and the ingredients necessary to maintain life. Therefore, knowledge and understanding of our planet is critical to our social well being and indeed, vital to our survival.

In recent years, media reports have made us increasingly aware of the geological forces at work in our physical environment. News stories graphically portray the violent force of a volcanic eruption, the devastation generated by a strong earthquake, and the large numbers left homeless by mudflows and flooding. Such events, and many others as well, are destructive to life and property, and we must be better able to understand and deal with them. To comprehend and prepare for such events requires an awareness of how science is done and the scientific principles that influence our planet, its rocks, mountains, atmosphere, and oceans.

The Sixth Edition of *Earth: An Introduction to Physical Geology*, like its predecessors, is a college-level text intended for both science majors and nonmajors taking their first course in geology. We have attempted to write a text that is not only informative and timely, but one that is highly usable as well.

Distinguishing Features

Readability

The language of this book is straightforward and *written to be understood*. Clear, readable discussions with a minimum of technical language are the rule. The frequent headings and subheadings help students follow discussions and identify the important ideas presented in each chapter. In the sixth edition, improved readability was achieved by examining chapter organization and flow, and writing in a more personal style. Large portions of the text were substantially rewritten in an effort to make the material more understandable.

Illustrations and Photographs

Geology is highly visual. Therefore, photographs and artwork are a very important part of an introductory book. *Earth, Sixth Edition*, contains dozens of new high-quality photographs that were carefully selected to aid understanding, add realism, and heighten the interest of the reader.

The illustrations in each new edition of *Earth* keep getting better and better. In the sixth edition more than 150 pieces of line art were redesigned. The new art illustrates ideas and concepts more clearly and realistically than ever before. The art program was carried out by Dennis Tasa, a gifted artist and respected Earth science illustrator.

Focus on Learning

New to the sixth edition: To assist student learning, every chapter now concludes with a *Chapter Summary*. When a chapter has been completed, three useful devices help students review. First, the *Chapter Summary* recaps all of the major points. Next is a checklist of *Key Terms* with page references. Learning the language of geology helps students learn the material. This is followed by *Review Questions* that help students examine their knowledge of significant facts and ideas.

Earth as a System

An important occurrence in modern science has been the realization that Earth is a giant multidimensional system. Our planet consists of many separate but interacting parts. A change in any one part can produce changes in any or all of the other parts—often in ways that are neither obvious nor immediately apparent. Although it is not possible to study the entire system at once, it is possible to develop an awareness and appreciation for the concept and for many of the system's important interrelationships. With this as a goal, a number of new headings and boxes have been added to the sixth edition of *Earth*.

Environmental Issues

Because knowledge about our planet and how it works is necessary to our survival and well-being, the treatment of environmental and resource topics has always been an important part of *Earth*. These issues serve to illustrate the relevance and application of geologic knowledge. The text integrates a great deal of information about the relationship between people and the physical environment and explores applications of geology to understanding and solving problems that arise from these interactions.

Maintaining a Focus on Basic Principles and Instructor Flexibility

The main focus of the Sixth Edition remains the same as in the first five—to foster a basic understanding of physical geology. As much as possible, we have attempted to provide the reader with a sense of the observational techniques and reasoning processes that constitute the discipline of geology.

The organization of the text remains intentionally traditional. Following the overview of geology in the introductory chapter, we turn to a discussion of Earth materials and the related processes of volcanism and weathering. Next, a discussion of a most basic topic, geologic time, is followed by an examination of the geological work of gravity, water, wind and ice in modifying and sculpturing landscapes. After this look at external processes, we examine Earth's internal structure and the processes that deform rocks and give rise to mountains. Finally, the text concludes with chapters on resources and the solar system. This particular organization was selected largely to accommodate the study of minerals and rocks in the laboratory, which usually comes early in the course.

Realizing that some instructors may prefer to structure their courses differently, we made each chapter self-contained, so that chapters may be taught in a different sequence. Thus, the instructor who wishes to discuss earthquakes, plate tectonics, and mountain building prior to dealing with erosional processes may do so without difficulty. We also chose to introduce plate tectonics in the first chapter so that this important theory could be incorporated in appropriate places throughout the text.

More About the Sixth Edition

In addition to adding chapter summaries, a new "Earth as a System" theme and making substantial changes to the photography and art programs, much more is new to the sixth edition.

The sixth edition of *Earth* represents a thorough revision. *Every* part of the book was examined carefully with the dual goals of keeping topics current and improving the clarity of text discussions. Based on feedback from reviewers and our students, we believe we have succeeded.

Moreover, fifteen of the special interest boxes are new to the sixth edition. Some highlight people—professional geologists and what they do, as well as contributions by some historical figures. Other new boxes focus on recent geologic events, such as the 1997 Red River floods, the volcanic crisis on Montserrat, and the *Pathfinder* mission to Mars. Additional boxes on environmental concerns, including debris-flow hazards and groundwater mining, are also part of the new edition.

Supplements

The authors and publisher have been pleased to work with a number of talented people to produce an excellent supplements package. This package includes the traditional supplements that students and professors have come to expect from authors and publishers, as well as some new kinds of supplements that involve electronic media.

For the Student

GEODe II CD-ROM: A revision of the popular *GEODe CD* by Dennis Tasa of Tasa Graphic Arts, Inc., Edward J. Tarbuck, and Frederick K. Lutgens. *GEODe II* is a dynamic program that reinforces key geologic concepts by using animations, tutorials, and interactive exercises. This *GEODe II* icon appears throughout the book wherever a text discussion has a corresponding *GEODe II* activity. A copy of *GEODe II* has been included with this text. This special offering gives students two valuable products (*GEODe II* and the textbook) for the price of one.

Internet Support: This site, specific to the text, contains numerous review exercises (from which students get immediate feedback), exercises to expand one's understanding of physical geology, and resources for further exploration. This web site provides an excellent platform from which to start using the Internet for the study of geology. Please visit the site at http://www.prenhall.com/tarbuck.

Geosciences on the Internet: A Student's Guide, by Andrew T. Stull and Duane Griffin, is a student's guide to the Internet and world wide web specific to geology. *Geosciences on the Internet* is available at no cost to qualified adopters of the text. Please contact your local Prentice Hall representative for details.

Study Guide: Written by experienced educator Richard Busch, the study guide helps students identify the important points from the text, and then provides them with review exercises, study questions, self-check exercises, and vocabulary review.

For the Professor

Transparency Set: More than 150 full-color acetates of illustrations from the text are available free of charge to qualified adopters.

Slides: More than 200 slides of images taken from the text, many of which were taken by the authors, are also available to qualified adopters.

Presentation Manager: This user-friendly navigation software enables professors to custom build multimedia presentations. *Prentice Hall Presentation Manager 3.0* contains several hundred images from the text. The CD-ROM allows professors to organize material in whatever order they choose; preview resources by chapter; search the digital library by keyword; integrate material from their hard drive, a network, or the Internet; or edit lecture notes and annotate images with an overlay tool. This powerful presentation tool is available at no cost to qualified adopters of the text.

 The New York Times Themes of the Times—Geology: This unique newspaper-format supplement features recent articles about geology from the pages of the *New York Times*. This supplement, available at no extra charge from your local Prentice Hall representative, encourages students to make connections between the classroom and the world around them.

Instructor's Manual: Written by Richard Mauger of East Carolina University, the instructor's manual is intended as a resource for both new and experienced instructors. It includes a variety of lecture outlines, additional source materials, teaching tips, advice about how to integrate visual supplements (including the web-based resources), and various other ideas for the classroom.

Test Item File: The test item file provides instructors with a wide variety of test questions.

PH Custom Test: Based on the powerful testing technology developed by Engineering Software Associates, Inc. (ESA), *Prentice Hall Custom Test* allows instructors to create and tailor exams to their own needs. With the online testing program, exams can also be administered online and data can then be automatically transferred for evaluation. The comprehensive desk reference guide is included along with online assistance.

Acknowledgments

Writing a college textbook requires the talents and cooperation of many individuals. Working with Dennis Tasa, who is responsible for all of the outstanding illustrations, is always special for us. We not only value his outstanding artistic talents and imagination, but his friendship as well. We benefited greatly from the skills of our Developmental Editor Carol Trueheart. Carol helped us make the sixth edition a more readable, user-friendly text. We are grateful to Professor Ken Pinzke at Belleville Area College. In addition to his many helpful suggestions regarding the manuscript, Ken helped in preparing the new chapter summaries. Thanks also to Nancy Lutgens who prepared the boxes on Louis Agassiz (chapter 12) and Inge Lehmann (chapter 17).

Our students remain our most effective critics. Their comments and suggestions continue to help us maintain our focus on readability and understanding.

Special thanks goes to those colleagues who prepared in-depth reviews. Their critical comments and thoughtful input helped guide our work and clearly strengthened the text. We wish to thank:

Molly T. Andries, *Brookhaven College*
R. Scott Babcock, *Western Washington University*
George Clark, *Kansas State University*
Paul F. Hudak, *University of North Texas*
Lawrence Kodosky, *Oakland Community College*
Erwin J. Mantei, *Southwest Missouri State University*
Barry Metz, *Delaware County Community College*
Karl A. Riggs, *Mississippi State University*
Veronica Stein, *Illinois Central College*
James D. Stewart, *Vincennes University*

We also want to acknowledge the team of professionals at Prentice Hall. Thanks to Editor in Chief Paul Corey. We sincerely appreciate his continuing strong support for excellence and innovation. Thanks also to our editor Dan Kaveney. His strong communication skills and energetic style contributed greatly to the project. The production team, led by Ed Thomas, has done an outstanding job. They are true professionals with whom we are very fortunate to be associated.

Edward J. Tarbuck
Frederick K. Lutgens

CHAPTER 1

An Introduction to Geology

Left Mount Agassiz, Mount Winchell, and Thunderbolt Peak, Kings Canyon National Park, California. Mount Agassiz is named for Louis Agassiz, a nineteenth-century scientist who was instrumental in developing the theory of the Ice Age. — *Photo by Carr Clifton*

Figure 1.1 At 6194 meters (20,320 feet), Mt. McKinley in Alaska's Denali National Park is the highest peak in North America. Geologists study the process that created this majestic peak. (Photo by Carr Clifton)

The spectacular eruption of a volcano, the terror brought by an earthquake, the magnificent scenery of a mountain valley, and the destruction created by a landslide are all subjects for the geologist (Figure 1.1). The study of geology deals with many fascinating and practical questions about our physical environment. What forces produce mountains? Will there soon be another great earthquake in California? What was the Ice Age like? Will there be another? What created this cave and the stone icicles hanging from its ceilings? Should we look for water here? Is strip mining practical in this area? Will oil be found if a well is drilled at that location? What if the landfill is located in the old quarry?

The subject of this text is **geology**, a word that literally means "the study of Earth." To understand Earth is not an easy task because our planet is not an unchanging mass of rock but rather a dynamic body with a long and complex history.

The science of geology is traditionally divided into two broad areas—physical and historical. **Physical geology**, which is the primary focus of this book, examines the materials composing Earth and seeks to understand the many processes that operate beneath and upon its surface (see Box 1.1). The aim of **historical geology**, on the other hand, is to understand the origin of Earth and its development through time. Thus, it strives to establish an orderly chronological arrangement of the multitude of physical and biological changes that have occurred in the geologic past. The study of physical geology logically precedes the study of Earth history because we must first understand how Earth works before we attempt to unravel its past.

Some Historical Notes about Geology

The nature of our Earth—its materials and processes—has been a focus of study for centuries. Writings about such topics as fossils, gems, earthquakes, and volcanoes date back to the early Greeks, more than 2300 years ago.

Certainly the most influential Greek philosopher was Aristotle. Unfortunately, Aristotle's explanations about the natural world were not based on keen observations and experiments. Instead, they were arbitrary pronouncements. He believed that rocks were created under the "influence" of the stars

Box 1.1

Geology: An Environmental Science

Environment refers to everything that surrounds and influences an organism. Some environmental conditions are biological and social: Others are *abiotic* (nonliving). The abiotic factors comprise our *physical environment* and include water, air, soil, and rock, and conditions such as temperature, humidity, and sunlight. Because geological phenomena and processes are basic to the physical environment, all geology is "environmental geology." However, the specific term *environmental geology* is usually reserved for the part of geologic science that focuses on the *relationships between people and the physical environment*. Environmental geology applies geologic principles to understanding and solving problems that arise from these human–environment interactions.

This book's primary focus is to present basic geologic principles, but we will explore may aspects of environmental geology along the way. Some involve *hazardous Earth processes* like volcanoes, earthquakes, landslides, floods, and coastal hazards (Figure 1.A). Each is responsible for significant losses of life and property each year. Of course, geologic hazards are simply natural processes. They become hazards only when people try to live where these processes occur.

Resources are another important focus of environmental geology. They include water and soil, a great variety of metallic and nonmetallic minerals, and energy. All form the very foundation of modern society. Geology must

Figure 1.A In late summer and fall 1997 a series of volcanic eruptions devastated the small Caribbean island of Montserrat. (Photo by Alain Buu/Gamma Liason)

deal not only with the *formation* and *occurrence* of these vital resources but also with *maintaining supplies*, and with the *environmental impact* of their extraction and use.

Complicating all environmental issues is rapid world population growth and everyone's aspiration to a better standard of living. Earth is now gaining about 100 million people each year. This means a ballooning demand for resources and that more people are being pushed to dwell in environments having geologic hazards.

We humans can dramatically influence geologic processes. For example,

river flooding is natural, but the magnitude and frequency of flooding can be changed significantly by human activities such as clearing forests, building cities, and constructing dams. Unfortunately, natural systems do not always adjust to artificial changes in ways that we can anticipate. Thus, an alteration to the environment that was intended to benefit society often has the opposite effect.

The best answer to environmental problems is to understand them. This is the goal of environmental geology—in fact, all of geology.

and that earthquakes occurred when air crowded into the ground, was heated by central fires, and escaped explosively. When confronted with a fossil fish, he explained that "a great many fishes live in the earth motionless and are found when excavations are made."

Although Aristotle's explanations may have been adequate for his day, they unfortunately continued to be expounded for many centuries, thus thwarting the acceptance of more up-to-date accounts. Frank D. Adams states in *The Birth and Development of the Geological Sciences* (New York:

Dover, 1938) that "throughout the Middle Ages Aristotle was regarded as the head and chief of all philosophers; one whose opinion on any subject was authoritative and final."

Catastrophism

During the seventeenth and eighteenth centuries the doctrine of **catastrophism** strongly influenced people's thinking about the dynamics of Earth. Briefly stated, catastrophists believed that Earth's landscapes had been shaped primarily by great catastrophes. Features such as mountains and canyons, which today we know take great periods of time to form, were explained as having been produced by sudden and often worldwide disasters produced by unknowable causes that no longer operate. This philosophy was an attempt to fit the rates of Earth processes to the then-current ideas on the age of Earth.

In the mid-1600s, James Ussher, Anglican Archbishop of Armagh, Primate of all Ireland, published a major work that had immediate and profound influences. A respected scholar of the Bible, Ussher constructed a chronology of human and Earth history in which he determined that Earth was only a few thousands of years old, having been created in 4004 B.C. Ussher's treatise earned widespread acceptance among Europe's scientific and religious leaders, and his chronology was soon printed in the margins of the Bible itself.

The relationship between catastrophism and the age of Earth has been summarized as follows:

> That the earth had been through tremendous adventures and had seen mighty changes during its obscure past was plainly evident to every inquiring eye; but to concentrate these changes into a few brief millenniums required a tailor-made philosophy, a philosophy whose basis was sudden and violent change.*

The Birth of Modern Geology

The late 1700s is generally regarded as the beginning of modern geology, for it was during this time that James Hutton, a Scottish physician and gentleman farmer, published his *Theory of the Earth* (Figure 1.2). In this work, Hutton put forth a principle that came to be known as the doctrine of **uniformitarianism**. Uniformitarianism is a fundamental principle of modern geology. It simply states that the *physical, chemical, and biological laws that operate today have also operated in the geologic past.* This means that the forces and processes that we observe presently shaping our planet have been at work for a very long time. Thus, to understand ancient rocks, we must first understand present-day processes and their results. This idea is commonly stated by saying "the present is the key to the past."

Prior to Hutton's *Theory of the Earth*, no one had effectively demonstrated that geological processes occur over extremely long periods of time. However, Hutton persuasively argued that forces which appear small could, over long spans of time, produce effects that were just as great as those resulting from sudden catastrophic events. Unlike his predecessors, Hutton carefully cited verifiable observations to support his ideas.

---

*H. E. Brown, V. E. Monnett, and J. W. Stovall, *Introduction to Geology* (New York: Blaisdell, 1958).

Figure 1.2 James Hutton, the eighteenth-century Scottish geologist who is often called the "father of modern geology." (Photo courtesy of the British Museum)

For example, when he argued that mountains are sculpted and ultimately destroyed by weathering and the work of running water, and that their wastes are carried to the oceans by processes that can be observed, Hutton said, "We have a chain of facts which clearly demonstrates . . . that the materials of the wasted mountains have traveled through the rivers"; and further, "There is not one step in all this progress . . . that is not to be actually perceived." He then went on to summarize this thought by asking a question and immediately providing the answer: "What more can we require? Nothing but time."

Hutton's literary style was cumbersome and difficult, so his work was not widely read nor easily understood. However, that began to change in 1802, when Hutton's friend and colleague, John Playfair, published *Illustrations of the Huttonian Theory*, a volume in which he presented Hutton's ideas in a much clearer and more attractive form. The following well-known passage from Playfair's work, which is a restatement of Hutton's basic principle, illustrates this style:

> Amid all the revolutions of the Globe, the economy of nature has been uniform and her laws are the only things which have resisted the general movement. The rivers and the rocks, the seas and the continents have been changed in all their parts; but the laws which direct those changes, and the rules to which they are subject, have remained invariably the same.

Although Playfair's book gave impetus to Hutton's ideas and aided the cause of modern geology, it is the English geologist Sir Charles Lyell (Figure 1.3) who is given the most credit for advancing the basic principles of modern geology. Between 1830 and 1872, he produced eleven editions of his great work, *Principles of Geology*. As was customary in those days, Lyell's book had a rather lengthy subtitle that outlined the main theme of the work: *Being an Attempt to Explain the Former Changes of the Earth's Surface, by Reference to Causes now in Operation.*

Lyell was able to show more convincingly than his predecessors that geologic processes observed today can be assumed to have operated in the past. Although the doctrine of uniformitarianism did not originate with Lyell, a fact that he openly acknowledged, he was most successful in interpreting and publicizing it for society at large.

Today the basic tenets of uniformitarianism are just as viable as in Lyell's day. Indeed, we realize more strongly than ever that the present gives us insight into the past and that the physical, chemical, and biological laws that govern geological processes remain unchanging through time. However, we also understand that the doctrine should not be taken too literally. To say that geological processes in the past were the same as those occurring today is not to suggest that they always had the same relative importance or that they operated at precisely the same rate. Although the same processes have prevailed through time, their rates have undoubtedly varied.*

The acceptance of uniformitarianism meant the acceptance of a very lengthy history for Earth. Although processes vary in their intensity, they still take a very long time to create or destroy major landscape features (Figure 1.4).

For example, geologists have established that mountains once existed in portions of present-day Minnesota, Wisconsin, and Michigan. Today the region consists of low hills and plains. Erosion (processes that wear land away) gradually destroyed these peaks. Estimates indicate that the North American continent is being lowered at a rate of about 3 centimeters per 1000 years. At this rate, it would take 100 million years for water, wind, and ice to lower mountains that were 3000 meters (10,000 feet) high.

But even this time span is relatively short on the time scale of Earth history, for the rock record contains evidence that shows Earth has experienced many

Figure 1.3 Charles Lyell. Lyell's book, *Principles of Geology*, did much to advance modern geology. (Courtesy of the Institute of Geological Sciences, London)

cycles of mountain building and erosion. Concerning the ever-changing nature of Earth through great expanses of geologic time, James Hutton made a statement that was to become his most famous. In concluding his classic 1788 paper published in the *Transactions of the Royal Society of Edingburgh*, he stated, "The results, therefore, of our present enquiry is, that we find no vestige of a beginning—no prospect of an end." A quote from William L. Stokes sums up the significance of Hutton's basic concept:

> In the sense that uniformitarianism implies the operation of timeless, changeless laws or principles, we can say that nothing in our incomplete but extensive knowledge disagrees with it.†

In the chapters that follow, we shall be examining the materials that compose our planet and the processes that modify it. It is important to remember that, although many features of our physical landscape may seem to be unchanging over the decades we might observe them, they are nevertheless changing, but on time scales of hundreds, thousands, or even many millions of years.

---

*It should be pointed out that during Earth's formative period, when our planet was very different from today, some processes were at work that are no longer operating.

†*Essentials of Earth History* (Englewood Cliffs, New Jersey: Prentice Hall, 1966), p. 34.

Figure 1.4 Weathering and erosion have gradually sculpted these striking rock spires, in Arizona's Monument Valley. Geologic processes often act so slowly that changes may not be visible during an entire human lifetime. (Photo by Carr Clifton)

Geologic Time

Although Hutton, Playfair, Lyell, and others recognized that geologic time is exceedingly long, they had no methods to accurately determine the age of Earth. However, with the discovery of radioactivity near the turn of the twentieth century and the continuing refinement of radiometric dating methods that were first attempted in 1905, geologists are now able to assign fairly accurate, specific dates to events in Earth's history.* Current estimates put the age of Earth at about 4.6 billion years.

Relative Dating and the Geologic Time Scale

During the nineteenth century, long before the advent of radiometric dating, a geologic time scale was developed using principles of relative dating. **Relative dating** means that events are placed in their proper sequence or order without knowing their absolute age in years. This is done by applying principles such as the **law of superposition**, which states that in an underformed sequence of sedimentary rocks or lava flows, each layer is older than the one above it and younger than the one below it (Figure 1.5). Today such a proposal appears to be elementary, but 300 years ago, it amounted to a major breakthrough in scientific reasoning by establishing a rational basis for relative time measurements. However, because no precise rate of deposition can be determined for most rock layers, the actual length of geologic time represented by any given layer is unknown.

Fossils, the remains or traces of prehistoric life, were also essential to the development of the geologic time scale (Figure 1.6). A fundamental principle in geology, one that was laboriously worked out over decades by collecting fossils from countless rock layers around the world, is known as the **principle of faunal succession.**[†] This principle states that fossil organisms succeed one another in a definite and determinable order, and therefore any time period can be

---

*A more complete discussion of this topic is found in Chapter 8.

[†]Because this principle is equally applicable to both plants and animals, the term *biotic succession* is preferred by some.

recognized by its fossil content. Once established, this principle allowed geologists to identify rocks of the same age in widely separated places and to build the geologic time scale as shown in Figure 1.7.

Notice that units having the same designations do not necessarily extend for the same number of years. For example, the Cambrian period lasted about 65 million years, whereas the Silurian period spanned just 30 million years. As we will emphasize again in Chapter 8, this situation exists because the basis for establishing the time scale was not the regular rhythm of a clock, but the changing character of life-forms through time. Specific dates were added long after the time scale was established. A glance at Figure 1.7 also reveals that the Phanerozoic eon is divided into many more units than earlier eons even though it encompasses only about 13 percent of Earth history. The meager fossil record for these earlier eons is the primary reason for the lack of detail on this portion of the time scale. Without abundant fossils, geologists lose their primary tool for subdividing geologic time.

The Magnitude of Geologic Time

The concept of geologic time is new to many nongeologists. People are accustomed to dealing with increments of time that are measured in hours, days, weeks, and years. Our history books often examine events over spans of centuries, but even a century is difficult to appreciate fully. For most of us, someone or something that is 90 years old is *very old*, and a 1000-year-old artifact is *ancient*.

By contrast, those who study geology must routinely deal with vast time periods—millions or billions

Figure 1.5 These rock layers were exposed by the downcutting of the Colorado River as it created the Grand Canyon. Relative ages of these layers can be determined by applying the law of superposition. The youngest rocks are on top and the oldest are at the bottom. (Photo by Tom Till)

A.

B.

Figure 1.6 Fossils are important tools for the geologist. In addition to being very important in relative dating, fossils can be useful environmental indicators. **A.** Natural cast of a trilobite. This diverse group of marine organisms was prominent during the Paleozoic era. **B.** This extinct coiled cephalopod, like its modern descendants, was a highly developed marine organism.

(thousands of millions) of years. When viewed in the context of Earth's 4.6-billion-year history, a geologic event that occurred 100 million years ago may be characterized as "recent" by a geologist, and a rock sample that has been dated at 10 million years may be called "young."

An appreciation for the magnitude of geologic time is important in the study of geology because many processes are so gradual that vast spans of time are needed before significant changes occur.

How long is 4.6 billion years? If you were to begin counting at the rate of one number per second and continued 24 hours a day, 7 days a week, and never stopped, it would take about two lifetimes (150 years) to reach 4.6 billion! Another interesting basis for comparison is as follows:

Compress, for example, the entire 4.6 billion years of geologic time into a single year. On that scale, the oldest rocks we know date from about mid-March. Living things first appeared in the sea in May. Land plants and animals emerged in late November and the widespread swamps that formed the Pennsylvanian coal deposits flourished for about four days in early December. Dinosaurs became dominant in mid-December, but disappeared on the 26th, at about the time the Rocky Mountains were first uplifted. Manlike creatures appeared sometime during the evening of December 31st, and the most recent continental ice sheets began to recede from the Great Lakes area and from northern Europe about 1 minute and 15 seconds before midnight on the 31st. Rome ruled the Western world for 5 seconds from 11:59:45 to 11:59:50. Columbus discovered America 3 seconds before midnight, and the science of geology was born with the writings of James Hutton just slightly more than one second before the end of our eventful year of years.*

The foregoing is just one of many analogies that have been conceived in an attempt to convey the magnitude of geologic time. Although helpful, all of them, no matter how clever, only begin to help us comprehend the vast expanse of Earth history.

The Nature of Scientific Inquiry

As members of a modern society, we are constantly reminded of the benefits derived from science. But what exactly is the nature of scientific inquiry?

All science is based on the assumption that the natural world behaves in a consistent and predictable manner. The overall goal of science is to discover the underlying patterns in the natural world and then to use this knowledge to make predictions about what should or should not be expected to happen given certain facts or circumstances (Box 1.2).

The development of new scientific knowledge involves some basic, logical processes that are universally

---

*Don L. Eicher, *Geologic Time*, 2nd ed. (Englewood Cliffs, New Jersey: Prentice Hall, 1978), pp. 18–19. Reprinted by permission.

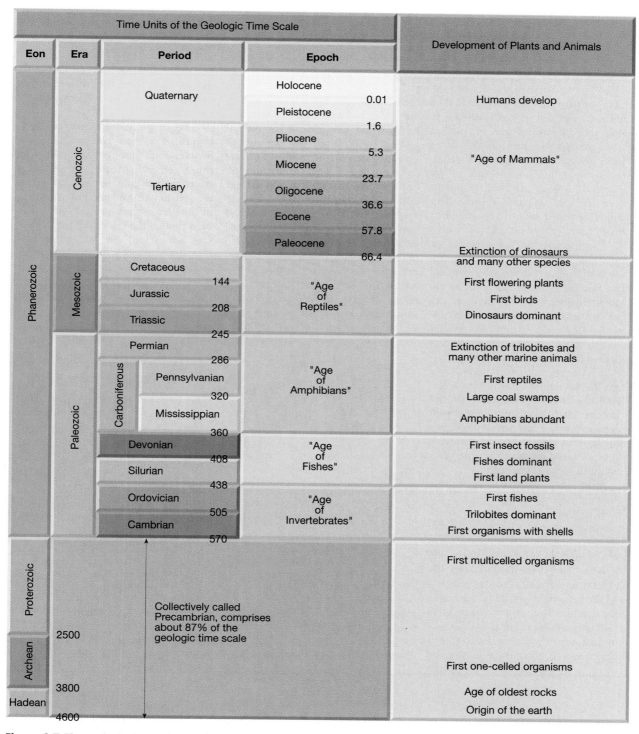

Figure 1.7 The geologic time scale. Numbers on the time scale represent time in millions of years before the present. These dates were added long after the time scale had been established using relative dating techniques. The Precambrian accounts for more than 85 percent of geologic time. (Data from Geological Society of America)

accepted. To determine what is occurring in the natural world, scientists collect scientific facts through observation and measurement (Figure 1.8). These data are essential to science and serve as the springboard for the development of scientific theories.

Hypothesis

Once facts have been gathered and principles have been formulated to describe a natural phenomenon, investigators try to explain how or why things happen

Box 1.2

Do Glaciers Move? An Application of the Scientific Method

Today we know that glaciers can do extraordinary erosional work and that in the past glacial ice has affected vast areas that are now ice-free. We understand much about how glaciers form and move, as well as how they erode and deposit. Such knowledge about glaciers has been acquired gradually over the past 200 years, yet scientists still have much more to learn about these moving masses of ice.

The study of glaciers provides an early application of the scientific method. High in the Alps of Switzerland and France, small glaciers exist in the upper portions of some valleys. In the late eighteenth and early nineteenth centuries, people who farmed and herded animals in these valleys suggested that glaciers in the upper reaches of the valleys had previously been much larger and had occupied downvalley areas. They based their explanation on the fact that the valley floors were littered with angular boulders and other rock debris that seemed identical to the materials that they could see in and near the glaciers at the heads of the valleys.

Although the explanation of these observations seemed logical, others did not accept the notion that masses of ice hundreds of meters thick were capable of movement. The disagreement was settled after a simple experiment was designed and carried out to test the hypothesis that glacial ice can move.

Markers were placed in a straight line completely across an alpine glacier. The position of the line was marked on the valley walls so that if the ice moved, the change in position could be detected. After a year or two the results were clear: The markers on the glacier had advanced down the valley, proving that glacial ice indeed moves. In addition, the experiment demonstrated that ice within a glacier does not move at a uniform rate, because the markers in the center advanced farther than did those along the margins. Although most glaciers move too slowly for direct visual detection, the experiment succeeded in demonstrating that movement nevertheless occurs. In the years that followed, this experiment was repeated many times with greater accuracy using more modern surveying techniques. Each time, the basic relationships established by earlier attempts were verified.

The experiment illustrated in Figure 1.B was carried out at Switzerland's Rhone Glacier later in the nineteenth century. It not only traced the movement of markers within the ice, but also mapped the position of the glacier's terminus. Notice that even though the ice within the glacier was advancing, the ice front was retreating. As often occurs in science, experiments and observations designed to test one hypothesis yield new information that requires further analysis and explanation.

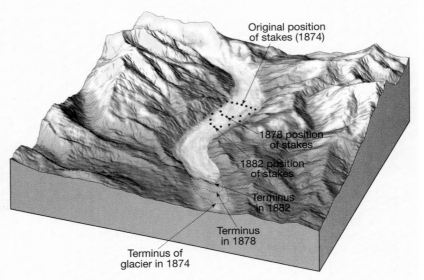

Figure 1.B Ice movement and changes in the terminus at Rhone Glacier, Switzerland. In this classic study of a valley glacier, the movement of stakes clearly showed that ice along the sides of the glacier moves slowest. Also notice that even though the ice front was retreating, the ice within the glacier was advancing.

in the manner observed. They can do this by constructing a tentative (or untested) explanation, which we call a scientific **hypothesis**. Often, several hypotheses are advanced to explain the same facts. Next, scientists think about what will occur or be observed if a hypothesis is correct and devise ways or methods to test the accuracy of predictions drawn from the hypothesis.

If a hypothesis cannot be tested, it is not scientifically useful, no matter how interesting it might seem. Testing usually involves making observations, developing models, and performing experiments. What if test results do not turn out as expected? One possibility is that there were errors in the observations or experiments. Of course, another possibility is that the hypothesis is not valid. Before rejecting the

Figure 1.8 This scientist is using a mass spectrometer to analyze rare gases for geophysics research. (Photo by Jean Miele/The Stock Market)

hypothesis, the tests might be repeated or new tests might be devised. The more tests the better. The history of science is littered with discarded hypotheses. One of the best known is the idea that Earth was at the center of the universe, a proposal that was supported by the apparent daily motion of the Sun, Moon, and stars around Earth.

Theory

When a hypothesis has survived extensive scrutiny and when competing hypotheses have been eliminated, a hypothesis might be elevated to the status of a scientific **theory**. In everyday language, we may say "that's only a theory." But a scientific theory is a well-tested and widely accepted view that scientists agree best explains certain observable facts. It is not enough for scientific theories to fit only the data that are already at hand. Theories must also fit additional observations that were not used to formulate them in the first place. Put another way, theories should have predictive power.

Scientific theories, like scientific hypotheses, are accepted only provisionally. It is always possible that a theory that has withstood previous testing may eventually be disproven. As theories survive more testing, they are regarded with higher levels of confidence. Theories that have withstood extensive testing, as, for example, the theory of plate tectonics and the theory of evolution, are held with a very high degree of confidence.

Scientific Methods

The process just described, in which scientists gather facts through observations and formulate scientific hypotheses and theories, is called the *scientific method*. Contrary to popular belief, the scientific method is not a standard recipe that scientists apply in a routine manner to unravel the secrets of our natural world. Rather, it is an endeavor that involves creativity and insight. Rutherford and Ahlgren put it this way: "Inventing hypotheses or theories to imagine how the world works and then figuring out how they can be put to the test of reality is as creative as writing poetry, composing music, or designing skyscrapers."*

Modern scientific endeavors frequently incorporate new technologies. With the development of high-speed computers, scientists began creating models that attempted to simulate what was happening in the "real" world. These models have become quite useful when dealing with processes that occur on very long time scales or are physically very large, or take place under extreme conditions. For example, it should be obvious that to design equipment capable of duplicating the conditions found deep within Earth and at the scale required would be virtually impossible. Although not without its own drawbacks, modeling helps narrow the gap between what we can measure and observe and the determination of the processes that produce those observations.

There is not a fixed path that scientists always follow that leads unerringly to scientific knowledge. Nevertheless, many scientific investigations involve the following steps: (1) the collection of scientific facts through observation and measurement (Figure 1.9); (2) the development of one or more working hypotheses to explain these facts; (3) development of observations and experiments to test the hypothesis; and (4) the acceptance, modification, or rejection of the hypothesis based on extensive testing. Other scientific discoveries represent purely theoretical ideas, which stand up to extensive examination. Still other scientific advancements have been made when a totally unexpected happening occurred during an experiment. These serendipitous discoveries are more than pure luck; for as Louis Pasteur said, "In the field of observation, chance favors only the prepared mind."

Scientific knowledge is acquired through several avenues, so it might be best to describe the nature of scientific inquiry as the methods of science, rather than the scientific method.

---

*F. James Rutherford and Andrew Ahlgren, *Science for All Americans* (New York: Oxford University Press, 1990), p. 7.

Figure 1.9 Temperature probe of active lava flow, Kilauea Caldera. (Photo by Norman Banks, U.S.G.S. Hawaiian Volcano Observatory)

 A View of Earth

Figure 1.10A is the view of Earth that greeted the *Apollo 8* astronauts as their spacecraft came from behind the Moon after circling it for the first time in December 1968. A view such as this provided the astronauts, as well as the people back on Earth, with a unique perspective of our planet. For the first time we were able to see Earth from the depths of space as a small, fragile-appearing sphere surrounded by the blackness of a vast universe. Such views were not only spectacular and exciting, but also humbling, for they showed us as never before what a tiny part of the universe our planet occupies.

As we look more closely at our planet from space, it becomes apparent that Earth is much more than rock and soil (Figure 1.10B). Indeed, the most conspicuous features are not the continents but the swirling clouds suspended above the surface and the vast global ocean. From such a vantage point we can appreciate why Earth's physical environment is traditionally divided into three major parts: the solid Earth; the water portion of our planet, the hydrosphere; and Earth's gaseous envelope, the atmosphere.

It should be emphasized that our environment is highly integrated. It is not dominated by rock, water, or air alone. Rather, it is characterized by continuous interactions as air comes in contact with rock, rock with water, and water with air. Moreover, the biosphere, which is the totality of all plant and animal life on our planet, interacts with each of the three physical realms and is an equally integral part of Earth (see Box 1.3).

The interactions among Earth's four spheres are continuous and uncountable. Figure 1.11 provides us with one easy-to-visualize example. The shoreline is an obvious meeting place for rock, water, and air. In this scene ocean waves that were created by the drag of air moving across the water are breaking against the rocky shore. The force of the water can be powerful and the erosional work that is accomplished can be great.

Hydrosphere

Earth is sometimes called the *blue* planet. Water more than anything else makes Earth unique. The **hydrosphere** is a dynamic mass of water that is continually on the move, evaporating from the oceans to the atmosphere, precipitating back to the land, and running back to the ocean again. The global ocean is certainly the most prominent feature of the hydrosphere, blanketing nearly 71 percent of Earth's surface and accounting for about 97 percent of Earth's water. However, the hydrosphere also includes the fresh water found in streams, lakes, and glaciers, as well as that found underground.

Although these latter sources constitute just a tiny fraction of the total, they are much more important than their meager percentage indicates.

A.

B.

Figure 1.10 **A.** View of Earth that greeted the *Apollo 8* astronauts as their spacecraft came from behind the Moon. **B.** A closer view of Earth from *Apollo 17*. (Both photos courtesy of NASA)

In addition to providing the fresh water that is so vital to life on land, rivers, glaciers, and groundwater are responsible for sculpting and creating many of our planet's varied landforms.

Atmosphere

Earth is surrounded by a life-giving gaseous envelope called the **atmosphere**. This blanket of air, very thin compared to Earth's diameter, is an integral part of

Box 1.3

Earth as a System

A *system* is a group of interrelated, interacting, or interdependent parts that form a complex whole. Most of us hear and use the term frequently. We may service our car's cooling *system*, make use of the city's transportation *system*, and be a participant in the political *system*. A news report might inform us of an approaching weather *system*.

We know that Earth is just a small part of a large system known as the *solar system*. As we study Earth, it also becomes clear that our planet can be viewed as a system with many separate but interacting parts or subsystems. The hydrosphere, atmosphere, biosphere, and solid Earth and all of their components can be studied separately. However, the parts are not isolated. Each is related in some way to the others to produce a complex and continuously interacting whole that we call the *Earth system*.

The parts of the Earth system are linked so that a change in one part can produce changes in any or all of the other parts. For example, when a volcano erupts, lava from Earth's interior may flow out at the surface and block a nearby valley. This new obstruction influences the region's drainage system by creating a lake or causing streams to change course.

Moreover, the large quantities of volcanic ash and gases that can be emitted during an eruption may be blown high into the atmosphere and influence the amount of solar energy that can reach the surface. The result could be a drop in air temperatures over the entire hemisphere. Where the surface is covered by lava flows or a thick layer of volcanic ash, existing soils are buried. This causes the soil-forming processes to begin anew to transform

the new surface material into soil (Figure 1.C). The soil that eventually forms will reflect the interaction among many parts of the Earth system. Of course, there would also be significant changes in the biosphere. Some organisms and their habitats would be eliminated by the lava and ash, while new settings for life, such as the lake, would be created. The potential climate change could also impact sensitive life-forms.

The Earth system is powered by energy from two sources. The Sun drives external processes that occur in the atmosphere, hydrosphere, and at Earth's surface. Weather and climate, ocean circulation, and erosional processes are driven by energy from the Sun. Earth's interior is the second source of energy. Heat remaining from when our planet formed and heat that is continuously

generated by radioactive decay powers the internal processes that produce volcanoes, earthquakes, and mountains.

Humans are *part of* the Earth system, a system in which the living and nonliving components are entwined and interconnected. Therefore, our actions produce changes in all of the other parts. When we burn gasoline and coal, build breakwaters along the shoreline, dispose of our wastes, and clear the land, we cause other parts of the system to respond, often in unforeseen ways. Throughout this book you will learn about many of Earth's subsystems: the hydrologic system, the tectonic (mountain-building) system, and the rock cycle, to name a few. Remember that these components *and we humans* are all part of the complex interacting whole we call the Earth system.

Figure 1.C When Mount St. Helens erupted in May 1980, the area shown here was buried by a volcanic mudflow. Now plants are reestablished and new soil is forming. (Photo by Jack Dykinga)

the planet. It not only provides the air that we breathe but also protects us from the Sun's intense heat and dangerous ultraviolet radiation. The energy exchanges that continually occur between the atmosphere and Earth's surface and between the atmosphere and space produce the effects we call weather.

If, like the Moon, Earth had no atmosphere, our planet would not only be lifeless but many of the processes and interactions that make the surface such a dynamic place could not operate. Without weathering and erosion, the face of our planet might more closely resemble the lunar surface,

Figure 1.11 The shoreline is one obvious meeting place for rock, water, and air. In this scene, ocean waves that were created by the force of moving air break against the rocky shore. The force of the water can be powerful and the erosional work that is accomplished can be great. Big Sur coast, California. (Photo by Mark Muench)

Solid Earth

Lying beneath the atmosphere and the oceans is the solid Earth (Figure 1.13). Much of our study of the solid Earth focuses on the more accessible surface features. Fortunately, these features represent the outward expressions of the dynamic behavior of the subsurface materials. By examining the most prominent surface features and their global extent, we can obtain clues to the dynamic processes that have shaped our planet.

Continent–Ocean Transition. The two principal divisions of Earth's surface are the continents and the ocean basins. It is important to realize that the present shoreline is not the boundary between these distinct regions. Rather, along most coasts a gently sloping platform of continental material, called the **continental shelf**, extends seaward from the shore. A glance at Figure 1.13 shows that there can be considerable variation in the extent of the continental shelf from one region to another.

For example, the shelf is broad along the East and Gulf coasts of the United States, but relatively narrow along the Pacific margin of the continent. The extent of the continental shelf also varies greatly from one time to another. For instance, during the most recent Ice Age, when more of the world's water was stored on land in the form of glacial ice, the level of the sea was about 150 meters (500 feet) lower than it is today. Consequently, during this period, more of Earth's surface was dry land. The boundary between the continents and the deep-ocean basins is perhaps best placed about halfway down the **continental slopes**, which are steep dropoffs that lead from the outer edge of the continental shelves to the deep-ocean basins. Using this as the dividing line, we find that about 60 percent of Earth's surface is represented by the ocean basins, and the remaining 40 percent exists as continental masses.

Elevations and Depths. The most obvious difference between the continents and the ocean basins is their relative levels. The average elevation of the continents above sea level is about 840 meters (2750 feet), whereas the average depth of the oceans is about 3800 meters (12,500 feet). Thus, the continents stand on the average 4640 meters (about 4.6 kilometers) above the level of the ocean floor.

The elevations of these crustal layers are largely reflections of their densities. The continental blocks are composed of material that has properties similar to those of granite, a common rock with a density about 2.7 times that of water. The crust of ocean basins, on the other hand, has a composition similar to that of basalt, a rock that is about 3 times denser than water.

which has not changed appreciably in nearly three billion years.

Biosphere

The **biosphere** includes all life on Earth. It is concentrated near the surface in a zone that extends from the ocean floor upward for several kilometers into the atmosphere. Plants and animals depend on the physical environment for the basics of life. However, organisms do more than just respond to their physical environment. Indeed, through countless interactions the biosphere powerfully influences the other three spheres. Without life, the makeup and nature of the solid Earth, hydrosphere, and atmosphere would be very different (Figure 1.12).

ATMOSPHERE

BIOSPHERE

HYDROSPHERE

SOLID EARTH

Figure 1.12 This schematic diagram illustrates the dynamic nature of Earth. The planet's four "spheres" constantly and vigorously interact with each other to produce a highly complex system. For more on Earth as a system, see Box 1.3.

This difference alone cannot account for the elevated positions of the continents.

The rocky material located below 100 kilometers is weak and capable of flow. Thus, the rigid outer layer can be thought of as floating on this weak layer, much like an ice cube floats on water. The continental blocks, which consist of thick slabs of less dense rock, float higher than the thinner, more dense oceanic materials.

Continents. Within these two diverse provinces, great variations in elevation exist. The most prominent features of the continents are linear mountain belts. Although the distribution of mountains appears to be random, this is not the case.

When the youngest mountains are considered, we find that they are located principally in two zones. The circum-Pacific belt includes the mountains of the western Americas and continues into the western Pacific in the from of volcanic island arcs. Island arcs are active mountainous regions composed largely of deformed volcanic rocks. Included in this group are the Aleutian Islands, Japan, the Philippines, and New Guinea.

The other major mountain belt extends eastward from the Alps through Iran and the Himalayas, and then dips southward into Indonesia. Careful examination of mountainous terrains reveals that most are places where thick sequences of rocks have been squeezed and highly deformed, as if placed in a gigantic vise.

Older mountains are also found on the continents. Examples include the Appalachians in the eastern United States and the Urals in Russia. Their once lofty peaks are now worn low, the result of millions of years of erosion. Still older are the stable continental interiors. Within these stable interiors are areas known as **shields**, extensive and relatively flat expanses composed largely of crystalline material. Radiometric dating of the shields has revealed that they are truly ancient regions. The ages of some samples exceed 3.8 billion years. Even these oldest known rocks exhibit evidence of enormous forces that have folded and deformed them.

Ocean Basins. At one time the ocean floor was thought to be a nondescript region with only an occasional volcanic structure emerging from the otherwise flat,

Figure 1.13 Major physical features of the continents and ocean basins. The diversity of features on the ocean floor is as varied as on the continents.

sediment-mantled depths. This perception of the ocean floor was incorrect.

The ocean basins are now known to contain the most prominent mountain ranges on Earth, the **oceanic ridge system**. In Figure 1.13 the Mid-Atlantic Ridge and the East Pacific Ridge are parts of this system. This broad elevated feature forms a continuous belt that winds for more than 70,000 kilometers (43,000 miles) around the globe in a manner similar to the seam of a baseball. Rather than consisting of highly deformed rock, such as most of the mountains found on the continents, the oceanic ridge system consists of layer upon layer of once molten rock that has been fractured and uplifted.

The ocean floor also contains extremely deep grooves that are occasionally more than 11,000 meters (36,000 feet) deep. Although these deep-ocean **trenches** are relatively narrow and represent only a small fraction of the ocean floor, they are nevertheless very significant features. Some trenches are located adjacent to young mountains that flank the continents. For example, in Figure 1.13 the Peru–Chile trench off the west coast of South America parallels the Andes Mountains. Other trenches parallel linear island chains called volcanic island arcs.

What is the connection, if any, between the young, active mountain belts and the oceanic trenches? What is the significance of the enormous ridge system that extends through all the world's oceans? What forces crumple rocks to produce majestic mountain ranges? These questions must be answered if we are to understand the dynamic processes that shape our planet.

Origin of Earth

Earth is one of nine planets that along with several dozen moons and numerous smaller bodies revolve around the Sun. The orderly nature of our solar system leads most astronomers to conclude that its members formed at essentially the same time and from the same primordial material as the Sun. This material formed a vast cloud of dust and gases called a *nebula*.

The **nebular hypothesis** suggests that the bodies of our solar system formed from an enormous nebular cloud composed mostly of hydrogen and helium with only a small percentage of the heavier elements.

About five billion years ago, this huge cloud of minute rocky fragments and gases began to contract under its own gravitational influence (Figure 1.14A). The contracting material somehow began to rotate. Like a spinning ice skater pulling in her arms, the cloud rotated faster and faster as it contracted. This rotation in turn caused the nebular cloud to flatten into a disk (Figure 1.14B). Within the rotating disk, smaller accumulations formed nuclei from which the planets eventually coalesced. However, the greatest concentration of material was pulled toward the center of this rotating mass. As it packed inward upon itself, it gravitationally heated, forming the hot *protosun* (sun in the making).

After the protosun formed, the temperature out in the rotating disk dropped significantly. This cooling caused substances with high melting points to condense into small particles, perhaps the size of sand grains. Iron and nickel solidified first. Next to condense were the elements of which rocky substances are composed. As these fragments collided over a few tens of millions of years, they accreted into the planets (Figure 1.14C,D). In the same manner, but on a lesser scale, the processes of condensation and accretion acted to form the moons and other small bodies of the solar system.

As the *protoplanets* (planets in the making) accumulated more and more debris, the space within the solar system began to clear. This removal of debris allowed sunlight to reach planetary surfaces unimpeded and to heat them. The resulting high surface temperatures of the inner planets (Mercury, Venus, Earth, Mars), coupled with their comparatively weak gravitational fields, meant that Earth and its neighbors were unable to retain appreciable amounts of the lighter

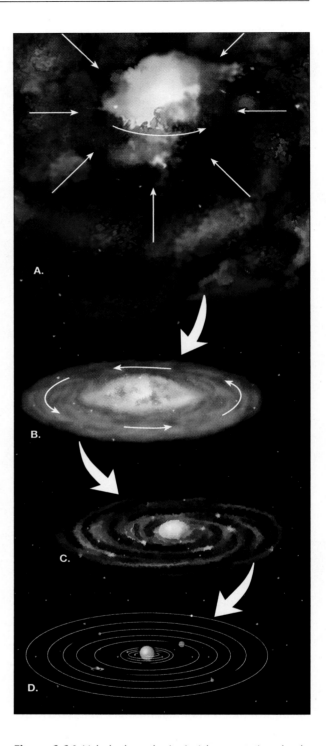

Figure 1.14 Nebular hypothesis. **A.** A huge rotating cloud of dust and gases (nebula) begins to contract. **B.** Most of the material is gravitationally swept toward the center, producing the Sun. However, owing to rotational motion some dust and gases remain orbiting the central body as a flattened disk. **C.** The planets begin to acrete from the material that is orbiting within the flattened disk. **D.** In time most of the remaining debris was either collected into the nine planets and their moons or swept out into space by the solar wind.

components of the primordial cloud. These light materials, which included hydrogen, helium, ammonia, methane, and water, vaporized from their surfaces and were eventually whisked from the inner solar system by streams of solar particles called the *solar winds*.

At distances beyond Mars, temperatures were much cooler. Consequently, the large outer planets (Jupiter, Saturn, Uranus, and Neptune) accumulated huge amounts of hydrogen and other light materials from the primordial cloud. The accumulation of these gaseous substances is thought to account for the comparatively large sizes and low densities of the outer planets.

 # Earth's Internal Structure

Shortly after Earth formed, heat released by colliding particles, coupled with the heat emitted by the decay of radioactive elements, caused some, or all, of Earth's interior to melt. Melting, in turn, allowed the heavier elements, principally iron and nickel, to sink, while the lighter rocky components floated upward.

An important consequence of this period of chemical differentiation is that large quantities of gaseous materials were allowed to escape from Earth's interior, as happens today during volcanic eruptions. By this process a primitive atmosphere gradually evolved. It is on this planet, with this atmosphere, that life as we know it came into existence.

In general, Earth's interior is characterized by a gradual increase in temperature, pressure, and density with depth. Estimates based on experimentation and modeling indicate that the temperature at 100 kilometers is between 1200°C and 1400°C, whereas the temperature at the core–mantle boundary is calculated to be about 4500°C and it may exceed 6700°C at Earth's center. Clearly, Earth's interior remains "hot"; however, energy is slowly, but continuously, flowing toward the surface where it is lost to space.

If temperature were the only factor that determined whether a material melted, our planet would be a molten ball covered with a thin, solid outer shell. However, pressure also increases with depth. Melting, which is accompanied by an increase in volume, occurs at higher temperatures at depth because of greater confining pressure. The increase in pressure with depth also causes a corresponding increase in density. Further, temperature and pressure greatly affect the *mechanical behavior* or *strength* of Earth materials. In particular, when a mineral approaches its melting temperature, its chemical bonds weaken and its mechanical strength (resistance to deformation) is greatly reduced.

Compositional Layers

The segregation of material, which began early in Earth's history, is still occurring, but on a much smaller scale. Because of this chemical differentiation, Earth's interior is not homogeneous. Rather, it consists of three major regions that have markedly different chemical compositions (Figure 1.15).

The principal divisions of Earth include (1) the **crust**, Earth's comparatively thin outer skin that ranges in thickness from 3 kilometers (2 miles) at the oceanic ridges to over 70 kilometers (40 miles) in some mountain belts such as the Andes and Himalayas; (2) the **mantle**, a solid rocky shell that extends to a depth of about 2885 kilometers (1800 miles); (3) the **core**, which can be further divided into the **outer core**, a molten metallic layer some 2270 kilometers (1410 miles) thick, and the **inner core**, a solid iron-rich sphere having a radius of 1216 kilometers (756 miles).

Crust. The crust, the rigid outermost layer of Earth, is divided into oceanic and continental crust (Figure 1.15). Typically, oceanic crust ranges from 3 to 15 kilometers in thickness and is composed of dark igneous rocks called *basalt*. By contrast, the upper continental crust consists of a large variety of rock types, which have an average composition of a *granitic rock* called *granodiorite* (see Chapter 3). The rocks of the oceanic crust are younger (180 million years or less) and more dense (about 3.0 g/cm^3) than continental rocks.* Continental rocks have an average density of about 2.7 g/cm^3 and some have been discovered that exceed 3.8 billion years in age.

Mantle. Over 82 percent of Earth's volume is contained in the mantle, a rocky shell about 2900 kilometers thick. The boundary between the crust and mantle reflects a change in composition. Although the mantle behaves like a solid when transmitting earthquake waves, mantle rocks are able to flow at an incredibly slow rate. The mantle is divided into the *lower mantle* or *mesosphere*, which extends from the core–mantle boundary to a depth of 660 kilometers, and the *upper mantle*, which continues to the base of the crust.

Core. The core is composed mostly of iron with lesser amounts of nickel and other elements. At the extreme pressure found in the core, this iron-rich material has an average density of about 11 g/cm^3 and approaches 14 times the density of water at Earth's center. The inner core and outer core are compositionally very similar, their division is based

*Liquid water has a density of 1 g/cm^3; therefore, the density of basalt is three times that of water.

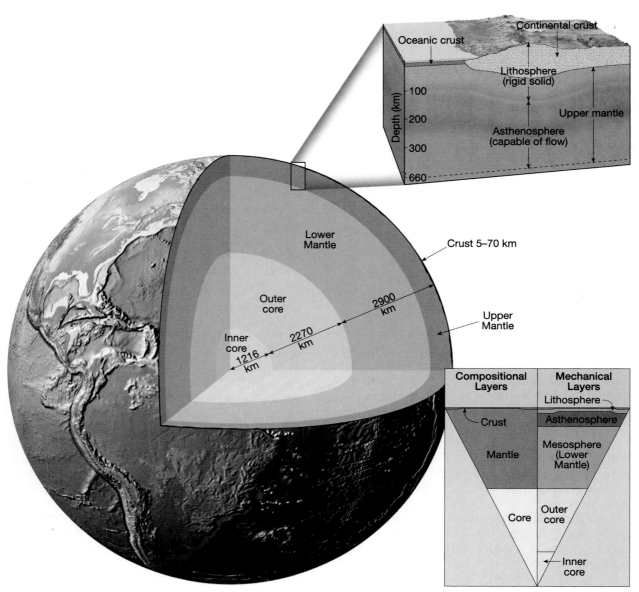

Figure 1.15 View of Earth's layered structure. **A.** The inner core, outer core, and mantle are drawn to scale, but the thickness of the crust is exaggerated by about five times. **B.** A blowup of Earth's outer shell. It shows the two types of crust (oceanic, continental), the rigid lithosphere, and weak asthenosphere.

on differences in mechanical strength. The outer core is a liquid that is capable of flow. It is the circulation within the core of our rotating planet that generates Earth's magnetic field. The inner core, despite its higher temperature, is stronger than the outer core and behaves like a solid.

Mechanical Layers

We now know that Earth's outer layer, including the uppermost mantle and crust, forms a relatively cool, rigid shell. This shell consists of materials with markedly different chemical compositions, but they act as a unit and behave similarly to mechanical deformation. This outermost rigid unit of Earth has been named the **lithosphere** ("sphere of rock"). Averaging about 100 kilometers in thickness, the lithosphere may be 250 kilometers or more in thickness below the older portions (shields) of the continents (Figure 1.15). Within the ocean basins the lithosphere is only a few kilometers thick along the oceanic ridges and increases to perhaps 100 kilometers in regions of older and cooler crustal rocks.

Beneath the lithosphere (to a depth of about 660 kilometers) lies a soft, relatively weak layer located in the upper mantle known as the **asthenosphere** ("weak sphere"). The region encompassing the upper 150 kilometers or so of the asthenosphere has a temperature/pressure regime in which a small amount of melting takes place (perhaps 1 to 5 percent). Within this very weak zone, the lithosphere is effectively detached from the asthenosphere located below. The result is that the lithosphere is able to move independently of the asthenosphere, a topic we will consider in the next section.

It is important to emphasize that the strength of various Earth materials is a function of both their composition and the temperature and pressure of their environment. You should not get the idea that the entire lithosphere is brittle, like the rocks found on the surface. Rather, the rocks of the lithosphere get progressively weaker (more easily deformed) with increasing depth. At the depth of the upper asthenosphere, the rocks are close enough to their melting temperature (some melting may occur) that they are easily deformed. Thus, the asthenosphere is weak because it is hot, just as hot wax is weaker than cold wax. However, in the material located below this weak zone, increased pressure offsets the effects of increased temperature. Therefore, these materials gradually stiffen with depth, forming the more rigid lower mantle. Despite their greater strength, the materials of the lower mantle are still capable of very gradual flow.

 Dynamic Earth

Earth is a dynamic planet! If we could go back in time a billion years or more, we would find a planet with a surface dramatically different from what it is today. There would be no Grand Canyon, Rocky Mountains, or Appalachian Mountains. Moreover, we would find continents with different shapes and located in different positions from today.

In contrast, a billion years ago the Moon's surface was almost the same as we now find it. In fact, if viewed telescopically from Earth, perhaps only a few craters would be missing. Thus, when compared to Earth, the Moon is a lifeless body wandering through space and time.

The processes that alter Earth's surface can be divided into two categories—destructive and constructive. *Destructive processes* are those that wear away the land, including weathering and erosion. Unlike the Moon, where weathering and erosion progress at infinitesimally slow rates, these processes are continually altering the landscape of Earth. In fact, these destructive forces would have long ago leveled the continents had it not been for opposing constructional processes. Included among *constructional processes* are volcanism and mountain building, which increase the

average elevation of the land. As we shall see, these forces depend on Earth's internal heat for their source of energy.

Plate Tectonics

Within the past decades, a great deal has been learned about the workings of our dynamic planet. In fact, this period has been an unequalled revolution in our knowledge about Earth. The revolution began in the early part of the twentieth century with the radical proposal of *continental drift*, the idea that the continents moved about the face of the planet. This proposal contradicted the established view that the continents and ocean basins are permanent and stationary features. For that reason, the notion was received with great skepticism and even ridicule. More than 50 years passed before enough data were gathered to transform this controversial hypothesis into a sound theory that weaved together the basic processes known to operate on Earth. The theory that finally emerged, called **plate tectonics**, provided geologists with the first comprehensive model of Earth's internal workings.*

According to the plate tectonics model, the lithosphere is broken into numerous segments called **plates**, which are in motion and are continually changing shape and size. As shown in Figure 1.16, seven major plates are recognized. They are the North American, South American, Pacific, African, Eurasian, Australian, and Antarctic plates. Intermediate-sized plates include the Caribbean, Nazca, Philippine, Arabian, Cocos, and Scotia plates. In addition, over a dozen smaller plates have been identified but are not shown in Figure 1.16. Note that several large plates include an entire continent plus a large area of seafloor (for example, the South American plate). However, none of the plates are defined entirely by the margins of a single continent.

The lithospheric plates move at very slow but continuous rates of a few centimeters a year. This movement is ultimately driven by the unequal distribution of heat within Earth. Hot material found deep in the mantle moves slowly upward and serves as one part of our planet's internal convective system. Concurrently, cooler, denser slabs of lithosphere descend back into the mantle, setting Earth's rigid outer shell in motion. Ultimately, the titanic, grinding movements of Earth's lithospheric plates generate earthquakes, create volcanoes, and deform large masses of rock into mountains.

---

*Plate tectonics can be defined as the composite of various ideas explaining the observed motion of Earth's lithosphere through the mechanisms of subduction and seafloor spreading, which, in turn, generate Earth's major features, including continents and ocean basins.

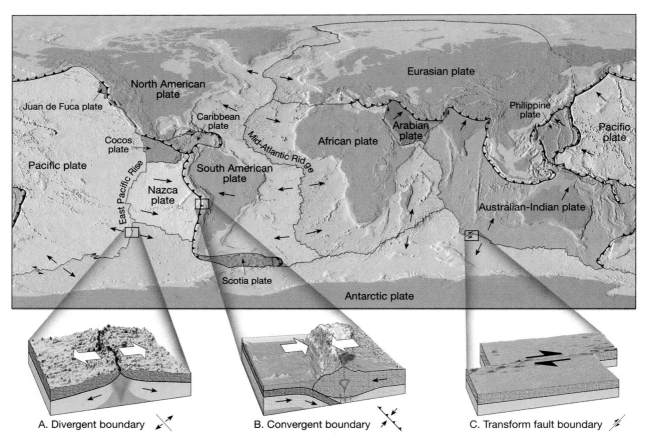

A. Divergent boundary

B. Convergent boundary

C. Transform fault boundary

Figure 1.16 Mosaic of rigid plates that constitute Earth's outer shell. (After W. B. Hamilton, U.S. Geological Survey)

Plate Boundaries

Plates move as coherent units relative to all other plates. Although the interiors of plates may be deformed, all major interactions among individual plates (and therefore most deformation) occurs along their *boundaries*. In fact, the first attempts to outline plate boundaries were made using locations of earthquakes. Later work showed that plates are bounded by three distinct types of boundaries, which are differentiated by the type of movement they exhibit. These boundaries are depicted at the bottom of Figure 1.16 and are briefly described here:

1. **Divergent boundaries**—where plates move apart, resulting in upwelling of material from the mantle to create new seafloor (Figure 1.16A).
2. **Convergent boundaries**—where plates move together, resulting in the subduction (consumption) of oceanic lithosphere into the mantle (Figure 1.16B).
3. **Transform fault boundaries**—where plates grind past each other without the production or destruction of lithosphere (Figure 1.16C).

Examine each large plate in Figure 1.16 and you can see that it is bounded by a combination of these boundaries. Movement along one boundary requires that adjustments be made at the others.

Divergent Boundaries. Plate spreading (divergence) occurs mainly at the mid-ocean ridge. As plates pull apart, the fractures created are immediately filled with molten rock that wells up from the asthenosphere below (Figure 1.17). This hot material slowly cools to hard rock, producing new slivers of seafloor. This happens again and again over millions of years, adding thousands of square kilometers of new seafloor.

This mechanism has created the floor of the Atlantic Ocean during the past 160 million years and is appropriately called **seafloor spreading**. A typical rate of seafloor spreading is 5 centimeters (2 inches) per year, although it varies considerably from one spreading center to another. This extremely slow rate of lithosphere production is nevertheless rapid enough so that all of Earth's ocean basins could have been generated within the last 200 million years. In fact, none of the ocean floor that has been dated exceeds 180 million years in age.

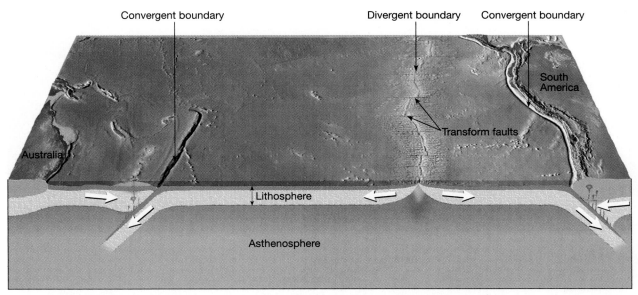

Figure 1.17 View of Earth showing the relationship between divergent and convergent plate boundaries.

Further, along divergent boundaries where molten rock emerges, the ocean floor is elevated. Worldwide, this ridge extends for over 70,000 kilometers through all major ocean basins. You can see parts of the mid-ocean ridge system in Figure 1.13.

As new lithosphere is formed along the oceanic ridge, it is slowly, yet continually, displaced away from the ridge axis. Thus, it begins to cool and contract, thereby increasing in density. This partially accounts for the greater depth of the older and cooler oceanic crust found in the deep ocean basins. In addition, cooling causes the mantle rocks below the oceanic crust to strengthen, thereby adding to the plate thickness. Stated another way, the thickness of oceanic lithosphere is age dependent. The older (cooler) it is, the greater its thickness.

Convergent Boundaries. Although new lithosphere is constantly being added at the oceanic ridges, the planet is not growing in size—its total surface area remains constant. To accommodate the newly created lithosphere, older oceanic plates return to the mantle along *convergent boundaries.* As two plates slowly converge, the leading edge of one slab is bent downward, allowing it to slide beneath the other. This is shown in Figure 1.17. The surface expression produced by the descending plate is an ocean *trench*, like the Peru–Chile trench illustrated in Figure 1.13.

The regions where oceanic crust is being consumed are called **subduction zones.** Here, as the subducted plate moves downward, it enters a high-pressure, high-temperature environment. Some subducted materials, as well as more voluminous amounts of the asthenosphere, melt and migrate upward into the overriding plate. Occasionally, this molten rock may reach the surface, where it gives rise to explosive volcanic eruptions like Mount St. Helens in 1980 and Soufriere Hills in 1997 (Figure 1.18).

Transform Fault Boundaries. Transform fault boundaries are located where plates grind past each other without either generating new lithosphere or consuming old lithosphere. These faults form in the direction of plate movement and were first discovered in association with offsets in oceanic ridges (Figure 1.17).

Although most transform faults are located along mid-ocean ridges, a few slice through the continents. The earthquake-prone San Andreas fault of California is a famous example. Along this fault the Pacific plate is moving toward the northwest, past the North American plate. The movement along this boundary does not go unnoticed. As these plates pass, strain builds in the rocks on opposite sides of the fault. Occasionally the rocks adjust, releasing energy in the form of a great earthquake of the type that devastated San Francisco in 1906.

Changing Boundaries. Although the total surface area of Earth does not change, individual plates may diminish or grow in area depending on the distribution of convergent and divergent boundaries. For example, the Antarctic and African plates are almost entirely bounded by spreading centers and hence are growing larger. By contrast, the Pacific plate is being subducted along its northern and western flanks and is therefore diminishing in size.

Figure 1.18 The August 20, 1997, eruption of Soufriere Hills volcano on the Caribbean island of Montserrat. This is one of several volcanoes in the region that owes its existence to the melting of a subducting plate. (Photo by Kevin West/Gamma Liaison)

Furthermore, new plate boundaries can be created in response to changes in the forces acting on these rigid slabs. For example, a relatively new divergent boundary is located in Africa, in a region known as the East African Rift Valleys. If spreading continues there, the African plate will split into two plates separated by a new ocean basin. At other locations, plates carrying continental crust are presently moving toward each other. Eventually, these continents may collide and be sutured together. Thus, the boundary that once separated two plates disappears as the plates become one. The result of such a continental collision is a majestic mountain range such as the Himalayas.

As long as temperatures deep within our planet remain significantly higher than those near the surface, material within Earth will continue to circulate. This internal flow, in turn, will keep the rigid outer shell of Earth in motion. Thus, while Earth's internal heat engine is operating, the positions and shapes of the continents and ocean basins will change, and Earth will remain a dynamic planet.

In the remaining chapters we will examine in more detail the workings of our dynamic planet in light of the plate tectonics theory.

Earth as a System: The Rock Cycle

Earth is a system. This means that our planet consists of many interacting parts that form a complex whole. Nowhere is this idea better illustrated than when we examine the rock cycle (Figure 1.19). The **rock cycle** allows us to view many of the interrelationships among different parts of the Earth system. It helps us understand the origin of igneous, sedimentary, and metamorphic rocks and to see that each type is linked to the others by the processes that act upon and within the planet. Learn the rock cycle well; you will be examining its interrelationships in greater detail throughout the book.

The Basic Cycle

Let us begin at the top of Figure 1.19. **Magma** is molten material that forms inside Earth. Eventually magma cools and solidifies. This process, called *crystallization*, may occur either beneath the surface or, following a volcanic eruption, at the surface. In either situation, the resulting rocks are called **igneous rocks**.

If igneous rocks are exposed at the surface, they will undergo *weathering*, in which the day-in and day-out

Figure 1.19 The rock cycle. Originally proposed by James Hutton, the rock cycle illustrates the role of the various geologic processes that act to transform one rock type into another. (Photos by J. D. Griggs, U.S.G.S. (A); E. J. Tarbuck (B, C, D); and Phil Dombrowski (E))

influences of the atmosphere slowly disintegrate and decompose rocks. The materials that result are often moved downslope by gravity before being picked up and

transported by any of a number of erosional agents— running water, glaciers, wind, or waves. Eventually these particles and dissolved substances, called **sediment**, are deposited. Although most sediment ultimately comes to rest in the ocean, other sites of deposition include river floodplains, desert basins, swamps, and dunes.

Next the sediments undergo *lithification*, a term meaning "conversion into rock." Sediment is usually lithified into **sedimentary rock** when compacted by the weight of overlying layers or when cemented as percolating water fills the pores with mineral matter.

If the resulting sedimentary rock is buried deep within Earth and involved in the dynamics of mountain building, or intruded by a mass of magma, it will be subjected to great pressures and/or intense heat. The sedimentary rock will react to the changing environment and turn into the third rock type, **metamorphic rock**. When metamorphic rock is subjected to additional pressure changes or to still higher temperatures, it will melt, creating magma, which will eventually crystallize into igneous rock.

Processes driven by heat from Earth's interior are responsible for creating igneous and metamorphic rocks. Weathering and erosion, external processes powered by a combination of energy from the Sun and gravity, produce the sediment from which sedimentary rocks form.

Alternative Paths

The paths shown in the basic cycle are not the only ones that are possible. To the contrary, other paths are just as likely to be followed as those described in the preceding section. These alternatives are indicated by the blue arrows in Figure 1.19.

Igneous rocks, rather than being exposed to weathering and erosion at Earth's surface, may remain deeply buried. Eventually these masses may be subjected to the strong compressional forces and high temperatures associated with mountain building. When this occurs, they are transformed directly into metamorphic rocks.

Metamorphic and sedimentary rocks, as well as sediment, do not always remain buried. Rather, overlying layers may be stripped away, exposing the once-buried rock. When this happens, the material is attacked by weathering processes and turned into new raw materials for sedimentary rocks.

Although rocks might seem to be unchanging masses, the rock cycle shows that they are not. The changes, however, take time—great amounts of time.

The Rock Cycle and Plate Tectonics

When the rock cycle was first proposed by James Hutton, very little was actually known about the processes by which one rock was transformed into another; only evidence for the transformation existed. In fact, it was not until the development of the theory of plate tectonics that a relatively complete picture of the rock cycle emerged.

Figure 1.20 illustrates the rock cycle in terms of the plate tectonics model. According to this model, weathered material from elevated landmasses is transported to the continental margins, where it is deposited in layers that collectively are thousands of meters thick. Once lithified, these sediments create a thick wedge of sedimentary rocks flanking the continents.

Eventually the relatively quiescent activity of sedimentation along a continental margin may be interrupted if the region becomes a convergent plate boundary. When this occurs, the oceanic lithosphere adjacent to the continent begins to inch downward into the asthenosphere beneath the continent. Along active continental margins such as this, convergence deforms the thick wedge of sedimentary rocks into linear belts of metamorphic rocks.

As the oceanic plate descends, some of the overlying sediments that were not crumpled into mountains are carried downward into the hot asthenosphere, where they too undergo metamorphism. Eventually some of these metamorphic rocks are transported to depths where the conditions may trigger some melting. The newly formed magma will then migrate upward through the lithosphere to produce igneous rocks. Some will crystallize prior to reaching the surface and the remainder will erupt and solidify at the surface. When igneous rocks are exposed at the surface, they are immediately attacked by the processes of weathering. Thus, the rock cycle begins anew.

Chapter Summary

- *Geology* literally means "the study of Earth." The two broad areas of the science of geology are (1) *physical geology*, which examines the materials composing Earth and the processes that operate beneath and upon its surface; and (2) *historical geology*, which seeks to understand the origin of Earth and its development through time.

- During the seventeenth and eighteenth centuries, *catastrophism* influenced the formulation of explanations about Earth. Catastrophism states that Earth's landscapes have been developed primarily by great catastrophes. By contrast, *uniformitarianism*, one of the fundamental principles of modern geology advanced by *James Hutton* in the late 1700s,

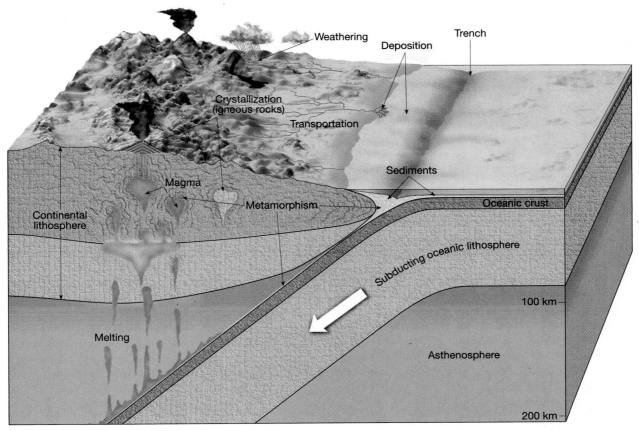

Figure 1.20 The rock cycle as it relates to the plate tectonics model.

states that the physical, chemical, and biological laws that operate today have also operated in the geologic past. The idea is often summarized as "the present is the key to the past." Hutton argued that processes that appear to be slow-acting could, over long spans of time, produce effects that were just as great as those resulting from sudden catastrophic events. *Sir Charles Lyell* (mid-1800s) is given the most credit for advancing the basic principles of modern geology with the publication of the eleven editions of his great work, *Principles of Geology*.

- Using the principles of *relative dating*, the placing of events in their proper sequence or order without knowing their absolute age in years, scientists developed a geologic time scale during the nineteenth century. Relative dates can be established by applying such principles as the *law of superposition* and the *principle of faunal succession*.

- All science is based on the assumption that the natural world behaves in a consistent and predictable manner. The process by which scientists gather facts and formulate scientific *hypotheses* and *theories* is called the *scientific method*. To determine what is occurring in the natural world, scientists often (1) collect facts, (2) develop a scientific hypothesis, (3) construct experiments to test the hypothesis, and (4) accept, modify, or reject the hypothesis on the basis of extensive testing. Other discoveries represent purely theoretical ideas that have stood up to extensive examination. Still other scientific advancements have been made when a totally unexpected happening occurred during an experiment.

- Earth's physical environment is traditionally divided into three major parts: the solid Earth; the water portion of our planet, the *hydrosphere*; and Earth's gaseous envelope, the *atmosphere*. In addition, the *biosphere*, the totality of life on Earth, interacts with each of the three physical realms and is an equally integral part of Earth.

- Two principal divisions of Earth's surface are the continents and ocean basins. The *continental shelf* and *continental slope* mark the continent–ocean basin transition. Major continental features include mountains and shields. Important zones on the ocean floor are *trenches* and the extensive *oceanic ridge system*.

- The *nebular hypothesis* describes the formation of the solar system. The planets and Sun began forming about five billion years ago from a large cloud of dust and gases. As the cloud contracted, it began to rotate and assume a disk shape. Material that was gravitationally pulled toward the center became the *protosun*. Within the rotating disk, small centers, called *protoplanets*, swept up more and more of the cloud's debris. Because of their high temperatures and weak gravitational fields, the inner planets were unable to accumulate and retain many of the lighter components. Because of the very cold temperatures existing far from the Sun, the large outer planets consist of huge amounts of lighter materials. These gaseous substances account for the comparatively large sizes and low densities of the outer planets.

- The solid Earth has several subdivisions. Compositionally, it is divided into a thin outer *crust*, a solid rocky *mantle*, and a dense *core*. The core, in turn, is divided into a liquid *outer core* and a solid *inner core*. Two important mechanical layers are the *lithosphere* (rigid outer shell averaging about 100 kilometers in thickness) and the *asthenosphere* (a relatively weak layer located in the mantle beneath the lithosphere).

- The theory of *plate tectonics* provides a comprehensive model of Earth's internal workings. It holds that Earth's rigid outer lithosphere consists of several segments called *plates* that are slowly and continually in motion relative to each other. Most earthquakes, volcanic activity, and mountain building are associated with the movements of these plates.

- The three distinct types of plate boundaries are (1) *divergent boundaries*—where plates move apart; (2) *convergent boundaries*—where plates move together, causing one to go beneath the other, or where plates collide, which occurs when the leading edges are made of continental crust; and (3) *transform fault boundaries*—where plates slide past each other.

- Earth is a system consisting of many interacting parts that form a complex whole. The *rock cycle* is an excellent example of this idea and is a means of viewing many of the interrelationships of geology. It illustrates the origin of the three basic rock types and the role of various geologic processes in transforming one rock type into another. *Igneous rock* forms from *magma* that cools and solidifies in a process called *crystallization*. *Sedimentary rock* forms when the products of *weathering*, called *sediment*, undergo *lithification*. *Metamorphic rock* forms from rock that has been subjected to great pressure and heat in a process called *metamorphism*.

Review Questions

1. Geology is traditionally divided into two broad areas. Name and describe these two subdivisions.

2. Briefly describe Aristotle's influence on the science of geology.

3. How did the proponents of catastrophism perceive the age of Earth?

4. Describe the doctrine of uniformitarianism. How did the advocates of this idea view the age of Earth?

5. Briefly describe the contributions of Hutton, Playfair, and Lyell.

6. About how old is Earth?

7. The geologic time scale was established without the aid of radiometric dating. What principles were used to develop the time scale?

8. Why are the Archean and Proterozoic eons not divided into as many subdivisions as the Phanerozoic eon?

9. How is a scientific hypothesis different from a scientific theory?

10. List and briefly describe the four "spheres" that constitute our environment.

11. The present shoreline is not the boundary between the continents and the ocean basin. Explain.

12. Briefly describe the events that led to the formation of the solar system.

13. List and briefly describe Earth's compositional divisions.

14. Contrast the asthenosphere and the lithosphere.

15. With which type of plate boundary is each of the following associated: subduction zone, San Andreas fault, seafloor spreading, and Mount St. Helens?

16. Using the rock cycle, explain the statement "one rock is the raw material for another."

Key Terms

asthenosphere (p. 20)
atmosphere (p. 13)
biosphere (p. 15)
catastrophism (p. 3)
continental shelf (p. 15)
continental slope (p. 15)
convergent boundary
  (p. 22)
core (p. 19)
crust (p. 19)
divergent boundary
  (p. 22)

faunal succession,
  principle of (p. 6)
geology (p. 2)
historical geology (p. 2)
hydrosphere (p. 12)
hypothesis (p. 10)
igneous rock (p. 24)
inner core (p. 19)
lithosphere (p. 20)
magma (p. 24)
mantle (p. 19)
metamorphic rock (p. 26)

nebular hypothesis (p. 18)
oceanic ridge system
  (p. 17)
outer core (p. 19)
physical geology (p. 2)
plate (p. 21)
plate tectonics (p. 21)
relative dating (p. 6)
rock cycle (p. 24)
seafloor spreading (p. 22)
sediment (p. 25)
sedimentary rock (p. 26)

shield (p. 16)
subduction zone (p. 23)
superposition, law of
  (p. 6)
theory (p. 11)
transform fault boundary
  (p. 22)
trench (p. 17)
uniformitarianism (p. 4)

Web Resources

The *Earth* Home Page provides on-line resources for this chapter on the World Wide Web. You will find review exercises, specific updates for items in the chapter, suggested reading, and links to interesting related pathways on the Internet. Visit the *Earth* Home Page at **http://www.prenhall.com/tarbuck.**

CHAPTER 2

Matter and Minerals

Left   Quartz (dark gray) and microcline feldspar crystals, Lake George, Colorado. — *Photo By Jeff Scovil*

Earth's crust and oceans are the source of a wide variety of useful and essential minerals (Figure 2.1). In fact, practically every manufactured product contains materials obtained from minerals. Most people are familiar with the common uses of many basic metals, including aluminum in beverage cans, copper in electrical wiring, and gold and silver in jewelry. But some people are not aware that pencil lead contains the greasy-feeling mineral graphite and that baby powder comes from a metamorphic rock made of the mineral talc. Moreover, many do not know that drill bits impregnated with diamonds are employed by dentists to drill through tooth enamel, or that the common mineral quartz is the source of silicon for computer chips. As the mineral requirements of modern society grow, the need to locate additional supplies of useful minerals also grows, and becomes more challenging as well.

In addition to the economic uses of rocks and minerals, all of the processes studied by geologists are in some way dependent on the properties of these basic Earth materials. Events such as volcanic eruptions, mountain building, weathering and erosion, and even earthquakes involve rocks and minerals. Consequently, a basic knowledge of Earth materials is essential to the understanding of all geologic phenomena.

Minerals: The Building Blocks of Rocks

We begin our discussion of Earth materials with an overview of **mineralogy** (the study of minerals), because minerals are the building blocks of rocks. Geologists define **minerals** as any naturally occurring inorganic solids that possess an orderly internal structure and a definite chemical composition. Thus, for any Earth material to be considered a mineral, it must exhibit the following characteristics:

1. It must occur naturally.
2. It must be inorganic.
3. It must be a solid.
4. It must possess an orderly internal structure, that is, its atoms must be arranged in a definite pattern.
5. It must have a definite chemical composition that may vary within specified limits.

When the term *mineral* is used by geologists, only those substances that meet these criteria are considered minerals. Consequently, synthetic diamonds, and a wide variety of other useful materials produced by chemists, are not considered minerals. Further, the gemstone *opal* is classified as a *mineraloid*, rather than a mineral, because it lacks an orderly internal structure.

Rocks, on the other hand, are more loosely defined. Simply, a **rock** is any solid mass of mineral, or mineral-like, matter that occurs naturally as part of our planet. A few rocks are composed almost entirely of one mineral. A common example is the sedimentary rock *limestone*, which is composed of impure masses of the mineral calcite. However, most rocks, like the common rock granite (shown in Figure 2.2), occur as aggregates of several kinds of minerals. Here, the term *aggregate* implies that the minerals are joined in such a way that the properties of each mineral are retained. Note that you can easily identify the mineral constituents of the sample of granite shown in Figure 2.2.

Figure 2.1 Native copper from Michigan's Keweenaw Peninsula. (Photo by E. J. Tarbuck)

Figure 2.2 Most rocks are aggregates of several kinds of minerals.

Granite
(Rock)

Quartz
(Mineral)

Hornblende
(Mineral)

Feldspar
(Mineral)

A few rocks are composed of nonmineral matter. These include the volcanic rocks *obsidian* and *pumice*, which are noncrystalline glassy substances, and *coal*, which consists of solid organic debris.

Although this chapter deals primarily with the nature of minerals, keep in mind that most rocks are simply aggregates of minerals. Because the properties of rocks are determined largely by the chemical composition and internal structure of those minerals contained within them, we will first consider these Earth materials. Subsequent chapters will consider the major rock types.

The Composition of Minerals

Each of Earth's nearly 4000 minerals is uniquely defined by its chemical composition and internal structure. In other words, every sample of the same mineral contains the same elements joined together in a consistent, repeating pattern. We will first review the basic building blocks of minerals, the *elements*, and

then examine how elements bond together to form mineral structures.

At present, 112 elements are known. Of these, only 92 are naturally occurring (Figure 2.3). Some minerals, such as gold and sulfur, are made entirely from one element. But most minerals are a combination of two or more elements, joined to form a chemically stable compound. To understand better how elements combine to form molecules and compounds, we must first consider the **atom**, the smallest part of matter that still retains the characteristics of an element. It is this extremely small particle that does the combining.

Atomic Structure

Two simplified models illustrating basic atomic structure are shown in Figure 2.4. Note that atoms have a central region, called the **nucleus**, which contains very dense **protons** (particles with positive electrical charges) and equally dense **neutrons** (particles with neutral electrical charges). Surrounding the nucleus are very light particles called **electrons**, which travel at high speeds and are negatively charged. For convenience, we often diagram atoms showing the electrons in orbits around the nucleus, like the orbits of

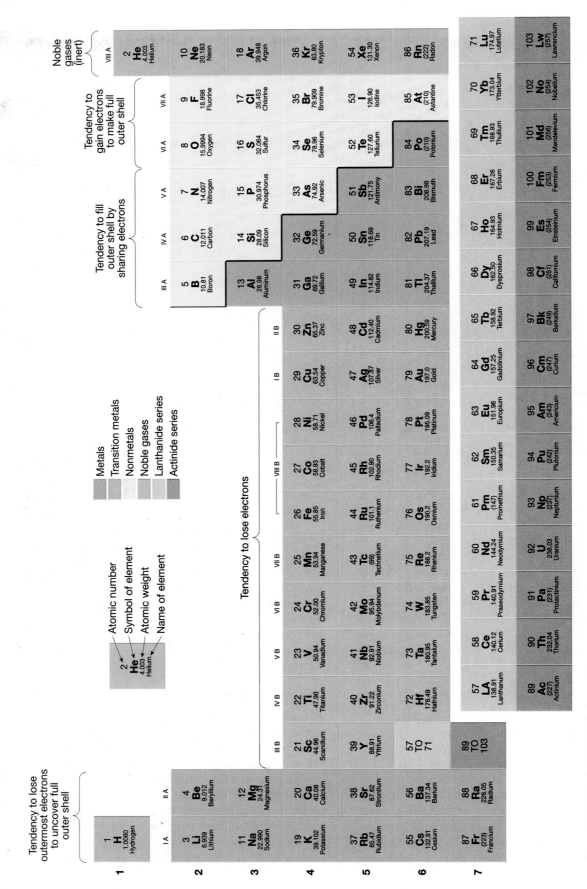

Figure 2.3 Periodic Table of the Elements.

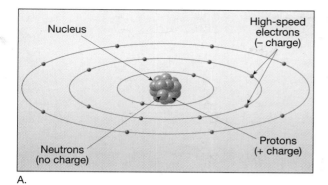

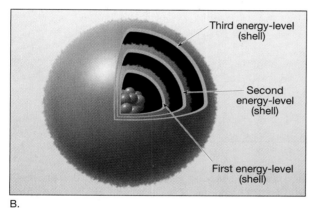

Figure 2.4 Two models of the atom. **A.** A very simplified view of the atom, which consists of a central nucleus, consisting of protons and neutrons, encircled by high-speed electrons. **B.** Another model of the atoms showing spherically shaped electron clouds (energy level shells). Note that these models are not drawn to scale. Electrons are minuscule in size compared to protons and neutrons, and the relative space between the nucleus and electron shells is much greater than illustrated.

the planets around the Sun. However, electrons *do not* travel in the same plane like planets. Further, because of their rapid motion, electrons create spherically shaped negatively charged zones around the nucleus called **energy levels**, or **shells**. Hence, a more realistic picture of the atom can be obtained by envisioning cloudlike shells of fast-moving electrons surrounding a central nucleus (Figure 2.4B). As we shall see, an important fact about these shells is that each can accommodate a specific number of electrons.

The number of protons found in an atom's nucleus determines the **atomic number** and name of the element. For example, all atoms with six protons are carbon atoms, those with eight protons are oxygen atoms, and so forth. Because atoms have the same number of electrons as protons, the atomic number also equals the number of electrons surrounding the nucleus (Table 2.1). Moreover, because neutrons have no charge, the positive charge of the protons is exactly balanced by the negative charge of the electrons. Consequently, atoms are electrically neutral. Thus, an

element is a large collection of electrically neutral atoms, all having the same atomic number.

The simplest element, hydrogen, is composed of atoms that have only one proton in the nucleus and one electron surrounding the nucleus. Each successively heavier atom has one more proton and one more electron, in addition to a certain number of neutrons (Table 2.1). Studies of electron configurations have shown that each electron is added in a systematic fashion to a particular energy level or shell. In general, electrons enter higher energy levels after lower energy levels have been filled to capacity.* The first principal shell holds a maximum of two electrons, while each of the higher shells holds eight or more electrons. As we shall see, it is generally the outermost electrons, also referred to as **valence electrons**, that are involved in chemical bonding.

Bonding

Elements combine with each other to form a wide variety of more complex substances. The strong attractive force linking atoms together is called a *chemical bond*. When chemical bonding joins two or more elements together in definite proportions, the substance is called a **compound**. Most minerals are chemical compounds.

Why do elements join together to form compounds? From experimentation it has been learned that the forces holding the atoms together are electrical. Further, it is known that chemical bonding results in a change in the electron configuration of the bonded atoms. As we noted earlier, it is the valence electrons (outer-shell electrons) that are generally involved in chemical bonding. Other than the first shell, which contains two electrons, *a stable configuration occurs when the valence shell contains eight electrons.* Only the so-called noble gases, such as neon and argon, have a complete outermost electron shell. Thus, the noble gases are the least chemically reactive, and thus their designation as "inert." However, all other atoms seek a valence shell containing eight electrons like the noble gases.

The octet rule, literally "a set of eight," refers to the concept of a complete outermost energy level. Simply, the **octet rule** states that atoms combine to form compounds and molecules in order to obtain the stable electron configuration of the noble gases. To satisfy the octet rule, an atom can either gain, lose, or share electrons with one or more atoms (see Figure 2.3). The result of this process is the formation of an electrical "glue" that bonds the atoms. In summary, *most atoms are chemically reactive and bond together in order to achieve the stable noble gas configuration while retaining overall electrical neutrality.*

---

*This principle holds for the first eighteen elements.

Table 2.1 Atomic Number and Distribution of Electrons

Element	Symbol	Atomic Number	Number of Electrons in Each Shell			
			1st	2nd	3rd	4th
Hydrogen	H	1	1			
Helium	He	2	2			
Lithium	Li	3	2	1		
Beryllium	Be	4	2	2		
Boron	B	5	2	3		
Carbon	C	6	2	4		
Nitrogen	N	7	2	5		
Oxygen	O	8	2	6		
Fluorine	F	9	2	7		
Neon	Ne	10	2	8		
Sodium	Na	11	2	8	1	
Magnesium	Mg	12	2	8	2	
Aluminum	Al	13	2	8	3	
Silicon	Si	14	2	8	4	
Phosphorus	P	15	2	8	5	
Sulfur	S	16	2	8	6	
Chlorine	Cl	17	2	8	7	
Argon	Ar	18	2	8	8	
Potassium	K	19	2	8	8	1
Calcium	Ca	20	2	8	8	2

Ionic Bonds. Perhaps the easiest type of bond to visualize is an **ionic bond**. In ionic bonding, one or more valence electrons are transferred from one atom to another. Simply, one atom gives up its valence electrons and the other uses them to complete its outer shell. A common example of ionic bonding is sodium (Na) and chlorine (Cl) joining to produce sodium chloride (common table salt). This is shown in Figure 2.5. Notice that sodium gives up its single outer electron to chlorine. As a result, sodium achieves a stable configuration having eight electrons in its outermost shell. By acquiring the electron that sodium loses, chlorine, which has seven valence electrons, completes its outermost shell. Thus, through the transfer of a single electron, both the sodium and chlorine atoms have acquired the stable noble gas configuration.

Once electron transfer takes place, atoms are no longer electrically neutral. By giving up one electron, a neutral sodium atom (11 protons/11 electrons)

becomes *positively charged* (11 protons/10 electrons). Similarly, by acquiring one electron, the neutral chlorine atom (17 protons/17 electrons) becomes *negatively charged* (17 protons/18 electrons). Atoms such as these, which have an electrical charge because of the unequal numbers of electrons and protons, are called **ions**. (An atom that picks up an extra electron and becomes negatively charged is called an *anion*. An atom that loses an electron and becomes positively charged is called a *cation*.)

We know that particles (ions) with like charges repel and those with unlike charges attract. Thus, an *ionic bond* is the attraction of oppositely charged ions to one another producing an electrically neutral compound. Figure 2.6 illustrates the arrangement of sodium and chloride ions in ordinary table salt. Notice that salt consists of alternating sodium and chloride ions, positioned in such a manner that each positive ion is attracted to and surrounded on all sides by negative

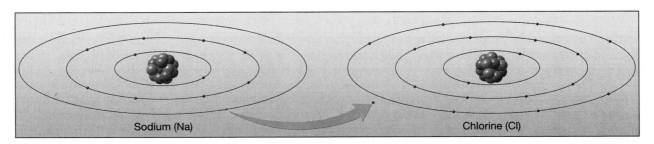

Figure 2.5 Chemical bonding of sodium and chlorine through the transfer of the lone outer electron from a sodium atom to a chlorine atom. The result is a positive sodium ion (Na⁺) and a negative chloride ion (Cl⁻). Bonding to produce sodium chloride (NaCl) is due to electrostatic attraction between the positive and negative ions. In this process note that both the sodium and chlorine atoms have achieved the stable noble gas configuration (eight electrons in their outer shell).

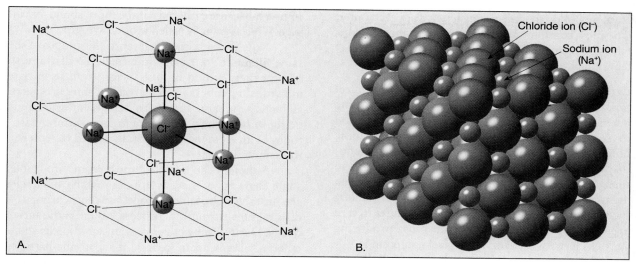

Figure 2.6 Schematic diagrams illustrating the arrangement of sodium and chloride ions in table salt. **A.** Structure has been opened up to show arrangement of ions. **B.** Actual ions are closely packed.

ions, and vice versa. This arrangement maximizes the attraction between ions with unlike charges while minimizing the repulsion between ions with like charges. Thus, *ionic compounds consist of an orderly arrangement of oppositely charged ions assembled in a definite ratio that provides overall electrical neutrality.*

The properties of a chemical compound are *dramatically different* from the properties of the elements comprising it. For example, chlorine is a green, poisonous gas that is so toxic it was used as a chemical weapon during World War I. Sodium is a soft, silvery metal that reacts vigorously with water and, if held in your hand, could burn it severely. Together, however, these atoms produce the compound sodium chloride (table salt), a clear crystalline solid that is essential for human life. This example also illustrates an important difference between a rock and a mineral. Most *minerals* are *chemical compounds* with unique properties that are very different from the elements that comprise them. A *rock*, on the other hand, is a *mixture* of minerals, with each mineral retaining its own identity.

Covalent Bonds. Not all atoms combine by transferring electrons to form ions. Other atoms *share* electrons. For example, the gaseous elements oxygen (O_2), hydrogen (H_2), and chlorine (Cl_2) exist as stable molecules consisting of two atoms bonded together, without a complete transfer of electrons.

Figure 2.7 illustrates the sharing of a pair of electrons between two chlorine atoms to form a molecule of chlorine gas (Cl_2). By overlapping their outer shells, these chlorine atoms share a pair of electrons. Thus, each chlorine atom has acquired, through cooperative action, the needed eight electrons to complete its outer shell. The bond produced by the sharing of electrons is called a **covalent bond.**

A common analogy may help you visualize a covalent bond. Imagine two people at opposite ends of a dimly lit room, each reading under a separate lamp. By moving the lamps to the center of the room, they are able to combine their light sources so each can see better. Just as the overlapping light beams meld, the shared electrons that provide the "electrical glue" in

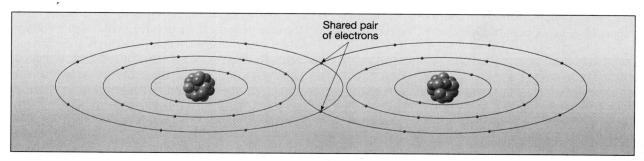

Figure 2.7 Illustration of the sharing of a pair of electrons between two chlorine atoms to form a chlorine molecule. Notice that by sharing a pair of electrons both chlorine atoms have eight electrons in their valence shell.

covalent bonds are indistinguishable from each other. The most common mineral group, the silicates, contains the element silicon, which readily forms covalent bonds with oxygen.

Other Bonds. As you might suspect, many chemical bonds are actually hybrids. They consist to some degree of electron sharing, as in covalent bonding, and to some degree of electron transfer, as in ionic bonding. Furthermore, both ionic and covalent bonds may occur within the same compound. This occurs in many silicate minerals, where silicon and oxygen atoms are covalently bonded to form the basic building block common to all silicates. These structures in turn are ionically bonded to metallic ions, producing various electrically neutral chemical compounds.

Another chemical bond exists in which valence electrons are free to migrate from one ion to another. The mobile valence electrons serve as the "electrical glue." This type of electron sharing is found in metals such as copper, gold, aluminum, and silver, and is called **metallic bonding**. Metallic bonding accounts for the high electrical conductivity of metals, the ease with which metals are shaped, and numerous other special properties of metals.

Isotopes and Radioactive Decay

Subatomic particles are so incredibly small that a special unit was devised to express their mass. A proton or a neutron has a mass just slightly more than one *atomic mass unit*, whereas an electron is only about one two-thousandth of an atomic mass unit. Thus, although electrons play an active role in chemical reactions, they do not contribute significantly to the mass of an atom.

The **mass number** of an atom is simply the total of its neutrons and protons in the nucleus. Atoms of the same element always have the same number of protons, but commonly have varying numbers of neutrons. This means that an element can have more than one mass number. These variants of the same element are called **isotopes** of that element.

For example, carbon has three well-known isotopes. One has a mass number of 12 (carbon-12), another has a mass number of 13 (carbon-13), and the third, carbon-14, has a mass number of 14. All atoms of the same element must have the same number of protons (atomic number), and carbon always has six. Hence, carbon-12 must have six protons plus *six* neutrons to give it a mass number of 12, whereas carbon-14 must have six protons plus *eight* neutrons to give it a mass number of 14. The *average* atomic mass of any random sample of carbon is much closer to 12 than

13 or 14, because carbon-12 is the more abundant isotope. This average is called **atomic weight.***

Note that in a chemical sense all isotopes of the same element are nearly identical. To distinguish among them would be like trying to differentiate individual members from a group of similar objects, all having the same shape, size, and color, with some being only slightly heavier. Further, different isotopes of an element are generally found together in the same mineral.

Although the nuclei of most atoms are stable, some elements do have isotopes in which the nuclei are unstable. Unstable isotopes, such as carbon-14, disintegrate through a process called **radioactive decay**. During radioactive decay unstable nuclei spontaneously break apart, giving off subatomic particles and/or electromagnetic energy similar to x-rays. The rate at which the unstable nuclei decay is steady and measurable, thus making such isotopes useful "clocks" for dating the events of Earth history. A discussion of radioactive decay and its applications in dating past geological events can be found in Chapter 8.

 ## The Structure of Minerals

A mineral is composed of an ordered array of atoms chemically bonded together to form a particular crystalline structure. This orderly packing of atoms is reflected in the regularly shaped objects we call crystals (see chapter-opening photo).

What determines the particular crystalline structure of a mineral? For those compounds formed by ions, the internal atomic arrangement is determined partly by the charges on the ions, but more importantly by the size of the ions involved. To form stable ionic compounds, each positively charged ion is surrounded by the largest number of negative ions that will fit, while maintaining overall electrical neutrality, and vice versa. Figure 2.8 shows some ideal arrangements for various-sized ions.

Let's examine the geometric arrangement of sodium and chloride ions in the mineral halite (Figure 2.9). We see in Figure 2.9A that the sodium and chloride ions pack together to form a cubic-shaped internal structure. Also note that the orderly arrangement of ions found at the atomic level is reflected on a much larger scale in the cubic shaped halite crystals shown in Figure 2.9B. Like halite, all samples of a particular mineral contain the same elements, joined together in the same orderly arrangement.

---

*The term *weight* as used here is a misnomer that has been sanctioned by long use. The correct term is atomic *mass*.

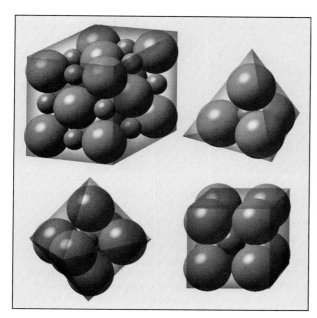

Figure 2.8 Ideal geometrical packing for various-sized positive and negative ions.

Although it is true that every specimen of the same mineral has the same internal structure, some *elements* are able to join together in more than one way. Thus, two minerals with totally different properties may have exactly the same chemical composition. Minerals of this type are said to be **polymorphs** (many forms). Graphite and diamond are particularly good examples of polymorphism because they both consist exclusively of carbon yet have drastically different properties. Graphite is the soft gray material of which pencil lead is made, whereas diamond is the hardest known mineral. The differences between these minerals can be attributed to the conditions under which they were formed. Diamonds form at depths approaching 200 kilometers, where extreme pressures produce the compact structure shown in Figure 2.10A. Graphite, on the other hand, consists of sheets of carbon atoms that are widely spaced and weakly held together (Figure 2.10B). Because these carbon sheets will easily slide past one another, graphite makes an excellent lubricant.

Scientists have learned that by heating graphite under high pressure they can produce diamonds. Although synthetic diamonds are generally not gem quality, because of their hardness they have many industrial uses. The transformation of one polymorph to another is called a *phase change*. In nature, certain minerals go through phase changes as they move from one environment to another. For example, when rocks are carried to greater depths by a subducting plate, the mineral *olivine* changes to a more complex form called *spinel*.

Two other minerals with identical chemical compositions ($CaCO_3$) but different crystal forms are calcite and aragonite. Calcite forms mainly through biochemical processes and is the major constituent of the sedimentary rock limestone. Aragonite is commonly deposited by hot springs and is also an important constituent of pearls and the shells of some marine organisms. Because aragonite changes to the more stable crystalline structure of calcite, it is rare in rocks older than 50 million years. Diamond is also somewhat unstable at Earth's surface, but (fortunately for jewelers) its rate of change to graphite is infinitesimal.

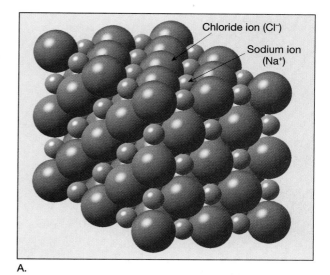

A.

B.

Figure 2.9 The structure of sodium chloride. **A.** The orderly arrangement of sodium and chloride ions in the mineral halite. **B.** The orderly arrangement at the atomic level produces regularly shaped crystals.

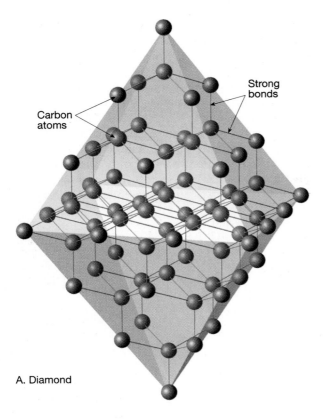

Carbon atoms

Strong bonds

A. Diamond

A. Diamond

Figure 2.10 Comparing the structures of diamond and graphite. Both are natural substances with the same chemical composition—carbon atoms. Nevertheless, their internal structure and physical properties reflect the fact that each formed in a very different environment. **A.** All carbon atoms in diamond are covalently bonded into a compact, three-dimensional framework, which accounts for the extreme hardness of the mineral. (Photo courtesy of Smithsonian Institution) **B.** In graphite the carbon atoms are bonded into sheets that are joined in a layered fashion by very weak electrical forces. These weak bonds allow the sheets of carbon to readily slide past each other, making graphite soft and slippery, and thus useful as a dry lubricant. (Photo by E. J. Tarbuck)

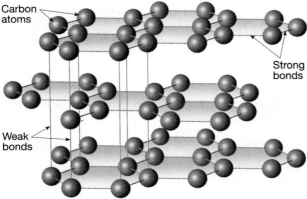

Carbon atoms

Strong bonds

Weak bonds

B. Graphite

B. Graphite

Physical Properties of Minerals

Minerals are solids formed by inorganic processes. Each mineral has an orderly arrangement of atoms (crystalline structure) and a definite chemical composition, which give it a unique set of physical properties. Because the internal structure and chemical composition of a mineral are difficult to determine without the aid of sophisticated tests and apparatus, the more easily recognized physical properties are frequently used in identification. A discussion of some diagnostic physical properties follows.

Crystal Form

Most people think of a crystal as a rare commodity, when in fact most inorganic solid objects are composed of crystals. The reason for this misconception is that most crystals do not exhibit their crystal form. The **crystal form** is the external expression of a mineral that reflects the orderly internal arrangement of atoms. Figure 2.11A illustrates the characteristic form of the iron-bearing mineral pyrite.

A.

B.

Figure 2.11 Crystal form is the external expression of a mineral's orderly internal structure. **A.** Pyrite, commonly known as "fool's gold," often forms cubic crystals. They may exhibit parallel lines (striations) on the faces. **B.** Quartz sample that exhibits well-developed hexagonal crystals with pyramidal-shaped ends. (Photo by Breck P. Kent)

Generally, whenever a mineral is permitted to form without space restrictions, it will develop individual crystals with well-formed crystal faces. Some crystals such as those of the mineral quartz have a very distinctive crystal form that can be helpful in identification (Figure 2.11B). However, most of the time crystal growth is interrupted because of competition for space, resulting in an intergrown mass of crystals, none of which exhibits its crystal form.

Luster

Luster is the appearance or quality of light reflected from the surface of a mineral. Minerals that have the appearance of metals, regardless of color, are said to have a *metallic luster*. Minerals with a *nonmetallic luster* are described by various adjectives, including vitreous (glassy), pearly, silky, resinous, and earthy (dull). Some minerals appear partially metallic in luster and are said to be submetallic.

Color

Although **color** is an obvious feature of a mineral, it is often an unreliable diagnostic property. Slight impurities in the common mineral quartz, for example, give it a variety of colors, including pink, purple (amethyst), white, and even black (see Figure 2.25, p. 52). When a mineral, such as quartz, exhibits a variety of colors, it is said to possess *exotic coloration*. Exotic coloration is usually caused by the inclusion of impurities, such as foreign ions, in the crystalline structure. Other minerals, for example, sulfur, which is yellow, and malachite, which is bright green, are said to have *inherent coloration*.

Streak

Streak is the color of a mineral in its powdered form and is obtained by rubbing the mineral across a piece of unglazed porcelain termed a *streak plate*. Although the color of a mineral may vary from sample to sample, the streak usually does not, and is therefore the more reliable property. Streak can also be an aid in distinguishing minerals with metallic lusters from those having nonmetallic lusters. Metallic minerals generally have a dense, dark streak, whereas minerals with nonmetallic lusters do not.

Hardness

One of the most useful diagnostic properties is **hardness**, a measure of the resistance of a mineral to abrasion or scratching. This property is determined by rubbing a mineral of unknown hardness against one of known hardness, or vice versa. A numerical value can be obtained by using the **Mohs scale** of hardness, which consists of ten minerals arranged in order from 1 (softest) to 10 (hardest), as shown in Table 2.2.

Any mineral of unknown hardness can be compared to these or to other objects of known hardness. For example, a fingernail has a hardness of 2.5, a copper penny 3, and a piece of glass 5.5. The mineral gypsum, which has a hardness of 2, can be easily scratched with your fingernail. On the other hand, the mineral calcite, which has a hardness of 3, will scratch your fingernail but will not scratch glass. Quartz, the hardest of the common minerals, will scratch glass.

Cleavage

In the crystal structure of a mineral, some bonds are weaker than others. These bonds are where a mineral will break when it is stressed. **Cleavage** is the tendency of a mineral to break along planes of weak bonding. Not all minerals have definite planes of weak bonding, but those that possess cleavage can be identified by the distinctive smooth surfaces that are produced when the mineral is broken.

The simplest type of cleavage is exhibited by the micas (Figure 2.12). Because the micas have weak bonds in one direction, they cleave to form thin, flat sheets. Some minerals have several cleavage planes, which produce smooth surfaces when broken, while others exhibit poor cleavage, and still others have no cleavage at all. When minerals break evenly in more than one direction, cleavage is described by the *number of planes* exhibited and the *angles at which they meet* (Figure 2.13).

Do not confuse cleavage with crystal form. When a mineral exhibits cleavage, it will break into pieces *that have the same geometry as each other.* By contrast, the quartz crystals shown in Figure 2.11B do not have cleavage. If broken, they fracture into shapes that do not resemble each other or the original crystals.

Fracture

Minerals that do not exhibit cleavage when broken, such as quartz, are said to **fracture**. Those that break into smooth curved surfaces resembling broken glass have a *conchoidal fracture* (Figure 2.14). Others break into splinters or fibers, but most minerals fracture irregularly.

Table 2.2 Mohs Scale of Hardness

Relative Scale		Mineral	Hardness of Some Common Objects
Hardest	10	Diamond	
	9	Corundum	
	8	Topaz	
	7	Quartz	
	6	Potassium Feldspar	
	5	Apatite	5.5 Glass, Pocketknife
	4	Fluorite	
	3	Calcite	3 Copper Penny
	2	Gypsum	2.5 Fingernail
Softest	1	Talc	

Figure 2.12 The thin sheets shown here were produced by splitting a mica (muscovite) crystal parallel to its perfect cleavage. (Photo by Breck P. Kent)

Figure 2.13 Smooth surfaces produced when a mineral with cleavage is broken. The sample on the left (fluorite) exhibits four planes of cleavage (eight sides), whereas the other two samples exhibit three planes of cleavage (six sides). Also notice that the mineral in the center (halite) has cleavage planes that meet at 90° angles, whereas the mineral on the right (calcite) has cleavage planes that meet at 75° angles. (Photo by E. J. Tarbuck)

Figure 2.14 Conchoidal fracture. The smooth curved surfaces result when minerals break in a glasslike manner. (Photo by E. J. Tarbuck)

Figure 2.15 Galena is lead sulfide and, like other metallic ores, has a relatively high specific gravity.

Specific Gravity

Specific gravity is a number representing the ratio of the weight of a mineral to the weight of an equal volume of water. For example, if a mineral weighs three times as much as an equal volume of water, its specific gravity is 3. With a little practice, you can estimate the specific gravity of minerals by hefting them in your hand. For example, if a mineral feels as heavy as the common rocks you have handled, its specific gravity will probably be somewhere between 2.5 and 3. Some metallic minerals have a specific gravity two or three times that of common rock-forming minerals. Galena, which is an ore of lead, has a specific gravity of roughly 7.5, whereas the specific gravity of 24-karat gold is approximately 20 (Figure 2.15).

Other Properties of Minerals

In addition to the properties already discussed, some minerals can be recognized by other distinctive properties. For example, halite is ordinary salt, so it is quickly identified with your tongue. Thin sheets of mica will bend and elastically snap back. Gold is malleable and can be easily shaped. Talc and graphite both have distinctive feels; talc feels soapy and graphite feels greasy. A few minerals, such as magnetite, have a high iron content and can be picked up with a magnet, while some varieties (lodestone) are natural magnets and will pick up small iron-based objects such as pins and paper clips.

Moreover, some minerals exhibit special optical properties. For example, when a transparent piece of calcite is placed over printed material, the letters appear twice. This optical property is known as *double refraction*. In addition, the streak of many sulfur-bearing minerals smells like rotten eggs.

One very simple chemical test involves placing a drop of dilute hydrochloric acid from a dropper bottle on a freshly broken mineral surface. Certain minerals, called carbonates, will effervesce (fizz) with hydrochloric acid. This test is useful in identifying the mineral calcite, which is a common carbonate mineral.

In summary, a number of special physical and chemical properties are useful in identifying certain minerals. These include taste, smell, elasticity, malleability, feel, magnetism, double refraction, and chemical reaction to hydrochloric acid. Remember that every one of these properties depends on the composition (elements) of a mineral and its structure (how the atoms are arranged).

 Mineral Groups

Nearly 4000 minerals have been named and about 40 to 50 new ones are being identified each year.* Fortunately, for students who are beginning to study minerals, no more than a few dozen are abundant! Collectively, these few make up most of the rocks of Earth's crust and as such, are classified as the *rock-forming minerals*. It is also interesting to note that *only eight elements* make up the bulk of these minerals and represent over 98 percent (by weight) of the continental crust (Table 2.3).

The two most abundant elements are silicon and oxygen, which combine to form the framework of the most common mineral group, the **silicates**. Perhaps the next most common mineral group is the carbonates, of which calcite is the most prominent member. Other common rock-forming minerals include gypsum and halite.

---

*Appendix C describes some of the common minerals of Earth's crust.

Table 2.3 Relative Abundance of the Most Common Elements in the Continental Crust

Element	Approximate Percentage by Weight
Oxygen (O)	46.6
Silicon (S)	27.7
Aluminum (Al)	8.1
Iron (Fe)	5.0
Calcium (Ca)	3.6
Sodium (Na)	2.8
Potassium (K)	2.6
Magnesium (Mg)	2.1
All others	1.7
Total	100

We will first discuss the most common mineral group, the silicates, and then consider some of the other prominent mineral groups.

The Silicates

Every silicate mineral contains the elements oxygen and silicon. Moreover, except for a few minerals such as quartz, every silicate mineral includes one or more additional elements that are needed to produce electrical neutrality. These additional elements give rise to the great variety of silicate minerals and their varied properties (Box 2.1).

The Silicon–Oxygen Tetrahedron. All silicates have the same fundamental building block, the **silicon–oxygen tetrahedron**. This structure consists of four oxygen ions surrounding a much smaller silicon ion (Figure 2.16). The silicon–oxygen tetrahedron is a complex ion (SiO_4^{4-}) with a charge of –4.

In nature, one of the simplest ways in which these tetrahedra join together to become neutral compounds is through the addition of positively charged

Figure 2.16 Two representations of the silicon–oxygen tetrahedron. **A.** The four large spheres represent oxygen ions, and the blue sphere represents a silicon ion. The spheres are drawn in proportion to the radii of the ions. **B.** An expanded view of the tetrahedron using rods to depict the bonds that connect the ions.

ions (Figure 2.17). In this way a chemically stable structure is produced, consisting of individual tetrahedra linked together by cations.

Other Silicate Structures. In addition to cations providing the opposite electrical charge needed to bind the tetrahedra, the tetrahedra may link with

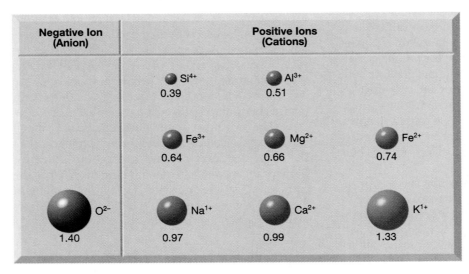

Negative Ion (Anion)	Positive Ions (Cations)		
	Si^{4+} 0.39	Al^{3+} 0.51	
	Fe^{3+} 0.64	Mg^{2+} 0.66	Fe^{2+} 0.74
O^{2-} 1.40	Na^{1+} 0.97	Ca^{2+} 0.99	K^{1+} 1.33

Figure 2.17 Relative sizes and electrical charges of ions of the eight most common elements in Earth's crust. These are the most common ions in rock-forming minerals. Ionic radii are expressed in angstroms (1 angstrom equals 10^{-8} cm).

themselves in a variety of configurations. For example, the tetrahedra may be joined to form *single chains*, *double chains*, or *sheet structures*, as shown in Figure 2.18. The joining of tetrahedra in each of these configurations results from the sharing of oxygen atoms between silicon atoms in adjacent tetrahedra.

To understand better how this sharing takes place, select one of the silicon ions (small blue spheres) near the middle of the single-chain structure shown in Figure 2.18. Notice that this silicon ion is completely surrounded by four larger oxygen ions (you are looking *through* one of the four, to see the blue silicon ion). Also notice that, of the four oxygen ions, two are joined to other silicon ions, whereas the other two are not shared in this manner. *It is the linkage across the shared oxygen ions that joins the tetrahedra into a chain structure.* Now, examine a silicon ion near the middle of the sheet structure and count the number of shared and unshared oxygen ions surrounding it. The increase in the degree of sharing accounts for the sheet structure. Other silicate structures exist, and the most common has all of the oxygen ions shared to produce a complex three-dimensional framework.

By now you can see that the ratio of oxygen ions to silicon ions differs in each of the silicate structures. In the isolated tetrahedron there are four oxygen ions for every silicon ion. In the single chain, the oxygen-to-silicon ratio is 3:1, and in the three-dimensional framework this ratio is 2:1. Consequently, as more of the oxygen ions are shared, the percentage of silicon in the structure increases. Silicate minerals are therefore described as having a "high" or "low" silicon content, based on their ratio of oxygen to silicon. This difference in silicon content is important, as we shall see in a later chapter when we consider the formation of igneous rocks.

Joining Silicate Structures. Most silicate structures, including single chains, double chains, or sheets, are not neutral chemical compounds. Thus, like the individual tetrahedra, they are all neutralized by the inclusion of metallic cations that bond them together into a variety of complex crystalline configurations. The cations that most often link silicate structures are those of the elements iron (Fe), magnesium (Mg), potassium (K), sodium (Na), aluminum (Al), and calcium (Ca).

Notice in Figure 2.17 that each of these cations has a particular atomic size and a particular charge. Generally, ions of approximately the same size are able to substitute freely for one another. For instance, ions of iron (Fe^{2+}) and magnesium (Mg^{2+}) are nearly the same size and substitute for each other without altering the mineral structure. This also holds true for calcium and sodium ions, which can occupy the same site in a crystalline structure. In addition, aluminum (Al) often substitutes for silicon in the silicon-oxygen tetrahedron.

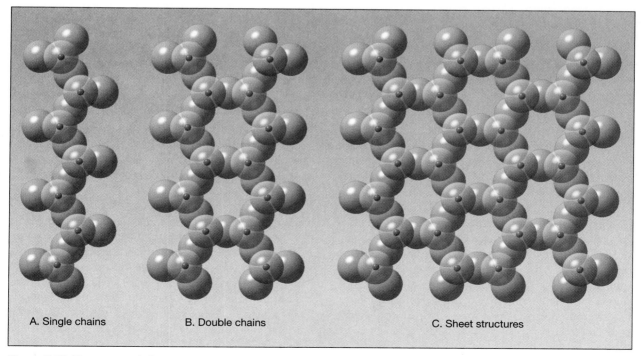

A. Single chains   B. Double chains   C. Sheet structures

Figure 2.18 Three types of silicate structures. **A.** Single chains. **B.** Double chains. **C.** Sheet structures.

Box 2.1

Asbestos: What Are the Risks?

Once considered safe enough to use in toothpaste, asbestos may have become the most feared contaminant on Earth. Although health concerns arose two decades earlier, the asbestos panic began in 1986 when the Environmental Protection Agency (EPA) instituted the Asbestos Hazard Emergency Response Act. It required inspection of all public and private schools for asbestos. This brought asbestos to public attention and raised parental fears that their children could contract asbestos-related cancers because of high levels of airborne fibers in schools. Since that time, billions of dollars have been spent on asbestos testing and removal.

What Is Asbestos?

Asbestos is not a single material. Rather, asbestos is a general term for a group of silicate minerals that *readily separate into thin, strong fibers* (Figure 2.A). Because these fibers are flexible, heat resistant, and relatively chemically inert, they have many uses. Asbestos has been widely used to strengthen concrete, make fireproof fabrics, and insulate boilers and hot pipes. It is a component in floor tiles and the major ingredient in automobile brake linings. Wall coatings rich in asbestos fibers were used extensively during the U.S. building boom of the 1950s and early 1960s.

Figure 2.A Chrysotile asbestos. This sample is a fibrous form of the mineral serpentine. (Photo by E. J. Tarbuck)

Most asbestos comes from three minerals. *Chrysotile* ("white asbestos") is a fibrous form of the mineral serpentine. It is the only type of asbestos mined in North America and once made up 95 percent of world production. *Crocidolite* ("blue asbestos") and *amosite* ("brown asbestos") are currently mined in South Africa and make up about 5 percent of world production.

Exposure and Risk

There is no question that prolonged exposure to air laden with certain types of asbestos dust in an unregulated workplace can be dangerous. When the thin, rodlike fibers are inhaled into the lungs, they are neither easily broken down nor easily expelled, but can remain *for life*. Three lung diseases can result: (1) *asbestosis*, a scarring of tissue, that

Because of the ability of silicate structures to readily accommodate different cations at a given bonding site, individual specimens of a particular mineral may contain varying amounts of certain elements. A mineral of this type is often expressed by a chemical formula that uses parentheses to show the variable component. A good example is the mineral olivine, $(Mg, Fe)_2SiO_4$, which is magnesium/iron silicate. As you can see from the formula, it is the iron (Fe^{2+}) and magnesium (Mg^{2+}) cations in olivine that freely substitute for each other. At one extreme, olivine may contain iron without any magnesium (Fe_2SiO_4, or iron silicate) and at the other, iron is totally lacking (Mg_2SiO_4, or magnesium silicate). Between these end members, any ratio of

iron to magnesium is possible. Thus, olivine, as well as many other silicate minerals, is actually a *family* of minerals with a range of composition between the two end members.

In certain substitutions, the ions that interchange do not have the same electrical charge. For instance, when calcium (Ca^{2+}) substitutes for sodium (Na^{1+}), the structure gains a positive charge. In nature, one way in which this substitution is accomplished, while still maintaining overall electrical neutrality, is the simultaneous substitution of aluminum (Al^{3+}) for silicon (Si^{4+}). This particular double substitution occurs in a mineral called plagioclase feldspar. It is a member of the most abundant family of minerals

decreases the lung's ability to absorb oxygen; (2) *mesothelioma*, a rare tumor that develops in the chest or gut; and (3) *lung cancer*.

Evidence that incriminates "blue asbestos" and "brown asbestos" comes from health studies at mines in South Africa and western Australia. Miners and mill workers showed an extremely high incidence of mesothelioma, sometimes after less than a year of exposure.

However, the U.S. Geological Survey concludes that the risks from the most widely used form of asbestos (chrysotile or "white asbestos") are minimal to nonexistent. They cite studies of miners of white asbestos in Canada and in northern Italy, where mortality rates from mesothelioma and lung cancer differ very little from the general public. Another study was conducted on miner's wives in the area of Thetford Mines, Quebec, once the largest chrysotile mine in the world. For many years there were no dust controls on mining and milling operations, so these women were exposed to extremely high levels of airborne asbestos. Nevertheless, they exhibited *below-normal* levels of the diseases thought to be associated with asbestos exposure.

The various types of asbestos fibers differ in their chemical composition, shape, and durability. The thin, rodlike "blue asbestos" and "brown asbestos" fibers, which can easily penetrate the lining of the lungs, are certainly pathogenic. But chrysotile fibers are curly and can be expelled more easily than can the rodlike fibers. Further, if inhaled, chrysotile fibers break down within a year. This is not the case for the other forms of asbestos, or for fiberglass, which is frequently used as a substitute for asbestos. These differences are thought to explain the fact that the mortality rates for chrysotile workers differ very little from the rates for the general population. Although the EPA has lifted its ban on asbestos, very little of this once exalted mineral is presently used in the United States.

What Risk?

Does asbestos present a risk to the nation's students? Available data indicate that the levels of airborne asbestos in schools are approximately 0.01 of the permissible exposure levels for the U.S. workplace. In addition, the indoor concentrations of the most biologically active fibers are comparable to outdoor levels. With few exceptions, the type of asbestos found in schools is chrysotile, which in low concentrations has been shown to be relatively harmless.

A comparison of the risk from asbestos exposure in schools to other risks in society is shown in Table 2.A. Playing high school football poses over 100 times the risk as exposure to airborne asbestos. These data clearly demonstrate that the asbestos panic was and is unwarranted. However, it has spawned an entire asbestos-removal industry that now has a life of its own. Misguided public policies, no matter how well intentioned they are, are expensive.

Table 2.A Estimates of Risk from Asbestos Exposure in Schools in Comparison with Other Risks

Cause	Annual Death Rate (per million)
Asbestos exposure in schools	0.005–0.093
Whooping cough vaccination (1970–80)	1–6
Aircraft accidents (1979)	6
High school football (1970–80)	10
Drowning (ages 5–14)	27
Motor vehicle accident, Pedestrian (ages 5–14)	32
Home accidents (ages 1–14)	60
Long-term smoking	1200

found in Earth's crust. The end members of this particular feldspar series are a calcium–aluminum silicate (anorthite, $CaAl_2Si_2O_8$), and a sodium-aluminum silicate (albite, $NaAlSi_3O_8$).

We are now prepared to review silicate structures in light of what we know about chemical bonding. An examination of Figure 2.17 shows that among the major constituents of the silicate minerals, only oxygen is an anion (negatively charged). Because oppositely charged ions attract (and similarly charged ions repel), the chemical bonds that hold silicate structures together form between oxygen and oppositely charged cations. Thus, cations arrange themselves so that they can be as close as possible to oxygen while remaining as far apart from each other as possible. Because of its small size and high charge (+4), the silicon (Si) cation forms the strongest bonds with oxygen. Aluminum (Al), although not as strongly bonded to oxygen as silicon is more strongly bonded than calcium (Ca), magnesium (Mg), iron (Fe), sodium (Na), or potassium (K). In many ways, aluminum plays a role similar to silicon by being the central ion in the basic tetrahedral structure.

Most silicate minerals consist of a basic framework composed of either a single silicon or aluminum cation surrounded by four negatively charged oxygen ions. These tetrahedra often link together to form a variety of other silicate structures (chains, sheets, etc.) through shared oxygen atoms. Finally, the other cations bond with the oxygen atoms of these silicate structures to create the more complex crystalline structures that characterize the silicate minerals.

Common Silicate Minerals

To reiterate, the silicates are the most abundant mineral group and have the silicate ion (SiO_4^{4-}) as their basic building block. The major silicate groups and common examples are given in Figure 2.19. The feldspars are by far the most plentiful silicate, comprising over 50 percent of Earth's crust. Quartz, the second most abundant mineral in the continental crust, is the only common mineral made completely of silicon and oxygen.

Notice in Figure 2.19 that each mineral *group* has a particular silicate *structure*, and that a relationship exists between the internal structure of a mineral and the *cleavage* it exhibits. Because the silicon–oxygen bonds are strong, silicate minerals tend to cleave between the silicon–oxygen structures rather than across them. For example, the micas have a sheet structure and thus tend to cleave into flat plates (see Figure 2.12, p. 42). Quartz, which has equally strong

Mineral		Idealized Formula	Cleavage	Silicate Structure
Olivine		$(Mg, Fe)_2SiO_4$	None	Single tetrahedron
Pyroxene group (Augite)		$(Mg,Fe)SiO_3$	Two planes at right angles	Single chains
Amphibole group (Hornblende)		$Ca_2(Fe,Mg)_5Si_8O_{22}(OH)_2$	Two planes at 60° and 120°	Double chains
Micas	Biotite	$K(Mg,Fe)_3AlSi_3O_{10}(OH)_2$	One plane	Sheets
	Muscovite	$KAl_2(AlSi_3O_{10})(OH)_2$		
Feld-spars	Orthoclase	$KAlSi_3O_8$	Two planes at 90°	Three-dimensional networks
	Plagioclase	$(Ca,Na)AlSi_3O_8$		
Quartz		SiO_2	None	(Expanded view)

Figure 2.19 Common silicate minerals. Note that the complexity of the silicate structure increases down the chart.

silicon–oxygen bonds in all directions, has no cleavage, but fractures instead.

Most silicate minerals form (crystallize) as molten rock is cooling. This cooling can occur at or near Earth's surface (low temperature and pressure) or at great depths (high temperature and pressure). The environment during crystallization and the chemical composition of the molten rock determine to a large degree which minerals are produced. For example, the silicate mineral olivine crystallizes at high temperatures, whereas quartz crystallizes at much lower temperatures.

In addition, some silicate minerals form at Earth's surface from the weathered products of older silicate minerals. Still other silicate minerals are formed under the extreme pressures associated with mountain building. Each silicate mineral, therefore, has a structure and a chemical composition that *indicate the conditions under which it formed*. Thus, by carefully examining the mineral constituents of rocks, geologists can often determine the circumstances under which the rocks formed.

We will now examine some of the most common silicate minerals, which we divide into two major groups on the basis of their chemical makeup.

Ferromagnesian (Dark) Silicates. The **dark** (or **ferromagnesian**) **silicates** are those minerals containing ions of iron (iron = *ferro*) and/or magnesium in their structure. Because of their iron content, ferromagnesian silicates are dark in color and have a greater specific gravity, between 3.2 and 3.6, than nonferromagnesian silicates. The most common dark silicate minerals are olivine, the pyroxenes, the amphiboles, the dark mica (biotite), and garnet.

Olivine is a family of high-temperature silicate minerals that are black to olive green in color and have a glassy luster and a conchoidal fracture. Rather than developing large crystals, olivine commonly forms small, rounded crystals that give rocks consisting largely of olivine a granular appearance. Olivine is composed of individual tetrahedra, which are bonded together by a mixture of iron and magnesium ions positioned so as to link the oxygen atoms and magnesium atoms together. Because the three-dimensional network generated in this fashion does not have its weak bonds aligned, olivine does not possess cleavage.

The *pyroxenes* are a group of complex minerals thought to be important components of Earth's mantle. The most common member, *augite*, is a black, opaque mineral with two directions of cleavage that meet at nearly a 90 degree angle. Its crystalline structure consists of single chains of tetrahedra bonded together by ions of iron and magnesium. Because the silicon–oxygen bonds are stronger than the bonds joining the silicate structures, augite cleaves parallel to the silicate chains. Augite is one of the dominant minerals in basalt, a common igneous rock of the oceanic crust and volcanic areas on the continents.

Hornblende is the most common member of a chemically complex group of minerals called *amphiboles* (Figure 2.20). Hornblende is usually dark green to black in color and except for its cleavage angles, which are about 60 degrees and 120 degrees, it is very similar in appearance to augite (Figure 2.21). The double chains of tetrahedra in the hornblende structure account for its particular cleavage. In a rock, hornblende often forms elongated crystals. This helps distinguish it from pyroxene, which forms rather blocky crystals. Hornblende is found predominantly in continental rocks, where it often makes up the dark portion of an otherwise light-colored rock.

Biotite is the dark iron-rich member of the mica family. Like other micas, biotite possesses a sheet structure that gives it excellent cleavage in one direction. Biotite also has a shiny black appearance that helps distinguish it from the other dark ferromagnesian minerals. Like hornblende, biotite is a common constituent of continental rocks, including the igneous rock granite.

Garnet is similar to olivine in that its structure is composed of individual tetrahedra linked by metallic ions. Also like olivine, garnet has a glassy luster, lacks cleavage, and possesses conchoidal fracture. Although the colors of garnet are varied, this mineral is most often brown to deep red. Garnet readily forms equidimensional crystals that are most commonly found in metamorphic rocks (Figure 2.22). When garnets are transparent, they may be used as gemstones (see Box 2.2).

Figure 2.20 Hornblende amphibole. Hornblende is a common dark silicate material having two cleavage directions that intersect at roughly 60° and 120°.

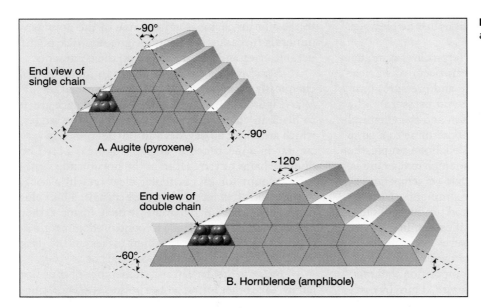

Figure 2.21 Cleavage angles for augite and hornblende.

Nonferromagnesian (Light) Silicates. As the name implies, the **light** (or **nonferromagnesian**) **silicates** are generally light in color and have a specific gravity of about 2.7, which is considerably less than the ferromagnesian silicates. As indicated earlier, these differences are mainly attributable to the presence or absence of iron and magnesium. The light silicates contain varying amounts of aluminum, potassium, calcium, and sodium, rather than iron and magnesium.

Muscovite is a common member of the mica family. It is light in color and has a pearly luster. Like other

Figure 2.22 A deep-red garnet crystal embedded in a light-colored, mica-rich metamorphic rock. (Photo by E. J. Tarbuck)

micas, muscovite has excellent cleavage in one direction (see Figure 2.12, p. 42). In thin sheets, muscovite is clear, a property that accounts for its use as window "glass" during the Middle Ages. Because muscovite is very shiny, it can often be identified by the sparkle it gives a rock. If you have ever looked closely at beach sand, you may have seen the glimmering brilliance of the mica flakes scattered among the other sand grains.

Feldspar, the most common mineral group, can form under a very wide range of temperatures and pressures, a fact that partially accounts for its abundance. All of the feldspars have similar physical properties. They have two planes of cleavage meeting at or near 90 degree angles, are relatively hard (6 on the Mohs scale), and have a luster that ranges from glassy to pearly. As one component in a rock, feldspar crystals can be identified by their rectangular shape and rather smooth shiny faces (Figure 2.23).

The structure of feldspar minerals is a three-dimensional framework formed when oxygen atoms are shared by adjacent silicon atoms. In addition, one-fourth to one-half of the silicon atoms in the feldspar structure are replaced by aluminum atoms. The difference in charge between aluminum (+3) and silicon (+4) is made up by the inclusion of one or more of the following ions into the crystal lattice: potassium (+1), sodium (+1), and calcium (+2). Because of the large size of the potassium ion as compared to the size of the sodium and calcium ions, two different feldspar structures exist. *Orthoclase feldspar* is a common member of a group of feldspar minerals that contains potassium

Box 2.2

Carrot, Karat, and Carat

Like many words in English, *carrot, karat,* and *carat* have the same sound but are different in meaning. Such words are called *homonyms*. We all know that a *carrot* is a crunchy, orange vegetable, but what about *karat* and *carat*?

Karat indicates the purity of gold (Figure 2.B). The purest gold is 24 karats. Gold less than 24 karats is actually an alloy (mixture) of gold and another metal, usually copper or silver. For example, 14-karat gold contains 14 parts of gold (by weight) mixed with 10 parts of other metals. Obviously, gold with a higher number is more expensive. Gold is much softer and more malleable than a gold–silver alloy, so pure gold is easiest to work with, but the alloy forms are more durable.

Carat is a unit of weight used for precious gems such as diamonds, emeralds, and rubies. The size of a carat has varied through history, but early in the twentieth century, it was established at 200 milligrams (or 0.2 gram). In everyday terms, a 1-ounce diamond is roughly 142 carats. Both *karat* and *carat* derive from the Greek word *keration,* meaning "carob bean," which early Greeks used as a weight standard.

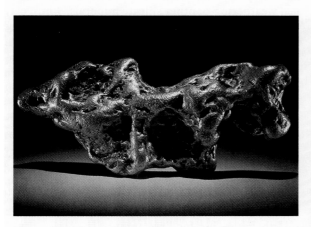

Figure 2.B The purity of gold, such as this 82-ounce nugget, is measured in units called karats. Pure gold is 24 karats. By comparison, the weight of gemstones is measured in carat weight. Each carat equals 0.2 gram. This diamond ring is 29.3 carats. (Courtesy of Smithsonian Institution)

Figure 2.23 Sample of the mineral orthoclase feldspar.

ions in its structure. The other group, called *plagioclase feldspar*, contains both sodium and calcium ions that freely substitute for one another depending on the environment during crystallization.

Orthoclase feldspar is usually light cream to salmon pink in color. The plagioclase feldspars, on the other hand, range in color from white to medium gray. However, color should not be used to distinguish these groups. The only sure way to distinguish the feldspars physically is to look for a multitude of fine parallel lines, called *striations*. Striations are found on some cleavage planes of plagioclase feldspar, but are not present on orthoclase feldspar (Figure 2.24).

Quartz is the only common silicate mineral consisting entirely of silicon and oxygen. As such, the term *silica* is applied to quartz, which has the chemical formula SiO_2. As the structure of quartz contains a ratio

← 2 cm →

Figure 2.24 These parallel lines, called striations, are a distinguishing characteristic of the plagioclase feldspars. (Photo by E. J. Tarbuck)

of two oxygen ions (O^{2-}) for every silicon ion (Si^{4+}), no other positive ions are needed to attain neutrality.

In quartz, a three-dimensional framework is developed through the complete sharing of oxygen by adjacent silicon atoms. Thus, all of the bonds in quartz are of the strong silicon–oxygen type. Consequently, quartz is hard, resistant to weathering, and does not have cleavage. When broken, quartz generally exhibits conchoidal fracture. In a pure form, quartz is clear and if allowed to solidify without interference, will form hexagonal crystals that develop pyramid-shaped ends (see Figure 2.11B, p. 41). However, like most other clear minerals, quartz is often colored by the inclusion of various ions (impurities) and forms without developing good crystal faces. The most common varieties of quartz are milky (white), smoky (gray), rose (pink), amethyst (purple), and rock crystal (clear) (Figure 2.25).

Clay is a term used to describe a variety of complex minerals that, like the micas, have a sheet structure. The clay minerals are generally very fine grained and can only be studied microscopically. Most clay minerals originate as products of the chemical weathering of other silicate minerals. Thus, clay minerals make up a large percentage of the surface material we call soil. Because of the importance of soil in agriculture, and because of its role as a supporting material for buildings, clay minerals are extremely important to humans.

One of the most common clay minerals is *kaolinite*, which is used in the manufacture of fine chinaware and in the production of high-gloss paper such as that used in this textbook. Further, some clay minerals absorb large amounts of water, which allows them to swell to several times their normal size. These clays have been used commercially in a variety of ingenious ways, including as an additive to thicken milkshakes in fast-food restaurants.

Figure 2.25 Quartz. Some minerals, such as quartz, occur in a variety of colors. These samples include crystal quartz (colorless), amethyst (purple quartz), citrine (yellow quartz), and smoky quartz (gray to black).

Table 2.4 Common Nonsilicate Mineral Groups

Group	Member	Formula	Economic Use
Oxides	Hematite	Fe_2O_3	Ore of iron, pigment
	Magnetite	Fe_3O_4	Ore of iron
	Corundum	Al_2O_3	Gemstone, abrasive
	Ice	H_2O	Solid form of water
	Chromite	$FeCr_2O_4$	Ore of chromium
	Ilmenite	$FeTiO_3$	Ore of titanium
Sulfides	Galena	PbS	Ore of lead
	Sphalerite	ZnS	Ore of zinc
	Pyrite	FeS_2	Sulfuric acid production
	Chalcopyrite	$CuFeS_2$	Ore of copper
	Bornite	Cu_5FeS_2	Ore of copper
	Cinnabar	HgS	Ore of mercury
Sulfates	Gypsum	$CaSO_4 \cdot 2H_2O$	Plaster
	Anhydrite	$CaSo_4$	Plaster
	Barite	$BaSO_4$	Drilling mud
Native elements	Gold	Au	Trade, jewelry
	Copper	Cu	Electrical conductor
	Diamond	C	Gemstone, abrasive
	Sulfur	S	Sulfa drugs, chemicals
	Graphite	C	Pencil lead, dry lubricant
	Silver	Ag	Jewelry, photography
	Platinum	Pt	Catalyst
Halides	Halite	$NaCl$	Common salt
	Fluorite	CaF_2	Used in steelmaking
	Sylvite	KCl	Fertilizer
Carbonates	Calcite	$CaCO_3$	Portland cement, lime
	Dolomite	$CaMg(CO_3)_2$	Portland cement, lime
	Malachite	$Cu_2(OH)_2CO_3$	Gemstone
	Azurite	$Cu_3(OH)_2(CO_3)_2$	Gemstone
Hydroxides	Limonite	$FeO(OH) \cdot nH_2O$	Ore of iron, pigments
	Bauxite	$Al(OH)_3 \cdot nH_2O$	Ore of aluminum
Phosphates	Apatite	$Ca_5(F,Cl,OH)(PO_4)_3$	Fertilizer
	Turquoise	$CuAl_6(PO_4)_4(OH)_8$	Gemstone

Important Nonsilicate Minerals

Other mineral groups can be considered scarce when compared to the silicates, although many are important economically. Table 2.4 lists examples of several nonsilicate mineral groups of economic value. A discussion of a few of the more common nonsilicate, rock-forming minerals follows.

The carbonate minerals are much simpler structurally than are the silicates. This mineral group is composed of the carbonate ion ($CO_2{}^{2-}$), and one or more kinds of positive ions. The two most common carbonate minerals are *calcite*, $CaCO_3$ (calcium carbonate), and *dolomite*, $CaMg(CO_3)_2$ (calcium/magnesium carbonate). Because these minerals are similar both physically and chemically, they are difficult to distinguish from one another. Both have a vitreous luster, a hardness between 3 and 4, and nearly perfect rhombic cleavage (see Figure 2.13, right side, p. 42). They can, however, be distinguished by using dilute hydrochloric acid. Calcite reacts vigorously with this acid, whereas dolomite reacts much more slowly. Calcite and dolomite are usually found together as the primary constituents in the sedimentary rocks limestone and dolostone. When calcite is the dominant mineral, the rock is called *limestone*, whereas *dolostone* results from a predominance of dolomite. Limestone has numerous economic uses, including uses as road aggregate, as building stone, and as the main ingredient in portland cement.

Two other nonsilicate minerals frequently found in sedimentary rocks are *halite* and *gypsum*. Both minerals are commonly found in thick layers, which are the last vestiges of ancient seas that have long since evaporated (Figure 2.26). Like limestone, both are important nonmetallic resources. Halite is the mineral name for common table salt ($NaCl$). Gypsum ($CaSO_4 \bullet 2H_2O$), which is calcium sulfate with water bound into the structure, is the mineral of which plaster and other similar building materials are composed.

Figure 2.26 Thick beds of halite (salt) at an underground mine near Grand Saline, Texas. (Photo by D. A. Humphreys)

In addition, a number of other minerals are prized for their economic value (see Table 2.4 and Box 2.3). Included in this group are the ores of metals, such as hematite (iron), sphalerite (zinc), and galena (lead); the native (free-occurring, not in compounds) elements, including gold, silver, carbon (diamonds); and a host of others, such as fluorite, corundum, and uraninite (a uranium source).*

*For more on the economic significance of these and other minerals, see Chapter 21.

 Box 2.3

Gemstones

Precious stones have been prized since antiquity. But misinformation abounds about gems and their mineral makeup. This stems partly from the ancient practice of grouping precious stones by color rather than mineral makeup. For example, *rubies* and red *spinels* are very similar in color, but they are completely different minerals. Classifying by color led to the more common spinels being passed off to royalty as rubies. Even today, with modern identification techniques, common *yellow quartz* is sometimes sold as the more valuable gemstone *topaz*.

Naming Gemstones

Most precious stones have common names that are different from their parent mineral. For example, *sapphire* is one of two names given to the mineral *corundum*. Minute amounts of foreign elements can produce vivid

Figure 2.C Australian sapphires showing variation in cuts and colors. (Photo by Fred Ward, Black Star)

sapphires of nearly every color (Figure 2.C). Traces of titanium and iron in corundum produce the most prized blue sapphires. When the mineral corundum contains a sufficient quantity of chromium, it exhibits a brilliant red color, and the gem is called *ruby*. Further, if a specimen is not suitable as a gem, it simply goes by the mineral name *corundum*. Because of its hardness, corundum that is not of gem quality is often crushed and sold as an abrasive.

To summarize, when corundum exhibits a red hue, it is called *ruby*, but if it exhibits any other color, the gem is called *sapphire*. Whereas corundum is the base mineral for two gems, quartz is the parent of more than a dozen gems. Table 2.B lists some well-known gemstones and their parent minerals.

What Constitutes a Gemstone?

When found in their natural state, most gemstones are dull and would be passed over by most people as "just another rock." Gems must be cut and polished by experienced professionals before their true beauty is displayed (Figure 2.C). Only those mineral specimens that are of such quality that they can command a price in excess of the cost of processing are considered gemstones.

Gemstones can be divided into two categories: precious and semiprecious. A *precious* gem has beauty, durability, size, and rarity, whereas a *semiprecious* gem generally has only one or two of these qualities. The gems traditionally in highest esteem are diamonds, rubies, sapphires, emeralds, and some varieties of opal (Table 2.B). All other gemstones are classified as semiprecious. However, large, high-quality specimens of semiprecious stones often command a very high price.

Table 2.B Important Gemstones

Gem	Mineral Name	Prized Hues
Precious		
Diamond	Diamond	Colorless, yellows
Emerald	Beryl	Greens
Opal	Opal	Brilliant hues
Ruby	Corundum	Reds
Sapphire	Corundum	Blues
Semiprecious		
Alexandrite	Chrysoberyl	Variable
Amethyst	Quartz	Purples
Cat's-eye	Chrysoberyl	Yellows
Chalcedony	Quartz (agate)	Banded
Citrine	Quartz	Yellows
Garnet	Garnet	Reds, greens
Jade	Jadeite or nephrite	Greens
Moonstone	Feldspar	Transparent blues
Peridot	Olivine	Olive greens
Smoky quartz	Quartz	Browns
Spinel	Spinel	Reds
Topaz	Topaz	Purples, reds
Tourmaline	Tourmaline	Reds, blue-greens
Turquoise	Turquoise	Blues
Zircon	Zircon	Reds

Today, translucent stones with evenly tinted colors are preferred. The most favored hues are red, blue, green, purple, rose, and yellow. The most prized stones are pigeon-blood rubies, blue sapphires, grass-green emeralds, and canary-yellow diamonds. Colorless gems are generally less than desirable except for diamonds that display "flashes of color" known as *brilliance*.

The durability of a gem depends on its hardness; that is, its resistance to abrasion by objects normally encountered in everyday living. For good durability, gems should be as hard or harder than quartz as defined by the Mohs scale of hardness. One notable exception is opal, which is comparatively soft (hardness 5 to 6.5) and brittle. Opal's esteem comes from its "fire," which is a display of a variety of brilliant colors including greens, blues, and reds.

It seems to be human nature to treasure that which is rare. In the case of gemstones, large, high-quality specimens are much rarer than smaller stones. Thus, large rubies, diamonds, and emeralds, which are rare in addition to being beautiful and durable, command the very highest prices.

Chapter Summary

- A *mineral* is a naturally occurring inorganic solid that possesses a definite chemical structure which gives it a unique set of physical properties. Most *rocks* are aggregates composed of two or more minerals.

- The building blocks of minerals are *elements*. An *atom* is the smallest particle of matter that still retains the characteristics of an element. Each atom has a *nucleus*, which contains *protons* (particles with positive electrical charges) and *neutrons* (particles with neutral electrical charges). Orbiting the nucleus of an atom in regions called *energy levels*, or *shells*, are *electrons*, which have negative electrical charges. The number of protons in an atom's nucleus determines its *atomic number* and the name of the element. An element is a large collection of electrically neutral atoms, all having the same atomic number.

- Atoms combine with each other to form more complex substances called *compounds*. Atoms bond together by either gaining, losing, or sharing electrons with other atoms. In *ionic bonding*, one or more electrons are transferred from one atom to another, giving the atoms a net positive or negative charge. The resulting electrically charged atoms are called *ions*. Ionic compounds consist of oppositely charged ions assembled in a regular, crystalline structure that allows for the maximum attraction of ions, given their sizes. Another type of bond, the *covalent bond*, is produced when atoms share electrons.

- *Isotopes* are variants of the same element, but with a different *mass number* (the total number of neutrons plus protons found in an atom's nucleus). Some isotopes are unstable and disintegrate naturally through a process called *radioactivity*.

- The properties of minerals include *crystal form, luster, color, streak, hardness, cleavage, fracture,* and *specific gravity*. In addition, a number of special physical and chemical properties (*taste, smell, elasticity, malleability, feel, magnetism, double refraction,* and *chemical reaction to hydrochloric acid*) are useful in identifying certain minerals. Each mineral has a unique set of properties that can be used for identification.

- Of the nearly 4000 minerals, no more than a few dozen make up most of the rocks of Earth's crust and, as such, are classified as *rock-forming minerals*.

Eight elements (oxygen, silicon, aluminum, iron, calcium, sodium, potassium, and magnesium) make up the bulk of these minerals and represent over 98 percent (by weight) of Earth's continental crust.

- The most common mineral group is the *silicates*. All silicate minerals have the negatively charged *silicon-oxygen tetrahedron* as their fundamental building block. In some silicate minerals the tetrahedra are joined in chains (the pyroxene and amphibole groups); in others, the tetrahedra are arranged into sheets (the micas, biotite and muscovite), or three-dimensional networks (the feldspars and quartz). The tetrahedra and various silicate structures are often bonded together by the positive ions of iron, magnesium, potassium, sodium, aluminum, and calcium. Each silicate mineral has a structure and a chemical composition that indicates the conditions under which it formed.

- The *nonsilicate* mineral groups, which contain several economically important minerals, include the *oxides* (e.g., the mineral hematite, mined for iron), *sulfides* (e.g., the mineral sphalerite, mined for zinc; and the mineral galena, mined for lead), *sulfates, halides,* and *native elements* (e.g., gold and silver). The more common nonsilicate rock-forming minerals include the *carbonate minerals*, calcite and dolomite. Two other nonsilicate minerals frequently found in sedimentary rocks are halite and gypsum.

Review Questions

1. Define the term *rock*.
2. List the three main particles of an atom and explain how they differ from one another.
3. If the number of electrons in a neutral atom is 35 and its mass number is 80, calculate the following:
   (a) The number of protons.
   (b) The atomic number.
   (c) The number of neutrons.
4. What is the significance of valence electrons?
5. Briefly distinguish between ionic and covalent bonding.
6. What occurs in an atom to produce an ion?
7. What is an isotope?
8. Although all minerals have an orderly internal arrangement of atoms (crystalline structure), most mineral samples do not exhibit their crystal form. Why?
9. Why might it be difficult to identify a mineral by its color?
10. If you found a glassy-appearing mineral while rock hunting and had hopes that it was a diamond, what simple test might help you make a determination?
11. Explain the use of corundum as given in Table 2.4 (p. 53) in terms of the Mohs hardness scale.
12. Gold has a specific gravity of almost 20. If a 25-liter pail of water weighs 25 kilograms, how much would a 25-liter pail of gold weigh?

13. Explain the difference between the terms *silicon* and *silicate*.

14. What do ferromagnesian minerals have in common? List examples of ferromagnesian minerals.

15. What do muscovite and biotite have in common? How do they differ?

16. Should color be used to distinguish between orthoclase and plagioclase feldspar? What is the best means of distinguishing between these two types of feldspar?

17. Each of the following statements describes a silicate mineral or mineral group. In each case, provide the appropriate name:

    (a) The most common member of the amphibole group.

    (b) The most common nonferromagnesian member of the mica family.

    (c) The only silicate mineral made entirely of silicon and oxygen.

    (d) A high-temperature silicate with a name that is based on its color.

    (e) Characterized by striations.

    (f) Originates as a product of chemical weathering.

18. What simple test can be used to distinguish calcite from dolomite?

Key Terms

atom (p. 33)
atomic number (p. 35)
atomic weight (p. 38)
cleavage (p. 42)
color (p. 41)
compound (p. 35)
covalent bond (p. 37)
crystal form (p. 40)
dark or ferromagnesian silicate (p. 49)
electron (p. 33)

element (p. 35)
energy-levels, or shells (p. 35)
fracture (p. 42)
hardness (p. 41)
ion (p. 36)
ionic bond (p. 36)
isotope (p. 38)
light or nonferromagnesian silicate (p. 50)
luster (p. 41)

mass number (p. 38)
metallic bond (p. 38)
mineral (p. 32)
mineralogy (p. 32)
Mohs scale (p. 41)
neutron (p. 33)
nucleus (p. 33)
octet rule (p. 35)
polymorph (p. 39)
proton (p. 33)
radioactive decay (p. 38)

rock (p. 32)
silicate mineral (p. 43)
silicon-oxygen tetrahedron (p. 44)
specific gravity (p. 43)
streak (p. 41)
valence electron (p. 35)

Web Resources

 The *Earth* Home Page provides on-line resources for this chapter on the World Wide Web. You will find review exercises, specific updates for items in the chapter, suggested reading, and links to interesting related pathways on the Internet. Visit the *Earth* Home Page at **http://www.prenhall.com/tarbuck**.

CHAPTER 3

Igneous Rocks

Left Devil's Tower, Wyoming. — *Photo by Carr Clifton*

Figure 3.1 Recent eruption of Hawaii's Kilauea Volcano. (Photo by Douglas Peebles/Westlight)

Igneous rocks make up the bulk of Earth's crust. In fact, with the exception of the liquid outer core, the remaining solid portion of our planet is basically a huge igneous rock partially mantled with a thin veneer of sedimentary rocks. Consequently, a basic knowledge of igneous rocks is essential to our understanding of the structure, composition, and internal workings of our planet.

In our discussion of the rock cycle, it was pointed out that **igneous rocks** (from the Latin *ignis*, or "fire") form as molten rock cools and solidifies. Abundant evidence supports the fact that the parent material for igneous rocks, called *magma*, is formed by a process called *partial melting*. Partial melting occurs at various levels within Earth's crust and upper mantle at depths that may exceed 200 kilometers. We will explore the origin of magma later in this chapter.

Once formed, a magma body buoyantly rises toward the surface because it is less dense than the surrounding rocks. On occasion molten rock breaks through, producing a spectacular volcanic eruption. Magma that reaches Earth's surface is called **lava**. The lava fountain shown in Figure 3.1 was produced when escaping gases propelled molten rock from a magma chamber. Sometimes blockage of a vent, coupled with a buildup of gas pressure, can produce catastrophic explosions. However, not all eruptions are violent; some volcanoes generate quiet outpourings of very fluid lava.

Igneous rocks that form when molten rock solidifies *at the surface* are classified as **extrusive**, or **volcanic**. Extrusive igneous rocks are abundant in western portions of the Americas, as well as on all other continents. In addition, many oceanic islands are composed almost entirely of extrusive igneous rocks.

Magma that loses it mobility before reaching the surface eventually crystallizes at depth. Igneous rocks that *form at depth* are termed **intrusive**, or **plutonic** (after Pluto, the god of the lower world in classical mythology). Intrusive igneous rocks would never be observed if portions of the crust were not uplifted and the overlying rocks stripped away by erosion.

Crystallization of Magma

Magma is molten rock that usually contains some suspended crystals and dissolved gases, principally water vapor, that are confined within the magma by the pressure of the surrounding rocks. The bulk of magma is composed of mobile ions of the eight most abundant elements of Earth's crust. These elements, which are also the major constituents of the silicate minerals, include silicon, oxygen, aluminum, potassium, calcium, sodium, iron, and magnesium. As magma cools, the random movements of these ions slow and they begin to arrange themselves into orderly crystalline structures. This process, called **crystallization**, generates mineral grains that are said to *precipitate* from the melt.

Before we examine how magma crystallizes, let us first examine how a simple crystalline solid melts. In any crystalline solid, the ions are arranged in a closely packed regular pattern. However, they are not without some motion. They exhibit a sort of restricted vibration about fixed points. As the temperature rises, the ions vibrate more and more rapidly and consequently collide with ever-increasing vigor with their neighbors. Thus, heating causes the ions to occupy more space causing the solid to expand and resulting in greater distance between ions.

When the ions become far enough apart and are vibrating rapidly enough to overcome the force of the chemical bonds, the solid begins to melt. At this stage, the ions are able to slide past one another, and their orderly crystalline structure disintegrates. Thus, melting converts a solid consisting of tight, uniformly packed ions into a liquid composed of unordered ions moving randomly about.

In the process of crystallization, cooling reverses the events of melting. As the temperature of the liquid drops, the ions pack closer together and begin to lose their freedom of movement. When cooled sufficiently, the forces of the chemical bonds will again confine the atoms to an orderly crystalline arrangement.

The crystallization of magma is much more complex than just described. Whereas a single compound, such as water, crystallizes at a specific temperature, solidification of magma with its diverse chemistry often spans a temperature range of 200°C. In addition, a magma may migrate to a new setting before crystallization is complete (or new magma may be introduced into the magma chamber), further complicating the process.

When magma cools, it is generally the silicon and oxygen atoms that link together first to form silicon–oxygen tetrahedra, the basic structure of the silicate minerals. As a magma continues to lose heat to its surroundings, the tetrahedra join with each other and with other ions to form embryonic crystal nuclei. Slowly, each nucleus grows as ions lose their mobility and join the crystalline network.

The earliest formed minerals have space to grow and tend to have better developed crystal faces than do the later ones that fill the remaining space. Eventually, all of the magma is transformed into a solid mass of interlocking silicate minerals that we call an *igneous rock* (Figure 3.2).

Because no two magmas are identical in composition, and because magma crystallizes in diverse environments, a great variety of igneous rocks exist. Nevertheless, it is possible to classify igneous rocks based on their mineral composition and the conditions under which they formed. Their environment during crystallization can be roughly inferred from the size and the arrangement of the mineral grains, a property called *texture*. Consequently, *igneous rocks are most often classified by their texture and mineral composition*. We will consider these two rock characteristics in the following sections.

 Igneous Textures

The term **texture** when applied to an igneous rock, is used to describe the overall appearance of the rock based on the size, shape, and arrangement of its interlocking crystals (Figure 3.3). Texture is an important characteristic because it reveals a great deal about the environment in which the rock formed. This fact

A.

B.

Figure 3.2 A. Close-up of interlocking crystals in a coarse-grained igneous rock. The largest crystals are about 1 centimeter in length. **B.** Photomicrograph of interlocking crystals in a coarse-grained igneous rock. (Photos by E.J. Tarbuck)

allows geologists to make inferences about a rock's origin while working in the field where sophisticated equipment is not available.

Factors Affecting Crystal Size

Three factors contribute to the textures of igneous rocks: (1) *the rate at which magma cools*; (2) *the amount of silica present*; and (3) *the amount of dissolved gases in the magma*. Of these, the rate of cooling is perhaps the most significant.

As a magma body loses heat to its surroundings, the mobility of its ions decreases. A very large magma body located at great depth will cool over a period of perhaps tens or hundreds of thousands of years. Initially, relatively few crystal nuclei form. Slow cooling permits ions to migrate over great distances where they may join with one of the few existing crystalline structures. Consequently, slow cooling promotes the growth of fewer but larger crystals.

On the other hand, when cooling occurs more rapidly—for example, in a thin lava flow—the ions quickly lose their mobility and readily combine. This results in the development of numerous embryonic nuclei, all of which compete for the available ions. The result is a solid mass of small intergrown crystals.

When molten material is quenched quickly, there may not be sufficient time for the ions to arrange into a crystalline network. Rocks that consist of unordered ions are referred to as **glass**.

A. Aphanitic

B. Phaneritic

C. Porphyritic

D. Glassy

Figure 3.3 Igneous rock textures. **A.** Aphanitic (fine-grained). **B.** Phaneritic (coarse-grained). **C.** Porphyritic (large crystals embedded in a matrix). **D.** Glassy (cooled too rapidly to form crystals). (Photos by E.J. Tarbuck)

Types of Igneous Textures

As we saw, the effect of cooling on rock textures is fairly straightforward. Slow cooling promotes the growth of large crystals, whereas rapid cooling tends to generate smaller crystals. We will consider the other two factors affecting crystal growth as we examine the major textural types.

Aphanitic (fine-grained) Texture. Igneous rocks that form at the surface or as small masses within the upper crust where cooling is relatively rapid possess a very fine-grained texture termed **aphanitic**. By definition, the crystals that make up aphanitic rocks are too small for individual minerals to be distinguished with the unaided eye (Figure 3.3A). Because mineral identification is not possible, we commonly characterize fine-grained rocks as being light, intermediate, or dark in color. Using this system of grouping, light-colored aphanitic rocks are those containing primarily light-colored nonferromagnesian silicate materials, and so forth (see the section titled "Common Silicate Minerals" in Chapter 2).

Commonly seen in many aphanitic rocks are the voids left by gas bubbles that escape as magma solidifies. These spherical or elongated openings are called **vesicles** and are most abundant in the upper portion of lava flows (Figure 3.4). It is in the upper zone of a lava flow that cooling occurs rapidly enough to "freeze" the lava, thereby preserving the openings produced by the expanding gas bubbles.

Phaneritic (coarse-grained) Texture. When large masses of magma slowly solidify far below the surface, they form igneous rocks that exhibit a coarse-grained texture described as **phaneritic**. These coarse-grained rocks consist of a mass of intergrown crystals, which are roughly equal in size and large enough so that the individual minerals can be identified with the unaided eye (Figure 3.3B). Because phaneritic rocks form deep within Earth's crust, their exposure at Earth's surface results only after erosion removes the overlying rocks that once surrounded the magma chamber.

Porphyritic Texture. A large mass of magma located at depth may require tens to hundreds of thousands of years to solidify. Because different minerals crystallize at different temperatures (as well as at differing rates) it is possible for some crystals to become quite large before others even begin to form. If magma containing some large crystals should change environments—for example, by erupting at the surface—the molten portion of the lava would cool quickly. The resulting rock, which has large crystals embedded in a matrix of smaller crystals, is said to have a **porphyritic texture** (Figure 3.3C). The large crystals in such a rock are referred to as **phenocrysts**, while the matrix

Figure 3.4 Scoria is a volcanic rock that exhibits a vesicular texture. Vesicles are small holes left by escaping gas bubbles. (Photo by E.J. Tarbuck)

of smaller crystals is called **groundmass**. A rock with such a texture is termed a **porphyry**.

Glassy Texture. During some volcanic eruptions, molten rock is ejected into the atmosphere where it is quenched quickly. Rapid cooling of this type may generate rocks having a **glassy texture**. As we indicated earlier, glass results when unordered ions are "frozen" before they are able to unite into an orderly crystalline structure. *Obsidian*, a common type of natural glass, is similar in appearance to a dark chunk of manufactured glass (Figure 3.3D).

Layers of obsidian (called obsidian flows) several tens of feet thick occur in some places (Figure 3.5). Thus, rapid cooling is not the only mechanism by which a glassy texture can form. As a general rule, magmas with a high silica content tend to form long, chainlike structures before crystallization is complete. These structures in turn impede ionic transport and increase the magma's viscosity. **Viscosity** is a measure of a fluid's resistance to flow.

Granitic magma, which is rich in silica, may be extruded as an extremely viscous mass that eventually solidifies as a glass. By contrast, basaltic magma, which is low in silica, forms very fluid lavas that upon cooling usually generate fine-grained crystalline rocks. However, the surface of basaltic lava may be quenched rapidly enough to form a thin, glassy skin. Moreover, Hawaiian volcanoes sometimes generate lava fountains, which spray basaltic lava tens of meters into the air. Such activity can produce strands of volcanic glass called *Pele's hair*, after the Hawaiian goddess of volcanoes.

Figure 3.5 This obsidian flow was extruded from a vent along the south wall of Newberry Caldera, Oregon. Note the road for scale. (Photo by E.J. Tarbuck)

Pyroclastic Texture. Some igneous rocks are formed from the consolidation of individual rock fragments that are ejected during a violent eruption. The ejected particles might be very fine ash, molten blobs, or large angular blocks torn from the walls of the vent during the eruption. Igneous rocks composed of these rock fragments are said to have a **pyroclastic texture** (Figure 3.6).

A common type of pyroclastic rock is composed of thin glass strands that remained hot enough during their flight to fuse together upon impact. Other pyroclastic rocks are composed of fragments that solidified

Figure 3.6 Pyroclastic texture. This volcanic rock consists of angular rock fragments embedded in a light-colored matrix of ash. (Photo by E.J. Tarbuck)

before impact and became cemented together at some later time. Because pyroclastic rocks are made of individual particles or fragments rather than interlocking crystals, their textures often appear to be more similar to sedimentary rocks than to other igneous rocks.

Pegmatitic Texture. Under special conditions, exceptionally coarse-grained igneous rocks, called **pegmatites**, may form. These rocks, which are composed of interlocking crystals all larger than a centimeter in diameter, are said to have a **pegamatitic texture**.

Most pegmatites form in veins near the margins of magma bodies during the last stage of crystallization. Because water and other volatiles do not crystallize within a magma body, these fluids make up a high percentage of the final melt (liquid portion of a magma excluding the solid components). Crystallization in such a fluid-rich environment, where ion migration is enhanced, results in the formation of the large crystals found in pegmatites. Pegamites are discussed in more detail in Box 3.1.

 Igneous Compositions

Igneous rocks are mainly composed of silicate minerals. Furthermore, the mineral makeup of a particular igneous rock is ultimately determined by the chemical composition of the magma from which it crystallizes. Recall that magma is composed largely of the eight elements that are the major constituents of the silicate minerals. Chemical analysis shows that silicon and oxygen (usually expressed as the silica [SiO_2] content of a magma) are by far the most abundant constituents of igneous rocks. These two elements, plus ions of aluminum (Al), calcium (Ca), sodium (Na), potassium (K), magnesium (Mg) and iron (Fe) make up roughly 98 percent by weight of most magmas. In addition, magma contains small amounts of many other elements including titanium and manganese, and trace amounts of much rarer elements such as gold, silver, and uranium.

As a magma cools and solidifies, these elements combine to form two major groups of silicate minerals. The *dark* (or *ferromagnesian*) *silicates* are minerals rich in iron and/or magnesium and typically low in silica. Olivine, pyroxene, amphibole, and biotite are the common ferromagnesian constituents of Earth's crust. By contrast, the light silicates contain greater amounts of potassium, sodium, and calcium rather than iron and magnesium. As a group these minerals are richer in silica than are the dark silicates. The light silicates include quartz, muscovite, and the most abundant mineral group, the feldspars. Igneous rocks can

Box 3.1

Pegmatites

Pegmatite is a name given to an igneous rock composed of abnormally large crystals (Figure 3.A). How large is *large*? Crystals in most pegmatite samples are more then 1 centimeter in diameter. In some specimens, crystals that are 1 meter or more across are common. Gigantic hexagonal crystals of muscovite measuring a few meters across have been found in Ontario, Canada. In the Black Hills of South Dakota, crystals as large as telephone poles of the lithium-bearing mineral spodumene have been mined. The largest of these was more than 12 meters (40 feet) long. Further, feldspar masses the size of houses have been quarried from a pegmatite located in North Carolina.

Most pegmatites have the composition of granite and contain unusually large crystals of quartz, feldspar, and muscovite. In addition to being an important source of excellent mineral specimens, large granitic pegmatites have been mined for their mineral constituents. Feldspar, for example, is used in the production of ceramics, and muscovite is used for electrical insulation and glitter. Although granitic pegmatites are most common, pegmatites with chemical makeups similar to those of other igneous rocks are also known. Further, pegmatites may contain significant amounts of some of the least abundant elements. Thus, in addition to the common silicates, pegmatites

Figure 3.A This pegmatite in the Black Hills of South Dakota was mined for its large crystals of spodumene, an important source of lithium. Arrows are pointing to impressions left by crystals. (Photo by James Kirchner)

with minerals containing the elements lithium, cesium, uranium, and the rare earths are known. In addition, semiprecious gems such as beryl, topaz, and tourmaline are occasionally found.

Most pegmatite bodies are located within large igneous masses or as veins that cut into the host rock that surrounds a pluton. Pegmatites form in the late stages of magma crystallization. Because water and other volatile

substances do not crystallize along with the bulk of the magma body, these fluids make up an unusually high percentage of the melt during the final phase of solidification. Crystallization in a fluid-rich environment where ion migration is enhanced results in the formation of crystals of abnormally large size. It is in this environment that the unusually large crystals of most pegmatites are thought to form.

be composed predominantly of dark or light silicates, or of members from both groups combined in various combinations and amounts.

Despite their great compositional diversity, igneous rocks can be divided into broad groups according to their proportions of light and dark minerals. Near one end of the continuum are rocks composed mainly of the light-colored silicates—quartz and feldspar. These so-called *granitic rocks* contain 70 percent silica and are major constituents of the continental crust. Rocks that contain abundant, dark (ferromagnesian) minerals and about 50 percent silica are said to have a *basaltic composition*. Basalt makes up the ocean floor as well as many of the volcanic islands located within the ocean basins. Basalt is also found on the continents, whereas granite is almost totally absent from the ocean basins. Igneous rocks with compositions between these major groups, as well as those totally devoid of either light or dark minerals, are also known.

Because such a large variety of igneous rocks exists, it is logical to assume that an equally large variety of magmas must also exist. However, geologists found that some volcanoes extrude lavas, or pyroclastic materials, exhibiting quite different compositions, particularly if an extensive period of time has separated the eruptions (Figure 3.7). Data of this type led them to look into the possibility that a single magma might have been the parent to a variety of igneous rocks. To explore this idea a pioneering investigation into the crystallization of magma was carried out by N.L. Bowen in the first quarter of the twentieth century.

Bowen's Reaction Series

In one laboratory study, Bowen demonstrated that as a basaltic magma cools, minerals tend to crystallize in a systematic fashion based on their melting points. As shown in Figure 3.8, the first mineral to crystallize from a basaltic magma is the ferromagnesian mineral olivine. Further cooling generates calcium-rich plagioclase feldspar as well as pyroxene, and so forth down the diagram.

During the crystallization process, the composition of the *melt* continually changes. For example, at the stage when about a third of the magma has solidified, the melt will be nearly depleted of iron, magnesium, and calcium because these elements are constituents of the earliest-formed minerals. The removal of these elements from the melt will cause it to become enriched in sodium, potassium, and aluminum. Further, because the original basaltic magma contained about 50 percent silica (SiO_2) the crystallization of the earliest-formed mineral, olivine, which is only about 40 percent silica, leaves the remaining

Figure 3.7 Ash and pumice ejected during a large eruption of Mt. Mazama (Crater Lake). Notice the gradation from light-colored, silica-rich ash near the base to dark-colored rocks at the top. It is likely that prior to this eruption the magma began to segregate as the less-dense, silica-rich magma migrated toward the top of the magma chamber. The zonation seen in the rocks resulted because a sustained eruption tapped deeper and deeper levels of the magma chamber. Thus, this rock sequence is an inverted representation of the compositional zonation in the magma body; that is, the magma from the top of the chamber erupted first and is found at the base of these ash deposits and vice versa. (Photo by E.J. Tarbuck)

melt richer in SiO_2. Thus, the silica component of the melt also becomes enriched as the magma evolves.

Bowen also demonstrated that if the solid components of a magma remain in contact with the remaining melt, they will chemically react and evolve into the next mineral in the sequence shown in Figure 3.8. For this reason, this arrangement of minerals became known as **Bowen's reaction series**. (As we shall see, in some natural settings the earliest-formed minerals are often separated from the melt, thus halting any further chemical reaction.)

Discontinuous Reaction Series. The upper left branch of Bowen's reaction series, shows that as a magma cools,

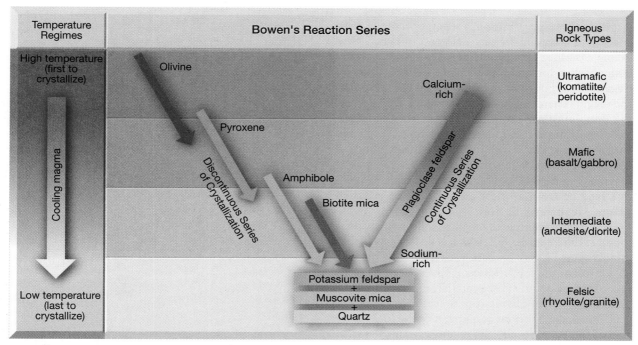

Figure 3.8 Bowen's reaction series shows the sequence in which minerals crystallize from a magma. Compare this figure to the mineral composition of the rock groups in Table 3.1 Note that each rock group consists of minerals that crystallize in the same temperature range.

olivine will react with the remaining melt to form the mineral pyroxene (Figure 3.8). In this reaction, olivine, which is composed of individual silica tetrahedra, incorporates more silica into its structure, thereby linking its tetrahedra into single-chain structures of the mineral pyroxene. As the magma body cools further, the pyroxene crystals will in turn react with the melt to generate the double-chain structure of amphibole. This reaction will continue until the last mineral in this series, biotite, is formed. Ordinarily, these reactions do not run to completion, so that various amounts of each of these minerals may exist at any given time and some minerals such as biotite may never form.

This branch of Bowen's reaction series is called a *discontinuous reaction series* because at each step a different silicate structure emerges. Olivine, the first mineral in the sequence to form, is composed of isolated tetrahedra, whereas pyroxene is composed of single chains, amphibole double chains, and biotite consists of sheet structures.

Continuous Reaction Series. The right branch of the reaction series, called the *continuous reaction series*, illustrates that calcium-rich plagioclase feldspar crystals react with the sodium ions in the melt to become progressively more sodium-rich (Figure 3.8). Here the sodium ions diffuse into the feldspar crystals and displace the calcium ions in the crystal lattice. Oftentimes, the rate of cooling occurs rapidly enough

to prohibit a complete replacement of the calcium ions by sodium ions. In these instances, the feldspar crystals will have calcium-rich interiors surrounded by zones that are progressively richer in sodium.

During the last stage of crystallization, after much of the magma has solidified, potassium feldspar forms. (Muscovite will form in pegmatites and other plutonic igneous rocks that crystalize at considerable depths.) Finally, if the remaining melt has excess silica, the mineral quartz will precipitate.

Bowen's reaction series illustrates the sequence in which minerals crystallize from a basaltic magma under laboratory conditions. Evidence that this crystallization model approximates what can happen in nature comes from the analysis of igneous rocks. In particular, we find that minerals which form in the same general temperature regime in Bowen's reaction series are found together in the same igneous rocks. For example, notice in Figure 3.8 that the minerals quartz, potassium feldspar, and muscovite, which are located in the same region of Bowen's diagram, are typically found together as major constituents of the plutonic igneous rock granite.

Magmatic Differentiation

Bowen demonstrated that minerals crystallize from magma in a systematic fashion. But how does Bowen's reaction series account for the great diversity of igneous rocks? It has been shown that, at one or more stages during crystallization, a separation of the solid and liquid components of a magma can occur. One example is called **crystal settling**. This process occurs if the earlier-formed minerals are denser (heavier) than the liquid portion and sink toward the bottom of the magma chamber, as shown in Figure 3.9A. When the remaining melt solidifies (either in place or in another location if it migrates into fractures in the surrounding rocks) it will form a rock with a chemical composition much different from the parent magma (Figure 3.9B). The formation of more than one magma from a single parent magma is called **magmatic differentiation**.

A classic example of magmatic differentiation is found in the Palisades Sill, which is a 300-meter-thick tabular mass of dark igneous rock exposed along the west bank of the lower Hudson River. Because of its great thickness and subsequent slow rate of solidification, crystals of olivine (the first mineral to form) sank and make up about 25 percent of the lower portion of the Palisades Sill. By contrast, near the top of this igneous body, where the last melt crystallized, olivine represents only 1 percent of the rock mass.[*]

At any stage in the evolution of a magma, the solid and liquid components can separate into two chemically distinct units. Further, continued magmatic differentiation within the secondary melt will generate additional chemically distinct fractions. Consequently, magmatic differentiation can produce several chemically diverse units and ultimately a variety of igneous rocks (see Figure 3.7).

Assimilation and Magma Mixing

Bowen successfully demonstrated that, through magmatic differentiation, a parent magma can generate several mineralogically different igneous rocks. However, more recent work indicates that this process alone cannot account for the great diversity of igneous rocks.

Once a magma body forms, its composition can change through the incorporation of foreign material. For example, as magma migrates upward it may incorporate some of the surrounding host rock, a process called **assimilation** (Figure 3.10). This process may operate in a near-surface environment where rocks are brittle. As the magma pushes upward, stress causes numerous cracks in the overlying rock. The force of the injected magma is often strong enough to dislodge blocks of "foreign" rock and incorporate them into

---

*Recent studies indicate that this igneous body was produced by multiple injections of magma and represents more than just a simple case of crystal settling.

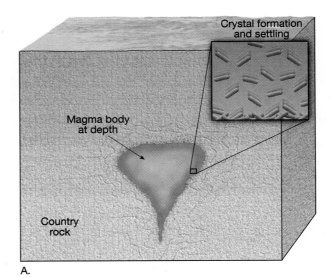

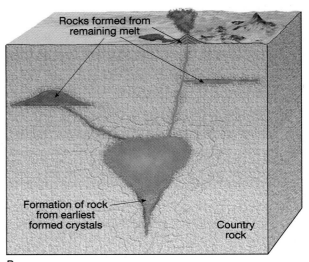

Figure 3.9 Separation of minerals by crystal settling. **A.** Illustration of how the earliest-formed minerals can be separated from a magma by settling. **B.** The remaining melt could migrate to a number of different locations and, upon further crystallization, generate rocks having a composition much different from that of the parent magma.

the magma body. In other environments, the magma may be hot enough to simply melt and assimilate some of the surrounding host rock.

Another means by which the composition of a magma body is altered is called **magma mixing**. This process occurs whenever one magma body intrudes another (Figure 3.10). Once combined, the two magmas generate a fluid with a different composition. Magma mixing may occur during ascent, as a more buoyant magma body overtakes a mass of magma that is rising more slowly.

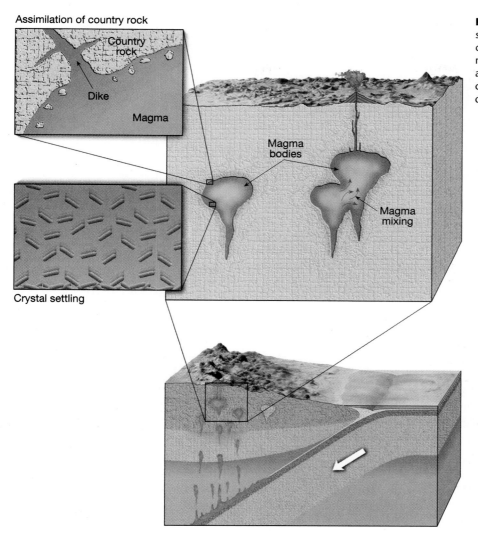

Figure 3.10 This illustration shows three ways that the composition of a magma body may be altered: magma mixing; assimilation of host rock; and crystal settling (magmatic differentiation).

Assimilation of country rock

Country rock

Dike

Magma

Crystal settling

Magma bodies

Magma mixing

 ## Naming Igneous Rocks

As we stated previously, igneous rocks are most often classified, or grouped, on the basis of their texture and mineral composition (Figure 3.11). The various igneous textures result mainly from different cooling histories, whereas the mineral composition of an igneous rock is the consequence of the chemical makeup of its parent magma (see Box 3.2). As we learned from Bowen's work, minerals that crystallize under similar conditions are often found together comprising the same igneous rock. Hence, the mineral composition categories used in the classification of igneous rocks closely correspond to Bowen's reaction series (compare Figure 3.8 and 3.11).

Igneous Rock Types

The first minerals to crystallize—olivine, pyroxene, and calcium-rich plagioclase—are high in iron, magnesium, or calcium, and low in silicon. As indicated earlier, basalt is a common rock with this mineral makeup—therefore, the term **basaltic** if often used to describe any rock having a similar mineral composition. Moreover, because basaltic rocks contain a high percentage of ferromagnesian minerals, geologists may also refer to them as **mafic** rocks (from *ma*gnesium and *fe*rrum, the Latin name for iron). Because of their iron content, mafic rocks are typically darker and denser than other igneous rocks commonly found at Earth's surface.

Among the last minerals to crystallize are potassium feldspar and quartz, the primary components of

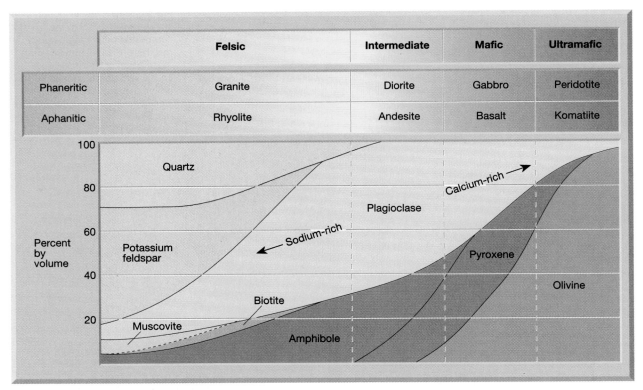

	Felsic	Intermediate	Mafic	Ultramafic
Phaneritic	Granite	Diorite	Gabbro	Peridotite
Aphanitic	Rhyolite	Andesite	Basalt	Komatiite

Figure 3.11 Mineralogy of the common igneous rocks. Phaneritic (coarse-grained) rocks are plutonic, solidifying deep underground. Aphanitic (fine-grained) rocks are volcanic, or solidify near Earth's surface. (After Dietrich)

the abundant rock *granite*. Igneous rocks in which these two minerals predominate are said to have a **granitic** composition. Geologists also refer to granitic rocks as being **felsic**, a term derived from *f*eldspar and *si*lica (quartz).

Intermediate igneous rocks contain minerals found near the middle of Bowen's reaction series. Amphibole and the intermediate plagioclase feldspars are the main constituents of this compositional group. We will refer to rocks that have a mineral makeup between that of granite and basalt as being **andesitic** (or **intermediate**), after the common volcanic rock *andesite*.

Although the rocks in each of these categories consist mainly of minerals located in a specific region of Bowen's reaction series, other constituents are usually present in lesser amounts. For example, granitic rocks are composed primarily of quartz and potassium feldspar, but they may also contain muscovite, biotite, amphibole, and sodium-rich plagioclase (see Table 3.1).

In this discussion we have identified three major rock categories, yet it is important to note that gradations among them exist (Figure 3.11). For example, an abundant intrusive igneous rock called *granodiorite* has a mineral composition between that of *granitic* (felsic) rocks and those with an *dioritic* (intermediate) composition.

Another important igneous rock, *peridotite*, contains mostly olivine and pyroxene and thus falls near the very beginning of Bowen's reaction series. Because peridotite is composed almost entirely of ferromagnesian minerals, its chemical composition is referred to as **ultramafic**. Although ultramafic rocks are uncommon at Earth's surface, peridotite is believed to be the main constituent of the upper mantle.

An important aspect of the chemical composition of igneous rocks is their silica (SiO_2) content. Recall that most of the minerals found in igneous rocks contain some silica. Typically, the silica content of crustal rocks ranges from a low of 50 percent in basaltic rocks to a high of over 70 percent in granitic rocks. The percentage of silica in igneous rocks actually varies in a systematic manner that parallels the abundance of the other elements. For example, rocks comparatively low in silica contain large amounts of calcium, iron, and magnesium. By contrast, rocks high in silica contain very small amounts of calcium, iron, and magnesium but contain relatively large concentrations of sodium and potassium. Consequently, the chemical makeup of an igneous rock can be inferred directly from its silica content.

Further, the amount of silica present in magma strongly influences its behavior. Granitic magma,

Box 3.2

Thin Sections and Rock Identification

Igneous rocks are classified on the basis of their mineral composition and texture. When analyzing specimens, geologists examine them closely to identify the minerals present and to determine the size and arrangement of the interlocking crystals. When out in the field, geologists use megascopic techniques to study rocks. The *megascopic* characteristics of rocks are those features that can be determined with the unaided eye or by using a low-magnification (x 10) hand lens. When practical to do so, geologists collect hand samples that can be taken back to the laboratory where *microscopic*, or high-magnification, methods can be employed. Microscopic examination is important to identify trace minerals, as well as those textural features that are too small to be visible with the unaided eye.

Because most rocks are not transparent, microscopic work requires the preparation of a very thin slice of rock known as a *thin section* (Figure 3.B). First, a saw containing diamonds embedded in its blade is used to cut a narrow slab from the sample. Next, one side of the slab is polished using grinding powder and then cemented to a microscope slide. Once the mounted sample is firmly in place, the other side of it is ground to a thickness of about 0.03 millimeter. When a slice of rock is that thin, it is usually transparent. Nevertheless, some metallic minerals, such as pyrite and magnetite, remain opaque.

Once produced, thin sections are examined under a specially designed microscope called a *polarizing microscope*. Such an instrument has a light source beneath the stage so that light can be transmitted upward through the thin section. Because minerals have crystalline structures that influence polarized light in a measurable way, this procedure allows for the identification of even the smallest components of a rock. Part C of Figure 3.B is a photomicrograph (photo taken through a microscope) of a thin section of granite

A. Hand sample of granite

B. Thin section

Quartz

Biotite

Feldspar

C. Photomicrograph taken with polarized light magnified about 27 times.

Figure 3.B Thin sections are very useful in identifying the mineral constituents in rocks. **A.** A slice of rock is cut from a hand sample using a diamond saw. **B.** This slice is cemented to a microscope slide and ground until it is transparent to light (about 0.03 millimeter thick). This very thin slice of rock is called a *thin section*. **C.** A thin section of granite viewed under polarized light. (Photos by E.J. Tarbuck)

shown under polarized light. The mineral constituents are identified by their unique optical properties. In addition to assisting in the study of igneous rocks, microscopic techniques are used with great success in analyzing sedimentary and metamorphic rocks as well.

Table 3.1   Classification of Igneous Rocks

	Felsic (Granitic)	Intermediate (Andesitic)	Mafic (Basaltic)	Ultramafic
Phaneritic (coarse-grained)	Granite	Diorite	Gabbro	Peridotite
Aphanitic (fine-grained)	Rhyolite	Andesite	Basalt	Komatiite (rare)
Mineral Composition	Quartz	Amphibole	Calcium feldspar	Olivine
	Potassium feldspar	Intermediate plagioclase	Pyroxene	Pyroxene
	Sodium feldspar			
Minor Mineral Constituents	Muscovite	Pyroxene	Olivine	Calcium feldspar
	Biotite	Amphibole	Amphibole	
	Amphibole	Biotite		
Rock color Based on % dark (mafic) minerals	Light-colored	Medium-colored	Dark gray to black	Dark-green to black
	Less than 15% dark minerals	15–40% dark minerals	More than 40% dark minerals	Nearly 100% dark minerals

which has a high silica content, is viscous and exists as a fluid at temperatures as low as 800°C. On the other hand, basaltic magmas are low in silica and generally more fluid. Further, basaltic magmas are largely crystalline below 950°C.

Felsic (Granitic) Rocks

Granite. *Granite* is perhaps the best known of all igneous rocks (Figure 3.12A). This is partly because of its natural beauty, which is enhanced when it is polished, and partly because of its abundance in the continental crust. Slabs of polished granite are commonly used for tombstones and monuments and as building stones.

Granite is a phaneritic rock composed of about 25 to 35 percent quartz and over 50 percent potassium feldspar and sodium-rich plagioclase. The quartz crystals, which are roughly spherical in shape, are often glassy and clear to light-gray in color. In contrast to quartz, feldspar crystals are not as glassy, are generally white to gray or salmon-pink in color, and exhibit a rectangular rather than a spherical shape.

Other common constituents of granite are muscovite and some dark silicates, particularly biotite and amphibole. Although the dark components generally make up less than 20 percent of most granites, dark minerals appear to be more prominent than their percentage would indicate.

A. Granite   Close up

B. Rhyolite   Close up

Figure 3.12 **A.** Granite, one of the most common phaneritic igneous rocks. **B.** Rhyolite, the aphanitic equivalent of granite, is less abundant. (Photos by E.J. Tarbuck)

When potassium feldspar is dominant and dark pink in color, granite appears almost reddish. This variety is popular as a building stone. However, most often the feldspar grains in granite are white to gray so that when mixed with lesser amounts of dark silicates the rock appears light gray in color.

Granite may also have a porphyritic texture. These specimens contain feldspar crystals of a centimeter or more in length that are scattered among a coarse-grained groundmass of quartz and amphibole.

Granite and other related crystalline rocks are often products of the processes that generate mountains. Because granite is a by-product of mountain building and is very resistant to weathering, it frequently forms the core of eroded mountains. For example, Pikes Peak in the Rockies, Mount Rushmore in the Black Hills, the White Mountains of New Hampshire, Stone Mountain in Georgia, and Yosemite National Park in the Sierra Nevada are all areas where large quantities of granite are exposed at the surface (Figure 3.13).

Granite is a very abundant rock. However, it has become common practice among geologists to apply the term *granite* to any coarse-grained intrusive rock composed predominantly of light silicate materials. We will follow this practice for the sake of simplicity. You should keep in mind that this use of the term *granite* covers rocks having a wider range of mineral compositions.

Rhyolite. Because igneous rocks are classified on the basis of their mineral composition and texture, two rocks may have the same mineral constituents but have

Figure 3.13 El Capitan, a large igneous monolith located in Yosemite National Park, California. Granite bedrock is exposed over much of the park. (Photo by Lewis Kemper/DRK Photo)

A. Obsidian flow.

←2 cm→

B. Hand sample of obsidian.

Figure 3.14 Obsidian is a dark-colored, glassy rock formed from silica-rich lava. (Photos by E. J. Tarbuck)

different textures and hence different names. For example, granite has a fine-grained volcanic equivalent called *rhyolite*. Although these rocks are mineralogically the same, they have different textures and do not look at all alike (Figure 3.12).

Like granite, rhyolite is composed primarily of the light-colored silicates (Figure 3.12B). This fact accounts for its color, which is usually buff to pink or occasionally very light gray. Rhyolite is usually aphanitic and frequently contains glassy fragments and voids indicating rapid cooling in a surface environment. In those instances when rhyolite contains phenocrysts, they are usually small and composed of either

quartz or potassium feldspar. In contrast to granite, rhyolite is rather uncommon. Yellowstone Park is one well-known exception. Here rhyolitic lava flows and ash deposits of similar composition are widespread.

Obsidian. *Obsidian* is a dark-colored, glassy rock that usually forms when silica-rich lava is quenched quickly (Figure 3.14). In contrast to the orderly arrangement of ions characteristic of minerals, *the ions in glass are unordered.* Consequently, glassy rocks such as obsidian are not composed of minerals in the same sense as most other rocks.

Although usually black or reddish-brown in color, obsidian has a high silica content (Figure 3.14). Thus, its composition is more akin to the light igneous rocks such as granite than to the dark rocks of basaltic composition. By itself, silica is clear like window glass; the dark color results from the presence of metallic ions. If you examine a thin edge of a piece of obsidian, it will be nearly transparent. Because of its excellent conchoidal fracture and ability to hold a sharp, hard edge, obsidian was a prized material from which Native Americans chipped arrowheads and cutting tools.

Pumice. *Pumice* is a volcanic rock that, like obsidian, has a glassy texture. Usually found with obsidian, pumice forms when large amounts of gas escape through lava to generate a gray, frothy mass (Figure 3.15). In some samples, the voids are quite noticeable, whereas in others, the pumice resembles fine shards of intertwined glass. Because of the large percentage of voids, many samples of pumice will float when placed in water. Oftentimes, flow lines are visible in pumice, indicating that some movement occurred before

←2 cm→

Figure 3.15 Pumice, a glassy rock containing numerous vesicles. (Photo by E. J. Tarbuck)

A.

B.

Figure 3.16 Andesite porphyry.
A. Hand sample of andesite porphyry, a common volcanic rock.
B. Photomicrograph of a thin section of andesite porphyry to illustrate texture. Notice that the few large crystals (phenocrysts) are surrounded by much smaller crystals (groundmass). (Photo by E.J. Tarbuck)

solidification was complete. Moreover, pumice and obsidian can often be found in the same rock mass, where they exist in alternating layers.

Intermediate (Andesitic) Rocks

Andesite. *Andesite* is a medium gray, fine-grained rock of volcanic origin. Its name comes from South America's Andes Mountains, where numerous volcanoes are composed of this rock type. In addition to the volcanoes of the Andes, many of the volcanic structures encircling the Pacific Ocean are of andesitic composition. Andesite commonly exhibits a porphyritic texture (Figure 3.16). When this is the case, the phenocrysts are often light, rectangular crystals of plagioclase feldspar or black, elongated hornblende crystals.

Diorite. *Diorite* is a coarse-grained intrusive rock that looks somewhat similar to gray granite. However, it can be distinguished from granite by the absence of visible quartz crystals. The mineral makeup of diorite is primarily sodium-rich plagioclase and amphibole, with lesser amounts of biotite. Because the light-colored feldspar grains and dark amphibole crystals are roughly equal in abundance, diorite has a "salt and pepper" appearance (Figure 3.17).

Mafic (Basaltic) Rocks

Basalt. *Basalt* is a very dark green to black, fine-grained volcanic rock composed primarily of pyroxene and calcium-rich plagioclase with lesser amounts of olivine and amphibole present (Figure 3.18A). When porphyritic, basalt commonly contains small, light-colored calcium feldspar phenocrysts or glassy-appearing olivine phenocrysts embedded in a dark groundmass.

Basalt is the most common extrusive igneous rock (Figure 3.19). Many volcanic islands, such as the

Hawaiian Islands and Iceland, are composed mainly of basalt. Further, the upper layers of the oceanic crust consist of basalt. In the United States, large portions of central Oregon and Washington were the sites of extensive basaltic outpourings (see Figure 4.18). At some locations these once-fluid basaltic flows have accumulated to thicknesses approaching 3 kilometers.

Gabbro. *Gabbro* is the intrusive equivalent of basalt (Figure 3.18B). Like basalt, it is very dark green to black in color and composed primarily of pyroxene

Close up

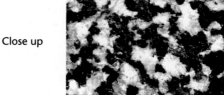

Figure 3.17 Diorite is a phaneritic igneous rock of intermediate composition. (Photo by E.J. Tarbuck)

A. Basalt

B. Gabbro

Close up

Close up

Figure 3.18 These dark-colored mafic rocks are composed primarily of pyroxene and calcium-rich plagioclase. **A.** Basalt is aphanitic and a very common extrusive rock. **B.** Gabbro, the phaneritic equivalent of basalt, is less abundant. (Photos by E.J. Tarbuck)

and calcium-rich plagioclase. Although gabbro is not a common constituent of the continental crust, it undoubtedly makes up a significant percentage of the oceanic crust. Here, large portions of the magma found in underground reservoirs that once fed basalt flows eventually solidified at depth to form gabbro.

Pyroclastic Rocks

Pyroclastic rocks are composed of fragments ejected during a volcanic eruption. One of the most common pyroclastic rocks, called *tuff*, is composed mainly of tiny ash-sized fragments that were later cemented together (Figure 3.20). In situations where the ash

Figure 3.19 Fluid basaltic lava moving down the slopes of Hawaii's Kilauea Volcano toward the sea. (Photo by Brad Lewis/Liaison International)

Figure 3.20 Outcrop of welded tuff interbedded with obsidian (black) near Shoshone, California. Tuff is composed mainly of ash-sized particles and may contain larger fragments of pumice or other volcanic rocks. (Photo by Breck P. Kent)

particles remained hot enough to fuse, the rock is called *welded tuff*. Although welded tuff consists mostly of tiny glass shards, it may contain walnut-size pieces of pumice and other rock fragments.

Welded tuffs blanket vast portions of once volcanically active areas of the western United States. Some of these tuff deposits are hundreds of feet thick and extend for tens of miles from their source. Most formed millions of years ago as volcanic ash spewed from large volcanic structures (calderas) in an avalanche style, spreading laterally at speeds approaching 100 kilometers per hour. Early investigators of these deposits incorrectly classified them as rhyolite lava flows. Today we know that this silica-rich lava is too viscous (thick) to flow more than a few miles from a vent.

Pyroclastic rocks composed mainly of particles larger than ash are called *volcanic breccia*. The particles in volcanic breccia can consist of streamlined fragments that solidified in air, blocks broken from the walls of the vent, crystals, and glass fragments.

Unlike some igneous rock names, such as granite and basalt, the terms *tuff* and *volcanic breccia* do not denote mineral composition. Thus, they are frequently used with a modifier, as, for example, rhyolite tuff.

Plate Tectonics and Igneous Rocks

The origin of magma has been a controversial topic in geology almost from the very beginning of the science. How do magmas of different compositions form? Why do volcanoes in the deep-ocean basins primarily extrude basaltic lava, whereas those on the continental margins adjacent to oceanic trenches extrude mainly andesitic lava? Why are basaltic rocks common on Earth's surface, whereas most granitic magma is emplaced at depth? Insights from the theory of plate tectonics are providing some answers.

Origin of Magma

Based on available scientific evidence, *Earth's crust and mantle are composed primarily of solid, not molten, rock.* Although the outer core is a fluid, its iron-rich material is very dense and remains deep within Earth. So, what is the source of magma that produces igneous activity?

Geologists conclude that magma originates when essentially solid rock, located in the crust and upper mantle, melts. The most obvious way to generate magma from solid rock is to raise the temperature above the rock's melting point. In a near-surface environment, silica-rich granitic rocks begin to melt at temperatures around 750°C, whereas basaltic rocks must be heated to temperatures above 1000°C before melting commences. In addition to raising a rock's temperature, rock that is near its melting point may begin to melt if the confining pressure drops, or if fluids (volatiles) are introduced. We will consider the role that heat, pressure, and volatiles play in generating magma.

Role of Heat. What source of heat is sufficient to melt rock? Workers in underground mines know that temperatures get higher as they go deeper. Although the rate of temperature change varies from place to place, it *averages* between 20°C and 30°C per kilometer in the *upper* crust. The change in temperature with depth is known as the **geothermal gradient** (Figure 3.21). From estimates of the geothermal gradient, the temperature at 100 kilometers ranges between 1200°

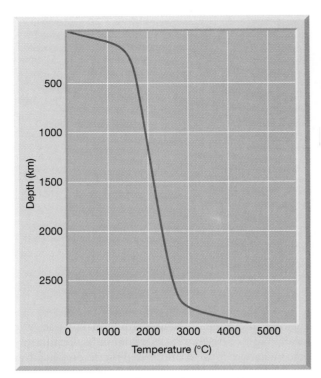

Figure 3.21 This graph illustrates the estimated temperature distribution for the crust and mantle. Notice that temperature increases significantly from the surface to the base of the lithosphere and that the temperature gradient (rate of change) is much less in the mantle. Because the temperature difference between the top and bottom of the mantle is relatively small, geologists conclude that slow convective flow (hot material rising, and cool mantle sinking) must occur in the mantle.

and 1400°C, whereas the temperature at the core–mantle boundary is calculated to be about 4500°C and it may exceed 6700°C at Earth's center.[*]

There are several ways that heat plays a role in generating magma. First, at subduction zones friction is thought to generate heat as rocks slide past one another. Second, rocks can be heated during subduction as they descend into a high temperature environment. Third, hot material at depth can rise and melt rocks located near the surface. Although all of these processes generate some magma, the quantities are usually small and the distribution is very localized. As we shall examine next, most magma is generated without the addition of heat.

Role of Pressure. If temperature were the only factor that determined whether or not rock melts, our planet would be a molten ball covered with a thin, solid outer shell. This, of course, is not the case. The reason is that pressure also increases with depth.

Melting which is accompanied by an increase in volume, *occurs at higher temperatures at depth* because of greater confining pressure (Figure 3.22). Consequently, an increase in confining pressure causes an increase in the rock's melting temperature. Conversely, reducing confining pressure lowers a rock's melting temperature. When confining pressure drops, melting is triggered. This may occur when rock *ascends* as a result of convective upwelling, thereby moving into zones of lower pressure. (Recall that even though the mantle is a *solid, it does flow* at very slow rates over time scales of millions of years.) This process is responsible for generating magma along ocean ridges where plates are rifting apart (Figure 3.23).

Role of Volatiles. Another important factor affecting the melting temperature of rock is its water content. Water and other volatiles act as salt does to melt ice. That is, volatiles cause rock to melt at lower temperatures. Further, the effect of volatiles is magnified by increased pressure. Consequently, "wet" rock buried at depth has a much lower melting temperature than does "dry" rock of the same composition and under the same confining pressure (Figure 3.22). Therefore, in addition to a rock's composition, its temperature, depth (confining pressure) and water content determine whether it exists as a solid or liquid.

Volatiles play an important role in generating magma in regions where cool slabs of oceanic lithosphere descend into the mantle (Figure 3.24). As an oceanic plate sinks, both heat and pressure drive water from the subducting crustal rocks. These volatiles, which are very mobile, migrate into the wedge of hot

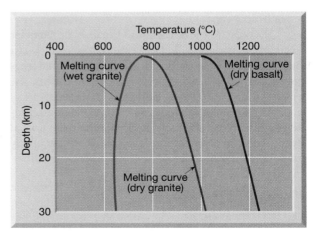

Figure 3.22 Idealized melting temperature curves. These curves portray the minimum temperatures required to melt rock within Earth's crust. Notice that dry granite and dry basalt melt at higher temperatures with increasing depth. By contrast, the melting temperature of wet granite actually decreases as the confining pressure increases.

---

[*]We will consider the heat sources for the geothermal gradient in Chapter 17.

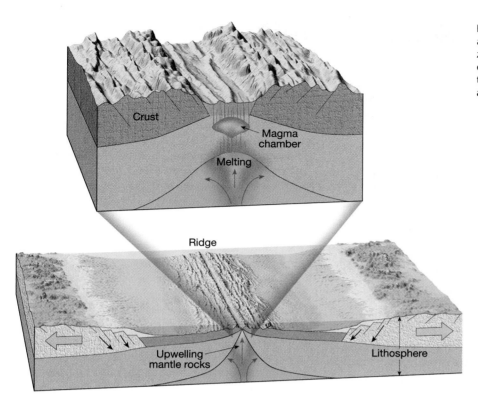

Figure 3.23 As hot mantle rock ascends, it continually moves into zones of lower pressure. This drop in confining pressure can trigger melting, even without additional heat.

mantle that lies above. This process is believed to lower the melting temperature of mantle rock sufficiently to generate some melt.

Once enough molten rock forms, it will buoyantly rise toward the surface. In a continental setting, the magma body may "pond" beneath crustal rocks, which are already near their melting temperature. This results in the generation of a secondary silica-rich magma.

In summary, magma can be generated under three sets of conditions: (1) *heat* may be added; for example, a magma body from a deeper source intrudes and melts crustal rock; (2) a *decrease in pressure* (without the addition of heat) can cause melting; and (3) the *introduction of volatiles* (principally water) can lower the melting temperature of mantle rock sufficiently to generate melt.

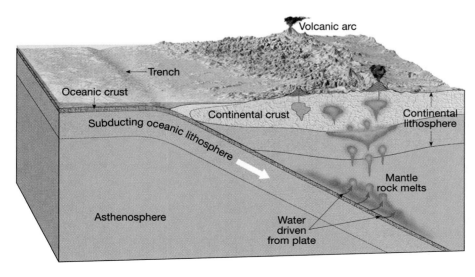

Figure 3.24 As an oceanic plate descends into the mantle, water and other volatiles are driven from the subducting crustal rocks. These volatiles lower the melting temperature of mantle rock sufficiently to generate melt.

Partial Melting and Magma Compositions

An important difference exists between the melting of a substance that consists of a single compound, such as ice, and melting igneous rocks, which are mixtures of several minerals. Ice melts at a specific temperature, whereas igneous rocks melt over a temperature range of about 200°C. As rock is heated, the minerals with the lowest melting points are the first to melt. Should melting continue, minerals with higher melting points begin to melt and the composition of the magma steadily approaches the overall composition of the rock from which it was derived. Most often, melting is not complete. This process, known as **partial melting**, produces most, if not all, magma.

An important consequence of partial melting is the *production of a melt with a higher silica content than the original rock*. Recall from the discussion of Bowen's reaction series that ultramafic rocks contain mostly high-melting-temperature minerals that are comparatively low in silica, whereas felsic rocks are composed primarily of low-melting-temperature silicates that are enriched in silica. Because silica-rich minerals melt first, magmas generated by partial melting are nearer to the felsic end of the compositional spectrum than are the rocks from which they formed (see Figure 3.8).

Formation of Mafic (Basaltic) Magma. Most mafic magmas originate from the partial melting of the ultramafic rock *peridotite*, the major constituent of the upper mantle. Laboratory studies confirm that partial melting of this dry, silica-poor rock produces magma having a basaltic composition. Because mantle rocks exist in environments that are characterized by high temperatures and pressures, melting often results from a reduction in confining pressure. This can occur, for example, where mantle rock ascends as part of slow-moving convective flow.

Because basaltic magmas form many kilometers below the surface, we might expect that most of this material would cool and crystallize before reaching the surface. However, as dry basaltic magma migrates upward, the confining pressure steadily diminishes and reduces the melting temperature. Most basaltic magmas ascend rapidly enough that, as they enter cooler environments, the heat loss is offset by a drop in the melting temperature. Consequently, large outpourings of basaltic magmas are common on Earth's surface.

Formation of Intermediate (Andesitic) Magma. If partial melting of mantle rocks generates mafic magmas, what is the source of the magma that generates andesitic and granitic rocks? Recall that intermediate and felsic magmas are not erupted from volcanoes in the deep-ocean basins; rather, they are found only within, or adjacent to, the continental margins (Figure 3.25). This is strong evidence that interactions between mantle-derived basaltic magmas and more felsic components of the crust generate these magmas. For example, as a basaltic magma migrates upward, it may melt and assimilate some of the felsic basement rock. The result is the formation of a magma of andesitic composition (intermediate between mafic and felsic).

Andesitic magma may also evolve from a basaltic magma by the process of magmatic differentiation. Recall from our discussion of Bowen's reaction series, that as a basaltic magma solidifies, it is the silica-poor ferromagnesian minerals that crystallize first. If these iron-rich components are separated from the liquid by crystal settling, the remaining melt, which is enriched in silica, will have an andesitic composition.

Formation of Felsic (Granitic) Magmas. Felsic magmas are too silica-rich to be produced directly from magmatic differentiation of mafic magmas. Most likely they are the end product of the crystallization of an andesitic magma, or the product of partial melting of silica-rich continental rocks.

The heat to melt crustal rocks often comes from hot mantle-derived mafic magmas that formed above a subducting plate and were emplaced beneath the crust. Here partial melting of wet, felsic rocks is thought to generate granitic magma.

As a wet felsic melt rises, the confining pressure decreases, which in turn *reduces* the effect of water in its role of lowering melting temperature. Further, granitic melts are higher in silica and thus more viscous (thicker) than other magmas. Therefore, in contrast to basaltic magmas that frequently produce vast outpourings of lava, granitic magmas usually lose their mobility before reaching the surface and tend to produce large intrusive features. On those occasions when silica-rich magmas do reach the surface, explosive pyroclastic eruptions, such as those from Mount St. Helens, are the rule.

Figure 3.25 1986 eruption of Mount Augustine, Cook Inlet, Alaska. Volcanoes that border the Pacific Ocean are fed largely by magmas that have intermediate or felsic compositions. These silica-rich magmas often erupt explosively, generating large plumes of volcanic dust and ash. (Photo by Steve Kaufman/DRK)

Chapter Summary

- *Igneous rocks* form when *magma cools* and solidifies. *Extrusive*, or *volcanic*, igneous rocks result when *lava* cools at the surface. Magma that solidifies at depth produces *intrusive*, or *plutonic*, igneous rocks.

- As magma cools, the ions that compose it arrange themselves into orderly patterns during a process called *crystallization*. Slow cooling results in the formation of rather large crystals. Conversely, when cooling occurs rapidly, the outcome is a solid mass consisting of tiny intergrown crystals. When molten material is quenched instantly, a mass of unordered atoms, referred to as *glass*, forms.

- Igneous rocks are most often classified by their *texture* and *mineral composition*.

- The texture of an igneous rock refers to the overall appearance of the rock based on the size and arrangement of its interlocking crystals. The most important factor affecting texture is the rate at which magma cools. Common igneous rock textures include *aphanitic*, with grains too small to

be distinguished with the unaided eye; *phaneritic*, with intergrown crystals that are roughly equal in size and large enough to be identified with the unaided eye; *porphyritic*, which has large crystals (*phenocrysts*) interbedded in a matrix of smaller crystals (*groundmass*); and *glassy*.

- The mineral makeup of an igneous rock is ultimately determined by the chemical composition of the magma from which it crystallizes. N.L. Bowen discovered that as magma cools in the laboratory, those minerals with higher melting points crystallize before minerals with lower melting points. *Bowen's reaction series* illustrates the sequence of mineral formation within basaltic magma. In the *discontinuous reaction series*, each mineral has a different crystalline structure that forms as the solid components (minerals) react with the remaining melt (liquid portion of a magma, excluding any solid material) and produce the next mineral in the sequence. The right branch of Bowen's reaction series, called the *continuous reaction series*, demonstrates that calcium-rich feldspar crystals react with the sodium ions contained in the melt to become progressively more sodium rich. During the last stage of crystallization, after most of the magma has solidified, the minerals muscovite, potassium feldspar, and quartz are generated.

- During the crystallization of magma, if the earlier-formed minerals are denser than the liquid portion, they will settle to the bottom of the magma chamber during a process called *crystal settling*. Owing to the fact that crystal settling removes the earlier-formed minerals, the remaining melt will form a rock with a chemical composition much different from the parent magma. The process of developing more than one magma type from a common magma is called *magmatic differentiation*.

- Once a magma body forms, its comparison can change through the incorporation of foreign material, a process termed *assimilation*, or by *magma mixing*.

- The mineral composition of an igneous rock is the consequence of the chemical makeup of the parent magma and the environment of crystallization. Hence, the classification of igneous rocks closely corresponds to Bowen's reaction series. *Felsic rocks* (e.g., granite and rhyolite) form from the last minerals to crystallize, potassium feldspar and quartz, and are light-colored. Rocks of *intermediate* composition, (e.g., andesite and diorite) form from plagioclase feldspar and amphibole minerals. *Mafic rocks* (e.g., basalt and gabbro) form from the first minerals to crystallize—olivine, pyroxene, and calcium-feldspar—are high in iron, magnesium, and calcium, low in silicon, and are dark-gray to black in color.

- Magma originates from essentially solid rock of the crust and mantle. In addition to a rock's composition, its temperature, depth (confining pressure), and water content determine whether it exists as a solid or liquid. Thus, magma can be generated by *raising a rock's temperature*, as occurs when a hot mantle plume "ponds" beneath crustal rocks. A *decrease in pressure* can cause rock to melt. Further, the *introduction of volatiles* (water) can lower a rock's melting point sufficiently to generate magma. Because melting is generally not complete, a process called *partial melting* produces a melt made of the lowest-melting-temperature minerals, which are higher in silica than the original rock. Thus, magmas generated by partial melting are nearer to the felsic end of the compositional spectrum than are the rocks from which they formed.

Review Questions

1. What is magma?
2. How does lava differ from magma?
3. How does the rate of cooling influence the crystallization process?
4. In addition to the rate of cooling, what two other factors influence the crystallization process?
5. The classification of igneous rocks is based largely on two criteria. Name these criteria.
6. The statements that follow relate to terms describing igneous rock textures. For each statement, identify the appropriate term.

   (a) Openings produced by escaping gases.

   (b) Obsidian exhibits this texture.

   (c) A matrix of fine crystals surrounding phenocrysts.

   (d) Crystals are too small to be seen with the unaided eye.

   (e) A texture characterized by two distinctly different crystal sizes.

   (f) Coarse grained, with crystals of roughly equal size.

   (g) Exceptionally large crystals exceeding 1 centimeter in diameter.

7. Why are the crystals in pegmatites so large?
8. What does a porphyritic texture indicate about igneous rock?
9. What is magmatic differentiation? How might this process lead to the formation of several different igneous rocks from a single magma?
10. Relate the classification of igneous rocks to Bowen's reaction series.
11. How are granite and rhyolite different? In what way are they similar? *p. 72–74*
12. Compare and contrast each of the following pairs of rocks: *p. 70*

   (a) Granite and diorite

   (b) Basalt and gabbro

   (c) Andesite and rhyolite

13. How do tuff and volcanic breccia differ from other igneous rocks such as granite and basalt?

14. What is the geothermal gradient?

15. Describe the three conditions that are thought to cause rock to melt.

16. What is partial melting?

17. How does the composition of a melt produced by partial melting compare with the composition of the parent rock?

18. How are most basaltic magmas generated?

19. Basaltic magma forms at great depth. Why doesn't most of it crystallize as it rises through the relatively cool crust?

20. Why are rocks of intermediate (andesitic) and felsic (granitic) composition generally *not* found in the ocean basins?

Key Terms

andesitic (p. 70)
aphanitic texture (p. 63)
assimilation (p. 68)
basaltic (p. 69)
Bowen's reaction series (p. 66)
crystallization (p. 61)
crystal settling (p. 68)
extrusive (p. 60)
felsic (p. 70)

geothermal gradient (p. 76)
glass (p. 62)
glassy texture (p. 63)
groundmass (p. 63)
granitic (p. 70)
igneous rocks (p. 60)
intermediate (p. 70)
intrusive (p. 60)
lava (p. 60)

mafic (p. 69)
magma (p. 60)
magma mixing (p. 68)
magmatic differentiation (p. 68)
partial melting (p. 80)
pegmatite (p. 64)
pegmatitic texture (p. 64)
phaneritic texture (p. 63)
phenocryst (p. 63)

plutonic (p. 60)
porphyritic texture (p. 63)
porphyry (p. 63)
pyroclastic texture (p. 64)
texture (p. 61)
ultramafic (p. 70)
vesicle (p. 63)
volcanic (p. 60)
viscosity (p. 63)

Web Resources

 The *Earth* Home Page provides on-line resources for this chapter on the World Wide Web. You will find review exercises, specific updates for items in the chapter, suggested reading, and links to interesting related pathways on the Internet. Visit the *Earth* Home Page at **http://www.prenhall.com/tarbuck.**

CHAPTER 4

Volcanic and Plutonic Activity

Left Eruption of Mount Ruapehu, Tongariro National Park, New Zealand, 1996. — *Photo by Tui De Roy/Minden Picture*

Figure 4.1 Before and after photographs show the transformation of Mount St. Helens caused by the May 18, 1980, eruption. The dark area in the "after" photo is debris-filled Spirit Lake, partially visible in the "before" photo. (Photos courtesy of U.S. Geological Survey)

On Sunday, May 18, 1980, the largest volcanic eruption to occur in North America in historic times transformed a picturesque volcano into a decapitated remnant (Figure 4.1). On this date in southwestern Washington State, Mount St. Helens erupted with tremendous force. The blast blew out the entire north flank of the volcano, leaving a gaping hole. In one brief moment, a prominent volcano whose summit had been more than 2900 meters (9500 feet) above sea level was lowered by more than 400 meters (1350 feet).

The event devastated a wide swath of timber-rich land on the north side of the mountain (Figure 4.2). Trees within a 400-square-kilometer area lay intertwined and flattened, stripped of their branches and appearing from the air like toothpicks strewn about. The accompanying mudflows carried ash, trees, and water-saturated rock debris 29 kilometers down the Toutle River. The eruption claimed 59 lives, some dying from the intense heat and the suffocating cloud of ash and gases, others from being hurled by the blast, and still others from entrapment in the mudflows.

The eruption ejected nearly a cubic kilometer of ash and rock debris. Following the devastating explosion, Mount St. Helens continued to emit great quantities of hot gases and ash. The force of the blast was so strong that some ash was propelled more than 18,000 meters (over 11 miles) into the stratosphere. During the next few days, this very fine-grained material was carried around Earth by strong upper-air winds. Measurable deposits were reported in Oklahoma and Minnesota, with crop damage into central Montana. Meanwhile, ash fallout in the immediate vicinity exceeded 2 meters in depth. The air over Yakima, Washington (130 kilometers to the east), was so filled with ash that residents experienced midnight-like darkness at noon.

Not all volcanic eruptions are as violent as the 1980 Mount St. Helens event. Some volcanoes, such as Hawaii's Kilauea volcano, generate relatively quiet outpourings of fluid lavas. These "gentle" eruptions are not without some fiery displays; occasionally fountains of incandescent lava spray hundreds of meters into the air (Figure 4.3). Such events, however, are typically short-lived and harmless, and the lava generally falls back into a lava pool.

Testimony to the quiet nature of Kilauea's eruptions is the fact that the Hawaiian Volcanoes Observatory has operated on its summit since 1912. This, despite the fact that Kilauea has had more than 50 eruptive phases since record keeping began in 1823. Further, the longest and largest of Kilauea's eruptions began in 1983 and remains active, although it has received only modest media attention.

Why do volcanoes like Mount St. Helens erupt explosively, whereas others like Kilauea are relatively quiet? Why do volcanoes occur in chains like the Aleutian Islands or the Cascade Range? Why do some volcanoes form on the ocean floor, while others occur on the continents? This chapter will deal with these and other questions as we explore the nature and movement of magma and lava.

Spirit Lake

The Nature of Volcanic Eruptions

Volcanic activity is commonly perceived as a process that produces a picturesque, cone-shaped structure that periodically erupts in a violent manner, like Mount St. Helens (Box 4.1). Although some eruptions may be very explosive, many are not. What determines whether a volcano extrudes magma violently or "gently"? The primary factors include the magma's *composition*, its *temperature*, and the amount of *dissolved gases* it contains. To varying degrees, these factors affect the magma's mobility, or **viscosity.** The more viscous the material, the greater its resistance to flow. (For example, syrup is more viscous than water.) The viscosity of magma associated with an explosive eruption may be five times more viscous than magma that is extruded in a quiescent manner.

Factors Affecting Viscosity

The effect of temperature on viscosity is easily seen. Just as heating syrup makes it more fluid (less viscous), the mobility of lava is strongly influenced by temperature. As lava cools and begins to congeal, its mobility decreases and eventually the flow halts.

A more significant factor influencing volcanic behavior is the chemical composition of the magma. This was discussed in Chapter 3 with the classification of igneous rocks. Recall that a major difference among various igneous rocks is their silica (SiO_2) content (Table 4.1). Magmas that produce mafic rocks such as basalt contain about 50 percent silica, whereas magmas that produce felsic rocks (granite and its extrusive equivalent, rhyolite) contain over 70 percent silica. The intermediate rock types, andesite and diorite, contain about 60 percent silica.

A magma's viscosity is directly related to its silica content. In general, the more silica in magma, the greater its viscosity. The flow of magma is impeded because silica structures link together into long chains, even before crystallization begins. Consequently, because of high silica content, felsic lavas are very viscous and tend to form comparatively short, thick flows. By contrast, mafic lavas, which contain less silica, tend to be quite fluid and have been known to travel distances of 150 kilometers (90 miles) or more before congealing.

Importance of Dissolved Gases

The gas content of a magma also affects its mobility. Dissolved gases tend to increase the fluidity of magma. Of far greater consequence is the fact that escaping gases provide enough force to propel molten rock from a volcanic vent.

Volcanoes inflate before an eruption, indicating a buildup in gas pressure directly below in a shallow magma chamber. When an eruption starts, gas-charged

Figure 4.2 Douglas fir trees were snapped off or uprooted by the lateral blast of Mount St. Helens on May 18, 1980. (Photo by Jeffrey Hutcherson/DRK)

Figure 4.3 Fluid basaltic lava erupting from Kilauea Volcano, Hawaii. (Photo by Douglas Peebles)

magma moves from the magma chamber and rises through the volcanic conduit or vent. As the magma nears the surface, the confining pressure is greatly reduced. This reduction in pressure allows the dissolved gases to be released suddenly, just as opening a warm soda bottle allows carbon dioxide gas bubbles to escape. At temperatures of 1000°C and low, near-surfaces pressures, these gases will expand to occupy hundreds of times their original volume.

Very fluid basaltic magmas allow the expanding gases to migrate upward and escape from the vent with relative ease. As they escape, the gases may propel incandescent lava hundreds of meters into the air, producing lava fountains (see Figure 4.3). Although spectacular, such fountains are mostly harmless and not generally associated with major explosive events that cause great loss of life and property. Rather, eruptions of fluid basaltic lavas, such as those that occur in Hawaii, are generally quiescent.

At the other extreme, highly viscous magmas explosively expel jets of hot ash-laden gases that evolve into buoyant plumes extending thousands of meters into the atmosphere (Figure 4.4). Prior to an explosive eruption, the upper portion of a magma body tends to become enriched in dissolved gases. As this magma moves up the volcanic vent toward the surface, these gases begin to collect as tiny bubbles. For reasons that are still poorly understood, at some height in the conduit this mixture is transformed into a gas jet containing tiny glass fragments that are explosively ejected from the volcano. This type of explosive eruption is exemplified by Mount Pinatubo in the Philippines (1991) and Mount St. Helens (1980).

Box 4.1

Mount St. Helens: Anatomy of the Eruption

The events leading to the May 18, 1980, eruption of Mount St. Helens began about two months earlier as a series of minor Earth tremors centered beneath the awakening mountain (Figure 4.A, part A). The tremors were caused by the upward movement of magma within the mountain. The first volcanic activity took place a week later, when a small amount of ash and steam rose from the summit. Over the next several weeks, sporadic eruptions of varied intensity occurred. Prior to the main eruption, the primary concern had been the potential hazard of mudflows. These moving lobes of saturated soil and rock are created as ice and snow melt from the heat emitted from magma within the volcano.

The only warning of a potential eruption was a bulge on the volcano's north flank (Figure 4.A, part B). Careful monitoring of this dome-shaped structure indicated a very slow but steady growth rate of a few meters per day. If the growth rate of the bulge changed appreciably, an eruption might quickly follow. Unfortunately, no such variation was detected prior to the explosion. In fact, the seismic activity decreased during the two days preceding the huge blast.

Dozens of scientists were monitoring the mountain when it exploded. "Vancouver, Vancouver, this is it!" was the only warning—and last words from one scientist—that preceded the unleashing of tremendous quantities of pent-up gases. The trigger was a medium-sized earthquake. Its vibrations sent the north slope of the cone plummeting into the Toutle River, removing the overburden that had trapped the magma below (Figure 4.A, part C). With the pressure reduced, the water in the magma vaporized and expanded, causing the mountainside to rupture like an overheated steam boiler. Because the eruption originated around the bulge, several hundred meters below the summit, the initial blast was directed laterally rather than vertically. Had the full force of the eruption been upward, far less destruction would have occurred.

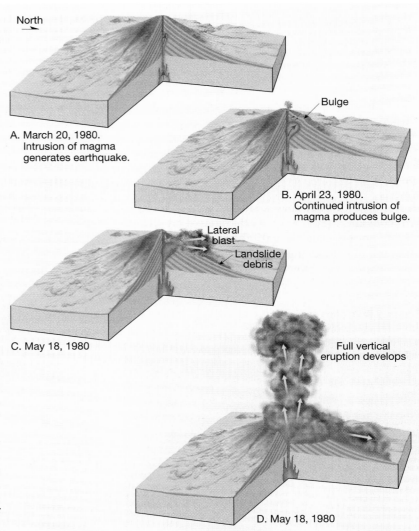

A. March 20, 1980.
Intrusion of magma
generates earthquake.

Bulge

B. April 23, 1980.
Continued intrusion of
magma produces bulge.

Lateral
blast
Landslide
debris

C. May 18, 1980

Full vertical
eruption develops

D. May 18, 1980

Figure 4.A Idealized diagrams showing the events in the May 18, 1980, eruption of Mount St. Helens. **A.** First, a sizable earthquake recorded on Mount St. Helens indicates that renewed volcanic activity is possible. **B.** Alarming growth of a bulge on the north flank suggests increasing magma pressure below. **C.** Triggered by an earthquake, a giant landslide relieved the confining pressure on the magma body and initiated an explosive lateral blast. **D.** Within seconds, a large vertical eruption sent a column of volcanic ash to an altitude of about 19 kilometers. This phase of the eruption continued for over 9 hours.

Mount St. Helens is one of 15 large volcanoes and innumerable smaller ones that comprise the Cascade Range, which extends from British Columbia to northern California. Eight of the largest volcanoes have been active in the past few hundred years. Of the remaining seven "active" volcanoes, the most likely to erupt again are Mount Baker and Mount Rainier in Washington, Mount Shasta and Lassen Peak in California, and Mount Hood in Oregon.

Table 4.1	Magmas Have Different Compositions, Which Cause Their Properties to Vary				
Composition	Silica Content	Viscosity	Gas Content	Tendency to Form Pyroclastics	Volcanic Landform
Mafic (Basaltic) magma	Least (~50%)	Least	Least (1–2%)	Least	Shield Volcanoes Basalt Plateaus Cinder Cones
Intermediate (Andesitic) magma	Intermediate (~60%)	Intermediate	Intermediate (3–4%)	Intermediate	Composite Cones
Felsic (Granitic) magma	Most (~70%)	Greatest	Most (4–6%)	Greatest	Volcanic Domes Pyroclastic Flows

As magma in the upper portion of the vent is ejected, the pressure on the molten rock directly below drops. Thus, rather than a single "bang," volcanic eruptions are really a series of explosions. This process might logically continue until the magma chamber is emptied, much like a geyser empties itself of water (see Chapter 11). However, this generally does not happen. The soluble gases in a viscous magma migrate upward quite slowly. Only within the uppermost portion of the magma body does the gas content build sufficiently to trigger explosive eruptions. Thus, an explosive event is commonly followed by the quiet emission of "degassed" lavas. However, once this eruptive phase ceases, the process of gas buildup begins anew. This time lag may explain the sporadic eruptive patterns of volcanoes that eject viscous lavas.

To summarize, the viscosity of magma, plus the quantity of dissolved gases and the ease with which they can escape, determines the nature of a volcanic eruption. We can now understand the "gentle" volcanic eruptions of hot, fluid lavas in Hawaii and the explosive and sometimes catastrophic eruptions of viscous lavas from volcanoes such as Mount St. Helens.

Figure 4.4 Eruption of Mount Augustine, Cook Inlet, Alaska, 1986. (Photo by Steve Kaufman/DRK)

Materials Extruded During an Eruption

Volcanoes extrude lava, large volumes of gas, and pyroclastic materials (broken rock, lava "bombs," fine ash, and dust). In this section we will examine each of these materials.

Lava Flows

Because of their low silica content, basaltic lavas are usually very fluid. They flow in thin, broad sheets or stream-like ribbons. On the island of Hawaii, such lavas have been clocked at speeds of 30 kilometers per hour down steep slopes, but flow rates of 10 to 300 meters per hour are more common. Further, basaltic lavas have been known to travel distances of 150 kilometers (90 miles) or more before congealing. In contrast, the movement of silica-rich (felsic) lava may be too slow to perceive.

When fluid basaltic lavas of the Hawaiian type congeal, they commonly form a relatively smooth skin that wrinkles as the still-molten subsurface lava continues to advance (Figure 4.5A). These are known as

A.

B.

Figure 4.5 A. Typical pahoehoe (ropy) lava flow, Kilauea, Hawaii. **B.** Typical slow-moving aa flow. (Photos by J.D. Griggs, U.S. Geological Survey)

pahoehoe flows (pronounced *pah-hoy-hoy*) and resemble the twisting braids in ropes.

Another common type of basaltic lava, called **aa** (pronounced *ah-ah*), has a surface of rough, jagged blocks with dangerously sharp edges and spiny projections (Figure 4.5B). Active aa flows are relatively cool and thick and advance at rates from 5 to 50 meters per hour. Further, escaping gases fragment the cool surface and produce numerous voids and sharp spines in the congealing lava. As the molten interior advances, the outer crust is broken further, giving the flow the appearance of an advancing mass of lava rubble.

The lava that flowed from the Mexican volcano Parícutin and buried the city of San Juan Parangaricutiro was of the aa type (see Figure 4.12). At times one of the flows from Parícutin moved only 1 meter per day, but it continued to advance day in and day out for more than 3 months.

Hardened lava flows commonly contain tunnels that once were horizontal conduits carrying lava from the volcanic vent to the flow's leading edge. These openings develop in the interior of a flow where temperatures remain high long after the surface congeals. Under these conditions, the still-molten lava within the conduits continues its forward motion, leaving behind the cavelike voids called **lava tubes** (Figure 4.6). Lava tubes can play an important role in allowing fluid lavas to advance great distances from their source.

When lava enters the ocean, or when outpourings of lava actually originate in the ocean basin, the flows' outer zones quickly congeal. However, the lava is usually able to move forward by breaking through the hardened surface. This process occurs over and over, generating a lava flow composed of elongated structures resembling large bed pillows stacked one upon the other. These structures, called **pillow lavas**, are useful in the reconstruction of Earth history (Figure 4.7). Whenever pillow lavas are located, they indicate that deposition occurred in an underwater environment.

Gases

Magmas contain varying amounts of dissolved gases held in the molten rock by confining pressure, just as carbon dioxide is held in soft drinks. As with soft drinks, as soon as the pressure is reduced, the gases begin to escape. Obtaining gas samples from an erupting volcano is difficult and dangerous, so geologists usually estimate the amount of gas originally contained within the magma.

The gaseous portion of most magmas makes up from 1 to 6 percent of the total weight, with most of this in the form of water vapor. Although the percentage may be small, the actual quantity of emitted gas can exceed thousands of tons per day.

The composition of volcanic gases is important because they contribute significantly to the gases that make up our planet's atmosphere. Analyses of samples taken during Hawaiian eruptions indicate that the gases are about 70 percent water vapor, 15 percent carbon dioxide, 5 percent nitrogen, 5 percent sulfur dioxide, and lesser amounts of chlorine, hydrogen, and argon. Sulfur compounds are easily recognized by their pungent odor. Volcanoes are a natural source of air pollution, including sulfur dioxide, which readily combines with water to form sulfuric acid.

In addition to propelling magma from a volcano, gases play an important role in creating the narrow conduit that connects the magma chamber to the surface. First, the intense heat from the magma body cracks the rock above. Then, hot blasts of high-pressure gases expand the cracks and develop a passageway to the surface. Once the passageway is completed,

A.

B.

Figure 4.6 Lava streams that flow in confined channels often develop a solid crust and become flows within lava tubes. **A.** Thurston Lava Tube, Hawaii Volcanoes National Park. **B.** View of an active lava tube as seen through the collapsed roof. (Photo **A** by Douglas Peebles and Photo **B** by Jeffrey B. Judd, U.S. Geological Survey)

the hot gases armed with rock fragments erode its walls, producing a larger conduit. Because these erosive forces are concentrated on any protrusion along the pathway, the volcanic pipes that are produced have a circular shape. As the conduit enlarges, magma moves upward to produce surface activity. Following an eruptive phase, the volcanic pipe often becomes choked with a mixture of congealed magma and debris that was not thrown clear of the vent. Before the next eruption, a new surge of explosive gases may again clear the conduit.

Figure 4.7 Pillow lavas exposed in Iceland's Eldgia fissure area. (Photo by Peter Kresan)

Pyroclastic Materials

When basaltic lava is extruded, dissolved gases escape quite freely and continually. These gases propel incandescent blobs of lava to great heights (see Figure 4.3). Some of this ejected material may land near the vent and build a cone-shaped structure, whereas smaller particles will be carried great distances by the wind. By contrast, viscous (felsic) magmas are highly charged with gases, and upon release they expand a thousand-fold as they blow pulverized rock, lava, and glass fragments from the vent. The particles produced in both of these situations are referred to as **pyroclastic materials** (meaning "fire fragments"). These ejected fragments range in size from very fine dust (less than 0.063 mm in diameter) and sand-sized volcanic ash (less than 2 mm in diameter) to pieces that weight more than a ton.

Ash and *dust* particles are produced from gas-laden, viscous magma during an explosive eruption (see Figure 4.4). As magma moves up in the vent, the gases rapidly expand, generating a froth of melt that might resemble froth which flows from a just-opened bottle of champagne. As the hot gases expand explosively, the froth is blown into very fine glassy fragments. When the hot ash falls, the glassy shards often fuse to form welded tuff. Sheets of this material, as well as ash deposits that later consolidate, cover vast portions of the western United States. Sometimes the frothlike lava is ejected as *pumice*, a material having so many voids (air spaces) that it often floats in water.

Also common are walnut-sized pyroclasts termed *lapilli* ("little stones"), and pea-sized particles called *cinders*. Particles larger than lapilli are called *blocks* when they are made of hardened lava and *bombs* when they are ejected as incandescent lava. Because

←2 cm→

Figure 4.8 Volcanic bombs. Ejected lava fragments take on a streamlined shape as they sail through the air. This bomb is approximately 10 centimeters in length. (Photo by E.J. Tarbuck)

bombs are semimolten upon ejection, they often take on a streamlined shape as they hurtle through the air (Figure 4.8). Because of their size, bombs and blocks usually fall on the slopes of a cone; however, they are occasionally propelled far from the volcano by the force of escaping gases.

Fine volcanic debris can be scattered great distances from its source. Dust in particular may be blasted high into the atmosphere, where it can remain for extended periods. When present, dust produces brilliant sunsets and has on occasion slightly lowered Earth's average temperature. The possible effects of volcanic eruptions on climate are discussed more thoroughly near the end of the chapter.

Volcanoes and Volcanic Eruptions

Successive eruptions from a central vent build a mountainous accumulation we call a **volcano.** Located at the summit of many volcanoes is a steep-walled **crater.** The crater is connected to a magma chamber via a circular conduit, or **pipe,** which terminates at an opening called a **vent.** Some volcanoes have unusually large summit depressions that exceed 1 kilometer in diameter and are known as **calderas*.**

When fluid Hawaiian-type lava leaves a conduit, it is often stored in the crater or caldera, until it overflows. Conversely, viscous lava forms a plug in the pipe, which rises slowly or is blown out, often enlarging the crater.

Lava does not always issue from a central crater. Sometimes the magma or escaping gases push through fissures located on the volcano's flanks. Continued

---

*The formation of calderas is discussed later in this chapter.

activity from a flank eruption may build a smaller **parasitic cone.** Mount Etna in Italy, for example, has more than 200 secondary vents. Some of these emit only gases and are appropriately called **fumaroles.**

The eruptive history of each volcano is unique, so volcanoes vary in size and form (see Box 4.2). Nevertheless, volcanologists recognize three general eruptive patterns and characteristic forms: shield volcanoes, cinder cones, and composite cones (stratovolcanoes).

Shield Volcanoes

When fluid Hawaiian-type lava is extruded, the volcano takes the shape of a broad, slightly domed structure called a **shield volcano** (Figure 4.9). They are so-called because they roughly resemble the shape of a warrior's shield. Shield volcanoes are built primarily of basaltic lava flows and contain only a small percentage of pyroclastic material.

Mauna Loa is one of five shield volcanoes that together make up the island of Hawaii. Its base rests on the ocean floor 5000 meters (16,400 feet) below sea level, and its summit is 4170 (13,677 feet) above the water, giving it a total height approaching 6 miles, greater than the height of Mount Everest. Nearly a million years and numerous eruptive cycles built this truly gigantic pile of volcanic rock. Many other volcanic structures, including Midway Island and the Galapagos Islands, have been built in a similar manner from the ocean's depths.

Perhaps the most active and intensively studied shield volcano is Kilauea, also on the island of Hawaii, on the flank of the larger Mauna Loa. Kilauea has erupted more than 50 times in recorded history and is still active today. Several months before an eruptive phase, Kilauea's summit inflates as magma rises from its source 60 kilometers or more below the surface. This molten rock gradually works its way upward and accumulates in smaller reservoirs 3 to 5 kilometers below the summit. For up to 24 hours in advance of each eruption, swarms of small earthquakes warn of the impending activity.

The longest and largest rift eruption ever recorded at Kilauea began in 1983 and continues as this is written. The eruption began along a 6.5-kilometer (4-mile) fissure in an inaccessible forested area east of the summit caldera (Figure 4.10). By the summer of 1986, the eruptions shifted 3 kilometers downslope. Here, smooth-surfaced pahoehoe lava formed a lava lake. Eventually, it overflowed, and the fast-moving pahoehoe destroyed nearly a hundred rural homes, covered a major roadway, and eventually flowed into the sea. Lava has been intermittently pouring into the ocean since that time, adding new land to the island.

Box 4.2

Volcano Sizes

In the section on the three basic volcano types (shield, cinder cone, composite cone), we often mentioned their sizes. It is useful to compare the three types side-by-side, as in Figure 4.B.

Smallest are the cinder cones. Parícutin rose from a Mexican cornfield during a period of just nine years. Because their eruption history is short, most cinder cones do not exceed 300 meters (1000 feet) in height.

In contrast to cinder cones, composite cones and shield volcanoes have recurring eruptions that can span a million years or longer. As shown in Figure 4.B, Mount Rainier, one of the largest composite cones in the Cascade Range, dwarfs a relatively large cinder cone, Sunset Crater in Arizona. Sunset Crater formed over a few years, but Mount Rainier gradually formed over the past 700,000 years.

Mauna Loa, one of five shield volcanoes that compose the island of Hawaii, is considered the world's largest active volcano (Figure 4.B).

This massive pile of basaltic lava has an estimated volume of 40,000 cubic kilometers that was extruded over approximately a million years. From its base on the floor of the Pacific Ocean to its summit, Mauna Loa approaches 9 kilometers in height.

Despite its enormous size, Mauna Loa is not the largest known volcano in the solar system. Olympus Mons, a huge shield volcano on Mars, is 25 kilometers high and 600 kilometers wide (see Chapter 22).

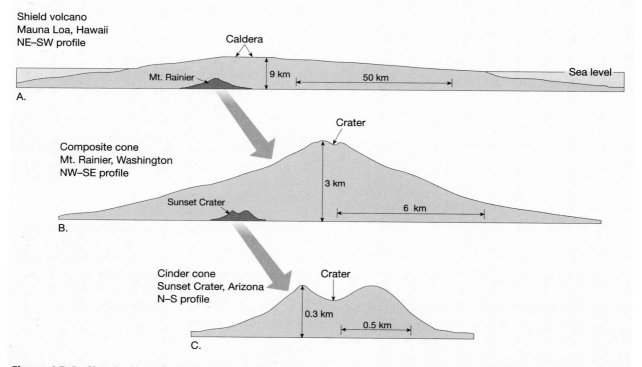

Figure 4.B Profiles of volcanic landforms. **A.** Profile of Mauna Loa, Hawaii, the largest shield volcano in the Hawaiian chain. Note size comparison with Mt. Rainier, Washington, a large composite cone. **B.** Profile of Mt. Rainier, Washington. Note how it dwarfs a typical cinder cone. **C.** Profile of Sunset Crater, Arizona, a typical steep-sided cinder cone.

Eruptions of this type are typical of shield volcanoes and have been occurring sporadically on the island of Hawaii for nearly a million years. The result is the formation of mountains with summits over 9000 meters (30,000 feet) above the seafloor. They are the tallest mountains on Earth (see Box 4.3).

There is now general agreement that the early stages of a shield volcano's formation consist of frequent eruptions of thin flows of very fluid basalts. As the structure enlarges, flank eruptions occur along with the summit eruptions. Collapse of the summit area frequently follows each eruptive phase. In the later stages of growth, the activity is more sporadic and pyroclastic ejections are more prevalent. Further, the lavas increase in viscosity, resulting in thicker, shorter flows. These activities tend to steepen the slope of the summit area.

Figure 4.9 Mauna Loa is one of five shield volcanoes that together make up the island of Hawaii. Shield volcanoes are built primarily of fluid basaltic lava flows and contain only a small percentage of pyroclastic materials. (Photo by Greg Vaughn)

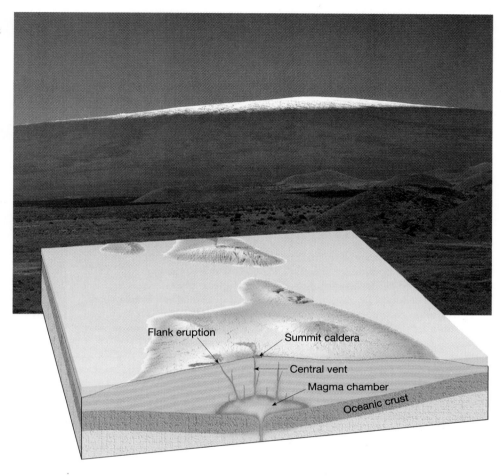

Flank eruption

Summit caldera

Central vent

Magma chamber

Oceanic crust

Figure 4.10 Lava extruded along the East Rift Zone, Kilauea, Hawaii. (Photo by Greg Vaughn)

Box 4.3

Loihi: The Next Hawaiian Island?

Loihi seamount is an undersea volcano (seamount) rising some 3500 meters (11,500 feet) above the floor of the Pacific Ocean. Although hidden beneath the waves, it is taller than Mount St. Helens prior to its explosive 1980 eruption. Situated 32 kilometers (20 miles) off the southeast coast of Hawaii's Big Island, Loihi was found to be active in the 1970s (Figure 4.C). Along with Kilauea, Loihi sits on the flank of Mauna Loa, a huge and still active shield volcano that makes up a substantial portion of the Big Island. All three derive lava from the same hot-spot source.

Loihi's most recent period of activity dates from July 1996 when the largest swarm of earthquakes ever recorded at any Hawaiian volcano (more than 4000) shook the seamount. Subsequent evidence of ongoing activity at Loihi was gathered in August and September 1997 during dives by the submersible vessel *Pisces*. It was at this time that scientists discovered high-temperature (200°C) hydrothermal (hot water) venting at the seamount's summit.

Will Loihi become the next Hawaiian Island? Any answer, of course, is speculative. If it is going to become an island, Loihi must build upward at least another 930 meters (3100 feet). Given the current rate at which the seamount is growing, it is likely to break the surface of the Pacific Ocean at some time in the distant future, most likely tens of thousands of years.

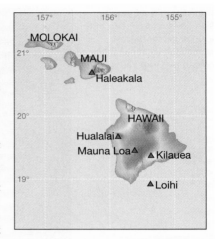

Figure 4.C Location map of Loihi seamount and other active volcanoes in the Hawaiian chain.

This explains why Mauna Kea, an older and inactive volcano to the north, has a steeper summit than does Mauna Loa.

Interestingly, shield volcanoes exist elsewhere in the solar system. Olympus Mons, a huge shield volcano on Mars, is the largest yet located. Its summit caldera is nearly large enough to contain the above-water portion of Mauna Loa.

Cinder Cones

As the name suggests, **cinder cones** are built from ejected lava fragments (Figure 4.11). Loose pyroclastic material has a high angle of repose (between 30 and 40 degrees), the steepest angle at which the material remains stable. Thus, volcanoes of this type have very steep slopes. Cinder cones are rather small, usually less than 300 meters (1000 feet) high, often forming near larger volcanoes, and often in groups.

One of the very few volcanoes observed by geologists from beginning to end is the cinder cone called Parícutin about 200 miles west of Mexico City. In 1943, it erupted in a cornfield owned by Dionisio Pulido, who witnessed the event as he prepared the field for planting.

For 2 weeks prior to the first eruption, numerous Earth tremors caused apprehension in the village of Parícutin about 3.5 kilometers away. Then sulfurous smoke began billowing from a small hole that had been in the cornfield for as long as Señor Pulido could remember. During the night, hot, glowing rock fragments thrown into the air from the hole produced a spectacular fireworks display. In one day the cone grew to 40 meters (130 feet) and by the fifth day it was over 100 meters (330 feet) high. Explosive eruptions threw hot fragments 1000 meters (3300 feet) above the crater rim. Larger fragments fell near the crater, some remaining incandescent as they rolled down the slope. These built an aesthetically pleasing cone, while finer ash fell over a much larger area, burning and eventually covering the village of Parícutin. Within 2 years the cone attained its final height of about 400 meters (1300 feet).

The first lava flow came from a fissure that opened just north of the cone, but after a few months flows began to emerge from the base of the cone itself. In June 1944, a clinkery aa flow 10 meters thick moved over much of the village of San Juan Parangaricutiro, leaving only the church steeple exposed (Figure 4.12). After 9 years, the activity ceased almost as quickly as it had begun. Today, Parícutin is just another one of the numerous quiet cinder cones dotting the landscape in this region of Mexico. Like the others, it will probably not erupt again.

Composite Cones

Earth's most picturesque volcanoes are composite cones. Most active composite cones are in a narrow zone that encircles the Pacific Ocean, appropriately named the *Ring of Fire*. In this region are Fujiyama (Mt. Fuji) in Japan, Mount Mayon in the Philippines, and the picturesque volcanoes of the Cascade Range in the

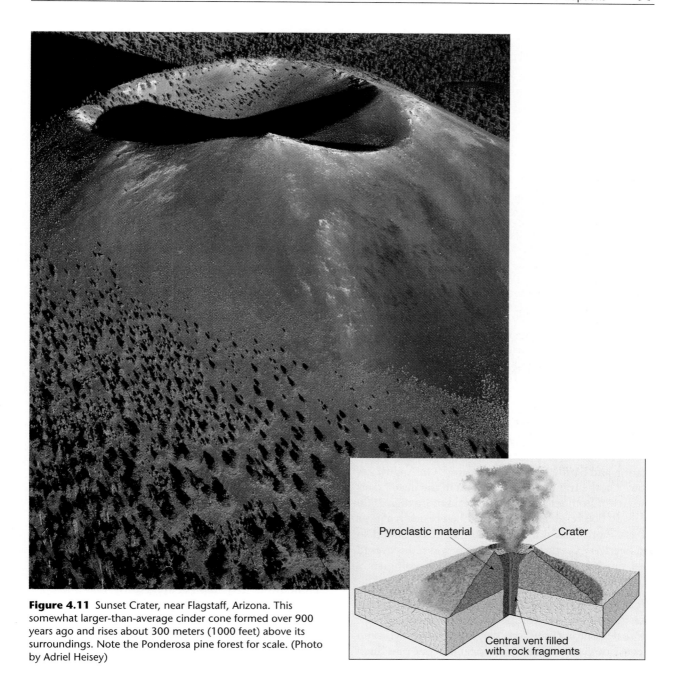

Figure 4.11 Sunset Crater, near Flagstaff, Arizona. This somewhat larger-than-average cinder cone formed over 900 years ago and rises about 300 meters (1000 feet) above its surroundings. Note the Ponderosa pine forest for scale. (Photo by Adriel Heisey)

Pyroclastic material

Crater

Central vent filled
with rock fragments

northwestern United States, including Mount St. Helens, Mount Ranier, and Mount Shasta (Figure 4.13).

A **composite cone** or **stratovolcano** is a large, nearly symmetrical structure composed of alternating lava flows and pyroclastic deposits, emitted mainly from a central vent. Just as shield volcanoes owe their shape to the highly fluid nature of the extruded lavas, so too do composite cones reflect the nature of the erupted material.

Composite cones are produced when relatively viscous lavas of andesitic composition are extruded. A composite cone may extrude viscous lava for long periods. Then, suddenly, the eruptive style changes and the volcano violently ejects pyroclastic material. Occasionally, both activities occur simultaneously. The resulting cone consists of alternating layers (strata) of lava and pyroclastic materials, giving rise to the name *stratovolcano*. Two of the most perfect cones, Mount Mayon in the Philippines and Fujiyama in Japan, exhibit the classic form of the stratovolcano with its steep summit and more gently sloping flanks.

Composite cones produce some of the most violent volcanic activity. Their eruption can be unexpected and devastating, as was the A.D. 79 eruption of

Figure 4.12 The village of San Juan Parangaricutiro engulfed by aa lava from Parícutin, shown in the background. Only the church towers remain. (Photo by Tad Nichols)

the Italian volcano we now call Vesuvius. Prior to this eruption, Vesuvius had been dormant for centuries. Although minor earthquakes probably warned of the events to follow, Vesuvius was covered with dense vegetation and hardly looked threatening. On August 24, however, the tranquility ended, and in the next 3 days the city of Pompeii (near Naples) and more than 2000 of its 20,000 residents were entombed beneath a layer of ash more than 6 meters thick. They remained so for nearly 17 centuries, until the city was rediscovered and much of it excavated, giving archaeologists a glimpse of ancient Roman life.

Nuée Ardente; A Deadly Pyroclastic Flow. Although the destruction of Pompeii was catastrophic, even more devastating eruptions may occur when a volcano ejects hot gases infused with incandescent ash. Such events produce a fiery pyroclastic flow called a **nuée ardente**. Also referred to as *glowing avalanches*, these turbulent steam clouds and companion ash flows race down steep volcanic slopes at speeds that can approach 200 kilometers (125 miles) per hour (Figure 4.14). The ground-hugging portions of glowing avalanches are rich in particulate matter, which is suspended by hot, buoyant gases. Thus, these flows, which can include larger rock fragments in addition to ash, travel downslope in a nearly frictionless environment cushioned by expanding volcanic gases. This explains why some nuée ardente deposits extend more than 100 kilometers (60 miles) from their source.

In 1902, a nuée ardente from Mount Pelée, a small volcano on the Caribbean island of Martinique, destroyed the port town of St. Pierre. The destruction happened in moments and was so devastating that almost all of St. Pierre's 28,000 inhabitants were killed. Only a shoemaker, a prisoner protected in a dungeon, and a few people on ships in the harbor were spared (Figure 4.15). Satis N. Coleman, in *Volcanoes, New and Old*, relates a vivid account of this event, which lasted less than 5 minutes.

I saw St. Pierre destroyed. The city was blotted out by one great flash of fire. Nearly 40,000 people were killed at once. Of eighteen vessels lying in the roads, only one, the British steamship *Roddam*, escaped and she, I hear, lost more than half of those on board. It was a dying crew that took her out. Our boat, the *Roraima*, arrived at St. Pierre early Thursday morning. For hours before entering the roadstead we could see flames and smoke rising from Mt. Pelée....The spectacle was magnificent. As we approached St. Pierre we could distinguish the rolling and leaping of red flames that belched from the mountain in huge volumes and gushed into the sky. Enormous clouds of black smoke hung over the volcano. There was a constant muffled roar. It was like the biggest oil refinery in the world burning up on the mountain top. There was a tremendous explosion about 7:45, soon after we

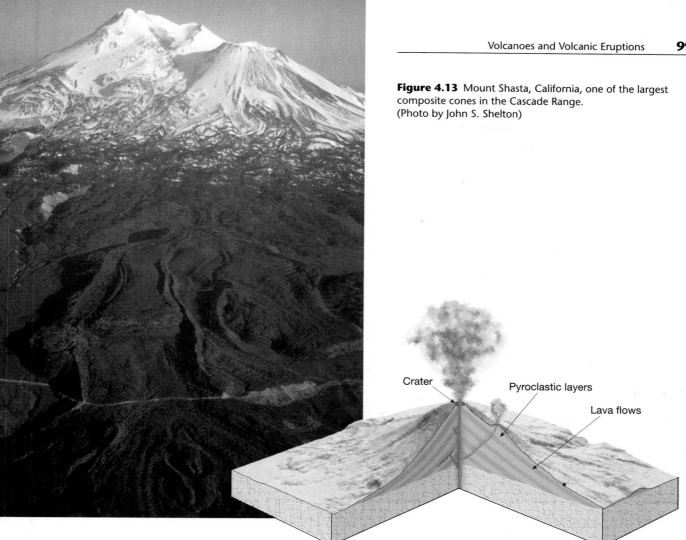

Figure 4.13 Mount Shasta, California, one of the largest composite cones in the Cascade Range. (Photo by John S. Shelton)

Crater   Pyroclastic layers   Lava flows

got in. The mountain was blown to pieces. There was no warning. The side of the volcano was ripped out and there was hurled straight toward us a solid wall of flame. It sounded like a thousand cannons.

The wave of fire was on us and over us like a flash of lightning. It was like a hurricane of fire. I saw it strike the cable steamship *Grappler* broadside on, and capsize her. From end to end she burst into flames and then sank. The fire rolled in mass straight down upon St. Pierre and the shipping. The town vanished before our eyes.

The air grew stifling hot and we were in the thick of it. Wherever the mass of fire struck the sea, the water boiled and sent up vast columns of steam.... The blast of fire from the volcano lasted only a few minutes. It shriveled and set fire to everything it touched. Thousands of casks of rum were stored in St. Pierre, and these were exploded by the terrific heat. The burning rum ran in streams down every street and out into the sea. This blazing rum set fire to the *Roraima* several times.... Before the volcano burst, the landings of St. Pierre were covered with people. After the explosion, not one living soul was seen on land.*

It is interesting to contrast the destruction of St. Pierre with that of Pompeii. In 3 days, Pompeii was completely buried, whereas St. Pierre was destroyed in moments and its remains were mantled by only a thin layer of volcanic debris. The structures of Pompeii remained intact, except for roofs that collapsed under the weight of the ash. In St. Pierre, masonry walls nearly a meter thick were knocked over like dominoes; large trees were uprooted and cannons were torn from their mounts. Clearly, volcanic hazards vary from one volcano to another.

*New York: John Day, 1946, pp. 80–81.

Figure 4.14 Nuée ardente races down the slope of Mount St. Helens on August 7, 1980, at speeds in excess of 100 kilometers (60 miles) per hour. (Photo by Peter W. Lipman, U.S. Geological Survey)

Lahar. In addition to their violent eruptions, large composite cones often generate a mudflow called by its Indonesian name **lahar**. These destructive flows occur when volcanic ash and debris become saturated with water and flow down steep volcanic slopes, generally following stream valleys. Some lahars are produced when rainfall saturates volcanic deposits, whereas others are triggered as large volumes of ice and snow melt during an eruption.

When Mount St. Helens erupted in 1980, several lahars formed. The flows and accompanying floods raced down the valleys of the north and south forks of the Toutle River at speeds exceeding 30 kilometers per hour. Water levels reached 4 meters above flood stage, and nearly all the homes and bridges along the river were destroyed or severely damaged. Fortunately, the affected area was not densely populated. This was not the case in 1985, when Nevado del Ruiz, a volcano in the Andes, erupted and generated a lahar that killed nearly 20,000 people (see the section entitled "Debris Flow" in Chapter 9).

Other Volcanic Landforms

The most obvious volcanic landform is a cone. But other distinctive landforms are also associated with volcanic activity.

Figure 4.15 St. Pierre as it appeared shortly after the eruption of Mount Pelée, 1902. (Reproduced from the collection of the Library of Congress)

Calderas and Pyroclastic Flows

Most volcanoes have a steep-walled *crater*. A crater is called a *caldera* when it exceeds 1 kilometer in diameter. Calderas are roughly circular and most form when the summit of a volcanic structure collapses into a partially emptied magma chamber below (Figure 4.16).

Crater Lake in Oregon, located in such a structure, is 10 kilometers (6 miles) at its widest and 1175 meters (over 3800 feet) deep. The formation of Crater Lake began about 7000 years ago when the volcano, later to be named Mount Mazama, violently extruded 50 to 70 cubic kilometers of pyroclastic material. With the loss of support, 1500 meters (nearly a mile) of this once-prominent 3600-meter cone collapsed. After the collapse, rainwater filled the caldera. Later volcanic activity built a small cinder cone in the lake called Wizard Island, which today provides a mute reminder of past activity (Figure 4.17).

Although most calderas are produced by *collapse following an explosive eruption*, some are not. For example, Hawaii's active shield volcanoes, Mauna Loa and Kilauea, have large calderas (3 to 5 kilometers, or 2 to 3 miles across, and nearly 200 meters, or 650 feet deep). They formed by gradual subsidence as magma slowly drained from the summit magma chambers during flank eruptions.

At least 138 calderas that exceed 5 kilometers in diameter are known. Unlike Crater Lake, these calderas are so large and irregular that many remained undetected until high-quality aerial, or satellite, images became available. One of these, LaGarita, located in the San Juan Mountains of southern Colorado, is about 32 kilometers wide and 80 kilometers long. Even today, with modern mapping techniques, the entire outline of this structure is still not known with certainty. The formation of a large caldera begins when a granitic (felsic) magma body is emplaced near the surface, upwarping the overlying rocks. Next, fracturing of the roof allows the highly viscous, gas-rich magma to reach the surface, where it explosively ejects huge volumes of pyroclastic materials, mainly ash and pumice fragments. Typically, this material, called a **pyroclastic flow**, avalanches across the landscape at speeds that may exceed 100 kilometers (60 miles) per hour, destroying most living things in its path. After coming to rest, these hot fragments often fuse to form a welded tuff that closely resembles a solidified lava flow. Finally, with the loss of support the roof collapses, generating a caldera. The cycle of caldera formation can repeat itself numerous times in the same location.

Perhaps the best-known caldera in the United States is located in the Yellowstone Plateau of northwestern Wyoming. Here a large viscous magma body exists a few kilometers below the surface. Several times over the past two million years, fracturing of

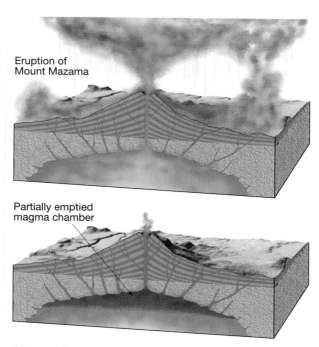

Eruption of Mount Mazama

Partially emptied magma chamber

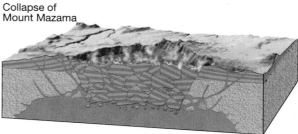

Collapse of Mount Mazama

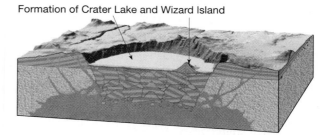

Formation of Crater Lake and Wizard Island

Figure 4.16 Sequence of events that formed Crater Lake, Oregon. About 7000 years ago, a violent eruption partly emptied the magma chamber, causing the summit of former Mount Mazama to collapse. Rainfall and groundwater contributed to form Crater Lake, the deepest lake in the United States. Subsequent eruptions produced the cinder cone called Wizard Island. (After H. Williams, *The Ancient Volcanoes of Oregon*, p. 47. Courtesy of the University of Oregon)

the rocks overlying the magma chamber has resulted in the ejection of huge quantities of pyroclastic material. One of the eruptions ejected 2500 cubic kilometers of magma in a devastating pyroclastic flow. (Fortunately, no eruption of this type has occurred in historic times.) In Yellowstone National Park, several fossil forests have been discovered, one above another. Following volcanic activity, a forest would develop

Figure 4.17 Crater Lake occupies a caldera about 10 kilometers (6 miles) in diameter. (Photo by Greg Vaughn/Tom Stack and Associates)

on a newly-formed volcanic surface, only to be covered by ash from the next eruptive phase. Other examples of large calderas that have formed in a similar manner are California's Long Valley Caldera and Valles Caldera located west of Los Alamos, New Mexico.

Fissure Eruptions and Lava Plateaus

We think of volcanic eruptions as building a cone or mountain from a central vent. But by far the greatest volume of volcanic material is extruded from fractures in the crust called **fissures**. Rather than building a cone, these long, narrow cracks pour forth a low-viscosity Hawaiian-type lava, blanketing a wide area.

The extensive Columbia Plateau in the northwestern United States was formed this way (Figure 4.18). Here, numerous **fissure eruptions** extruded very fluid basaltic lava. Successive flows, some 50 meters thick, buried the existing landscape as they built a lava plateau that is nearly a mile thick in places (Figure 4.19). The fluidity is evident, because some lava remained molten long enough to flow 150 kilometers (90 miles) from its source. The term **flood basalts** appropriately describes these flows. For a more complete discussion of the origin of flood basalt provinces, see Box 19.2, "Hot Spots and Flood Basalts."

Although a few large continental areas are covered with basalt flows, the greatest activity of this type occurs hidden from view on the ocean floor. Along oceanic ridges, where seafloor spreading is active, fissure eruptions generate new seafloor.

Iceland, which is located astride the Mid-Atlantic Ridge, has experienced numerous fissure eruptions of this type. The largest Icelandic eruptions in historic times occurred in 1783 and were named the Laki eruptions. A rift 25 kilometers long generated over 20 separate vents, which initially extruded sulfurous gases and ash deposits that built several small cinder cones. This activity was followed by huge outpourings of very fluid basaltic lava. The total volume of lava extruded by the Laki eruptions was in excess of 12 cubic kilometers. Volcanic gases stunted grasslands and directly killed most of Iceland's livestock. The ensuing famine caused 10,000 deaths.

Lava Domes

In contrast to mafic lavas, silica-rich lavas near the felsic end of the compositional spectrum are so viscous they hardly flow. As the thick lava is "squeezed" out of the vent, it may produce a bulbous mass of congealed lava called a **lava dome** (see Box 4.4). Because domes develop from viscous magma, most are composed of rhyolite, and many consist of obsidian.

Most volcanic domes develop following an explosive eruption of a gas-rich magma. This is exemplified by the volcanic dome that continues to "grow" from the vent that produced the 1980 eruption of Mount St. Helens (Figure 4.20). Although most volcanic domes form in association with preexisting composite cones, some form independently, such as the line of rhyolitic and obsidian domes at Mono Craters, California.

Figure 4.18 Volcanic areas in the northwestern United States. The Columbia River basalts cover an area of nearly 200,000 square kilometers (80,000 square miles). Activity here began about 17 million years ago as lava began to pour out of large fissures, eventually producing a basalt plateau with an average thickness of more than 1 kilometer. (After U.S. Geological Survey)

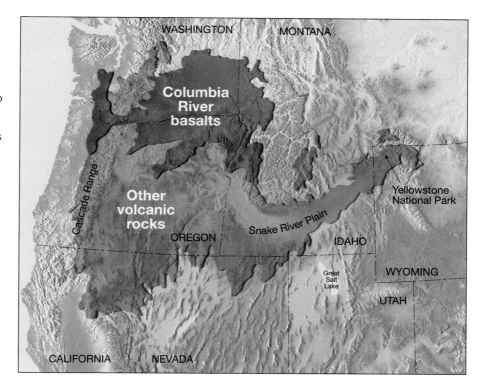

Volcanic Pipes and Necks

Most volcanoes are fed magma through short conduits, called *pipes*, that connect a magma chamber to the surface. In rare circumstances, pipes may extend tubelike to depths exceeding 200 kilometers. When this occurs, the ultramafic magmas that migrate up these structures produce rocks that are thought to be samples of the mantle that have undergone very little alteration during their ascent. Geologists consider these unusually deep conduits to be "windows" into Earth, for they allow us to view rock normally found only at great depth.

The best-known volcanic pipes are the diamond-bearing structures of South Africa. Here, the rocks filling the pipes originated at depths of at least 150 kilometers (90 miles), where pressure is high enough to generate diamonds and other high-pressure minerals. The task of transporting essentially unaltered magma (along with diamond inclusions) through 150 kilometers of solid rock is exceptional. This fact accounts for the scarcity of natural diamonds.

Volcanoes on land are continually being lowered by weathering and erosion. Cinder cones are easily eroded, because they are composed of unconsolidated materials. However, all volcanoes will eventually succumb to relentless erosion over geologic time. As erosion progresses, the rock occupying the volcanic pipe is often more resistant and may remain standing above the surrounding terrain long after most of the cone has vanished. Shiprock, New Mexico, is such a feature, and

Figure 4.19 Basalt flows near Idaho Falls. (Photo by John S. Shelton)

Box 4.4

Volcanic Crisis on Montserrat

The Caribbean's Lesser Antilles are mostly volcanic in origin and extend from near the northeast coast of South America in an arc toward Puerto Rico and the Virgin Islands (Figure 4.D). Near the beginning of the twentieth century, devastating eruptions occurred when volcanoes on Martinique (Mt. Pelée) and St. Vincent (Soufrière) killed more than 30,000 people. As the twentieth century comes to a close, the Caribbean is once again the focus of volcanologists. This time their attention is on the island of Montserrat.

This small island is dominated by the Soufriére Hills volcano, which began erupting in July 1995, after thousands of years of inactivity. The volcano, like most in the Caribbean, erupts viscous lava that oozes out at the surface forming a lava dome. Such domes have the potential to produce devastating explosions of pulverized rock, ash, and gases known as pyroclastic flows. Because there may be little warning, such eruptions are extremely dangerous.

The activity at Soufriére Hills volcano included many large pyroclastic flows that eventually covered large parts of the island. Moreover, plumes of volcanic ash were sometimes erupted to heights of 6000 meters (20,000 feet) or more. By January 1998, many of the island's nearly 12,000 residents had been evacuated to neighboring islands. To say the least, the erupting volcano caused serious hardship and economic distress to the people of Montserrat. On the positive side, loss of life was small.

Since the onset of eruptive activity, Soufriére Hills has become one of the most closely monitored volcanoes in the world. Almost immediately after the unexpected activity began, the Montserrat Volcano Observatory was established. Staffed by scientists from the University of the West Indies and the British Geological Survey, the mountain was wired with seismometers, tiltmeters, and gas analyzers. The valuable data being gathered may one day contribute to a reliable method of predicting volcanic eruptions.

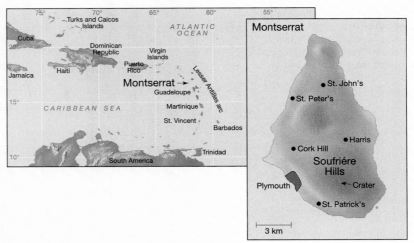

Figure 4.D Map of Caribbean and Lesser Antilles arc showing location of Montserrat and Soufriére Hills volcano.

Figure 4.E Homes in Plymouth, the largest town on Montserrat, buried in pyroclastic debris from Soufriére Hills Volcano, August 1997. (Photo by Alain Buu/GAMMA)

---

is called a **volcanic neck** (Figure 4.21). This structure, higher than many skyscrapers, is but one of many such landforms that protrude conspicuously from the red desert landscapes of the American Southwest.

 Plutonic Igneous Activity

Although volcanic eruptions can be among the most violent and spectacular events in nature and therefore worthy of detailed study, most magma is emplaced at

Figure 4.20 Following the May, 1980, eruption of Mount St. Helens, a lava dome began to develop. (Photo by David Falconer/DRK Photo)

Figure 4.21 Shiprock, New Mexico, is a volcanic neck. This structure, which stands over 420 meters (1380 feet) high, consists of igneous rock, which crystallized in the vent of a volcano that has long since been eroded away. (Photo by Adriel Heisey)

depth. Thus, an understanding of intrusive igneous activity is as important to geologists as the study of volcanic events.

The structures that result from the emplacement of igneous material at depth are called **plutons**, named for Pluto, the god of the lower world in classical mythology. Because all plutons form out of view beneath Earth's surface, they can be studied only after uplifting and erosion have exposed them. The challenge lies in reconstructing the events that generated these structures millions or even hundreds of millions of years ago.

For the sake of clarity, we have separated our discussions of volcanism and plutonic activity. Keep in mind, however, that these diverse processes occur simultaneously and involve basically the same materials.

Nature of Plutons

Plutons are known to occur in a great variety of sizes and shapes. Some of the most common types are illustrated in Figure 4.22. Notice that some of these structures have a tabular (tabletop) shape, whereas others are quite massive. Also, observe that some of these bodies cut across existing structures, such as layers of sedimentary rock; others form when magma is injected between sedimentary layers. Because of these differences, intrusive igneous bodies are generally classified according to their shape as either **tabular** (sheetlike) or **massive** and by their orientation with respect to the host rock. Plutons are said to be **discordant** if they cut across existing structures and **concordant** if they form parallel to features such as sedimentary strata. As you can see in Figure 4.22A, plutons are closely associated with volcanic activity. Many of the largest intrusive bodies are the remnants of magma chambers that once fed ancient volcanoes.

Dikes. **Dikes** are tabular discordant bodies that are produced when magma is injected into fractures. The force exerted by the emplaced magma can be great enough to separate the walls of the fracture further. Once crystallized, these sheetlike structures have thicknesses ranging from less than a centimeter to more than a kilometer. The largest have lengths of hundreds of kilometers. Most dikes, however, are a few meters thick and extend laterally for no more than a few kilometers.

Dikes are often found in groups that once served as vertically oriented pathways followed by molten rock that fed ancient lava flows. The parent pluton is generally not observable. Some dikes are found radiating, like spokes on a wheel, from an eroded volcanic neck. In these situations, the active ascent of magma is thought to have generated fissures in the volcanic cone out of which lava flowed.

Dikes often weather more slowly than the surrounding rock. When exposed by erosion, these dikes have the appearance of a wall, as shown in Figure 4.23.

Sills. **Sills** are tabular plutons formed when magma is injected along sedimentary bedding surfaces (Figure 4.24). Horizontal sills are the most common, although all orientations, even vertical, are known to exist. Because of their relatively uniform thickness and large areal extent, sills are likely the product of very fluid magmas. Magmas having a low silica content are more fluid, so most sills are composed of the rock basalt.

The emplacement of a sill requires that the overlying sedimentary rock be lifted to a height equal to the thickness of the sill. Although this is a formidable task, in shallow environments it often requires less energy than forcing the magma up the remaining distance to the surface. Consequently, sills form only at shallow depths, where the pressure exerted by the weight of overlying rock layers is low. Although sills are intruded between layers, they can be locally discordant. Large sills frequently cut across sedimentary layers and resume their concordant nature at a higher level.

One of the largest and most studied of all sills in the United States is the Palisades Sill. Exposed for 80 kilometers along the west bank of the Hudson River in southeastern New York and northeastern New Jersey, this sill is about 300 meters thick. Because of its resistant nature, the Palisades Sill forms an imposing cliff that can be seen easily from the opposite side of the Hudson.

In many respects, sills closely resemble buried lava flows. Both are tabular and often exhibit columnar jointing (Figure 4.25). **Columnar joints** form as igneous rocks cool and develop shrinkage fractures that produce elongated, pillarlike columns. Further, because sills generally form in near-surface environments and may be only a few meters thick, the emplaced magma often cools quickly enough to generate an aphanitic texture.

When attempts are made to reconstruct the geologic history of a region, it becomes important to differentiate between sills and buried lava flows. Fortunately, under close examination these two phenomena can be readily distinguished. The upper portion of a buried lava flow usually contains voids produced by escaping gas bubbles. Further, only the rocks beneath a lava flow show evidence of metamorphic alteration. Sills, on the other hand, form when magma has been forcefully intruded between sedimentary layers. Thus, fragments of the overlying rock can be found only in sills. Lava flows, conversely, are extruded before the overlying strata are deposited. Further, "baked" zones in the rock above and below are trademarks of a sill.

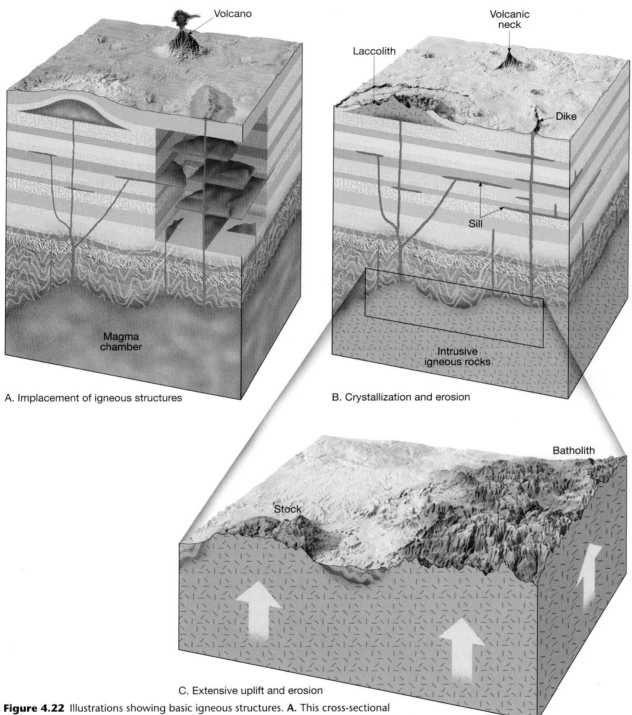

Figure 4.22 Illustrations showing basic igneous structures. **A.** This cross-sectional view shows the relationship between volcanism and intrusive igneous activity. **B.** This view illustrates the basic intrusive igneous structures, some of which have been exposed by erosion long after their formation. **C.** After millions of years of uplifting and erosion a stock and batholith are exposed at the surface.

Laccoliths. **Laccoliths** are similar to sills because they form when magma is intruded between sedimentary layers in a near-surface environment. However, the magma that generates laccoliths is more viscous. This less fluid magma collects as a lens-shaped mass that arches the overlying strata upward (see Figure 4.22). Consequently, a laccolith can occasionally be detected because of the dome-shaped bulge it creates at the surface.

Figure 4.23 The vertical structure in the foreground is a dike, which is more resistant to weathering than the surrounding rock. This dike is located west of Granby, Colorado, near Arapaho National Forest. (Photo by P. Jay Fleisher)

Sill

Figure 4.24 Salt River Canyon, Arizona. The dark, essentially horizontal band is a sill of basaltic composition that intruded horizontal layers of sedimentary rock. (Photo by E. J. Tarbuck)

Most large laccoliths are probably not much wider than a few kilometers. The Henry Mountains in southeastern Utah are largely composed of several laccoliths believed to have been fed by a much larger magma body emplaced nearby (Figure 4.26).

Batholiths. By far the largest intrusive igneous bodies are **batholiths**. Most often, batholiths occur in groups that form linear structures several hundreds of kilometers long and up to 100 kilometers wide, as shown in Figure 4.27. The Idaho batholith, for example, encompasses an area of more than 40,000 square kilometers and consists of many plutons. Indirect evidence gathered from gravitational studies indicates that batholiths are also very thick, possibly extending dozens of kilometers into the crust. Based on the amount exposed by erosion, some batholiths are at least several kilometers thick.

By definition, a plutonic body must have a surface exposure greater than 100 square kilometers (40 square miles) to be considered a batholith. Smaller plutons of this type are termed **stocks**. Many stocks appear to be portions of batholiths that are not yet fully exposed.

Batholiths usually consist of rock types having chemical compositions toward the granitic end of the spectrum, although diorite is commonly found. Smaller batholiths can be rather simple structures composed almost entirely of one rock type. However, studies of large batholiths have shown that they consist of several distinct plutons that were intruded over a period of millions of years. The plutonic activity that created the Sierra Nevada batholith, for example, occurred nearly continuously over a 130-million-year period that ended about 80 million years ago during the Cretaceous period.

Batholiths may compose the core of mountain systems. Here uplifting and erosion have removed the surrounding rock, thereby exposing the resistant igneous body. Some of the highest peaks in the Sierra Nevada, such as Mount Whitney, are carved from such a granitic mass.

Large expanses of granitic rock also occur in the stable interiors of the continents, such as the Canadian Shield of North America. These relatively flat exposures are the remains of ancient mountains that have long since been leveled by erosion. Thus, the rocks that make up the batholiths of youthful mountain ranges, such as the Sierra Nevada, were generated near the top of a magma chamber, whereas in shield areas, the roots of former mountains and, thus, the lower portions of batholiths, are exposed. In Chapter 20, we will further consider the role of igneous activity as it relates to mountain building.

Figure 4.25 Columnar jointing in basalt, Giants Causeway National Park, Northern Ireland. These five-to-seven-sited columns are produced by contraction and fracturing that results as a lava flow or sill gradually cools. (Photo by Tom Till)

Equally spaced centers of contraction

Figure 4.26 Laccolith exposed in the Henry Mountains of southern Utah. This dark, dome-shaped structure is one of several laccoliths in the Henry Mountains. Notice the upturned sedimentary beds that flank this intrusion of mafic igneous rocks. (Photo by John S. Shelton)

Emplacement of Batholiths

An interesting problem that faced geologists was trying to explain how large granitic batholiths form within largely unaltered sedimentary rocks. One group of geologists supported the idea that batholiths originated from magma that formed at depth and then migrated upward to its present position. This idea, however, presents a space problem. What happened to the rock originally in the location now occupied by these igneous masses? Further, the problem of explaining how magma is able to force its way through several kilometers of solid rock also plagued those supporting the magmatic origin of batholiths.

The group opposing this hypothesis suggested that granite batholiths originate when hot ion-rich

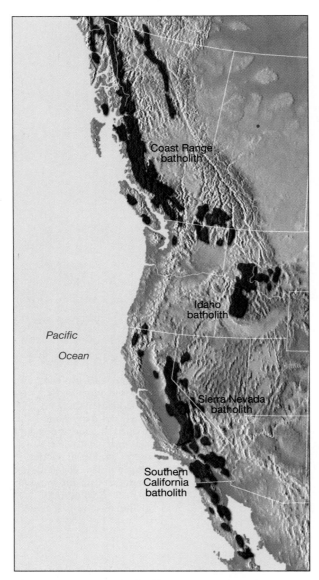

Figure 4.27 Granitic batholiths that occur along the western margin of North America. These gigantic, elongated bodies consist of numerous plutons that were emplaced during the last 150 million years of Earth history.

fluids and gases migrated through sedimentary rock and chemically altered the rock's composition. This essentially metamorphic process of converting rock into granite without passing through a molten stage is called *granitization*. Although granitization undoubtedly generates small quantities of granite, this process is clearly not capable of generating these large intrusive bodies.

This controversy was resolved when careful studies were made of structures called *salt domes*. These structures are of economic importance because they are found in close association with major oil-producing areas in the Gulf Coast states and in the Persian

Gulf region. Salt domes are produced where extensive salt deposits were buried by thousands of meters of sediment. Salt, which is less dense than overlying sediments, migrates very slowly upward. This is possible because salt behaves like a mobile fluid when it is subjected to stress over a long period of time. Because salt beds are not perfectly uniform, the zone of upward movement is thought to originate at a high spot. As the salt moves slowly upward, the stress exerted on the overlying sediments causes them to mobilize and be pushed aside (Figure 4.28A). Occasionally the salt breaches the surface, where it begins to flow outward, not unlike a very thick lava flow.

It is now generally accepted that batholiths are emplaced in a manner similar to salt domes (Figure 4.28B). Because magma is less dense than the surrounding rock, its buoyancy propels it upward. At depths of several kilometers, the overlying rocks are subjected to very high temperatures and pressures; thus, they deform plastically as the rising magma forcibly makes room for itself. As the magma continues to move upward, some of the host rock that was shouldered aside will fill in the space left by the magma body as it passes.*

As a magma body nears the surface, it encounters relatively cool, brittle rocks that resist deformation. Further upward movement, if any, is accomplished by a process called *stoping*, where the injected magma dislodges blocks of host rock from the roof of the magma chamber and incorporates them into the magma body. Evidence for this process are inclusions called **xenoliths**, the unmelted remnants of the host rock (Figure 4.29).

Magma gradually cools during its ascent, which reduces its rate of upward mobility. Further, because movement through the brittle uppermost crust is greatly restricted, most magma accumulates in chambers several kilometers below the surface. As a result, roughly 10 to 1000 times more magma is emplaced at depth than is involved in volcanic output.

Plate Tectonics and Igneous Activity

For many years geologists have realized that the global distribution of igneous activity is not random, but rather exhibits a definite pattern. In particular, volcanoes that extrude mainly intermediate to felsic lavas

---

*An analogous situation occurs when a can of oil-base paint is left in storage. The oil paint is less dense than the pigments used for coloration; thus, oil collects into drops that slowly migrate upward while the heavier pigments settle to the bottom.

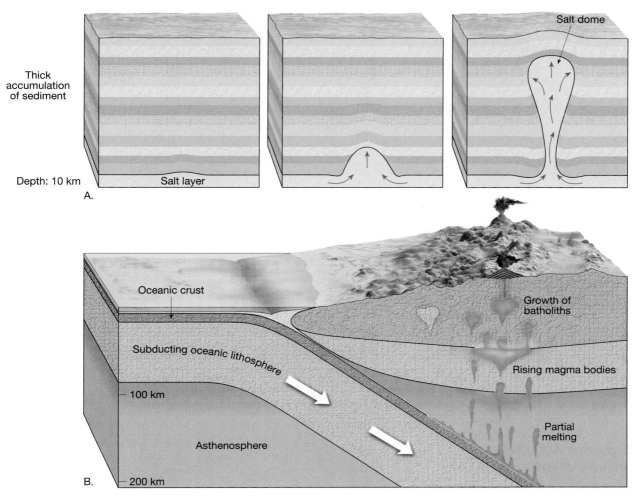

Figure 4.28 Many geologists suggest that the emplacement of a salt dome (Part **A**) is analogous to the processes involved in the emplacement of large magma bodies (Part **B**).

are confined largely to continental margins. By contrast, most volcanoes located within the ocean basins, such as those in Hawaii and Iceland, extrude lavas that are of mafic composition. Moreover, basaltic rocks are common in both oceanic and continental settings, whereas granitic rocks are rarely observed in the oceans. This pattern puzzled geologists until the development of the plate tectonics theory, which greatly clarified the picture.

Many of the more than 800 known active volcanoes are located along continental margins adjacent to oceanic trenches (Figure 4.30). Further, volcanic activity occurs along the oceanic ridge system. This later activity, although extensive, is hidden from view by the world's oceans.

In this section we will examine three zones of igneous activity and relate them to global tectonics. These active areas are along the oceanic ridges (spreading centers), adjacent to ocean trenches (subduction zones), and within the plates themselves.

Figure 4.29 The dark xenolith is an unmelted remnant of host rock that was incorporated into the igneous rocks that compose this batholith. (Photo by Tom Bean)

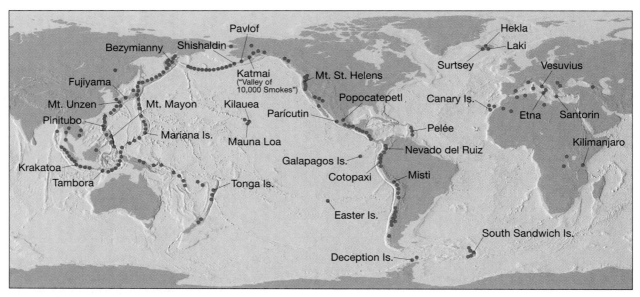

Figure 4.30 Locations of some of Earth's major volcanoes.

These three settings are shown in Figure 4.31, which highlights two examples of each. Please refer to Figure 4.31 as you read the following.

Igneous Activity at Spreading Centers

The greatest volume of volcanic rock is produced along the oceanic ridge system, where seafloor spreading is active (Figure 4.31). As plates of rigid lithosphere pull apart, pressure on the underlying rocks is lessened. This reduced pressure, in turn, lowers the melting temperature of the mantle rocks. Partial melting of these mantle materials (primarily peridotite) produces large quantities of basaltic magma that move upward to fill the newly formed cracks between the diverging plates. In this way new slivers of oceanic crust are generated.

Some of the molten basalt reaches the ocean floor, where it produces extensive lava flows or occasionally grows into a volcanic cone. Sometimes this activity produces a volcanic structure large enough to rise above sea level, such as the islands of Vestmann off the south coast of Iceland (Figure 4.32). Numerous submerged volcanoes also dot the flanks of the ridge system and the adjacent deep-ocean floor. Many of these formed along the ridge crest and gradually moved away as new oceanic crust was created by the seemingly unending process that generates new seafloor.

Igneous Activity at Subduction Zones

Recall that deep-ocean trenches are sites where slabs of water-rich oceanic crust are bent and descend into the mantle (Figure 4.33). As a slab sinks, volatiles are driven from the oceanic crust and migrate upward into the wedge-shaped piece of mantle located directly above. At a depth of 100 to 150 kilometers, these water-rich fluids reduce the melting point of the mantle's peridotite sufficiently to promote partial melting. In some environments silica and other components of the subducted sediments may become incorporated into the magma body. This process generates basaltic magma, and in some cases, small amounts of andesitic magma.

After a sufficient quantity of magma has accumulated, it slowly migrates upward because it is less dense than the surrounding rock. When igneous activity occurs along a subduction zone in the ocean, a chain of volcanoes called a *volcanic island arc* is produced. These structures, which usually form 200 to 300 kilometers from the oceanic trench, include such island chains as the Aleutians, the Tongas, and the Marianas.

When subduction of oceanic crust occurs beneath a continent, the magma that forms may become contaminated with silica-rich rocks as it moves through the crust. As you saw in Chapter 3, magmatic differentiation and the assimilation of crustal fragments into the ascending mantle-derived basaltic magma will change it into one exhibiting an

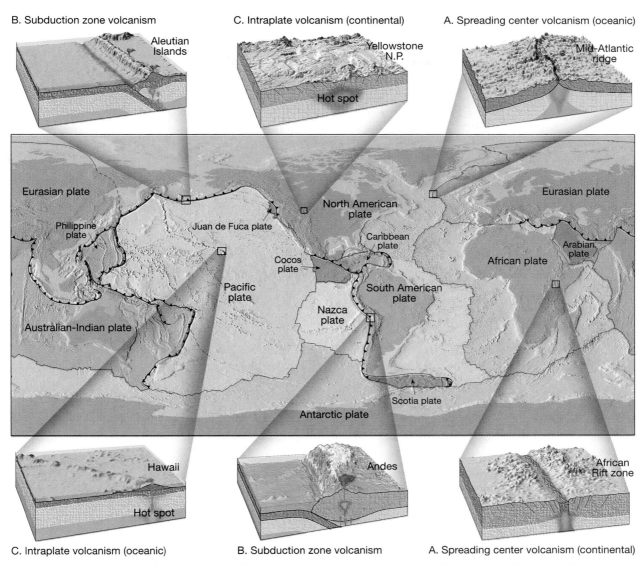

B. Subduction zone volcanism

Aleutian Islands

C. Intraplate volcanism (continental)

Yellowstone N.P.

Hot spot

A. Spreading center volcanism (oceanic)

Mid-Atlantic ridge

Eurasian plate

Philippine plate

Juan de Fuca plate

North American plate

Caribbean plate

Cocos plate

Pacific plate

South American plate

Nazca plate

Eurasian plate

Arabian plate

African plate

Australian-Indian plate

Scotia plate

Antarctic plate

C. Intraplate volcanism (oceanic)

Hawaii

Hot spot

B. Subduction zone volcanism

Andes

A. Spreading center volcanism (continental)

African Rift zone

Figure 4.31 Three zones of volcanism. Two of these zones are plate boundaries, and the third is the interior area of the plates.

Figure 4.32 Houses destroyed in the fishing port of Heimaey, off the south coast of Iceland, 1973. Cinders and ash covered much of the town and a thick aa flow nearly closed off the harbor entrance. (Photo by Wolfgang Kaehler)

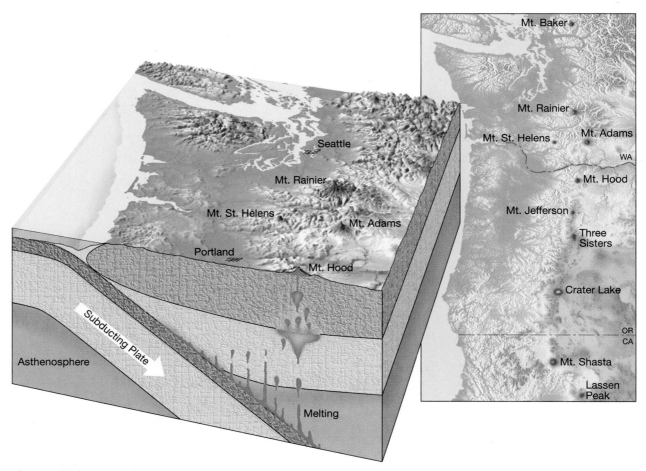

Figure 4.33 Locations of several of the larger composite cones that comprise the Cascade Range.

andesitic to granitic composition. The *continental margin volcanic arc* consisting of South America's Andes Mountains is one place where andesitic magma having such an origin is extruded.

Many subduction-zone volcanoes border the Pacific Basin. Because of this pattern, the region has come to be called the *Ring of Fire*. Here volcanism is associated with subduction of the Pacific seafloor. As oceanic plates sink, they carry sediments and oceanic crust containing abundant water to great depths. The presence of water contributes to the high gas content and explosive nature of volcanoes that make up the Ring of Fire. The volcanoes of the Cascade Region in the northwestern United States, including Mount St. Helens, Mount Rainier, and Mount Shasta, are all of this type (Figure 4.33).

Intraplate Igneous Activity

We know why igneous activity occurs along plate boundaries. But why do eruptions occur in the middle of plates? Equally perplexing is the activity in Yellowstone National Park where felsic pumice and ash flows actually overlap basalt flows that cover vast portions of the Pacific Northwest.

Because basalts having relatively similar compositions are found in the ocean basins as well as on the continents, partial melting of mantle rocks is the most probable source for many of these rocks. The source of some intraplate basaltic magma are hot **mantle plumes**, which may originate at the core–mantle boundary. Upon reaching the crust, these structures begin to spread laterally. The result is localized volcanic regions a few hundred kilometers across called **hot spots**. More than 100 hot spots have been identified and most appear to have persisted for tens of millions of years. One hot spot is situated beneath the island of Hawaii. Another may be responsible for the large outpourings of lava that make up Iceland.

With few exceptions, lavas and ash of felsic composition are restricted to vents located landward of the continental margins. This suggests that remelting of the continental crust may be the mechanism responsible for the formation of these silica-rich magmas. But what mechanism causes large quantities of continental material to be melted? One explanation is that a thick segment of continental crust occasionally becomes situated over a rising plume of hot mantle material. Rather than producing vast outpourings of basaltic lava as occurs at oceanic sites such as Hawaii, the magma from the rising plume is emplaced at depth. Here melting and assimilation of the surrounding host rock, coupled with magmatic differentiation, result in the formation of a highly evolved *secondary magma.* As this buoyant magma of intermediate to felsic composition slowly migrates upward, continued hot-spot activity supplies heat to the rising mass, thereby aiding its ascent. Volcanism in the Yellowstone region may have resulted from this process.

Although the plate tectonics theory has answered many questions regarding the distribution of igneous activity, many new questions have arisen: Why does seafloor spreading occur in some areas but not others? How do mantle plumes and associated hot spots originate? These and other questions are the subject of continuing geologic research.

Volcanoes and Climate

The idea that explosive volcanic eruptions change Earth's climate was first proposed many years ago and still appears to explain some aspects of climatic variability. Explosive eruptions emit huge quantities of gases and fine-grained debris into the atmosphere (Figure 4.34). The greatest eruptions inject material high into the stratosphere, where it spreads around the globe and remains for months or even years. The basic premise is that this suspended volcanic material (most importantly, droplets of sulfuric acid) will filter out a portion of the incoming solar radiation, and this, in turn, will lower air temperatures worldwide.

Perhaps the most notable cool period linked to a volcanic event is the "year without a summer" that followed the 1815 eruption of Mount Tambora in Indonesia. The abnormally cold spring and summer of 1816 in many parts of the Northern Hemisphere, including New England, are believed to have been caused by the cloud of volcanic debris and gases ejected from Tambora (Figure 4.34).

When Mount St. Helens erupted in 1980, there was almost immediate speculation: Can an eruption such as this change our climate? Although spectacular,

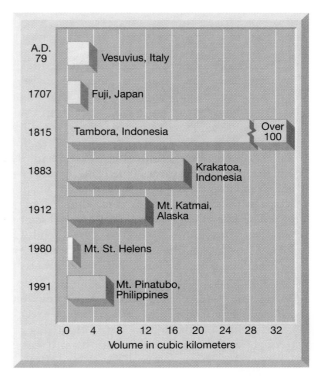

Figure 4.34 Approximate volume of volcanic debris emitted during some well-known eruptions. The 1815 eruption of Tambora, the largest known eruption in historic time, ejected over 100 times more ash than did Mount St. Helens in 1980.

a single explosive volcanic eruption of the magnitude of Mount St. Helens occurs somewhere in the world every 2 to 3 years. Studies of these events indicate that a very slight cooling of the lower atmosphere does occur. However, the cooling is so slight, less than one-tenth of one degree Celsius, as to be inconsequential.

The huge eruption of the Philippines' Mount Pinatubo in 1991 gave scientists another opportunity to test the connection between volcanism and climatic change. In 1991, this volcano emitted huge quantities of ash and 25 million to 30 million tons of sulfur dioxide gas. The gaseous component turned into a haze of tiny sulfuric acid droplets. Because this material was superhot, it rose to heights of 20 to 30 kilometers and quickly encircled the entire globe. This sun-blocking haze was probably the most massive since the eruption of the Indonesian volcano Krakatau in 1883. Studies indicated that the average global surface temperature declined by about 0.6°C (1°F) when compared to the average of the year before. Scientists hope to learn more about the volcano/climate connection as subsequent data are analyzed.

Chapter Summary

- The primary factors that determine the nature of volcanic eruptions include the magma's *composition*, its *temperature*, and the *amount of dissolved gases* it contains. As lava cools, it begins to congeal, and, as *viscosity* increases, its mobility decreases. The *viscosity of magma is directly related to its silica content*. Granitic (felsic) lava, with its high silica content (over 70 percent), is very viscous and forms short, thick flows. Basaltic (mafic) lava, with a lower silica content (about 50 percent), is more fluid and may travel a long distance before congealing. Dissolved gases tend to increase the fluidity of magma and, as they expand, provide the force that propels molten rock from the vent of a volcano.

- The materials associated with a volcanic eruption include (1) *lava flows* (*pahoehoe* flows, which resemble twisted braids, and *aa* flows, consisting of rough jagged blocks; both form from basaltic lavas); (2) *gases* (primarily *water vapor*); and (3) *pyroclastic material* (pulverized rock and lava fragments blown from the volcano's vent, which include *ash*, *pumice*, *lapilli*, *cinders*, *blocks*, and *bombs*.)

- Successive eruptions of lava from a central vent result in a montainous accumulation of material known as a *volcano*. Located at the summit of many volcanoes is a steep-walled depression called a *crater*. *Shield cones* are broad, slightly domed volcanoes built primarily of fluid, basaltic lava. *Cinder cones* have very steep slopes composed of pyroclastic material. *Composite cones*, or *stratovolcanoes*, are large, nearly symmetrical structures built of interbedded lavas and pyroclastic deposits. Composite cones produce some of the most violent volcanic activity. Often associated with a violent eruption is a *nuée ardente*, a fiery cloud of hot gases infused with incandescent ash that races down steep volcanic slopes. Large composite cones may also generate a type of mudflow known as a *lahar*.

- Most volcanoes are fed by conduits called *pipes*. As erosion progresses, the rock occupying the pipe is often more resistant and may remain standing above the surrounding terrain as a *volcanic neck*. The summits of some volcanoes have large,

nearly circular depressions that exceed 1 kilometer in diameter called *calderas*. Although volcanic eruptions from a central vent are the most familiar, by far the largest amounts of volcanic material are extruded from cracks in the crust called *fissures*. The term *flood basalts* describes the fluid, waterlike, basaltic lava flows that cover an extensive region in the northwestern United States known as the Columbia Plateau. When silica-rich magma is extruded, *pyroclastic flows* consisting largely of ash and pumice fragments usually result.

- Intrusive igneous bodies are classified according to their *shape* and by their *orientation with respect to the host rock*, generally sedimentary rock. The two general shapes are *tabular* (sheetlike) and *massive*. Intrusive igneous bodies that cut across existing sedimentary beds are said to be *discordant*; those that form parallel to existing sedimentary beds are *concordant*.

- *Dikes* are tabular, discordant igneous bodies produced when magma is injected into fractures that cut across rock layers. Tabular, concordant bodies, called *sills*, form when magma is injected along the bedding surfaces of sedimentary rocks. In many respects sills closely resemble buried lava flows. *Laccoliths* are similar to sills but form from less-fluid magma that collects as a lens-shaped mass that arches the overlying strata upward. *Batholiths*, the largest intrusive igneous bodies with surface exposures of more than 100 square kilometers (40 square miles), frequently make up the cores of mountains.

- *Most active volcanoes are associated with plate boundaries*. Active areas of volcanism are found along the oceanic ridges (*spreading center volcanism*), adjacent to ocean trenches (*subduction-zone volcanism*), as well as the interiors of plates themselves (*intraplate volcanism*).

- Explosive volcanic eruptions are regarded as an explanation for some aspects of earth's climatic variability. The basic premise is that suspended volcanic material will filter out a portion of the incoming solar radiation, which, in turn, will lower air temperature in the lower atmosphere.

Review Questions

1. What triggered the May 18, 1980, eruption of Mount St. Helens? (see Box 4.1)

2. List three factors that determine the nature of a volcanic eruption? What role does each play?

3. Why is a volcano fed by highly viscous magma likely to be a greater threat than a volcano supplied with very fluid magma?

4. Describe pahoehoe and aa lava.

5. List the main gases released during a volcanic eruption. Why are gases important in eruptions?

6. How do volcanic bombs differ from blocks of pyroclastic debris?

7. Compare a volcanic crater to a caldera.

8. Compare and contrast the main types of volcanoes (size, shape, eruptive style, and so forth).

9. Name a prominent volcano for each of the three types.

10. Briefly compare the eruptions of Kilauea and Parícutin.

11. Contrast the destruction of Pompeii with the destruction of St. Pierre.

12. Describe the formation of Crater Lake. Compare it to the caldera formed during the eruption of Kilauea.

13. What is Shiprock, New Mexico, and how did it form?

14. How do the eruptions that created the Columbia Plateau differ from eruptions that create volcanic peaks?

15. Where are fissure eruptions most common?

16. Extensive pyroclastic flow deposits are most often associated with which volcanic structures?

17. Describe each of the four intrusive features discussed in the text.

18. Why might a laccolith be detected at Earth's surface before being exposed by erosion?

19. What is the largest of all intrusive igneous bodies? Is it tabular or massive? Concordant or discordant?

20. Spreading center volcanism is associated with which rock type? What causes rocks to melt in regions of spreading center volcanism?

21. What is the Ring of Fire?

22. Are volcanoes in the Ring of Fire generally described as quiescent or violent? Name a volcano that would support your answer.

23. Describe the situation that generates magma in subduction-zone volcanism.

24. The Hawaiian Islands and Yellowstone are associated with which of the three zones of volcanism?

Key Terms

aa flow (p. 91)
batholith (p. 108)
caldera (p. 93)
cinder cone (p. 96)
columnar joint (p. 106)
composite cone (strato-volcano) (p. 97)
concordant (p. 106)
crater (p. 93)
discordant (p. 106)
dike (p. 106)

fissure (p. 102)
fissure eruption (p. 102)
flood basalt (p. 102)
fumarole (p. 93)
hot spot (p. 114)
laccolith (p. 107)
lahar (p. 100)
lava dome (p. 102)
lava tube (p. 91)
mantle plume (p. 114)
massive (p. 106)

nuée ardente (p. 98)
pahoehoe flow (p. 91)
parasitic cone (p. 93)
pillow lava (p. 91)
pipe (p. 93)
pluton (p. 106)
pyroclastic flow (p. 101)
pyroclastic material (p. 92)
shield volcano (p. 93)
sill (p. 106)

stock (p. 108)
tabular (p. 106)
vent (p. 93)
viscosity (p. 87)
volcanic neck (p. 104)
volcano (p. 93)
xenolith (p. 110)

Web Resources

The *Earth* Home Page provides on-line resources for this chapter on the World Wide Web. You will find review exercises, specific updates for items in the chapter, suggested reading, and links to interesting related pathways on the Internet. Visit the *Earth* Home Page at **http://www.prenhall.com/tarbuck**.

CHAPTER 5

Weathering and Soil

Weathering

Mechanical Weathering
Frost Wedging
Unloading
Thermal Expansion
Biological Activity

Chemical Weathering
Dissolution
Oxidation
Hydrolysis
Alterations Caused by Chemical Weathering

Rates of Weathering
Rock Characteristics
Climate
Differential Weathering

Soil
An Interface in the Earth System
What Is Soil?

Controls of Soil Formation
Parent Material
Time
Climate
Plants and Animals
Slope

The Soil Profile

Soil Types
Pedalfer
Pedocal
Laterite

Soil Erosion
How Soil Is Eroded
Rates of Erosion
Sedimentation and Chemical Pollution

Left Differential weathering exhibited by Delicate Arch, Arches National Park, Utah. *Photo by David Muench*

The surface of Earth is surprisingly dynamic, continuously changing over time. Two hundred years ago most people believed that mountains, lakes, and deserts were permanent features of an Earth that was thought to be no more than a few thousand years old. Today, however, we know that weathering and erosion gradually wear down mountains, lakes fill with sediment and vegetation or are drained by streams, and deserts come and go as climates change.

Earth is indeed a dynamic body. Volcanic forces and other mountain-forming processes elevate portions of Earth's surface, while opposing processes continually move material from higher elevations to lower elevations. The latter processes include:

1. **Weathering**—the physical breakdown (disintegration) and chemical alteration (decomposition) of rocks at or near Earth's surface.

2. **Mass wasting**—the transfer of rock and soil downslope under the influence of gravity.

3. **Erosion**—the physical removal of material by mobile agents such as water, wind, or ice.

In this chapter we will focus on rock weathering and the products generated by this activity. However, weathering cannot be easily separated from mass wasting and erosion because as weathering breaks rocks apart, erosion and mass wasting remove the rock debris (Figure 5.1). This transport of material by erosion and mass wasting further disintegrates and decomposes the rock.

Weathering

All materials are susceptible to weathering. Let's consider, for example, the fabricated product concrete, which closely resembles a sedimentary rock called conglomerate. A newly poured concrete sidewalk has a smooth, fresh, unweathered look. However, not many years later the same sidewalk will appear chipped, cracked, and rough, with pebbles exposed at the surface. If a tree is nearby, its roots may heave and buckle the concrete as well. The same natural processes that eventually break apart a concrete sidewalk also act to disintegrate rock.

Weathering occurs when rock is mechanically fragmented (disintegrated) and/or chemically altered (decomposed). **Mechanical weathering** is accomplished by physical forces that break rock into smaller and smaller pieces without changing the rock's mineral composition. **Chemical weathering** involves a chemical transformation of rock into one or more new compounds. These two concepts can be illustrated with a piece of paper. The paper can be disintegrated by tearing it into smaller and smaller pieces, whereas decomposition occurs when the paper is set afire and burned.

Why does rock weather? Simply, weathering is the response of Earth materials to a changing environment. For instance, after millions of years of uplift and erosion, the rocks overlying a large intrusive igneous body may be removed, exposing it at the surface. This mass of crystalline rock—formed deep below ground where temperatures and pressures are high—is now subjected to a very different and comparatively hostile surface environment. In response, this rock mass will gradually change. This transformation of rock is what we call weathering.

In the following sections we will discuss the various modes of mechanical and chemical weathering. Although we will consider these two processes separately, keep in mind that they usually work simultaneously in nature.

Mechanical Weathering

When a rock undergoes *mechanical weathering* it is broken into smaller and smaller pieces, each retaining the characteristics of the original material. The end result is many small pieces from a single large one. Figure 5.2 shows that breaking a rock into smaller pieces increases the surface area available for chemical attack. An analogous situation occurs when sugar is added to a liquid. In this situation, a cube of sugar will dissolve much slower than an equal volume of sugar granules because the cube has much less surface area available for dissolution. Hence, by breaking rocks into smaller pieces, mechanical weathering increases the amount of surface area available for chemical weathering.

In nature, four important physical processes lead to the fragmentation of rock: frost wedging, expansion resulting from unloading, thermal expansion, and biological activity. In addition, although the work of erosional agents such as wind, glacial ice, and running water is usually considered separately from mechanical weathering, it is nevertheless important. As these mobile agents move rock debris, they relentlessly disintegrate these materials.

Frost Wedging

Repeated cycles of freezing and thawing represent an important process of mechanical weathering. Liquid water has the unique property of expanding about 9 percent upon freezing, because water molecules in the regular crystalline structure of ice are farther apart than they are in liquid water near the freezing point. As a result, water freezing in a confined space exerts tremendous outward pressure on the walls of its container. To verify this, consider a tightly sealed glass jar filled with water. As the water freezes, the container is cracked.

Figure 5.1 Bryce Canyon, National Park, Utah. When weathering accentuates differences in rocks, spectacular landforms are sometimes created. As the rock gradually disintegrates and decomposes, mass wasting and erosion remove the products of weathering. (Photo by Barbara Gerlach/DRK Photo)

In nature, water works its way into cracks in rock and, upon freezing, expands and enlarges these openings. After many freeze–thaw cycles, the rock is broken into angular fragments. This process is appropriately called **frost wedging** (Figure 5.3). Frost wedging is most pronounced in mountainous regions where a daily freeze–thaw cycle often exists. Here, sections of rock are wedged loose and may tumble into large piles called **talus slopes** that often form at the base of steep rock outcrops (Figure 5.4).

Frost wedging also causes great destruction to highways in the northern United States, particularly in the early spring when the freeze–thaw cycle is well established. Roadways acquire numerous potholes and are occasionally heaved and buckled by this destructive force.

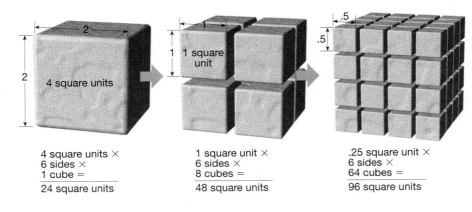

Figure 5.2 Chemical weathering can occur only to those portions of a rock that are exposed to the elements. Mechanical weathering breaks rock into smaller and smaller pieces, thereby increasing the surface area available for chemical attack.

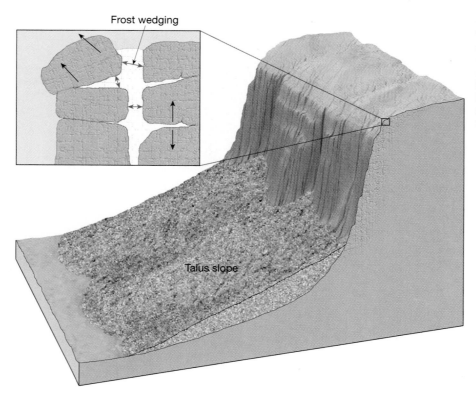

Figure 5.3 Frost wedging. As water freezes it expands, exerting a force great enough to break rock. When frost wedging occurs in a setting such as this, the broken rock fragments fall to the base of the cliff and create a cone-shaped accumulation known as a talus slope.

Unloading

When large masses of igneous rock, particularly granite, are exposed by erosion, concentric slabs begin to break loose. The process generating these onionlike layers is called **sheeting**. It is thought that this occurs, at least in part, because of the great reduction in pressure when the overlying rock is eroded away, a process called *unloading*. Accompanying this unloading, the outer layers expand more than the rock below, and thus separate from the rock body (Figure 5.5). Continued weathering eventually causes the slabs to separate and spall off, creating **exfoliation domes**. Excellent examples of exfoliation domes are Stone Mountain, Georgia, and Half Dome and Liberty Cap in Yosemite National Park (Figure 5.6).

Deep underground mining provides us with another example of how rocks behave once the confining pressure is removed. Large rock slabs have been known to explode off the walls of newly cut mine tunnels because of the abruptly reduced pressure. Evidence of this type, plus the fact that fracturing occurs parallel to the floor of a quarry when large

Figure 5.4 After frost wedging loosens rock fragments on a steep rock exposure, the angular pieces fall to the base of the cliff and produce a talus slope. Endicott Mountains, Brooks Range, Gates of the Arctic National Park, Alaska. (Photo by Tom Bean)

Figure 5.5 Sheeting is caused by the expansion of crystalline rock as erosion removes the overlying material. Fractures that roughly parallel the surface topography are common in large, intrusive granitic masses. Olmstead Point, Yosemite National Park, California. (Photo by Jeff Gnass/The Stock Market)

blocks of rock are removed, strongly supports the process of unloading as the cause of sheeting.

Although many fractures are created by expansion, others are produced by contraction during the crystallization of magma, and still others by tectonic forces during mountain building. Fractures produced by these activities generally form a definite pattern and are called *joints* (Figure 5.7). Joints are important rock structures that allow water to penetrate to depth and start the process of weathering long before the rock is exposed.

Thermal Expansion

The daily cycle of temperature may weaken rocks, particularly in hot deserts where daily variations may exceed 30°C. Heating a rock causes expansion and cooling causes contraction. Repeated swelling and shrinking of minerals with different expansion rates should logically exert some stress on the rock's outer shell.

Although this process was once thought to be of major importance in the disintegration of rock, laboratory experiments have not substantiated this. In one test, unweathered rocks were heated to temperatures much higher than those normally experienced on Earth's surface and then cooled. This procedure was repeated many times to simulate hundreds of years of weathering, but the rocks showed little apparent change.

Nevertheless, pebbles in desert areas do show evidence of shattering that may have been caused by temperature changes (Figure 5.8). A proposed solution to this dilemma suggests that rocks must first be weakened by chemical weathering before they can be broken down by thermal activity. Further, this process may be aided by the rapid cooling of a desert rainstorm.

Additional data are needed before a definite conclusion can be reached as to the impact of temperature variation on rock disintegration.

Biological Activity

Weathering is also accomplished by the activities of organisms, including plants, burrowing animals, and humans. Plant roots in search of nutrients and water grow into fractures, and as the roots grow, they wedge

Figure 5.6 Summit of Half Dome, an exfoliation dome in Yosemite National Park. (Photo by Breck Kent)

Figure 5.7 Aerial view of nearly parallel joints near Moab, Utah. (Photo by Michael Collier)

the rock apart (Figure 5.9). Burrowing animals further break down rock by moving fresh material to the surface, where physical and chemical processes can more effectively attack it. Decaying organisms also produce acids that contribute to chemical weathering. Where rock has

Figure 5.8 These stones were once rounded stream gravels; however, long exposure in a hot desert climate disintegrated them. (Photo by C. B. Hunt, U.S. Geological Survey)

been blasted in search of minerals or for road construction, the impact of humans is particularly noticeable, but on a worldwide scale, humans probably rank behind burrowing animals in Earth-moving accomplishments.

 # Chemical Weathering

Chemical weathering involves the complex processes that break down rock components and internal structures of minerals. Such processes convert the constituents to new minerals or release them to the surrounding environment. During this transformation, the original rock decomposes into substances that are stable in the surface environment. Consequently, the products of chemical weathering will remain essentially unchanged as long as they remain in an environment similar to the one in which they formed.

Water is by far the most important agent of chemical weathering. Pure water alone is a good solvent, and small amounts of dissolved materials result in increased chemical activity for weathering solutions. The major processes of chemical weathering are dissolution, oxidation, and hydrolysis. Water plays a leading role in each.

Dissolution

Perhaps the easiest type of decomposition to envision is the process of **dissolution**. Just as sugar dissolves in water, so too do certain minerals. One of the most water-soluble minerals is halite (common salt), which,

Figure 5.9 Root wedging widens fractures in rock and aids the process of mechanical weathering. (Photo by Tom Bean)

as you may recall, is composed of sodium and chloride ions. Halite readily dissolves in water because, although this compound maintains overall electrical neutrality, the individual ions retain their respective charges.

Moreover, the surrounding water molecules are polar; that is, the oxygen end of the molecule has a small residual negative charge whereas the end with hydrogen has a small positive charge. As the water molecules come in contact with halite, their negative ends approach sodium ions and their positive ends cluster about chloride ions. This disrupts the attractive forces in the halite crystal and releases the ions to the water solution (Figure 5.10).

Although most minerals are, for all practical purposes, insoluble in pure water, the presence of even a small amount of acid dramatically increases the corrosive force of water. (An acidic solution contains the reactive hydrogen ion, H^+.) In nature, acids are produced by a number of processes. For example, carbonic acid is created when carbon dioxide in the atmosphere dissolves in raindrops. As acidic rainwater soaks into the ground, carbon dioxide in the soil may increase the acidity of the weathering solution. Various organic acids are also released into the soil as organisms decay, and sulfuric acid is produced by the weathering of pyrite and other sulfide minerals.

Regardless of the source of the acid, this highly reactive substance readily decomposes most rocks and produces certain products that are water soluble. For example, the mineral calcite, $CaCO_3$, which composes the common building stones marble and limestone, is easily attacked by even a weakly acidic solution:

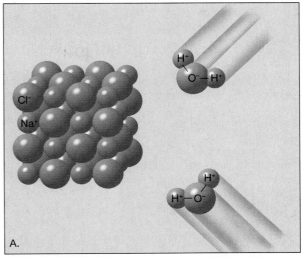

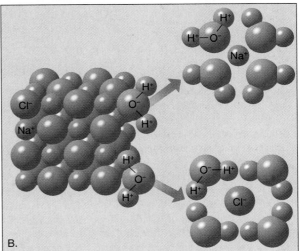

Figure 5.10 Illustration of halite dissolving in water. **A.** Sodium and chloride ions are attacked by the polar water molecules. **B.** Once removed, these ions are surrounded and held by a number of water molecules as shown.

$$\underset{\text{calcium carbonate}}{CaCO_3} \quad + \quad \underset{\text{aqueous acid}}{2[H^+(H_2)O]} \quad \rightarrow$$

$$\underset{\substack{\text{calcium ion}\\(\text{soluable})}}{Ca^{2+}} \quad + \quad \underset{\substack{\text{carbon}\\\text{dioxide}}}{CO_2\uparrow} \quad + \quad \underset{\text{water}}{3(H_2)O}$$

During this process, the insoluble calcium carbonate is transformed into soluble products. In nature, over periods of thousands of years, large quantities of limestone are dissolved and carried away by underground water. This activity is clearly evidenced by the large number of subsurface caverns found in every one of the contiguous 48 states. Monuments and buildings made of limestone or marble are also subjected to the corrosive work of acids, particularly in industrial areas that have smoggy, polluted air (see Box 5.1).

The soluble ions from reactions of this type are retained in our underground water supply. It is these dissolved ions that are responsible for the so-called hard water found in many locales. Simply, hard water is undesirable because the active ions react with soap to produce an insoluble material that renders soap nearly useless in removing dirt. To solve this problem a water softener can be used to remove these ions, generally by replacing them with others that do not chemically react with soap.

Oxidation

Everyone has seen iron and steel objects that rusted when exposed to water (Figure 5.11). The same thing can happen to iron-rich minerals. The process of rusting occurs when oxygen combines with iron to form iron oxide as follows:

$$\underset{\text{iron}}{4Fe} \quad + \quad \underset{\text{oxygen}}{3O_2} \quad \rightarrow \quad \underset{\text{iron oxide (hematite)}}{2Fe_2O_3}$$

Box 5.1

Acid Precipitation: A Human Impact on the Earth System

Humans are part of the complex interacting whole we call the Earth system (see Box 1.3, p. 14). As such, our actions cause changes to all the other parts of the system. For example, by going about our normal routine, we humans modify the composition of the atmosphere. These atmospheric modifications, in turn, cause unintended and unwanted changes to occur in the hydrosphere, biosphere, and solid Earth. Acid precipitation is one small but significant example.

Decomposed stone monuments and structures are common sights in many cities (Figure 5.A). Although we expect rock to gradually decompose, many of these monuments have succumbed prematurely. An important cause for this accelerated chemical weathering is acid precipitation.

Rain is naturally somewhat acidic. When carbon dioxide from the atmosphere dissolves in water, the product is weak carbonic acid. However, the term *acid precipitation* refers to precipitation that is much more acidic than natural, unpolluted rain and snow.

As a consequence of burning large quantities of fossil fuels like coal and petroleum products, nearly 40 million tons of sulfur and nitrogen oxides are released into the atmosphere each year in the United States. The major sources of these emissions include power-generating plants, industrial processes, such as ore smelting and petroleum refining, and vehicles of all kinds. Through a series of complex chemical reactions, some of these pollutants are converted into acids that then fall to Earth's surface as rain or snow. Another portion is deposited in dry form and subsequently converted into acid after coming in contact with precipitation, dew, or fog.

Figure 5.A Acid rain accelerates the chemical weathering of stone monuments and structures, including this building facade in Leipzig, Germany. (Photo by Doug Plummer)

Northern Europe and eastern North America have experienced widespread acid rain for some time. Studies have also shown that acid rain occurs in many other regions, including western North America, Japan, China, Russia, and South America. In addition to local pollution sources, a portion of the acidity found in the northeastern United States and eastern Canada originates hundreds of kilometers away in industrialized regions to the south and southwest. This situation occurs because many pollutants remain in the atmosphere as long as five days, during which time they may be transported great distances.

The damaging environmental effects of acid rain are thought to be considerable in some areas and imminent in others. The best-known effect is an increased acidity in thousands of lakes in Scandinavia and eastern North America. Accompanying this have been substantial increases in dissolved aluminum that is leached from the soil by the acidic water and that, in turn, is toxic to fish. Consequently, some lakes are virtually devoid of fish, and other are approaching this condition. Ecosystems are characterized by many interactions at many levels of organization, which means that evaluating the effects of acid precipitation on these complex systems is difficult and expensive, and far from complete.

In addition to the many lakes that can no longer support fish, research indicates that acid precipitation may also reduce agricultural crop yields and impair the productivity of forests. Acid rain not only harms the foliage, but also damages roots and leaches nutrient minerals from the soil. Finally, acid precipitation promotes the corrosion of metals and contributes to the destruction of stone structures.

This type of chemical reaction, called **oxidation,**[*] occurs when electrons are lost from one element during the reaction. In this case, we say that iron was oxidized because it lost electrons to oxygen. Although the oxidation of iron progresses very slowly in a dry environment, the addition of water greatly speeds the reaction.

Oxidation is important in decomposing such ferromagnesian minerals as olivine, pyroxene, and hornblende. Oxygen readily combines with the iron in

[*] The reader should note that *oxidation* is a term referring to any chemical reaction in which a compound or radical loses electrons The element oxygen is not necessarily present.

Figure 5.11 Iron reacts with oxygen to form iron oxide as seen on these rusted barrels. (Photo by Stephen J. Krasemann/DRK Photo)

these minerals to form the reddish-brown iron oxide called *hematite* (Fe_2O_3) or in other cases a yellowish-colored rust called *limonite* [$FeO(OH)$]. These products are responsible for the rusty color on the surfaces of dark igneous rocks, such as basalt, as they begin to weather. However, oxidation can occur only after iron is freed from the silicate structure by another process called hydrolysis.

Another important oxidation reaction occurs when sulfide minerals such as pyrite decompose. Sulfied minerals are major constituents in many metallic ores, and pyrite is frequently associated with coal deposits as well. In a moist environment, chemical weathering of pyrite (FeS_2) yields sulfuric acid (H_2SO_4) and iron oxide [$FeO(OH)$]. In many mining locales this weathering process creates a serious environmental hazard, particularly in humid areas where abundant rainfall infiltrates spoil banks (waste material left after coal or other minerals are removed). This so-called *mine acid* eventually makes its way to streams, killing aquatic organisms and degrading aquatic habitats.

Hydrolysis

The most common mineral group, the silicates, is decomposed primarily by the process of **hydrolysis**, which basically is the reaction of any substance with

water. Ideally, the hydrolisis of a mineral could take place in pure water as some of the water molecules dissociate to form the very reactive hydrogen (H^+) and hydroxyl (OH^-) ions. It is the hydrogen ion that attacks and replaces other positive ions found in the crystal lattice. With the introduction of hydrogen ions into the crystalline structure, the original orderly arrangement of atoms is destroyed and the mineral decomposes.

In nature, water usually contains other substances that contribute additional hydrogen ions, thereby greatly accelerating hydrolysis. The most common of these substances is carbon dioxide, CO_2, which dissolves in water to form carbonic acid, H_2CO_3. Rain dissolves some carbon dioxide in the atmosphere, and additional amounts, released by decaying organic matter, are acquired as the water percolates through the soil.

In water, carbonic acid ionizes to form hydrogen ions (H^+) and bicarbonate ions (HCO_3^-). To illustrate how a rock undergoes hydrolysis in the presence of carbonic acid, let's examine the chemical weathering of granite, a common continental rock. Recall that granite consists mainly of quartz and potassium feldspar. The weathering of the potassium feldspar component of granite is as follows:

$$2KAlSi_3O_8 \; + \; 2(H^+ + HCO_3^-) \; + \; H_2O \; \rightarrow$$

potassium feldspar   carbonic acid   water

$$Al_2Si_2O_5(OH)_4 \; + \; 2K^+ \; + \; 2HCO_3^- + \; 4SiO_2$$

kaolinite (residual clay)   potassium ion   bicarbonate ion   silica

in solution

In this reaction, the hydrogen ions (H^+) attack and replace potassium ions (K^+) in the feldspar structure, thereby disrupting the crystalline network. Once removed, the potassium is available as a nutrient for plants or becomes the soluble salt potassium bicarbonate ($KHCO_3$), which may be incorporated into other minerals or carried to the ocean.

The most abundant product of the chemical breakdown of potassium feldspar is the clay mineral kaolinite. Clay minerals are the end products of weathering and are very stable under surface conditions. Consequently, clay minerals make up a high percentage of the inorganic material in soils. Moreover, the most abundant sedimentary rock, shale, contains a high proportion of clay minerals.

In addition to the formation of clay minerals during the weathering of potassium feldspar, some silica is removed from the feldspar structure and carried away by groundwater. This dissolved silica will eventually precipitate, producing nodules of chert or flint, or it will fill in the pore spaces between grains of sediment, or it will be carried to the ocean, where microscopic animals will remove it from the water to build hard silica shells.

To summarize, the weathering of potassium feldspar generates a residual clay mineral, a soluble salt (potassium bicarbonate), and some silica, which enters into solution.

Quartz, the other main component of granite, is very resistant to chemical weathering; it remains substantially unaltered when attacked by weak acidic solutions. As a result, when granite weathers, the feldspar crystals dull and slowly turn to clay, releasing the once-interlocked quartz grains, which still retain their fresh, glassy appearance. Although some of the quartz remains in the soil, much is eventually transported to the sea or to other sites of deposition, where it becomes the main constituent of such features as sandy beaches and sand dunes. In time these quartz grains may be lithified to form the sedimentary rock sandstone.

Table 5.1 lists the weathered products of some of the most common silicate minerals. Remember that silicate minerals make up most of Earth's crust and that these minerals are essentially composed of only eight elements. When chemically weathered, the silicate minerals yield sodium, calcium, potassium, and magnesium ions that form soluble products, which may be removed from groundwater. The element iron combines with oxygen, producing relatively insoluble iron oxides, most notably hematite and limonite, which give soil a reddish-brown or yellowish color. Under most conditions the three remaining elements, aluminum, silicon, and oxygen, join with water to produce residual clay minerals. However, even the highly insoluble clay minerals are very slowly removed by subsurface water.

Alterations Caused by Chemical Weathering

As noted earlier, the most significant result of chemical weathering is the decomposition of unstable minerals and the generation or retention of those materials that are stable at Earth's surface. This accounts for the predominance of certain minerals in the surface material we call soil.

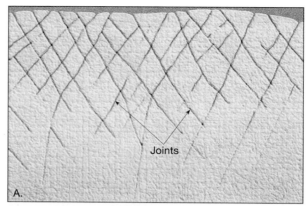
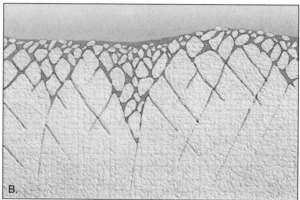
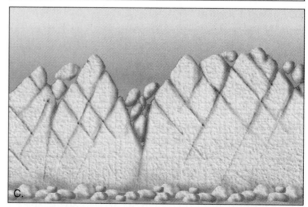

Figure 5.12 Spheroidal weathering of extensively jointed rock. Water moving through the joints begins to enlarge them. Because the rocks are attacked more on the corners and edges, they take on a spherical shape. The photo shows spheroidal weathering in Joshua Tree National Monument, California. (Photo by E. J. Tarbuck)

Table 5.1 Products of Weathering

Mineral	Residual Products	Material in Solution
Quartz	Quartz grains	Silica
Feldspars	Clay minerals	Silica, K^+, Na^+, Ca^{2+}
Amphibole (hornblende)	Clay minerals, Limonite, Hematite	Silica, Ca^{2+}, Mg^{2+}
Olivine	Limonite, Hematite	Silica, Mg^{2+}

In addition to altering the internal structure of minerals, chemical weathering causes physical changes as well. For instance, when angular rock fragments are attacked by water flowing through joints, the fragments tend to take on a spherical shape. The gradual rounding of the corners and edges of angular blocks is illustrated in Figure 5.12. The corners are attacked most readily because of the greater surface area for their volume as compared to the edges and faces. This process, called **spheroidal weathering**, gives the weathered rock a more rounded or spherical shape (Figure 5.12).

Sometimes during the formation of spheroidal boulders, successive shells separate from the rock's main body (Figure 5.13). Eventually the outer shells break off, allowing the chemical weathering activity to penetrate deeper into the boulder. This spherical scaling results because, as the minerals in the rock weather to clay, they increase in size through the addition of water to their structure. This increased bulk exerts an outward force that causes concentric layers of rock to break loose and fall off.

Hence, chemical weathering does produce forces great enough to cause mechanical weathering. This type of spheroidal weathering in which shells spall off should not be confused with the phenomenon of sheeting discussed earlier. In sheeting, the fracturing occurs as a result of unloading, and the rock layers that separate from the main body are largely unaltered at the time of separation.

Rates of Weathering

Several factors influence the type and rate of rock weathering. We have already seen how mechanical weathering affects the rate of weathering. By breaking rock into smaller pieces, the amount of surface area exposed to chemical weathering is increased. Other important factors examined here include the roles of rock characteristics and climate.

Rock Characteristics

Rock characteristics encompass all of the chemical traits of rocks, including mineral composition and solubility. In addition, any physical features such as joints (cracks) can be important because they influence the ability of water to penetrate rock.

The variations in weathering rates due to the mineral constituents can be demonstrated by comparing old headstones made from different rock types. Headstones of granite, which is composed of silicate minerals, are relatively resistant to chemical weathering. We can see this by examining the inscriptions on the headstones shown in Figure 5.14. In contrast, the marble headstone shows signs of extensive chemical alteration over a relatively short period. Marble is composed of calcite (calcium carbonate), which readily dissolves even in a weakly acidic solution.

The most abundant mineral group, the silicates, weathers in the order shown in Figure 5.15. This arrangement of minerals is identical to Bowen's reaction series. The order in which the silicate minerals weather is essentially the same as their order of crystallization. The explanation for this is related to the crystalline structure of silicate minerals. The strength of silicon–oxygen bonds is great. Because quartz is composed entirely of these strong bonds, it is very resistant to weathering. By contrast, olivine has far fewer silicon–oxygen bonds and is not nearly as resistant to chemical weathering.

Figure 5.13 Successive shells are loosened as the weathering process continues to penetrate ever deeper into the rock. (Photo by Martin Schmidt, Jr.)

Figure 5.14 An examination of headstones reveals the rate of chemical weathering on diverse rock types. The granite headstone (left) was erected a few years after the marble headstone (right). The inscription date of 1885 on the marble monument is nearly illegible. (Photos by E. J. Tarbuck)

Climate

Climatic factors, particularly temperature and moisture, are crucial to the rate or rock weathering. One important example from mechanical weathering is that the frequency of freeze–thaw cycles greatly affects the amount of frost wedging. Temperature and moisture also exert a strong influence on rates of chemical weathering and on the kind and amount of vegetation present. Regions with lush vegetation generally have a thick mantle of soil rich in decayed organic matter from which chemically active fluids such as carbonic acid and humic acids are derived.

The optimum environment for chemical weathering is a combination of warm temperatures and abundant moisture. In polar regions chemical weathering is ineffective because frigid temperatures keep the available moisture locked up as ice, whereas in arid regions there is insufficient moisture to foster rapid chemical weathering.

Human activities can influence the composition of the atmosphere, which, in turn, can impact the rate of chemical weathering. Box 5.1 examines one well-known example—acid rain.

Differential Weathering

Many factors influence the type and rate of rock weathering for a given place. There is generally enough variation, even within a relatively small area, for the rocks to exhibit some differential weathering. **Differential weathering** simply relates to the fact that rocks exposed at Earth's surface usually do not weather at the same rate. Because of variations in such

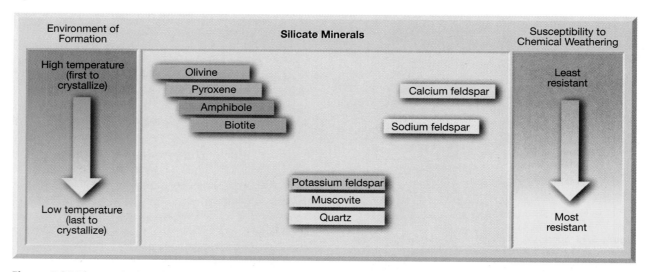

Figure 5.15 The weathering of common silicate minerals. The order in which the silicate minerals chemically weather is essentially the same as their order of crystallization.

factors as mineral makeup, degree of jointing, and exposure to the elements, significant differences occur. Consequently, differential weathering and subsequent erosion create many unusual and often spectacular rock formations and landforms. Included are features such as the arches in Arches National Park (see chapter-opening photo) and the sculpted rock pinnacles in Bryce Canyon National Park (see Figure 5.1, p. 121).

Soil

Soil covers most land surfaces. Along with air and water, it is one of our most indispensable resources (Figure 5.16). Also like air and water, soil is taken for granted by many of us. The following quote helps put this vital layer in perspective.

> Science, in recent years, has focused more and more on the Earth as a planet, one that for all we know is unique—where a thin blanket of air, a thinner film of water, and the thinnest veneer of soil combine to support a web of life of wondrous diversity in continuous change.*

*Jack Eddy, "A Fragile Seam of Dark Blue Light," in *Proceedings of the Global Change Research Forum*. U.S. Geological Survey Circular 1086, 1993, p. 15.

Figure 5.16 Soil is an essential resource that we often take for granted. Soil is not a living entity, but it contains a great deal of life. Moreover, this complex medium supports nearly all plant life, which, in turn, supports animal life. (Photo by James E. Patterson)

Soil has accurately been called "the bridge between life and the inanimate world." All life—the entire biosphere—owes its existence to a dozen or so elements that must ultimately come from Earth's crust. Once weathering and other processes create soil, plants carry out the intermediary role of assimilating the necessary elements and making them available to animals including humans.

An Interface in the Earth System

When Earth is viewed as a system, soil is referred to as an *interface*—a common boundary where different parts of a system interact. This is an appropriate designation because soil forms where the solid Earth, the atmosphere, the hydrosphere, and the biosphere meet. Soil is a material that develops in response to complex environmental interactions among different parts of the Earth system. Over time, soil gradually evolves to a state of equilibrium or balance with the environment. Soil is dynamic and sensitive to almost every aspect of its surroundings. Thus, when environmental changes occur, such as climate, vegetative cover, and animal (including human) activity, the soil responds. Any such change produces a gradual alteration of soil characteristics until a new balance is reached. Although thinly distributed over the land surface, soil functions as a fundamental interface, providing an excellent example of the integration among many parts of the Earth system.

What Is Soil?

With few exceptions, Earth's land surface is covered by **regolith**, the layer of rock and mineral fragments produced by weathering. Some would call this material soil, but soil is more than an accumulation of weathered debris. **Soil** is a combination of mineral and organic matter, water, and air—that portion of the regolith that supports the growth of plants. Although the proportions of the major components in soil vary, the same four components always are present to some extent (Figure 5.17). About one-half of the total volume of a good quality surface soil is a mixture of disintegrated and decomposed rock (mineral matter) and **humus**, the decayed remains of animal and plant life (organic matter). The remaining half consists of pore spaces among the solid particles where air and water circulate.

Although the mineral portion of the soil is usually much greater than the organic portion, humus is an essential component. In addition to being an important source of plant nutrients, humus enhances the soil's ability to retain water. Because plants require air and water to live and grow, the portion of the soil consisting of pore spaces that allow for the circulation of these fluids is as vital as the solid soil constituents.

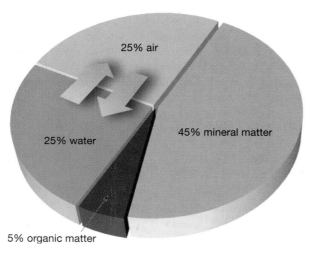

Figure 5 17 Composition (by volume) of a soil in good condition for plant growth. Although the percentages vary, each soil is composed of mineral and organic matter, water, and air.

Soil water is far from "pure" water; instead, it is a complex solution containing many soluble nutrients. Soil water not only provides the necessary moisture for the chemical reactions that sustain life; it also supplies plants with nutrients in a form they can use. The pore spaces not filled with water contain air. This air is the source of necessary oxygen and carbon dioxide for most microorganisms and plants that live in the soil.

Controls of Soil Formation

Soil is the product of the complex interplay of several factors, including parent material, time, climate, plants and animals, and slope. Although all these factors are interdependent, their roles will be examined separately.

Parent Material

The source of the weathered mineral matter from which soils develop is called the **parent material** and is a major factor influencing a newly forming soil. Gradually it undergoes physical and chemical changes as the processes of soil formation progress. Parent material can either be the underlying bedrock or a layer of unconsolidated deposits. When the parent material is bedrock, the soils are termed **residual soils**. By contrast, those developed on unconsolidated sediment are called **transported soils** (Figure 5.18). It should be pointed out that transported soils form *in place* on parent materials that have been carried from elsewhere and deposited by gravity, water, wind, or ice.

The nature of the parent material influences soils in two ways. First, the type of parent material will affect the rate of weathering, and thus the rate of soil formation. Also, because unconsolidated deposits are already partly weathered, soil development on such material will likely progress more rapidly than when

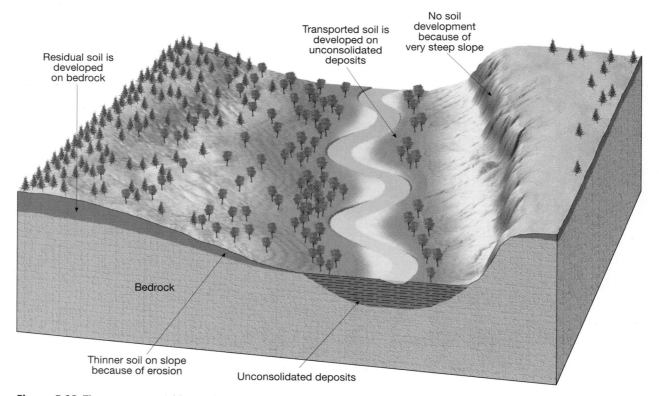

Figure 5.18 The parent material for residual soils is the underlying bedrock, whereas transported soils form on unconsolidated deposits. Also note that as slopes become steeper, soil becomes thinner.

bedrock is the parent material. Second, the chemical makeup of the parent material will affect the soil's fertility. This influences the character of the natural vegetation the soil can support.

At one time the parent material was believed to be the primary factor causing differences among soils. Today soil scientists realize that other factors, especially climate, are more important. In fact, it has been found that similar soils are often produced from different parent materials and that dissimilar soils have developed from the same parent material. Such discoveries reinforce the importance of other soil-forming factors.

Time

Time is an important component of every geological process, and soil formation is no exception. The nature of soil is strongly influenced by the length of time processes have been operating. If weathering has been going on for a comparatively short time, the character of the parent material determines to a large extent the characteristics of the soil. As weathering processes continue, the influence of parent material on soil is overshadowed by the other soil-forming factors, especially climate. The amount of time required for various soils to evolve cannot be listed because the soil-forming processes act at varying rates under different circumstances. However, as a rule the longer a soil has been forming, the thicker it becomes and the less it resembles the parent material.

Climate

Climate is considered to be the most influential control of soil formation. Temperature and precipitation are the elements that exert the strongest impact on soil formation. Variations in temperature and precipitation determine whether chemical or mechanical weathering will predominate and also greatly influence the rate and depth of weathering. For instance, a hot, wet climate may produce a thick layer of chemically weathered soil in the same amount of time that a cold, dry climate produces a thin mantle of mechanically weathered debris. Also, the amount of precipitation influences the degree to which various materials are leached from the soil, thereby affecting soil fertility. Finally, climatic conditions are an important control on the type of plant and animal life present.

Plants and Animals

Plants and animals play a vital role in soil formation. The types and abundance of organisms present have a strong influence on the physical and chemical properties of a soil. In fact, for well-developed soils in many regions, the significance of natural vegetation in influencing soil type is frequently implied in the description used by soil scientists. Such phrases as *prairie soil*, *forest soil*, and *tundra soil* are common.

Plants and animals furnish organic matter to the soil. Certain bog soils are composed almost entirely of organic matter, whereas desert soils might contain as little as a small fraction of 1 percent. Although the quantity of organic matter varies substantially among soils, it is the rare soil that completely lacks it.

The primary source of organic matter in soil is plants, although animals and an infinite number of microorganisms also contribute. When organic matter is decomposed, important nutrients are supplied to plants, as well as to animals and microorganisms living in the soil. Consequently, soil fertility is in part related to the amount of organic matter present. Furthermore, the decay of plant and animal remains causes the formation of various organic acids. These complex acids hasten the weathering process. Organic matter also has a high water-holding ability and thus aids water retention in a soil.

Mircoorganisms, including fungi, bacteria, and single-celled protozoa, play an active role in the decay of plant and animal remains. The end product is humus, a material that no longer resembles the plants and animals from which it is formed. In addition, certain microorganisms aid soil fertility because they have the ability to convert atmospheric nitrogen into soil nitrogen.

Earthworms and other burrowing animals act to mix the mineral and organic portions of a soil. Earthworms, for example, feed on organic matter and thoroughly mix soils in which they live, often moving and enriching many tons per acre each year. Burrows and holes also aid the passage of water and air through the soil.

Slope

The slope of the land can vary greatly over short distances. Such variations, in turn, can lead to the development of a variety of localized soil types. Many of the differences exist because slope has a significant impact on the amount of erosion and the water content of soil.

On steep slopes, soils are often poorly developed. In such situations the quantity of water soaking in is slight; as a result, the moisture content of the soil may not be sufficient for vigorous plant growth. Further, because of accelerated erosion on steep slopes, the soils are thin, or in some cases nonexistent (Figure 5.18).

In contrast, poorly drained and water-logged soils found in bottomlands have a much different character. Such soils are usually thick and dark. The dark color results from the large quantity of organic matter that accumulates because saturated conditions retard the decay of vegetation. The optimum terrain for soil development is a flat-to-undulating upland surface. Here we find good drainage, minimum erosion, and sufficient infiltration of water into the soil.

Slope orientation, or the direction the slope is facing, is another factor we should note. In the mid-latitudes of the Northern Hemisphere, a south-facing slope will receive a great deal more sunlight than a north-facing slope. In fact, a steep north-facing slope may receive no direct sunlight at all. The difference in the amount of solar radiation received will cause differences in soil temperature and moisture, which in turn may influence the nature of the vegetation and the character of the soil.

Although this section dealt separately with each of the soil-forming factors, remember that all of them work together to form soil. No single factor is responsible for a soil's character; rather, it is the combined influence of parent material, time, climate, plants and animals, and slope that determines this character.

The Soil Profile

Because soil-forming processes operate from the surface downward, variations in composition, texture, structure, and color gradually evolve at varying depths. These vertical differences, which usually become more pronounced as time passes, divide the soil into zones or layers known as **horizons**. If you were to dig a trench in soil, you would see that its walls are layered. Such a vertical section through all of the soil horizons constitutes the **soil profile** (Figure 5.19).

Figure 5.20 presents an idealized view of a well-developed soil profile in which five horizons are identified. From the surface downward, they are designated as *O, A, E, B,* and *C*. These five horizons are common to soils in temperate regions. The characteristics and extent of development of horizons vary in different environments. Thus, different localities exhibit soil profiles that can contrast greatly with one another.

The *O* soil horizon consists largely of organic material. This is in contrast to the layers beneath it, which consist mainly of mineral matter. The upper portion of the *O* horizon is primarily plant litter such as loose leaves and other organic debris that are still recognizable. By contrast, the lower portion of the *O* horizon is made up of partly decomposed organic matter (humus) in which plant structures can no longer be identified. In addition to plants, the *O* horizon is teeming with microscopic life including bacteria, fungi, algae, and insects. All of these organisms contribute oxygen, carbon dioxide, and organic acids to the developing soil.

Underlying the organic-rich *O* horizon is the *A* horizon. This zone is largely mineral matter, yet biological activity is high and humus is generally present—up to 30 percent in some instances. Together

Figure 5.19 A soil profile is a vertical cross-section from the surface down to the parent material. Well-developed soils show distinct layers called horizons. (Photo by E. J. Tarbuck)

the *O* and *A* horizons make up what is commonly called the *topsoil*. Below the *A* horizon, the *E* horizon is a light-colored layer that contains little organic material. As water percolates downward through this zone, finer particles are carried away. This washing out of fine soil components is termed **eluviation**. Water percolating downward also dissolves soluble inorganic soil components and carries them to deeper zones. This depletion of soluble materials from the upper soil is termed **leaching**.

Immediately below the *E* horizon is the *B* horizon, or *subsoil*. Much of the material removed from the *E* horizon by eluviation is deposited in the *B* horizon, which is often referred to as the *zone of accumulation*. The accumulation of the fine clay particles enhances water retention in the subsoil. However, in extreme cases, clay accumulation can form a very compact and impermeable layer called *hardpan*. The *O, A, E,* and *B* horizons together constitute the **solum**, or "true soil." It is in the solum that the soil-forming processes are active and that living roots and other plant and animal life are largely confined.

Below the solum and above the unaltered parent material is the *C* horizon, a layer characterized by partially altered parent material. Whereas the *O, A, E,* and *B* horizons bear little resemblance to the parent material, it is easily identifiable in the *C* horizon.

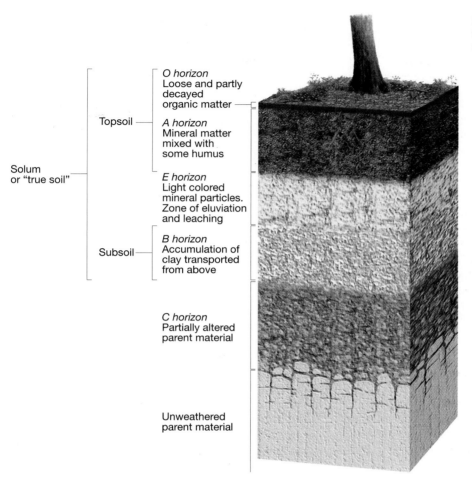

Figure 5.20 Idealized soil profile from a humid climate in the middle latitudes.

Solum or "true soil"

Topsoil
- *O horizon* Loose and partly decayed organic matter
- *A horizon* Mineral matter mixed with some humus

Subsoil
- *E horizon* Light colored mineral particles. Zone of eluviation and leaching
- *B horizon* Accumulation of clay transported from above

C horizon Partially altered parent material

Unweathered parent material

Although this material is undergoing changes that will eventually transform it into soil, it has not yet crossed the threshold that separates regolith from soil.

The characteristics and extent of development can vary greatly among soils in different environments. The boundaries between soil horizons may be sharp, or the horizons may blend gradually from one to another. Consequently, a well-developed soil profile indicates that environmental conditions have been relatively stable over an extended time span and that the soil is *mature*. By contrast, some soils lack horizons altogether.

Such soils are called *immature* because soil building has been going on for only a short time. Immature soils are also characteristic of steep slopes where erosion continually strips away the soil, preventing full development.

Soil Types

In the following discussion, we will briefly examine some common soil types. As you read, notice that the characteristics of each soil type primarily depend on the prevailing climatic conditions. A summary of the soil characteristics discussed in this section is provided in Table 5.2.

In cold or dry climates soils are generally very thin and poorly developed. The reasons for this are fairly obvious. Chemical weathering progresses very slowly in such climates, and the scanty plant life yields very little organic matter.

Pedalfer

The term **pedalfer** gives a clue to the basic characteristic of this soil type. The word is derived from the Greek **ped**on, meaning "soil," and the chemical symbols **Al** (aluminum) and **Fe** (iron). Pedalfers are characterized by an accumulation of iron oxides and aluminum-rich clays in the *B* horizon. In mid-latitude areas where the annual rainfall exceeds 63 centimeters (25 inches), most of the soluble materials, such as calcium carbonate, are leached from the soil and carried away by underground water. The less soluble iron oxides and clays are carried from the *E* horizon and deposited in the *B* horizon, giving it a

Table 5.2 Summary of Soil Types

Climate	Temperate humid (>63 cm rainfall)	Temperate dry (<63 cm rainfall)	Tropical (heavy rainfall)	Extreme arctic or desert
Vegetation	Forest	Grass and brush	Grass and trees	Almost none, so no humus develops
Typical Area	Eastern U.S.	Western U.S.		
Soil Type	Pedalfer	Pedocal	Laterite	
Topsoil	Sandy, light colored; acid	Commonly enriched in calcite; whitish color	Zones not developed: Enriched in iron (and aluminum); brick red color	No real soil forms because there is no organic material. Chemical weathering is very slow
Subsoil	Enriched in aluminum, iron and clay; brown color	Enriched in calcite; whitish color	All other elements removed by leaching	
Remarks	Extreme development in conifer forests, because abundant humus makes groundwater very acidic. Produces light gray soil because of removal of iron	*Caliche* is name applied to the accumulation of calcite	Apparently bacteria destroy humus, so no acid is available to remove iron	

brown to red-brown color. These soils are best developed under forest vegetation where large quantities of decomposing organic matter provide the acid conditions necessary for leaching. In the United States pedalfers are found east of a line extending from northwest Minnesota to south-central Texas.

Pedocal

Pedocal is derived from the Greek **ped***on*, meaning "soil," and the first three letter of **cal***cite* (calcium carbonate). As the name implies, pedocals are characterized by an accumulation of calcium carbonate. This soil type is found in the drier western United States in association with grassland and brush vegetation. Because chemical weathering is less intense in drier areas, pedocals generally contain a smaller percentage of clay materials than do pedalfers.

In the arid and semiarid western states a calcite-enriched layer called **caliche** may be present in the soils. In these areas not much rainfall penetrates to great depths. Rather, the rainwater is held by the soil particles near the surface until it evaporates. As a result, the soluble materials, chiefly calcium carbonate, are removed from the uppermost layer and redeposited below, forming the caliche layer.

Laterite

In the hot, wet climates of the tropics, soils called **laterites** may develop. Because chemical weathering is intense under such climate conditions, these soils are usually deeper than soils developing over a similar period in the mid-latitudes. Not only does leaching remove the soluble materials such as calcite, but the great quantities of percolating water also remove much of the silica, with the result that oxides of iron and aluminum become concentrated in the soil. The iron gives the soil a distinctive red color. As bacterial activity is very high in the tropics, laterites contain practically no humus. This fact, coupled with the highly leached and bricklike nature of these soils, makes laterites poor for growing crops. The infertility of these soils has been demonstrated repeatedly in tropical countries where cultivation has been expanded into such areas (see Box 5.2).

Soil Erosion

Soils are just a tiny fraction of all Earth materials, yet they are a vital resource. Because soils are necessary for the growth of rooted plants, they are the very foundation of the human life-support system. Just as human ingenuity can increase the agricultural productivity of soils through fertilization and irrigation, soils can be damaged or destroyed by careless activities. Despite their basic role in providing food, fiber, and other basic materials, soils are among our most abused resources.

Perhaps this neglect and indifference has occurred because a substantial amount of soil seems to remain even where soil erosion is serious. Nevertheless, although the loss of fertile topsoil may not be obvious to the untrained eye, it is a growing problem as human activities expand and disturb more and more of Earth's surface.

Box 5.2

Laterites and the Clearing of the Rain Forest

Laterites are thick red soils that form in the wet tropics and subtropics. They are the end product of extreme chemical weathering. Because lush tropical rain forests have laterite soils, we might assume the soils are fertile, with great potential for agriculture. However, just the opposite is true: Laterites are among the poorest soils for farming. How can this be?

Because laterites develop under rain-forest conditions of high temperature and heavy rainfall, they are severely leached. Leaching destroys fertility because most plant nutrients are removed by the large volume of downward-percolating water. Therefore, even though the vegetation may be dense and luxuriant, the soil itself contains few available nutrients.

Most nutrients that support the rain forest are locked up in the trees themselves. As vegetation dies and decomposes, the roots of the rainforest trees quickly absorb the nutrients before they are leached from the soil. The nutrients are continuously recycled as trees die and decompose.

Therefore, when forests are cleared to provide land for farming or to harvest the timber, most of the nutrients are removed as well (Figure 5.B). What remains is a soil that contains little to nourish planted crops.

The clearing of rain forests not only removes plant nutrients but also accelerates erosion. When vegetation is present, its roots anchor the soil and its leaves and branches provide a canopy that protects the ground by deflecting the full force of the frequent heavy rains.

The removal of vegetation also exposes the ground to strong direct sunlight. When baked by the Sun, laterites can harden to a bricklike consistency and become practically impenetrable to water and crop roots. In only a few years, lateritic soils in a freshly cleared area may no longer be cultivable.

The term *laterite* is derived from the Latin word *latere*, meaning "brick," and was first applied to the use of this material for brick making in India and

Figure 5.B Clearing the rain forest near the Barum River in Malaysia. The thick lateric soil is highly leached. (Photo by David Hiser Photographers)

Figure 5.C This ancient temple at Angor Wat, Cambodia, was built of bricks made of laterite. (Photo by R. Ian Lloyd/The Stock Market)

Cambodia. Laborers simply excavated the soil, shaped it, and allowed it to harden in the sun. Ancient but still well-preserved structures built of laterite remain standing today in the wet tropics (Figure 5.C). Such structures have withstood centuries of weathering because all of the original soluble materials were already removed from the soil by chemical weathering. Laterites are therefore virtually insoluble and thus very stable.

In summary, we have seen that laterites are highly leached soils that are the products of extreme chemical weathering in the warm, wet tropics. Although they may be associated with lush tropical rain forests, these soils are unproductive when vegetation is removed. Moreover, when cleared of plants, laterites are subject to accelerated erosion and can be baked to bricklike hardness by the Sun.

Figure 5.21 When it is raining, millions of water drops are falling at velocities approaching 10 meters per second (35 kilometers per hour). When water drops strike an exposed surface, soil particles may splash as high as 1 meter into the air and land more than a meter away from the point of raindrop impact. Soil dislodged by splash erosion is more easily moved by sheet erosion. (Photo courtesy of U.S. Department of Agriculture)

How Soil Is Eroded

Soil erosion is a natural process; it is part of the constant recycling of Earth materials that we call the *rock cycle*. Once soil forms, erosional forces, especially water and wind, move soil components from one place to another. Every time it rains, raindrops strike the land with surprising force (Figure 5.21). Each drop acts like a tiny bomb, blasting movable soil particles out of their positions in the soil mass. Then, water flowing across the surface carries away the dislodged soil particles. Because the soil is moved by thin sheets of water, this process is termed *sheet erosion*.

After flowing as a thin, unconfined sheet for a relatively short distance, threads of current typically develop and tiny channels called *rills* begin to form. Still deeper cuts in the soil, known as *gullies*, are created as rills enlarge (Figure 5.22). When normal farm cultivation cannot eliminate the channels, we know the rills have grown large enough to be called gullies. Although most dislodged soil particles move only a short distance during each rainfall, substantial quantities eventually leave the fields and make their way downslope to a stream. Once in the stream channel, these soil particles, which can now be called *sediment*, are transported downstream and eventually deposited.

Rates of Erosion

We know that soil erosion is the ultimate fate of practically all soils. In the past, erosion occurred at slower rates than it does today because more of the land surface was covered and protected by trees, shrubs, grasses, and other plants. However, human activities such as farming, logging, and construction, which remove or disrupt the natural vegetation, have greatly accelerated the rate of soil erosion. Without the stabilizing effect of plants, the soil is more easily swept away by the wind or carried downslope by sheet wash.

Natural rates of soil erosion vary greatly from one place to another and depend on soil characteristics as well as such factors as climate, slope, and type of vegetation. Over a broad area, erosion caused by surface runoff may be estimated by determining the sediment loads of the streams that drain the region. When studies of this kind were made on a global scale they indicated that, prior to the appearance of humans, sediment transport by rivers to the ocean amounted to

Figure 5.22 Gully erosion in poorly protected soil. (Photo by James E. Patterson)

just over 9 billion metric tons per year. By contrast, the amount of material currently transported to the sea by rivers is about 24 billion metric tons per year, or more than two and one-half times the earlier rate.

It is more difficult to measure the loss of soil due to wind erosion. However, the removal of soil by wind is generally much less significant than erosion by flowing water except during periods of prolonged drought. When dry conditions prevail, strong winds can remove large quantities of soil from unprotected fields. Such was the case in the 1930s in the portions of the Great Plains that came to be called the Dust Bowl (see Box 5.3).

In many regions the rate of soil erosion is significantly greater than the rate of soil formation. This means that a renewable resource has become nonrenewable in these places. At present, it is estimated that topsoil is eroding faster than it forms on more than one-third of the world's croplands. The result is lower productivity, poorer crop quality, reduced agricultural income, and an ominous future.

Sedimentation and Chemical Pollution

Another problem related to excessive soil erosion involves the deposition of sediment. Each year in the United States hundreds of millions of tons of eroded soil are deposited in lakes, reservoirs, and streams. The detrimental impact of this process can be significant. For example, as more and more sediment is deposited in a reservoir, the capacity of the reservoir is diminished, limiting its usefulness for flood control, water supply, and/or hydroelectric power generation. In addition, sedimentation in streams and other waterways can restrict navigation and lead to costly dredging operations.

In some cases soil particles are contaminated with pesticides used in farming. When these chemicals are introduced into a lake or reservoir, the quality of the water supply is threatened and aquatic organisms may be endangered. In addition to pesticides, nutrients found naturally in soils as well as those added by agricultural fertilizers make their way into streams and lakes, where they stimulate the growth of plants. Over a period of time, excessive nutrients accelerate the process by which plant growth leads to the depletion of oxygen and an early death of the lake.

The availability of good soils is critical if the world's rapidly growing population is to be fed. On every continent unnecessary soil loss is occurring because appropriate conservation measures are not being used. Although it is a recognized fact that soil erosion can never be completely eliminated, soil conservation programs can substantially reduce the loss of this basic resource. Windbreaks (rows of trees), terracing, and plowing along the contours of hills are some of the effective measures, as are special tillage practices and crop rotation.

Chapter Summary

- External processes that continually remove materials from higher elevations and move them to lower elevations include (1) *weathering*—the disintegration and decomposition of rock at or near Earth's surface; (2) *mass wasting*—the transfer of rock material downslope under the influence of gravity; and (3) *erosion*—the removal of material by a mobile agent, usually water, wind, or ice.

- *Mechanical weathering* is the physical breaking up of rock into smaller pieces. Rocks can be broken into smaller fragments by *frost wedging* (where water works its way into cracks or voids in rock, and, upon freezing, expands and enlarges the openings), *unloading* (expansion and breaking due to a great reduction in pressure when the overlying rock is eroded away), *thermal expansion* (weakening of rock as the result of expansion and contraction as it heats and cools), and *biological activity* (by humans, burrowing animals, plant roots, etc.).

- *Chemical weathering* alters a rock's chemistry, changing it into different substances. Water is by far the most important agent of chemical weathering. *Dissolution* occurs when water-soluble minerals such as halite become dissolved in water. Oxygen dissolved in water will *oxidize* iron-rich minerals. When carbon dioxide (CO_2) is dissolved in water it forms *carbonic acid*, which accelerates the decomposition of silicate minerals by *hydrolysis*. The chemical weathering of silicate minerals frequently produces (1) soluble products containing sodium, calcium, potassium, and magnesium ions, and silica in solution; (2) insoluble iron oxides; and (3) clay minerals.

- The rate at which rock weathers depends on such factors as (1) *particle size*—small pieces generally weather faster than large pieces; (2) *mineral makeup*—calcite readily dissolves in mildly acidic solutions, and silicate minerals that form first

Box 5.3

Dust Bowl: Soil Erosion in the Great Plains

During a span of dry years in the 1930s, large dust storms plagued the Great Plains. Because of the size and severity of these storms, the region came to be called the "Dust Bowl," and the time period, the "dirty thirties." The heart of the Dust Bowl was nearly 100 million acres in the panhandles of Texas and Oklahoma and adjacent parts of Colorado, New Mexico, and Kansas (Figure 5.D). To a lesser extent, dust storms were also a problem over much of the Great Plains, from North Dakota to west-central Texas.

At times dust storms were so severe that they were called "black blizzards" and "black rollers" because visibility was reduced to only a few feet. Numerous storms lasted for hours and stripped huge volumes of topsoil from the land.

In the spring of 1934, a windstorm that lasted for a day and a half created a dust cloud 2000 kilometers (1200 miles) long. As the sediment moved east, New York had "muddy rains" and Vermont had "black snows." Another storm carried dust more than 3 kilometers (2 miles) into the atmosphere and transported it 3000 kilometers from its source in Colorado to create "midday twilight" in New England and New York.

What caused the Dust Bowl? Clearly, the fact that portions of the Great Plains experienced some of North America's strongest winds is important. However,

Figure 5.D An abandoned farmstead shows the disastrous effects of wind erosion and deposition during the "Dust Bowl" period. This photo of a previously prosperous farm was taken in Oklahoma in 1937. Also see Figure 13.8, p. 336. (Photo courtesy of Soil Conservation Service, U.S. Department of Agriculture)

it was the expansion of agriculture that set the stage for the disastrous period of soil erosion. Mechanization allowed the rapid transformation of the grass-covered prairies of this semiarid region into farms. Between the 1870s and 1930, cultivation expanded nearly tenfold, from about 10 million acres to more than 100 million acres.

As long as precipitation was adequate, the soil remained in place. However, when a prolonged drought struck in the 1930s, the unprotected fields were vulnerable to the wind. The result was severe soil loss, crop failure, and economic hardship.

Beginning in 1939, a return to rainier conditions brought relief. New farming practices that reduced soil loss by wind were instituted. Although dust storms are less numerous and not as severe as in the "dirty thirties," soil erosion by strong winds still occurs periodically whenever the combination of drought and unprotected soil exists.

---

from magma are least resistant to chemical weathering; and (3) *climatic factors*, particularly temperature and moisture. Frequently, rocks exposed at Earth's surface do not weather at the same rate. This *differential weathering* of rocks is influenced by such factors as mineral makeup, degree of jointing, and exposure to the elements.

- *Soil* is a combination of mineral and organic matter, water, and air—that portion of the *regolith* (the layer of rock and mineral fragments produced by weathering) that supports the growth of plants. About one-half of the total volume of a good-quality soil is a mixture of disintegrated and decomposed rock (mineral matter) and *humus* (the decayed remains of animal and plant life); the remaining half consists of pore spaces, where air and water circulate. The most important factors that control soil formation are *parent material*, *time*, *climate*, *plants* and *animals*, and *slope*.

- Soil-forming processes operate from the surface downward and produce zones or layers in the soil that are called *horizons*. From the surface downward, the soil horizons are respectively designated as *O* (largely organic matter), *A* (largely mineral matter), *E* (where the fine soil components and soluble materials have been removed by *eluviation* and *leaching*), *B* (or *subsoil*, often referred to as the *zone of accumulation*), and *C* (partially altered parent material). Together the *O* and *A* horizons make up what is commonly called the *topsoil*.

- Although there are hundreds of soil types and sub-types worldwide, the three very generic types are (1) *pedalfer*—characterized by an accumulation of iron oxides and aluminum-rich clays in the *B* horizon; (2) *pedocal*—characterized by an accumulation of calcium carbonate; and (3) *laterite*—deep soils that develop in the hot, wet tropics that are poor for growing crops because they are highly leached.

- Soil erosion is a natural process; it is part of the constant recycling of Earth materials that we call the rock cycle. Once in a stream channel, soil particles are transported downstream and eventually deposited. *Rates of soil erosion* vary from one place to another and depend on the soil's characteristics as well as such factors as climate, slope, and type of vegetation.

Review Questions

1. Differentiate between the products of mechanical weathering and chemical weathering.

2. In what type of environment is frost wedging most effective?

3. Describe the processes of sheeting and spheroidal weathering. How are they different and how are they similar?

4. How does mechanical weathering add to the effectiveness of chemical weathering?

5. Granite and basalt are exposed at the surface in a hot, wet region.

    (a) Which type of weathering will predominate?
    (b) Which of these rocks will weather most rapidly? Why?

6. Heat speeds up a chemical reaction. Why then does chemical weathering proceed slowly in a hot desert?

7. How is carbonic acid (H_2CO_3) formed in nature? What results when this acid reacts with potassium feldspar?

8. What is the difference between soil and regolith?

9. What factors might cause different soils to develop from the same parent material, or similar soils to form from different parent materials?

10. Which of the controls of soil formation is most important? Explain.

11. How can slope affect the development of soil? What is meant by the terms *slope orientation*?

12. List the characteristics associated with each of the horizons in a well-developed soil profile. Which of the horizons constitute the solum? Under what circumstances do soils lack horizons?

13. Distinguish between pedalfers and pedocals.

14. Briefly describe the conditions that led to the Dust Bowl of the 1930s (see Box 5.3).

15. Soils formed in the humid tropics and the Arctic contain little organic matter. Do both lack humus for the same reasons?

16. What soil type is associated with tropical rain forests? As this soil is associated with luxuriant natural vegetation, is it also excellent for growing crops? Briefly explain.

17. List three detrimental effects of soil erosion, other than the loss of topsoil from croplands.

Key Terms

caliche (p. 136)
chemical weathering
    (p. 120)
dissolution (p. 124)
eluviation (p. 134)
erosion (p. 120)
exfoliation dome (p. 122)
frost wedging (p. 121)

horizon (p. 134)
humus (p. 131)
hydrolysis (p. 127)
laterite (p. 136)
leaching (p. 134)
mass wasting (p. 120)
mechanical weathering
    (p. 120)

oxidation (p. 126)
parent material (p. 132)
pedalfer (p. 135)
pedocal (p. 136)
regolith (p. 131)
residual soil (p. 132)
sheeting (p. 122)
soil (p. 131)

soil profile (p. 134)
solum (p. 134)
spheroidal weathering
    (p. 129)
talus slope (p. 121)
transported soil (p. 132)
weathering (p. 120)

Web Resources

The *Earth* Home Page provides on-line resources for this chapter on the World Wide Web. You will find review exercises, specific updates for items in the

chapter, suggested reading, and links to interesting related pathways on the Internet. Visit the *Earth* Home Page at **http://www.prenhall.com/tarbuck.**

Sedimentary Rocks

Left   This eroded sandstone monument in Arizona's Vermillion Cliffs Wilderness was once a sand dune.
— *Photo by Jack W. Dykinga*

Figure 6.1 Sedimentary rocks in Arizona's Monument Valley are very colorful. In the background, harder, more resistant sandstone stands above weaker, crumbling shale. Sedimentary rocks are exposed at the surface more than igneous and metamorphic rocks. Because they contain fossils and other clues about our geologic past, sedimentary rocks are important in the study of Earth history. (Photo by Carr Clifton)

Chapter 5 gave you the background needed to understand the origin of sedimentary rocks. Recall that weathering of existing rocks begins the process. Next, erosional agents such as running water, wind, waves, and ice remove the products of weathering and carry them to a new location where they are deposited. Usually the particles are broken down further during the transport phase. Following deposition, this material, which is now called **sediment,** becomes lithified. In most cases, the sediment is lithified into solid sedimentary rock by the processes of *compaction* and *cementation*.

Thus, the products of mechanical and chemical weathering constitute the raw materials for sedimentary rocks. The word *sedimentary* indicates the nature of these rocks, for it is derived from the Latin *sedimentum*. which means "settling," a reference to solid material settling out of a fluid (water or air). Most, but not all, sediment is deposited in this fashion. Weathered debris is constantly being swept from bedrock, carried away, and eventually deposited in lakes, river valleys, seas, and countless other places. The particles in a desert sand dune, the mud on the floor of a swamp, the gravel in a stream bed, and even household dust are examples of this never-ending process. Because the weathering of bedrock and the transport and deposition of the weathering products are continuous, sediment is found almost everywhere. As piles of sediment accumulate, the materials near the bottom are compacted. Over long periods, these sediments become cemented together by mineral matter deposited in the spaces between particles, forming solid rock.

Geologists estimate that sedimentary rocks account for only about 5 percent (by volume) of Earth's outer 16 kilometers (10 miles). However, the importance of this group of rocks is far greater than this percentage would imply. If we were to sample the rocks exposed at the surface, we would find that the great majority are sedimentary. Indeed, about 75 percent of all rock outcrops on the continents are sedimentary (Figure 6.1). Therefore, we may think of sedimentary rocks as comprising a relatively thin and somewhat discontinuous layer in the uppermost portion of the crust. This fact is readily understood when we consider that sediment accumulates at the surface.

Because sediments are deposited at Earth's surface, the rock layers that they eventually form contain evidence of past events that occurred at the surface. By their very nature, sedimentary rocks contain within them indications of past environments in which their particles were deposited and, in some cases, clues to the mechanisms involved in their transport. Furthermore, it is sedimentary rocks that contain fossils, which are vital tools in the study of the geologic past. Thus, it is largely from this group of rocks that geologists must reconstruct the details of Earth history.

Finally, it should be mentioned that many sedimentary rocks are very important economically. Coal, which is burned to provide a significant portion of U.S. electrical energy, is classified as a sedimentary rock. Our other major energy sources, petroleum and natural gas, are associated with sedimentary rocks. So are major sources of iron, aluminum, manganese, and fertilizer, plus numerous materials essential to the construction industry.

Types of Sedimentary Rocks

Sediment has two principal sources. First, sediment may be an accumulation of material that originates and is transported as solid particles derived from both mechanical and chemical weathering. Deposits of this type are termed *detrital* and the sedimentary rocks that they form are called **detrital sedimentary rocks**. The second major source of sediment is soluble material produced largely by chemical weathering. When these dissolved substances are precipitated by either inorganic or organic processes, the material is known as chemical sediment and the rocks formed from it are called **chemical sedimentary rocks**.

We will now look at each type of sedimentary rock, and some examples of each.

Detrital Sedimentary Rocks

Though a wide variety of minerals and rock fragments may be found in detrital rocks, clay minerals and quartz are the chief constituents of most sedimentary rocks in this category. Recall from Chapter 5 that clay minerals are the most abundant product of the chemical weathering of silicate minerals, especially the feldspars. Clays are fine-grained minerals with sheet-like crystalline structures similar to the micas. The other common mineral, quartz, is abundant because it is extremely durable and very resistant to chemical weathering. Thus, when igneous rocks such as granite are attacked by weathering processes, individual quartz grains are freed.

Other common minerals in detrital rocks are feldspars and micas. Because chemical weathering rapidly transforms these minerals into new substances, their presence in sedimentary rocks indicates that erosion and deposition were fast enough to preserve some of the primary minerals from the source rock before they could be decomposed.

Particle size is the primary basis for distinguishing among various detrital sedimentary rocks. Table 6.1 presents the size categories for particles making up detrital rocks. Note that in this context the term *clay* refers only to a particle size and not to the minerals of the same name. Although most clay minerals are of clay size, not all clay-sized sediment consists of clay minerals.

Particle size is not only a convenient method of dividing detrital rocks; the sizes of the component grains also provide useful information about environments of deposition. Currents of water or air sort the

Table 6.1 Particle Size Classification for Detrital Rocks

Size Ranges (millimeters)	Particle Name	Common Sediment Name	Detrital Rock
>256	Boulder		
64–256	Cobble	Gravel	Conglomerate or breccia
4–64	Pebble		
2–4	Granule		
1/16–2	Sand	Sand	Sandstone
1/256–1/16	Silt	mud	Shale or mudstone
<1/256	Clay		

particles by size; the stronger the current, the larger the particle size carried. Gravels, for example, are moved by swiftly flowing rivers as well as by landslides and glaciers. Less energy is required to transport sand; thus, it is common to such features as windblown dunes and some river deposits and beaches. Very little energy is needed to transport clay, so it settles very slowly. Accumulation of these tiny particles are generally associated with the quiet water of a lake, lagoon, swamp or certain marine environments.

Common detrital sedimentary rocks, in order of increasing particle size, are shale, sandstone, and conglomerate or breccia. We will now look at each type and how it forms.

Shale

Shale is a sedimentary rock consisting of silt- and clay-sized particles (Figure 6.2). These fine-grained detrital rocks account for well over half of all sedimentary rocks. The particles in these rocks are so small that they cannot be readily identified without great magnification

← 5 cm →

Figure 6.2 Shale is a fine-grained detrital rock that is by far the most abundant of all sedimentary rocks. Dark shales containing plant remains are relatively common. (Photo by E. J. Tarbuck)

and for this reason make shale more difficult to study and analyze than most other sedimentary rocks.

Much of what can be learned is based on particle size. The tiny grains in shale indicate that deposition occurs as the result of gradual settling from relatively quiet, nonturbulent currents. Such environments include lakes, river floodplains, lagoons, and portions of the deep-ocean basins. Even in these "quiet" environments, there is usually enough turbulence to keep clay-sized particles suspended almost indefinitely. Consequently, much of the clay is deposited only after the individual particle coalesce to form larger aggregates.

Sometimes the chemical composition of the rock provides additional information. One example is black shale, which is black because it contains abundant organic matter (carbon). When such a rock is found, it strongly implies that deposition occurred in an oxygen-poor environment such as a swamp, where organic materials do not readily oxidize and decay.

As silt and clay accumulate, they tend to from thin layers, which are commonly referred to as *laminea*. Initially the particles in the laminae are oriented randomly. This disordered arrangement leaves a high percentage of open space (called *pore space*) that is filled with water. However, this situation usually changes with time as additional layers of sediment pile up and compact the sediment below.

During this phase the clay and silt particles take on a more nearly parallel alignment and become tightly packed. This rearrangement of grains reduces the size of the pore spaces and forces out much of the water. Once the grains are pressed closely together, the tiny spaces between particles do not readily permit solutions containing cementing material to circulate. Therefore, shales are often described as being weak because they are poorly cemented and therefore not well lithified.

The inability of water to penetrate its microscopic poor spaces explains why shale often forms barriers to the subsurface movement of water and petroleum. Indeed, rock layers that contain groundwater are commonly underlain by shale beds that block further downward movement. The opposite is true for underground reservoirs of petroleum. They are often capped by shale beds that effectively prevent oil and gas from escaping to the surface.*

It is common to apply the term *shale* to all fine-grained sedimentary rocks, especially in a nontechnical context. However, be aware that there is a more restricted use of the term. In this narrower usage, shale must exhibit the ability to split into thin layers along well-developed closely space planes. This property is termed **fissility**. If the rock breaks into chunks or blocks, the name *mudstone* is applied. Another fine-grained sedimentary rock that, like mudstone, is often grouped with shale but lacks fissility is *siltstone*. As its name implies, siltstone is composed largely of silt-sized particles and contains less clay-sized material than shale and mudstone.

Although shale is far more common than other sedimentary rocks, it does not usually attract as much notice as other less abundant members of this group. The reason is that shale does not form prominent outcrops as sandstone and limestone often do. Rather, shale crumbles easily and usually forms a cover of soil that hides the unweathered rock below. This is illustrated nicely in the Grand Canyon, where the gentler slopes of weathered shale are quite inconspicuous and overgrown with vegetation, in sharp contrast with the bold cliffs produced by more durable rocks (Figure 6.3).

Although shale beds may not form striking cliffs and prominent outcrops, some deposits have economic value. Certain shales are quarried to obtain raw material for pottery, brick, tile, and china. Moreover, when mixed with limestone, shale is used to make portland cement. In the future, one type of shale, called oil shale, may become a valuable energy resource. This possibility will be explored in Chapter 21.

Sandstone

Sandstone is the name given rocks in which sand-sized grains predominate (Figure 6.4). After shale, sandstone is the most abundant sedimentary rock, accounting for approximately 20 percent of the entire group. Sandstones form in a variety of environments and often contain significant clues about their origin, including sorting, particle shape, and composition.

Sorting is the degree of similarity in particle size in a sedimentary rock. For example, if all the grains in a sample of sandstone are about the same size, the sand is considered *well sorted*. Conversely, if the rock contains mixed large and small particles, the sand is said to be *poorly sorted*. By studying the degree of sorting, we can learn much about the depositing current. Deposits of windblown sand are usually better sorted than deposits sorted by wave activity (Figure 6.5). Particles washed by waves are commonly better sorted than materials deposited by streams. Sediment accumulations that exhibit poor sorting usually result when particles are transported for only a relatively short time and then rapidly deposited. For example, when a turbulent stream reaches the gentler slopes at the base of a steep mountain, its velocity is quickly reduced and poorly sorted sands and gravels are deposited.

The shapes of sand grains can also help decipher the history of a sandstone. When streams, winds, or

---

*The relationship between impermeable beds and the occurrence and movement of groundwater is examined in Chapter 11. Shale beds as cap rocks in oil traps are discussed in Chapter 21.

Figure 6.3 Sedimentary rock layers exposed in the walls of the Grand Canyon, Arizona. Beds of resistant sandstone and limestone produce bold cliffs. By contrast, weaker, poorly cemented shale crumbles and produces a gentler slope of weathered debris in which some vegetation is growing. (Photo by Tom Till)

waves move sand and other sedimentary particles, the grains lose their sharp edges and corners and become more rounded as they collide with other particles during transport. Thus, rounded grains likely have been airborne or waterborne. Further, the degree of rounding indicates the distance or time involved in the transportation of sediment by currents of air or water. Highly rounded grains indicate that a great deal of abrasion and hence a great deal of transport has occurred.

Very angular grains, on the other hand, imply two things: that the materials were transported only a short distance before they were deposited, and that some other medium may have transported them. For example, when glaciers move sediment, the particles are usually mad more irregular by the crushing and grinding action of the ice.

In addition to affecting the degree of rounding and the amount of sorting that particles undergo, the length of transport by turbulent air and water currents also influences the mineral composition of a sedimentary deposit. Substantial weathering and long transport lead to the gradual destruction of weaker and less stable minerals, including the feldspars and ferromagnesians. Because quartz is very durable, it is usually the mineral that survives the long trip in a turbulent environment.

The preceding discussion has shown that the origin and history of sandstone can often be deduced by examining the sorting, roundness and mineral composition of its constituent grains. Knowing this information allows us to infer that a well-sorted, quartz-rich sandstone consisting of highly rounded grains must be the result of a great deal of transport. Such a rock, in fact, may represent several cycles of weathering, transport, and deposition. We may also conclude that a sandstone containing significant amount of feldspar and angular grains of ferromagnesian minerals underwent little chemical weathering and transport and was probably deposited close to the source area of the particles.

Owing to its durability, quartz is the predominant mineral in most sandstones. When this is the case, the rock may simply be called *quartz sandstone*. When a sandstone contains appreciable quantities of feldspar, the rock is called *arkose*. In addition to feldspar, arkose usually contains quartz and sparkling bits of mica. The mineral composition of arkose indicates that the grains were derived from granitic source rocks. The particles are generally poorly sorted and angular, which suggests short-distance transport, minimal chemical weathering in a relatively dry climate, and rapid deposition and burial.

2 cm

Figure 6.4 Quartz sandstone. After shale, sandstone is the most abundant sedimentary rock. This thick layer of sandstone, called the Navajo Sandstone, is exposed in and near Utah's Zion National Park. The hand sample comes from this layer.

Close up

A third variety of sandstone is known as *graywacke*. Along with quartz and feldspar, this dark-colored rock contains abundant rock fragments and matrix. *Matrix* refers to the silt- and clay-sized particles found in spaces between larger sand grains. More than 15 percent of graywacke's volume is matrix. The poor sorting and angular grains characteristic of graywacke suggest that the particles were transported only a relatively short distance from their source area and then rapidly deposited. Before the sediment could be reworked and sorted further, it was buried by additional layers of material. Graywacke is frequently associated with submarine deposits made by dense sediment-choked torrents called turbidity currents.*

Conglomerate and Breccia

Conglomerate consists largely of gravels (Figure 6.6). As Table 6.1 indicates, these particles can range in size from large boulders to particles as small as garden peas. The particles are commonly large enough to be identified as distinctive rock types; thus, they can be valuable in identifying the source areas of sediments. More often than not conglomerates are poorly sorted because the opening between the large gravel particles contain sand or mud.

---

*More on these currents and the *graded beds* they create may be found in the section on "Submarine Canyons and Turbidity Currents" in Chapter 18.

Figure 6.5 Sorting is the degree of similarity in particle size. The wind-transported sand grains in this dune are well sorted because they are all practically the same size. Mesquite Flat Dunes, Death Valley National Monument, California. (Photo by David Muench)

Gravels accumulate in a variety of environments and usually indicate the existence of steep slopes or very turbulent currents. The coarse particles in a conglomerate may reflect the action of energetic mountain streams or result from strong wave activity along a rapidly eroding coast. Some glacial and landslide deposits also contain plentiful gravel.

If the large particles are angular rather the rounded, the rock is called *breccia* (Figure 6.7). Because large particles abrade and become rounded very rapidly during transport, the pebbles and cobbles in a breccia indicate that they did not travel far from their source area before they were deposited. Thus, as with many sedimentary rocks, conglomerates and breccias contain clues to their history. Their particle sizes reveal the strength of the currents that transported them, whereas the degree of rounding indicates how far the particles traveled. The fragments within a sample identify the source rocks that supplied them.

⊙ Chemical Sedimentary Rocks

In contrast to detrital rocks, which form from the solid products of weathering, chemical sediments derive from material that is carried *in solution* to lakes and seas. This material does not remain dissolved in the water indefinitely, however. Some of it precipitates to form chemical sediments. These become rocks such as limestone, chert, and rock salt.

This precipitation of material occurs in two ways. *Inorganic* processes such as evaporation and chemical activity can produce chemical sediments. *Organic* (life)

Figure 6.6 Conglomerate is composed primarily of rounded gravel-sized particles.

Close up

processes of water-dwelling organisms also form chemical sediments, said to be of **biochemical** origin.

One example of a deposit resulting from inorganic chemical processes is the dripstone that decorates many caves (Figure 6.8). Another is the salt left behind as a body of seawater evaporates. In contrast, many water-dwelling animals and plants extract dissolved mineral matter to form shells and other hard

Close up

Figure 6.7 When the gravel-sized particles in a detrital rock are angular, the rock is called breccia. (Photo by E. J. Tarbuck)

Figure 6.8 Because many cave deposits are created by the seemingly endless dripping of water over long time spans, they are commonly called *dripstone*. The material being deposited is calcium carbonate ($CaCO_3$) and the rock is a form of limestone called *travertine*. The calcium carbonate is precipitated when some dissolved carbon dioxide escapes from a water drop. (Photo by Clifford Stroud, National Park Service)

parts. After the organisms die, their skeletons collect by the millions on the floor of a lake or ocean as biochemical sediment (Figure 6.9).

Limestone

Representing about 10 percent of the total volume of all sedimentary rocks, *limestone* is the most abundant chemical sedimentary rock. It is composed chiefly of the mineral calcite ($CaCO_3$) and forms either by inorganic means or as the result of biochemical processes. Regardless of its origin, the mineral composition of all limestone is similar, yet many different types exist. This is true because limestones are produced under a variety of conditions. Those forms having a marine biochemical origin are by far the most common.

Coral Reefs. Corals are one important example of organisms that are capable of creating large quantities of marine limestone. These relatively simple invertebrate animals secrete a calcareous (calcite-rich) external skeleton. Although they are small, corals are capable of creating massive structures called *reefs* (Figure 6.10A). Reefs consist of coral colonies made up of great numbers of individuals that live side by side on a calcite structure secreted by the animals. In addition, calcium carbonate-secreting algae live with the corals and help cement the entire structure into a solid mass. A wide variety of other organisms also live in and near the reefs.

Certainly the best-known modern reef is Australia's Great-Barrier Reef, 2000 kilometers long,

but many lesser reefs also exist. They develop in the shallow, warm waters of the tropics and subtropics equatorward of about 30° latitude. Striking examples exist in the Bahamas and Florida Keys.

Close up

Figure 6.9 This rock, called coquina, consists of shell fragments; therefore, it has a biochemical origin. (Photo by E. J. Tarbuck)

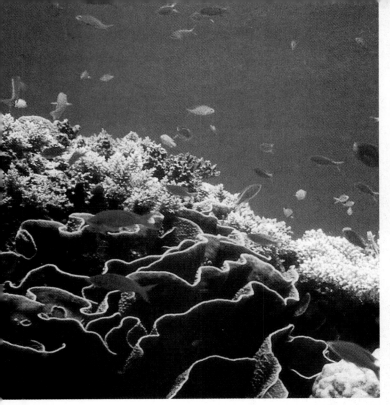

A.   B.

Figure 6.10 **A.** This modern coral reef is at Bora Bora in French Polynesia. (Photo by Nancy Sefton/Photo Researchers)
B. El Capitan Peak, a massive limestone cliff in Guadalupe Mountains National Park, Texas. The rocks here are an exposed portion of
a large reef that formed during the Permian period. (Photo by Steve Elmore/The Stock Market)

Of course, not only modern corals build reefs. Corals have been responsible for producing vast quantities of limestone in the geologic past as well. In the United States, reefs of Silurian age are prominent features in Wisconsin, Illinois, and Indiana. In west Texas and adjacent southeastern New Mexico, a massive reef complex formed during the Permian period is strikingly exposed in Guadalupe Mountains National Park (Figure 6.10B)

Coquina and Chalk. Although most limestone is the product of biological processes, this origin is not always evident, because shells and skeletons may undergo considerable change before becoming lithified into rock. However, one easily identified biochemical limestone is *coquina*, a coarse rock composed of poorly cemented shells and shell fragments (see Figure 6.9). Another less obvious, but nevertheless familiar example, is *chalk*, a soft , porous rock made up almost entirely of the hard parts of microscopic marine organisms. Among the most famous chalk deposits are those exposed along the southeast coast of England (Figure 6.11).

Inorganic Limestones. Limestones having an inorganic origin form when chemical changes or high water temperatures increase the concentration of calcium carbonate to the point that it precipitates. *Travertine*, the type of limestone commonly seen in caves, is an example (see Figure 6.8). When travertine is deposited in caves, groundwater is the source of the calcium carbonate. As water droplets become exposed to the air in a cavern, some of the carbon dioxide dissolved in the water escapes, causing calcium carbonate to precipitate.

Another variety of inorganic limestone is *oolitic limestone*. It is a rock composed of small spherical grains called *ooids*. Ooids form in shallow marine waters as tiny "seed" particles (commonly small shell fragments) are moved back and forth by currents. As the grains are rolled about in the warm water, which is supersaturated with calcium carbonate, they become coated with layer upon layer of the precipitate.

Dolostone

Closely related to limestone is *dolostone*, rock composed of the calcium–magnesium carbonate mineral dolomite. Although dolostone can form by direct precipitation from seawater, most dolostone probably originates when magnesium in seawater replaces some of the calcium in limestone. The latter hypothesis is reinforced by the fact that there is practically no young dolostone. Rather, most dolostones are ancient rocks in which there was ample time for magnesium to replace calcium.

Chert

Chert is a name used for a number of very compact and hard rocks made of microcrystalline silica (SiO_2). One well-known form is *flint*, whose dark color results from the organic matter it contains. *Jasper*, a red variety, gets

Figure 6.11 The White Cliffs of Dover. This prominent chalk deposit underlies large portions of southern England as well as parts of northern France. (Photo by Laguna Photo/Liaison International)

its bright color from the iron oxide it contains. The banded form is usually referred to as *agate* (Figure 6.12).

Chert deposits are commonly found in one of two situations: as irregularly shaped nodules in limestone and as layers of rock. The silica composing many chert nodules may have been deposited directly from water. Such nodules have an inorganic origin. However, it is unlikely that a very large percentage of chert layers was precipitated directly from seawater, because seawater is seldom saturated with silica. Hence, beds of chert are thought to have originated largely as biochemical sediment.

Most water-dwelling organisms that produce hard parts make them of calcium carbonate. But some, such as diatoms and radiolarians, produce glasslike silica skeletons. These tiny organisms are able to extract silica even though seawater contains only tiny quantities. It is from their remains that most chert beds are believed to originate.

Some bedded cherts occur in association with lava flows and layers of volcanic ash. For these occurrences it is probable that the silica was derived from the decomposition of the volcanic ash and not from biochemical sources. Note that when a hand specimen of chert is being examined, there are few reliable criteria by which the mode of origin (inorganic versus biochemical) can be determined.

Like glass, most chert has a conchoidal fracture. Its hardness, ease of chipping, and ability to hold a sharp edge made chert a favorite of Native Americans for fashioning "points" for spears and arrows. Because of chert's durability and extensive use, "arrowheads" are found in many parts of North America.

Evaporites

Very often evaporation is the mechanism triggering deposition of chemical precipitates. Minerals commonly precipitated in this fashion include halite (sodium chloride, NaCl), the chief component of *rock salt*, and gypsum (hydrous calcium sulfate, $CaSO_4 \cdot 2H_2O$), the main ingredient of *rock gypsum*. Both have significant importance. Halite is familiar to everyone as the common salt used in cooking and

seasoning foods (see Box 6.1). Of course, it has many other uses, from melting ice on roads to making hydrochloric acid, and has been considered important enough that people have sought, traded, and fought over it for much of human history. Gypsum is the basic ingredient of plaster of Paris. This material is used most extensively in the construction industry for wallboard and interior plaster.

In the geologic past, many areas that are now dry land were basins, submerged under shallow arms of a sea that had only narrow connections to the open ocean. Under these conditions, seawater continually moved into the bay to replace water lost by evaporation. Eventually the waters of the bay became saturated and salt deposition began. Such deposits are called **evaporites**.

When a body of seawater evaporates, the minerals that precipitate do so in a sequence that is determined by their solubility. Less-soluble minerals precipitate first and more-soluble minerals precipitate later as salinity increases. For example, gypsum precipitates when about two-thirds to three-quarters of the seawater has evaporated, and halite settles out when nine-tenths of the water has been removed. During the last stages of this process, potassium and magnesium salts precipitate. One of these last-formed salts, the mineral *sylvite*, is mined as a significant source of potassium ("potash") for fertilizer.

Figure 6.12 Agate is the banded form of chert. (Photo by Jeff Scovil)

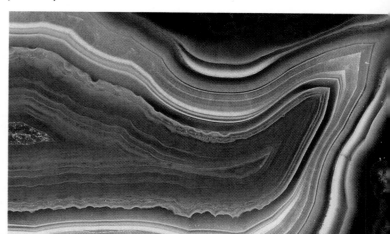

Box 6.1

Harvesting Salt from the Sea

Each year about 30 percent of the world's supply of salt is extracted from seawater. In this process, salt water is held in shallow ponds while energy from the Sun evaporates the water. The nearly pure salt deposits that eventually form are essentially artificial evaporite deposits.

Although seawater is certainly plentiful, there are only a few places where the solar evaporation process is feasible. In addition to abundant salt water, other requirements include long, uninterrupted periods of sunshine; little rainfall; ample air movement; flat, impervious soil; and access to a major market or to inexpensive transportation. In the United States, these conditions exist at the Great Salt Lake in Utah and near San Diego and San Francisco in California.

At the southern end of San Francisco Bay, it takes nearly 38,000 liters (10,000 gallons) of water to produce 900 kilograms (1 ton) of salt. Each year about 1 million tons of salt are harvested in the Bay Area. The process involves a large amount of land and 560 kilometers (350 miles) of earthen dikes to confine the evaporating water in ponds of the right size and depth. The cycle, from bay water to harvested salt, requires about 5 years.

Bay water is admitted to the first of the concentrating ponds in early summer when the discharge of the rivers entering the bay is reduced and the water in the bay is at its highest salinity. Once in the ponds, the brine is moved from pond to pond as it reaches certain preestablished levels of salinity. Evaporation continues until the solution is concentrated enough to crystallize virtually pure sodium chloride (NaCl).

At this point, the saturated brine is pumped into carefully cleaned 20- to 60-acre crystallization ponds to a depth of about 0.5 meter. Within a few months, a 12- to 15-centimeter-thick layer of pure salt crystals has collected on the bottom. The remaining fluid, called *bittern*, is rich in magnesium, bromine, and potassium. The bittern is drained, and the salt is ready for harvest, 5 years after the seawater first entered the process (Figure 6.A).

Figure 6.A Harvesting salt produced by the evaporation of seawater. (Photo by William E. Townsend)

On a smaller scale, evaporite deposits can be seen in such places as Death Valley, California. Here, following rains or periods of snowmelt in the mountains, streams flow from the surrounding mountains into an enclosed basin. As the water evaporates, **salt flats** form when dissolved materials are precipitated as a white crust on the ground (Figure 6.13).

Coal

Coal is quite different from other rocks. Unlike limestone and chert, which are calcite- and silica-rich, coal is made of organic matter. Close examination of coal under a magnifying glass often reveals plant structures such as leaves, bark, and wood that have been chemically altered but are still identifiable. This supports the conclusion that coal is the end product of large amounts of plant material, buried for millions of years (Figure 6.14).

The initial stage in coal formation is the accumulation of large quantities of plant remains. However, special conditions are required for such accumulations, because dead plants readily decompose when exposed to the atmosphere or other oxygen-rich environments. One important environment that allows for the buildup of plant material is a swamp (Figure 6.15).

Stagnant swamp water is oxygen-deficient, so complete decay (oxidation) of the plant material is not possible. Instead, the plants are attacked by certain bacteria that partly decompose the organic material and liberate oxygen and hydrogen. As these elements escape, the percentage of carbon gradually increases. The bacteria are not able to finish the job of decomposition because they are destroyed by acids liberated from the plants.

The partial decomposition of plant remains in an oxygen-poor swamp creates a layer of *peat*, a soft, brown

Figure 6.13 These salt flats in Utah are examples of evaporite deposits and are common in basins located in the arid West. (Photo by Scott T. Smith)

material in which plant structures are still easily recognized. With shallow burial, peat slowly changes to *lignite*, a soft, brown coal. Burial increases the temperature of sediments as well as the pressure on them.

The higher temperatures bring about chemical reactions within the plant materials and yield water and organic gases (volatiles). As the load increases from more sediment on top of the developing coal, the water and volatiles are pressed out and the proportion of *fixed carbon* (the remaining solid combustible material) increases. The greater the carbon content, the greater the coal's energy ranking as a fuel. During burial, the coal also becomes increasingly compact. For example, deeper burial transforms lignite into a harder, more compacted black rock called *bituminous* coal. Compared to the peat from which it formed, a bed of bituminous coal may be only one-tenth as thick.

Lignite and bituminous coals are sedimentary rocks. However, when sedimentary layers are subjected to the folding and deformation associated with mountain building, the heat and pressure cause a further loss of volatiles and water, thus increasing the concentration of fixed carbon. This metamorphoses bituminous coal into *anthracite*, a very hard, shiny black *metamorphic* rock. Although anthracite is a clean-burning fuel, only a relatively small amount is mined. Anthracite is not widespread and is more difficult and expensive to extract than the relatively flat-lying layers of bituminous coal.

Coal is a major energy resource. Its role as a fuel and some of the problems associated with burning coal are discussed in Chapter 21.

Turning Sediment into Sedimentary Rock

Having examined the general types of sedimentary rocks, let us look at how sediment becomes rock.

Lithification refers to the process by which unconsolidated sediments are transformed into solid sedimentary rocks. One of the most common processes affecting sediments is **compaction**. As sediments accumulate through time, the weight of overlying material compresses the deeper sediments. As the grains are pressed closer and closer, there is a considerable reduction in pore space. For example, when clays are buried beneath several thousand meters of material, the volume of clay may be reduced by as much as 40 percent. Because sands and other coarse sediments are only slightly compressible, compaction is most significant as a lithification process in fine-grained sedimentary rocks such as shale.

Cementation is the most important process by which sediments are converted to sedimentary rocks. The cementing materials are carried in solution by water percolating through the open spaces between particles. Through time, the cement precipitates onto the sediment grains, fills the open spaces, and joins the particles.

Calcite, silica, and iron oxide are the most common cements. It is often a relatively simple matter to identify the cementing material. Calcite cement will effervesce with dilute hydrochloric acid. Silica is the hardest cement and thus produces the hardest sedimentary rocks. An orange or dark red color in a sedimentary rock means that iron oxide is present.

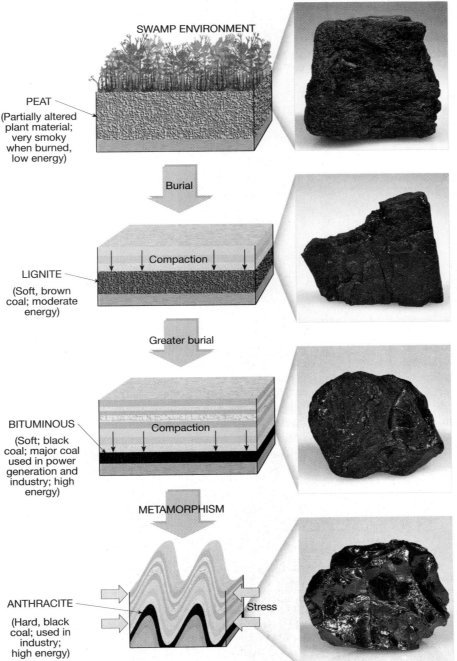

Figure 6.14 Successive stages in the formation of coal.

SWAMP ENVIRONMENT

PEAT
(Partially altered plant material; very smoky when burned, low energy)

Burial

Compaction

LIGNITE
(Soft, brown coal; moderate energy)

Greater burial

Compaction

BITUMINOUS
(Soft; black coal; major coal used in power generation and industry; high energy)

METAMORPHISM

ANTHRACITE
(Hard, black coal; used in industry; high energy)

Stress

Most sedimentary rocks are lithified by means of compaction and cementation. However, certain chemical sedimentary rocks, such as the evaporites, initially form as solid masses of intergrown crystals, rather than beginning as accumulations of separate particles that later become solid. Other crystalline sedimentary rocks do not begin that way but are transformed into masses of interlocking crystals sometime after the sediment is deposited.

For example, with time and burial, loose sediment consisting of delicate calcareous skeletal debris may be recrystallized into a relatively dense crystalline limestone. Because crystals grow until they fill all the available space, pore spaces are frequently lacking in crystalline sedimentary rocks. Unless the rocks later develop joints and fractures, they will be relatively impermeable to fluids like water and oil.

Figure 6.15 Because stagnant swamp water does not allow plants to completely decay, large quantities of plant material can accumulate in a swamp. (Photo by Carr Clifton)

Classification of Sedimentary Rocks

The classification scheme in Table 6.2 divides sedimentary rocks into two major groups: detrital and chemical. Further, we can see that the main criterion for subdividing the detrital rocks is particle size, whereas the primary basis for distinguishing among different rocks in the chemical group is their mineral composition.

As is the case with many (perhaps most) classifications of natural phenomena, the categories presented in Table 6.2 are more rigid than the actual state of nature. In reality, many of the sedimentary rocks classified into the chemical group also contain at least small quantities of detrital sediment. Many limestones, for example, contain varying amounts of mud or sand, giving them a "sandy" or "shaly" quality. Conversely, because practically all detrital rocks are cemented with

Table 6.2 Classification of Sedimentary Rocks

DETRITAL ROCKS

Texture	Sediment Name and Particle Size	Comments	Rock Name
Clastic	Gravel (>2 mm)	Rounded rock fragments	Conglomerate
		Angular rock fragments	Breccia
	Sand (1/16–2 mm)	Quartz predominates	Quartz sandstone
		Quartz with considerable feldspar	Arkose
		Dark color; quartz with considerable feldspar, clay, and rocky fragments	Graywacke
	Mud (<1/16 mm)	Splits into thin layers	Shale
		Breaks into clumps or blocks	Mudstone

CHEMICAL ROCKS

Group	Texture	Composition	Rock Name
Inorganic	Clastic or nonclastic	Calcite, $CaCo_3$	Limestone
	Nonclastic	Dolomite, $CaMg(CO_3)_2$	Dolostone
	Nonclastic	Microcrystalline quartz, SiO_2	Chert
	Nonclastic	Halite, NaCl	Rock salt
	Nonclastic	Gypsum, $CaSO_4 \cdot 2H_2O$	Rock gypsum
Biochemical	Clastic or nonclastic	Calcite, $CaCO_3$	Limestone
	Nonclastic	Microcrystalline quartz, SiO_2	Chert
	Nonclastic	Altered plant remains	Coal

material that was originally dissolved in water, they too are far from being "pure."

As was the case with the igneous rocks examined in Chapter 3, *texture* is a part of sedimentary rock classification. There are two major textures used in the classification of sedimentary rocks: clastic and nonclastic. The term **clastic** is taken from a Greek word meaning "broken." Rocks that display a clastic texture consist of discrete fragments and particles that are cemented and compacted together. Although cement is present in the spaces between particles, these openings are rarely filled completely. Table 6.2 shows that *all* detrital rocks have a clastic texture. The table also shows that some chemical sedimentary rocks exhibit this texture, too. For example, coquina, the limestone composed of shells and shell fragments, is obviously as clastic as a conglomerate or sandstone. The same applies for some varieties of oolitic limestone.

Some chemical sedimentary rocks have a **nonclastic** texture in which the minerals form a pattern of interlocking crystals. The crystals may be microscopically small or large enough to be visible without magnification. Common examples of rocks with nonclastic textures are those deposited when seawater evaporates (Figure 6.16). The materials that make up many other nonclastic rocks may actually have originated as detrital deposits. In these instances, the particles probably

consisted of shell fragments and other hard parts rich in calcium carbonate or silica. The clastic nature of the grains was subsequently obliterated or obscured because the particles recrystallized when they were consolidated into limestone or chert.

Nonclastic rocks consist of intergrown crystals, and some may resemble igneous rocks, which are also crystalline. The two rock types are usually easy to distinguish because the minerals contained in nonclastic sedimentary rocks are quite unlike those found in most igneous rocks. For example, rock salt, rock gypsum, and some forms of limestone consist of intergrown crystals, but the minerals within these rocks (halite, gypsum, and calcite) are seldom associated with igneous rocks.

 ## Sedimentary Environments

As stated at the beginning of the chapter, sedimentary rocks are important in the interpretation of Earth history. By understanding the conditions under which sedimentary rocks form, geologists can often deduce the history of a rock, including information about the origin of its component particles, the method and length of its transport, and the nature of the place where the grains eventually came to rest; that is, the environment of deposition.

Sediments are deposited at Earth's surface. Thus, they hold many clues about the physical, chemical, and biological conditions that existed in the areas where the materials accumulated (see Box 6.2). By applying a thorough knowledge of present-day conditions, geologists attempt to reconstruct the ancient environments and geographical relationships of an area at the time a particular set of sedimentary layers were deposited. Such analyses often lead to the creation of maps, which depict the distribution of land and sea, mountains and plains, deserts and glaciers, and other environments of deposition.

Sedimentary environments are commonly placed into one of two broad categories: *terrestrial* (continental) or *marine*. Because the shore zone exhibits characteristics of both, it can be considered transitional between land and sea. Figure 6.17 divides the two broad categories of terrestrial and marine into several major sedimentary environments. Chapters 10 through 14, as well as portions of Chapter 18, will describe these environments in detail. Each is an area where sediment accumulates and where organisms live and die. Each produces a characteristic sedimentary rock or assemblage that reflects prevailing conditions.

When a series of sedimentary layers is studied, we can see the successive changes in environmental

Figure 6.16 Like other evaporites, this sample of rock salt is said to have a nonclastic texture because it is composed of intergrown crystals.

 Close up

Box 6.2

Seafloor Sediments and Climatic Change

Reliable climate records go back only a couple of hundred years, at best. How do scientists learn about climates and climatic changes prior to that time? The obvious answer is that they must reconstruct past climates from *indirect evidence*; that is, they must examine and analyze phenomena that respond to and reflect changing atmospheric conditions. An interesting and important technique for analyzing Earth's climatic history is the study of sediments from the ocean floor.

Although seafloor sediments are of many types, most contain the remains of organisms that once lived near the sea surface (the ocean-atmosphere interface). When such near-surface organisms die, their shells slowly settle to the ocean floor where they become part of the sedimentary record. One reason that seafloor sediments are useful recorders of worldwide climatic changes is that the numbers and types of organisms living near the sea surface change as the climate changes. This principle is explained by Richard Foster Flint as follows:

...we would expect that in any area of the ocean/atmosphere interface the average annual temperature of the surface water of the ocean would approximate that of the contiguous atmosphere. The temperature equilibrium established between surface seawater and the air above it should mean that...changes in climate should be reflected in changes in organisms living near the surface of the deep sea....

When we recall that the seafloor sediments in vast areas of the ocean consist mainly of shells of pelagic foraminifers, and that these animals are sensitive to variations in water temperature, the connection between such sediments and climatic changes become obvious.*

Thus, in seeking to understand climatic change as well as other environmental transformations, scientists are tapping the huge reservoir of data in seafloor sediments. The sediment cores

Glacial and Quaternary Geology (New York: Wiley, 1971), p. 718.

gathered by drilling ships and other research vessels have provided invaluable data that have significantly expanded our knowledge and understanding of past climates (Figure 6.B).

One notable example of the importance of seafloor sediments to our understanding of climate change relates to unraveling the fluctuating atmospheric conditions of the Ice Age. The record of temperature changes contained in cores of sediment from the ocean floor have proven critical to our present understanding of this recent span of Earth history.†

†For more on this topic, see "Causes of Glaciation" in Chapter 12.

Figure 6.B The *JOIDES Resolution*, drilling ship of the Ocean Drilling Program. The seafloor sediments recovered by this and other research vessels provide scientists with data that allow them to reconstruct past climates. (Photo courtesy of the Ocean Drilling Program)

conditions that occurred at a particular place with the passage of time. Changes in past environments may also be seen when a single unit of sedimentary rock is traced laterally. This is true because at any one time many different depositional environments can exist over a broad area. For example, when sand is accumulating in a beach environment, finer muds are often being deposited in quieter offshore waters. Still farther out, perhaps in a zone where biological activity is high and land-derived sediments are scarce, the deposits consist largely of the calcareous remains of small organisms. In this example, different sediments are accumulating adjacent to one another at the same time. Each unit possesses a distinctive set of characteristics reflecting the conditions in a particular environment. To describe such sets of sediments, the term **facies** is used. When a sedimentary unit is examined in cross-section from

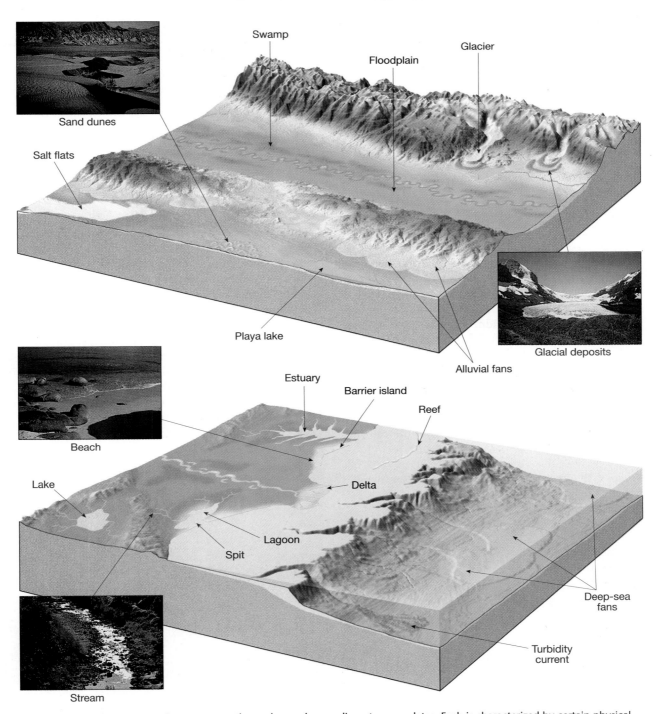

Figure 6.17 Sedimentary environments are those places where sediment accumulates. Each is characterized by certain physical, chemical, and biological conditions. Because each sediment contains clues about the environment in which it was deposited, sedimentary rocks are important in the interpretation of Earth history. A number of important terrestrial, shoreline (transitional), and marine sedimentary environments are represented in these idealized diagrams.

one end to the other, each facies grades laterally into another that formed at the same time but which exhibits different characteristics (Figure 6.18).

Commonly, the merging of adjacent facies tends to be a gradual transition rather than a sharp boundary, but abrupt changes do sometimes occur.

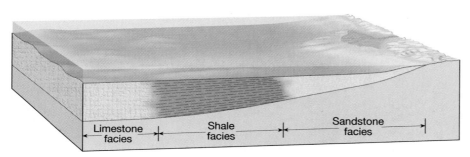

Figure 6.18 When a single sedimentary layer is traced laterally, we may find that it is made up of several different rock types. This can occur because many sedimentary environments can exist at the same time over a broad area. The term *facies* is used to describe such sets of sedimentary rocks. Each facies grades laterally into another that formed at the same time but in a different environment.

Sedimentary Structures

In addition to variations in grain size, mineral composition, and texture, sediments exhibit a variety of structures. Some, such as graded beds, are created when sediments are accumulating and are a reflection of the transporting medium. Others, such as *mud cracks*, form after the materials have been deposited and result from processes occurring in the environment. When present, sedimentary structures provide additional information that can be useful in the interpretation of Earth history.

Sedimentary rocks form as layer upon layer of sediment accumulates in various depositional environments.

These layers, called **strata**, or **beds**, are probably *the single most common and characteristic feature of sedimentary rocks*. Each stratum is unique. It may be a coarse sandstone, a fossil-rich limestone, a black shale, and so on. When you look at Figure 6.19, or look back through this chapter at Figure 6.1 (p. 144), 6.3 (p. 147), and 6.10B (p. 151), you will see many such layers, each different from the others. The variations in texture, composition, and thickness reflect the different conditions under which each layer was deposited.

The thickness of beds ranges from microscopically thin to tens of meters thick. Separating the strata are **bedding planes**, flat surfaces along which rocks tend to separate or break. Changes in the grain size or

Figure 6.19 This outcrop of sedimentary strata illustrates the characteristic layering of this group of rocks. (Photo by Tom Till)

A.

Figure 6.20 **A.** The cut-away section of this sand dune shows cross-bedding. (Photo by John S. Shelton) **B.** The cross-bedding of this sandstone indicates it was once a sand dune. (Photo by David Muench)

B.

in the composition of the sediment being deposited can create bedding planes. Pauses in deposition can also lead to layering because chances are slight that newly deposited material will be exactly the same as previously deposited sediment. Generally each bedding plane marks the end of one episode of sedimentation and the beginning of another.

Because sediments usually accumulate as particles that settle from a fluid, most strata are originally deposited as horizontal layers. There are circumstances, however, when sediments do not accumulate in horizontal beds. Sometimes when a bed of sedimentary rock is examined, we see layers within it that are inclined to the horizontal. When this occurs, it is called **cross-bedding** and is most characteristic of sand dunes, river deltas, and certain stream channel deposits (Figure 6.20).

Graded beds represent another special types of bedding. In this case the particles within a single sedimentary layer gradually change from coarse at the bottom to fine at the top (Figure 6.21). Graded beds are most characteristic of rapid deposition from water containing sediment of varying sizes. When a current experiences a rapid energy loss, the largest particles settle first, followed by successively smaller grains. The deposition of a graded bed is most often associated with a turbidity current, a mass of sediment-choked water that is denser than clear water and that moves downslope along the bottom of a lake or ocean.*

---

*More on these currents and graded beds may be found in the section on "Submarine Canyons and Turbidity Currents" in Chapter 18.

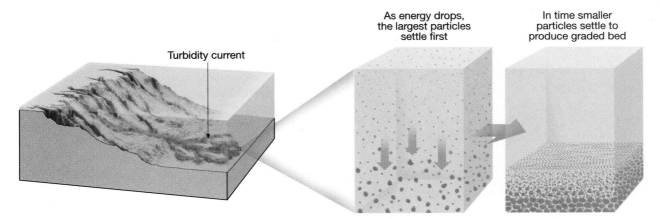

Turbidity current

As energy drops, the largest particles settle first

In time smaller particles settle to produce graded bed

Figure 6.21 Graded beds. Each layer grades from coarse at its base to fine at the top.

As geologists examine sedimentary rocks, much can be deduced. A conglomerate, for example, may indicate a high-energy environment, such as a surf zone or rushing stream, where only coarse materials settle out and finer particles are kept suspended (Figure 6.22). If the rock is arkose, it may signify a dry climate where little chemical alteration of feldspar is possible. Carbonaceous shale is a sign of a low-energy, organic-rich environment, such as a swamp or lagoon.

Other features found in some sedimentary rocks also give clues to past environments. Ripple marks are such a feature. **Ripple marks** are small waves of sand that develop on the surface of a sediment layer by the action of moving water or air (Figure 6.23A). The ridges form at right angles to the direction of motion. If the ripple marks were formed by air or water moving in essentially one direction, their form will be asymmetrical. These *current ripple marks* will have steeper sides in the downcurrent direction and more gradual slopes on the upcurrent side. Ripple marks produced by a stream flowing across a sandy channel or by wind blowing over a sand dune are two common examples of current ripples. When present in solid rock, they may be used to determine the direction of movement of ancient wind or water currents. Other ripple marks have a symmetrical form. These features, called *oscillation ripple marks*, result from the back-and-forth movement of surface waves in a shallow nearshore environment.

Figure 6.22 In a turbulent stream channel, only large particles settle out. Finer sediments remain suspended and continue their downstream journey. (Photo by E. J. Tarbuck)

Mud cracks (Figure 6.23B) indicate that the sediment in which they were formed was alternately wet and dry. When exposed to air, wet mud dries out and shrinks, producing cracks. Mud cracks are associated with such environments as shallow lakes and desert basins.

Fossils: Evidence of Past Life

Fossils, the remains or traces of prehistoric life, are important inclusions in sediment and sedimentary rocks (see Box 6.3). They are important tools for interpreting the geologic past. Knowing the nature of the life-forms that existed at a particular time helps researchers understand past environmental conditions. Further, fossils are important time indicators and play a key role in correlating rocks of similar ages that are from different places.*

Only a tiny fraction of the organisms that lived during the geologic past have been preserved as fossils. Normally, the remains of an animal or plant are totally destroyed. Under what circumstances are they preserved? Two special conditions seem to be necessary: rapid burial and the possession of hard parts.

Usually when an organism perishes, its remains are quickly eaten by scavengers or decomposed by bacteria. Occasionally, however, the remains are buried by sediment. When this occurs the remains are removed from the environment where destructive forces operate most effectively. Rapid burial therefore is an important condition favoring presentation.

In addition, organisms have a much better chance of being preserved as part of the fossil record if they have hard parts. Although traces and imprints of soft-bodied animals such as jellyfish, worms, and insects exist, they are far less common. Flesh usually decays so rapidly that the chance of preservation is exceedingly unlikely. Hard parts like shells, bones, and teeth predominate in the record of past life.

Because preservation is contingent on special conditions, the record of life in the geologic past is biased. The fossil remains of those organisms with hard parts that lived in areas of sedimentation are quite abundant. However, we get only an occasional glimpse of the vast array of other life-forms that did not meet the special conditions favoring preservation.

---

*The section entitled "Correlation" in Chapter 8 contains a more detailed discussion of the role of fossils in the interpretation of Earth history.

A.

B.

Figure 6.23 A. Ripple marks can be produced by currents of water or wind. (Photo by Stephen Trimble)
B. Mud cracks form when wet mud or clay dries out and shrinks. (Photo by Gary Yeowell/Tony Stone Images)

Box 6.3

Types of Fossilization

Fossils are of many types. The remains of relatively recent organisms may not have been altered at all; teeth, bones, and shells are common examples. Far less common are entire animals, flesh included, that have been preserved because of rather unusual circumstances. Remains of prehistoric elephants called *mammoths* that were frozen in the Arctic tundra of Siberia and Alaska are examples, as are the mummified remains of sloths preserved in a dry cave in Nevada.

Given enough time, the remains of an organism are likely to be modified. Often fossils become *petrified* (literally, "turned into stone"), meaning that the small internal cavities and pores of the original structure are filled with precipitated mineral matter (Figure 6.C). In other instances, *replacement* may occur, in which cell walls and other solid material are removed and replaced with mineral matter. Sometimes the microscopic details of the replaced structure are faithfully retained.

Molds and casts constitute another common class of fossils. When a shell or other structure is buried in sediment and then dissolved away by underground water, a *mold* is left behind. The mold faithfully reflects only the shape and surface markings of the organism; it does not reveal any information concerning its internal structure. If these hollow spaces are subsequently filled with mineral matter, *casts* are created (Figure 6.D).

A type of fossilization called *carbonization* is particularly effective in preserving leaves and delicate animal forms such as insects. It occurs when fine sediment encases the remains of an organism. As time passes, pressure squeezes out the liquid and gaseous components, leaving behind only a thin residue of carbon (Figure 6.E). Black shales deposited as organic-rich mud in oxygen-poor environments often contain abundant carbonized remains. If the film of carbon is lost from a fossil preserved in fine-grained sediment, a replica of the surface, called an *impression*, may still show considerable detail (Figure 6.F).

Delicate organisms, such as insects, are difficult to preserve and consequently are quite rare in the fossil record. Not only must they be protected from decay; they also must not be subjected to any pressure that would crush them. One way in which some insects have been preserved is in *amber*, the hardened resin of ancient trees. The insect in Figure 6.G was preserved after being trapped in a drop of sticky resin. Resin sealed off the insect from the atmosphere and protected the remains from damage by water and air. As the resin hardened, a protective, pressure-resistant case was formed.

In addition to the fossils already mentioned, there are numerous other types, many of them only traces of prehistoric life. Examples of such indirect evidence include:

1. *Tracks*—footprints made by animals in soft sediment that was later lithified (Figure 6.H).

2. *Burrows*—tubes in sediment, wood, or rock made by an animal. These holes may later become filled with mineral matter and preserved. Some of the oldest-known fossils are believed to be worm burrows.

3. *Coprolites*—fossil dung and stomach contents that can provide useful information pertaining to food habits of organisms.

4. *Gastroliths*—highly polished stomach stones that were used in the grinding of food by some extinct reptiles.

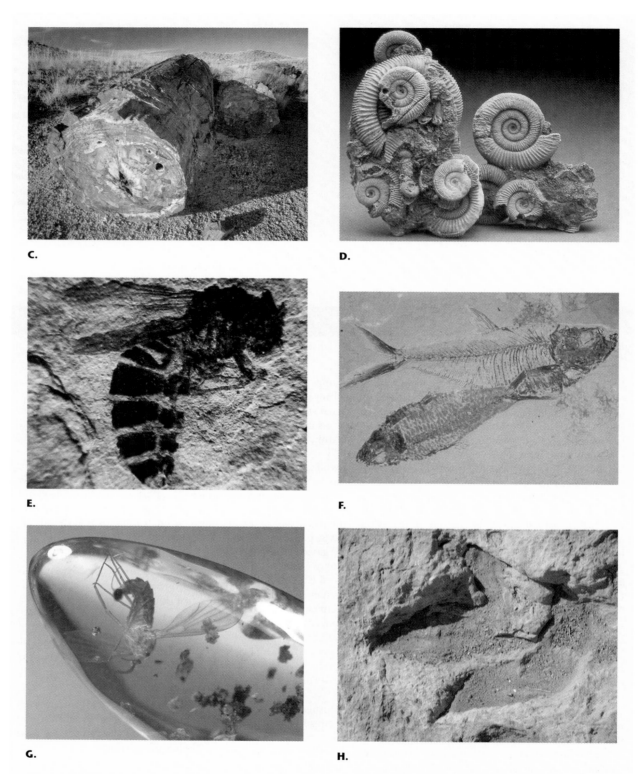

Figures 6.C—6.H There are many types of fossilization. Six examples are shown here. **C.** Petrified wood in Petrified Forest National Park, Arizona. **D.** Natural casts of shelled invertebrates. **E.** A fossil bee preserved as a thin carbon film. **F.** Impressions are common fossils and often show considerable detail. **G.** Insect in amber. **H.** Dinosaur footprint in fine-grained limestone near Tuba City, Arizona. (Photo C by David Muench; Photos D,F, and H by E. J. Tarbuck; Photo E courtesy of the National Park Service; Photo G by Breck P. Kent)

Chapter Summary

- *Sedimentary rock* consists of *sediment* that, in most cases, has been *lithified* into solid rock by the processes of *compaction* and *cementation*. Sediment has two principal sources: (1) as *detrital material*, which originates and is transported as solid particles from both mechanical and chemical weathering, which, when lithified, forms detrital sedimentary rocks; and (2) from soluble material produced largely by chemical weathering, which, when precipitated, forms *chemical sedimentary rocks*.

- *Particle size* is the primary basis for distinguishing among various detrital sedimentary rocks. The size of the particles in a detrital rock indicates the energy of the medium that transported them. For example, gravels are moved by swiftly flowing rivers, whereas less energy is required to transport sand. Common detrital sedimentary rocks include *shale* (silt-and clay-sized particles), *sandstone*, and *conglomerate* (rounded gravel-sized particles) or *breccia* (angular gravel-sized particles).

- Precipitation of chemical sediments occurs in two ways: (1) by *inorganic processes*, such as evaporation and chemical activity; or by (2) *organic processes* of water-dwelling organisms that produce sediments of *biochemical origin*. Limestone, the most abundant chemical sedimentary rock, consists of the mineral calcite ($CaCO_3$) and forms either by inorganic means or as the result of biochemical processes. Inorganic limestones include *travertine*, which is commonly seen in caves, and *oolitic limestone*, consisting of small spherical grains of calcium carbonate. Other common chemical sedimentary rocks include *dolostone* (composed of the calcium-magnesium carbonate mineral dolomite), *chert* (made of microcrystalline quartz), *evaporites* (such as rock salt and rock gypsum), and *coal* (lignite and bituminous).

- *Lithification* refers to the processes by which unconsolidated sediments are transformed into solid sedimentary rock. Most sedimentary rocks are lithified by means of *compaction* and/or *cementation*. Compaction occurs when the weight of overlying materials compresses the deeper sediments. Cementation, the most important process by which sediments are converted to sedimentary rocks, occurs when soluble cementing materials, such as *calcite, silica,* and *iron oxide*, are precipitated onto sediment grains, fill open spaces, and join the particles. Although most sedimentary rocks are lithified by compaction or cementation, certain chemical rocks, such as the evaporites, initially form as solid masses of intergrown crystals.

- Sedimentary rocks can be divided into two main groups: *detrital* and *chemical* All detrital rocks have a *clastic texture*, which consists of discrete fragments and particles that are cemented and compacted together. The main criterion for subdividing the detrital rocks is particle size. Common detrital rocks include *conglomerate, sandstone,* and *shale.* The primary basis for distinguishing among different rocks in the chemical group is their mineral composition. Some chemical rocks, such as those deposited when seawater evaporates, have a *nonclastic texture* in which the minerals form a pattern of interlocking crystals. However, in reality, many of the sedimentary rocks classified into the chemical group also contain at least small quantities of detrital sediment. Common chemical rocks include *limestone, rock gypsum,* and *coal* (e.g., lignite and bituminous).

- Sedimentary environments are those places where sediment accumulates. They are grouped into terrestrial (continental), transitional (shoreline), and marine environments. Each is characterized by certain physical, chemical, and biological conditions. Because sediment contains clues about the environment in which it was deposited, sedimentary rocks are important in the interpretation of Earth's history.

- Sedimentary rocks are particularly important in interpreting Earth's history because, as layer upon layer of sediment accumulates, each records the nature of the environment at the time the sediment was deposited. These layers, called *strata*, or *beds*, are probably the single most characteristic feature of sedimentary rocks. Other features found in some sedimentary rocks, such as *ripple marks, mud cracks, cross-bedding,* and *fossils*, also give clues to part environments.

Review Questions

1. How does the volume of sedimentary rocks in Earth's crust compare with the volume of igneous rocks in the crust? Are sedimentary rocks evenly distributed throughout the crust?

2. What minerals are most common in detrital sedimentary rocks? Why are these minerals so abundant?

3. What is the primary basis for distinguishing among various detrital sedimentary rocks?

4. The term *clay* can be used in two different ways. Describe the two meanings.

5. Why does shale usually crumble easily?

6. How are the degree of sorting and the amount of rounding related to the transportation of sand grains?

7. Distinguish between conglomerate and breccia.

8. Distinguish between the two categories of chemical sedimentary rocks.

9. What are evaporite deposits? Name a rock that is an evaporite.

10. When a body of seawater evaporates, minerals precipitate in a certain order. What determines this order?

11. Each of the following statements describes one or more characteristics of a particular sedimentary rock. For each statement, name the sedimentary rock that is being described.

    (a) An evaporite used to make plaster.
    (b) A fine-grained detrital rock that exhibits *fissility*.
    (c) Dark-colored sandstone containing angular rock particles as well as clay, quartz, and feldspar.
    (d) The most abundant chemical sedimentary rock.
    (e) A dark-colored, hard rock made of microcrystalline quartz.
    (f) A variety of limestone composed of small spherical grains.

12. How is coal different from other biochemical sedimentary rocks?

13. Compaction is an important lithification process with which sediment size?

14. List three common cements for sedimentary rocks. How might each be identified?

15. What is the primary basis for distinguishing among different chemical sedimentary rocks?

16. Distinguish between clastic and nonclastic textures. What type of texture is common to all detrital sedimentary rocks?

17. Some nonclastic sedimentary rocks closely resemble igneous rocks. How might the two be distinguished easily?

18. Why are seafloor sediments useful in studying climates of the past? (See Box 6.2, p. 158.)

19. What is probably the single most characteristic feature of sedimentary rocks?

20. Distinguish between cross-bedding and graded bedding.

21. How do current ripple marks differ from oscillation ripple marks?

22. List two conditions that favor the preservation of organisms as fossils.

23. What type of fossilization is indicated by each of the following statements? Which one is an example of indirect evidence? (See Box 6.3, p. 163.)

    (a) A leaf preserved as a thin carbon film.
    (b) Small internal cavities and pores of a log are filled with mineral matter.
    (c) Fossil dung.
    (d) This is created when a mold is filled with mineral matter.

Key Terms

bedding plane (p. 160)	clastic (p. 159)	facies (p. 158)	nonclastic (p. 157)
beds (strata) (p. 160)	compaction (p. 154)	fissility (p. 146)	ripple mark (p. 162)
biochemical (p. 149)	cross-bedding (p. 161)	fossil (p. 162)	salt flat (p. 153)
cementation (p. 154)	detrital sedimentary rock	graded bed (p. 161)	sediment (p. 144)
chemical sedimentary	(p. 145)	lithification (p. 154)	sorting (p. 146)
rock (p. 145)	evaporite deposit (p. 152)	mud crack (p. 162)	strata (beds) (p. 160)

Web Resources

The *Earth* Home Page provides on-line resources for this chapter on the World Wide Web. You will find review exercises, specific updates for items in the chapter, suggested reading, and links to interesting related pathways on the Internet. Visit the *Earth* Home Page at **http://www.prenhall.com/tarbuck.**

CHAPTER 7

Metamorphic Rocks

Left Uplifting and erosion exposed this metamorphic rock, Brooks Range, Gates of the Arctic National Park, Alaska. — *Photo by Jeff Gnass/The Stock Market*

Figure 7.1 Deformed strata; north end of Cottonwood Mountains, Death Valley, California. (Photo by Michael Collier)

Consider the conditions necessary to fold and distort the rock shown in Figure 7.1. It takes an enormous amount of directed pressure and temperatures hundreds of degrees above surface conditions acting for, typically, thousands to millions of years to produce the deformation shown. Under these extreme conditions rocks respond by folding and flowing. This chapter looks at the tectonic forces that forge metamorphic rocks, and how these rocks change in appearance and mineral makeup.

Extensive areas of metamorphic rocks are exposed on every continent in the relatively flat regions known as *shields* (see Figure 7.22). These areas include eastern Canada, Brazil, much of Africa, India, half of Australia, and Greenland. Moreover, metamorphic rocks are an important component of many mountain belts, where they make up a large portion of a mountain's crystalline

core. Even those portions of the stable continental interiors that are covered by sedimentary rocks are underlain by metamorphic basement rocks. In all of these settings the metamorphic rocks are usually highly deformed and are often intruded by igneous masses. Indeed, significant parts of Earth's continental crust are composed of metamorphic and associated igneous rocks.

Unlike some igneous and sedimentary processes that take place in surface or near-surface environments, metamorphism almost always occurs deep within Earth beyond our direct observation. Notwithstanding this significant obstacle, geologists have developed techniques that have allowed them to learn a great deal about the conditions under which metamorphic rocks form. Thus, metamorphic rocks provide important clues about the geologic processes that operate within Earth's crust.

Metamorphic Environments

Recall from the section on the rock cycle in Chapter 1 that metamorphism is the transformation of one rock type into another. Metamorphic rocks can be transformed from igneous, sedimentary, or even from other metamorphic rocks. **Metamorphism** is a very appropriate name for this process because it literally means to "change form." The agents of metamorphism include heat, pressure (stress), and chemically active fluids. The changes that occur are both textural and mineralogical.

Metamorphism occurs incrementally, from slight change (low grade) to dramatic change (high grade). For example, under low-grade metamorphism, the common sedimentary rock *shale* becomes the more compact metamorphic rock called *slate*. Hand samples of these rocks are sometimes difficult to distinguish.

In other cases, high-grade metamorphism causes a transformation so complete that the identity of the original rock cannot be determined. In high-grade metamorphism, such features as bedding planes, fossils, and vesicles that may have existed in the parent rock are obliterated. Further, when rocks at depth are subjected to uneven pressure (stress) they slowly flow and bend into intricate folds (Figure 7.2). In the most extreme metamorphic environments, the temperatures approach those at which rocks melt. However, during metamorphism some material must remain solid, for if complete melting occurs, we have entered the realm of igneous activity.

Metamorphism takes place *where rock is subjected to conditions unlike those in which it formed* (see Box 7.1). In response to these new conditions, the unstable rock gradually changes until a state of equilibrium with the new environment is reached. Most metamorphic changes occur at the elevated temperatures

and pressures that exist in the zone extending from a few kilometers below Earth's surface to the crust–mantle boundary.

Metamorphism most often occurs in one of three settings:

1. *When rock is near or touching a mass of magma,* **contact metamorphism** takes place. Here the changes are caused primarily by the high temperatures of the molten material, which in effect "bake" the surrounding rock.

2. The least common type of metamorphism occurs along *fault zones* and is called **cataclastic metamorphism**. Here rock is broken and pulverized as crustal blocks on opposite sides of a fault grind past one another.

3. *During mountain building,* great quantities of rock are subjected to directed pressures and high temperatures associated with large-scale deformation called **regional metamorphism**. The end result may be extensive areas of metamorphic rocks.

By far the greatest volume of metamorphic rock is produced during regional metamorphism in conjunction with mountain building. Here, large segments of Earth's crust are highly deformed by folding and faulting. Further, in the most intense zones of metamorphism, magma may be generated. Thus, areas affected by regional metamorphism frequently display zones of contact metamorphism, as well as cataclastic metamorphism. After examining the agents of metamorphism and some common metamorphic rocks, we will return to the topic of metamorphic environments.

 ## Agents of Metamorphism

As stated earlier, the agents of metamorphism include *heat, pressure (stress),* and *chemically active fluids.* During metamorphism, rocks are often subjected to all three metamorphic agents simultaneously. However, the degree of metamorphism and the contribution of each agent vary greatly from one environment to another. In low-grade metamorphism, rocks are subjected to temperatures and pressures only slightly greater than those associated with the lithification of sediments. High-grade metamorphism, on the other hand, involves extreme tectonic forces and temperatures close to those at which rocks melt.

In addition, the mineral makeup of the parent rock determines, to a large extent, the degree to which each metamorphic agent will cause change. For example, when intruding magma forces its way into existing rock, hot, ion-rich fluids (mostly water) circulate through the host rock. If the host rock is quartz

Figure 7.2 Deformed metamorphic rocks exposed in a road cut in the Eastern Highland of Connecticut. (Photo by Phil Dombrowski)

Box 7.1

Impact Metamorphism and Tektites

It now is clear that comets and asteroids have collided with Earth far more frequently than once thought. The evidence: More than a hundred giant impact structures called *astroblemes* have been identified to date. Many had been thought to result from some poorly understood volcanic process. Most astroblemes are too old to look like an impact crater, but a notable exception is the very fresh-looking Meteor Crater in Arizona.

One earmark of astroblemes is impact or shock metamorphism. When high-velocity projectiles (comets, asteroids) impact Earth's surface, pressures reach millions of atmospheres and temperatures exceed 2000°C momentarily. The result is pulverized, shattered, and melted rock. The products of these impacts, called *impactiles* include fused mixtures of fragmented rock and melted material, plus glass-rich ejecta that resemble volcanic bombs. In some cases, a very dense form of quartz and minute diamonds are found. These high-pressure minerals indicate shock metamorphism. The best-known impactite structures in North America are Meteor Crater, Arizona, and astroblemes located in Indiana and Ohio (Figure 7.A).

Where impact craters are relatively fresh, shock-melted ejecta and rock fragments ring the impact site. Although most material is deposited close to its source, some ejecta can travel great distances. One example is *tektites*, beads of silica-rich glass, some of which have been aerodynamically shaped like teardrops during flight (Figure 7.B). Most tektites are no more than a few centimeters across and are jet-black to dark green or yellowish. In Australia, millions of tektites are strewn over an area seven times the size of Texas. Several such tektite groupings have been identified worldwide, one stretching nearly halfway around the globe.

Figure 7.A Meteor Crater, located west of Winslow, Arizona. (Photo by Michael Collier)

No tektite falls have been observed, so their origin is now known with certainty. Because tektites are much higher in silica than volcanic glass (obsidian), a volcanic origin is unlikely. Most researchers agree that tektites are the result of impacts of large projectiles.

One hypothesis suggests an extraterrestrial origin for tektites. Asteroids may have struck the Moon with such force that ejecta "splashed" outward fast enough to escape the Moon's gravity.

Others argue that tektites are terrestrial, but an objection is that some groupings, such as Australia's, lack an identifiable impact crater. However, the object that produced the Australian tektites might have struck the continental shelf, leaving the remnant crater out of sight below sea level. Evidence supporting a terrestrial origin includes tektite falls in western Africa that appear to be the same age as a crater in the same region.

Figure 7.B Tektites recovered from Nullarbor Plain, Australia. (Photo by Bill Mason/ Smithsonian Institution)

---

sandstone, very little alteration may take place. On the other hand, if the host rock is limestone, the impact of these fluids can be dramatic and the effects of metamorphism might extend for several kilometers from the magma body.

Heat as a Metamorphic Agent

Perhaps the most important agent of metamorphism is *heat* because it provides the energy to drive the chemical changes that result in the recrystallization of minerals. Rocks formed near Earth's surface may be subjected to intense heat when they are intruded by molten material rising from below. The effects of this contact metamorphism are most apparent when it occurs at or near the surface where the temperature contrast between the magma and the host rock is most pronounced. Here the adjacent host rock is "baked" by the emplaced magma. In this high-temperature and low-pressure environment, the boundary that forms between the intruding magma and the altered rocks is usually quite distinct.

In addition to magma rising and metamorphosing near-surface rocks, rocks near the surface may be slowly thrust downward to become metamorphosed at depth. As we discussed earlier, Earth materials are continually being transported to great depths at convergent plate boundaries. Recall that temperatures increase with depth at a rate known as the *geothermal gradient*. In the upper crust, this increase in temperature averages between 20°C and 30°C per kilometer. When buried to a depth of only a few kilometers, certain minerals, such as clay, become unstable and begin to recrystallize into minerals, such as muscovite (mica), that are stable in this environment. Other minerals, particularly those found in crystalline igneous rocks, are stable at relatively high temperatures and pressures and therefore may require burial to 20 kilometers or more before metamorphism will occur.

Pressure and Stress as Metamorphic Agents

Pressure, like temperatures, also increases with depth. Buried rocks are subjected to the force, or **stress**, exerted by the load above (Figure 7.3A). This **confining pressure** is analogous to water pressure where the force is applied equally in all directions. The deeper you go in the ocean, the greater the pressure on all sides. The same is true for rocks at depth.

In addition to the confining pressure exerted by the load of material above, rocks are also subjected to directional tectonic forces during mountain building (Figure 7.3B). These forces, which are unequal in different directions, are called **differential stresses**. Most often these differential forces are *compressional* and act to shorten a rock body. In some environments, however, the stresses are *tensional* and tend to elongate, or pull apart, the rock mass (Figure 7.4).

Differential stress can also cause a rock to **shear**. Shearing is similar to the slippage that occurs between individual cards when you hold a deck flat between your hands and slide your hands in opposite directions, shearing the deck. In near-surface environments, shearing results when relatively brittle rock breaks into thin slabs that are forced to slide past one another. This deformation grinds and pulverizes the original mineral grains into small fragments. By contrast, because

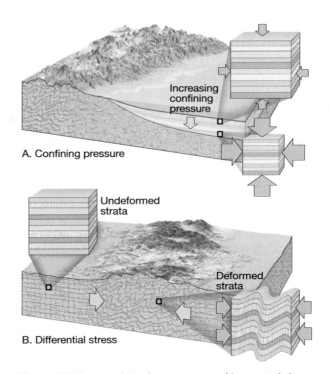

Figure 7.3 Pressure (stress) as a metamorphic agent. **A.** In a depositional environment, as confining pressure increases, rocks deform by decreasing in volume. **B.** During mountain building, differential stress shortens and deforms rock strata.

rocks located at great depths are warmer, and under greater confining pressure, they tend to behave plastically during deformation. This accounts for their ability to flow and bend into intricate folds when subjected to shearing (see Figure 7.2).

Figure 7.4 Metaconglomerate or stretched pebble conglomerate. These once nearly spherical pebbles have been heated and stretched into elongated structures. (Photo by E. J. Tarbuck)

Chemical Activity as a Metamorphic Agent

Chemically active fluids also enhance the metamorphic process. Most commonly, the fluid is water containing ions in solution. Water is plentiful, because some water is contained in the pore spaces of virtually every rock. In addition, many minerals are *hydrated* (have water bound chemically) and thus contain water within their crystalline structures.

When deep burial occurs, rocks become more compact, reducing the amount of pore space. Thus, water is forced out of the rock and becomes available to aid in chemical reactions. Further, heating causes the dehydration of minerals and the release of water. Water that surrounds the crystals acts as a catalyst by aiding ion migration. In some instances water promotes the recrystallization of minerals, which form more stable configurations. In other cases, ion exchange among minerals results in the formation of completely new minerals.

Complete alteration of rock by hot, mineral-rich water has been observed in the near-surface environment of Yellowstone National Park. On a much larger scale similar activity occurs along the mid-ocean ridge system. Here, seawater circulates through the still-hot basaltic rocks, transforming iron- and magnesium-rich minerals into metamorphic minerals such as serpentine and talc.

How Metamorphism Alters Rocks

The metamorphic process causes many changes in rocks, including increased density, growth of larger crystals, reorientation of the mineral grains into a layered or banded texture, and the transformation of low-temperature minerals into high-temperature minerals. Further, the introduction of ions generates new minerals, some of which are economically important. Thus, the grade of metamorphism is reflected in the *texture* and *mineralogy* (mineral makeup) of metamorphic rocks.

Textural Changes

When rocks are subjected to low-grade metamorphism, they become more compact and thus more dense. A common example is conversion of the sedimentary rock *shale* into the metamorphic rock *slate*. When shale is subjected to temperatures and pressures only slightly greater than those of the sedimentary processes that formed it, slate results. In this case, the directed pressure causes the microscopic clay minerals in shale to align into the more compact arrangement found in slate. This realignment of particles that occurs when shale is converted to slate gives slate a distinctive **texture**. (Recall that texture is the size, shape, and distribution of particles that constitute a rock.)

Foliated Textures. Under more extreme conditions, pressure causes mineral grains in a rock to do much more than just realign. Pressure can cause certain minerals to *recrystallize*. In general, recrystallization encourages the growth of larger crystals. Consequently, many metamorphic rocks consist not of microscopic crystals, but of visible crystals, much like coarse-grained igneous rocks.

The crystals of some minerals, such as micas (platy minerals) and hornblende (needlelike minerals), will recrystallize with a *preferred orientation*. The new orientations will be essentially perpendicular to the direction of the compressional force as shown on the right side in Figure 7.5. The resulting mineral alignment usually gives the rock a layered or banded texture termed **foliation** (Figure 7.5). Simply put, *a foliated texture results whenever the minerals and structural features of a metamorphic rock are forced into parallel alignment.*

Various types of foliation exist, depending largely upon the grade of metamorphism and the mineralogy of the parent rock. We will look at three: rock or slaty cleavage; schistosity; and gneissic texture.

Rock or Slaty Cleavage. During the transformation of shale to slate, clay minerals (stable at the surface) recrystallize into minute mica flakes (stable at much higher temperatures and pressures). Further, these platy mica crystals become aligned so that their flat surfaces are nearly parallel. Consequently, slate can be split easily along these layers of mica grains into rather flat slabs. This property is called **rock cleavage** or **slaty cleavage**, to differentiate it from the type of cleavage exhibited by minerals (Figure 7.6). Because the mica flakes composing slate are minute, slate is often not visibly foliated. But slate is considered foliated because it can be split easily into slabs, evidence that its minerals are aligned.

Schistosity. Under more extreme temperature–pressure regimes, the very fine mica grains in slate will grow many times larger. These mica crystals, which are up to a centimeter in diameter, give the rock a scaly appearance. This type of foliation is called **schistosity**, and a rock having this texture is called *schist*. Many types of schist exist depending on the original parent rock, and they are named according to their mineral constituents—mica schist, talc schist, and so on. By far the most abundant are the mica schists.

Gneissic Texture. During high-grade metamorphism, ion migrations can be extreme enough to cause minerals to segregate. An example is Figure 7.5, lower right. Notice that the dark and light silicate minerals

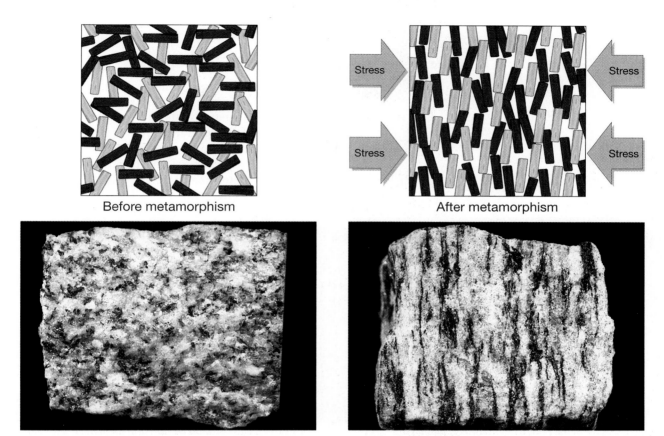

Figure 7.5 Under directed pressure, flat or needle-shaped minerals (top left) become reoriented or recrystallized so that they are aligned at right angles to the stress (top right). The resulting parallel orientation of mineral grains gives the rock a foliated texture. If the coarse-grained igneous rock (granite, bottom left) underwent intense metamorphism, it could end up closely resembling the metamorphic rock on the bottom right (gneiss). (Photos by E. J. Tarbuck)

have separated, giving the rock a banded appearance called **gneissic texture**. Metamorphic rocks with this texture are called *gneiss* (pronounced "nice") and are quite common. Gneiss often forms from the metamorphism of granite or diorite but can form from gabbro or even the high-grade metamorphism of shale. Although foliated, gneiss will not usually split parallel to the crystals as easily as will slate.

Nonfoliated Texture. Not all metamorphic rocks have a foliated texture. Those that do not are **nonfoliated**. Metamorphic rocks composed of only one mineral that exhibits equidimensional crystals usually are not visibly foliated. For example, when a fine-grained limestone (made of a single mineral, calcite) is metamorphosed, the small calcite crystals combine to form relatively large interlocking crystals. The resulting rock, *marble*, has a texture similar to that of a coarse-grained igneous rock. Although most marbles are nonfoliated, microscopic investigation of marble may reveal some flattening and parallelism of the grains.

Further, some limestones contain thin layers of clay minerals that may become distorted during metamorphism. The "impurities" will often appear as curved bands of dark material flowing through the marble, a clear indication of metamorphism.

Mineralogical Changes

In the metamorphism of shale to slate, you saw that clay minerals recrystallize to form mica crystals. During most recrystallization, including this example, the chemical composition of the rock does not change (except for the loss of water and carbon dioxide). Rather, the existing minerals and available ions in the water will recombine to form minerals that are stable in the new environment. A common example is when limestone ($CaCO_3$), containing abundant quartz (SiO_2), is heated during contact metamorphism. The calcite and quartz crystals chemically react to form wollastonite ($CaSiO_3$), and carbon dioxide is liberated.

In some environments, however, new materials are actually introduced during the metamorphic process. For example, rock adjacent to a large magma body would acquire new elements from **hydrothermal solutions** (hot water). Many metallic ore deposits are formed by the deposition of minerals from hydrothermal solutions.

Figure 7.6 Slaty cleavage is the type of rock cleavage exhibited by this metamorphic rock, California. The parallel mineral alignment in this rock allows it to split easily in the flat plates visible in the photo. (Photo by E. J. Tarbuck)

With the development of plate tectonics, it became clear that some metal-rich hydrothermal deposits originate along ancient spreading centers (mid-ocean ridges). As seawater percolates through newly formed oceanic crust, it dissolves metallic sulfides from the basaltic rocks. The hot, metal-rich fluids rise along fractures and gush from the seafloor as particle-filled clouds called *black smokers*. Upon mixing with the cold seawater, the sulfides precipitate to form massive metallic deposits. This is the origin of the copper ores mined today on the island of Cyprus. Some of Earth's richest copper deposits have formed in this manner.

Common Metamorphic Rocks

As you learned, metamorphism causes many changes in rocks, including increased density, growth of larger crystals, reorientation of the mineral grains into a layered or banded appearance known as foliation, and the transformation of low-temperature minerals into high-temperature minerals. Further, the introduction of ions generates new minerals, some of which are economically important. We will now examine some of the most common rocks generated by diverse metamorphic processes (Table 7.1).

Foliated Rocks

Slate. To review, *slate* is a very fine-grained foliated rock composed of minute mica flakes. The most noteworthy characteristic of slate is its excellent rock cleavage, or tendency to break into flat slabs. This property traditionally made slate a most useful rock for roof and floor tile, blackboards, and billiard tables (Figure 7.7).

Slate is most often generated by the low-grade metamorphism of shale, although less frequently it is metamorphosed from volcanic ash. Slate's color depends on its mineral constituents. Black (carbonaceous) slate contains organic material (carbon-bearing), red slate gets its color from iron oxide, and green slate usually contains chlorite, a mica-like mineral formed by the metamorphism of iron-rich silicates.

Because slate forms during low-grade metamorphism, evidence of shale's original bedding planes is often preserved. However, the orientation of slate's rock cleavage is at a pronounced angle to the original sedimentary layering (Figure 7.8). Thus, unlike shale, which splits along bedding planes, slate splits across them.

Phyllite. *Phyllite* represents a gradation in metamorphism between slate and schist. Its constituent platy minerals are larger than those in slate, but not yet large enough to be clearly identifiable. Although phyllite appears similar

Table 7.1 Common Metamorphic Rocks

Metamorphic Rock	Texture	Parent Rock	Comments
Slate	Foliated	Shale	Very fine-grained
Phyllite	Foliated	Shale	Fine- to medium-grained
Schist	Foliated	Shale, granitic and volcanic rocks	Coarse-grained micaceous minerals
Gneiss	Foliated	Shale, granitic and volcanic rocks	Coarse-grained (nonmicaceous)
Marble	Nonfoliated	Limestone, dolostone	Composed of interlocking calcite grains
Quartzite	Nonfoliated	Quartz sandstone	Composed of interlocking quartz grains
Hornfels	Nonfoliated	Any fine-grained material	Fine-grained
Migmatite	Weakly foliated	Mixture of granitic and mafic rocks	Composed of contorted layers
Mylonite	Weakly foliated	Any material	Hard, fine-grained rock
Metaconglomerate	Weakly foliated	Quartz-rich conglomerate	Strongly stretched pebbles
Amphibolite	Weakly foliated	Mafic volcanic rocks	Coarse-grained

Figure 7.7 Slate used for roofing material on a house in Switzerland. (Photo by E. J. Tarbuck)

to slate, it can be easily distinguished from slate by its glossy sheen (Figure 7.9). Phyllite usually exhibits rock cleavage and is composed mainly of very fine crystals of either muscovite or chlorite.

Schist. *Schists* are strongly foliated rocks that can be readily split into thin flakes or slabs. By definition, schists contain more than 50 percent platy and elongated minerals that commonly include mica (muscovite, biotite) and amphibole. Like slate, the parent material from which many schists originate is shale, but to form schist, the metamorphism is more intense. In addition, most schists are products of major mountain-building episodes.

The term *schist* describes the texture of a rock. To denote the composition, the mineral names are used as well. For example, schists composed primarily of the micas muscovite and biotite are called *mica schists*. Depending upon the degree of metamorphism and composition of the parent rock, mica schists often contain *accessory minerals* unique to metamor-

phic rocks. Some common accessory minerals include *garnet, staurolite*, and *sillimanite*, in which case the rock is called *garnet-mica schist, staurolite-mica schist* and so forth (Figure 7.10). Some schists contain another accessory mineral, *graphite*, which is used as pencil "lead," graphite fibers (used in fishing rods), and lubricant (commonly for locks).

In addition, schists may be composed largely of the minerals chlorite or talc, in which case they are called *chlorite schist* and *talc schist*, respectively. Both chlorite and talc schists can form when rocks with a basaltic composition undergo metamorphism.

Gneiss. *Gneiss* is the term applied to banded metamorphic rocks that contain mostly elongated and granular (as opposed to platy) minerals. The most common minerals in gneiss are quartz, potassium feldspar, and sodium feldspar. Lesser amounts of muscovite, biotite, and hornblende are common. The segregation of light and dark silicates is developed in gneisses, giving them a characteristic banded appearance. Thus, most gneisses consist of alternating bands of white or reddish feldspar-rich zones and layers of dark ferromagnesian minerals (see Figure 7.5, lower right). These banded gneisses are often deformed by folding and flowing while in a plastic state (Figure 7.11). Some gneisses will split along the layers of platy minerals, but most break in an irregular fashion.

Those gneisses that have a composition similar to that of granite are probably derived from granite or its aphanitic equivalent. However, they may also form from the high-grade metamorphism of shale. In this instance, gneiss represents the last rock in the sequence of shale, slate, phyllite, schist, and gneiss.

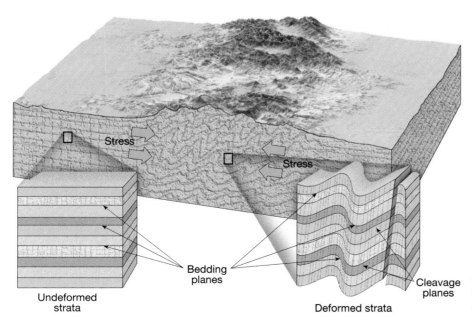

Undeformed strata

Stress

Stress

Bedding planes

Deformed strata

Cleavage planes

Figure 7.8 Illustration showing the relationship between slaty cleavage and bedding planes.

Figure 7.9 Phyllite (left) can be distinguished from slate (right) by its glossy sheen. (Photo by E. J. Tarbuck)

← 5 cm →

Close up

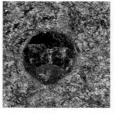

Figure 7.10 Garnet-mica schist. The dark-red garnet crystals and the light-colored mica matrix formed during the metamorphism of shale. (Photo by E. J. Tarbuck)

Like schists, gneisses may also include large crystals of accessory minerals such as garnet and staurolite. Gneisses can also be made up primarily of dark minerals such as those that compose basalt. For example, an amphibole-rich rock that exhibits a gneissic texture is called *amphibolite gneiss*.

Nonfoliated Rocks

Marble. *Marble* is a coarse, crystalline rock whose parent rock was limestone or dolostone (Figure 7.12). Pure marble is white and composed essentially of the mineral calcite. Because of its attractive color and relative softness (hardness of 3), marble is a popular building stone in banks and government buildings. White marble is particularly prized as a stone from which to carve monuments and statues, such as the famous statue of David by Michelangelo. Unfortunately, because marble is basically calcium carbonate, it is readily attacked by acid rain. Some historic monuments and tombstones already show severe chemical weathering.

Figure 7.11 Deformed and folded gneiss. (Photo by E. J. Tarbuck)

Often the limestone from which marble forms contains impurities that color the marble. Thus, marble can be pink, gray, green, or even black. Also, when impure limestone is metamorphosed, the resulting marble may contain a variety of accessory minerals (chlorite, mica, garnet, and commonly wollastonite). When marble forms from limestone interbedded with shales, it will appear banded. Under extreme deformation, such banded marble may become highly contorted and give the rock a rather artistic design.

Quartzite. *Quartzite* is a very hard metamorphic rock most often formed from quartz sandstone (Figure 7.13). Under moderate-to-high-grade metamorphism, the quartz grains in sandstone fuse like chips of glass melting together (inset in Figure 7.13). The recrystallization is so complete that when broken, quartzite will not split between the original quartz grains, but through them. In some instances, such sedimentary features as crossbedding are preserved and give the rock a banded appearance. Quartzite is typically white, but iron oxide may produce reddish or pinkish stains. Dark mineral grains may impart a gray color.

Contact Metamorphism

Contact metamorphism occurs when magma invades cooler rock. Here, a zone of alteration called an **aureole** (or halo) forms around the emplaced magma (Figure 7.14). Small intrusive magma bodies that generate dikes and sills have aureoles only a few centimeters thick. By contrast, large magma bodies that crystallize to form batholiths may create zones of metamorphic rock several kilometers thick (Figure 7.14). These large aureoles often consist of distinct *zones of metamorphism*. Near the magma body, high-temperature minerals such as garnet may form, whereas farther away such low-grade minerals as chlorite are produced.

In addition to the size of the intrusive magma body, the mineral composition of the host rock and the availability of water greatly affect the size of the aureole produced. In chemically active rock such as limestone, the zone of alteration can extend 10 kilometers or more from the magma body. Here the occurrence of minerals such as garnet and wollastonite mark the areas of metamorphism.

Close up

Figure 7.12 Marble, a crystalline rock formed by the metamorphism of limestone. (Photo by E. J. Tarbuck)

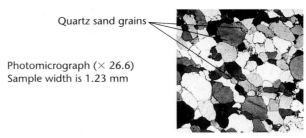

Quartz sand grains

Photomicrograph (× 26.6)
Sample width is 1.23 mm

Figure 7.13 Quartzite is a nonfoliated metamorphic rock formed from quartz sandstone. The photomicrograph shows the interlocking quartz grains typical of quartzite. (Photos by E. J. Tarbuck)

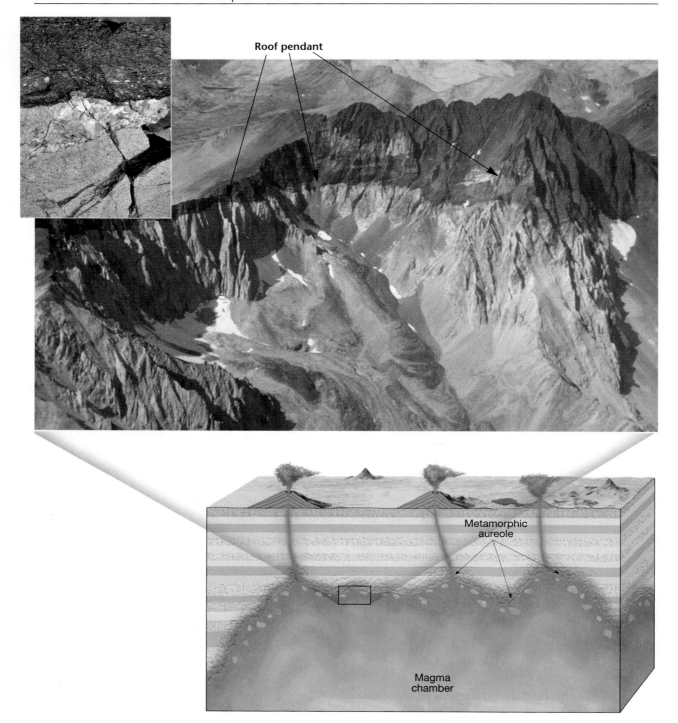

Figure 7.14 Contact metamorphism produces a zone of alteration called an *aureole* around an intrusive igneous body. In the photo, the dark layer, called a *roof pendant*, consists of metamorphosed host rock adjacent to the upper part of the light-colored igneous pluton. The term *roof pendant* implies that the rock was once the roof of a magma chamber. Sierra Nevada, near Bishop, California. The inset photo is a closeup of a contact between an igneous pluton and metamorphosed host rock. (Photo by John S. Shelton)

Most contact metamorphic rocks are fine-grained, dense, tough, and of various chemical compositions. For example, during contact metamorphism, clay minerals are baked, as if placed in a kiln, and can generate a very hard, fine-grained rock resembling porcelain. Because directional pressure is not a major factor in forming these rocks, they generally are not foliated. *Hornfels* is the name applied to the wide variety of rather hard, nonfoliated metamorphic rocks formed during contact metamorphism.

When large igneous plutons cool, hot, ion-rich fluids (*hydrothermal solutions*) that do not crystallize are expelled. These solutions percolate through the surrounding host rock, chemically reacting with it and promoting the metamorphic process. In addition, hydrothermal solutions are the source of a variety of metallic ore deposits that can be profitably extracted from metamorphic rocks. Such deposits include ores of copper, zinc, lead, iron, and gold.[*]

Contact metamorphism is easily recognized only when it occurs at the surface or in a near-surface environment where the temperature contrast between the magma and host rock is great. Undoubtedly, contact metamorphism is also an active process at great depth. However, its effect is blurred owing to the general alteration caused by regional metamorphism.

Metamorphism Along Fault Zones

Near the surface, rock behaves like a brittle solid. Consequently, movement along a fault zone fractures and pulverizes rock. In some cases, rock may even be milled into very fine components. The result is a loosely coherent rock called *fault breccia* that is composed of broken and crushed rock fragments (Figure 7.15). Displacements along California's San Andreas fault have created a zone of fault breccia and related rock types over 1000 kilometers long and up to 3 kilometers wide. This type of localized metamorphism, which involves purely mechanical forces that pulverize individual mineral grains, is called *cataclastic metamorphism.*

Much of the intense deformation associated with fault zones occurs at great depth. In this environment the rocks deform by ductile flow, which generates elongated grains that often give the rock a foliated or lineated appearance. Rocks formed in this manner are termed *mylonites.* Worldwide, the quantity of metamorphic rock generated solely by faulting is small when compared to the other processes. Nevertheless, in some areas these granulated rocks are quite abundant.

Regional Metamorphism

As indicated earlier, the greatest quantity of metamorphic rock is produced during regional metamorphism in association with mountain building. During these dynamic events, large segments of Earth's crust are intensely squeezed and become highly deformed (Figure 7.16). As the rocks are folded and faulted, the crust is shortened and thickened, like a rumpled carpet. This general thickening of the crust results in terrains that are lifted high above sea level.

Figure 7.15 Fault breccia consisting of large angular fragments. This outcrop, located in Titus Canyon, Death Valley, California, was produced along a fault zone. The largest dark fragments are about 2 to 3 meters across. (Photo by E. J. Tarbuck)

Although material is obviously elevated to great heights during mountain building, an equally voluminous quantity of rock is forced downward, where it experiences high temperatures and pressures. Here in the "roots" of mountains, the most intense metamorphic activity occurs. On occasion, the deformed rock is heated enough to melt. Once enough magma forms, it will buoyantly rise toward the surface. Magmas emplaced in a near-surface environment will cause contact metamorphism within the zone of regional metamorphism. Thus, areas affected by regional metamorphism frequently display zones of contact metamorphism, as well as cataclastic metamorphism.

Consequently, the cores of many mountain ranges consist of folded and faulted metamorphic rocks often intermixed with igneous rocks. As these deformed rock masses are uplifted, erosion removes the overlying material to expose the igneous and metamorphic rocks that comprise the central core of the mountain range.

Zones of Regional Metamorphism

In regional metamorphism, there usually exists a gradation in intensity. As we shift from areas of low-grade metamorphism to areas of high-grade metamorphism, changes in mineralogy and rock texture can be observed.

For a simplified example, we will begin with the sedimentary rock shale. Under low-grade metamorphism it yields the metamorphic rock slate (Figure 7.17). In high-temperature, high-pressure environments, slate will turn into phyllite and then into mica schist. Under more extreme conditions, the micas in schist will recrystallize into minerals such as feldspar and hornblende, and eventually generate a gneiss.

You can see this transition by approaching the Appalachian Mountains from the west. Beds of shale, which once extended over large areas of the eastern

---

*For more on hydrothermal deposits, see Chapter 21.

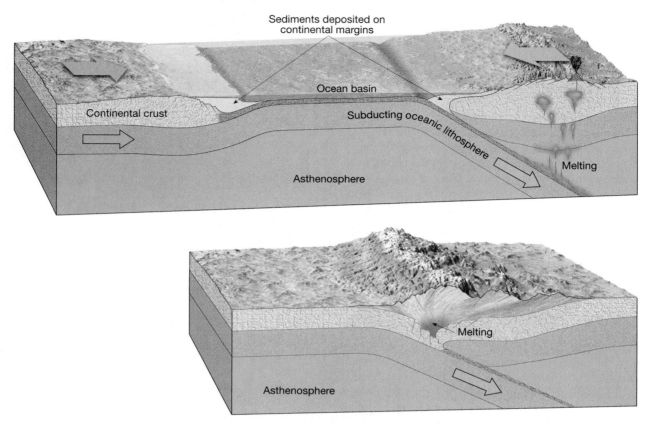

Figure 7.16 Regional metamorphism occurs where rocks are squeezed between two converging plates during mountain building.

United States, still occur as nearly flat-lying strata in Ohio. However, in the broadly folded Appalachians of central Pennsylvania, these beds are inclined and composed of low-grade slate.

As we move farther eastward to the intensely deformed crystalline Appalachians, we find large out-crops of schist and gneiss, some of which are perhaps remnants of once flat-lying shale beds. The most intense zones of metamorphism are found in Vermont and New Hampshire, often near igneous intrusions.

Index Minerals. In addition to textural changes, we encounter changes in mineralogy as we progress from low-grade to high-grade metamorphism. An idealized transition in mineralogy that results from the regional metamorphism of shale is shown in Figure 7.18. The first new mineral to be produced in the formation of slate is chlorite. Moving toward the region of high-grade metamorphism, chlorite is replaced by ever-greater amounts of muscovite and biotite. Mica schists form under more extreme

Figure 7.17 Idealized illustration of progressive regional metamorphism. From left to right, we progress from low-grade metamorphism (slate) to high-grade metamorphism (gneiss). (Photos by E. J. Tarbuck)

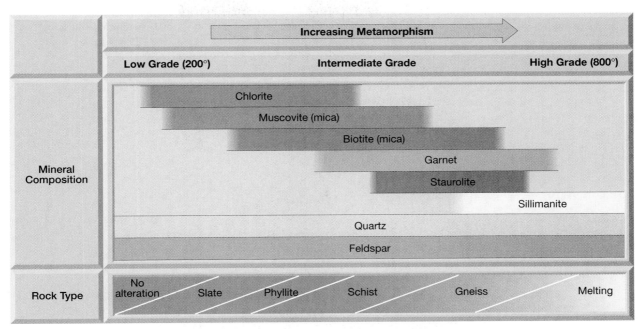

Figure 7.18 The typical transition in mineralogy that results from progressive metamorphism of shale.

conditions and may also contain garnet and staurolite crystals. At temperatures and pressures approaching the melting point of rock, sillimanite forms. Sillimanite is a high-temperature metamorphic mineral used to make refractory porcelains such as those used in spark plugs.

Through the study of metamorphic rocks and laboratory experiments, researchers have learned that certain minerals are good indicators of the metamorphic environment in which they formed. Using these **index minerals**, geologists distinguish among different zones of regional metamorphism. For example, the mineral chlorite is produced when temperatures are relatively low, about 200°C (Figure 7.18). By contrast, the mineral sillimanite forms in very extreme environments where temperatures exceed 600°C. By mapping the occurrences of index minerals, geologists in effect map zones of varying metamorphic intensities. Figure 7.19 outlines the zones that outcrop in New England.

Migmatites. In the most extreme environments, even the highest-grade metamorphic rocks are subjected to change. For example, in a near-surface environment where temperatures approach 750°C, a schist or gneiss having a chemical composition similar to granite will begin to melt. However, recall from our discussion of igneous rocks that not all minerals melt at the same temperature. The light-colored silicates, usually quartz and potassium feldspar, will melt first, whereas the mafic silicates, such as amphibole and biotite, will remain solid. If this partially melted rock cools, the light bands will be made of crystalline igneous rock while the dark bands will consist of unmelted metamorphic material. Rocks

of this type fall into a transitional zone somewhere between "true" igneous rocks and "true" metamorphic rocks. They are called **migmatites** (Figure 7.20).

Metamorphism and Plate Tectonics

Most of our knowledge of metamorphism supports what we know about the dynamic behavior of Earth as outlined by plate tectonics theory. In this model, most deformation and associated metamorphism occur in the vicinity of *convergent plate boundaries* where slabs of lithosphere are moving toward one another. Along some convergent zones, continental blocks collide to form mountainous structures as is illustrated earlier in Figure 7.16. In these settings compressional forces squeeze and generally deform the edges of converging plates, as well as the sediments that have accumulated along the margins of continents. Many of Earth's major mountain belts, including the Alps, Himalayas, and Appalachians, formed in this manner. All of these mountain systems are composed (to varying degrees) of deformed and metamorphosed rocks that were squeezed between two converging plates.

Large-scale metamorphism also occurs along subduction zones where oceanic plates are descending into the mantle. A close examination of Figure 7.21 shows that several metamorphic environments exist along this type of convergent boundary. Near the trenches, slabs of cold oceanic lithosphere are being subducted to great depths. As the lithosphere descends,

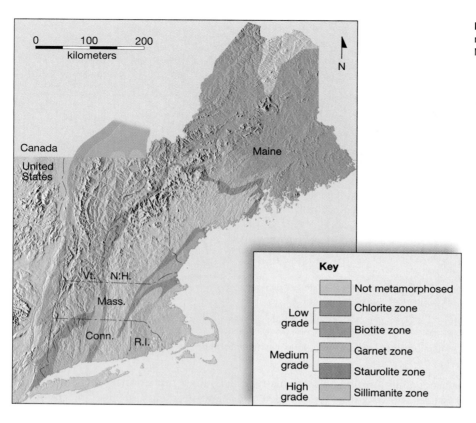

Figure 7.19 Zones of metamorphic intensities in New England.

Key

	Not metamorphosed
Low grade	Chlorite zone
	Biotite zone
Medium grade	Garnet zone
	Staurolite zone
High grade	Sillimanite zone

sediments and crustal rocks are subjected to steadily increasing temperatures and pressures (Figure 7.21). However, temperatures in the slab remain cooler than the surrounding mantle because rock is a poor heat conductor. Rock formed in this high-pressure, low-temperature environment is called *blueschist*, because of the presence of the blue-colored amphibole glauco-phane. The rocks of the Coast Range of California were formed in this manner. Here, highly deformed rocks that were once deeply buried have been uplifted because of a change in the plate boundary.

Subduction zones are also an important site of magma generation (Figure 7.21). Recall from Chapter 3 that as an oceanic plate sinks, heat and pressure drives

Figure 7.20 Migmatite. The lightest colored layers are igneous rock composed of quartz and feldspar, while the darker layers have a metamorphic origin. (Photo by Hal Roepke)

water from the subducting crustal rocks. These volatiles migrate into the wedge of hot material above and lower the melting temperature of these mantle rocks suffi-ciently to generate magma. Once enough molten rock forms, it buoyantly rises toward the surface, baking and further deforming the strata it intrudes. Thus, in near-surface zones landward of trench areas, high-tempera-ture and low-pressure contact metamorphism is common (Figure 7.21). The Sierra Nevada, which con-sists of numerous igneous intrusions and associated metamorphic rocks, exemplify this type of environment (see Figure 7.14).

Thus, mountainous terrains that form along sub-duction zones are generally composed of two distinct linear belts of metamorphic rocks. Nearest the trench we find a high-pressure, low-temperature metamorphic regime similar to that of the Coast Range of California. Farther inland, in the region of igneous intrusions, metamorphism is dominated by high temperatures and low pressures; that is, environments similar to those associated with the Sierra Nevada Batholith.

Ancient Metamorphic Environments

In addition to the linear belts of metamorphic rocks that are found in the axes of most mountain belts, even larger expanses of metamorphic rocks exist within the stable continental interiors (Figure 7.22). These relatively flat expanses of metamorphic rocks and associated igneous

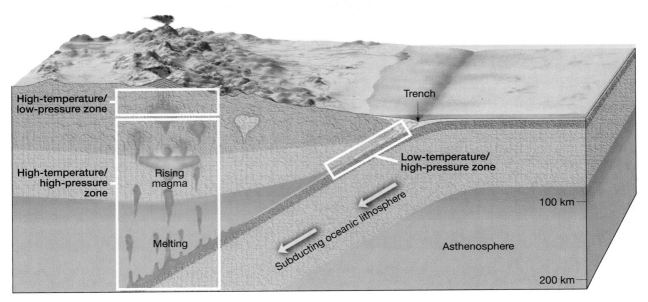

Figure 7.21 Metamorphic environments according to the plate tectonics model.

plutons are called **shields**. One such structure, the Canadian Shield, has very little topographic expression and forms the bedrock over much of central Canada, extending from Hudson Bay to northern Minnesota. Radiometric dating of the Canadian Shield indicates that it is composed of rocks that range in age from 1.8 billion to 3.8 billion years. Because shields are ancient, and their rock structure is similar to that found in the cores of younger deformed mountainous terrains, they are assumed to be remnants of much earlier periods of mountain building (Figure 7.23). This evidence strongly supports the generally accepted view that Earth has been a dynamic planet throughout most of its history. Studies of these vast areas of metamorphism in the context of the plate tectonics model have given geologists new insights into the problem of discerning just how the continents came to exist. We will consider this topic further in Chapter 20.

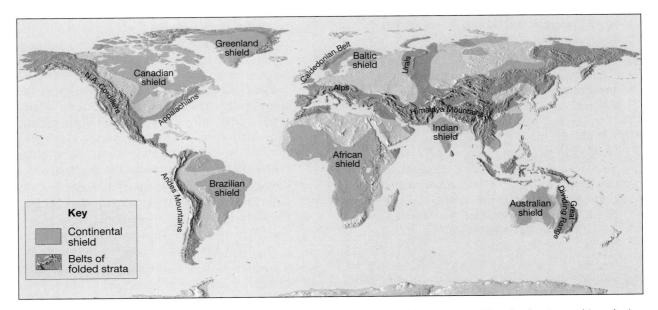

Figure 7.22 Occurrences of metamorphic rocks. The continental shields of the world are composed largely of metamorphic rocks. In addition, the deformed portions of many mountain belts are also metamorphic. The remaining area shown in this map is the stable continental interior, which generally consists of undeformed sedimentary beds that overlie metamorphic and igneous basement rocks.

Figure 7.23 Ancient metamorphic rocks compose this portion of the Greenland Shield. These 3.8-billion-year-old rocks are among the most ancient on Earth. (Photo by Kevin Schafer/Peter Arnold, Inc.)

Chapter Summary

- *Metamorphism* is the transformation of one rock type into another. *Metamorphic rocks* form from preexisting rocks (either igneous, sedimentary, or other metamorphic rocks) that have been altered by the agents of metamorphism, which include *heat, pressure,* and *chemically active fluids.* During metamorphism some of the material must remain solid. The changes that occur in the rocks are textural as well as mineralogical.

- Metamorphism most often occurs in one of three settings: (1) when rock is in contact with or near a mass of magma, *contact metamorphism* occurs; (2) during mountain building, where extensive areas of rock undergo *regional metamorphism;* or (3) *along fault zones.* The greatest volume of metamorphic rock is produced during regional metamorphism.

- The three agents of metamorphism include heat, pressure, and chemically active fluids. The mineral makeup of the parent rock determines, to a large extent, the degree to which each metamorphic agent will cause change. Heat is perhaps the most important agent because it provides the energy to drive chemical reactions that result in the recrystallization of minerals. Pressure, like temperature, also increases with depth. When subjected to *stress* exerted by the load above or during mountain building, rock located at great depth is warmed and behaves plastically. Chemically active fluids, most commonly water containing ions in solution, also enhance the metamorphic process by acting as a catalyst and aiding ion migration.

- *The grade of metamorphism is reflected in the texture and mineralogy of metamorphic rocks.* During the metamorphic process, rocks become more dense and, as the degree of metamorphism intensifies, recrystallization can cause mineral crystals to grow larger. Minerals with a sheet or elongated structure often become oriented essentially perpendicular to the direction of compressional force, giving the rock a layered or banded appearance termed *foliation.* Foliation can result in a fine crystalline rock having *slaty cleavage,* such as slate, or in rocks with larger crystals, a scaly appearance called *schistosity.* During high-grade metamorphism, ion migrations can cause minerals to segregate into bands. Metamorphic rocks with a banded texture are called *gneiss.* Metamorphic rocks composed of only one mineral forming equidimensional crystals are often *nonfoliated.* Most *marble* (metamorphosed limestone) is nonfoliated. Further, metamorphism can cause the transformation of low-temperature minerals into high-temperature minerals and, through the introduction of ions from *hydrothermal solutions,* generate new minerals, some of which form economically important metallic ore deposits.

- Common foliated metamorphic rocks include *slate, phyllite,* various types of *schists* (e.g., garnet-mica schist), and *gneiss.* Nonfoliated rocks include *marble* (parent rock—limestone) and *quartzite* (most often formed from quartz sandstone).

- The three geologic environments in which metamorphism commonly occurs are (1) in contact with igneous rocks (*contact metamorphism*), (2) during dynamic episodes associated with mountain building called *regional metamorphism,* or (3) along fault zones called *cataclastic metamorphism.* Contact metamorphism occurs when rocks are in contact with igneous bodies and a zone of alteration called an *aureole* forms around the emplaced magma. Most contact metamorphic rocks are fine-grained, dense, tough rocks of various chemical compositions and, because directional pressure is not a major factor, are not generally foliated. Regional metamorphism takes place at considerable depths over an extensive area and is associated with the process of mountain building. A gradation in the intensity of metamorphism usually exists in regional metamorphism, with the intensity of metamorphism (low- to high-grade) reflected in the texture and mineralogy of the rock. In the most extreme metamorphic environments, rocks, called *migmatites*, that fall into a transition zone somewhere between "true" igneous rocks and "true" metamorphic rocks may form. During metamorphism along fault zones, rocks are deformed by brittle fracture in low-pressure environments and by ductile flow at depth that generates elongated grains that often give the rock a foliated or lineated appearance. Compared to the other two processes, the quantity of metamorphic rock generated solely by faulting is very small.

Review Questions

1. What is metamorphism? What are the agents that change rocks?

2. What is foliation? Distinguish between *slaty cleavage* and *schistosity*.

3. List some changes that might occur to a rock in response to metamorphic processes.

4. Slate and phyllite resemble each other. How might you distinguish one from another?

5. Each of the following statements describes one or more characteristics of a particular metamorphic rock. For each statement, name the metamorphic rock that is being described.

   (a) Calcite-rich and nonfoliated.

   (b) Foliated and composed mainly of granular minerals.

   (c) Represents a grade of metamorphism between slate and schist.

   (d) Very fine-grained and foliated; excellent rock cleavage.

   (e) Foliated and composed of more than 50 percent platy minerals.

   (f) Often composed of alternating bands of light and dark silicate minerals.

   (g) Hard, nonfoliated rock resulting from contact metamorphism.

6. Distinguish between contact metamorphism and regional metamorphism. Which creates the greatest quantity of metamorphic rock?

7. What feature would easily distinguish schist and gneiss from quartzite and marble?

8. Briefly describe the textural and mineralogical differences among slate, mica schist, and gneiss. Which one of these rocks represents the highest degree of metamorphism?

9. Are migmatites associated with high-grade or low-grade metamorphism?

10. With what type of plate boundary is regional metamorphism associated?

Key Terms

aureole (p. 179)
cataclastic metamorphism (p. 171)
confining pressure (p. 173)
contact metamorphism (p. 171)

differential stresses (p. 173)
foliation (p. 174)
gneissic texture (p. 175)
hydrothermal solution (p. 175)
index mineral (p. 183)

metamorphism (p. 171)
migmatite (p. 183)
nonfoliated (p. 175)
regional metamorphism (p. 171)
rock cleavage (p. 174)
schistosity (p. 174)

shear (p. 173)
shield (p. 185)
slaty cleavage (p. 174)
stress (p. 173)
texture (p. 174)

Web Resources

The *Earth* Home Page provides on-line resources for this chapter on the World Wide Web. You will find review exercises, specific updates for items in the chapter, suggested reading, and links to interesting related pathways on the Internet. Visit the *Earth* Home Page at **http://www.prenhall.com/tarbuck**.

CHAPTER 8

Geologic Time

Left Clearing winter storm, South Rim, Grand Canyon National Park, Arizona. — *Photo by Carr Clifton*

In the late 1700s, James Hutton recognized that Earth is very old. But how old? For many years there was no reliable method to determine the age of Earth or the dates of various events in the geologic past. Rather, a geologic time scale was developed that showed the sequence of events based on relative dating principles. What are these principles? What part do fossils play? With the discovery of radioactivity and radiometric dating techniques, geologists now can assign fairly accurate dates to many of the events in Earth history. What is radioactivity? Why is it a good "clock" for dating the geologic past?

Geology Needs a Time Scale

In 1869 John Wesley Powell, who was later to head the U.S. Geological Survey, led a pioneering expedition down the Colorado River and through the Grand Canyon (Figure 8.1). Writing about the rock layers that were exposed by the downcutting of the river, Powell noted that "the canyons of this region would be a Book of Revelations in the rock-leaved Bible of geology." He was undoubtedly impressed with the millions of years of Earth history exposed along the walls of the Grand Canyon (see chapter-opening photo).

Powell realized that the evidence for an ancient Earth is concealed in its rocks. Like the pages in a long and complicated history book, rocks record the geological events and changing life-forms of the past. The book, however, is not complete. Many pages, especially in the early chapters, are missing. Others are tattered, torn, or smudged. Yet, enough of the book remains to allow much of the story to be deciphered.

Interpreting Earth history is a prime goal of the science of geology. Like a modern-day sleuth, the geologist must interpret the clues found preserved in the rocks. By studying rocks, especially sedimentary rocks, and the features they contain, geologists can unravel the complexities of the past.

Geological events by themselves, however, have little meaning until they are put into a time perspective. Studying history, whether it be the Civil War or the Age of Dinosaurs, requires a calendar. Among geology's major contributions to human knowledge is the geologic time scale and the discovery that Earth history is exceedingly long.

Relative Dating— Key Principles

During the late 1800s and early 1900s, attempts were made to determine Earth's age. Although some of the methods appeared promising at the time, none of these early efforts proved to be reliable (see Box 8.1). What these scientists were seeking was an **absolute date.** Such dates pinpoint the time in history when something took place. Today radiometric dating allows

A.

B.

Figure 8.1 **A.** Start of the expedition from Green River station. A drawing from Powell's 1875 book. **B.** Major John Wesley Powell, pioneering geologist and the second director of the U.S. Geological Survey. (Courtesy of the U.S. Geological Survey)

us to accurately determine absolute dates for rock units that represent important events in Earth's distant past. However, prior to the discovery of radioactivity, geologists had no reliable method of absolute dating and had to rely solely on relative dating (we will address radiometric dating later in this chapter).

Relative dating means that rocks are placed in their proper *sequence of formation*. Relative dating cannot tell us how long ago something took place, only that it followed one event and preceded another. The relative dating techniques that were developed are still widely used. Absolute dating methods did not replace these techniques; they simply supplemented them. To establish a relative time scale, a few simple principles or rules had to be discovered and applied. Although they may seem obvious to us today, they were major breakthroughs in thinking at the time, and their discovery was an important scientific achievement.

Law of Superposition

Nicolaus Steno, a Danish anatomist, geologist, and priest (1638–1686), is credited with being the first to recognize a sequence of historical events in an outcrop of sedimentary rock layers. Working in the mountains of western Italy, Steno applied a very simple rule that has come to be the most basic principle of relative dating—the **law of superposition.** The law simply states that in an undeformed sequence of sedimentary rocks, each bed is older than the one above and younger than the one below. Although it may seem obvious that a rock layer could not be

deposited with nothing beneath it for support, it was not until 1669 that Steno clearly stated this principle.

This rule also applies to other surface-deposited materials such as lava flows and beds of ash from volcanic eruptions. Applying the law of superposition to the beds exposed in the upper portion of the Grand Canyon (Figure 8.2), we can easily place the layers in their proper order. Among those that are pictured, the sedimentary rocks in the Supai Group are the oldest, followed in order by the Hermit Shale, Coconino Sandstone, Toroweap Formation, and Kaibab Limestone.

Principle of Original Horizontality

Steno is also credited with recognizing the importance of another basic principle, called the **principle of original horizontality**. Simply stated, it means that layers of sediment are generally deposited in a horizontal position. Thus, if we observe rock layers that are flat, they have not been disturbed and still have their *original* horizontality. But if they are folded or inclined at a steep angle they must have been moved into that position by crustal disturbances sometime *after* their deposition (Figure 8.3).

Principle of Cross-Cutting Relationships

When a fault cuts through other rocks, or when magma intrudes and crystallizes, we can assume that the fault or intrusion is younger than the rocks

A.

B.

Figure 8.2 Applying the law of superposition to these layers exposed in the upper portion of the Grand Canyon, the Supai Group is oldest and the Kaibab Limestone is youngest (Photo by E. J. Tarbuck)

Box 8.1

Early Methods of Dating Earth

Today, thanks to reliable radiometric dating methods, we know that the age of Earth is about 4.6 billion years. However, this great age is a relatively recent discovery. In the late eighteenth century, James Hutton recognized the immensity of Earth history and the importance of time as a component in all geological processes. In the nineteenth century, Sir Charles Lyell and others effectively demonstrated that Earth had experienced many episodes of mountain building and erosion, which must have required great spans of geologic time. Although these pioneering scientists understood that Earth was very old, they had no way of knowing its true age. Was it tens of millions, hundreds of millions, or even billions of years old? During the latter half of the nineteenth and the early twentieth centuries, solutions to this problem were sought and several methods were subsequently devised.

Rate of Sedimentation

One method that was attempted several times involved the rate at which sediment is deposited. Some reasoned that if they could determine the rate that sediment accumulates, and could further ascertain the total thickness of sedimentary rock that had been deposited during Earth history, they could estimate the length of geologic time. All that was necessary was to divide the rate of sediment accumulation into the total thickness of sedimentary rock. Unfortunately, this method was riddled with difficulties, some of which are as follows:

1. Different sediments accumulate at different rates under varying conditions. Thus, determining an overall rate of sediment accumulation is difficult. Further, if such a rate is determined, it does not necessarily mean that the same rate can be applied to the past.
2. Since no single locality has a complete geologic column, estimates of the total thickness of sedimentary

rocks had to be compiled by adding together the maximum known thickness of rocks of each age. These estimates had to be revised each time a thicker section was discovered.
3. Sediment compacts when it is lithified; thus, a correction for compaction had to be made.

Needless to say, estimates of Earth's age varied considerably each time this method was attempted. The figure representing the maximum thickness of sedimentary rock ranged from 9600 meters (32,000 feet) to over 100,500 meters (330,000 feet). The amount of time for 0.3 meter (1 foot) of sediment to accumulate varied from 100 years to over 8600 years. The age of Earth as calculated by this method therefore ranged from 3 million to 1.5 billion years!

Ocean Salinity

Another method for dating Earth involved the salinity of the oceans, which were assumed to originally have been fresh water. Those who advocated this method felt that if they could accurately estimate the quantity of salt being carried to the ocean each year by rivers and the total amount of salt currently in the oceans, they could determine the length of geologic time by dividing the latter figure by the former.

Near the turn of the twentieth century, John Joly calculated the age of Earth at about 90 million years using this method. Joly, however, had no accurate notion of the amount of salt lost from the oceans because of deposition and winds blowing salt inland. It is also probable that the rate of salt accumulation has not always been constant. Thus, Joly's estimate for the age of Earth was not accurate. However, both of the methods for dating Earth that have just been described indicated that Earth was considerably older than the 6000 years given it by Archbishop Ussher.*

_____
*See the section on catastrophism in Chapter 1.

Lord Kelvin's Calculations

Very influential estimates of the age of Earth were compiled by the highly respected physicist Lord Kelvin in the latter part of the nineteenth century. Because Kelvin's estimates required few assumptions and were based on precise measurements, they were widely accepted for a time.

One of Kelvin's methods was founded on the widely held assumption that Earth had originally been molten and had cooled to its present condition. Although his data and calculations were limited, Kelvin still made it obvious that Earth could not be more than 100 million years old, and likely much less.

The second of Kelvin's estimates was based on the assumption that the source of the Sun's tremendous output of energy was of a conventional nature (nuclear fusion and radioactivity had not yet been discovered). His calculations indicated that the Sun could have illuminated Earth for only a few tens of millions of years. Furthermore, he said that in the past it had been much hotter and in the future it would become much cooler. He believed that Earth was inhabitable for organisms for a period of only 20–40 million years. Kelvin's apparently irrefutable estimates had a rather profound impact:

> Evolutionists found it virtually impossible to accept these figures, but all they had were educated guesses in the face of Kelvin's potent mathematics. Darwin and others compromised their original theories in their later years in an effort to reconcile evolution and uniformitarianism with the physicist's estimates. Eventually, however, they were vindicated.[†]

_____
[†]Leigh W. Mintz, *Historical Geology: The Science of a Dynamic Earth*, 2nd ed. (Columbus, Ohio: Merrill/Macmillan 1977), pp. 84–85.

Figure 8.3 Most layers of sediment are deposited in a nearly horizontal position. Thus, when we see rock layers that are inclined, we can assume that they must have been moved into that position by crustal disturbances after their deposition. Hartland Quay, Devon, England. (Photo by Tom Bean/DRK Photo)

affected. For example, in Figure 8.4, the faults and dikes clearly must have occurred *after* the sedimentary layers were deposited.

This is the **principle of cross-cutting relationships**. By applying the cross-cutting principle, you can see that fault A occurred *after* the sandstone layer was deposited because it "broke" the layer. Likewise, fault A occurred *before* the conglomerate was laid down because that layer is unbroken.

We can also state that dike B and its associated sill are older than dike A because dike A cuts the sill. In the same manner, we know that the batholith was emplaced after movement occurred along fault B but before dike B was formed. This is true because the batholith cuts across fault B while dike B cuts across the batholith.

Inclusions

Sometimes inclusions can aid the relative dating process. **Inclusions** are fragments of one rock unit that have been enclosed within another. The basic principle is logical and straightforward. The rock mass adjacent to the one containing the inclusions must have been there first in order to provide the rock fragments. Therefore, the rock mass containing inclusions is the younger of the two. Figure 8.5 provides an example. Here the inclusions of granite in the adjacent sedimentary layer indicate that the sedimentary layer was deposited on top of

a weathered mass of granite rather than being intruded from below by magma that later crystallized.

Unconformities

When we observe layers of rock that have been deposited essentially without interruption, we call them **conformable**. Particular sites exhibit conformable beds representing certain spans of geologic time. However, no place on Earth has a complete set of conformable strata. Throughout Earth history, the deposition of sediment has been interrupted over and over again. All such breaks in the rock record are termed unconformities. An **unconformity** represents a long period during which deposition ceased, erosion removed previously formed rocks, and then deposition resumed. In each case uplift and erosion are followed by subsidence and renewed sedimentation. Unconformities are important features because they represent significant geologic events in Earth history. Moreover, their recognition helps us identify what intervals of time are not represented by strata.

The rocks exposed in the Grand Canyon of the Colorado River represent a tremendous span of geologic history. It is a wonderful place to take a trip through time. The canyon's colorful strata record a long history of sedimentation in a variety of environments—advancing seas, rivers, and deltas, tidal flats, and sand dunes. But the

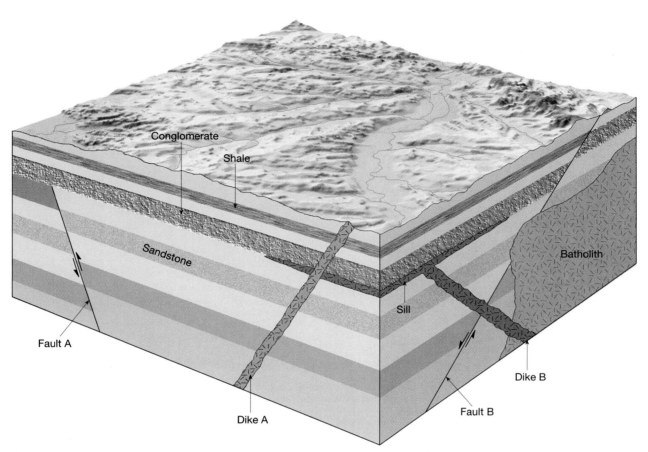

Figure 8.4 Cross-cutting relationships represent one principle used in relative dating. An intrusive rock body is younger than the rocks it intrudes. A fault is younger than the rock layers it cuts.

record is not continuous. Unconformities represent vast amounts of time that have not been recorded in the canyon's layers. Figure 8.6 is a geologic cross-section of the Grand Canyon. Refer to it as you read about the three basic types of unconformities: angular unconformities, disconformities, and nonconformities.

Angular Unconformity. Perhaps the most easily recognized unconformity is an **angular unconformity**. It consists of tilted or folded sedimentary rocks that are overlain by younger, more flat-lying strata. An angular unconformity indicates that during the pause in deposition, a period of deformation (folding or tilting) and erosion occurred (Figure 8.7).

 When James Hutton and John Playfair studied an angular unconformity in Scotland more than 200 years ago, it was clear to them that it represented a major episode of geologic activity.* They also appreciated the immense time span implied by such relationships. When Playfair later wrote of their visit to this site he stated that, "The mind seemed to grow giddy by looking so far into the abyss of time."

*These two pioneering geologists are discussed in the section on the birth of modern geology in Chapter 1.

Disconformity. When contrasted with angular unconformities, **disconformities** are more common, but usually far less conspicuous because the strata on either side are essentially parallel. Many disconformities are difficult to identify because the rocks above and below are similar and there is little evidence of erosion. Such a break often resembles an ordinary bedding plane. Other disconformities are easier to identify because the ancient erosion surface is cut deeply into the older rocks below.

Nonconformity. The third basic type of unconformity is a **nonconformity**. Here the break separates older metamorphic or intrusive igneous rocks from younger sedimentary strata (Figures 8.5 and 8.6). Just as angular unconformities and disconformities imply crustal movements, so too do nonconformities. Intrusive igneous masses and metamorphic rocks originate far below the surface. Thus, for a nonconformity to develop, there must be a period of uplift and the erosion of overlying rocks. Once exposed at the surface, the igneous or metamorphic rocks are subjected to weathering and erosion prior to subsidence and the renewal of sedimentation.

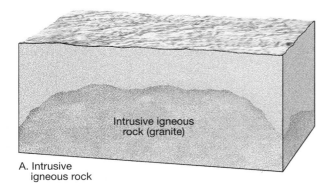

Intrusive igneous
rock (granite)

A. Intrusive
igneous rock

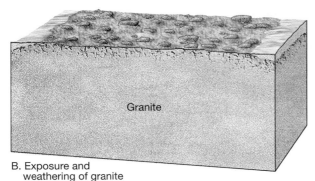

Granite

B. Exposure and
weathering of granite

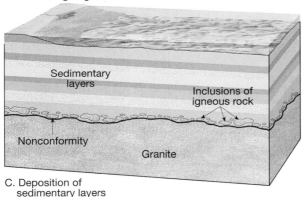

Sedimentary
layers

Inclusions of
igneous rock

Nonconformity

Granite

C. Deposition of
sedimentary layers

Figure 8.5 Because pieces of granite are contained within the overlying sedimentary bed, we know the granite must be older. When older intrusive igneous rocks are overlain by younger sedimentary layers, a type of unconformity termed a nonconformity is said to exist.

Using Relative Dating Principles

If you apply the principles of relative dating to the hypothetical geologic cross-section in Figure 8.8, you can place in proper sequence the rocks and the events they represent. The statements within the figure summarize the logic used to interpret the cross-section.

In this example, we establish a relative time scale for the rocks and events in the area of the cross-section. Remember that this method gives us no idea of how many years of Earth history are represented, for we have no absolute dates. Nor do we know how this area compares to any other.

Correlation of Rock Layers

To develop a geologic time scale that is applicable to the entire Earth, rocks of similar age in different regions must be matched up. Such a task is referred to as **correlation**.

Correlation by Physical Criteria

Within a limited area, correlating rocks of one locality with those of another may be done simply by walking along the outcropping edges. However, this may not be possible when the rocks are mostly concealed by soil and vegetation. Correlation over short distances is often achieved by noting the position of a bed in a sequence of strata. Or, a layer may be identified in another location if it is composed of distinctive or uncommon minerals.

By correlating the rocks from one place to another, a more comprehensive view of the geologic history of a region is possible. Figure 8.9, for example, shows the correlation of strata at three sites on the Colorado Plateau. No single local exhibits the entire sequence, but correlation reveals a more complete picture of the sedimentary rock record.

Many geologic studies involve relatively small areas. Although they are important in their own right, their full value is realized only when they are correlated with other regions. Although the methods just described are sufficient to trace a rock formation over relatively short distances, they are not adequate for matching up rocks that are separated by great distances. When correlation between widely separated areas or between continents is the objective, the geologist must rely on fossils.

Fossils and Correlation

Although the existence of fossils had been known for centuries, it was not until the late 1700s and early 1800s that their significance as geologic tools was made evident. During this period an English engineer and canal builder, William Smith, discovered that each rock formation in the canals he worked on contained fossils unlike those in the beds either above or below. Further, he noted that sedimentary strata in widely separated areas could be identified—and correlated—by their distinctive fossil content.

Based on Smith's classic observations and the findings of many geologists who followed, one of the most important and basic principles in historical geology was formulated: *fossil organisms succeed one another in a definite and determinable order, and therefore any time period can be recognized by its fossil content*. This has come

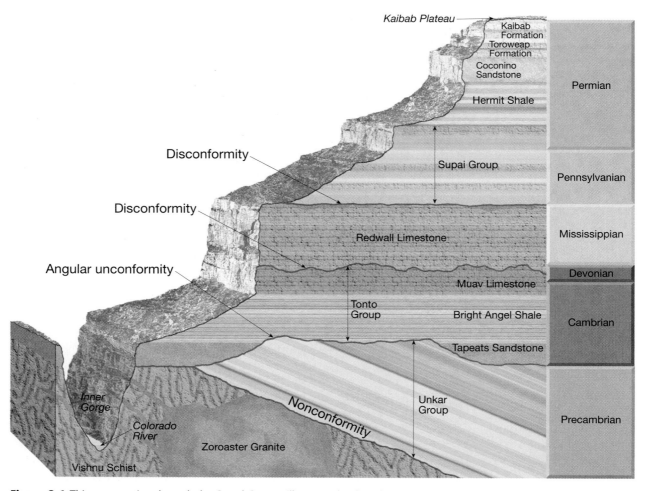

Figure 8.6 This cross-section through the Grand Canyon illustrates the three basic types of unconformities. An angular unconformity can be seen between the tilted Precambrian Unkar Group and the Cambrian Tapeats Sandstone. Two disconformities are marked, above and below the Redwall Limestone. A nonconformity occurs between the igneous and metamorphic rocks of the Inner Gorge and the sedimentary strata of the Unkar Group.

to be known as the **principle of fossil succession**. In other words, when fossils are arranged according to their age, they do not present a random or haphazard picture. To the contrary, fossils document the evolution of life through time.

For example, an Age of Trilobites is recognized quite early in the fossil record. Then, in succession, paleontologists recognize an Age of Fishes, an Age of Coal Swamps, an Age of Reptiles, and an Age of Mammals. These "ages" pertain to groups that were especially plentiful and characteristic during particular time periods. Within each of the "ages" there are many subdivisions based, for example, on certain species of trilobites, and certain types of fish, reptiles, and so on. This same succession of dominant organisms, never out of order, is found on every major landmass.

When fossils were found to be time indicators, they became the most useful means of correlating rocks of similar age in different regions. Geologists pay particular attention to certain fossils called **index fossils**. These fossils are widespread geographically and are limited to a short span of geologic time, so their presence provides an important method of matching rocks of the same age. Rock formations, however, do not always contain a specific index fossil. In such situations, groups of fossils are used to establish the age of the bed. Figure 8.10 illustrates how an assemblage of fossils may be used to date rocks more precisely than could be accomplished by the use of any one of the fossils.

In addition to being important and often essential tools for correlation, fossils are important environmental indicators. Although much can be deduced about past environments by studying the nature and characteristics of sedimentary rocks, a close examination of the fossils present can usually provide a great deal more information. For example, when the remains of certain clam shells are found in limestone, the geologist quite reasonably assumes that the region was once covered by a shallow sea. Also, by using what we know of

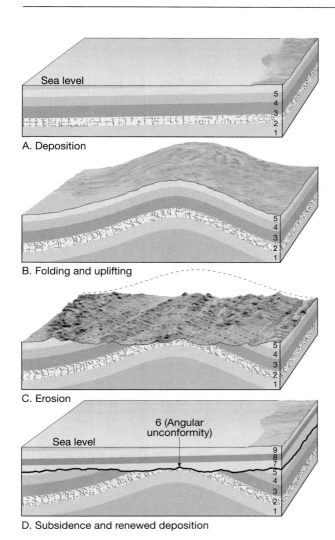

A. Deposition

B. Folding and uplifting

C. Erosion

6 (Angular unconformity)

Sea level

D. Subsidence and renewed deposition

E.

Figure 8.7 Formation of an angular unconformity. An angular unconformity represents an extended period during which deformation and erosion occurred. Part E shows an angular unconformity at Siccar Point, Scotland that was first described by James Hutton and John Playfair more than 200 years ago. (Photo by Edward Hay)

living organisms, we can conclude that fossil animals with thick shells capable of withstanding pounding and surging waves inhabited shorelines.

On the other hand, animals with thin, delicate shells probably indicate deep, calm offshore waters. Hence, by looking closely at the types of fossils, the approximate position of an ancient shoreline may be identified. Further, fossils can be used to indicate the former temperature of the water. Certain kinds of present-day corals must live in warm and shallow tropical seas like those around Florida and the Bahamas. When similar types of coral are found in ancient limestones, they indicate the marine environment that must have existed when they were alive. These examples illustrate how fossils can help unravel the complex story of Earth history.

Absolute Dating with Radioactivity

In addition to establishing relative dates by using the principles described in the preceding sections, it is also possible to obtain reliable absolute dates for events in the geologic past. For example, we know that Earth is about 4.6 billion years old and that the dinosaurs became extinct about 66 million years ago. Dates that are expressed in millions and billions of years truly stretch our imagination because our personal calendars involve time measured in hours, weeks, and years. Nevertheless, the vast expanse of geologic time is a reality and it is radiometric dating that allows us to measure it accurately. In this section you will learn about radioactivity and its application in radiometric dating.

Recall from Chapter 2 that each atom has a *nucleus* containing protons and neutrons, and that the nucleus is orbited by electrons. *Electrons* have a negative electrical charge, and *protons* have a positive charge. A *neutron* is actually a proton and an electron combined, so it has no charge (it is neutral).

The *atomic number* (each element's identifying number) is the number of protons in the nucleus. Every element has a different number of protons, and thus a different atomic number (hydrogen = 1, carbon = 6, oxygen = 8, uranium = 92, etc.). Atoms of the same element always have the same number of protons, so the atomic number stays constant.

Practically all of an atom's mass (99.9%) is in the nucleus, indicating that electrons have virtually no mass at all. So, by adding the protons and neutrons in an atom's nucleus, we derive the atom's *mass number*. The number of neutrons can vary, and these variants, or *isotopes*, have different mass numbers.

To summarize with an example, uranium's nucleus always has 92 protons, so its atomic number

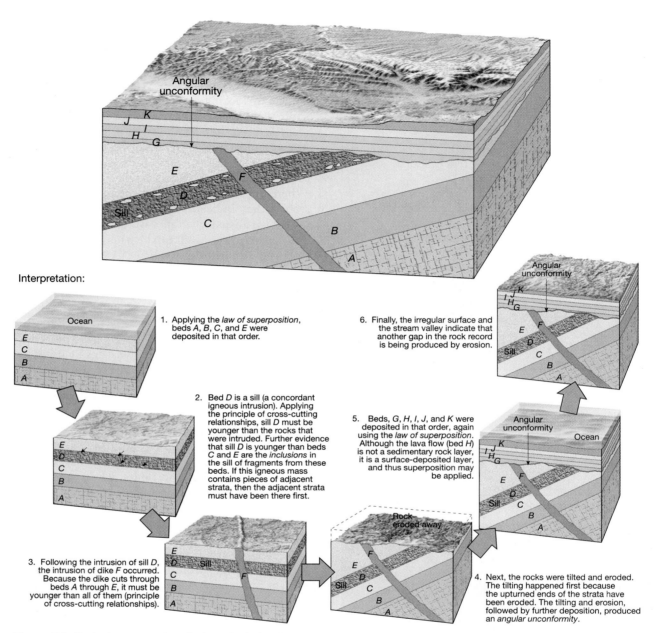

Figure 8.8 Geologic cross-section of a hypothetical region.

always is 92. But its neutron population varies, so uranium has three isotopes: uranium-234 (protons + neutrons = 234), uranium-235, and uranium-238. All three isotopes are mixed in nature. They look the same and behave the same in chemical reactions.

Radioactivity

The forces that bind protons and neutrons together in the nucleus usually are strong. However, in some isotopes, the nuclei are unstable because the forces binding protons and neutrons together are not strong enough. As a result, the nuclei spontaneously break apart, or decay, a process called **radioactivity**.

What happens when unstable nuclei break apart? Three common types of radioactive decay are illustrated in Figure 8.11 and can be summarized as follows:

1. *Alpha particles* (α particles) may be emitted from the nucleus. An alpha particle is composed of 2 protons and 2 neutrons. Thus, the emission of an alpha particle means that the mass number of the isotope is reduced by 4 and the atomic number is lowered by 2.

2. When a *beta particle* (β particle), or electron, is given off from a nucleus, the mass number remains unchanged, because electrons have

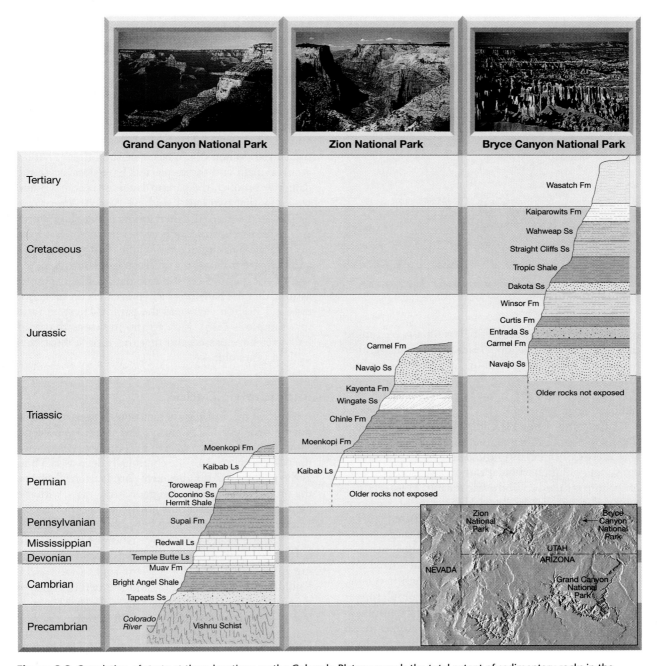

Figure 8.9 Correlation of strata at three locations on the Colorado Plateau reveals the total extent of sedimentary rocks in the region. (After U.S. Geological Survey; Photos by E. J. Tarbuck)

practically no mass. However, because the electron has come from a neutron (remember, a neutron is a combination of a proton and an electron), the nucleus contains one more proton than before. Therefore, the atomic number increases by 1.

3. Sometimes an electron is captured by the nucleus. The electron combines with a proton and forms a neutron. As in the last example, the mass number remains unchanged. However,

since the nucleus now contains one less proton, the atomic number decreases by 1.

An unstable radioactive isotope is referred to as the *parent*, and the isotopes resulting from the decay of the parent are termed the *daughter products*. Figure 8.12 provides an example of radioactive decay. Here it can be seen that when the radioactive parent, uranium-238 (atomic number 92, mass number 238) decays, it follows a number of steps, emitting 8 alpha

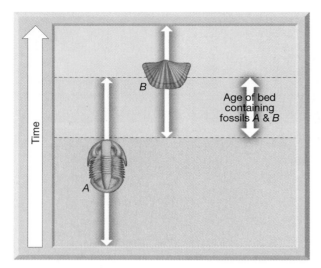

Figure 8.10 Overlapping ranges of fossils help date rocks more exactly than using a single fossil.

particles and 6 beta particles before finally becoming the stable daughter product lead-206 (atomic number 82, mass number 206). One of the unstable daughter products produced during this decay series is radon. Box 8.2 examines the hazards associated with this radioactive gas.

Certainly among the most important results of the discovery of radioactivity is that it provided a reliable means of calculating the ages of rocks and minerals that contain particular radioactive isotopes. The procedure is called **radiometric dating.** Why is radiometric dating reliable? Because the rates of decay for many isotopes have been precisely measured and do not vary under the physical conditions that exist in Earth's outer layers. Therefore, each radioactive isotope used for dating has been decaying at a fixed rate since the formation of the rocks in which it occurs, and the products of decay have been accumulating at a corresponding rate. For example, when uranium is incorporated into a mineral that crystallizes from magma, there is no lead (the stable daughter product) from previous decay. The radiometric "clock" starts at this point. As the uranium in this newly formed mineral disintegrates, atoms of the daughter product are trapped, and measurable amounts of lead eventually accumulate.

Half-Life

The time required for one-half of the nuclei in a sample to decay is called the **half-life** of the isotope. Half-life is a common way of expressing the rate of radioactive disintegration. Figure 8.13 illustrates what occurs when a radioactive parent decays directly into its stable daughter product. When the quantities of parent and daughter are equal (ratio 1:1), we know that one half-life has transpired. When one-quarter of the original parent atoms remain and three-quarters have decayed to the daughter product, the parent/daughter ratio is 1:3 and we know that two half-lives have passed. After three half-lives, the ratio of parent atoms to daughter atoms is 1:7 (one parent atom for every seven daughter atoms).

If the half-life of a radioactive isotope is known and the parent/daughter ratio can be determined, the age of the sample can be calculated. For example, assume that the half-life of a hypothetical unstable isotope is 1 million years and the parent/daughter ratio in a sample is 1:15. Such a ratio indicates that four half-lives have passed and that the sample must be 4 million years old.

Radiometric Dating

Notice that the *percentage* of radioactive atoms that decay during one half-life is always the same: 50 percent. However, the *actual number* of atoms that decay with the passing of each half-life continually decreases. Thus, as the percentage of radioactive parent atoms declines, the proportion of stable daughter atoms rises, with the increase in daughter atoms just matching the drop in parent atoms. This fact is the key to radiometric dating.

Of the many radioactive isotopes that exist in nature, five have proven particularly useful in providing radiometric ages for ancient rocks (Table 8.1). Rubidium-87, thorium-232, and the two isotopes of uranium are used only for dating rocks that are millions of years old, but potassium-40 is more versatile.

Potassium-Argon. Although the half-life of potassium-40 is 1.3 billion years, analytical techniques make possible the detection of tiny amounts of its stable daughter product, argon-40, in some rocks that are younger than 100,000 years. Another important reason for its frequent use is that potassium is an abundant

Table 8.1. Isotopes Frequently Used in Radiometric Dating

Radioactive Parent	Stable Daughter Product	Currently Accepted Half-life Values
Uranium-238	Lead-206	4.5 billion years
Uranium-235	Lead-207	713 million years
Thorium-232	Lead-208	14.1 billion years
Rubidium-87	Strontium-87	47.0 billion years
Potassium-40	Argon-40	1.3 billion years

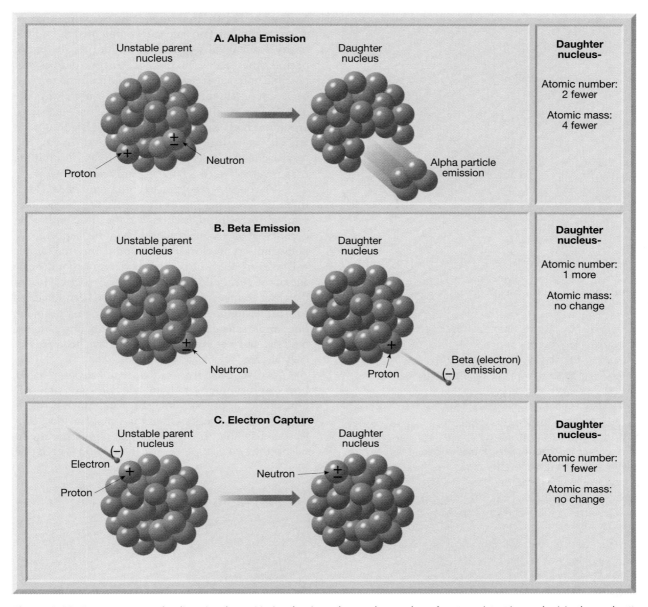

Figure 8.11 Common types of radioactive decay. Notice that in each case the number of protons (atomic number) in the nucleus changes, thus producing a different element.

constituent of many common minerals, particularly micas and feldspars.

Although potassium (K) has three natural isotopes K^{39}, K^{40}, and K^{41}, only K^{40} is radioactive. When K^{40} decays, it does so in two ways. About 11 percent changes to argon-40 (Ar^{40}) by means of electron capture (see Figure 8.11C). The remaining 89 percent of K^{40} decays to calcium-40 (Ca^{40}) by beta emission (see Figure 8.11B). The decay of K^{40} to Ca^{40}, however, is not useful for radiometric dating, because the Ca^{40} produced by radioactive disintegration cannot be distinguished from calcium that may have been present when the rock formed.

The potassium-argon clock begins when potassium-bearing minerals crystallize from a magma or form within a metamorphic rock. At this point the new minerals will contain K^{40} but will be free of Ar^{40}, because this element is an inert gas that does not chemically combine with other elements. As time passes, the K^{40} steadily decays by electron capture. The Ar^{40} produced by this process remains trapped within the mineral's crystal lattice. Because no Ar^{40} was present when the mineral formed, all of the daughter atoms trapped in the mineral must have come from the decay of K^{40}. To determine a sample's age, the K^{40}/Ar^{40} ratio is measured precisely and the known half-life for K^{40} applied.

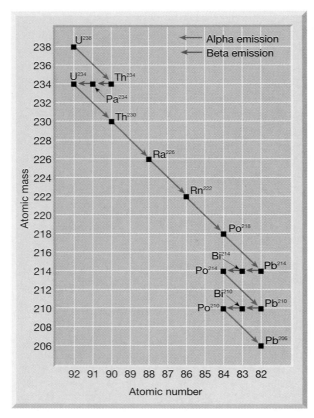

Figure 8.12 The most common isotope of uranium (U-238) is an example of a radioactive decay series. Before the stable end product (Pb-206) is reached, many different isotopes are produced as intermediate steps.

Sources of Error. It is important to realize that an accurate radiometric date can be obtained only if the mineral remained a closed system during the entire period since its formation. A correct date is not possible unless there was neither the addition nor loss of parent or daughter isotopes. This is not always the case. In fact, an important limitation of the potassium-argon method arises from the fact that argon is a gas and it may leak from minerals, throwing off measurements. Indeed, losses can be significant if the rock is subjected to relatively high temperatures.

Of course, a reduction in the amount of Ar^{40} leads to an underestimation of the rock's age. Sometimes temperatures are high enough for a sufficiently long period that all argon escapes. When this happens, the potassium-argon clock is reset and dating the sample will give only the time of thermal resetting, not the true age of the rock. For other radiometric clocks, a loss of daughter atoms can occur if the rock has been subjected to weathering or leaching. To avoid such a problem, one simple safeguard is to use only fresh, unweathered material and not samples that may have been chemically altered.

If parent/daughter ratios are not always reliable, how can meaningful radiometric dates be obtained? One common precaution against unknown errors is the use of cross checks. Often this simply involves subjecting a sample to two different radiometric methods. If the two dates agree, the likelihood is high that

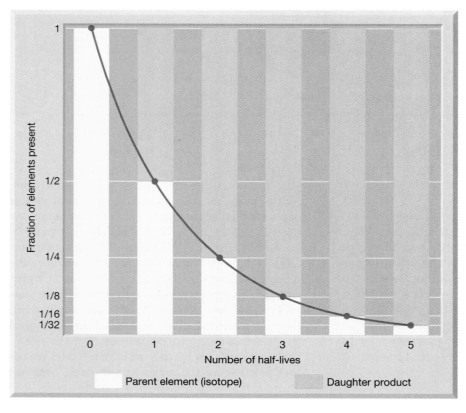

Figure 8.13 The radioactive decay curve shows change that is exponential. Half of the radioactive parent remains after one half-life. After a second half-life one-quarter of the parent remains, and so forth.

Box 8.2

Radon

Richard L. Hoffman*

Radioactivity is defined as the spontaneous emission of atomic particles and/or electromagnetic waves from unstable atomic nuclei. For example, in a sample of uranium-238, unstable nuclei decay and produce a variety of radioactive progeny or "daughter" products as well as energetic forms of radiation (Table 8.A). One of its radioactive decay products is radon—a colorless, odorless, invisible gas.

Radon gained public attention in 1984 when a worker in a Pennsylvania nuclear power plant set off radiation alarms—not when he left work, but as he entered. His clothing and hair were contaminated with radon decay products. Investigation revealed that his basement at home had a radon level 2800 times the average level in indoor air. The home was located along a geological formation known as the Reading Prong—a mass of uranium-bearing rock that runs from near Reading, Pennsylvania, to near Trenton, New Jersey.

Originating in the radio-decay of traces of uranium and thorium found in almost all soils, radon isotopes (Rn-222 and Rn-220) are continually renewed in an ongoing, natural process. Geologists estimate that the top six feet of soil from an average acre of land contains about fifty pounds of uranium (about 2 to 3 parts per million); some types of rocks contain more. Radon is continually generated by the gradual decay of this uranium. Because uranium has a half-life of about 4.5 billion years, radon will be with us forever.

Radon itself decays, having a half-life of only about four days. Its decay products (except lead-206) are all radioactive solids that adhere to dust particles, many of which we inhale. During prolonged exposure to a radon-contaminated environment, some decay will occur while the gas is in the lungs, thereby placing the radioactive radon progeny in direct contact with delicate lung tissue. Steadily accumulating evidence indicates radon to be a significant cause of lung cancer second only to smoking.

A house with a radon level of 4.0 picocuries per liter of air has about 8 to 9 atoms of radon decaying every minute in every liter of air. The EPA suggests indoor radon levels be kept below this level. EPA risk estimates are conservative—they are based on an assumption that one would spend 75 percent of a 70-year time span (about 52 years) in the contaminated space, which most people would not.

Once radon is produced in the soil, it diffuses throughout the tiny spaces between soil particles. Some radon ultimately reaches the soil surface where it dissipates into the air. Radon enters buildings and homes through holes and cracks in basement floors and walls. Radon's density is greater than air, so it tends to remain in basements during its short decay cycle.

The source of radon is as enduring as its generation mechanism within Earth; radon will never go away. However, cost-effective mitigation strategies are available to reduce radon to acceptable levels, generally without great expense.

*Dr. Hoffmann is Professor of Chemistry, Illinois Central College.

Table 8.A Decay Products of Uranium-238

Some Decay Products of Uranium-238	Decay Particle Produced	Half-Life
Uranium-238	alpha	4.5 billion years
Radium-226	alpha	1600 years
Radon-222	**alpha**	**3.82 days**
Polonium-218	alpha	3.1 minutes
Lead-214	beta	26.8 minutes
Bismuth-214	beta	19.7 minutes
Polonium-214	alpha	1.6×10^{-4} second
Lead-210	beta	20.4 years
Bismuth-210	beta	5.0 days
Polonium-210	alpha	138 days
Lead-206	none	stable

the date is reliable. If, on the other hand, there is an appreciable difference between the two dates, other cross checks must be employed to determine which, if either, is correct.

Carbon-14. To date very recent events, carbon-14 is used. Carbon-14 is the radioactive isotope of carbon. The process is often called **radiocarbon dating.** Because the half-life of carbon-14 is only 5730 years, it can be used for dating events from the historic past as well as those from recent geologic history. In some cases carbon-14 can be used to date events as far back as 75,000 years.

Carbon-14 is continuously produced in the upper atmosphere as a consequence of cosmic ray bombardment. Cosmic rays (high-energy nuclear particles) shatter the nuclei of gas atoms, releasing neutrons. Some of the neutrons are absorbed by nitrogen atoms (atomic number 7, mass number 14), causing each nucleus to emit a proton. As a result, the atomic number decreases by 1 (to 6), and a different element, carbon-14, is created (Figure 8.14A). This isotope of carbon quickly becomes incorporated into carbon dioxide, which circulates in the atmosphere and is absorbed by living matter. As a result, all organisms contain a small amount of carbon-14, including yourself.

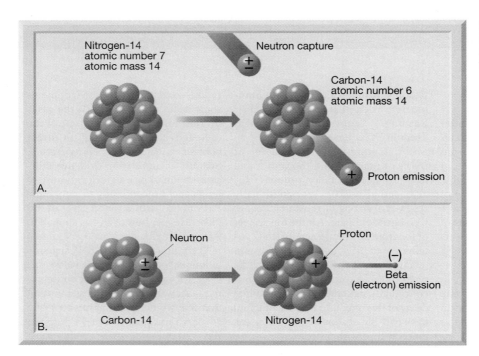

Figure 8.14 A. Production and B. decay of carbon-14. These sketches represent the nuclei of the respective atoms.

As long as an organism is alive, the decaying radiocarbon is continually replaced, and the proportions of carbon-14 and carbon-12 remain constant. Carbon-12 is the stable and most common isotope of carbon. However, when any plant or animal dies, the amount of carbon-14 gradually decreases as it decays to nitrogen-14 by beta emission (Figure 8.14B). By comparing the proportions of carbon-14 and carbon-12 in a sample, radiocarbon dates can be determined.

Although carbon-14 is only useful in dating the last small fraction of geologic time, it has become a very valuable tool for anthropologists, archaeologists, and historians, as well as for geologists who study very recent Earth history. In fact, the development of radiocarbon dating was considered so important that the chemist who discovered this application, Willard F. Libby, received a Nobel Prize in 1960.

Bear in mind that, although the basic principle of radiometric dating is simple, the actual procedure is quite complex. The analysis that determines the quantities of parent and daughter must be painstakingly precise. In addition, some radioactive materials do not decay directly into the stable daughter product as was the case with our hypothetical example, a fact that may further complicate the analysis. In the case of uranium-238, there are thirteen intermediate unstable daughter products formed before the fourteenth and last daughter product, the stable isotope lead-206, is produced (see Figure 8.12).

Value of Radiometric Dating. Radiometric dating methods have produced literally thousands of dates for events in Earth history. Rocks from several localities have been dated at more than 3 billion years, and geologists realize that still-older rocks exist. For example, a granite from South Africa has been dated at 3.2 billion years and contains inclusions of quartzite. Quartzite, a metamorphic rock, was originally the sedimentary rock sandstone. Sandstone, in turn, is the product of the lithification of sediments produced by the weathering of existing rocks. Thus, we have a positive indication that even older rocks existed.

Radiometric dating has vindicated the ideas of Hutton, Darwin, and others who over 150 years ago inferred that geologic time must be immense. Indeed, radiometric dating has proven that there has been enough time for the processes we observe to have accomplished tremendous tasks.

 The Geologic Time Scale

Geologists have divided the whole of geologic history into units of varying magnitude. Together, they comprise the **geologic time scale** of Earth history (Figure 8.15). The major units of the time scale were delineated during the nineteenth century, principally by workers in western Europe and Great Britain. Because absolute dating was unavailable at that time, the entire time scale was created using methods of relative dating. It has only been in this century that radiometric methods permitted absolute dates to be added.

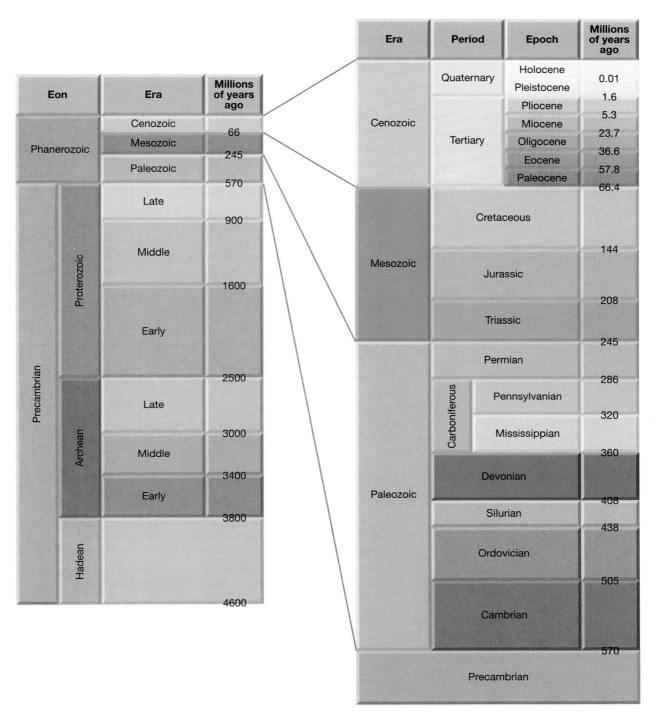

Figure 8.15 The geologic time scale. The absolute dates were added long after the time scale had been established using relative dating techniques. (Data from Geological Society of America)

Structure of the Time Scale

The geologic time scale subdivides the 4.6-billion-year history of Earth into many different units and provides a meaningful time frame within which the events of the geologic past are arranged. As shown in Figure 8.15, **eons** represent the greatest expanses of time. The eon that began about 570 million years ago is the **Phanerozoic,** a term derived from Greek words meaning *visible life*. It is an appropriate description because the rocks and deposits of the Phanerozoic eon contain abundant fossils that document major evolutionary trends.

Another glance at the time scale reveals that the Phanerozoic eon is divided into **eras.** The three eras within the Phanerozoic are the **Paleozoic** ("ancient life"), the **Mesozoic** ("middle life"), and the **Cenozoic** ("recent life"). As the names imply, the eras are bounded by profound worldwide changes in life-forms (see Box 8.3).

Each era is subdivided into time units known as **periods.** The Paleozoic has seven, the Mesozoic three, and the Cenozoic two. Each of these dozen periods is characterized by a somewhat less profound change in life-forms as compared with the eras. The eras and periods of the Phanerozoic, with brief explanations of each, are shown in Table 8.2.

Finally, each of the twelve periods is divided into still smaller units called **epochs.** As you can see in Figure 8.15, seven epochs have been named for the periods of the Cenozoic. The epochs of other periods usually are simply termed *early*, *middle*, and *late*.

Precambrian Time

Notice that the detail of the geologic time scale does not begin until about 570 million years ago, the date for the beginning of the Cambrian period. The more than four billion years prior to the Cambrian are divided into three eons, the **Hadean**, the **Archean**, and the **Proterozoic**. It is also common for this vast expanse of time to simply be referred to as the **Precambrian**. Although it represents more than 85 percent of Earth

history, the Precambrian is not divided into nearly as many smaller time units as the Phanerozoic eon.

Why is the huge expanse of Precambrian time not divided into numerous eras, periods, and epochs? The reason is that Precambrian history is not known in great enough detail. The quantity of information geologists have deciphered about Earth's past is somewhat analogous to the detail of human history. The farther back we go, the less that is known. Certainly more data and information exist about the past 10 years than for the first decade of the twentieth century; the events of the nineteenth century have been documented much better than the events of the first century A.D.; and so on. So it is with Earth history. The more recent past has the freshest, least disturbed, and more observable record. The farther back in time the geologist goes, the more fragmented the record and clues become. There are other reasons to explain our lack of a detailed time scale for this vast segment of Earth history:

1. The first abundant fossil evidence does not appear in the geologic record until the beginning of the Cambrian period. Prior to the Cambrian, simple life-forms such as algae, bacteria, fungi, and worms predominated. All of these organisms lack hard parts, an important condition favoring preservation. For this reason, there is only a meager Precambrian fossil record. Many exposures of Precambrian rocks

Table 8.2 Major Divisions of Geologic Time

Cenozoic Era (Age of Recent Life)	**Quaternary period**	The several geologic eras were originally named Primary, Secondary, Tertiary, and Quaternary. The first two names are no longer used; Tertiary and Quaternary have been retained but used as period designations.
	Tertiary period	
Mesozoic Era (Age of Middle Life)	**Cretaceous period**	Derived from Latin word for chalk (creta) and first applied to extensive deposits that form white cliffs along the English Channel (see Figure 6.11).
	Jurassic period	Named for the Jura Mountains, located between France and Switzerland, where rocks of this age were first studied.
	Triassic period	Taken from word "trias" in recognition of the threefold character of these rocks in Europe.
Paleozoic Era (Age of Ancient Life)	**Permian period**	Named after the province of Perm, Russia, where these rocks were first studied.
	Pennsylvanian period*	Named for the state of Pennsylvania where these rocks have produced much coal.
	Mississippian period*	Named for the Mississippi River valley where these rocks are well exposed.
	Devonian period	Named after Devonshire County, England, where these rocks were first studied.
	Silurian period	Named after Celtic tribes, the Silures and the Ordovices, that lived in Wales during the Roman Conquest.
	Ordovician period	
	Cambrian period	Taken from Roman name for Wales (Cambria), where rocks containing the earliest evidence of complex forms of life were first studied.
Precambrian		The time between the birth of the planet and the appearance of complex forms of life. More than 85 percent of Earth's estimated 4.6 billion years fall into this span.

SOURCE: U.S. Geological Survey.

*Outside of North America, the Mississippian and Pennsylvanian periods are combined into the Carboniferous period.

Box 8.3

The KT Extinction

The boundaries between divisions on the geologic time scale represent times of significant geological and/or biological change. Of special interest is the boundary between the Mesozoic era ("middle life") and Cenozoic era ("recent life"), about 66 million years ago. Around this time, more than half of all plant and animal species died out in a *mass extinction*. This boundary marks the end of the era in which dinosaurs and other reptiles were prominent and the beginning of the era when mammals became very important (Figure 8.A).

Because the last period of the Mesozoic is the Cretaceous (abbreviated K to avoid confusion with other "C" periods), and the first period of the Cenozoic is Tertiary (abbreviated T), the time of this mass extinction is called the Cretaceous-Tertiary or *KT boundary*.

The dinosaurs met their demise around the KT boundary, along with large numbers of other animal and plant groups, both terrestrial and marine. More important, of course, is that many species survived. Human beings are descended from these survivors. Perhaps this explains why an event that occurred 66 millions years ago has captured the interest of so many people.

The extinction of the great reptiles is generally attributed to this group's inability to adapt to some radical change in environmental conditions. What event could have triggered the rapid extinction of the dinosaurs—the most successful group of land animals ever to have lived?

One view proposes that, approximately 66 million years ago, a large asteroid or comet about 10 kilometers in diameter collided with Earth. The impact of such a body would have produced a dust cloud thousands of times larger than that released during most volcanic eruptions. For many months the dust would have greatly restricted the sunlight reaching Earth's surface. With insufficient sunlight for photosynthesis, delicate food chains would collapse. Large, plant-eating dinosaurs would be affected more adversely than would smaller life-forms because of the tremendous volume of vegetation they

Figure 8.A Dinosaurs dominated the Mesozoic landscape until their extinction at the close of the Cretaceous period. This skeleton of *Triceratops* is on display in London's Natural History Museum. (Photo by Natural History Museum, London)

consumed. When the sunlight returned, more than half of the species on Earth, including numerous marine organisms, had become extinct.

What evidence points to such a catastrophic collision 66 million years ago? First, a thin layer of sediment nearly 1 centimeter thick has been discovered at the KT boundary, worldwide. This sediment contains a high level of the element *iridium*, rare in Earth's crust but found in high proportions in stony meteorites. Could this layer be the scattered remains of an asteroid that was responsible for the environmental changes that led to the demise of many reptile groups?

The second piece of evidence is that this period of mass extinction appears to have affected all land animals larger than dogs. Supporters of this catastrophic-event scenario suggest that small, ratlike mammals could survive a breakdown of food chains lasting perhaps several months. Large animals, proponents argue, could not have survived such an event.

Although the impact hypothesis is widely held, some scientists disagree. They claim that what appears to be a

mass extinction over a short period in fact occurred over a much broader time span. Based on the fossil record at the KT boundary, these geologists conclude that the decline of the dinosaurs was gradual.

Those who disagree with the impact hypothesis have suggested an upsurge in volcanism. This is based on enormous outpourings of basaltic lavas in India approximately 65 million years ago. Volcanism advocates argue that the consequences of extensive volcanism would be quite similar to those from an asteroid impact: large quantities of dust and ash in the atmosphere that reduced sunlight, causing food chains to collapse. Further, some eruptions emit large quantities of sulfur gases that could form toxic sulfuric acid rains.

Volcano advocates point to the 1783 eruption at Laki, Iceland. Although small, this activity killed 75 percent of all livestock and ultimately 24 percent of the inhabitants of Iceland.

Whatever the cause, the KT extinction provided habitat vacancies for the mammals, allowing mammals to attain dominance during the Cenozoic era.

have been studied in some detail, but correlation is often difficult when fossils are lacking.

2. Because Precambrian rocks are very old, most have been subjected to a great many changes. Much of the Precambrian rock record is composed of highly distorted metamorphic rocks. This makes the interpretation of past environments difficult, because many of the clues present in the original sedimentary rocks have been destroyed.

Radiometric dating has provided a partial solution to the troublesome task of dating and correlating Precambrian rocks. But untangling the complex Precambrian record still remains a daunting task.

◉ Difficulties in Dating the Geologic Time Scale

Although reasonably accurate dates have been worked out for the periods of the geologic time scale (Figure 8.15), the task is not without difficulty. The primary difficulty in assigning absolute dates to units of time is the fact that not all rocks can be dated by radiometric methods. Recall that for a radiometric date to be useful, all the minerals in the rock must have formed at about the same time. For this reason, radioactive isotopes can be used to determine when minerals in an igneous rock crystallized and when pressure and heat created new minerals in a metamorphic rock.

However, samples of sedimentary rock can only rarely be dated directly by radiometric means. Although a detrital sedimentary rock may include particles that contain radioactive isotopes, the rock's age cannot be accurately determined because the grains composing the rock are not the same age as the rock in which they occur. Rather, the sediments have been weathered from rocks of diverse ages.

Radiometric dates obtained from metamorphic rocks may also be difficult to interpret, because the age of a particular mineral in a metamorphic rock does not necessarily represent the time when the rock initially formed. Instead, the date might indicate any one of a number of subsequent metamorphic phases.

If samples of sedimentary rocks rarely yield reliable radiometric ages, how can absolute dates be assigned to sedimentary layers? Usually the geologist must relate the strata to datable igneous masses, as in Figure 8.16. In this example, radiometric dating has determined the ages of the volcanic ash bed within the Morrison Formation and the dike cutting the Mancos Shale and Mesaverde Formation. The sedimentary beds below the ash are obviously older than the ash, and all the layers above the ash are younger. The dike is younger than the Mancos Shale and the Mesaverde Formation but older than the Wasatch Formation because the dike does not intrude the Tertiary rocks.

From this kind of evidence, geologists estimate that a part of the Morrison Formation was deposited about 160 million years ago as indicated by the ash, bed. Further, they conclude that the Tertiary period began after the intrusion of the dike, 66 millions years ago. This is one example of literally thousands that illustrate how datable materials are used to bracket the various episodes in Earth history within specific time periods. It shows the necessity of combining laboratory dating methods with field observations of rocks.

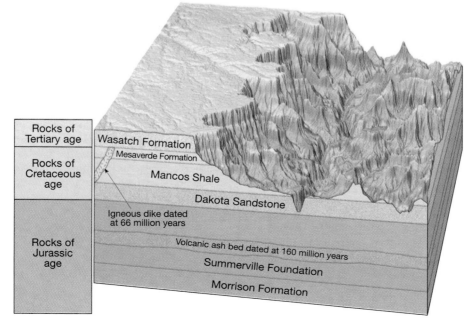

Figure 8.16 Absolute dates for sedimentary layers are usually determined by examining their relationship to igneous rocks. (After U.S. Geological Survey)

Chapter Summary

- The two types of dates used by geologists to interpret Earth history are (1) *relative dates*, which put events in their *proper sequence of formation*, and (2) *absolute dates*, which pinpoint the *time in years* when an event occurred.

- Relative dates can be established using the *law of superposition* (in an underformed sequence of sedimentary rocks or surface-deposited igneous rocks, each bed is older than the one above, and younger than the one below), *principle of original horizontality* (most layers are deposited in a horizontal position), *principle of cross-cutting relationships* (when a fault or intrusion cuts through another rock, the fault or intrusion is younger than the rocks cut through), and *inclusions* (the rock mass containing the inclusion is younger than the rock that provided the inclusion).

- *Unconformities* are gaps in the rock record. Each represents a long period during which deposition ceased, erosion removed previously formed rocks, and then deposition resumed. The three basic types of unconformities are *angular unconformities* (tilted or folded sedimentary rocks that are overlain by younger, more flat-lying strata), *disconformities* (the strata on either side of the unconformity are essentially parallel), and *nonconformities* (where a break separates older metamorphic or intrusive igneous rocks from younger sedimentary strata).

- *Correlation*, the matching up of two or more geologic phenomena in different areas, is used to develop a geologic time scale that applies to the whole Earth.

- Fossils are used to *correlate* sedimentary rocks that are from different regions by using the rocks' distinctive fossil content and applying the *principle of fossil succession*. The principle of fossil succession, which is based on the work of *William Smith* in the late 1700s, states that fossil organisms succeed one another in a definite and determinable order, and therefore any time period can be recognized by its fossil content. The use of *index fossils*, those that are wide-spread geographically and are limited to a short span of geologic time, provides an important method for matching rocks of the same age.

- Each atom has a nucleus containing *protons* (positively charged particles) and *neutrons* (neutral particles). Orbiting the nucleus are negatively charged *electrons*. The *atomic number* of an atom is the number of protons in the nucleus. The *mass number* is the number of protons plus the number of neutrons in an atom's nucleus. *Isotopes* are variants of the same atom, but with a different number of neutrons, and hence a different mass number.

- *Radioactivity* is the spontaneous breaking apart (decay) of certain unstable atomic nuclei. Three common forms of radioactive decay are (1) emission of *alpha particles* from the nucleus, (2) emission of *beta particles* from the nucleus, and (3) *capture of an electron* by the nucleus.

- An unstable *radioactive isotope*, called the *parent*, will decay and form *daughter products*. The length of time for one-half of the nuclei of a radioactive isotope to decay is called the *half-life* of the isotope. Using a procedure called *radiometric dating*, if the half-life of the isotope is known, and the parent/daughter ratio can be measured, the age of a sample can be calculated. An accurate radiometric date can only be obtained if the mineral containing the radioactive isotope remained in a closed system during the entire period since its formation.

- The *geologic time scale* divides Earth's history into units of varying magnitude. It is commonly presented in chart form, with the oldest time and event at the bottom and the youngest at the top. The principle subdivisions of the geologic time scale, called *eons*, include the *Hadean, Archean, Proterozoic* (together, these three eons are commonly referred to as the *Precambrian*), and, beginning about 570 million years ago, the *Phanerozoic*. The Phanerozoic (meaning "visible life") eon is divided into the following *eras: Paleozoic* ("ancient life"), *Mesozoic* ("middle life"), and *Cenozoic* ("recent life").

- One problem in assigning absolute dates is that *not all rocks can be radiometrically dated*. A sedimentary rock may contain particles of many ages that have been weathered from different rocks that formed at various times. One way geologists assign absolute dates to sedimentary rocks is to relate them to datable igneous masses, such as volcanic ash beds.

Review Questions

1. Distinguish between absolute and relative dating.

2. Describe two early methods for dating Earth. How old was Earth thought to be according to these estimates? List a weakness of each method. (See Box 8.1, p. 191.)

3. What is the law of superposition? How are cross-cutting relationships used in relative dating?

4. Refer to Figure 8.4 (p. 194) and answer the following questions:

    (a) Is fault A older or younger than the sandstone layer?

    (b) Is dike A older or younger than the sandstone layer?

    (c) Was the conglomerate deposited before or after fault A?

    (d) Was the conglomerate deposited before or after fault B?

    (e) Which fault is older, A or B?

    (f) Is dike A older or younger than the batholith?

5. When you observe an outcrop of steeply inclined sedimentary layers, what principle allows you to assume that the beds were tilted after they were deposited?

6. A mass of granite is in contact with a layer of sandstone. Using a principle described in this chapter, explain how you might determine whether the sandstone was desposited on top of the granite, or whether the granite was intruded from below after the sandstone was deposited?

7. Distinguish among angular unconformity, disconformity, and nonconformity.

8. What is meant by the term *correlation?*

9. Describe William Smith's important contribution to the science of geology.

10. Why are fossils such useful tools in correlation?

11. Figure 8.17 is a block diagram of a hypothetical area in the American Southwest. Place the lettered features in the proper sequence, from oldest to youngest. Identify an angular unconformity and a nonconformity.

12. If a radioactive isotope of thorium (atomic number 90, mass number 232) emits 6 alpha particles and 4 beta particles during the course of radioactive decay, what are the atomic number and mass number of the stable daughter product?

13. Why is radiometric dating the most reliable method of dating the geologic past?

14. A hypothetical radioactive isotope has a half-life of 10,000 years. If the ratio of radioactive parent to stable daughter product is 1:3, how old is the rock containing the radioactive material?

15. Why is potassium-40 used more frequently in radiometric dating than other isotopes?

16. Why is the ratio between potassium-40 and calcium-40 not used for radiometric dating?

17. In order to provide a reliable radiometric date, a mineral must remain a closed system from the time of its formation until the present. Why is this true?

18. What precautions are taken to ensure reliable radiometric dates?

19. To make calculations easier, let us round the age of Earth to five billion years.

    (a) What fraction of geologic time is represented by recorded history (assume 5000 years for the length of recorded history)?

    (b) The first abundant fossil evidence does not appear until the beginning of the Cambrian period (570 million years ago). What percent of geologic time is represented by abundant fossil evidence?

20. What subdivisions make up the geologic time scale?

21. Explain the lack of a detailed time scale for the vast span known as the Precambrian.

22. Briefly describe the difficulties in assigning absolute dates to layers of sedimentary rock.

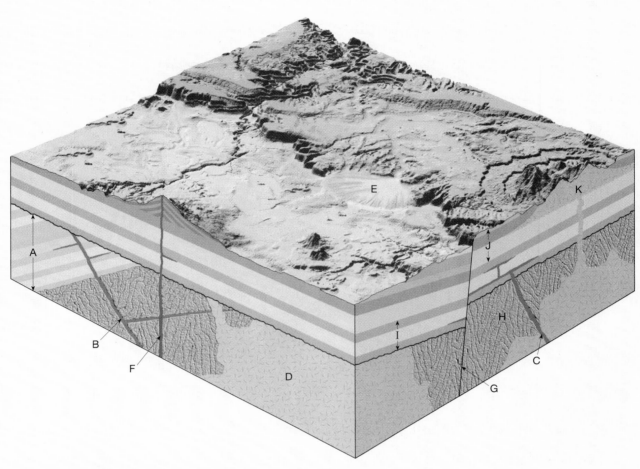

Figure 8.17 Use this block diagram in conjunction with Review Question 11.

Key Terms

absolute date (p. 190)
angular unconformity (p. 194)
Archean eon (p. 206)
Cenozoic era (p. 206)
conformable (p. 193)
correlation (p. 195)
cross-cutting relationships, principle of (p. 193)

disconformity (p. 194)
eon (p. 205)
epoch (p. 206)
era (p. 206)
fossil succession, principle of (p. 196)
geologic time scale (p. 204)
Hadean eon (p. 206)
half-life (p. 200)

inclusions (p. 193)
index fossil (p. 196)
Mesozoic era (p. 206)
nonconformity (p. 194)
original horizontality, principle of (p. 191)
Paleozoic era (p. 206)
period (p. 206)
Phanerozoic eon (p. 205)
Precambrian (p. 206)

Proterozoic eon (p. 206)
radioactivity (p. 198)
radiocarbon dating (p. 193)
radiometric dating (p. 200)
relative dating (p. 191)
superposition, law of (p. 191)
unconformity (p. 193)

Web Resources

The *Earth* Home Page provides on-line resources for this chapter on the World Wide Web. You will find review exercises, specific updates for items in the

chapter, suggested reading, and links to interesting related pathways on the Internet. Visit the *Earth* Home Page at **http://www.prenhall.com/tarbuck.**

CHAPTER 9
Mass Wasting

Left The landslide that damaged this house was triggered by the January 1994 Northridge, California, earthquake. — *Photo by Chromo Sohm/The Stock Market*

arth's surface is never perfectly flat but instead consists of slopes of many different varieties. Some are steep and precipitous; others are moderate or gentle. Some are long and gradual; others are short and abrupt. Slopes can be mantled with soil and covered by vegetation or consist of barren rock and rubble. Taken together, slopes are the most common elements in our physical landscape. Although most slopes may appear to be stable and unchanging, the force of gravity causes material to move downslope. At one extreme, the movement may be gradual and practically imperceptible. At the other extreme, it may consist of a thundering rockfall or avalanche.

We periodically hear news reports relating the terrifying and often grim details of landslides. On May 31, 1970, one such event occurred when a gigantic rock avalanche buried more than 20,000 people in Yungay and Ranrahirca, Peru (Figure 9.1). There was little warning of the impending disaster; it began and ended in just a matter of minutes. The avalanche started 14 kilometers from Yungay, near the summit of the 6700-meter (22,000-foot) Nevado Huascaran, the loftiest peak in the Peruvian Andes. Triggered by the ground motion from a strong offshore earthquake, a huge mass of rock and ice broke free from the precipitous north face of the mountain. After plunging nearly a kilometer, the material pulverized on impact and immediately began rushing down the mountainside, made fluid by trapped air and melted ice.

The falling debris ripped loose millions of tons of additional debris as it roared downhill. Hurricane-speed winds generated as compressed air escaped from beneath the avalanche mass created thunderlike noise and stripped nearby hillsides of vegetation. Although the material followed a previously eroded gorge, a portion of the debris jumped a 200–300-meter bedrock ridge that had protected Yungay from similar events in the past and buried the entire city. After inundating another town in its path, Ranrahirca, the mass of debris finally reached the bottom of the valley. There, its momentum carried it across the Rio Santa and tens of meters up the valley wall on the opposite side.

This was not the first such disaster in the region and will probably not be the last. Just 8 years earlier, a less spectacular, but nevertheless devastating, rock avalanche took the lives of an estimated 3500 people on the heavily populated valley floor at the base of the mountain. Fortunately, mass movements such as the one just described are infrequent and only occasionally affect large numbers of people.

Mass Wasting and Landform Development

Landslides are spectacular examples of a basic geologic process called **mass wasting.** Mass wasting refers to the downslope movement of rock, regolith, and soil under the direct influence of gravity. It is distinct from the erosional processes that are examined in subsequent chapters because mass wasting does not require a transporting medium.

In the evolution of most landforms, mass wasting is the step that follows weathering. By itself, weathering does not produce significant landforms. Rather, landforms develop as the products of weathering are removed from the places where they originate. Once weathering weakens and breaks rock apart, mass wasting transfers the debris downslope, where a stream, acting as a conveyor belt, usually carries it away. Although there may be many intermediate stops along the way, the sediment is eventually transported to its ultimate destination, the sea.

The combined effects of mass wasting and running water produce stream valleys, which are the most common and conspicuous of Earth's landforms. If streams alone were responsible for creating the valleys in which they flow, the valleys would be very narrow features. However, the fact that most river valleys are much wider than they are deep is a strong indication of

Figure 9.1 This Peruvian valley was devastated by a rock avalanche that was triggered by an off-shore earthquake in May 1970. **A.** Before. **B.** After the rock avalanche. (Photos courtesy of Iris Lozier)

A.

B.

the significance of mass wasting processes in supplying material to streams. This is illustrated by the Grand Canyon (Figure 9.2). The walls of the canyon extend far from the Colorado River owing to the transfer of weathered debris downslope to the river and its tributaries by mass wasting processes. In this manner, streams and mass wasting combine to modify and sculpture the surface. Of course, glaciers, groundwater, waves, and wind are also important agents in shaping landforms and developing landscapes.

Controls and Triggers of Mass Wasting

Gravity is the controlling force of mass wasting, but several factors play an important role in overcoming inertia and triggering downslope movements. Among these factors are saturation of material with water, oversteepening of slopes, removal of anchoring vegetation, and ground vibrations from earthquakes.

The Role of Water

When the pores in sediment become filled with water, the cohesion among particles is destroyed, allowing them to slide past one another with relative ease. For example, when sand is slightly moist, it sticks together quite well. However, if enough water is added to fill the openings between the grains, the sand will ooze out in all directions. Thus, saturation reduces the internal resistance of materials, which are then easily set in motion by the force of gravity. When clay is wetted, it becomes very slick—another example of the "lubricating" effect of water. Water also adds considerable weight to a mass of material. The added weight in itself may be enough to cause the material to slide or flow downslope.

Oversteepened Slopes

Oversteepening of slopes is another cause of many mass movements. There are many situations in nature where oversteepening takes place. A stream undercutting a valley wall and waves pounding against the base of a cliff are but two familiar examples. Furthermore, through their activities, people often create oversteepened and unstable slopes that become prime sites for mass wasting (see Box 9.1).

Unconsolidated, granular particles (sand-sized or coarser) assume a stable slope called the **angle of repose**. This is the steepest angle at which material remains stable. Depending on the size and shape of the particles, the angle varies from 25 to 40 degrees. The larger, more angular particles maintain the steepest

Figure 9.2 The walls of the Grand Canyon extend far from the channel of the Colorado River. This results primarily from the transfer of weathered debris downslope to the river and its tributaries by mass wasting processes. (Photo by Michael Collier)

Box 9.1

The Vaiont Dam Disaster

A massive rock avalanche in Peru is described at the beginning of this chapter. As with most occurrences of mass wasting, this tragic episode was triggered by a natural event—in this case, an earthquake. However, disasters also result from the mass movement of surface material triggered by the actions of humans.

In 1960, a large dam, almost 265 meters tall, was built across Vaiont Canyon in the Italian Alps. It was engineered without good geological input, and the result was a disaster only 3 years later.

The bedrock in Vaiont Canyon slanted steeply downward toward the lake impounded behind the dam. The bedrock was weak, highly fractured limestone strata with beds of clay and numerous solution cavities. As the reservoir filled behind the completed dam, rocks became saturated and the clays became swollen and more plastic. The rising water reduced the internal friction that had kept the rock in place.

Measurements made shortly after the reservoir was filled hinted at the problem, because they indicated that a portion of the mountain was slowly creeping downhill at the rate of 1 centimeter per week. In September 1963, the rate increased to 1 centimeter per day, then 10–20 centimeters per day, and eventually as much as 80 centimeters on the day of the disaster.

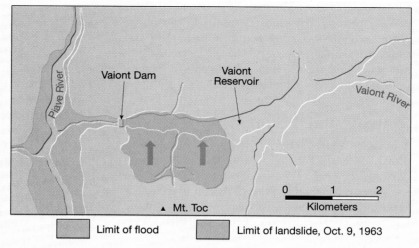

Figure 9.A Sketch map of the Vaiont River area showing the limits of the landslide, the portion of the reservoir that was filled with debris, and the extent of flooding downstream. (After G.A. Kiersh, "Vaiont Reservoir Disaster," *Civil Engineering* 34 (1964): 32–39.)

Finally, the mountainside let loose. In just an instant, 240 million cubic meters of rock and rubble slid down the mountainside and filled nearly 2 kilometers of the gorge to heights of 150 meters above the reservoir level (Figure 9.A). This pushed the water completely over the dam in a wave more than 90 meters high. More than 1.5 kilometers downstream, the wall of water was still 70 meters high, destroying everything in its path.

The entire event lasted less than 7 minutes, yet it claimed an estimated 2600 lives. Although this is known as the worst dam disaster in history, the Vaiont Dam itself remained intact. Although the catastrophe was triggered by human interference with the Vaiont River, the slide would have eventually occurred on its own; however, the effects would not have been nearly as tragic.

---

slopes. If the angle is increased, the rock debris will adjust by moving downslope.

Oversteepening is not just important because it triggers movements of unconsolidated granular materials. Oversteepening also produces unstable slopes and mass movements in cohesive soils, regolith, and bedrock. The response will not be immediate, as with loose, granular material, but sooner or later, one or more mass wasting processes will eliminate the oversteepening and restore stability to the slope.

Vegetation

Plants protect against erosion and contribute to the stability of slopes because their root systems bind soil and regolith together. Where plants are lacking, mass wasting is enhanced, especially if slopes are steep and water is plentiful. When anchoring vegetation is removed by forest fires or by people (for timber, farming, or development), surface materials frequently move downslope.

An unusual example occurred several decades ago on steep slopes near Menton, France. Farmers replaced olive trees, which have deep roots, with a more profitable, but shallow-rooted crop, carnations. When the less stable slope failed, the landslide took 11 lives.

Earthquakes as Triggers

Conditions that favor mass wasting may exist in an area for a long time without movement occurring. An additional factor is sometimes necessary to trigger the movement. Among the more important and dramatic triggers are earthquakes. An earthquake and its aftershocks can dislodge enormous volumes of rock

Figure 9.3 Various forms of mass wasting can be triggered by earthquakes. This home in Pacific Palisades, California, was destroyed by a landslide triggered by the January 1994 Northridge earthquake. In some cases, damages from earthquake-induced mass wasting are greater than damages caused directly by an earthquake's ground vibrations. (Photo by Chromo Sohm/The Stock Market)

Classification of Mass Wasting Processes

There is a broad array of different processes that geologists call mass wasting. Four processes are illustrated in Figure 9.4. Generally, the different types are classified based on the type of material involved, the kind of motion displayed, and the velocity of the movement.

Type of Material

The classificaiton of mass wasting processes on the basis of the material involved in the movement depends upon whether the descending mass began as unconsolidated material or as bedrock. If soil and regolith dominate, terms such as "debris," "mud," or "earth" are used in the description. In contrast, when a mass of bedrock breaks loose and moves downslope, the term "rock" may be part of the description.

Type of Motion

In addition to characterizing the type of material involved in a mass wasting event, the way in which the material moves may also be important. Generally, the kind of motion is described as either a fall, a slide, or a flow.

and unconsolidated material. The event in the Peruvian Andes described at the beginning of the chapter is one tragic example. In many areas that are jolted by earthquakes, it is not ground vibrations directly, but landslides and ground subsidence triggered by the vibrations that cause the greatest damage. Figure 9.3 illustrates this effect.

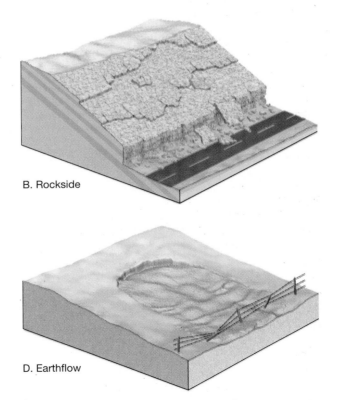

A. Slump

B. Rockside

C. Debris flow

D. Earthflow

Figure 9.4 The four processes illustrated here are all considered to be relatively rapid forms of mass wasting. Because material in slumps **A.** and rockslides **B.** move along well-defined surfaces, they are said to move by sliding. By contrast, when material moves downslope as a viscous fluid, the movement is described as a flow. Debris flow **C.** and earthflow **D.** advance downslope in this manner.

Fall. When the movement involves the free-fall of detached individual pieces of any size, it is termed a **fall**. Fall is a common form of movement on slopes that are so steep that loose material cannot remain on the surface. The rock may fall directly to the base of the slope or move in a series of leaps and bounds over other rocks along the way. Many falls result when freeze and thaw cycles and/or the action of plant roots loosen rock to the point that gravity takes over. Although signs along bedrock cuts on highways warn of falling rock few of us have actually witnessed such an event. However, as Figure 9.5 illustrates, they do indeed occur. In fact, this is the primary way in which *talus slopes* are built and maintained (Figure 9.6). Sometimes falls may trigger other forms of downslope movement. For example, recall that the Yungay disaster described at the beginning of the chapter was initiated by a mass of freefalling material that broke from the nearly vertical summit of Nevado Huascaran.

Slide. Many mass wasting processes are described as **slides**. Slides occur whenever material remains fairly coherent and moves along a well-defined surface. Sometimes the surface is a joint, a fault, or a bedding plane that is approximately parallel to the slope. However, in the case of the movement called slump, the descending material moves en masse along a curved surface of rupture.

A note of clarification is appropriate at this point. Sometimes the word *slide* is used as a synonym for the word *landslide*. It should be pointed out that

Figure 9.6 Talus is a slope built of angular rock fragments. Mechanical weathering, especially frost wedging, loosens the pieces of bedrock, which then fall to the base of the cliff. With time, a series of steep, cone-shaped accumulations build up at the base of the vertical slope. (Photo by Wolfgang Kaehler)

although many people, including geologists, use the term, the word *landslide* has no specific definition in geology. Rather, it should be considered as a popular nontechnical term used to describe all perceptible forms of mass wasting, including those in which sliding does not occur.

Flow. The third type of movement common to mass wasting processes is termed **flow**. Flow occurs when material moves downslope as a viscous fluid. Most flows are saturated with water and typically move as lobes or tongues.

Rate of Movement

The event described at the beginning of this chapter clearly involved rapid movement. The rock and debris moved downslope at speeds well in excess of

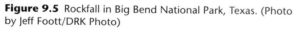

Figure 9.5 Rockfall in Big Bend National Park, Texas. (Photo by Jeff Foott/DRK Photo)

200 kilometers (125 miles) per hour. This most rapid type of mass movement is termed a **rock avalanche**. Many researchers believe that rock avalanches, such as the one that produced the scene in Figure 9.7, must literally "float on air" as they move downslope. That is, high velocities result when air becomes trapped and compressed beneath the falling mass of debris, allowing it to move as a buoyant, flexible sheet across the surface.

Most mass movements, however, do not move with the speed of a rock avalanche. In fact, a great deal of mass wasting is imperceptibly slow. One process that we will examine later, termed *creep*, results in particle movements that are usually measured in millimeters or centimeters per year. Thus, as you can see, rates of movement can be spectacularly sudden or exceptionally gradual. Although various types of mass wasting are often classified as either rapid or slow, such a distinction is highly subjective because a wide range of rates exists between the two extremes. Even the velocity of a single process at a particular site can vary considerably.

Slump

Slump refers to the downward sliding of a mass of rock or unconsolidated material moving as a unit along a curved surface (Figure 9.8). Usually the slumped material does not travel spectacularly fast nor very far. This is a common form of mass wasting, especially in thick accumulations of cohesive materials such as clay. The rupture surface is characteristically spoon-shaped and concave upward or outward. As the movement occurs, a crescent-shaped scarp is created at the head and the block's upper surface is sometimes tilted backwards. Although slump may involve a single mass, it often consists of multiple blocks. Sometimes water collects between the base of the scarp and the top of the tilted block. As this water percolates downward along the surface of rupture, it may promote further instability and additional movement.

Slump commonly occurs because a slope has been oversteepened. The material on the upper portion of a slope is held in place by the material at the bottom of the slope. As this anchoring material at the

Figure 9.7 This 4-kilometer long tongue of rubble was deposited atop Alaska's Sherman Glacier by a rock avalanche. The event was triggered by a tremendous earthquake in March 1964. (Photo by Austin Post, U.S. Geological Survey)

Rock Avalanche
Debris

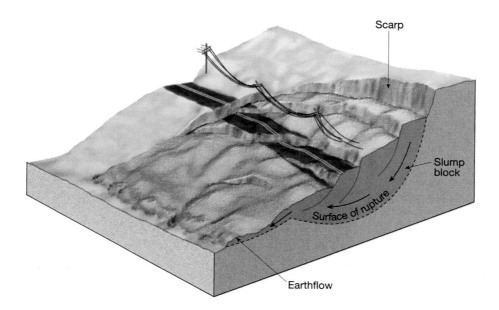

Scarp

Slump block

Surface of rupture

Earthflow

Figure 9.8 Slump occurs when material slips downslope en masse along a curved surface of rupture. Earthflows frequently form at the base of the slump.

base is removed, the material above is made unstable and reacts to the pull of gravity. One relatively common example is a valley wall that becomes oversteepened by a meandering river. Figure 9.9 provides another example in which a coastal cliff has been undercut by wave action at its base. Slumping may also occur when a slope is overloaded, causing internal stress on the material below. This type of slump often occurs where weak, clay-rich material underlies layers of stronger, more resistant rock such as sandstone. The seepage of water through the upper layers reduces the strength of the clay below and slope failure results.

Rockslide

Rockslides occur when blocks of bedrock break loose and slide down a slope (see Figure 9.4B, p. 217). If the material involved is largely unconsolidated, the term **debris slide** is used instead. Such events are among the fastest and most destructive mass movements. Usually rockslides take place in a geologic setting where the rock strata are inclined, or where joints and fractures exist parallel to the slope. When such a rock unit is undercut at the base of the slope, it loses support and the rock eventually gives way. Sometimes the rockslide

Figure 9.9 Slump at Point Fermin, California. Slump is often triggered when slopes become oversteepened by erosional processes such as wave action. (Photo by John S. Shelton)

Scarp

Scarp

Slump block

Figure 9.10 On August 17, 1959, an earthquake triggered a massive rockslide in the canyon of Montana's Madison River. An estimated 27 million cubic meters of debris slid down the canyon wall, forming a dam that created Earthquake Lake. (Photo by John Montagne)

is triggered when rain or melting snow lubricates the underlying surface to the point that friction is no longer sufficient to hold the rock unit in place. As a result, rockslides tend to be more common during the spring, when heavy rains and melting snow are most prevalent.

Earthquakes can trigger rockslides and other mass movements. The 1811 earthquake at New Madrid, Missouri, for example, caused slides in an area of more than 13,000 square kilometers (5000 square miles) along the Mississippi River valley. A more recent example occurred on August 17, 1959, when a severe earthquake west of Yellowstone National Park triggered a massive slide in the canyon of the Madison River in southwestern Montana (Figure 9.10). In a matter of moments an estimated 27 million cubic meters of rock, soil, and trees slid into the canyon. The debris dammed the river and buried a campground and highway. More than 20 unsuspecting campers perished.

Not far from the site of the Madison Canyon slide, the classic Gros Ventre rockslide occurred 34 years earlier. The Gros Ventre River flows west from the northernmost part of the Wind River Range in northwestern Wyoming, through Grand Teton National Park, and eventually empties into the Snake River. On June 23, 1925, a massive rockslide took place in its valley, just east of the small town of Kelly. In the span of just a few minutes a great mass of sandstone, shale, and soil crashed down the south side of the valley, carrying with it a dense pine forest. The volume of debris, estimated at 38 million cubic meters (50 million cubic yards), created a 70-meter-high dam on the Gros Ventre River. Because the river was completely blocked, a lake was formed. It filled so quickly that a house that

had been 18 meters (60 feet) above the river was floated off its foundation 18 hours after the slide. In 1927, the lake overflowed the dam, partially draining the lake and resulting in a devastating flood downstream.

Why did the Gros Ventre rockslide take place? Figure 9.11 shows a diagrammatic cross-sectional view of the geology of the valley. Notice the following points: (1) the sedimentary strata in this area dip (tilt) 15–21 degrees; (2) underlying the bed of sandstone is a relatively thin layer of clay; and (3) at the bottom of the valley the river had cut through much of the sandstone layer. During the spring of 1925, water from heavy rains and melting snow seeped through the sandstone, saturating the clay below. Because much of the sandstone layer had been cut through by the Gros Ventre River, the layer had virtually no support at the bottom of the slope. Eventually the sandstone could no longer hold its position on the wetted clay, and gravity pulled the mass down the side of the valley. The circumstances at this location were such that the event was inevitable.

Debris Flow

Debris flow is a relatively rapid type of mass wasting that involves a flow of soil and regolith containing a large amount of water (Figure 9.4C, p. 217). Debris flows, which are also called *mudflows*, are most characteristic of semiarid mountainous regions and are also common on the slopes of some volcanoes. Because of their fluid properties, debris flows frequently follow canyons and stream channels. In populated areas, debris flows can pose a significant hazard to life and property (see Box 9.2).

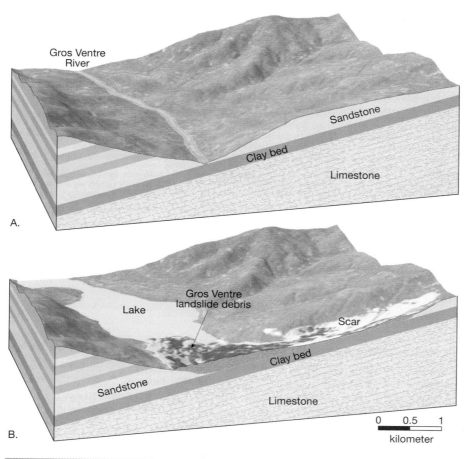

A.

B.

Figure 9.11 Parts A and B show a before and after cross-sectional view of the Gros Ventre rockslide. The slide occurred when the tilted and undercut sandstone bed could no longer maintain its position atop the saturated bed of clay. As the photo in part C illustrates, even though the Gros Ventre rockslide occurred in 1925, the scar left on the side of Sheep Mountain is still a prominent feature. (Parts A and B after W.C. Alden, "Landslide and Flood at Gros Ventre, Wyoming," Transactions (AIME) 76 (1928); 348. Part C photo by Stephen Trimble)

C.

Debris Flows in Semiarid Regions

When a cloudburst or rapidly melting mountain snows create a sudden flood in a semiarid region, large quantities of soil and regolith are washed into nearby stream channels because there is usually little vegetation to anchor the surface material. The end product is a flowing tongue of well-mixed mud, soil, rock, and water. Its consistency may range from that of wet concrete to a soupy mixture not thicker than muddy water. The rate of flow therefore depends not only on the slope but also on the water content. When dense, debris flows are capable of carrying or pushing large boulders, trees, and even houses with relative ease.

Box 9.2

Reducing Debris-Flow Hazards in the San Francisco Bay Region*

Debris flows are among the most common and dangerous types of mass wasting in the San Francisco Bay region. When prolonged, intense rains fall on steep hillsides, the saturated soils can become unstable and move rapidly downslope. Such flows are capable of destroying homes, washing out roads and bridges, knocking down trees, and obstructing streams and roadways with thick deposits of mud and rocks. An especially destructive event occurred in 1982 when thousands of debris flows caused nearly $70 million in damages and took 25 lives. Since then, several serious but less severe events have occurred. As more and more people build in the hills around the Bay region, the potential impact of debris flows on life and property is increasing.

The Debris-Flow Warning System

Debris flows can begin suddenly, often with little warning. Loss of lives during the intense 1982 storm prompted the U.S. Geological Survey (USGS) and the National Weather Service (NWS) to develop a debris-flow warning system for the San Francisco Bay area. Established in 1985, the system first issued warnings during severe storms in February 1986, successfully predicting the occurrence of debris flows in hilly parts of the Bay region. Today, the system is being refined with the aim of providing a prototype system for other parts of the nation.

During the rainy season (October through April), the USGS measures rainfall using more than 50 radio-telemetered rain gauges coordinated by the NWS, called the ALERT network (Figure 9.B). Early in each rainy season, these rainfall measurements, along with measurements of soil moisture from a study site in the hills south of San Francisco, are used to estimate the moisture level of soils throughout the Bay region. Soils must reach a sufficient moisture level each year before slopes become susceptible to debris flows during intense rainstorms. Once soils reach this moisture level, the USGS monitors weather forecasts and uses up-to-the-minute data from the ALERT network to determine the potential for imminent debris flows during each subsequent rainstorm. Warnings are then broadcast by the National Weather Service.

How Much Rain Is Needed to Trigger Debris Flows?

Once soils have reached sufficient moisture levels during a rainy season, it is the rainfall rate, rather than total rainfall amount, that is most important for determining whether debris flows will occur. For example, 4 inches of rain in 24 hours is generally not sufficient to trigger debris flows in the San Francisco Bay region. However, 4 inches of rain in 6 hours generally will trigger numerous debris flows.

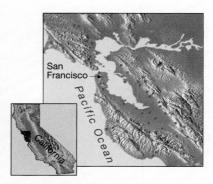

Figure 9.B San Francisco Bay region, showing hilly areas where debris flows are possible. Dots show locations of ALERT rain gauges. During storms, these gauges radio-transmit rainfall data to the U.S. Geological Survey and the National Weather Service. Scientists analyze the data to determine debris-flow danger. If danger is high, a Debris-Flow Watch or Warning is issued.

The USGS has developed thresholds that describe the minimum rainfall rates needed to trigger debris flows on natural slopes in the Bay area. On burned slopes that have lost their anchoring vegetation, and altered slopes, such as road cuts, greater caution is needed because debris flows can be triggered by less severe rainfall conditions.

*Based on material prepared by the U.S. Geological Survey

Debris flows pose a serious hazard to development in relatively dry mountainous areas such as southern California. The construction of homes on canyon hillsides and the removal of native vegetation by brush fires and other means have increased the frequency of these destructive events (Figure 9.12). Moreover, when a debris flow reaches the end of a steep, narrow canyon, it spreads out, covering the area beyond the mouth of the canyon with a mixture of wet debris. This material contributes to the buildup of fanlike deposits at canyon mouths.* The

fans are relatively easy to build on, often have nice views, and are close to the mountains; in fact, like the nearby canyons, many have become preferred sites for development. Because debris flows occur only sporadically, the public is often unaware of the potential hazard of such sites.

Lahars

Debris flows are also common on the slopes of some volcanoes, in which case they are termed **lahars**. The word originated in Indonesia, a volcanic region that has experienced many of these often destructive

---

*These structures are called *alluvial fans* and will be discussed in greater detail in Chapters 10 and 13.

Figure 9.12 On January 25, 1997, a mudflow literally buried this one-story home in Mill Valley, California. Heavy rains from a powerful Pacific storm triggered the event. (AP Photo/Justin Sullivan)

events. Lahars result when highly unstable layers of ash and debris become saturated with water and flow down steep volcanic slopes. These flows generally follow existing stream channels. Heavy rainfalls eroding volcanic deposits often trigger these flows. Others are initiated when large volumes of ice and snow are suddenly melted by heat flowing to the surface from within the volcano or by the hot gases and near-molten debris emitted during a violent eruption.

When Mount St. Helens erupted in May 1980, several lahars were created. The flows and accompanying floods raced down the valleys of the north and south forks of the Toutle River at speeds that were often in excess of 30 kilometers per hour. Fortunately, the affected area was not densely settled. Nevertheless, more than 200 homes were destroyed or severely damaged (Figure 9.13). Most bridges met a similar fate. According to the U.S. Geological Survey:

Even after traveling many tens of miles from the volcano and mixing with cold waters, the mudflows maintained temperatures in the range of about 84° to 91° C; they undoubtedly had higher temperatures closer to the eruption source.... Locally the mudflows surged up the valley walls as much as 360 feet and over hills as

Figure 9.13 A house damaged by a lahar along the Toutle River, west-northwest of Mount St. Helens. The end section of the house was torn free and lodged against trees. (Photo by D.R. Crandell, U.S. Geological Survey)

high as 250 feet. From the evidence left by the "bathtub-ring" mudlines, the larger mudflows at their peak averaged from 33 to 66 feet deep.*

Eventually the lahars in the Toutle River drainage area carried more than 50 million cubic meters of material to the lower Cowlitz and Columbia rivers. The deposits temporarily reduced the water-carrying capacity of the Cowlitz River by 85 percent, and the depth of the Columbia River navigational channel was decreased from 12 meters to less than 4 meters.

In November 1985, lahars were produced during the eruption of Nevado del Ruiz, a 5300-meter (17,400-foot) volcano in the Andes Mountains of Colombia. The eruption melted much of the snow and ice that capped the uppermost 600 meters of the peak, producing torrents of hot viscous mud, ash, and debris. The lahars moved outward from the volcano, following the valleys of three rain-swollen rivers that radiate from the peak. The flow that moved down the valley of the Lagunilla River was the most destructive. It devastated the town of Armero, 48 kilometers from the mountain. Most of the more than 25,000 deaths caused by the event occurred in this once-thriving agricultural community.

Death and property damage due to the lahars also occurred in 13 other villages within the 180-square-kilometer disaster area. Although a great deal of pyroclastic material was explosively ejected from Nevado del Ruiz, it was the lahars triggered by this eruption that made this such a devastating natural disaster. In fact, it was the worst volcanic disaster since 28,000 people died following the 1902 eruption of Mount Pelée on the Caribbean island of Martinique.[†]

Earthflow

We have seen that debris flows are frequently confined to channels in semiarid regions. In contrast, **earthflows** most often form on hillsides in humid areas during times of heavy precipitation or snowmelt (see Figure 9.4D, p. 217). When water saturates the soil and regolith on a hillside, the material may break away, leaving a scar on the slope and forming a tongue- or teardrop-shaped mass that flows downslope (Figure 9.14).

The materials most commonly involved are rich in clay and silt and contain only small proportions of sand and coarser particles. Earthflows range in size from bodies a few meters long, a few meters wide, and less than a meter deep to masses more than 1 kilometer

Figure 9.14 This small, tongue-shaped earthflow occurred on a newly formed slope along a recently constructed highway. It formed in clay-rich material following a period of heavy rain. Notice the small slump at the head of the earthflow. (Photo by E.J. Tarbuck)

long, several hundred meters wide, and more than 10 meters deep. Because earthflows are quite viscous, they generally move at slower rates than the more fluid debris flows described in the preceding section. They are characterized by a slow and persistent movement and may remain active for periods ranging from days to years. Depending on the steepness of the slope and the material's consistency, measured velocities range from less than a millimeter a day up to several meters a day. Over the time span that earthflows are active, movement is typically faster during wet periods than during drier times. In addition to occurring as isolated hillside phenomena, earthflows commonly take place in association with large slumps. In this situation, they may be seen as tonguelike flows at the base of the slump block (see Figure 9.8, p. 220).

A special type of earthflow, known as *liquefaction*, sometimes occurs in association with earthquakes. Porous, clay- to sand-sized sediments that are saturated with water are most vulnerable. When shaken suddenly, the grains lose cohesion and the ground flows. Liquefaction can cause buildings to sink or tip on their sides and underground storage tanks and sewer lines to float upward (see Figure 16.21, p. 418). To say the least, damage can be substantial.

Slow Movements

Movements such as rockslides, rock avalanches, and lahars are certainly the most spectacular and catastrophic forms of mass wasting. As these events have been known to kill thousands, they deserve intensive

---

*Robert I. Tilling, *Eruptions of Mount St. Helens: Past, Present, and Future*. Washington, DC: U.S. Government Printing Office, 1987.

[†]A discussion of the Mount Pelée eruption can be found in the section on composite cones in Chapter 4.

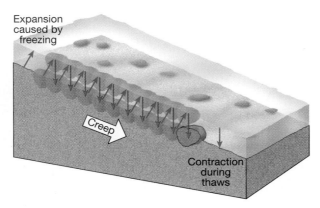

Figure 9.15 The repeated expansion and contraction of the surface material causes a net downslope migration of rock particles—a process called *creep*.

study so that, through more effective prediction, timely warnings and better controls can help save lives. However, because of their large size and spectacular nature, they give us a false impression of their importance as a mass wasting process. Indeed, sudden movements are responsible for moving less material than the slower and far more subtle action of creep. Whereas rapid types of mass wasting are characteristic of mountains and steep hillsides, creep can take place on gentle slopes and is thus much more widespread.

Creep

Creep is a type of mass wasting that involves the gradual downhill movement of soil and regolith. One of the primary causes of creep is the alternate expansion and contraction of surface material caused by freezing and thawing or wetting and drying. As shown in Figure 9.15, freezing or wetting lifts particles at right angles to the slope, and thawing or drying allows the particles to fall back to a slightly lower level. Each cycle therefore moves the material a short distance downhill. Creep may also be initiated if the ground becomes saturated with water. Following a heavy rain or snowmelt, a waterlogged soil may lose its internal cohesion, allowing gravity to pull the material downslope. Because creep is imperceptibly slow, the process cannot be observed in action. What can be observed, however, are the effects of creep. Creep causes fences and utility poles to tilt and retaining walls to be displaced (Figure 9.16).

Solifluction

Solifluction is a form of mass wasting that is common in regions underlain by **permafrost**. Permafrost refers to the permanently frozen ground that occurs in association with Earth's harsh tundra and ice cap climates (see Box 9.3). Solifluction may be regarded as a form of creep in which unconsolidated, water-saturated material gradually moves downslope. Solifluction occurs in a zone above the permafrost called the *active layer*, which thaws in summer and refreezes in winter. During the summer season, water is unable to percolate into the impervious permafrost layer below. As a result, the active layer becomes saturated and slowly flows. The process can occur on slopes as gentle as 2–3 degrees. Where there is a well-developed mat of vegetation, a solifluction sheet may move in a series of well-defined lobes or as a series of partially overriding folds (Figure 9.17).

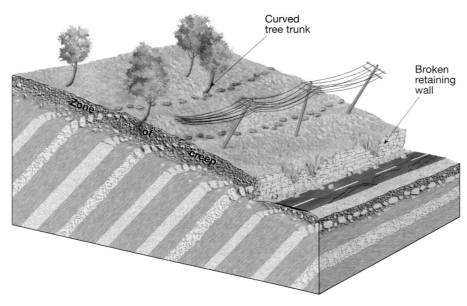

Figure 9.16 Although creep is an imperceptibly slow movement, its effects are often visible.

Figure 9.17 Solifluction lobes northeast of Fairbanks, Alaska. Solifluction occurs when the active layer thaws in summer. (Photo by James E. Patterson)

Chapter Summary

- *Mass wasting* refers to the downslope movement of rock, regolith, and soil under the direct influence of gravity. In the evolution of most landforms, mass wasting is the step that follows weathering. The combined effects of mass wasting and erosion by running water produce stream valleys.

- *Gravity is the controlling force of mass wasting.* Other factors that play an important role in overcoming inertia and triggering downslope movements are saturation of the material with water and oversteepening of slopes beyond the *angle of repose*.

- The various processes included under the name of mass wasting are divided and described on the basis of (1) the type of material involved (debris, mud, earth, or rock); (2) the type of motion (fall, slide, or flow); and (3) the rate of movement (rapid or slow).

- The more rapid forms of mass wasting include *slump*, the downward sliding of a mass of rock or unconsolidated material moving as a unit along a curved surface; *rockslide*, blocks of bedrock breaking loose and sliding downslope; *debris flow*, a relatively rapid flow of soil and regolith containing a large amount of water; and *earthflow*, an unconfined flow of saturated, clay-rich soil that most often occurs on a hillside in a humid area following heavy precipitation or snowmelt.

- The slowest forms of mass wasting include *creep*, the gradual downhill movement of soil and regolith, and *solifluction*, a form of mass wasting that is common in regions underlain by *permafrost* (permanently frozen ground associated with tundra and ice cap climates).

Review Questions

1. Describe how mass wasting processes contribute to the development of stream valleys.

2. How did the building of a dam contribute to the Vaiont Canyon disaster? Was the disaster avoidable? (See Box 9.1, p. 216).

3. What is the controlling force of mass wasting?

4. How does water affect mass wasting processes?

5. Describe the significance of the angle of repose.

6. Distinguish among fall, slide, and flow.

7. Why can rock avalanches move at such great speeds?

8. Both slump and rockslide move by sliding. In what ways do these processes differ?

9. What factors led to the massive rockslide at Gros Ventre, Wyoming?

10. Compare and contrast mudflow and earthflow.

11. Describe the mass wasting that occurred at Mount St. Helens during its active period in 1980 and at Nevado del Ruiz in 1985.

12. Because creep is an imperceptibly slow process, what evidence might indicate that this phenomenon is affecting a slope? Describe the mechanism that creates this slow movement.

13. Why is solifluction only a summertime process?

14. What is permafrost? What portion of Earth's land surface is affected? (See Box 9.3.)

Box 9.3

The Sensitive Permafrost Landscape

Many of the mass wasting disasters described in this chapter had sudden and disastrous impacts on people. When the activities of people cause ice contained in permanently frozen ground to melt, the impact is more gradual and less deadly. Nevertheless, because permafrost regions are sensitive and fragile landscapes, the scars resulting from poorly planned actions can remain for generations.

Permanently frozen ground, known as *permafrost*, occurs where summers are too cool to melt more than a shallow surface layer. Deeper ground remains frozen year-round. Strictly speaking, permafrost is defined only on the basis of temperature; that is, it is ground with temperatures that have remained below 0° C (32° F) continuously for 2 years or more. The degree to which ice is present in the ground strongly affects the behavior of the surface material. Knowing how much ice is present and where it is located is very important when it comes to constructing roads, buildings, and other projects in areas underlain by permafrost.

Permafrost underlies about 20 percent of Earth's land area. In addition to occurring in Antarctica and in some high mountain areas, permafrost is extensive in the lands surrounding the Arctic Ocean. It covers more than 80 percent of Alaska, about 50 percent of Canada, and a substantial portion of northern Siberia (Figure 9.C). Near the southern margins of the region, the permafrost consists of relatively thin, isolated masses. Farther north, the area and thickness gradually increase to the point where the permafrost is essentially continuous and its thickness may approach or even exceed 500 meters. In the discontinuous zone, land-use planning is

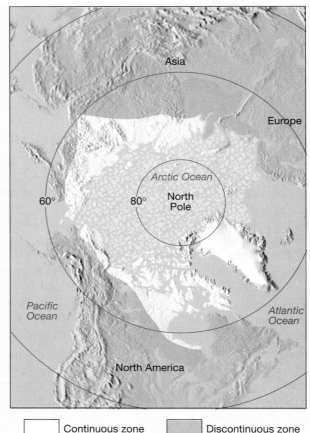

Figure 9.C
Distribution of permafrost in the Northern Hemisphere. More than 80 percent of Alaska and about 50 percent of Canada are underlain by permafrost. Two zones are recognized. In the continuous zone, the only ice-free areas are beneath deep lakes or rivers. In the higher-latitude portions of the discontinuous zone, there are only scattered islands of thawed ground. Moving southward, the percentage of unfrozen ground increases until all the ground is unfrozen. (After the U.S. Geological Survey)

☐ Continuous zone   ▦ Discontinuous zone

frequently more difficult than in the continuous zone farther north because the occurrences of permafrost are patchy and difficult to predict.

When people disturb the surface, such as by removing the insulating vegetation mat or by building roads and buildings, the delicate thermal balance is disturbed, and the permafrost can thaw (Figure 9.D). Thawing produces unstable ground that may slide, slump, subside, and undergo severe frost heaving.

As Figure 9.E illustrates, when a heated structure is built directly on permafrost that contains a high proportion of ice, thawing creates soggy material into which a building can sink. One solution is to place buildings and other structures on piles, like stilts. Such piles allow subfreezing air to circulate between the floor of the building and the soil and thereby keep the ground frozen.

When oil was discovered on Alaska's North Slope, many people were concerned about the building of a pipeline linking the oil fields at Prudhoe Bay to the ice-free port of Valdez 1300 kilometers to the south. There was serious concern that such a massive project might damage the sensitive permafrost environment. Many also worried about possible oil spills.

Because oil must be heated to about 60°C to flow properly, special engineering procedures had to be developed to isolate this heat from the permafrost. Methods included insulating the pipe, elevating portions of the pipeline above ground level, and even placing cooling devices in the ground to keep it frozen. The Alaska pipeline is clearly one of the most complex and costly projects ever built in the Arctic tundra. Detailed studies and careful engineering helped minimize adverse effects resulting from the disturbance of frozen ground.

Figure 9.D When a rail line was built across this permafrost landscape in Alaska, the ground subsided. (Photo by Lynn A. Yehle, U.S. Geological Survey)

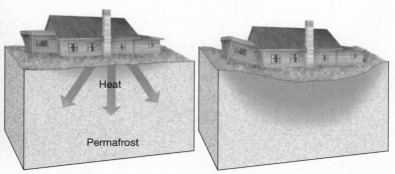

Figure 9.E This building, located south of Fairbanks, Alaska, subsided because of thawing permafrost. Notice that the right side, which was heated, settled much more than the unheated porch on the left.

Key Terms

angle of repose (p. 215)	earthflow (p. 225)	mass wasting (p. 214)	slide (p. 218)
creep (p. 226)	fall (p. 218)	permafrost (p. 226)	slump (p. 219)
debris flow (p. 221)	flow (p. 218)	rock avalanche (p. 219)	solifluction (p. 226)
debris slide (p. 220)	lahar (p. 223)	rockslide (p. 220)	

Web Resources

The *Earth* Home Page provides on-line resources for this chapter on the World Wide Web. You will find review exercises, specific updates for items in the chapter, suggested reading, and links to interesting related pathways on the Internet. Visit the *Earth* Home Page at **http://www.prenhall.com/tarbuck**.

CHAPTER 10

Running Water

Left: The Little Missouri River in Theodore Roosevelt National Park, North Dakota. — *Photo by Carr Clifton*

ivers are very important to people. We use them as highways for moving goods, as sources of water for irrigation, and as an energy source. Their fertile floodplains have been cultivated since the dawn of civilization. When viewed as part of the Earth system, rivers and streams represent a basic link in the constant cycling of the planet's water. Moreover, running water is the dominant agent of landscape alteration, eroding more terrain and transporting more sediment than any other process. Because so many people live near rivers, floods are among the most destructive of all geologic hazards. Despite huge investments in levees and dams, rivers cannot always be controlled (Figure 10.1).

Earth as a System: The Hydrologic Cycle

All the rivers run into the sea;
yet the sea is not full;
unto the place from whence the rivers come,
thither they return again. (Ecclesiastes 1:7)

As the perceptive writer of Ecclesiastes indicated, water is continually on the move, from the ocean to the land and back again in an endless cycle.

The water found in each of the reservoirs depicted in Figure 10.2 does not remain in these places indefinitely. Water can readily change from one state of matter (solid, liquid, or gas) to another at the temperatures and pressures that occur at Earth's surface. Therefore, water is constantly moving among the hydrosphere, the atmosphere, the solid Earth, and the biosphere. This unending circulation of Earth's water supply is called the **hydrologic cycle.** The cycle shows us many critical interrelationships among different parts of the Earth system.

The hydrologic cycle is a gigantic worldwide system powered by energy from the Sun in which the atmosphere provides the vital link between the oceans and continents (Figure 10.3). Water evaporates into the atmosphere from the ocean and to a much lesser extent from the continents. Winds transport this moisture-laden air, often great distances, until conditions cause the moisture to condense into clouds, and precipitation to fall. The precipitation that falls into the

Figure 10.1 A helicopter rescues a man stranded on the roof of his Olivehurst, California home after a levee along Feather River broke. Huge floods occurred in California's Central Valley in January, 1997. Prolonged heavy rains melted snow from a Sierra Nevada snowpack that was nearly double the average. Rain runoff combined with snowmelt created such high river levels that more than 30 breaks occurred in major levees and caused one of the biggest floods of the century for northern California. Floodwaters covered nearly 650 square kilometers (250 square miles) and destroyed or damaged 16,000 homes. Forty-three counties were declared disaster areas. Eight people died and damages were estimated at $1.6 billion. (Photo by John Trotter/*Sacramento Bee*)

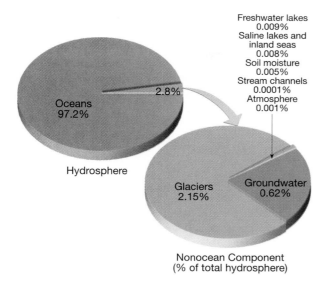

Figure 10.2 Distribution of Earth's water. The amount of water on Earth is immense: an estimated 1.36 billion cubic kilometers (326 million cubic miles). Of this total, the vast bulk—97.2 percent—is part of the world ocean. Ice sheets and glaciers account for another 2.15 percent, leaving only 0.65 percent to be divided among lakes, streams, subsurface water, and the atmosphere. Although the percentage of Earth's total water found in each of the latter sources is but a small fraction of the total inventory, the absolute quantities are great.

ocean has completed its cycle and is ready to begin another. The water that falls on land, however, must make its way back to the ocean.

What happens to precipitation once it has fallen on the land? A portion of the water soaks into the ground (called **infiltration**) moving downward, then laterally, and finally seeping into lakes, streams, or directly into the ocean. When the rate of rainfall is greater than the land's ability to absorb it, the additional water flows over the surface into lakes and streams, a process called **runoff**. Much of the water that infiltrates or runs off eventually finds its way back to the atmosphere because of evaporation from the soil, lakes, and streams. Also, some of the water that infiltrates the ground is absorbed by plants, which later release it into the atmosphere. This process is called **transpiration.**

Each year a field of crops may transpire an amount of water equivalent to a layer 60 centimeters deep over the entire field. The same area of trees may pump twice this amount into the atmosphere. Because we cannot clearly distinguish between the amount of water that is evaporated and the amount that is transpired by plants, the term **evapotranspiration** is often used for the combined effect.

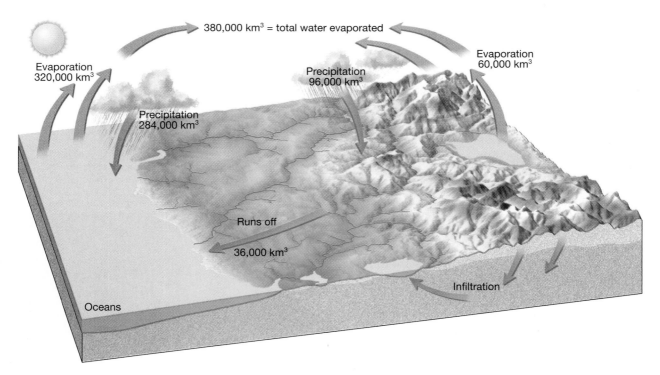

Figure 10.3 Earth's water balance. Each year, solar energy evaporates about 320,000 cubic kilometers of water from the oceans, while evaporation from the land (including lakes and streams) contributes 60,000 cubic kilometers of water. Of this total of 380,000 cubic kilometers of water, about 284,000 cubic kilometers fall back to the ocean, and the remaining 96,000 cubic kilometers fall on the land surface. Of that 96,000 cubic kilometers, only 60,000 cubic kilometers of water evaporate from the land, leaving 36,000 cubic kilometers of water to erode the land during the journey back to the oceans.

When precipitation falls in very cold areas—at high elevations or high latitudes—the water may not immediately soak in, run off, or evaporate. Instead it may become part of a snowfield or a glacier. In this way glaciers store large quantities of water on land. If present-day glaciers were to melt and release their stored water, sea level would rise by several tens of meters worldwide and submerge many heavily populated coastal areas. As we shall see in Chapter 12, over the past two million years, huge ice sheets have formed and melted on several occasions, each time changing the balance of the hydrologic cycle.

Figure 10.3 also shows Earth's overall *water balance*, or the volume of water that passes through each part of the cycle annually. The amount of water vapor in the air is just a tiny fraction of Earth's total water supply. But the absolute quantities that are cycled through the atmosphere over a one-year period are immense—some 380,000 cubic kilometers.

It is important to know that the hydrologic cycle is balanced. Because the total water vapor in the atmosphere remains about the same, average annual precipitation over Earth must be equal to the quantity of water evaporated. However, for all of the continents taken together, precipitation exceeds evaporation. Conversely, over the oceans, evaporation exceeds precipitation. As the level of the world ocean is not dropping, the system must be in balance.

The erosional work accomplished by the 36,000 cubic kilometers of water that flows annually from the land to the ocean is enormous. Arthur Bloom effectively described it as follows:

> The average continental height is 823 meters above sea level....If we assume that the 36,000 cubic kilometers of annual runoff flow downhill an average of 823 meters, the potential mechanical power of the system can be calculated. Potentially, the runoff from all lands would continuously generate almost 9×10^9 kW. If all this power were used to erode the land, it would be comparable to having...one horse-drawn scraper or scoop at work on each 3-acre piece of land, day and night, year round. Of course, a large part of the potential energy of runoff is wasted as frictional heat by turbulent flow and splashing of water.*

Although only a small percentage of the energy of running water is used to erode the surface, running water nevertheless is *the single most important agent sculpturing Earth's land surface.*

Geomorphology: A Systematic Analysis of Late Cenozoic Landforms (Englewood Cliffs, N.J.: Prentice Hall, 1978), p. 97.

To summarize, the hydrologic cycle represents the continuous movement of water from the oceans to the atmosphere, from the atmosphere to the land, and from the land back to the sea. The wearing down of Earth's land surface is largely attributable to the last of these steps and is the primary focus of the remainder of this chapter.

Running Water

Although we have always depended to a great extent on running water, its source eluded us for centuries. Not until the sixteenth century did we realize that streams were supplied by surface runoff and underground water, which ultimately had their sources as rain and snow.

Runoff initially flows in broad, thin sheets across the ground, appropriately termed **sheet flow**. The amount of water that runs off in this manner rather than sinking into the ground depends upon the **infiltration capacity** of the soil. Infiltration capacity is controlled by many factors, including: (1) the intensity and duration of the rainfall, (2) the prior wetted condition of the soil, (3) the soil texture, (4) the slope of the land, and (5) the nature of the vegetative cover. When the soil becomes saturated, sheet flow, commences as a layer only a few millimeters thick. After flowing as a thin, unconfined sheet for only a sort distance, threads of current typically develop and tiny channels called **rills** begin to form and carry the water to a stream.

To some, the term *stream* implies relative size. That is to say, streams are thought of as being larger than creeks or brooks but smaller than rivers. In geology, however, this is not the case. Here the word **stream** is used to denote channelized flow of any size,

Figure 10.4 Most streamflow is turbulent, although it is usually not as rough as that experienced by these rafters on the Colorado River. (Photo by Tom Bean)

from the smallest trickle to the mightiest river. It should be pointed out, however, that although the terms *river* and *stream* are used interchangeably, the term *river* is often preferred when describing a main stream into which several tributaries flow.

The remainder of this chapter will concentrate on that part of the hydrologic cycle in which the water moves in stream channels. The discussion will deal primarily with streams in humid regions. Streams are also important in arid landscapes, but we will examine that in Chapter 13, "Deserts and Winds."

 Streamflow

Water may flow in one of two ways, either as **laminar flow** or **turbulent flow**. When the movement is laminar, the water particles flow in straight-line paths that are parallel to the channel. The water particles move steadily downstream without mixing. By contrast, when the flow is turbulent, the water moves in a confused and erratic fashion that is often characterized by swirling, whirlpool-like eddies (Figure 10.4).

The stream's velocity is a primary factor that determines whether the flow is laminar or turbulent. Laminar flow is possible only when water is moving very slowly through a smooth channel. If the velocity increases or the channel becomes rough, laminar flow changes to turbulent flow. The movement of water in streams is usually fast enough that flow is turbulent. The multidirectional movement of turbulent flow is very effective both in eroding a stream's channel and in keeping sediment suspended within the water so that it can be transported downstream.

Flowing water makes its way to the sea under the influence of gravity. Some sluggish streams flow at less than 1 kilometer per hour, whereas a few rapid ones may exceed 30 kilometers per hour. Velocities are determined at gauging stations where measurements are taken at several locations across the stream channel and then averaged. This is done because the rate of water movement is not uniform within a stream channel. When the channel is straight, the highest velocities occur in the center just below the surface (Figure 10.5). It is here that friction is least. Minimum velocities occur along the sides and bottom (bed) of the channel where friction is always greatest. When a stream channel is crooked or curved, the fastest flow is not in the center. Rather, the zone of maximum velocity shifts toward the outside of each bend. As we shall see later, this shift plays an important part in eroding the stream's channel on that side.

The ability of a stream to erode and transport material is directly related to its velocity. Even slight variations in velocity can lead to significant changes in the load of sediment that water can transport. Several

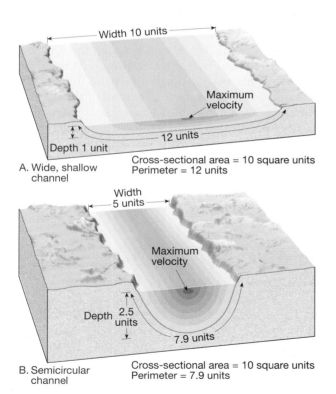

A. Wide, shallow channel
Cross-sectional area = 10 square units
Perimeter = 12 units

B. Semicircular channel
Cross-sectional area = 10 square units
Perimeter = 7.9 units

C. Gauging station

Figure 10.5 Influence of channel shape on velocity. **A.** The stream in this wide, shallow channel moves more slowly than does water in the semicircular channel because of greater frictional drag. **B.** The cross-sectional area of this semicircular channel is the same as the one in part A, but it has less water in contact with its channel and therefore less frictional drag. Thus, water will flow more rapidly in channel B, all other factors being equal. **C.** Continuous records of stage and discharge are collected by the U.S. Geological Survey at more than 7000 gauging stations in the United States. Average velocities are determined by using measurements from several spots across the stream. This station is on the Rio Grande south of Taos, New Mexico. (Photo by E. J. Tarbuck)

factors determine the velocity of a stream and therefore control the amount of erosional work a stream may accomplish. These factors include (1) the gradient, (2) the shape, size, and roughness of the channel, and (3) the discharge.

Gradient and Channel Characteristics

Certainly one of the most obvious factors controlling stream velocity is the **gradient**, or slope, of a stream channel. Gradient is typically expressed as the vertical drop of a stream over a fixed distance. Gradients vary considerably from one stream to another as well as along the course of a given stream.

Portions of the lower Mississippi River, for example, have gradients of 10 centimeters per kilometer and less. By way of contrast, some steep mountain stream channels decrease in elevation at a rate of more than 40 meters per kilometer, 400 times more abruptly than the lower Mississippi. The steeper the gradient, the more energy available for streamflow. If two streams were identical in every respect except gradient, the stream with the higher gradient would obviously have the greater velocity.

The *cross-sectional shape* of a channel determines the amount of water in contact with the channel and hence affects the frictional drag. The most efficient channel is one with the least perimeter for its cross-sectional area. Figure 10.5 compares two channel shapes. Although the cross-sectional area of both is identical, the semicircular shape has less water in contact with the channel and therefore less frictional drag. As a result, if all other factors are equal, the water will flow more rapidly in the semicircular channel.

The size and roughness of the channel also affect the amount of friction. An increase in the size of a channel reduces the ratio of perimeter to cross-sectional area

and therefore increases the efficiency of flow. The effect of roughness is obvious. A smooth channel promotes a more uniform flow, whereas an irregular channel filled with boulders creates enough turbulence to significantly retard the stream's forward motion.

Discharge

The **discharge** of a stream is the amount of water flowing past a certain point in a given unit of time. This is usually measured in cubic meters per second or cubic feet per second. Discharge is determined by multiplying a stream's cross-sectional area by its velocity:

discharge (m³/second) =

channel width (meters) × **channel depth (meters)**

× **velocity (meters/second)**

Table 10.1 lists the world's largest rivers in terms of discharge. The largest river in North America, the Mississippi, discharges an average 17,300 cubic meters per second. Although this is a huge quantity of water, it is nevertheless dwarfed by the mighty Amazon, the world's largest river. Draining an area that is nearly three-quarters the size of the conterminous United States and that averages about 200 centimeters of rain per year, the Amazon discharges 12 times more water than the Mississippi. In fact, it has been estimated that the flow of the Amazon accounts for about 15 percent of all the fresh water discharged into the ocean by all of the world's rivers. Just one day's discharge would supply the water needs of New York City for 9 years!

The discharges of most rivers are far from constant (see Box 10.1). This is true because of such variables as rainfall and snowmelt. When discharge changes, then the factors noted earlier must also

Table 10.1 World's Largest Rivers Ranked by Discharge

Rank	River	Country	Drainage Area		Average Discharge	
			Square kilometers	Square miles	Cubic meters per second	Cubic feet per second
1	Amazon	Brazil	5,778,000	2,231,000	212,400	7,500,000
2	Congo	Zaire	4,014,500	1,550,000	39,650	1,400,000
3	Yangtze	China	1,942,500	750,000	21,800	770,000
4	Brahmaputra	Bangladesh	935,000	361,000	19,800	700,000
5	Ganges	India	1,059,300	409,000	18,700	660,000
6	Yenisei	Russia	2,590,000	1,000,000	17,400	614,000
7	Mississippi	United States	3,222,000	1,244,000	17,300	611,000
8	Orinoco	Venezuela	880,600	340,000	17,000	600,000
9	Lena	Russia	2,424,000	936,000	15,500	547,000
10	Parana	Argentina	2,305,000	890,000	14,900	526,000

Box 10.1

The Effect of Urbanization on Discharge

When rains occur, stream discharge increases. If the rains are sufficiently heavy, the ability of the channel to contain the discharge is exceeded, and water spills over the banks as a flood. Floods are natural events that should be expected. However, when cities are built, the magnitude and frequency of flooding increases.

The top portion of Figure 10.A is a hypothetical hydrograph that shows the time relationship between a rainstorm and the occurrence of flooding. Notice that the water level in the stream does not rise at the onset of precipitation because time is needed for water to move from the place where it fell to the stream. This time span is called the *lag time*.

When an area changes from being predominantly rural to largely urban, streamflow is affected. The effect of urbanization on streamflow is illustrated by the bottom hydrograph in Figure 10.A. Notice that after urbanization the peak discharge during a flood is greater, and that the lag time between precipitation and flood peak is shorter than before urbanization. The explanation for this effect is relatively simple. The construction of streets, parking lots, and buildings covers over the ground that once soaked up water. Thus, less water infiltrates the ground, and the rate and amount of runoff increase.

Further, because much less water soaks into the ground, the low-water (dry-season) flow in urban streams, which is maintained by the seepage of groundwater into the channel, is greatly reduced. As one might expect, the magnitude of these effects is a function of the percentage of land that is covered by impermeable surfaces.

Urbanization is just one example of human interference with streams. There are many other ways that land use inadvertently influences the flow of streams and the work they carry out. Moreover, there are also many ways by which people intentionally attempt to manipulate and control streams. Some of these are discussed at appropriate points in this chapter.

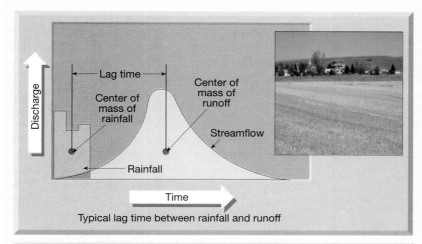

Typical lag time between rainfall and runoff

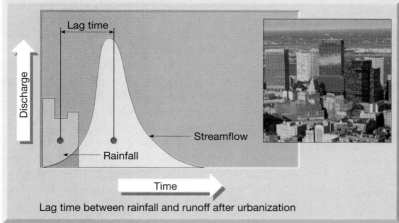

Lag time between rainfall and runoff after urbanization

Figure 10.A When an area changes from rural to urban, the lag time between rainfall and flood peak is shortened. The flood peak is also higher following urbanization. (After L. B. Leopold, U.S. Geological Survey)

change. When discharge increases, the width or depth of the channel must increase or the water must flow faster, or some combination of these factors must change. Indeed, measurements show that when the amount of water in a stream increases, the width, depth, and velocity all increase in an orderly fashion (Figure 10.6). To handle the additional water, the stream will increase the size of its channel by widening and deepening it. As we saw earlier, when the size of the channel increases, proportionally less of the water is in contact with the bed and banks of the channel. This means that friction, which acts to retard the flow, is reduced. The less friction, the more swiftly the water will flow.

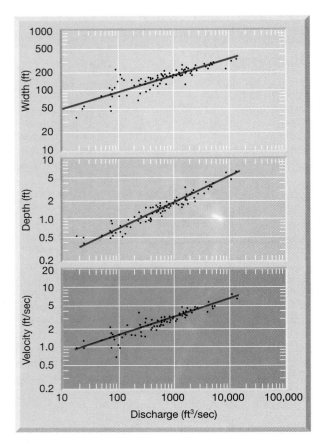

Figure 10.6 Relationship of width, depth, and velocity to discharge of the Powder River at Locate, Montana. As discharge increases, width, depth, and velocity all increase in an orderly fashion. (From L.B. Leopold and Thomas Maddock, Jr., U.S. Geological Survey Professional Paper 252, 1953)

Changes Downstream

One useful way of studying a stream is to examine its **longitudinal profile**. Such a profile is simply a cross-sectional view of a stream from its source area (called the **head** or **headwaters**) to its **mouth**, the point downstream where the river empties into another water body. By examining Figure 10.7, you can see that the most obvious feature of a typical longitudinal profile is a constantly decreasing gradient from the head to the mouth. Although many local irregularities may exist, the overall profile is a smooth concave-upward curve.

The longitudinal profile shows that the gradient decreases downstream. To see how other factors change in a downstream direction, observations and measurements must be made. When data are collected from successive gauging stations along a river, they show that discharge increases toward the mouth. This should come as no surprise because, as we move downstream, more and more tributaries contribute water to the main channel. In the case of the Amazon, for example, about 1000 tributaries join the main river along its 6500-kilometer course across South America.

Furthermore, in most humid regions, additional water is continually being added from the groundwater supply. Because this is the case, the width, depth, and velocity all must change in response to the increased volume of water carried by the stream. Indeed, the downstream changes in these variables have been shown to vary in a manner similar to what occurs when discharge increases at one place; that is, width, depth, and velocity all increase systematically.

The observed increase in average velocity that occurs downstream contradicts our intuitive impressions concerning wild, turbulent, mountain streams and wide, placid, rivers in the lowlands. A mental picture that we may have of "old man river just rollin' along" is really correct. The mountain stream has much higher instantaneous, turbulent velocities, but the water moves vertically, laterally, and actually upstream in some cases. Thus, the average flow velocity may be lower than in a wide, placid river just "rollin' along" very efficiently with far less turbulence.

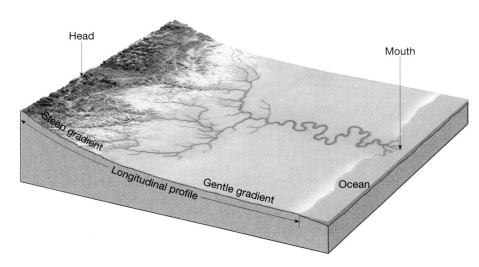

Figure 10.7 A longitudinal profile is a cross-section along the length of a stream. Note the concave-upward curve of the profile, with a steeper gradient upstream and a gentler gradient downstream.

In the headwaters region where the gradient is steepest, the water must flow in a relatively small and often boulder-strewn channel. The small channel and rough bed create great drag and inhibit movement by sending water in all directions with almost as much backward motion as forward motion. However, as one progresses downstream, the material on the bed of the stream becomes much smaller, offering less resistance to flow, and the width and depth of the channel increase to accommodate the greater discharge. These factors, especially the wider and deeper channel, permit the water to flow more freely and hence more rapidly.

In summary, we have seen that there is an inverse relationship between gradient and discharge. Where the gradient is high, the discharge is small, and where the discharge is great, the gradient is small. Stated another way, a stream can maintain a higher velocity near its mouth even though it has a lower gradient than upstream because of the greater discharge, larger channel, and smoother bed.

Base Level and Graded Streams

In 1875, John Wesley Powell, the pioneering geologist who first explored the Grand Canyon and later headed the U.S. Geological Survey, introduced the concept that there is a downward limit to stream erosion, which he called **base level**. Although the idea is relatively straightforward, it is nevertheless a key concept in the study of stream activity. Base level is defined as the lowest elevation to which a stream can erode its channel. Essentially this is the level at which the mouth of a stream enters the ocean, a lake, or another stream. Base level accounts for the fact that most stream profiles have low gradients near their mouths, because the streams are approaching the elevation below which they cannot erode their beds. Powell recognized that two types of base level exist:

> We may consider the level of the sea to be a grand base level, below which the dry lands cannot be eroded; but we may also have, for local and temporary purposes, other base levels of erosion.*

Sea level, which Powell called "grand base level," is now referred to as **ultimate base level. Local** or **temporary base levels** include lakes, resistant layers of rock, and main streams which act as base levels for their tributaries. All have the capacity to limit a stream at a certain level.

For example, when a stream enters a lake, its velocity quickly approaches zero and its ability to erode ceases. Thus, the lake prevents the stream from eroding below its level at any point upstream from the lake. However, because the outlet of the lake can cut downward and drain the lake, the lake is only a temporary hindrance to the stream's ability to downcut its channel. In a similar manner, the layer of resistant rock at the lip of the waterfall in Figure 10.8

Exploration of the Colorado River of the West (Washington, D.C.: Smithsonian Institution, 1875), p. 203.

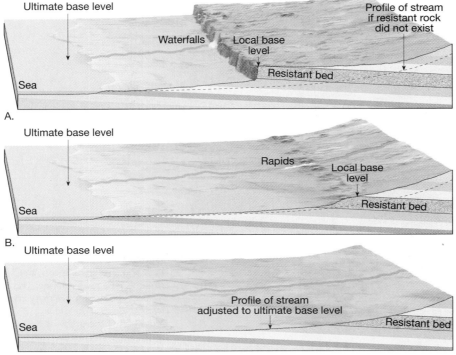

A.

B.

C.

Figure 10.8 A resistant layer of rock can act as a local (temporary) base level. Because the durable layer is eroded more slowly, it limits the amount of downcutting upstream.

acts as temporary base level. Until the ledge of hard rock is eliminated, it will limit the amount of downcutting upstream.

Any change in base level will cause a corresponding readjustment of stream activities. When a dam is built along a stream course, the reservoir that forms behind it raises the base level of the stream (Figure 10.9). Upstream from the reservoir the stream gradient is reduced, lowering its velocity and, hence, its sediment-transporting ability. The stream, now unable to transport all of its load, will deposit material, thereby building up its channel. This process continues until the stream again has a gradient sufficient to carry its load. The profile of the new channel would be similar to the old, except that it would be somewhat higher.

If, on the other hand, the base level should be lowered, either by uplifting of the land or by a drop in sea level, the stream would again readjust. The stream, now above base level, would have excess energy and

downcut its channel to establish a balance with its new base level. Erosion would first progress near the mouth, then work upstream until the stream profile was adjusted along its full length.

The observation that streams adjust their profiles to changes in base level led to the concept of a graded stream. A **graded stream** has the correct slope and other channel characteristics necessary to maintain just the velocity required to transport the material supplied to it. On the average, a graded system is neither eroding nor depositing material but is simply transporting it. Once a stream has reached this state of equilibrium, it becomes a self-regulating system in which a change in one characteristic causes an adjustment in the others to counteract the effect. Referring again to our example of a stream adjusting to a lowering of its base level, the stream would not be graded while it was cutting its new channel but would achieve this state after downcutting had ceased.

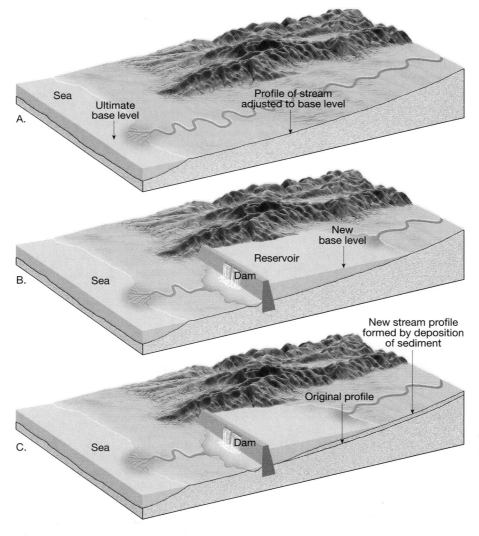

Figure 10.9 When a dam is built and a reservoir forms, the stream's base level is raised. This reduces the stream's velocity and leads to deposition and a reduction of the gradient upstream from the reservoir.

manner, the force of running water swiftly erodes poorly consolidated materials on the bed and banks of the stream. The stronger the current, the more effectively the stream will lift particles. In some instances, water is forced into cracks and bedding planes with sufficient strength to actually pry up pieces of rock from the bed of the channel.

Observing a muddy stream demonstrates that currents of water can lift and carry debris. However, it is not as obvious that a stream is capable of eroding solid rock in a manner similar to sandpapering. Just as the particles of grit on sandpaper can wear down a piece of wood, so too can the sand and gravel carried by a stream abrade a bedrock channel. Many steep-sided gorges cut through solid rock by the ceaseless bombardment of particles against the bed and banks of a channel serve as testimony to this erosional strength (Figure 10.10). In addition, the individual sediment grains are also abraded by their many impacts with the channel and with one another. Thus, by scraping, rubbing, and bumping, abrasion erodes a bedrock channel and simultaneously smoothes and rounds the abrading particles.

Common features on some river beds are rounded depressions known as **potholes** (Figure 10.11). They are created by the abrasive action of particles swirling in fast-moving eddies. The rotational motion of the sand and pebbles acts like a drill to bore the holes.

Figure 10.11 Potholes in the bed of a small stream in Cataract Falls State Park, Indiana. The rotational motion of swirling pebbles acts like a drill to create potholes. (Photo by Tom Till)

Figure 10.10 This steep-sided gorge in Paria Canyon Wilderness Area, Arizona–Utah, has been cut through solid rock by stream erosion. Canyons such as this are narrow due to rapid downcutting by the stream and very slow weathering of the canyon walls. (Photo by Tom Bean)

 ## Stream Erosion

Streams erode their channels by lifting loosely consolidated particles, by abrasion, and by dissolution. The last of these is by far the least significant. Although some erosion results from the dissolution of soluble bedrock and channel debris, most dissolved material in a stream is contributed by inflows of groundwater.

As we learned earlier, when the flow of water is turbulent, the water whirls and eddies. When an eddy is sufficiently strong, it can dislodge particles from the channel and lift them into the moving water. In this

As the particles wear down to nothing, they are replaced by new ones that continue to drill the stream bed. Eventually, smooth depressions several meters across and just as deep may result.

Transport of Sediment by Streams

Streams are Earth's most important erosional agent. Not only do they have the ability to downcut their channels, but streams also have the capacity to transport the enormous quantities of sediment produced by weathering. Although erosion by running water in a stream's channel contributes significant amounts of material for transport, by far the greatest quantity of sediment carried by a stream is derived from the products of weathering. Weathering produces tremendous amounts of material that are delivered to the stream by sheet flow, mass wasting, and groundwater.

Streams transport their loads of sediment in three ways: (1) in solution (**dissolved load**), (2) in suspension (**suspended load**), and (3) along the bottom of the channel (**bed load**). Let us now look at each of these.

Dissolved Load

The greatest portion of the dissolved load transported by most streams is supplied by groundwater. As water percolates through the ground, it first acquires soluble soil compounds. As the water seeps deeper through cracks and pores in the bedrock below, it may dissolve additional mineral matter. Eventually, much of this mineral-rich water finds its way into streams.

The velocity of stream flow has essentially no effect on a stream's ability to carry its dissolved load. Once material is in solution, it goes wherever the stream goes, regardless of velocity. Precipitation occurs only when the chemistry of the water changes.

The quantity of material carried in solution is highly variable and depends on such factors as climate and the geological setting. Usually the dissolved load is expressed as parts of dissolved material per million parts of water (parts per million, or ppm). Although some rivers may have a dissolved load of 1000 ppm or more, the average figure for the world's rivers is estimated at between 115 and 120 ppm. Almost 4 billion metric tons of dissolved mineral matter are supplied to the oceans each year by streams.

Suspended Load

Most streams (but not all) carry the largest part of their load in *suspension*. Indeed, the visible cloud of sediment suspended in the water is the most obvious portion of a stream's load. Usually only fine sand-, silt-, and clay-sized particles can be carried this way, but during flood stage, larger particles are carried as well. Also during flood stage, the total quantity of material carried in suspension increases dramatically, as can be verified by people whose homes have become sites for the deposition of this material. During flood stage, the Hwang Ho (Yellow River) of China is reported to carry an amount of sediment equal in weight to the water that carries it. Rivers like this are appropriately described as "too thick to drink but too thin to cultivate."

The type and amount of material carried in suspension are controlled by two factors: the velocity of the water and the settling velocity of each sediment grain. **Settling velocity** is defined as the speed at which a particle falls through a still fluid. The larger the particle, the more rapidly it settles toward the stream bed. In addition to size, the shape and specific gravity of particles also influence settling velocity. Flat grains sink through water more slowly than do spherical grains, and dense particles fall toward the bottom more rapidly than do less dense particles. The slower the settling velocity and the stronger the turbulence, the longer a sediment particle will stay in suspension and the farther it will be carried downstream with the flowing water.

Bed Load

A portion of a stream's load of solid material consists of sediment that is too large to be carried in suspension. These coarser particles move along the bottom of the stream and constitute the *bed load* (Figure 10.12). In terms of the erosional work accomplished by a downcutting stream, the grinding action of the bed load is of great importance.

The particles that make up the bed load move along the bottom by rolling, sliding, and saltation. Sediment moving by **saltation** appears to jump or skip along the stream bed. This occurs as particles are propelled upward by collisions or lifted by the current and then carried downstream a short distance until gravity pulls them back to the bed of the stream. Particles that are too large or heavy to move by saltation either roll or slide along the bottom, depending on their shapes.

Unlike the suspended and dissolved loads, which are constantly in motion, the bed load is in motion only intermittently, when the force of the water is sufficient to move the larger particles. The bed load usually does not exceed 10 percent of a stream's total load, although it may constitute up to 50 percent of the total load of a few streams. For example, consider the distribution of the 750 million tons of material carried to the Gulf of Mexico by the Mississippi River each year. Of this total, it is estimated that approximately 67 percent is carried in suspension, 26 percent in solution, and the remaining

Figure 10.12 Although the bed load of many rivers consists of sand, the bed load of this stream is made up of boulders and is easily seen during periods of low water. During floods, the seemingly immovable rocks in this channel are rolled along the bed of the stream. The maximum-size particle a stream can move is determined by the velocity of the water. (Photo by E.J. Tarbuck)

7 percent as bed load. Estimates of a stream's bed load should be viewed cautiously, however, because this fraction of the load is very difficult to measure accurately. Not only is the bed load more inaccessible than the suspended and dissolved loads, but it moves primarily during periods of flooding when the bottom of a stream channel is most difficult to study.

Capacity and Competence

A stream's ability to carry solid particles is typically described using two criteria. First, the maximum load of solid particles that a stream can transport is termed its **capacity**. The greater the amount of water flowing in a stream (discharge), the greater the stream's capacity for hauling sediment. Second, the **competence** of a stream indicates the maximum particle size that a stream can transport. The stream's velocity determines its competence; the stronger the flow, the larger the particles it can carry in suspension and as bed load.

It is a general rule that the competence of a stream increases as the square of its velocity. Thus, if the velocity of a stream doubles, the impact force of the water increases four times; if the velocity triples, the force increases nine times, and so forth. Hence, the large boulders that are often visible during a low-water stage and that seem immovable can, in fact, be transported during flood stage because of the stream's increased competence (Figure 10.12).

By now it should be clear why the greatest erosion and transportation of sediment occur during floods (Figure 10.13). The increase in discharge results in greater capacity; the increased velocity produces greater competency. With rising velocity the water becomes more turbulent, and larger and larger particles are set in motion. In the course of just a few days, or perhaps just a few hours, a stream in flood stage can erode and transport more sediment than it does during months of normal flow (see Box 10.2, p.245).

Deposition of Sediment by Streams

Whenever a stream's velocity slows, its competence is reduced, and particles of sediment are deposited in definite order by size. As streamflow drops below the critical settling velocity of a certain particle size, sediment in that category begins to settle out. Thus, stream transport provides a mechanism by which solid particles of various sizes are separated. This process, called **sorting**, explains why particles of similar size are deposited together.

The well-sorted material typically deposited by a stream is called **alluvium**, a general term for all stream-deposited sediment. Many different depositional features are composed of alluvium. Some of these features may be found within stream channels, some occur on the valley floor adjacent to the channel, and some exist at the mouth of the stream.

Channel Deposits

As a river transports sediment toward the sea, some material may be deposited within the channel. Channel deposits are most often composed of sand and gravel, the coarser components of a stream's load, and are commonly referred to as **bars**. Such features, however, are only temporary, for the material will be picked up again by the running water and be transported farther downstream. Eventually, most of the material will be carried to its ultimate destination, the ocean.

Sand and gravel bars can form in a number of situations. For example, they are common where

Figure 10.13 During floods both capacity and competency increase. Therefore the greatest erosion and sediment transport occur during these high-water periods. Here we see the sediment-filled floodwaters of Kenya's Mara River. (Photo by Tim Davis/Tony Stone Images)

streams flow in a series of bends, called *meanders*. As a stream flows around a bend, the velocity of the water on the outside increases, leading to erosion at that site. At the same time, the water on the inside of the meander slows, which causes some of the sediment load to settle out. As these deposits occur on the inside "point" of the bend, they are called **point bars** (Figure 10.14). More accurately, these deposits would be better described as "crescent-shaped accumulations of sand and gravel."

Sometimes a stream deposits materials on the floor of its channel. If these accumulations become thick enough to choke the channel, they force the stream to split and follow several paths. What results is a complex network of converging and diverging channels that thread their way among the bars. Because such channels have a interwoven appearance, the stream is said to be **braided** (Figure 10.15). Braided patterns most often form when the load supplied to a stream exceeds its competency or capacity. This can happen under several circumstances: (1) If a steeper, turbulently flowing tributary enters a main stream, its rocky bed load may be deposited at the junction because the velocity abruptly decreases. (2) Excessive load may also be provided when debris from barren slopes is flushed into a channel during a heavy downpour. (3) Excessive load may occur at the end of a glacier where ice-eroded sediment is dumped into a meltwater stream flowing away from the glacier.

Braided streams also form when there is an abrupt decrease in gradient or a decrease in the stream's discharge. The latter situation could result from a drop in rainfall in the area drained by the stream. It also commonly occurs when a stream leaves a humid area with many tributaries and enters a dry region with few tributaries. In this case, the loss of water to evaporation and seepage into the channel results in a diminished discharge.

Floodplain Deposits

As its name implies, a **floodplain** is that part of a valley that is inundated during a flood. Most streams are flanked by floodplains. Although some are impressive features that are many kilometers across, others have modest widths of just a few meters. If we were to sample the alluvium that covers a floodplain, we would find that some of it consists of coarse sands and gravels that were originally deposited as point bars by meanders shifting laterally across the valley floor. Other sediments would be composed of fine sands, silts, and clays that were spread across the floodplain whenever water overflowed the channel during flood stage.

Rivers that occupy valleys with broad, flat floors sometimes create landforms called **natural levees** that flank the stream channel. Natural levees are built by successive floods over many years. When a stream overflows its banks onto the floodplain, the water flows over the surface as a broad sheet. Because such

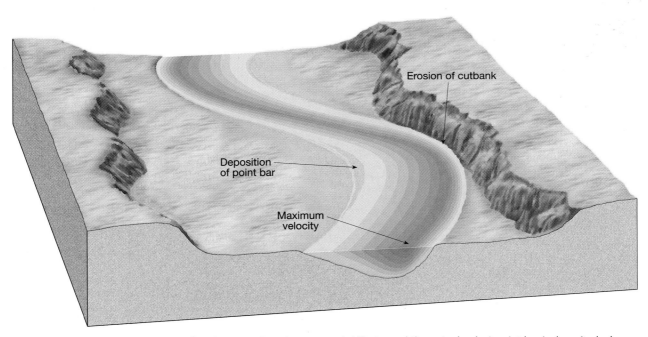

Figure 10.14 When a stream meanders, its zone of maximum speed shifts toward the outer bank. A point bar is deposited when the water on the inside of a meander slows. By eroding its outer bank and depositing material on the inside of the bend, a stream is able to shift its channel.

Figure 10.15 Braided stream choked with sediment near the edge of a melting glacier. (Photo by Bradford Washburn)

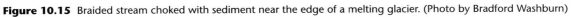

Box 10.2

The 1997 Red River Floods

The Red River of the North originates near the point where South Dakota, North Dakota, and Minnesota meet. It flows northward, finally emptying into Manitoba's Lake Winnepeg (Figure 10.B). Along most of its course in the United States, the river serves as the North Dakota–Minnesota boundary.

In April 1997, Grand Forks, North Dakota, and other communities along the Red River experienced disastrous flooding—the worst ever recorded in the region (Figure 10.C). The flood was preceded by an especially snowy winter. As April began, the winter snows were melting and flooding seemed imminent. Then on April 5 and 6 a final winter storm pounded the northern Great Plains. The blizzard rebuilt once-shrinking snowdrifts back to heights of 6 meters (20 feet) in some places. However, the drifts did not last long because rapidly rising temperatures melted the snow in a matter of days, causing rapid runoff into the Red River.

The discharge at Grand Forks jumped from about 70 cubic meters (2500 cubic feet) per second on April 1 to

Figure 10.C Floodwaters surround Harwood, North Dakota, during the record-breaking 1997 floods. (Photo by Dan Koeck/Gamma Liaison)

Figure 10.B The Red River flows north into Manitoba's Lake Winnepeg.

nearly 4200 cubic meters (150,000 cubic feet) per second three weeks later. The floodwaters at Grand Forks crested at 17.7 meters (58.1 feet), almost 8 meters (more than 26 feet) above flood stage! This exceeded the previous record flood stage set in 1897 by 1.2 meters (4 feet). Roughly 4.5 million acres of land were under water, and most of Grand Fork's nearly 50,000 inhabitants had to evacuate. The losses in Grand Forks alone were estimated to exceed $1 billion.

The 1997 Red River flood was truly a record-breaking event—a 500-year flood (that is, the statistical probability of such a flood is once in 500 years). Although such an extraordinary flood is rare, less severe flooding of broad areas along the Red River of the North is common. Why do floods here occur so frequently and over such large areas? Four factors seem to be especially important.*

1. The Red River flows northward. When the spring thaw takes place, it proceeds in the same direction the river flows, from south to north. Therefore, runoff from the southern portion of the drainage basin progressively joins with newly created snowmelt from more northerly localities. If the progressive south-to-north thaw is perfectly synchronized, extensive floods occur in the northern portion of the valley.

2. Ice derived from the southern valley progressively meets with freshly broken ice in the central and northern valley. Such ice concentrations can slow or block the water flow, causing river levels to swell.

3. The Red River flows across a broad and extremely flat plain, the former floor of a now-abandoned lake. Therefore, when the river floods onto this plain, the area covered can be dramatic.

4. The river's gradient is very low and decreases downstream. Near Fargo the gradient averages about 8 centimeters per kilometer (5 inches per mile). Farther north, near the Canadian border, the gradient drops to just 2.5 centimeters per kilometer (1.5 inches per mile). During floods in this border region, the river's lack of slope causes it to form what is essentially a massive shallow lake.

*Based on material prepared by Professor Donald P. Schwert, North Dakota State University.

a flow pattern significantly reduces the water's velocity and turbulence, the coarser portion of the suspended load is deposited in strips bordering the channel. As the water spreads out over the floodplain, a lesser amount of finer sediment is laid down over the valley floor. This uneven distribution of material produces the gentle, almost imperceptible slope of the natural levee (Figure 10.16).

The natural levees of the lower Mississippi River rise 6 meters above the lower portions of the valley floor. The area behind the levee is characteristically poorly drained for the obvious reason that water cannot flow up over the levee and into the river. Marshes called **back swamps** often result. When a tributary stream enters a valley having substantial natural levees, it may not be able to make its way into the main channel. As a consequence, the tributary may flow through the back swamp parallel to the main river for many kilometers before crossing the natural levee and joining the main river. Such streams are called **yazoo tributaries**, after the Yazoo River, which parallels the lower Mississippi River for more than 300 kilometers.

Alluvial Fans and Deltas

Two of the most common landforms composed of alluvium are alluvial fans and deltas. They are sometimes similar in shape and are deposited for essentially the same reason: an abrupt loss of competence in a stream. The key distinction between them is that alluvial fans are deposited on land whereas deltas are deposited in a body of water. In addition, alluvial fans can be quite steep, but deltas are relatively flat, barely protruding above the level surface of the ocean or lake in which they formed.

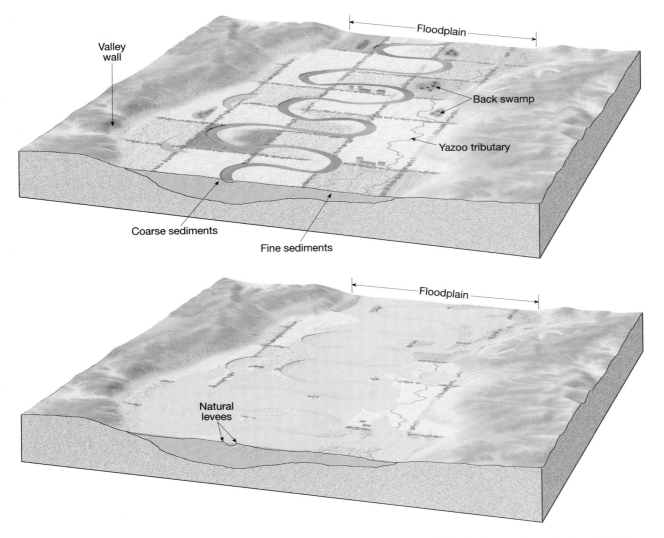

Figure 10.16   Natural levees are gently sloping structures that are created by repeated floods. Because the ground next to the stream channel is higher than the adjacent floodplain, back swamps and yazoo tributaries may develop.

Figure 10.17 Alluvial fans develop where the gradient of a stream changes abruptly from steep to flat. Such a situation exists in Death Valley, California, where streams emerge from the mountains into a flat basin. As a result, Death Valley has many large alluvial fans. (Photo by Michael Collier)

Alluvial Fans. **Alluvial fans** typically develop where a high-gradient stream leaves a narrow valley in mountainous terrain and comes out suddenly onto a broad, flat plain or valley floor (Figure 10.17). Alluvial fans form in response to the abrupt drop in gradient combined with the change from a narrow channel of a mountain stream to less confined channels at the base of the mountains. The sudden drop in velocity causes the stream to dump its load of sediment quickly in a distinctive cone- or fan-shaped accumulation. As illustrated by Figure 10.17, the surface of the fan slopes outward in a broad arc from an apex at the mouth of the steep valley. Usually, coarse material is dropped near the apex of the fan, while finer material is carried toward the base of the deposit. As we learned in Chapter 9, steep canyons in dry regions are prime locations for debris flows. Therefore, it should be expected that many alluvial fans in arid areas have debris flow deposits interbedded with the alluvium.

Deltas. In contrast to an alluvial fan, a **delta** forms when a stream enters an ocean or a lake. Figure 10.18A depicts the structure of a simple delta that might form in the relatively quiet waters of a lake. As the stream's forward motion is checked upon entering the lake, the dying current deposits its load of sediments. These deposits occur in three types of beds.

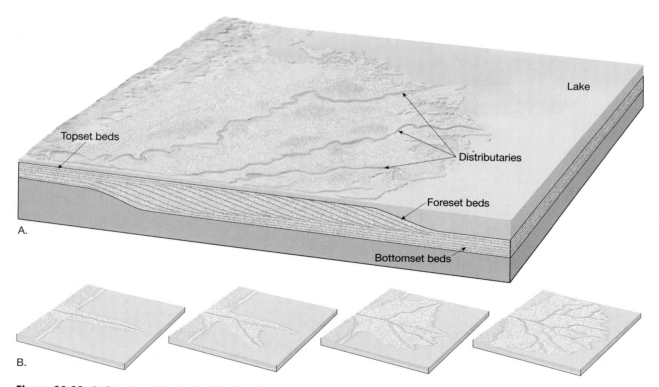

Figure 10.18 A. Structure of a simple delta that forms in the relatively quiet waters of a lake. **B.** Growth of a simple delta. As a stream extends its channel, the gradient is reduced. Frequently, during flood stage the river is diverted to a higher-gradient route, forming a new distributary. Old abandoned distributaries are gradually invaded by aquatic vegetation and fill with sediment. (After Ward's Natural Science Establishment, Inc., Rochester, N.Y.)

Forest beds are composed of coarser particles that drop almost immediately upon entering the lake to form layers that slope downcurrent from the delta front. The forest beds are usually covered by thin, horizontal *topset beds* that are deposited during flood stage. The finer silts and clays settle out some distance from the mouth in nearly horizontal layers called *bottomset beds*.

As the delta grows outward, the stream's gradient continually lessens. This circumstance eventually causes the channel to become choked with sediment from the slowing water. As a consequence, the river seeks a shorter, higher-gradient route to base level, as illustrated in Figure 10.18B. This illustration shows the main channel dividing into several smaller ones, called **distributaries**. Most deltas are characterized by these shifting channels that act in an opposite way to that of tributaries. Rather than carrying water into the main channel, distributaries carry water away from the main channel in varying paths to base level. After numerous shifts of the channel, the simple delta may grow into the triangular shape of the Greek letter delta (Δ), for which it was named. Note, however, that many deltas do not exhibit this shape. Differences in the configurations of shorelines, and variations in the nature and strength of wave activity, result in many different shapes (Figure 10.19).

Although deltas that form in the ocean generally exhibit the same basic form as the simple lake-deposited feature just described, most large marine deltas are far more complex and have foreset beds that are inclined at a much lower angle than those depicted in Figure 10.18A.

Indeed, many of the world's great rivers have created massive deltas, each with its own peculiarities and none as simple as the one illustrated in Figure 10.18B.

The Mississippi Delta. Many large rivers have deltas that extend over thousands of square kilometers. The delta of the Mississippi River is one such feature. It resulted from the accumulation of huge quantities of sediment derived from the vast region drained by the river and its tributaries. Today New Orleans rests where there was ocean less than 5000 years ago. Figure 10.20 shows that portion of the Mississippi delta which has been built over the past 5000 to 6000 years. As the figure illustrates, the delta is actually a series of seven coalescing subdeltas. Each was formed when the river left its existing channel to find a shorter, more direct path to the Gulf of Mexico. The individual subdeltas interfinger and partially cover one another to produce a very complex structure. It is also apparent from Figure 10.20 that, after each portion was abandoned, coastal erosion modified the features. The present subdelta, called a *bird-foot* delta because of the configuration of its distributaries, has been built by the Mississippi in the last 500 years.

At present this active bird-foot delta has extended about as far as natural forces will allow. In fact, for many years the river has been struggling to cut through a narrow neck of land and shift its course to that of the Atchafalaya River (see inset in Figure 10.20). If this were to happen, the Mississippi would abandon its lowermost 500-kilometer path in favor of

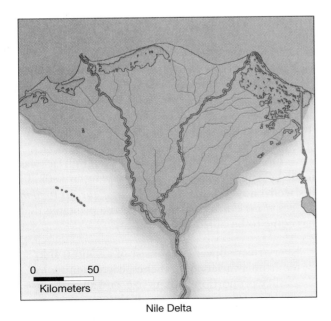

Nile Delta

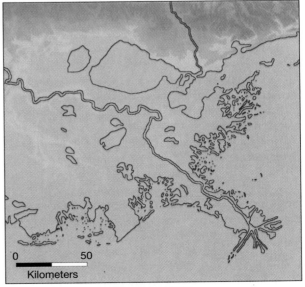

Mississippi Delta

Figure 10.19 The shapes of deltas vary and depend on such factors as a river's sediment load and the strength and nature of shoreline processes. The triangular shape of the Nile delta was the basis for naming this feature. The present Mississippi delta is called a *bird-foot delta*.

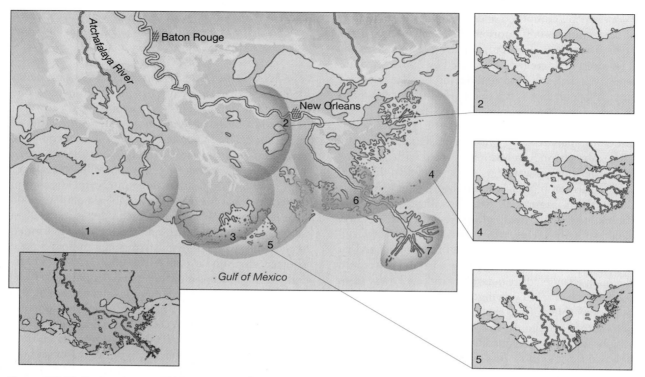

Figure 10.20 During the past 5000 to 6000 years, the Mississippi River has built a series of seven coalescing subdeltas. The numbers indicate the order in which the subdeltas were deposited. The present bird-foot delta (number 7) represents the activity of the past 500 years. Without ongoing human efforts, the present course will shift and follow the path of the Atchafalaya River. The inset on left shows the point where the Mississippi may someday break through (arrow) and the shorter path it would take to the Gulf of Mexico. (After C.R. Kolb and J.R. Van Lopik)

the Atchafalaya's much shorter 225-kilometer route. From the early 1940s until the 1950s, an increasing portion of the Mississippi's discharge was diverted to this new path, indicating that the river was ready to shift and begin building a new subdelta.

To prevent such an event, and to keep the Mississippi following its present course, a damlike structure was erected at the site where the channel was trying to break through. Floods in the early 1970s weakened the control structure, and the river again threatened to shift until a massive auxiliary dam was completed in the mid-1980s. For the time being, at least, the inevitable has been avoided, and the Mississippi River will continue to flow past Baton Rouge and New Orleans on its way to the Gulf of Mexico.

Although deltas are deposited by many large rivers, not all rivers create these features. Even streams that transport large loads of sediment may lack deltas because powerful currents and waves quickly redistribute the material as soon as it is deposited. The Columbia River in the Pacific Northwest is an example. In other cases, rivers do not carry sufficient quantities of sediment to build a delta. The St. Lawrence River, for example, has little opportunity to pick up much sediment between Lake Ontario and its mouth in the Gulf of St. Lawrence.

Stream Valleys

Valleys are the most common landforms on Earth's surface. In fact, they exist in such large numbers that they have never been counted except in limited areas used for study. Prior to the turn of the nineteenth century, it was generally believed that valleys were created by catastrophic events that pulled the crust apart and created troughs in which streams could flow. Today, however, we know that with a few exceptions, streams create their own valleys.

One of the first clear statements of this fact was made by English geologist, John Playfair, in 1802. In his well-known work, *Illustrations of the Huttonian Theory of the Earth*, Playfair stated the principle that has come to be called **Playfair's law**:

Every river appears to consist of a main trunk, fed from a variety of branches, each running in a valley proportioned to its size, and all of them together forming a system of valleys, communicating with one another, and having such a nice adjustment of their declivities, that none of them join the principal valley, either on too high or too low a level; a circumstances that

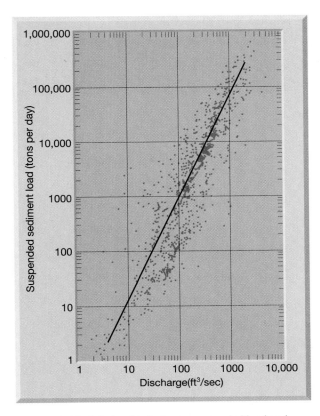

Figure 10.21 Relationship between suspended load and discharge on the Powder River at Arvada, Wyoming. (From L.B. Leopold and Thomas Maddock, Jr., U.S. Geological Survey Professional Paper 252, 1953)

Figure 10.21 shows the relationship between suspended load and discharge at a gauging station on the Powder River in Wyoming. Notice that as the discharge increases, the quantity of suspended sediment increases. In fact, the increase is exponential; that is to say, if the discharge at the gauging station increases tenfold, the suspended load may increase by a factor of 100 or more. Measurements and calculations have shown that stream channel erosion during periods of increased discharge can account for only a portion of the additional sediment transported by a stream. Therefore, much of the increased load must be delivered to the stream by sheet flow and mass wasting.

A narrow, V-shaped valley indicates that the primary work of the stream has been downcutting toward base level. The most prominent features in such a valley are **rapids** and **waterfalls** (Figure 10.22). Both occur

Figure 10.22 V-shaped valley of the Yellowstone River. The rapids and waterfalls indicate that the river is vigorously downcutting. (Photo by Art Wolfe)

would be indefinitely improbable, if each of these valleys were not the work of the stream that flows in it.

Not only were Playfair's observations essentially correct but they were written in a style that is seldom achieved in scientific prose.

Stream valleys can be divided into two general types: narrow, V-shaped valleys and wide valleys with flat floors. These exist as the ideal forms, with many gradations between.

Narrow Valleys

In some arid regions, where downcutting is rapid and weathering is slow, and in places where rock is particularly resistant, narrow valleys may have nearly vertical walls (see Figure 10.10, p. 241). However, most valleys, even those that are narrow at the base, are much broader at the top than the width of the channel at the bottom. This would not be the case if the only agent responsible for eroding valleys were the streams flowing through them.

The sides of most valleys are shaped primarily as the result of weathering, sheet flow, and mass wasting. Consider the following example of this process.

Lockport
Dolostone

Shale

Dolostone

Mudstone

Sandstone

Shale

Figure 10.23 The smaller American Falls at Niagara Falls. The river plunges over the falls and erodes the shale beneath the more resistant Lockport Dolostone. As a section of dolostone is undercut, it loses support and breaks off. (Photo by David Ball/Tony Stone Images)

where the stream profile drops rapidly, a situation usually caused by variations in the erodibility of the bedrock into which the stream channel is cutting. A resistant bed produces a rapids by acting as a temporary base level upstream while allowing downcutting to continue downstream. Once erosion has eliminated the resistant rock, the stream profile smoothes out again.

Waterfalls are places where the stream makes a vertical drop. One type of waterfall is exemplified by Niagara Falls (Figure 10.23). Here, the falls are supported by a resistant bed of dolostone that is underlain by a less resistant shale. As the water plunges over the lip of the falls it erodes the less resistant shale, undermining a section of dolostone, which eventually breaks off. In this manner the waterfall retains its vertical cliff while slowly but continually retreating upstream. Over the past 12,000 years, Niagara Falls has retreated more than 11 kilometers (7 miles) upstream.

Wide Valleys

Once a stream has cut its channel closer to base level, it approaches a graded condition, and downward erosion becomes less dominant. At this point more of the

stream's energy is directed from side to side. The reason for this change is not fully understood, but the reduced gradient probably is an important factor. Nevertheless, it does occur, and the result is a widening of the valley as the river erodes first one bank and then the other (Figure 10.24). In this manner the flat valley floor, or floodplain, is produced. This is an appropriate name because the river is confined to its channel except during flood stage, when it overflows its banks and inundates the floodplain.

When a river erodes laterally and creates a floodplain as just described, it is called an *erosional floodplain*. However, floodplains can be depositional as well. *Depositional floodplains* are produced by a major fluctuation in conditions, such as a change in base level. The floodplain in California's Yosemite Valley is one such feature; it was produced when a glacier gouged the former stream valley about 300 meters (1000 feet) deeper than it had been. After the glacial ice melted, the stream readjusted itself to its former base level by refilling the valley with alluvium.

Streams that flow upon floodplains, whether erosional or depositional, move in sweeping bends called

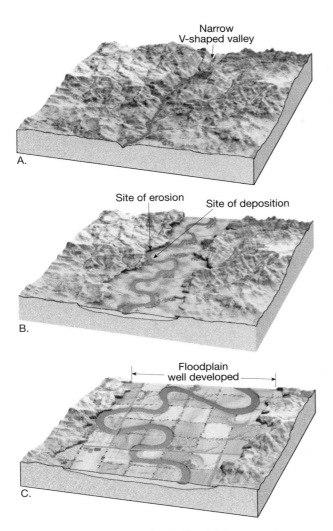

Narrow
V-shaped valley

A.

Site of erosion   Site of deposition

B.

Floodplain
well developed

C.

Figure 10.24 Stream eroding its floodplain.

meanders. The term is derived from a river in western Turkey, the Menderes, which has a very sinuous course. Once a bend in a channel begins to form, it grows larger. Erosion occurs on the outside of the meander where velocity and turbulence are greatest. Most often, the outside bank is undermined, especially during periods of high water. As the bank becomes oversteepened, it fails by slumping into the channel. Because the outside of a meander is a zone of active erosion, it is often referred to as the **cut bank** (Figure 10.25). Much of the debris detached by the stream at the cut bank moves downstream and is soon deposited as point bars in zones of decreased velocity on the insides of meanders. In this manner, meanders migrate laterally, while keeping the same cross-sectional area, by eroding the outside of the bends and depositing on the inside (see Figure 10.14, p. 78). Growth ceases when the meander reaches a critical size that is determined by the size of the stream. The larger the stream, the larger its meanders can be.

Owing to the slope of the channel, erosion is more effective on the downstream side of a meander. Therefore, in addition to growing laterally, the bends also gradually migrate down the valley. Sometimes the downstream migration of a meander is slowed when it reaches a more resistant portion of the floodplain. This allows the next meander upstream to "catch up." Gradually the neck of land between the meanders is narrowed. When they get close enough, the river may erode through the narrow neck of land to the next loop (Figure 10.26). The new, shorter channel segment is called a **cutoff** and, because of its shape, the abandoned bend is called an **oxbow lake** (Figure 10.27). Over a period of time, the oxbow lake fills with sediment to create a **meander scar**.

A.

B.

Figure 10.25 Erosion of a cut bank along the Newaukum River, Washington. **A.** January 1965. **B.** March 1965. (Photos by P.A. Glancy, U.S. Geological Survey)

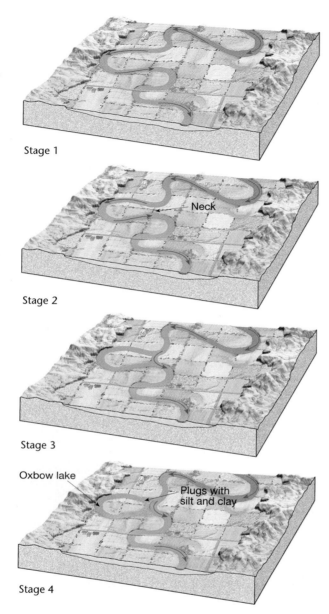

Stage 1

Stage 2

Neck

Stage 3

Stage 4

Oxbow lake

Plugs with
silt and clay

Figure 10.26 Formation of a cutoff and oxbow lake.

The process of meander cutoff formation has the effect of shortening the river and was described humorously by Mark Twain in *Life on the Mississippi*.

In the space of one hundred and seventy-six years the lower Mississippi has shortened itself two hundred and forty-two miles. This is an average of a trifle over one mile and a third per year. Therefore, any calm person, who is not blind or idiotic, can see that in the Old Oolitic Silurian Period, just a million years ago next November, the Lower Mississippi River was upwards of one million three hundred thousand miles long, and stuck out over the Gulf of Mexico like a fishing

rod. And by the same token any person can see that seven hundred and forty-two years from now the Lower Mississippi will be only a mile and three quarters long, and Cairo and New Orleans will have joined their streets together, and be plodding comfortably along under a single mayor and a mutual board of aldermen. One gets such wholesale returns of conjecture out of such a trifling investment of fact.

Although the data used by Mark Twain may be reasonably accurate, he intentionally forgot to include the fact that the Mississippi also created many new meanders, thus lengthening its course by a similar amount. In fact, with the growth of its delta, the Mississippi is actually getting longer, not shorter.

Incised Meanders and Stream Terraces

We usually expect a stream with a highly meandering course to be on a floodplain in a wide valley. However, certain rivers exhibit meandering channels that flow in

Figure 10.27 Oxbow lakes occupy abandoned meanders. As they fill with sediment, oxbow lakes gradually become swampy meander scars. Aerial view of oxbow lake created by the meandering Green River near Bronx, Wyoming. (Photo by Michael Collier)

A.

B.

Figure 10.28
A. This high-altitude image shows incised meanders of the Delores River in western Colorado. (Courtesy of USDA-ASCS)
B. A close-up view of incised meanders of the Colorado River in Canyonlands National Park, Utah. (Photo by Michael Collier)
In both places, meandering streams began downcutting because of the uplift of the Colorado Plateau.

steep, narrow valleys. Such meanders are called **incised meanders** (Figure 10.28). How do such features form?

Originally the meanders probably developed on the floodplain of a stream that was relatively near base level. Then, a change in base level caused the stream to begin downcutting. One of two events could have occurred. Either base level dropped or the land upon which the river was flowing was uplifted.

An example of the first circumstance happened during the Ice Age when large quantities of water were withdrawn from the ocean and locked up in glaciers on land. The result was that sea level (ultimate base

level) dropped, causing rivers flowing into the ocean to begin to downcut (see Box 12.1). Of course, this activity ceased at the close of the Ice Age when the glaciers melted and the ocean rose to its former level.

Regional uplift of the land, the second cause for incised meanders, is exemplified by the Colorado Plateau in the southwestern United States. Here, as the plateau was gradually uplifted, numerous meandering rivers adjusted to being higher above base level by downcutting (Figure 10.28).

After a river has adjusted to a relative drop in base level by downcutting, it may once again produce

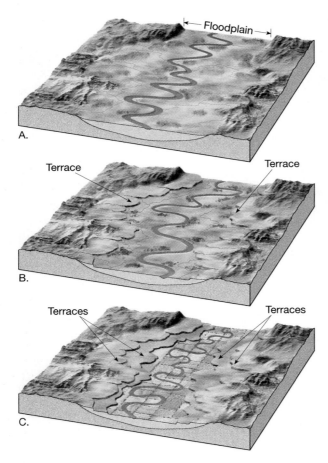

 Drainage Networks

a floodplain at a level below the old one. The remnants of a former floodplain are sometimes present in the form of flat surfaces called **terraces** (Figure 10.29).

A stream is just a small component of a larger system. Each system consists of a **drainage basin**, the land area that contributes water to the stream. The drainage basin of one stream is separated from another by an imaginary line called a **divide** (Figure 10.30). Divides range in size from a ridge separating two small gullies to continental divides, which split continents into enormous drainage basins. The Mississippi River has the largest drainage basin in North America (Figure 10.31). Extending between the Rocky Mountains in the West and the Appalachian Mountains in the East, the Mississippi River and its tributaries collect water from more than 3.2 million square kilometers (1.2 million square miles) of the continent.

Drainage Patterns

All drainage systems are made up of an interconnected network of streams that together form particular patterns. The nature of a drainage pattern can vary greatly from one type of terrain to another, primarily in response to the kinds of rock on which the streams developed or the structural pattern of faults and folds.

The most commonly encountered drainage pattern is the **dendritic pattern** (Figure 10.32A). This pattern is characterized by irregular branching of tributary

Figure 10.29 Terraces can form when a stream downcuts through previously deposited alluvium. This may occur in response to a lowering of base level or as a result of regional uplift.

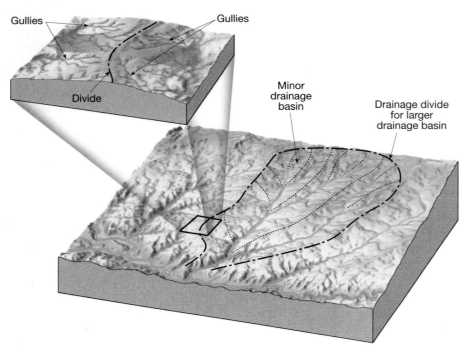

Figure 10.30 A *drainage basin* is the land area drained by a stream and its tributaries. *Divides* are the boundaries separating drainage basins.

Figure 10.31 The drainage basin of the Mississippi River, North America's largest river, covers about 3 million square kilometers. *Divides* are the boundaries that separate drainage basins from each other. Drainage basins and divides exist for all streams.

streams that resembles the branching pattern of a deciduous tree. In fact, the word *dendritic* means "treelike." The dendritic pattern forms where underlying bedrock is relatively uniform, such as flat-lying sedimentary strata or massive igneous rocks. Because the underlying material is essentially uniform in its resistance to erosion, it dose not control the pattern of streamflow. Rather, the pattern is determined chiefly by the direction of slope of the land.

When streams diverge from a central area like spokes from the hub of a wheel, the pattern is said to be **radial** (Figure 10.32B). This pattern typically develops on isolated volcanic cones and domal uplifts.

Figure 10.32C illustrates a **rectangular pattern**, with many right-angle bends. This pattern develops when the bedrock is crisscrossed by a series of joints and faults. Because these structures are eroded more easily than unbroken rock, their geometric pattern guides the directions of streams as they carve their valleys.

Figure 10.32D illustrates a **trellis drainage pattern**, a rectangular pattern in which tributary streams are nearly parallel to one another and have the appearance of a garden trellis. This pattern forms in areas underlain by alternating bands of resistant and less resistant rock and is particularly well displayed in the folded Appalachian Mountains, where both weak and strong strata outcrop in nearly parallel belts (see Box 10.3).

Headward Erosion and Stream Piracy

We have seen that a stream can lengthen its course by building a delta at its mouth. A stream also lengthens its course by **headward erosion**; that is, by extending the head of its valley upslope. As sheet flow converges and becomes concentrated at the head of a stream channel, its velocity, and hence its power to erode, increases. The result can be vigorous erosion at the head of the valley. Thus, through headward erosion, the valley becomes extended into previously undissected terrain (Figure 10.33).

Headward erosion by streams plays a major role in the dissection of upland areas. In addition, an understanding of this process helps explain changes that take place in drainage patterns. One cause for changes that occur in the pattern of streams is **stream piracy**, the diversion of the drainage of one stream because of the headward erosion of another stream. Piracy can occur, for example, if a stream on one side of a divide has a steeper gradient than a stream on the other side. Because the stream with the steeper gradient has more energy, it can extend its valley headward, eventually breaking down the divide and capturing part or all of the drainage of the slower stream. In Figure 10.34, the flow of stream *A* was captured when a more swiftly

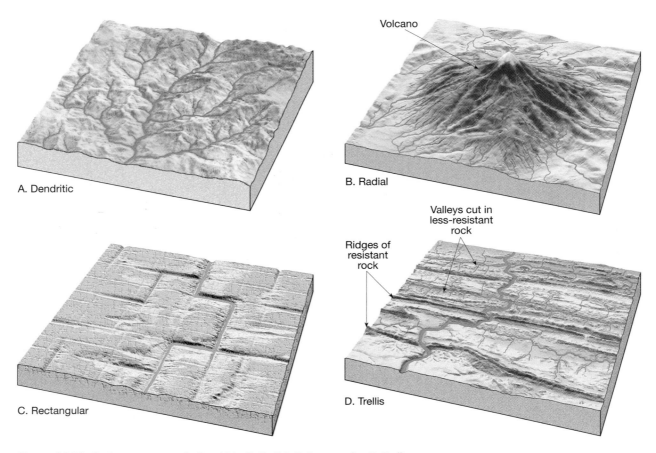

Figure 10.32 Drainage patterns. **A.** Dendritic. **B.** Radial. **C.** Rectangular. **D.** Trellis.

Figure 10.33 By headward erosion valleys extend into previously undissected terrain. The San Rafael River is shown above its confluence with the Green River in Utah. (Photo by Michael Collier)

Box 10.3

Formation of a Water Gap

Sometimes to understand fully the pattern of streams in an area, we must understand the history of the streams. For example, in many places a river valley can be seen cutting through a ridge or mountain that lies across its path. The steep-walled notch followed by the river through the structure is called a *water gap* (Figure 10.D).

Why does a stream cut across such a structure and not flow around it? One possibility is that the stream existed before the ridge or mountain was formed. In this situation, the stream, called an *antecedent stream*, would have to keep pace downcutting while the uplift progressed. That is, the stream would maintain its course as folding or faulting raised an area of the crust across the path of the stream.

A second possibility is that the stream was *superposed*, or let down, upon the structure (Figure 10.E). This can occur when a ridge or mountain is buried beneath layers of relatively flat-lying sediments or sedimentary strata. Streams originating on this cover would establish their courses without regard to the structures below. Then, as the valley was deepened and the structure was encountered, the river would continue to cut its valley into it. The folded Appalachians provide some good examples. Here, a number of major rivers, such as the Potomac and the Susquehanna, cut across the folded strata on their way to the Atlantic.

Figure 10.D Harpers Ferry gap at the confluence of the Shenandoah and Potomac rivers near the West Virginia–Maryland border. Water gaps such as this one are common in parts of the Appalachians. (Photo by John S. Shelton)

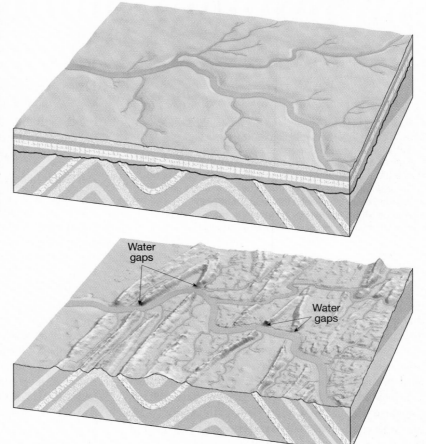

Figure 10.E Development of a superposed stream. The river establishes its course on relatively uniform strata; it then encounters and cuts through the underlying structure.

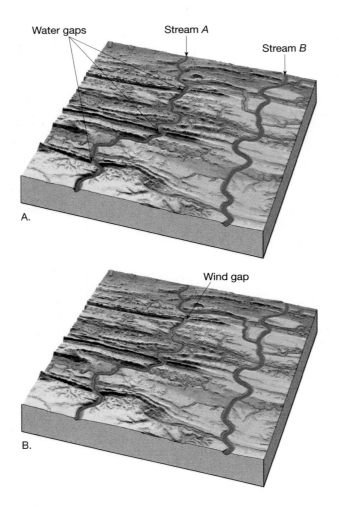

Water gaps   Stream *A*   Stream *B*

A.

Wind gap

B.

Figure 10.34 Stream piracy and the formation of wind gaps. A tributary of stream *B* erodes headward until it eventually captures and diverts stream *A*. A water gap through which stream *A* flowed is abandoned because of the piracy. As a result, this feature is now a wind gap. In this valley and ridge-type setting, the softer rocks in the valleys are eroded more easily than the resistant ridges. Consequently, as the valleys are lowered, the ridges and wind gaps become elevated relative to the valleys.

flowing tributary of stream *B* breached the divide at its head and diverted stream *A*.

Stream piracy also explains the existence of narrow, steep-sided gorges that have no active streams running through them. These abandoned water gaps (called *wind gaps*) form when the stream that cut the notch has its course changed by a pirate stream. In Figure 10.34, one water gap that had been created by stream *A* became a wind gap as a result of stream piracy.

Floods and Flood Control

When the discharge of a stream becomes so great that it exceeds the capacity of its channel, it overflows its banks as a **flood**. Floods are the most common and most destructive of all geologic hazards. They are, nevertheless, simply part of the *natural* behavior of streams.

Causes of Floods

Rivers flood because of the weather. Rapid melting of snow in the spring, and/or major storms that bring heavy rains over a large region, cause most floods. The extensive 1997 flood along the Red River of the North (described in Box 10.2) is a recent example of an event triggered by rapid snowmelt. Exceptional rains caused the devastating floods in the upper Mississippi River Valley during the summer of 1993 (Figure 10.35).

Human interference with the stream system can worsen or even cause floods. One example is the transformation of a rural landscape to an urban one. Pavement and storm sewers increase runoff and speed water to a nearby river (see Box 10.1, p. 69). A second example is the failure of a dam. The bursting of a dam in 1889 on the Little Conemaugh River caused the devastating Johnstown, Pennsylvania, flood that took some 3000 lives. A second dam failure occurred there again in 1977 and caused 77 fatalities.

Flood Control

Several strategies have been devised to eliminate or lessen the catastrophic effects of floods. Engineering efforts include the construction of artificial levees, the building of flood-control dams, and river channelization.

Artificial Levees. Artificial levees are earthen mounds built on the banks of a river to increase the volume of water the channel can hold. These most common of stream-containment structures have been used since ancient times and continue to be used today.

Artificial levees are usually easy to distinguish from natural levees because their slopes are much steeper. When a river is confined by levees during periods of high water, it frequently deposits material in its channel as the discharge diminishes. This is sediment that otherwise would have been dropped on the floodplain. Thus, each time there is a high flow, deposits are left on the river bed and the bottom of the channel is built up. With the buildup of the bed, less water is required to overflow the original levee. As a result, the height of the levee may have to be raised periodically to protect the floodplain. Moreover, many artificial levees are not built to withstand periods of

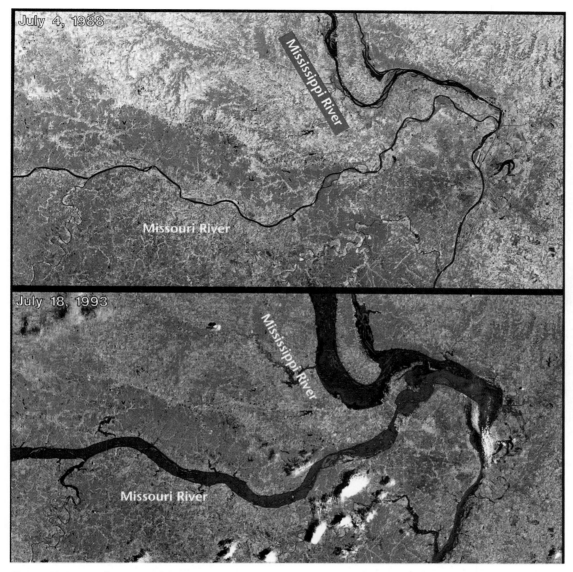

Figure 10.35 Satellite views of the Missouri River flowing into the Mississippi River. St. Louis is just south of their confluence. The upper image shows the rivers during a drought that occurred in summer 1988. The lower image depicts the peak of the record-breaking 1993 flood. Exceptional rains produced the wettest spring and early summer of the twentieth century in the upper Mississippi River basin. In all, nearly 14 million acres were inundated, displacing at least 50,000 people. (Photos by Earth Observation Satellite Company)

extreme flooding. For example, levee failures were numerous in the Midwest during the summer of 1993, when the upper Mississippi and many of its tributaries experienced record floods (Figure 10.36).

Flood-Control Dams. *Flood-control dams* are built to store floodwater and then let it out slowly. This lowers the flood crest by spreading it out over a longer time span. Since the 1920s, thousands of dams have been built on nearly every major river in the United States. Many dams have significant nonflood-related functions

such as providing water for irrigated agriculture and for hydroelectric power generation. Many reservoirs are also major regional recreational facilities.

Although dams may reduce flooding and provide other benefits, building these structures also has significant costs and consequences. For example, reservoirs created by dams may cover fertile farmland, useful forests, historic sites, and scenic valleys. Of course, dams trap sediment. Therefore, deltas and floodplains downstream erode because they are no longer replenished with silt during floods. Large dams

River

Break in Levee

Figure 10.36 Water rushes through a break in an artificial levee in Monroe County, Illinois. During the record-breaking 1993 Midwest floods, many artificial levees could not withstand the force of the floodwaters. Sections of many weakened structures were overtopped or simply collapsed. (Photo by James A. Finley/AP/Wide World Photos)

can also cause significant ecological damage to river environments that took thousands of years to establish.

Building a dam is not a permanent solution to flooding. Sedimentation behind a dam means that the volume of its reservoir will gradually diminish, reducing the effectiveness of this flood-control measure.

Channelization. *Channelization* involves altering a stream channel in order to speed the flow of water to prevent it from reaching flood height. This may simply involve clearing a channel of obstructions or dredging a channel to make it wider and deeper.

A more radical alteration involves straightening a channel by creating *artificial cutoffs*. The idea is that by shortening the stream, the gradient and hence the velocity are increased. By increasing velocity, the larger discharge associated with flooding can be dispersed more rapidly.

Since the early 1930s, the Army Corps of Engineers has created many artificial cutoffs on the Mississippi for the purpose of increasing the efficiency of the channel and reducing the threat of flooding. In all, the river has been shortened more than 240 kilometers (150 miles). The program has been somewhat successful in reducing the height of the river in flood stage. However, because the river's tendency toward meandering still exists, preventing the river from returning to its previous course has been difficult.

Artificial cutoffs increase a stream's velocity and may also accelerate erosion of the bed and banks of the channel. A case in point is the Blackwater River in Missouri, whose meandering course was shortened in 1910. Among the many effects of this project was a dramatic increase in the width of the channel caused by the increased velocity of the stream. One particular bridge over the river collapsed because of bank erosion in 1930. Over the next 17 years the same bridge was replaced on three more occasions, each time with a longer span.

A Nonstructural Approach. All of the flood-control measures described so far have involved structural solutions aimed at "controlling" a river. These solutions are expensive and often give people residing on the floodplain a false sense of security.

Today many scientists and engineers advocate a nonstructural approach to flood control. They suggest that an alternative to artificial levees, dams, and channelization is sound floodplain management. By identifying high-risk areas, appropriate zoning regulations can be implemented to minimize development and promote more appropriate land use.

Chapter Summary

- The *hydrologic cycle* describes the continuous interchange of water among the oceans, atmosphere, and continents. Powered by energy from the Sun, it is a global system in which the atmosphere provides the link between the oceans and continents. The processes involved in the hydrologic cycle include *precipitation*, *evaporation*, *infiltration* (the movement of water into rocks or soil through cracks and pore spaces), *runoff* (water that flows over the land), and *transpiration* (the release of water vapor to the atmosphere by plants). *Running water is the single most important agent sculpturing Earth's land surface.*

- The amount of water running off the land rather than sinking into the ground depends upon the *infiltration capacity* of the soil. Initially, runoff flows as broad, thin sheets across the ground, appropriately termed *sheet flow*. After a short distance, threads of current typically develop and tiny channels called *rills* form.

- The factors that determine a stream's *velocity* are *gradient* (slope of the stream channel), *cross-sectional shape*, *size* and *roughness* of the channel, and the stream's *discharge* (amount of water passing a given point per unit of time, frequently measured in cubic meters or cubic feet per second). Most often, the gradient and roughness of a stream decrease downstream, while width, depth, discharge, and velocity increase.

- The two general types of *base level* (the lowest point to which a stream may erode its channel) are (1) *ultimate base level* (sea level) and (2) *temporary*, or *local*, *base level*. Any change in base level will cause a stream to adjust and establish a new balance. Lowering base level will cause a stream to erode, whereas raising base level results in deposition of material in the channel.

- Streams transport their load of sediment in solution (*dissolved load*), in suspension (*suspended load*), and along the bottom of the channel (*bed load*). Much of the dissolved load is contributed by groundwater. Most streams carry the greatest part of the load in suspension. The bed load moves only intermittently and usually represents the smallest portion of a stream's load.

- A stream's ability to transport solid particles is described using two criteria: *capacity* (the maximum load of solid particles a stream can carry) and *competence* (the maximum particle size a stream can transport). Competence increases as the square of stream velocity, so if velocity doubles, water's force increases fourfold.

- Streams deposit sediment when velocity slows and competence is reduced. This results in *sorting*, the process by which like-sized particles are deposited together. Stream deposits are called *alluvium* and may occur as channel deposits called *bars*, as floodplain deposits, which include *natural levees*, and as *deltas* or *alluvial fans* at the mouths of streams.

- Although many gradations exist, the two general types of stream valleys are (1) *narrow V-shaped valleys* and (2) *wide valleys with flat floors*. Because the dominant activity is downcutting toward base level, narrow valleys often contain *waterfalls* and *rapids*. When a stream has cut its channel closer to base level, its energy is directed from side-to-side, and erosion produces a flat valley floor, or *floodplain*. Streams that flow upon floodplains often move in sweeping bends called *meanders*. Widespread meandering may result in shorter channel segments, called *cutoffs*, and/or abandoned bends, called *oxbow lakes*.

- The land area that contributes water to a stream is called a *drainage basin*. Drainage basins are separated by an imaginary line called a *divide*. Common *drainage patterns* (the form of a network of streams) produced by a main channel and its tributaries include (1) *dendritic*, (2) *radial*, (3) *rectangular*, and (4) *trellis*.

- *Headward erosion* lengthens a stream course by extending the head of its valley upslope. This process can lead to *stream piracy* (the diversion of the drainage of one stream by another). Former water gaps called *wind gaps* can result from stream piracy.

- *Floods* are triggered by heavy rains and/or snowmelt. Sometimes human interference can worsen or even cause floods. Flood-control measures include the building of *artificial levees* and dams, as well as *channelization*, which could involve creating *artificial cutoffs*. Many scientists and engineers advocate a nonstructural approach to flood control that involves more appropriate land use.

Review Questions

1. Describe the movement of water through the hydrologic cycle. Once precipitation has fallen on land, what paths are available to it?

2. Over the oceans, evaporation exceeds precipitation. Why does sea level not drop?

3. List several factors that influence infiltration capacity.

4. "Water in streams moves primarily in laminar flow." Briefly explain whether this statement is true or false.

5. A stream originates at 2000 meters above sea level and travels 250 kilometers to the ocean. What is its average gradient in meters per kilometer?

6. Suppose that the stream mentioned in Question 5 developed extensive meanders so that its course was lengthened to 500 kilometers. Calculate this new gradient. How does meandering affect gradient?

7. When the discharge of a stream increases, what happens to the stream's velocity?

8. What typically happens to channel width, channel depth, velocity, and discharge from the point where a stream begins to the point where it ends? Briefly explain why these changes take place.

9. When an area changes from being predominantly rural to largely urban, how is streamflow affected? (See Box 10.1, p. 237)

10. Define *base level*. Name the main river in your area. For what streams does it act a base level? What is the base level for the Mississippi River?

11. Why do most streams have low gradients near their mouths?

12. Describe three ways in which a stream may erode its channel. Which one of these is responsible for creating potholes?

13. If you were to collect a jar of water from a stream, what part of the load would settle to the bottom of the jar? What portion would remain in the water? What part of a stream's load would probably not be present in your sample?

14. What is settling velocity? What factors influence settling velocity?

15. Distinguish between capacity and competency.

16. What factors promote the frequent flooding of broad areas along the Red River of the North? (See Box 10.2, p. 246)

17. Describe a situation that might cause a stream channel to become braided.

18. Briefly describe the formation of a natural levee. How is this feature related to backswamps and yazoo tributaries?

19. In what way is a delta similar to an alluvial fan? In what way are they different?

20. Why does a river flowing across a delta eventually change course?

21. Each of the following statements refers to a particular drainage pattern. Identify the pattern.

    (a) Streams diverging from a central high area such as a dome.

    (b) Branching, "treelike" pattern.

    (c) A pattern that develops when bedrock is crisscrossed by joints and faults.

22. Describe how a water gap might form. (See Box 10.3, p. 259)

23. List and briefly describe three basic flood-control strategies. What are some drawbacks of each?

Key Terms

alluvial fan (p. 248)
alluvium (p. 243)
backswamp (p. 247)
bar (p. 244)
base level (p. 239)
bed load (p. 242)
braided stream (p. 244)
capacity (p. 243)
competence (p. 243)
cut bank (p. 253)
cutoff (p. 253)
delta (p. 248)
dendritic pattern (p. 256)
discharge (p. 236)
dissolved load (p. 242)
distributary (p. 249)
divide (p. 256)

drainage basin (p. 256)
evapotranspiration
  (p. 233)
flood (p. 260)
floodplain (p. 244)
graded stream (p. 240)
gradient (p. 236)
head (headwaters) (p. 238)
headward erosion (p. 257)
hydrologic cycle (p. 232)
incised meander (p. 255)
infiltration (p. 233)
infiltration capacity
  (p. 234)
laminar flow (p. 235)
local (temporary) base
  level (p. 239)

longitudinal profile
  (p. 238)
meander (p. 253)
meander scar (p. 253)
mouth (p. 238)
natural levee (p. 244)
oxbow lake (p. 253)
Playfair's law (p. 250)
point bar (p. 244)
pothole (p. 241)
radial pattern (p. 257)
rectangular pattern
  (p. 257)
rills (p. 234)
runoff (p. 233)
saltation (p. 243)

settling velocity (p. 243)
sheet flow (p. 234)
sorting (p. 243)
stream (p. 234)
stream piracy (p. 257)
suspended load (p. 242)
temporary (local) base
  level (p. 239)
terrace (p. 256)
transpiration (p. 233)
trellis drainage pattern
  (p. 257)
turbulent flow (p. 235)
ultimate base level
  (p. 239)
waterfall (p. 251)
yazoo tributary (p. 247)

Web Resources

The *Earth* Home Page provides on-line resources for this chapter on the World Wide Web. You will find review exercises, specific updates for items in the chapter, suggested reading, and links to interesting related pathways on the Internet. Visit the *Earth* Home Page at **http://www.prenhall.com/tarbuck.**

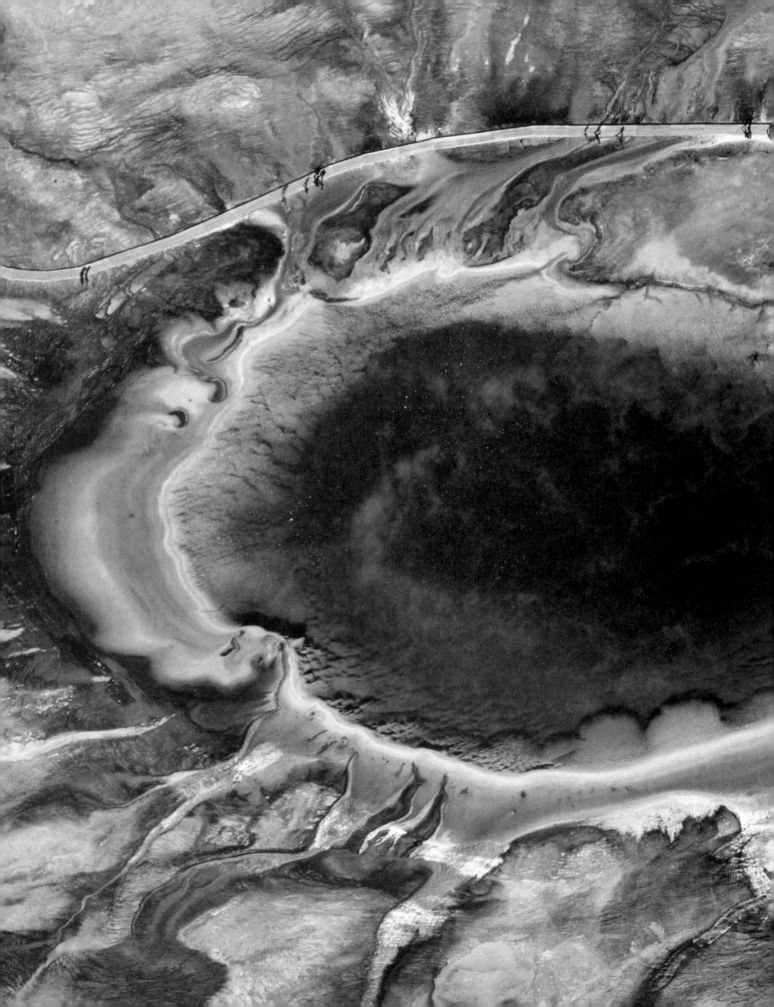

CHAPTER 11

Groundwater

Left Aerial view of Grand Prismatic Hot Springs in Yellowstone National Park, Wyoming. Note the people on the trail for scale. — *Photo by Paul Chesley*

Worldwide, wells and springs provide water for cities, crops, livestock and industry. In the United States, groundwater is the source of about 40 percent of the water used for all purposes (except hydroelectric power generation and power-plant cooling). Groundwater is the drinking water for more than 50 percent of the population, is 40 percent of the water used for irrigation, and provides more than 25 percent of industry's needs. In some areas, however, overuse of this basic resource has resulted in water shortage, stream-flow depletion, land subsidence, contamination by saltwater, increased pumping cost, and groundwater pollution.

Importance of Underground Water

Groundwater is one of our most important and widely available resources, yet people's perceptions of the subsurface environment from which it comes are often unclear and incorrect. The reason is that the groundwater environment is largely hidden from view except in caves and mines, and the impressions people gain from these subsurface openings are misleading. Observations on the land surface give and impression that Earth is "solid." This view remains when we enter a cave and see water flowing in a channel that appears to have been cut into solid rock.

Because of such observations, many people believe that groundwater occurs only in underground "rivers." In reality, most of the subsurface environment is not "solid" at all. It includes countless tiny *pore spaces* between grains of soil and sediment, plus narrow joints and fractures in bedrock. Together, these spaces add up to an immense volume. It is in these small openings that groundwater collects and moves.

Considering the entire hydrosphere, or all of Earth's water, only about six-tenths of 1 percent occurs underground. Nevertheless, this small percentage, stored in the rocks and sediments beneath Earth's surface, is a vast quantity. When the oceans are excluded

and only sources of freshwater are considered, the significance of groundwater becomes more apparent.

Table 11.1 contains estimates of the distribution of fresh water in the hydrosphere. Clearly the largest volume occurs as glacial ice. Second in rank is groundwater, with slightly more than 14 percent of the total. However, when ice is excluded and just liquid water is considered, more than 94 percent of all fresh water is groundwater. Without question, *groundwater represents the largest reservoir of fresh water that is readily available to humans*. Its value in terms of economics and human well-being is incalculable.

Geologically, goundwater is important as an erosional agent. The dissolving action of groundwater slowly removes soluble rock such as limestone, allowing surface depressions known as *sinkholes* to form as well as creating subterranean caverns (Figure 11.1). Groundwater is also an equalizer of stream flow. Much of the water that flows in rivers is not direct runoff from rain and snowmelt. Rather, a large percentage of precipitation soaks in and then moves slowly underground to stream channels. Groundwater is thus a form of storage that sustains streams during periods when rain does not fall. The information in Table 11.1 reinforces this point. Here we see that the rate of exchange for groundwater is 280 years. This figure represents the amount of time required to replace the water now stored underground. By contrast, the rate of water exchange for rivers is just slightly more than 11 days: If the groundwater supply to a river were cut off and no rain fell, the river would run dry in just over 11 days. Thus, when we see water flowing in a river during a dry period, it represents rain that fell at some earlier time and was stored underground.

Distribution of Underground Water

When rain falls, some of the water runs off, some evaporates, and the remainder soaks into the ground. This last path is the primary source of practically all

Table 11.1 Fresh Water of the Hydrosphere

Parts of the Hydrosphere	Volume of Freshwater (km³)	Share of Total Volume of Freshwater (percent)	Rate of Water Exchange
Ice sheets and glaciers	24,000,000	84.945	8000 years
Groundwater	4,000,000	14.158	280 years
Lakes and reservoirs	155,000	0.549	7 years
Soil moisture	83,000	0.294	1 year
Water vapor in the atmosphere	14,000	0.049	9.9 days
River water	1,200	0.004	11.3 days
Total	28,253,200	100.000	

SOURCE: U.S. Geological Survey Water Supply Paper 2220, 1987

Figure 11.1 The Papoose Room at Carlsbad Caverns National Park, New Mexico, is the handiwork of groundwater. (Photo by Larry Ulrich/DRK Photo)

underground water. The amount of water that takes each of these paths, however, varies greatly both in time and space. Influential factors include steepness of slope, nature of surface material, intensity of rainfall, and type and amount of vegetation. Heavy rains falling on steep slopes underlain by impervious materials will obviously result in a high percentage of the water running off. Conversely, if rain falls steadily and gently upon more gradual slopes composed of materials that are easily penetrated by the water, a much larger percentage of water soaks into the ground.

Some if the water that soaks in does not travel far, because it is held by molecular attraction as a surface film on soil particles. A portion of this moisture evaporates back into the atmosphere. Much of the remainder is used by plants between rains. Water that is not held in this **belt of soil moisture** penetrates downward until it reaches a zone where all of the open spaces in sediment and rock are completely filled with water (Figure 11.2). This is the **zone of saturation**. Water within it is called **groundwater**. The upper limit of this zone is known as the **water table**. Extending upward from the water table is the **capillary fringe**. Here groundwater is held by surface tension in tiny passages between grains of soil or sediment. The area above the water table that includes the capillary fringe and the belt of soil moisture is called the **zone of aeration**. The open spaces are unsaturated and filled mainly with air.

 ## The Water Table

The water table, the upper limit of the zone of saturation, is a very significant feature of the groundwater system. The water table level is important in predicting the productivity of wells, explaining the changes in the flow of springs and streams, and accounting for fluctuations in the levels of lakes.

Although we cannot observe the water table directly, its elevation can be mapped and studied in detail where wells are numerous because the water level in wells coincides with the water table. Such maps reveal that the water table is rarely level, as we might expect a table to be. Instead, its shape is usually a subdued replica of the surface topography, reaching its highest elevations beneath hills and then descending toward valleys (Figure 11.2). Where a wetland (swamp) is encountered, the water table is right at the surface. Lakes and streams generally occupy areas low enough that the water table is above the land surface.

Several factors contribute to the irregular surface of the water table. The most important cause is that groundwater moves very slowly and at varying rates under different conditions. Because of this, water tends to "pile up" beneath high areas between stream valleys. If rainfall were to cease completely, these water table "hills" would slowly subside and gradually approach the level of the valleys. However, new supplies of rainwater are usually added frequently enough to prevent this. Nevertheless, in times of extended drought, the water table may drop enough to dry up shallow wells (Figure 11.2). Other causes for the uneven water table are variations in rainfall and permeability from place to place.

The relationship between the water table and a stream in a humid region is illustrated in Figure 11.3A. Even during dry periods, the movement of groundwater into the channel maintains a flow in the stream. In situations such as this, streams are said to be **effluent**.

By contrast, in arid regions, where the water table is far below the surface, groundwater cannot contribute to stream flow. Therefore, the only permanent streams in such areas are those that originate in wet regions and then happen to traverse the desert. (Examples are the Nile River in Egypt and the Colorado River in the American Southwest.) Under these conditions the zone of saturation beneath the valley floor is supplied by downward seepage from the stream channel, which, in turn, produces an upward bulge in the water table. Streams that provide water to the water table in this manner are called **influent streams** (Figure 11.3B).

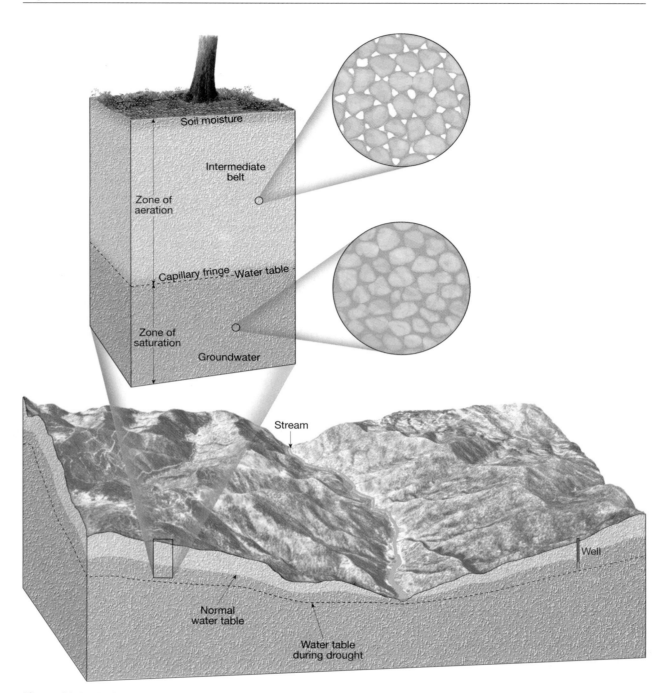

Figure 11.2 Distribution of underground water. The shape of the water table is usually a subdued replica of the surface topography. During periods of drought, the water table falls, reducing streamflow and drying up some wells.

Factors Influencing the Storage and Movement of Groundwater

The nature of subsurface materials strongly influences the rate of groundwater movement and the amount of groundwater that can be stored. Two factors are especially important—porosity and permeability.

Porosity

Water soaks into the ground because bedrock, sediment, and soil contain countless voids or openings. These openings are similar to those of a sponge and are often called pore spaces. The quantity of groundwater that can be stored depends on the **porosity** of the material, which is the percentage of the total volume of rock or sediment that consists of pore spaces. Voids

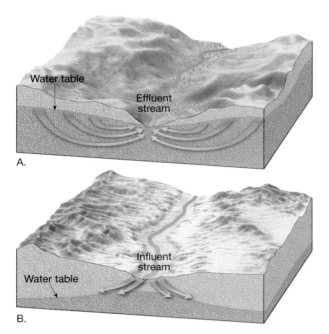

Figure 11.3 **A.** Effluent streams are characteristic of humid areas and are supplied by water from the zone of saturation. **B.** Influent streams are found in deserts. Seepage from such streams produces an upward bulge in the water table.

most often are spaces between sedimentary particles, but also common are joints, faults, cavities formed by the dissolving of soluble rock such as limestone, and vesicles (voids left by gases escaping from lava).

Variations in porosity can be great. Sediment is commonly quite porous, and open spaces may occupy 10 to 50 percent of the sediment's total volume. Pore space depends on the size and shape of the grains, how they are packed together, the degree of sorting, and in sedimentary rocks, the amount of cementing material. For example, clay may have a porosity as high as 50 percent,

whereas some gravels may have only 20 percent voids.

Where sediments of various sizes are mixed, the porosity is reduced because the finer particles tend to fill the openings among the larger grains. Most igneous and metamorphic rocks, as well as some sedimentary rocks, are composed of tightly interlocking crystals so the voids between the grains may be negligible. In these rocks, fractures must provide the voids.

Permeability, Aquitards, and Aquifers

Porosity alone cannot measure a material's capacity to yield groundwater. Rock or sediment may be very porous yet, still not allow water to move through it. The pores must be *connected* to allow water flow, and they must be *large enough* to allow flow. Thus, the **permeability** of a material, its ability to *transmit* a fluid, is also very important.

Groundwater moves by twisting and turning through small interconnected openings. The smaller the pore spaces, the slower the water moves. This idea is clearly illustrated by examining the information about the water-yielding potential of different materials in Table 11.2. Here groundwater is divided into two categories: (1) that portion which will drain under the influence of gravity (called *specific yield*) and (2) that part which is retained as a film on particle and rock surfaces and in tiny openings (called *specific retention*). Specific yield indicates how much water is actually available for use, whereas specific retention indicates how much water remains bound in the material. For example, clay's ability to store water is great owing to its high porosity, but its pore spaces are so small that water is unable to move through it. Thus, clay's porosity is high but because its permeability is poor, clay has a very low specific yield.

Table 11.2 Selected Values of Porosity, Specific Yield, and Specific Retention*

Material	Porosity	Specific Yield	Specific Retention
Soil	55	40	15
Clay	50	2	48
Sand	25	22	3
Gravel	20	19	1
Limestone	20	18	2
Sandstone (semiconsolidated)	11	6	5
Granite	0.1	0.09	0.01
Basalt (fresh)	11	8	3

*Values in percent by volume
SOURCE: U.S. Geological Survey Water Supply Paper 2220, 1987

Impermeable layers that hinder or prevent water movement are termed **aquitards**. Clay is a good example. On the other hand, larger particles, such as sand or gravel, have larger pore spaces. Therefore, the water moves with relative ease. Permeable rock strata or sediment that transmit groundwater freely are called **aquifers**. Sands and gravels are common examples.

In summary, you have seen that porosity is not always a reliable guide to the amount of groundwater that can be produced, and permeability is significant in determining the rate of groundwater movement and the quantity of water that might be pumped from a well.

Movement of Groundwater

We noted the common misconception that groundwater occurs in underground rivers that resemble surface streams. Although subsurface streams do exist, they are not common. Rather, as you learned in the preceding sections, groundwater exists in the pore spaces and fractures in rock and sediment. Thus, contrary to any impressions of rapid flow that an underground river might evoke, the movement of most groundwater is exceedingly slow, from pore to pore. By exceedingly slow, we mean typical rates of a few centimeters each day.

The energy that makes groundwater move is provided by the force of gravity. In response to gravity, water moves from areas where the water table is high to zones where the water table is lower. This means that water gravitates toward a stream channel, lake, or spring. Although some water takes the most direct path down the slope of the water table, much of the water follows long curving paths toward the zone of discharge.

Figure 11.4 shows how water percolates into a stream from all possible directions. Some paths actually turn upward, apparently against the force of gravity, and enter through the bottom of the channel. This is easily explained: The deeper you go into the zone of saturation, the greater is the water pressure. Thus, the looping curves followed by water in the saturated zone may be thought of as a compromise between the downward pull of gravity and the tendency of water to move toward areas of reduced pressure. As a result, water at any given height is under greater pressure beneath a hill than beneath a stream channel, and the water tends to migrate toward points of lower pressure.

The modern concepts of groundwater movement were formulated in the middle of the nineteenth century. During this period Henry Darcy, a French engineer studying the water supply of the city of Dijon in east-central France, formulated a law that now bears his name and is basic to an understanding of groundwater movement. Darcy found that if permeability remains

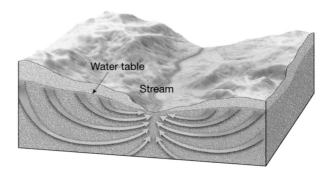

Figure 11.4 Arrows indicate groundwater movement through uniformly permeable material. The looping curves may be thought of as a compromise between the downward pull of gravity and the tendency of water to move toward areas of reduced pressure.

uniform, the velocity of groundwater will increase as the slope of the water table increases. The water table slope, known as the **hydraulic gradient**, is determined by dividing the vertical difference between the recharge and discharge points (a quantity known as the **head**) by the length of flow between these points. **Darcy's Law** can be expressed by the following formula:

$$V = K \frac{h}{l}$$

where V represents velocity, h the head, l the length of flow, and K a coefficient that accounts for a material's permeability.

Rates of groundwater movement have been determined directly in a number of ways. In one experiment dye is introduced into a well and the time is measured until the coloring agent appears in another well at a known distance from the first. Measurements of groundwater movements have also been made by applying radiometric dating techniques, specifically by using the radioactive isotope of carbon, carbon-14. Upon entering the ground the carbon dioxide dissolved in rainwater contains a characteristic amount of carbon-14. As the water gradually percolates through the ground, the radioactive carbon decays. The rate of movement is determined by measuring the distance between the well where the water was withdrawn and the recharge area, and dividing this by the radiometrically determined age.*

Experiments such as those just described have shown that the rate of groundwater movement is highly variable. Although a typical rate for many aquifers is about 15 meters per year (about 4 centimeters per day), velocities more than 15 times this figure have been measured in exceptionally permeable materials.

---

*A discussion of radiometric dating appears in Chapter 8.

⊙ Springs

Springs have aroused the curiosity and wonder of people for thousands of years. The fact that springs were, and to some people still are, rather mysterious phenomena is not difficult to understand, for here is water flowing freely from the ground in all kinds of weather in seemingly inexhaustible supply, but with no obvious source.

Not until the middle of the seventeenth century did the French physicist Pierre Perrault invalidate the age-old assumption that precipitation could not adequately account for the amount of water emanating from springs and flowing in rivers. Over several years, Perrault computed the quantity of water that fell on France's Seine River basin. He then calculated the mean annual runoff by measuring the river's discharge. After allowing for the loss of water by evaporation, he showed that there was sufficient water remaining to feed the springs. Thanks to Perrault's pioneering efforts and the measurements by many afterward, we now know that the source of springs is water from the zone of saturation and that the ultimate source of this water is precipitation.

Whenever the water table intersects Earth's surface, a natural outflow of groundwater results, which we call a **spring** (Figure 11.5). Springs such as the one pictured in Figure 11.5 form when an aquitard blocks the downward movement of groundwater and forces it to move laterally. Where the permeable bed outcrops, a spring results. Another situation leading to the formation of a spring is illustrated in Figure 11.6. Here an aquitard is situated above the main water

Figure 11.5 Spring in Arizona's Marble Canyon. (Photo by Michael Collier)

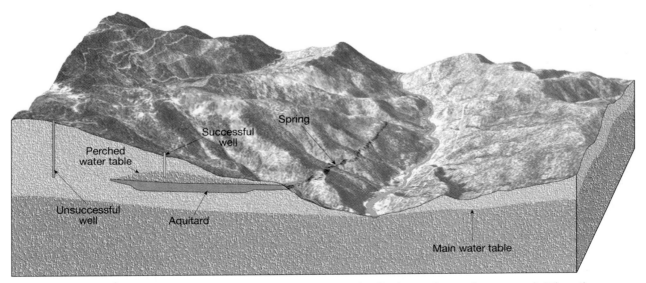

Figure 11.6 When an aquitard is situated above the main water table, a localized zone of saturation may result. Where the perched water table intersects the side of the valley, a spring flows. The perched water table also caused the well on the right to be successful, whereas the well on the left will be unsuccessful unless it is drilled to a greater depth.

table. As water percolates downward, a portion of it is intercepted by the aquitard, thereby creating a localized zone of saturation and a **perched water table**.

Springs, however, are not confined to places where a perched water table creates a flow at the surface. Many geological situations lead to the formation of springs because subsurface conditions vary greatly from place to place. Even in areas underlain by impermeable crystalline rocks, permeable zones may exist in the form of fractures or solution channels. If these openings fill with water and intersect the ground surface along a slope, a spring will result.

Hot Springs and Geysers

By definition, the water in **hot springs** is 6–9°C (10–15°F) warmer than the mean annual air temperature for the localities where they occur. In the United States alone, there are over 1000 such springs (Figure 11.7).

Temperatures in deep mines and oil wells usually rise with increasing depth, an average of about 2°C per 100 meters (1°F per 100 feet). Therefore, when groundwater circulates at great depths, it becomes heated. If it rises to the surface, the water may emerge as a hot spring. The water of some hot springs in the eastern United States is heated in this manner. However, the great majority (over 95 percent) of the hot springs (and geysers) in the United States are found in the West (Figure 11.7). The reason for such a distribution is that the source of heat for most hot springs is cooling igneous rock, and it is in the West that igneous activity has occurred more recently.

Geysers are intermittent hot springs or fountains in which columns of water are ejected with great force at various intervals, often rising 30–60 meters (100–200 feet) into the air. After the jet of water ceases, a column of steam rushes out, usually with a thunderous roar. Perhaps the most famous geyser in the world is Old Faithful in Yellowstone National Park, which erupts about once each hour (Figure 11.8). The great abundance, diversity, and spectacular nature of Yellowstone's geysers and other thermal features undoubtedly was the primary reason for its becoming the first national park in the United States. Geysers are also found in other parts of the world, notably New Zealand and Iceland. In fact, the Icelandic word *geysa*, to gush, gives us the name *geyser*.

Geysers occur where extensive underground chambers exist within hot igneous rocks. How they operate is shown in Figure 11.9. As relatively cool groundwater enters the chambers, it is heated by the surrounding rock. At the bottom of the chambers, the water is under great pressure because of the weight of the overlying water. This great pressure prevents the water from boiling at the normal surface temperature of 100°C (212°F). For example, water at the bottom of a 300-meter (1000-foot) water-filled chamber must attain nearly 230°C before it will boil. The heating causes the water to expand, with the result that some is forced out at the surface. This loss of water reduces the pressure on the remaining water in the chamber, which lowers the boiling point. A portion of the water deep within the chamber quickly turns to steam and the geyser erupts (Figure 11.9). Following eruption, cool groundwater again seeps into the chamber and the cycle begins anew.

When groundwater from hot springs and geysers flows out at the surface, material in solution is often precipitated, producing an accumulation of chemical sedimentary rock. The material deposited at any given place

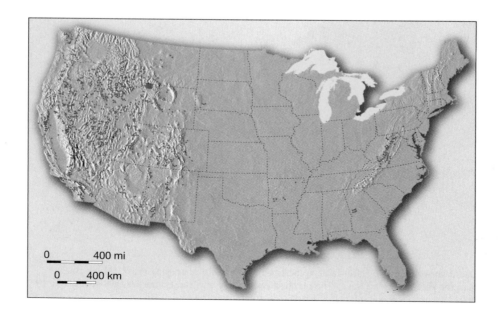

0 400 mi
0 400 km

Figure 11.7 Distribution of hot springs and geysers in the United States. Note the concentration in the West, where igneous activity has been most recent. (After G.A. Waring, U.S. Geological Survey Professional Paper 492, 1965)

Figure 11.8 Old Faithful, one of the world's most famous geysers, emits as much as 45,000 liters (almost 12,000 gallons) of hot water and steam about once each hour. (Photo by R. Scott Dunham)

commonly reflects the chemical makeup of the rock through which the water circulated. When the water contains dissolved silica, a material called *siliceous sinter* or *geyserite* is deposited around the spring. When the water contains dissolved calcium carbonate, a form of limestone called *travertine* or *calcareous tufa* is deposited. The latter term is used if the material is spongy and porous.

The deposits at Mammoth Hot Springs in Yellowstone National Park are more spectacular than most (Figure 11.10). As the hot water flows upward through a series of channels and then out at the surface, the reduced pressure allows carbon dioxide to separate and escape from the water. The loss of carbon dioxide causes the water to become supersaturated with calcium carbonate, which then precipitates. In addition to containing dissolved silica and calcium carbonate, some hot springs contain sulfur, which gives water a poor taste and unpleasant odor. Undoubtedly Rotten Egg Spring, Nevada, is such a situation.

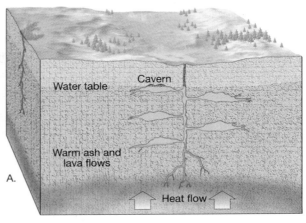
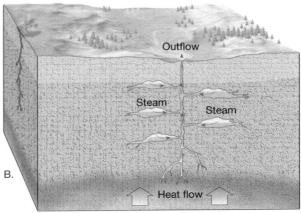
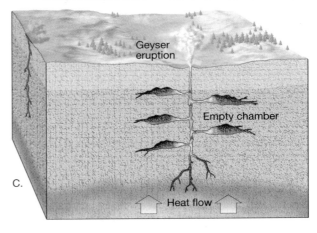

Figure 11.9 Idealized diagrams of a geyser. A geyser can form if the heat is not distributed by convection. **A.** In this figure, the water near the bottom is heated to near its boiling point. The boiling point is higher there than at the surface because the weight of the water above increases the pressure. **B.** The water higher in the geyser system is also heated; therefore, it expands and flows out at the top, reducing the pressure on the water at the bottom. **C.** At the reduced pressure on the bottom, boiling occurs. Some of the bottom water flashes into steam, and the expanding steam causes an eruption.

Figure 11.10 Mammoth Hot Springs at Yellowstone National Park. Although most of the deposits associated with geysers and hot springs in Yellowstone Park are silica-rich geyserite, the deposits at Mammoth Hot Springs consist of a form of limestone called travertine. (Photo by Stephen Trimble)

 ## Wells

The most common method for removing groundwater is the **well**, a hole bored into the zone of saturation. Wells serve as small reservoirs into which groundwater migrates and from which it can be pumped to the surface. The use of wells dates back may centuries and continues to be an important method of obtaining water today. By far the single greatest use of this water in the United States is irrigation for agriculture. More than 65 percent of the groundwater used each year is for this purpose. Industrial uses rank a distant second, followed by the amount used in city water systems and rural homes.

The water table level may fluctuate considerably during the course of a year, dropping during the dry seasons and rising following periods of rain. Therefore, to ensure a continuous supply of water, a well must penetrate below the water table. Whenever water is withdrawn from a well, the water table around the well is lowered. This effect, termed **drawdown**, decreases with increasing distance from the well. The result is a depression in the water table, roughly conical in shape, known as a **cone of depression** (Figure 11.11). Because the cone of depression increases the hydraulic gradient near the well, groundwater will flow more rapidly toward the opening. For most smaller domestic wells, the cone of depression is negligible. However, when wells are heavily pumped for irrigation or for industrial purposes, the withdrawal of water can be great enough to create a very wide and steep cone of depression. This may substantially lower the water table in an area and cause nearby shallow wells to become dry. Figure 11.11 illustrates this situation.

Digging a successful well is a familiar problem for people in areas where groundwater is the primary source of supply (see Box 11.1). One well may be successful at a depth of 10 meters (33 feet) whereas a neighbor may have to go twice as deep to find an adequate supply. Still others may be forced to go deeper or try a

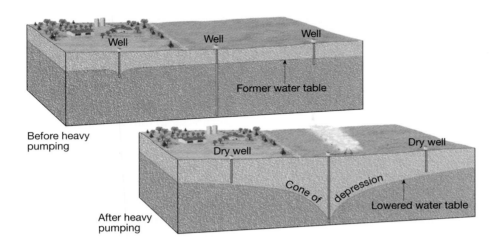

Figure 11.11 A cone of depression in the water table often forms around a pumping well. If heavy pumping lowers the water table, the shallow wells may be left dry.

Artesian Wells

different site altogether. When subsurface materials are heterogenous, the amount of water a well is capable of providing may vary a great deal over short distances. For example, when two nearby wells are drilled to the same level and only one is successful, it may be caused by the presence of a perched water table beneath one of them. Such a case is shown in Figure 11.6. Massive igneous and metamorphic rocks provide a second example. These crystalline rocks are usually not very permeable except where they are cut by many intersecting joints and fractures. Therefore, when a well drilled into such rock does not intersect an adequate network of fractures, it is likely to be unproductive.

In most wells, water cannot rise on its own. If water is first encountered at 30 meters depth, it remains at that level, fluctuating perhaps a meter or two with seasonal wet and dry periods. However, in some wells, water rises, sometimes overflowing at the surface. Such wells are abundant in the *artois* region of northern France, and so we call these self-rising wells *artesian*.

To many people the term *artesian* is applied to any well drilled to great depths. This use of the term is incorrect. Others believe that an artesian well must flow freely at the surface (Figure 11.12). Although this is a more correct notion than the first, it represents too narrow a definition. The term **artesian** is applied to *any* situation in which groundwater under pressure rises above the level of the aquifer. As we shall see, this does not always mean a free-flowing surface discharge.

For an artesian system to exist, two conditions must be met (Figure 11.13): (1) water must be confined to an aquifer that is inclined so that one end can receive water; and (2) aquitards, both above and below the aquifer, must be present to prevent the water from

escaping. When such a layer is tapped, the pressure created by the weight of the water above will force the water to rise. If there were no friction, the water in the well would rise to the level of the water at the top of the aquifer. However, friction reduces the height of the pressure surface. The greater the distance from the recharge area (area where water enters the inclined aquifer), the greater the friction and the less the rise of water.

In Figure 11.13, well 1 is a **nonflowing artesian well**, because at this location the pressure surface is below ground level. When the pressure surface is above the ground and a well is drilled into the aquifer, a **flowing artesian well** is created (well 2, Figure 11.13). It is important to realize that not all artesian systems are wells. *Artesian springs* also exist. Here groundwater reaches the surface by rising through a natural fracture rather than through an artificially produced hole.

Artesian systems act as conduits, often transmitting water great distances from remote areas of recharge to points of discharge. A well-known artesian system in South Dakota is a good example of this. In the western part of the state, the edges of a series of sedimentary layers have been bent up to the surface along the flanks of the Black Hills. One of these beds, the permeable Dakota Sandstone, is sandwiched between impermeable strata and gradually dips into the ground toward the east. When the aquifer was first tapped, water poured from the ground surface, creating fountains many meters high (Figure 11.14). In some places, the force of the water was sufficient to power waterwheels. Scenes such as the one pictured in Figure 11.14, however, can no longer occur, because thousands of additional wells now tap the same aquifer. This depleted the reservoir and the water table in the recharge area was lowered. As a consequence, the pressure dropped to the point where many wells stopped flowing altogether and had to be pumped.

Box 11.1

Dowsing for Water*

"Water dowsing" refers in general to the practice of using a forked stick, rod, pendulum, or similar device to locate underground water, minerals, or other hidden or lost substances, and has been a subject of discussion and controversy for hundreds of years.

Although tools and methods vary widely, most dowsers (also called diviners or water witches) probably still use the traditional forked stick, which may come from a variety of trees, including the willow, peach, and witch hazel (Figure 11.A). Other dowsers may use keys, wire coat hangers, pliers, wire rods, pendulums, or various kinds of elaborate boxes and electrical instruments.

In the classic method of using a forked stick, one fork is held in each hand with the palms upward and the bottom of the "Y" is pointed skyward. The dowser then walks back and forth over the area to be tested. When a source of water is detected, the butt end of the stick is supposed to rotate or be attracted downward.

According to dowsers, the attraction of the water may be so great that the bark peels off as the rod twists in the hands. Some dowsers are said to have suffered blistered or bloody hands from the twisting.

The exact origin of the divining rod in Europe is not known. The first detailed description of it is in Johannes Agricola's *De Re Metallica* (1556), a description of German mines and mining methods. The device was introduced into England during the reign of Elizabeth I (1558–1603) to locate mineral deposits, and soon afterward it was adopted as a water finder throughout Europe.

Water dowsing seems to be a mainly European cultural phenome-non, completely unknown to New World Indians and Eskimos. It was carried to America by some of the earliest settlers from England and Germany. Although the published record was very slight at first, water dowsing or witching began to be mentioned after 1675 in connection with witches and witchcraft. Two articles condemning it appeared in the 1821 and 1826 issues of the *American Journal of Science* and were among the first in a long line of treatises on water witching.

Despite almost unanimous condemnation by geologists and technicians, the practice of water dowsing has spread throughout America. Thousands of dowsers are currently active in the United States.

Case histories and demonstrations of dowsers may seem convincing, but when dowsing is exposed to scientific examination, it presents a very different picture. A study in Australia compared geologist's successes at locating groundwater with those of dowsers. The dowsers caused twice as many dry holes to be dug as the geologists. At Iowa State University, dowsers were given the task of finding water along a prescribed path on campus. They were not able to "discover" the water mains directly beneath them.

What does it mean to say that a dowser is successful? The dowser may find water, but how much? And of what quality? At what rate can it be withdrawn? For how long and with what impact on other wells and on nearby streams?

The natural explanation of "successful" water dowsing is that in many areas water would be hard to miss. The dowser commonly implies that the spot indicated by the rod is the *only* one where water could be found, but this is not necessarily true. In a region of adequate rainfall and favorable geology, it is difficult *not* to drill and find water!

Some water exists under the surface almost everywhere. This explains why many dowsers appear to be successful. To locate groundwater accurately, however, as to depth, quantity, and quality, a number of techniques must be used. Hydrologic, geologic, and geophysical knowledge is needed to determine the depths and extent of the different water-bearing strata and the quantity and quality of water found in each. The area must be thoroughly tested and studied to determine these facts.

*Much of this information is based on material prepared by the U.S. Geological Survey.

Figure 11.A Most dowsers still use the traditional forked stick.

On a different scale, city water systems can be considered examples of artificial artesian systems (Figure 11.15). The water tower, into which water is pumped, would represent the area of recharge, the pipes the confined aquifer, and the faucets in homes the flowing artesian wells.

Problems Associated with Groundwater Withdrawal

As with many of our valuable natural resources, groundwater is being exploited at an increasing rate. In some areas, overuse threatens the groundwater supply. In

Figure 11.12 Sometimes water flows freely at the surface when an artesian well is developed. However, for most artesian wells the water must be pumped to the surface. (Photo by James E. Patterson)

other places, groundwater withdrawal has caused the ground and everything resting on it to sink. Still other localities are concerned with the possible contamination of the groundwater supply.

Treating Groundwater as a Nonrenewable Resource

Many natural systems tend to establish a condition of equilibrium. The groundwater system is no exception. The water table's height reflects a balance between

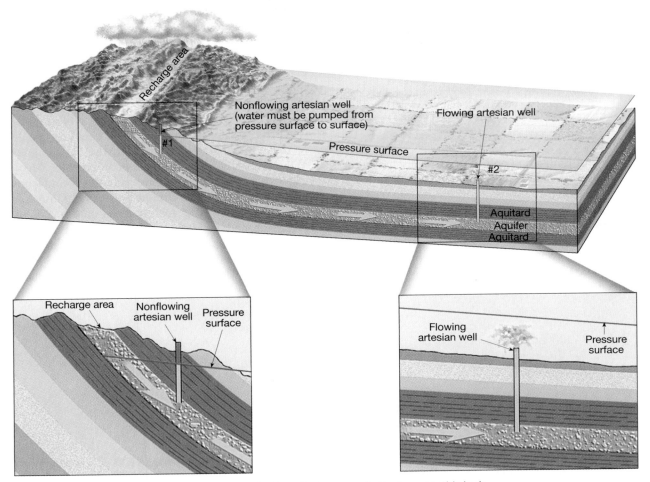

Figure 11.13 Artesian systems occur when an inclined aquifer is surrounded by impermeable beds.

Figure 11.14 A "gusherlike" flowing artesian well in South Dakota in the early part of the century. Thousands of additional wells now tap the same confined aquifer; thus, the pressure has dropped to the point that many wells stopped flowing altogether and have to be pumped. (Photo by N. H. Darton, U.S. Geological Survey)

the rate of infiltration and the rate of discharge and withdrawal. Any imbalance will either raise or lower the water table. Long-term imbalances can lead to a significant drop in the water table if there is either a decrease in recharge due to prolonged drought, or an increase in groundwater discharge or withdrawal.

For many, groundwater appears to be and endlessly renewable resource, because it is continually replenished by rainfall and melting snow. But in some regions, groundwater has been and continues to be treated as a *nonrenewable* resource. Where this occurs, the water available to recharge the aquifer falls significantly short of the amount being withdrawn.

The High Plains provides one example. Here an extensive agricultural economy is largely dependent on irrigation (Figure 11.16). In some parts of the region, where intense irrigation has been practiced for an extended period, depletion of groundwater has been severe. Under these circumstances, it can be said that the groundwater is literally being "mined." Even if pumping were to cease immediately, it could take hundreds or thousands of years for the groundwater to be fully replenished. Box 11.2 explores this issue more fully.

Subsidence

As you shall see later in this chapter, surface subsidence can result from natural processes related to groundwater. However, the ground may also sink when water is pumped from wells faster than natural recharge processes can replace it. This effect is particularly pronounced in areas underlain by thick layers of unconsolidated sediments. As the water is withdrawn, the water pressure drops and the weight of the overburden is transferred to the sediment. The greater pressure packs the sediment grains tightly together and the ground subsides.

Many areas may be used to illustrate land subsidence caused by the excessive pumping of groundwater from relatively loose sediment. A classic example in the United States occurred in the San Joaquin Valley of California and is discussed in Box 11.3. Many other cases of land subsidence due to groundwater pumping exist in the United States including Las Vegas, Nevada, New Orleans and Baton Rouge, Louisiana, and the Houston–Galveston area of Texas. In the low-lying coastal area between Houston and Galveston, land subsidence ranges from 1.5 to 3 meters (5 to 9 feet). The result is that about 78 square kilometers (30 square miles) are permanently flooded.

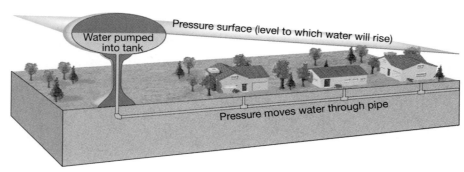

Figure 11.15 City water systems can be considered to be artificial artesian systems.

Figure 11.16 In some agricultural regions, water is pumped from the ground faster than it is replenished. In such instances, groundwater is being treated as a nonrenewable resource. (Photo by Bill Kamin/Visuals Unlimited)

Outside the United States, one of the most spectacular examples of subsidence occurred in Mexico City, which is built on a former lake bed. In the first half of the twentieth century thousands of wells were sunk into the water-saturated sediments beneath the city. As water was withdrawn, portions of the city subsided by as much as 6 to 7 meters. In some places buildings have sunk to such a point that access to them from the street is at what used to be the second-floor level!

Saltwater Contamination

In many coastal areas the groundwater resource is being threatened by the encroachment of salt water. To understand this problem, we must examine the relationship between fresh groundwater and salt groundwater. Figure 11.17A is a diagrammatic cross-section that illustrates this relationship in a coastal area underlain by permeable homogenous materials. Fresh water is less dense than salt water, so it floats on the salt water and forms a large, lens-shaped body that may extend to considerable depths below sea level. In such a situation, if the water table is 1 meter above sea level, the base of the freshwater body will extend to a depth of about 40 meters below sea level. Stated another way, the depth of the fresh water below sea level is about 40 times greater than the elevation of the water table above sea level. Thus, when excessive pumping lowers the water table by a certain amount, the bottom of the freshwater zone will rise by 40 times that amount. Therefore, if groundwater

withdrawal continues to exceed recharge, there will come a time when the elevation of the salt water will be sufficiently high to be drawn into wells, thus contaminating the freshwater supply (Figure 11.17B). Deep wells and wells near the shore are usually the first to be affected.

In urbanized coastal areas, the problems created by excessive pumping are compounded by a decrease in the rate of natural recharge. As more and more of the surface is covered by streets, parking lots, and buildings, infiltration into the soil is diminished.

In an attempt to correct the problem of saltwater contamination of groundwater resources, a network of recharge wells may be used. These wells allow waste water to be pumped back into the groundwater system. A second method of correction is accomplished by building large basins. These basins collect surface drainage and allow it to seep into the ground. On New York's Long Island, where the problem of saltwater contamination was recognized more than 40 years ago, both of these methods have been employed with considerable success.

Contamination of freshwater aquifers by salt water is primarily a problem in coastal areas, but it can also threaten noncoastal locations. Many ancient sedimentary rocks of marine origin were deposited when the ocean covered places that are now far inland.

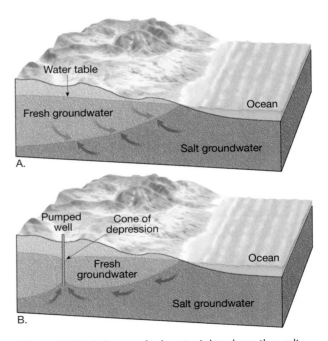

Figure 11.17 A. Because fresh water is less dense than salt water, it floats on the salt water and forms a lens-shaped body that may extend to considerable depths below sea level. **B.** When excessive pumping lowers the water table, the base of the freshwater zone will rise by 40 times that amount. The result may be saltwater contamination of wells.

Box 11.2

The Ogallala Aquifer—How Long Will the Water Last?

The High Plains extend from the western Dakotas south to Texas. Despite being a land of little rain, this is one of the most important agricultural regions in the United States. The reason is a vast endowment of groundwater that makes irrigation possible throughout most of the region's 450,000 square kilometers (175,000 square miles). The source of most of this water is the Ogallala Formation, the largest aquifer in the United States (Figure 11.B).

Geologically, the Ogallala Formation is young, consisting of a number of sandy and gravelly rock layers of late Tertiary and Quaternary age (see geologic time scale Figure 8.15, p. 205). The sediments were derived from the erosion of the Rocky Mountains and carried eastward by sluggish streams. Erosion has removed much of the formation from eastern Colorado, severing the Ogallala's connection to the Rockies.

The Ogallala Formation averages 60 meters (200 feet) thick and in places it is as much as 180 meters (600 feet) thick. Groundwater in the aquifer originally came downslope from the Rockies as well as from surface precipitation that soaked into the ground over thousands of years. Because of its high porosity and great size, the Ogallala accumulated huge amounts of groundwater—enough fresh water to fill Lake Huron! Today, with the connection between the aquifer and the Rockies severed, all of the Ogallala's recharge must come from the region's meager rainfall.

The Ogallala aquifer was first used for irrigation in the late 1800s, but its use was limited by the capacity of the pumps available at the time. In the 1920s, with the development of large-capacity irrigation pumps, High Plains farmers, particularly in Texas, began tapping the Ogallala for irrigation. Then in the 1950s, improved technology brought large-scale exploitation of the aquifer. Today, nearly 170,000 wells are being used to irrigate more than 65,000 square kilometers (16 million acres) of land.

With the increase in irrigation has come a drastic drop in the Ogallala's water table, especially in the southern High Plains. Declines in the water table of 3 to 15 meters (10 to 50 feet) are common, and in places the water table is now 60 meters (200 feet) below its level prior to irrigation.

Beginning in the 1980s, irrigated acreage has declined in the High Plains. One important reason has been higher energy costs. As water levels have dropped, the costs of pumping groundwater to the surface have increased.

Although the water table decline has slowed in parts of the southern High Plains, substantial pumping continues, often in excess of recharge. The future of irrigated farming in this region is clearly in jeopardy.

The southern High Plains represent an area of the United States that will return, sooner or later, to dry-land farming. The transition will come sooner and with fewer ecological and economic crises if the agricultural industry is weaned gradually from its dependence on groundwater irrigation. If nothing is done until all the accessible water in the Ogallala aquifer has been removed, the transition will be ecologically dangerous and economically dreadful.*

*National Research Council. *Solid-Earth Sciences and Society*. Washington, DC: National Academy Press, 1993, p. 148.

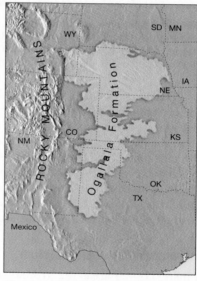

Figure 11.B The Ogallala Formation underlies about 450,000 square kilometers (175,000 square miles) of the High Plains, making it the largest aquifer in the United States.

In some instances, significant quantities of seawater were trapped and still remain in the rock. These strata sometimes contain quantities of fresh water and may be pumped for use by people. However, if fresh water is removed more rapidly than it is replenished, saline water may encroach and render the wells unusable. Such a situation threatened users of a deep (Cambrian age) sandstone aquifer in the Chicago area. To counteract this, water from Lake Michigan was allocated to the affected communities to offset the rate of withdrawal from the aquifer.

Groundwater Contamination

The pollution of groundwater is a serious matter, particularly in areas where aquifers provide a large part of the water supply. One common source of groundwater pollution is sewage. Its sources include an ever increasing number of septic tanks, as well as inadequate or broken sewer systems and farm wastes.

If sewage water that is contaminated with bacteria enters the groundwater system, it may become

Box 11.3

Land Subsidence in the San Joaquin Valley

The San Joaquin Valley is a broad structural basin that contains a thick fill of sediments. The size of Maryland, it constitutes the southern two-thirds of California's Central Valley, a flatland separating two mountain ranges, the Coast Ranges to the west and the Sierra Nevada to the east (Figure 11.C). The valley's aquifer system is a mixture of alluvial materials derived from the surrounding mountains. Sediment thicknesses average about 870 meters (about half a mile). The valley's climate is arid to semiarid, with average annual precipitation ranging from 12 to 35 centimeters (5 to 14 inches).

The San Joaquin Valley has a strong agricultural economy that requires large quantities of water for irrigation. For many years up to 50 percent of this need was met by groundwater. In addition, nearly every city in the region uses groundwater as its principal source for homes and industry.

Although development of the valley's groundwater for irrigation began in the late 1800s, land subsidence did not begin until the mid-1920s when withdrawals were substantially increased. By the early 1970s, water levels had declined up to 120 meters (400 feet). The resulting ground subsidence exceeded 8.5 meters (29 feet) at one place in the region (Figure 11.D). At that time, areas within the valley were subsiding faster than 0.3 meter (1 foot) per year.

Then, because surface water was being imported and groundwater pumping was being decreased, water levels in the aquifer recovered and subsidence ceased. However, during a drought in 1976–1977, heavy groundwater pumping led to renewed subsidence. This time, water levels dropped much faster because of the reduced storage capacity caused by earlier compaction of sediments. In all, one-half the entire valley was affected by subsidence. According to the U. S. Geological Survey:

> Subsidence in the San Joaquin Valley probably represents one of the greatest single manmade alterations in the configuration of the Earth's surface....It has caused serious and costly problems in construction and maintenance of water-transport structures, highways, and highway structures; also many millions of dollars have been spent on the repair or replacement of deep-water wells. Subsidence, besides changing the gradient and course of valley creeks and streams, has caused unexpected flooding, costing farmers many hundreds of thousands of dollars in recurrent land leveling.*

Similar effects have been documented in the San Jose area of the Santa

Figure 11.D The marks on this utility pole indicate the level of the surrounding land in preceding years. Between 1925 and 1975 this part of the San Joaquin Valley subsided almost 9 meters because of the withdrawal of groundwater and the resulting compaction of sediments. (Photo courtesy of U.S. Geological Survey)

Clara Valley, California, where, between 1916 and 1966, subsidence approached 4 meters. Flooding of lands bordering the southern part of San Francisco Bay was one of the results. As was the case in the San Joaquin Valley, the subsidence stopped when imports of surface water were increased, allowing groundwater withdrawal rates to be decreased.

*R. L. Ireland, J. F. Poland, and F. S. Riley, *Land Subsidence in the San Joaquin Valley, California, as of 1980*, U.S. Geological Survey Professional Paper 437-I (Washington, DC: U.S. Government Printing Office, 1984), p. 11.

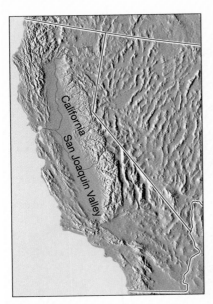

Figure 11.C The shaded area shows California's San Joaquin Valley.

purified through natural processes. The harmful bacteria may be mechanically filtered by the sediment through which the water percolates, destroyed by chemical oxidation, and/or assimilated by other organisms. For purification to occur, however, the aquifer must be of the correct composition. For example, extremely permeable aquifers (such as highly fractured crystalline rock, coarse gravel, or cavernous

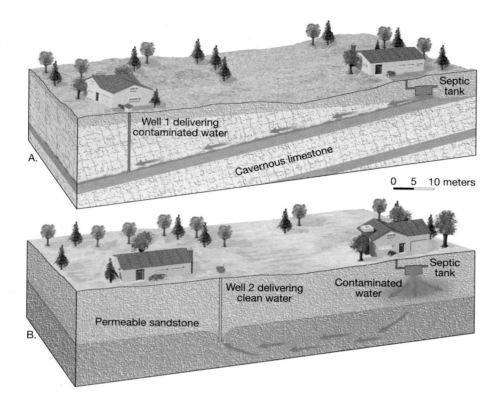

Figure 11.18 A. Although the contaminated water has traveled more than 100 meters before reaching well 1, the water moves too rapidly through the cavernous limestone to be purified. **B.** As the discharge from the septic tank percolates through the permeable sandstone, it is purified in a relatively short distance.

limestone) have such large openings that contaminated groundwater may travel long distances without being cleansed. In this case, the water flows too rapidly and is not in contact with the surrounding material long enough for purification to occur. This is the problem at well 1 in Figure 11.18A.

On the other hand, when the aquifer is composed of sand or permeable sandstone, it can sometimes be purified after traveling only a few dozen meters through it. The openings between sand grains are large enough to permit water movement, yet the movement of the water is slow enough to allow ample time for its purification (well 2, Figure 11.18B).

Sometimes sinking a well can lead to groundwater pollution problems. If the well pumps a sufficient quantity of water, the cone of depression will locally increase the slope of the water table. In some instances, the original slope may even be reversed. This could lead to the contamination of wells that yielded unpolluted water before heavy pumping began (Figure 11.19). Also recall that the rate of groundwater movement increases as the slope of the water table steepens. This could produce problems because a faster rate of movement allows less time for the water to be purified in the aquifer before it is pumped to the surface.

Other sources and types of contamination also threaten groundwater supplies (Figure 11.20). These include widely used substances such as highway salt, fertilizers that are spread across the land surface, and

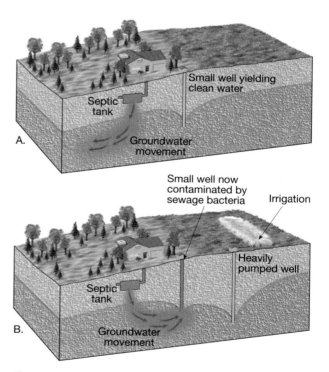

Figure 11.19 A. Originally the outflow from the septic tank moved away from the small well. **B.** The heavily pumped well changed the slope of the water table, causing contaminated groundwater to flow toward the small well.

Figure 11.20 Sometimes, agricultural chemicals **A.** and materials leached from landfills **B.** find their way into the groundwater. These are two of the potential sources of groundwater contamination. (Photo A by Roy Morsch/The Stock Market; Photo B by F. Rossotto/The Stock Market)

pesticides. In addition, a wide array of chemicals and industrial materials may leak from pipelines, storage tanks, landfills, and holding ponds. Some of these pollutants are classified as *hazardous*, meaning that they are either flammable, corrosive, explosive, or toxic. In land disposal, potential contaminants are heaped onto mounds or spread directly over the ground. As rainwater oozes through the refuse, it may dissolve a variety of organic and inorganic materials. If the leached material reaches the water table, it will mix with the groundwater and contaminate the supply. Similar problems may result from leakage of shallow excavations called holding ponds into which a variety of liquid wastes are disposed.

Because groundwater movement is usually slow, polluted water can go undetected for a long time. In fact, most contamination is discovered only after drinking water has been affected and people become ill. By this time, the volume of polluted water may be very large, and even if the source of contamination is removed immediately, the problem is not solved. Although the sources of groundwater contamination are numerous, the solutions are relatively few.

Once the source of the problem has been identified and eliminated, the most common practice is simply to abandon the water supply and allow the pollutants to be flushed away gradually. This is the least costly and easiest solution, but the aquifer must remain unused for many years. To accelerate this process, polluted water is sometimes pumped out and treated. Following removal of the tainted water, the aquifer is allowed to recharge naturally or, in some cases, the treated water

or other fresh water is pumped back in. This process is costly, time-consuming, and it may be risky because there is no way to be certain that all of the contamination has been removed. Clearly, the most effective solution to groundwater contamination is prevention.

The Geologic Work of Groundwater

Groundwater dissolves rock. This fact is the key to understanding how caverns and sinkholes form. Because soluble rocks, especially limestone, underlie millions of square kilometers of Earth's surface, it is here that the groundwater carries on its important role as an erosional agent. Limestone is nearly insoluble in pure water, but is quite easily dissolved by water containing small quantities of carbonic acid, and most groundwater contains this acid. It forms because rainwater readily dissolves carbon dioxide from the air and from decaying plants. Therefore, when groundwater comes in contact with limestone, the carbonic acid reacts with the calcite (calcium carbonate) in the rocks to form calcium bicarbonate, a soluble material that is then carried away in solution.

Caverns

The most spectacular results of groundwater's erosional handiwork are limestone **caverns**. In the United States alone about 17,000 caves have been discovered and new ones are being found every year. Although most are relatively small, some have spectacular dimensions. Mammoth Cave in Kentucky and Carlsbad Caverns in southeastern New Mexico are famous examples. The Mammoth Cave system is the most extensive in the world, with more than 540 kilometers of interconnected passages. The dimensions at Carlsbad Caverns are impressive in a different way. Here we find the largest and perhaps most spectacular single chamber. The Big Room at Carlsbad Caverns has an area equivalent to fourteen football fields and enough height to accommodate the U.S. Capitol Building.

Most caverns are created at or just below the water table in the zone of saturation. Here acidic groundwater follows lines of weakness in the rock, such as joints and bedding planes. As time passes, the dissolving process slowly creates cavities and gradually enlarges them into caverns. Material that is dissolved by the groundwater is eventually discharged into streams and carried to the ocean.

In many caves, development has occurred at several levels, with the current cavern-forming activity occurring at the lowest elevation. This situation reflects the close relationship between the formation of major subterranean passages and the river valleys into which they drain. As streams cut their valleys deeper, the water table drops as the elevation of the river drops. Consequently, during periods when surface streams are rapidly downcutting, surrounding groundwater levels drop rapidly and cave passages are abandoned by the water while the passages are still relatively small in cross-sectional area. Conversely, when the entrenchment of streams is slow or negligible, there is time for large cave passages to form.

Certainly the features that arouse the greatest curiosity for most cavern visitors are the stone formations that give some caverns a wonderland appearance. These are not erosional features, like the cavern itself, but depositional features created by the seemingly endless dripping of water over great spans of time. The calcium carbonate that is left behind produces the limestone we call travertine. These cave deposits, however, are also commonly called *dripstone*, an obvious reference to their mode of origin. Although the formation of caverns takes place in the zone of saturation, the deposition of dripstone is not possible until the caverns are above the water table in the zone of aeration. As soon as the chamber is filled with air, the stage is set for the decoration phase of cavern building to begin.

The various dripstone features found in caverns are collectively called **speleothems**, no two of which are exactly alike (Figure 11.21). Perhaps the most familiar speleothems are **stalactites**. These icicle-like pendants hang from the ceiling of the cavern and form where water seeps through cracks above. When the water reaches air in the cave, some of the dissolved carbon dioxide escapes from the drop and calcite precipitates. Deposition occurs as a ring around the edge of the water drop. As drop after drop follows, each leaves an infinitesimal trace of calcite behind, and a hollow limestone tube is created. Water then moves through the tube, remains suspended momentarily at the end, contributes a tiny ring of calcite, and falls to the cavern floor. The stalactite just described is appropriately called a *soda straw* (Figure 11.22). Often the hollow tube of the soda straw becomes plugged or its

Figure 11.21 Speleothems are of many types, including stalactites, stalagmites, and columns. Chinese Theater, Carlsbad Caverns National Park. (Photo by David Muench Photography)

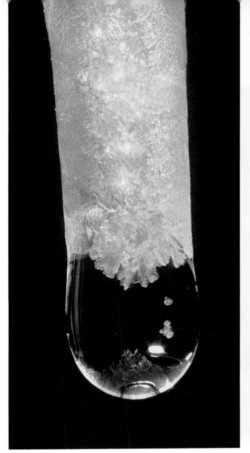

A.

B.

Figure 11.22 A. A "live" solitary soda straw stalactite. (Photo by Clifford Stroud, National Park Service) B. A soda straw "forest" in Carlsbad Caverns. (Photo courtesy of the National Park Service)

supply of water increases. In either case, the water is forced to flow, and hence deposit, along the outside of the tube. As deposition continues, the stalactite takes on the more common conical shape.

Speleothems that form on the floor of a cavern and reach upward toward the ceiling are called **stalagmites**. The water supplying the calcite for stalagmite growth falls from the ceiling and splatters over the surface. As a result, stalagmites do not have a central tube and are usually more massive in appearance and rounded on their upper ends than stalactites. Given enough time, a downward-growing stalactite and an upward-growing stalagmite may join to form a *column*.

Karst Topography

Many areas of the world have landscapes that to a large extent have been shaped by the dissolving power of groundwater. Such areas are said to exhibit **karst topography**, named for the Kras Plateau in Slovenia (formerly a part of Yugoslavia) located along the northeastern shore of the Adriatic Sea where such topography is strikingly developed. In the United States, karst landscapes occur in many areas that are underlain by limestone, including portions of Kentucky, Tennessee, Alabama, southern Indiana, and central and northern Florida. Generally, arid and semi-arid areas are too dry

to develop karst topography. When solution features exist in such regions, they are likely to be remnants of a time when rainier conditions prevailed.

Karst areas typically have irregular terrain punctuated with many depressions, called **sinkholes** or **sinks**. In the limestone areas of Florida, Kentucky, and southern Indiana, there are literally tens of thousands of these depressions varying in depth from just a meter of two to a maximum of more than 50 meters (Figure 11.23).

Sinkholes commonly form in two ways. Some develop gradually over many years without any physical disturbance to the rock. In these situations, the limestone immediately below the soil is dissolved by downward-sweeping rainwater that is freshly charged with carbon dioxide. With time, the bedrock surface is lowered and the fractures into which the water seeps are enlarged. As the fractures grow in size, soil subsides into the widening voids, from which it is removed by groundwater flowing in the passages below. These depressions are usually shallow and have gentle slopes.

By contrast, sinkholes can also form abruptly and without warning when the roof a cavern collapses under its own weight. Typically, the depressions created in this manner are steep-sided and deep. When they form in populous areas, they may represent a serious geologic hazard. Such a case is described in Box 11.4.

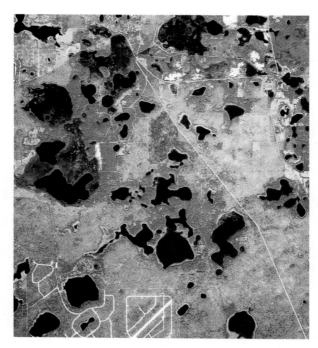

A.

B.

Figure 11.23 A. This high-altitude infrared image shows an area of karst topography in central Florida. The numerous lakes occupy sinkholes. (Courtesy of USDA–ASCS) **B.** This small sinkhole formed suddenly in 1991 when the roof of a cavern collapsed, destroying this home in Frostproof, Florida. (Photo by *St. Petersburg Times*/Gamma Liaison)

In addition to a surface pockmarked by sinkholes, karst regions characteristically show a striking lack of surface drainage (streams). Following a rainfall, the runoff is quickly funneled below ground through sinks. It then flows through caverns until it finally reaches the water table. Where streams do exist at the surface, their paths are usually short. The names of such streams often give a clue to their fate. In the Mammoth Cave area of Kentucky, for example, there is Sinking Creek, Little Sinking Creek, and Sinking Branch. Some sinkholes become plugged with clay and debris, creating small lakes or ponds. The development of a karst landscape is depicted in Figure 11.24.

Chapter Summary

- As a resource, *groundwater* represents the largest reservoir of freshwater that is readily available to humans. Geologically, the dissolving action of groundwater produces *caves* and *sinkholes*. Groundwater is also an equalizer of stream flow.

- Groundwater is that water which completely fills the pore spaces in sediment and rock in the subsurface *zone of saturation*. The upper limit of this zone is the *water table*. The *zone of aeration* is above the water table where the soil, sediment, and rock are not saturated.

- Materials with very small pore spaces (such as clay) hinder or prevent groundwater movement and are called *aquitards*. *Aquifers* consist of materials with larger pore spaces (such as sand) that are permeable and transmit groundwater freely.

- Groundwater moves in looping curves that are a compromise between the downward pull of gravity and the tendency of water to move toward areas of reduced pressure.

- *Springs* occur whenever the water table intersects the land surface and a natural flow of groundwater results. *Wells*, openings bored into the zone of saturation, withdraw groundwater and create roughly conical depressions in the water table known as *cones of depression*. *Artesian wells* occur when water rises above the level at which it was initially encountered.

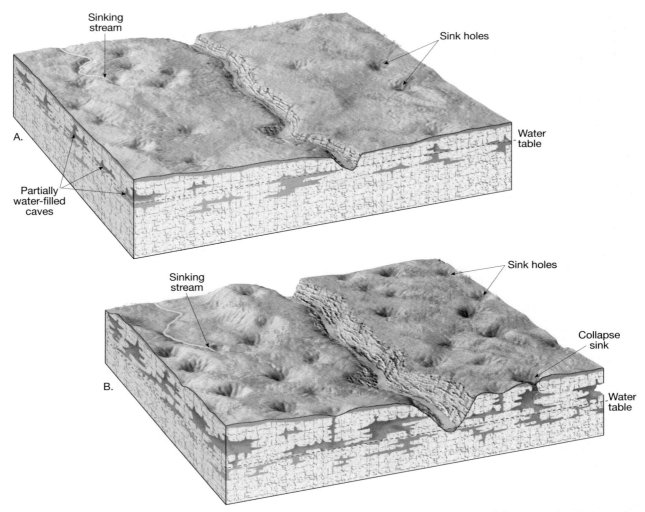

Figure 11.24 Development of a karst landscape. During early stages, groundwater percolates through limestone along joints and bedding planes. Solution activity creates and enlarges caverns at and below the water table. **A.** In this view, sinkholes are well developed and surface streams are funneled below ground. **B.** With the passage of time, caverns grow larger and the number and size of sinkholes increase. Collapse of caverns and coalescence of sinkholes form larger, flat-footed depressions. Eventually solution activity may remove most of the limestone from the area, leaving only isolated remnants.

- When groundwater circulates at great depths, it becomes heated. If it rises, the water may emerge as a *hot spring*. *Geysers* occur when groundwater is heated in underground chambers, expands, and some water quickly changes to steam, causing the geyser to erupt. The source of heat for most hot springs and geysers is hot igneous rock.

- Some of the current environmental problems involving groundwater include (1) *overuse* by intense irrigation, (2) *land subsidence* caused by groundwater withdrawal, (3) *saltwater contamination*, and (4) contamination by pollutants.

- Most *caverns* form in limestone at or below the water table when acidic groundwater dissolves rock along lines of weakness, such as joints and bedding planes. The various *dripstone* features found in caverns are collectively called *speleothems*. Landscapes that to a large extent have been shaped by the dissolving power of groundwater exhibit *karst topography*, an irregular terrain punctuated with many depressions, called *sinkholes* or *sinks*.

Box 11.4

The Winter Park Sinkhole

The craterlike sinkhole in Figure 11.E began forming in Winter Park, Florida, on May 8, 1981. Newspaper accounts, such as the one that follows, were front-page news and made this sinkhole one of the most publicized ever.

Sinkhole Nibbles Away at Florida City

Winter Park, Fla. (AP)—A giant sinkhole—already several hundred feet wide after swallowing a three-bedroom bungalow, half a swimming pool and six Porsches —nibbled away at a side street yesterday and threatened a main thoroughfare.

"It has slowed down, but it hasn't quit," said Winter Park Fire Capt. Gus LaGarde.

The crater, estimated at between 450 and 600 feet wide and 125 to 170 feet deep, grew by eight to 10 feet yesterday and was filling with water, authorities said.

It developed Friday night and opened rapidly Saturday, when it gulped the wood-frame cottage, cars and part of a foreign car lot and wrecked the city's $150,000 municipal swimming pool.

It was slowly eating its way west yesterday, leaving a group of businesses, their backs lost in Saturday's slide, hanging at the edge of the pit, LaGarde said.

The hole devoured most of a side street yesterday and was about 50 feet from one main thoroughfare and moving closer to several others, he said.

"We're still losing some of the perimeter," he said, "It doesn't appear to get any deeper...it continues to eat up the roadway, power poles, anything that gets in the way."

Residents and owners of nearby homes and businesses were warned on Saturday to leave until the sinkhole stopped growing. Some people rented trucks and began moving furniture and other property."*

Sinkhole formation is not uncommon in northern and central Florida. In fact, the Winter Park event was just one of three that occurred in the area over a two-week span. In each case the collapse at the surface was probably triggered by a lowering of the water table brought about by a severe drought. As the water table dropped, the roofs of the underground cavities lost support and fell into the voids below.

*Courtesy of Associated Press.

Figure 11.E The Winter Park, Florida, sinkhole that grew rapidly for 3 days, swallowing part of a community swimming pool as well as several businesses, houses, and automobiles. (Photo by Leif Skoogfors/Woodfin Camp and Associates)

Review Questions

1. What percentage of fresh water is groundwater? If glacial ice is excluded and only liquid fresh water is considered, about what percentage is groundwater?

2. Geologically, groundwater is important as an erosional agent. Name another significant geological role for groundwater.

3. Compare and contrast the zones of aeration and saturation. Which of these zones contains groundwater?

4. Although we usually think of tables as being flat, the water table generally is not. Explain.

5. What is an effluent stream? How does an influent stream differ?

6. Distinguish between porosity and permeability.

7. What is the difference between an aquitard and an aquifer?

8. Under what circumstances can a material have a high porosity but not be a good aquifer?

9. As illustrated in Figure 11.4 (p. 272), groundwater moves in looping curves. What factors cause the water to follow such paths?

10. Briefly describe the important contribution to our understanding of groundwater movement made by Henry Darcy.

11. When an aquitard is situated above the main water table, a localized saturated zone may be created. What term is applied to such a situation?

12. What is the source of heat for most hot springs and geysers? How is this reflected in the distribution of these features?

13. Two neighbors each dig a well. Although both wells penetrate to the same depth, one neighbor is successful and the other is not. Describe a circumstance that might explain what happened.

14. What is meant by the term *artesian*?

15. In order for artesian wells to exist, two conditions must be present. List these conditions.

16. When the Dakota Sandstone was first tapped, water poured freely from many artesian wells. Today these wells must be pumped. Explain.

17. What problem is associated with the pumping of groundwater for irrigation in the southern part of the High Plains? (See Box 11.2, p. 282.)

18. Briefly explain what happened in the San Joaquin Valley as the result of excessive groundwater withdrawal. (See Box 11.3, p. 283.)

19. In a particular coastal area the water table is 4 meters above sea level. Approximately how far below sea level does the fresh water reach?

20. Why does the rate of natural groundwater recharge decrease as urban areas develop?

21. Which aquifer would be most effective in purifying polluted groundwater: coarse gravel, sand, or cavernous limestone?

22. What is meant when a groundwater pollutant is classified as hazardous?

23. Name two common speleothems and distinguish between them.

24. Speleothems form in the zone of saturation. True or False? Briefly explain your answer.

25. Areas whose landscapes largely reflect the erosional work of groundwater are said to exhibit what kind of topography?

26. Describe two ways in which sinkholes are created.

Key Terms

aquifer (p. 272)
aquitard (p. 272)
artesian (p. 277)
belt of soil moisture (p. 269)
capillary fringe (p. 269)
cavern (p. 285)
cone of depression (p. 276)
Darcy's law (p. 272)

drawdown (p. 276)
effluent stream (p. 269)
flowing artesian well (p. 277)
geyser (p. 274)
groundwater (p. 269)
head (p. 272)
hot spring (p. 274)
hydraulic gradient (p. 272)

influent stream (p. 269)
karst topography (p. 287)
nonflowing artesian well (p. 277)
perched water table (p. 274)
permeability (p. 271)
porosity (p. 270)
sinkhole (sink) (p. 287)
speleothem (p. 286)

spring (p. 273)
stalactite (p. 286)
stalagmite (p. 287)
water table (p. 269)
well (p. 276)
zone of aeration (p. 269)
zone of saturation (p. 269)

Web Resources

The *Earth* Home Page provides on-line resources for this chapter on the World Wide Web. You will find review exercises, specific updates for items in the chapter, suggested reading, and links to interesting related pathways on the Internet. Visit the *Earth* Home Page at **http://www.prenhall.com/tarbuck.**

CHAPTER 12

Glaciers and Glaciation

Left Icebergs form when portions of a glacier's leading edge break off into an adjacent body of water. — *Photo by Liaison International*

Today glaciers cover nearly 10 percent of Earth's land surface; however, in the recent geologic past ice sheets were three times more extensive, covering vast areas with ice thousand of meters thick. Many regions still bear the mark of these glaciers (Figure 12.1). The basic character of such diverse places as the Alps, Cape Cod, and Yosemite Valley was fashioned by now vanished masses of glacial ice. Moreover, Long Island, the Great Lakes, and the fiords of Norway and Alaska all owe their existence to glaciers. Glaciers, of course, are not just a phenomenon of the geologic past. As we shall see, they are still sculpting and depositing debris in many regions today.

Glaciers: A Part of the Hydrologic Cycle

Recall the hydrologic cycle in Chapter 10: Time and time again, the same water evaporates from the oceans into the atmosphere, precipitates upon the land, and flows in rivers and underground back to the sea. However, when precipitation falls at high elevations or high latitudes, the water may not immediately make its way toward the sea. Instead, it may become part of a

glacier. Although the ice will eventually melt, allowing the water to continue its path to the sea, water can be stored as glacial ice for many tens, hundreds, or even thousands of years (see Box 12.1).

A **glacier** is a thick ice mass that originates on land from the accumulation, compaction, and recrystallization of snow. Because glaciers are agents of erosion, they must also *flow*. Indeed, like running water, groundwater, wind, and waves, glaciers are dynamic agents that accumulate, transport and deposit sediment. Although glaciers are found in many parts of the world today, most are located in remote areas.

Valley (Alpine) Glaciers

Literally thousands of relatively small glaciers exist in lofty mountain areas, where they usually follow valleys that were originally occupied by streams. Unlike the rivers that previously flowed in these valleys, the glaciers advance slowly, perhaps only a few centimeters per day. Because of their setting, these moving ice masses are termed **valley glaciers** or **alpine glaciers** (Figure 12.2). Each glacier actually is a stream of ice, bounded by precipitous rock walls, that flows downvalley from

Figure 12.1 Glacier National Park, Montana, presents a landscape carved by glaciers. St. Mary Lake occupies a glacially eroded valley (glacial trough) and exists because of glacial deposits that act as a dam. The sharp ridges on the left (arêtes) and the pointed peak in the background (a horn) were also sculptured by glacial ice. (Photo by James Blank/The Stock Market)

Figure 12.2 Aerial view of South Cascade Glacier, a valley glacier about 3 kilometers long in the North Cascade Range, Washington. The snowline and cracks called crevasses are clearly visible. (Photo by Austin Post, U.S. Geological Survey)

an accumulation center near its head. Like rivers, valley glaciers can be long or short, wide or narrow, single or with branching tributaries. Generally, the widths of alpine glaciers are small compared to their lengths. Some extend for just a fraction of a kilometer, whereas others go on for may tens of kilometers. The west branch of the Hubbard Glacier, for example, runs through 112 kilometers of mountainous terrain in Alaska and the Yukon Territory.

Ice Sheets

In contrast to valley glaciers, **ice sheets** exist on a much larger scale. The low total annual solar radiation reaching the poles makes these regions eligible for great ice accumulations. Although many ice sheets have existed in the past, just two achieve this status at present (Figure 12.3). In the area of the North Pole, Greenland is covered by an imposing ice sheet that occupies 1.7 million square kilometers, or about 80 percent of this large island. Averaging nearly 1500 meters thick, in places the ice extends 3000 meters above the island's bedrock floor.

In the South Polar realm, the huge Antarctic Ice Sheet attains a maximum thickness of nearly 4300 meters and covers an area of more than 13.9 million square kilometers. Because of the proportions of these huge features, they often are called *continental ice sheets*. Indeed, the combined areas of present-day continental ice sheets represent almost 10 percent of Earth's land area.

These enormous masses flow out in all directions from one or more snow-accumulation centers and completely obscure all but the highest areas of underlying terrain. Even sharp variations in the topography beneath the glacier usually appear as relatively subdued undulations on the surface of the ice. Such topographic differences, however, do affect the behavior of the ice sheets, especially near their margins, by guiding flow in certain directions and creating zones of faster and slower movement .

Along portions of the Antarctic coast, glacial ice flows into bays, creating features called **ice shelves**. They are large, relatively flat masses of floating ice that extend seaward from the coast but remain attached to the land along one or more sides. The shelves are thickest on their landward sides and they become thinner seaward. They are sustained by ice from the adjacent ice sheet as well as being nourished by snowfall and the freezing of seawater to their bases. Antarctica's ice shelves extend over nearly 1.4 million square kilometers. The Ross and Filchner ice shelves are the largest, with the Ross Ice Shelf alone covering an area nearly the size of Texas (Figure 12.3).

Other Types of Glaciers

In addition to valley glaciers and ice sheets, other types of glaciers are also identified. Covering some uplands and plateaus are masses of glacial ice called **ice caps**. They resemble ice sheets but are much smaller than the continental-scale features. Ice caps occur in many places, including Iceland and several of the large islands in the Arctic Ocean.

Often ice caps and ice sheets feed **outlet glaciers**. These tongues of ice flow down valleys extending

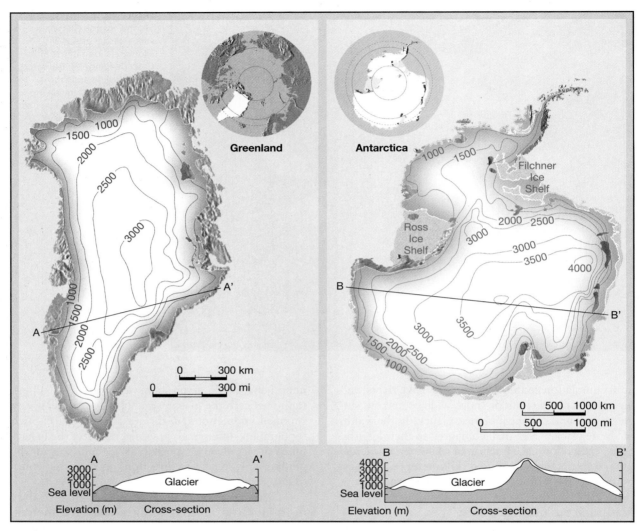

Figure 12.3 The only present-day continental ice sheets are those covering Greenland and Antarctica. Their combined areas represent almost 10 percent of Earth's land area. Greenland's ice sheet occupies 1.7 million square kilometers, or about 80 percent of the island. The area of the Antarctic Ice Sheet is almost 14 million square kilometers. Ice shelves occupy an additional 1.4 million square kilometers adjacent to the Antarctic Ice Sheet. The contours indicate the thickness of the ice in meters.

outward from the margins of these larger ice masses. The tongues are essentially valley glaciers that are avenues for ice movement from an ice cap or ice sheet through mountainous terrain to the sea. Where they encounter the ocean, some outlet glaciers spread out as floating ice shelves. Often large numbers of icebergs are produced.

Piedmont glaciers occupy broad lowlands at the bases of steep mountains and form when one or more alpine glaciers emerge from the confining walls of mountain valleys. Here the advancing ice spreads out to from a broad sheet. The size of individual piedmont glaciers varies greatly. Among the largest is the broad Malaspina Glacier along the coast of southern Alaska. It covers more than 5000 square kilometers of the flat coastal plain at the foot of the lofty St. Elias range.

Formation of Glacial Ice

Snow is the raw material from which glacial ice originates; therefore, glaciers form in areas where more snow falls in winter than melts during the summer. Before a glacier is created, snow must be converted into glacial ice. This transformation is shown in Figure 12.4.

When temperatures remain below freezing following a snowfall, the fluffy accumulation of delicate hexagonal crystals soon changes. As air infiltrates the spaces between the crystals, the extremities of the crystals evaporate and the water vapor condenses near the centers of the crystals. In this manner snowflakes become smaller, thicker, and more spherical, and the large pore spaces disappear.

Box 12.1

What if the Ice Melted?

How much water is stored as glacial ice? Estimates by the U.S. Geological Survey indicate that only slightly more than 2 percent of the world's water is tied up in glaciers. But even 2 percent of a vast quantity is still large. The total volume of just valley glaciers is about 210,000 cubic kilometers, comparable to the combined volume of the world's largest saline and freshwater lakes.

As for ice sheets, the Antarctic ice sheet includes 80 percent of the world's ice and nearly two-thirds of Earth's fresh water, and covers almost one and one-half times the area of the United States. If this ice melted, sea level would rise an estimated 60 to 70 meters, and the ocean would inundate many densely populated coastal areas (Figure 12.A).

The hydrologic importance of the Antarctic ice can be illustrated in another way. If the ice sheet were melted at a uniform rate, it could feed (1) the Mississippi River for more than 50,000 years, (2) all the rivers in the United States for about 17,000 years, (3) the Amazon River for approximately

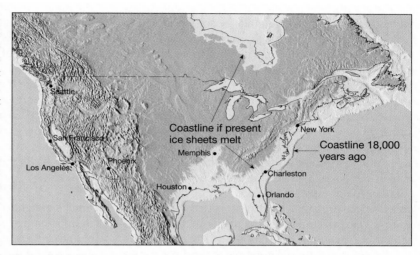

Figure 12.A This map of a portion of North America shows the present-day coastline compared to the coastline that existed during the last ice age maximum (18,000 years ago) and the coastline that would exist if present ice sheets in Greenland and Antarctica melted. (After R. H. Dott, Jr., and R. L. Battan, *Evolution of the Earth*, New York: McGraw Hill, 1971. Reprinted by permission of the publisher.)

5000 years, or (4) all the rivers of the world for about 750 years.

Although the quantity of ice on Earth today is truly immense, present glaciers occupy only about one-third the area they did in the very recent geologic past.

By this process air is forced out and what was once light, fluffy snow is recrystallized into a much denser mass of small grains having the consistency of coarse sand. This granular recrystallized snow is called **firn** and is commonly found making up old snowbanks near the end of winter. As more snow is added, the pressure on the lower layers increases, compacting the ice grains at depth. Once the thickness of ice and snow exceeds 50 meters, the weight is sufficient to fuse firn into a solid mass of interlocking ice crystals. Glacial ice has now been formed.

Movement of a Glacier

The movement of glacial ice is generally referred to as *flow*. The fact that glacial movement is described in this way seems paradoxical—how can a solid flow? The way in which ice flows is complex and is of two basic types. The first of these, **plastic flow**, involves movement *within* the ice. Ice behaves as a brittle solid until the pressure upon it is equivalent to the weight of about 50 meters (165 feet) of ice. Once that load is surpassed,

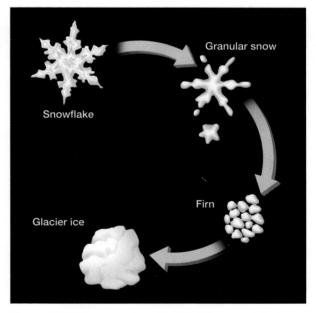

Figure 12.4 The conversion of freshly fallen snow into dense, crystalline glacial ice.

ice behaves as a plastic material and flow begins. Such flow occurs because of the molecular structure of ice. Glacial ice consists of layers of molecules stacked one upon the other. The bonds between layers are weaker than those within each layer. Therefore, when a stress exceeds the strength of the bonds between the layers, the layers remain intact and slide over one another.

A second, and often equally important, mechanism of glacial movement consists of the entire ice mass slipping along the ground. With the exception of some glaciers located in polar regions where the ice is probably frozen to the solid bedrock floor, most glaciers are thought to move by this sliding process, called **basal slip**. In this process, meltwater probably acts as a hydraulic jack and perhaps as a lubricant helping the ice over the rock. The source of the liquid water is related in part to the fact that the melting point of ice decreases as pressure increases. Therefore, deep within a glacier the ice may be at the melting point even though its temperature is below 0°C.

In addition, other factors may contribute to the presence of meltwater deep within the glacier. Temperatures may be increased by plastic flow (an effect similar to frictional heating), by heat added from the Earth below, and by the refreezing of meltwater that has seeped down from above. This last process relies on the property that, as water changes state from liquid to solid, heat (termed latent heat of fusion) is released.

Figure 12.5 illustrates the effects of these two basic types of glacial motion. This vertical profile through a glacier also shows that not all the ice flows forward at the same rate. Frictional drag with the bedrock floor causes the lower portions of the glacier to move more slowly.

In contrast to the lower portion of the glacier, the upper 50 meters or so is not under sufficient pressure to exhibit plastic flow. Rather, the ice in this uppermost zone is brittle and is appropriately referred to as the **zone of fracture**. The ice in the zone of fracture is carried along "piggy-back" style by the ice below. When the glacier moves over irregular terrain, the zone of fracture is subjected to tension, resulting in cracks called **crevasses** (see Figure 12.2 and 12.6). These gaping cracks can make travel across glaciers dangerous and may extend to depths of 50 meters. Below this depth, plastic flow seals them off.

Rates of Glacial Movement

Unlike streamflow, the movement of glaciers is not readily apparent to the casual observer. If we could watch an alpine glacier move, we would see that, like the water in a river, all of the ice in the valley does not move down-valley at an equal rate. Just as friction with the bedrock floor slows the movement of the ice at the bottom of the glacier, the drag created by the valley walls leads to the flow being greatest in the center of the glacier.

This was first demonstrated by experiments in the nineteenth century, in which markers were carefully placed in a straight line across the top of a valley glacier. Periodically, the positions of the stakes were recorded, revealing the type of movement just described (see Box 1.2, page 10).

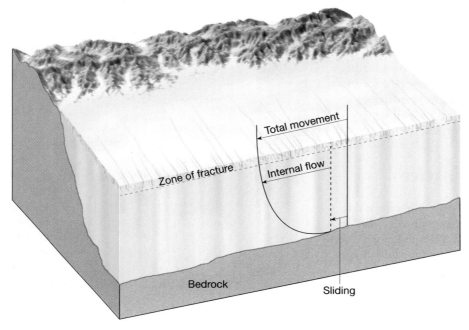

Figure 12.5 Vertical cross-section through a glacier to illustrate ice movement. Glacial movement is divided into two components. Below about 50 meters, ice behaves plastically and flows. In addition, the entire mass of ice may slide along the ground. The ice in the zone of fracture is carried along "piggyback" style. Notice that the rate of movement is slowest at the base of the glacier where frictional drag is greatest.

Figure 12.6 This climber is trying to negotiate a deep crevasse on Mt. McKinley, Alaska. Crevasses form in the brittle ice of the zone of fracture. They do not continue down into the zone of flow. (Photo by Robert Kaufman Photographers)

How rapidly does glacial ice move? Average velocities vary considerably from one glacier to another. Some move so slowly that trees and other vegetation may become well established in the debris that has accumulated on the glacier's surface, whereas others may move at rates of up to several meters per day. For example, Byrd Glacier, an outlet glacier in Antarctica that was the subject of a 10-year study using satellite images, moved at an average rate of 750 to 800 meters per year (about 2 meters per day). Other glaciers in the study advanced at one-fourth that rate.

The advance of some glaciers is characterized by periods of extremely rapid movements called **surges**. Glaciers that exhibit such movement may flow along in an apparently normal manner, then speed up for a relatively short time before returning to the normal rate again. The flow rates during surges are as much as 100 times the normal rate. Evidence indicates that many glaciers may be of the surging type (Box 12.2).

Budget of a Glacier

Snow is the raw material from which glacial ice originates; therefore, glaciers form in areas where more snow falls in winter than melts during the summer. Glaciers are constantly gaining and losing ice. Snow accumulation and ice formation occur in the **zone of accumulation**. Its outer limits are defined by the **snowline**. The elevation of the snowline varies greatly. In polar regions, it may be sea level, whereas in tropical areas, the snowline exists only high in mountain areas, often at altitudes exceeding 4500 meters. Above the snowline, in the zone of accumulation, the addition of snow thickens the glacier and promotes movement. Beyond the snowline is the **zone of wastage**. Here there is a net loss to the glacier as all of the snow from the previous winter melts, as does some of the glacial ice (Figure 12.7).

In addition to melting, glaciers also waste as large pieces of ice break off the front of the glacier in a process called **calving**. Calving creates *icebergs* in places where the glacier has reached the sea or a lake (Figure 12.8). Because icebergs are just slightly less dense than seawater, they float very low in the water, with more than 80 percent of their mass submerged. Along the margins of Antarctica's ice shelves, calving is the primary means by which these masses lose ice. The relatively flat icebergs produced here can be several kilometers across and 600 meters thick. By comparison, thousands of irregularly shaped icebergs are produced by outlet glaciers flowing from the margins of the Greenland Ice Sheet. Many drift southward and find their way into the North Atlantic, where they can pose a hazard to navigation.

Whether the margin of a glacier is advancing, retreating, or remaining stationary depends on the budget of the glacier. The **glacial budget** is the balance, or lack of balance, between accumulation at the upper end of the glacier, and loss at the lower end. This loss is termed **ablation**. If ice accumulation exceeds ablation, the glacial front advances until the two factors balance. When this happens, the terminus of the glacier stationary.

If a warming trend increases ablation and/or if a drop in snowfall decreases accumulation, the ice front will retreat. As the terminus of the glacier retreats, the extent of the zone of wastage diminishes. Therefore, in time a new balance will be reached between accumulation and wastage, and the ice front will again become stationary.

Whether the margin of a glacier is advancing, retreating, or stationary, the ice within the glacier continues to flow forward. In the case of a receding glacier, the ice still flows forward, but not rapidly enough to offset ablation. This point is illustrated well in Figure 1.C on page 10. As the line of stakes within Rhone Glacier continued to move downvalley, the terminus of the glacier slowly retreated upvalley.

Box 12.2

Surging Glaciers

The copious snows that fall in the mountains of southeastern Alaska and adjacent Canada nurture an extensive system of valley glaciers. Because these glaciers are relatively accessible and exhibit a diversity of behaviors, they are of great interest to scientists. The ice lobes are used as natural laboratories for the study of glacial processes and climatic change.

Although many of these glaciers are retreating, others are advancing, and a few exhibit the comparatively rapid movements called *surges*. Hubbard Glacier is one example. From about 1900, when records of its movements were begun, until 1986, the ice within Hubbard Glacier moved forward at a relatively steady rate of 100 meters or so each year. Because ablation did not entirely offset this movement, the front of the ice gradually advanced. Then, unexpectedly, during the winter of 1986, Hubbard Glacier experienced a period of surging. At times the ice

moved as fast as 14 meters a day. A tributary also surged forward at rates exceeding 30 meters a day.

Certainly the most studied glacier exhibiting such surging movement is the 24-kilometer-long Variegated Glacier northwest of Juneau, Alaska (Figure 12.B). Unlike Hubbard Glacier, Variegated Glacier is known to surge at regular intervals. Since 1906, surges have occurred about once every 17 to 20 years. The most recent and thoroughly investigated event took place in 1982–1983. During this surge, the ice advanced as fast as 54 meters (180 feet) per day.

It is not yet clear whether the mechanism that triggers these rapid movements is the same for each surging-type glacier. However, researchers studying the Variegated Glacier have determined that the surges of this ice mass take the form of a rapid increase in basal sliding that is caused by increases in water pressure beneath the ice. The increased water

pressure at the base of the glacier acts to reduce friction between the underlying bedrock and the moving ice. The pressure buildup, in turn, is related to changes in the system of passageways that conduct water along the glacier's bed and deliver it as outflow to the terminus.

Geologists study the surge phenomenon not only because it is an unsolved problem, but also because an understanding of the mechanism that triggers surges may have wider significance. This broader interest exists,

because of the possibilities that glacier surges may impinge upon works of man and that surging of the Antarctic ice sheet may be a factor in the initiation of ice ages and in cyclic variations of sea level.*

*Barclay Kamb et al., "Glacier Surge Mechanism: 1982–1983 Surge of Variegated Glacier, Alaska," *Science* 227 (February 1985): 469.

August 1964

August 1965

Figure 12.B The surge of Variegated Glacier, a valley glacier near Yakutat, Alaska, is captured in these two aerial photographs taken one year apart. During a surge, ice velocities in Variegated Glacier are 20 to 50 times greater than during a quiescent phase. (Photos by Austin Post, U.S. Geological Survey)

Glacial Erosion

Glaciers are capable of great erosion. For anyone who has observed the terminus of an alpine glacier, the evidence of its erosive force is clear. The observer can

witness firsthand the release of rock material of various sizes from the ice as it melts. All signs lead to the conclusion that the ice has scraped, scoured, and torn rock from the floor and walls of the valley and carried it downvalley. Indeed, as a medium of sediment transport, ice has no equal.

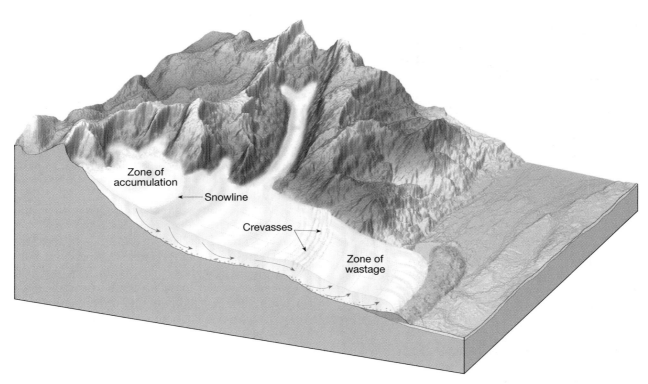

Figure 12.7 The snowline separates the zone of accumulation and the zone of wastage. Above the snowline, more snow falls each winter than melts each summer. Below the snowline, the snow from the previous winter completely melts as does some of the underlying ice. Whether the margin of a glacier advances, retreats, or remains stationary depends on the balance between accumulation and wastage.

Figure 12.8 Icebergs are created when large pieces calve from the front of a glacier after it reaches a water body. These icebergs are in the Weddell Sea off the coast of Antarctica. Only about 20 percent of an iceberg protrudes above the water line. (Photo by Frans Lanting/Minden Pictures)

Figure 12.9 A large, glacially transported boulder near Hensler, North Dakota. (Photo by Tom Bean/DRK Photo)

When the ice at the bottom of a glacier contains large rock fragments, long scratches and grooves called **glacial striations** may even be cut into the bedrock (Figure 12.10). These linear grooves provide clues to the direction of ice flow. By mapping the striations over large areas, patterns of glacial flow can often be reconstructed. On the other hand, not all abrasive action produces striations. The rock surfaces over which the glacier moves may also become highly polished by the ice and its load of finer particles. The broad expanses of smoothly polished granite in Yosemite National Park provide an excellent example.

As is the case with other agents of erosion, the rate of glacial erosion is highly variable. This differential erosion by ice is largely controlled by four factors: (1) rate of glacial movement; (2) thickness of the ice; (3) shape, abundance, and hardness of the rock fragments contained in the ice at the base of the glacier; and (4) the erodibility of the surface beneath the glacier. Variations in any or all of these factors from time to time and/or from place to place mean that the features, effects, and degree of landscape modification in glaciated regions can vary greatly.

Figure 12.10 Glacial abrasion created the scratches and grooves in this bedrock. Glacier Bay National Park, Alaska. (Photo by Carr Clifton)

Once rock debris is acquired by the glacier, the enormous competency of ice will not allow the debris to settle out like the load carried by a stream or by the wind. Consequently, glaciers can carry huge blocks that no other erosional agent could possibly budge (Figure 12.9). Although today's glaciers are of limited importance as erosional agents, many landscapes that were modified by the widespread glaciers of the most recent Ice Age still reflect to a high degree the work of ice.

Glaciers erode the land primarily in two ways—plucking and abrasion. First, as a glacier flows over a fractured bedrock surface, it loosens and lifts blocks of rock and incorporates them into the ice. This process, known as **plucking**, occurs when meltwater penetrates the cracks and joints of bedrock beneath a glacier and freezes. As the water expands, it exerts tremendous leverage that pries the rock loose. In this manner sediment of all sizes becomes part of the glacier's load.

The second major erosional process is **abrasion** (Figure 12.10). As the ice and its load of rock fragments slide over bedrock, they function like sandpaper to smooth and polish the surface below. The pulverized rock produced by the glacial "grist mill" is appropriately called **rock flour**. So much rock flour may be produced that meltwater streams flowing out of a glacier often have the grayish appearance of skim milk and offer visible evidence of the grinding power of ice.

Landforms Created by Glacial Erosion

The erosional effects of valley glaciers and ice sheets are quite different. A visitor to a glaciated mountain region is likely to see a sharp and angular topography (see Figure 12.1, p.294). The reason is that as alpine glaciers move downvalley, they tend to accentuate the irregularities of the mountain landscape by creating steeper canyon walls and making bold peaks even more jagged. By contrast, continental ice sheets generally override the terrain and hence subdue rather than accentuate the irregularities they encounter. Although the erosional potential of ice sheets is enormous, landforms carved by these huge ice masses usually do not inspire the same wonderment and awe as do the erosional features created by valley glaciers. Much of the rugged mountain scenery so celebrated for its majestic beauty is the product of erosion by alpine glaciers. Figure 12.11 shows a mountain area before, during, and after glaciation.

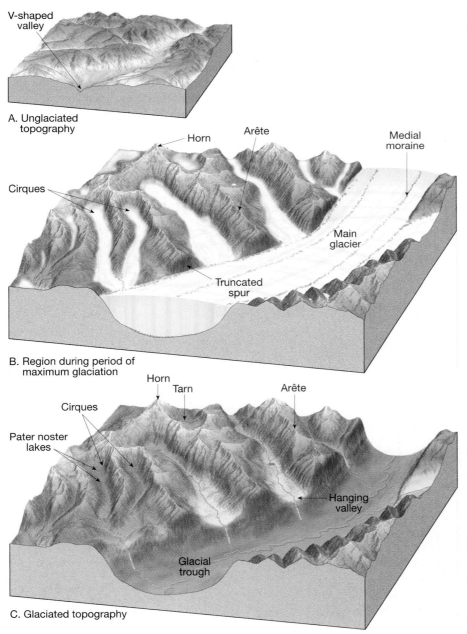

A. Unglaciated topography

B. Region during period of maximum glaciation

C. Glaciated topography

Figure 12.11 Erosional landforms created by alpine glaciers. The unglaciated landscape in part **A** is modified by valley glaciers in part **B**. After the ice recedes, in part **C**, the terrain looks very different than before glaciation.

Glaciated Valleys

A hike up a glaciated valley reveals a number of striking ice-created features. The valley itself is often a dramatic sight. Unlike streams, which create their own valleys, glaciers take the path of least resistance by following the course of existing stream valleys. Prior to glaciation, mountain valleys are characteristically narrow and V-shaped because streams are well above base level and are therefore downcutting. However, during glaciation these narrow valleys undergo a transformation as the glacier widens and deepens them, creating a U-shaped **glacial trough** (Figure 12.11C and Figure 12.12). In addition to producing a broader and deeper valley, the glacier also straightens the valley. As ice flows around sharp curves, its great erosional force removes the spurs of land that extend into the valley. The results of this activity are triangular-shaped cliffs called **truncated spurs** (Figure 12.11B).

The intensity of glacial erosion depends in part upon the thickness of the ice. Consequently, main glaciers, also called trunk glaciers, cut their valleys deeper than do their smaller tributary glaciers. Thus, when the glaciers eventually recede, the valleys of tributary glaciers are left standing above the main glacial trough, and are termed **hanging valleys** (Figure

12.11C). Rivers flowing through hanging valleys may produce spectacular waterfalls, such as those in Yosemite National Park (Figure 12.13).

As hikers walk up a glacial trough, they may pass a series of bedrock depressions on the valley floor that were probably created by plucking and then scoured by the abrasive force of the ice. If these depressions are filled with water, they are called **pater noster lakes** (Figure 12.12). The Latin name means "our Father," and is a reference to a string of rosary beads.

At the head of a glacial valley is a very characteristic and often imposing feature called a **cirque**. As Figure 12.14 illustrates, these bowl-shaped depressions have precipitous walls on three sides but are open on the downvalley side. The cirque is the focal point of the glacier's growth, because it is the area of snow accumulation and ice formation. Cirques begin as irregularities in the mountainside that are subsequently enlarged by the frost wedging and plucking that occur along the sides and bottom of the glacier. After the glacier has melted away, the cirque basin is often occupied by a small lake called a **tarn** (Figure 12.15).

Sometimes when two glaciers exist on opposite sides of a divide, each flowing away from the other, the dividing ridge between their cirques is largely

Figure 12.12 Prior to glaciation, a mountain valley is typically narrow and V-shaped. During glaciation, an alpine glacier widens, deepens, and straightens the valley, creating the U-shaped glacial trough seen here. The string of lakes is called pater noster lakes. This valley is in Glacier National Park, Montana. (Photo by John Montagne)

Figure 12.13 Bridalveil Falls in Yosemite National Park cascades from a hanging valley into the glacial trough below. (Photo by E. J. Tarbuck)

eliminated as plucking and frost action enlarge each one. When this occurs, the two glacial troughs come to intersect, creating a gap or pass from one valley into the other. Such a feature is termed a **col**. Some important and well-known mountain passes that are cols include St. Gotthard Pass in the Swiss Alps, Tioga Pass in California's Sierra Nevada, and Berthoud Pass in the Colorado Rockies.

Before leaving the topic of glacial troughs and their associated featured, one more rather well-known feature should be discussed—fiords. **Fiords** are deep, often spectacular, steep-sided inlets of the sea that are present at high latitudes where mountains are adjacent to the ocean (Figure 12.16). They are drowned glacial troughs that became submerged as the ice left the valley and sea level rose following the Ice Age. The depths of fiords may exceed 1000 meters. However, the great depths of these flooded troughs is only partly explained by the post–Ice Age rise in sea level. Unlike the situation governing the downward erosional work of rivers, sea level does not act as base level for glaciers. As a consequence, glaciers are capable of eroding their beds far below the surface of the sea. For example, a 300-meter-thick glacier can carve its valley floor more than 250 meters below sea level before downward erosion ceases and the ice begins to float. Norway, British Columbia, Greenland, New Zealand, Chile, and Alaska all have coastlines characterized by fiords.

Figure 12.14 Aerial view of cirques in Utah's Uinta Range. These bowl-shaped depressions are found at the heads of glacial valleys and represent a focal point of ice formation. (Photo by John S. Shelton)

Figure 12.15 A small lake occupying a cirque is called a tarn. Tombstone Mountains, Yukon Territory, Canada. (Photo by Stephen J. Krasemann)

Arêtes and Horns

A visit to the Alps, and the Northern Rockies, or many other scenic mountain landscapes carved by valley glaciers reveals not only glacial troughs, cirques, pater noster lakes, and the other related features just discussed. You also are likely to see sinuous, sharp-edged ridges called **arêtes** (French for *knife-edge*) and sharp,

pyramid-like peaks called **horns** projecting above the surroundings. Both features can originate from the same basic process, the enlargement of cirques produced by plucking and frost action (Figure 12.11B, C, p. 303). In the case of the spires of rock called horns, groups of cirques around a single high mountain are

Figure 12.16 Like other fiords, Muir Inlet, Alaska, is a drowned glacial trough The south coast of Alaska has many fiords. (Photo by Tom Bean)

responsible. As the cirques enlarge and converge, an isolated horn is produced. The most famous example is the Matterhorn in the Swiss Alps (Figure 12.17).

Arêtes can be formed in a similar manner, except that the cirques are not clustered around a point but rather exist on opposite sides of a divide. As the cirques grow, the divide separating them is reduced to a narrow knife-like partition. An arête, however, may also be created in another way. When two glaciers occupy parallel valleys, an arête can form when the divide separating the moving tongues of ice is progressively narrowed as the glaciers scour and widen their adjacent valleys.

Roches Moutonnées

In many glaciated landscapes, but most frequently where continental ice sheets have modified the terrain, the ice carves small streamlined hills from protruding bedrock knobs. Such an asymmetrical knob of bedrock is called a **roche moutonnée** (French for *sheep rock*). They are formed when glacial abrasion smoothes the gentle slope facing the oncoming ice sheet and plucking steepens the opposite side as the ice rides over the knob (Figure 12.18). Roches moutonnées indicate the direction of glacial flow, because the gentler slope is generally on the side from which the ice advanced.

Figure 12.17 Horns are sharp, pyramid-like peaks that are fashioned by alpine glaciers. This example is the famous Matterhorn in the Swiss Alps. (Photo by E. J. Tarbuck)

 # Glacial Deposits

Glaciers pick up and transport a huge load of debris as they slowly advance across the land. Ultimately these materials are deposited when the ice melts. In regions where glacial sediment is deposited, it can play a truly significant role in forming the physical landscape. For example, in many areas once covered by the continental ice sheets of the recent Ice Age, the bedrock is rarely exposed because glacial deposits that are tens or even hundreds of meters thick completely mantle the terrain. The general effect of these deposits is to reduce the local relief and thus level the topography. Indeed, rural country scenes that are familiar to many of us—rocky pastures in New England, wheat fields in the Dakotas, rolling farmland in the Midwest—result directly from glacial deposition.

Long before the theory of an extensive Ice Age was ever proposed, much of the soil and rock debris covering portions of Europe was recognized as coming from somewhere else. At the time, these "foreign" materials were believed to have been "drifted" into their present positions by floating ice during an ancient flood. As a consequence, the term *drift* was applied to this sediment. Although rooted in an incorrect concept, this term was so well established by the time the true glacial origin of the debris became widely recognized that it remained in the basic glacial vocabulary. Today **glacial drift** is an all-embracing term for sediments of glacial origin, no matter how, where, or in what shape they were deposited.

One of the features distinguishing drift from sediments laid down by other erosional agents is that glacial deposits consist primarily of mechanically weathered rock debris that underwent little or no chemical weathering prior to deposition. Thus, minerals that are notably prone to chemical decomposition, such as hornblende and the plagioclase feldspars, are often conspicuous components in glacial sediments.

Glacial drift is divided by geologists into two distinct types: (1) materials deposited directly by the glacier, which are known as **till**; and (2) sediments laid down by glacial meltwater, called **stratified drift**. We will now look at landforms made of each type.

 # Landforms Made of Till

Till is deposited as glacial ice melts and drops its load of rock fragments. Unlike moving water and wind, ice cannot sort the sediment it carries; therefore, deposits of till are characteristically unsorted mixtures of many particle sizes (Figure 12.19). A close examination of this sediment shows that many of the pieces are scratched and

Figure 12.18 Roche
Moutonnée, in Yosemite National
Park, California. The gentle slope
was abraded and the steep side
was plucked. The ice moved from
right to left. (Photo by E. J.
Tarbuck)

polished as the result of being dragged along by the glacier. Such pieces help distinguish till from other deposits that are a mixture of different sediment sizes, such as material from a debris flow or a rockslide.

Boulders found in the till or lying free on the surface are called **glacial erratics**, if they are different from the bedrock below (see Figure 12.9, p. 302). Of course, this means that they must have been derived from a source outside the area where they are found. Although the locality of origin for most erratics is unknown, the origin of some can be determined. In many cases, boulders were transported as far as 500 kilometers from their source area and, in a few instances, more than 1000 kilometers. Therefore, by studying glacial erratics as well as the mineral composition of the remaining till, geologists are sometimes able to trace the path of a lobe of ice.

In portions of New England and other areas, erratics dot pastures and farm fields. In fact, in some places these large rocks were cleared from fields and piled to make fences and walls (Figure 12.20). Keeping the fields clear, however, is an ongoing chore because each spring newly exposed erratics appear. Wintertime frost heaving has lifted them to the surface.

End and Ground Moraines

The most common term for landforms made of glacial deposits is *moraine*. Originally this term was used by French peasants when referring to the ridges and embankments of debris found near the margins of glaciers in the French Alps. Today, however, moraine, has a broader meaning, because it is applied to a number of landforms, all of which are composed primarily of till.

An **end moraine** is a ridge of till that forms at the terminus of a glacier. These relatively common landforms are deposited when a state of equilibrium is attained between ablation and ice accumulation. That is, the end moraine forms when the ice is melting and evaporating near the end of the glacier at a rate equal to the forward advance of the glacier from its region of nourishment. Although the terminus of the glacier is now stationary, the ice continues to flow forward, delivering a continuous supply of sediment in the same manner a conveyor belt delivers goods to the end of a production line. As the ice melts, the till is dropped and the end moraine grows. The longer the ice front remains stable, the larger the ridge of till will become.

Eventually, the time comes when ablation exceeds nourishment. At this point, the front of the

Close up
of cobble

Figure 12.19 Glacial till is an unsorted mixture of many different sediment sizes. A close examination often reveals cobbles that have been scratched as they were dragged along by the glacier. (Photos by E. J. Tarbuck)

glacier begins to recede in the direction from which it originally advanced. However, as the ice front retreats, the conveyor-belt action of the glacier continues to provide fresh supplies of sediment to the terminus. In this manner a large quantity of till is deposited as the ice melts away, creating a rock-strewn, undulating

plain. This gently rolling layer of till laid down as the ice front recedes is termed **ground moraine**. Ground moraine has a leveling effect, filling in low spots and clogging old stream channels, often leading to a derangement of the existing drainage system. In areas where this layer of till is still relatively fresh, such as the northern Great Lakes region, poorly drained swampy lands are quite common.

Periodically, a glacier will retreat to a point where ablation and nourishment once again balance. When this happens, the ice front stabilizes and a new end moraine forms.

The pattern of end moraine formation and ground moraine deposition may be repeated many times before the glacier has completely vanished. Such a pattern is illustrated by Figure 12.21. It should be pointed out that the outermost end moraine marks the limit of the glacial advance. Because of its special status, this end moraine is also called the **terminal moraine**. On the other hand, the end moraines that were created as the ice front occasionally stabilized during retreat are termed **recessional moraines**. Note that both terminal and recessional moraines are essentially alike; the only difference between them is their relative positions.

End moraines deposited by the most recent major stage of Ice Age glaciation are prominent features in many parts of the Midwest and Northeast. In Wisconsin, the wooded, hilly terrain of the Kettle Moraine near Milwaukee is a particularly picturesque example. A well-known example in the Northeast is Long Island. This linear strip of glacial sediment that extends northeastward from New York City is part of an end moraine complex that stretches from eastern Pennsylvania to Cape Cod, Massachusetts (Figure 12.22). The end moraines that make up Long Island represent materials that were deposited by a continental ice sheet in the relatively shallow waters off the coast and built up many meters above sea level. Long Island Sound, the narrow body of water separating the island and the mainland, was not built up as much by glacial deposition and was therefore subsequently flooded by the rising sea following the Ice Age.

Lateral and Medial Moraines

Alpine glaciers produce two types of moraines that occur exclusively in mountain valleys. The first of these is called a **lateral moraine** (Figure 12.23). As we learned earlier, when an alpine glacier moves down-valley, the ice erodes the sides of the valley with great efficiency. In addition, large quantities of debris are added to the glacier's surface as rubble falls or slides from higher up on the valley walls and collects on the edges of the moving ice. When the ice eventually

Figure 12.20 Land cleared of glacial erratics, which were then piled atop one another to build this stone wall near West Bend, Wisconsin. (Photo by Tom Bean)

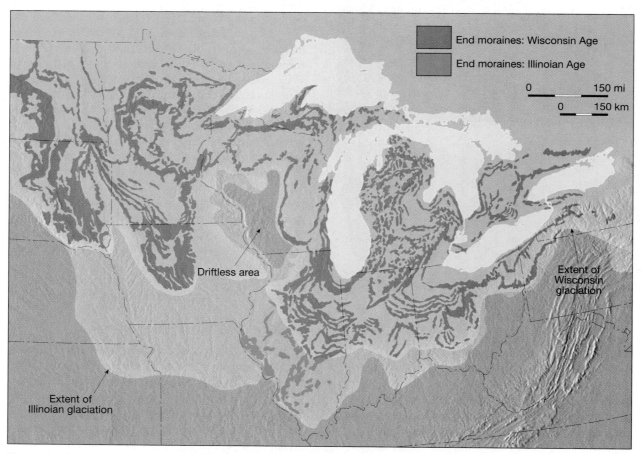

Figure 12.21 End moraines of the Great Lakes region. Those deposited during the most recent (Wisconsinan) stage are most prominent.

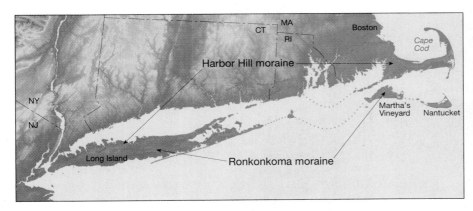

Figure 12.22 End moraines make up substantial parts of Long Island, Cape Code, Martha's Vineyard, and Nantucket. Although portions are submerged, the Ronkonkoma moraine (a terminal moraine) extends through central Long Island, Martha's Vineyard, and Nantucket. It was deposited about 20,000 years ago. The recessional Harbor Hill moraine, which formed about 14,000 years ago, extends along the north shore of Long Island, through southern Rhode Island and Cape Cod.

melts, this accumulation of debris is dropped next to the valley walls. These ridges of till paralleling the sides of the valley constitute the lateral moraines.

The second type of moraine that is unique to alpine glaciers is the **medial moraine** (Figure 12.24). Medial moraines are created when two alpine glaciers coalesce to form a single ice stream. The till that was once carried along the sides of each glacier joins to form a single dark stripe of debris within the newly enlarged glacier. The creation of these dark stripes within the ice stream is one obvious proof that glacial

ice moves, because the moraine could not form if the ice did not flow downvalley. It is quite common to see several medial moraines within a single large alpine glacier, because a streak will from whenever a tributary glacier joins the main valley.

Drumlins

Moraines are not the only landforms deposited by glaciers. In some areas that were once covered by continental ice sheets, a special variety of glacial landscape exists—one characterized by smooth, elongate, parallel

Figure 12.23 Well-developed lateral moraines deposited by the shrinking Athabaska Glacier in Canada's Jasper National Park. (Photo by David Barnes/The Stock Market)

Figure 12.24 Medial moraines form when the lateral moraines of merging valley glaciers join. Saint Elias National Park, Alaska. (Photo by Tom Bean)

hills called **drumlins** (Figure 12.25). Certainly one of the best-known drumlins is Bunker Hill in Boston, the site of the famous Revolutionary War battle in 1775.

An examination of Bunker Hill or other less famous drumlins would reveal that drumlins are streamlined asymmetrical hills composed largely of till. They range in height from about 15 to 50 meters and may be up to one kilometer long. The steep side of the hill faces the direction from which the ice advanced, whereas the gentler, longer slope points

in the direction the ice moved. Drumlins are not found as isolated landforms, but rather occur in clusters called *drumlin fields* (Figure 12.26). One such cluster, east of Rochester New York, is estimated to contain about 10,000 drumlins. Although drumlin formation is not fully understood, their streamlined shape indicates that they were molded in the zone of plastic flow within an active glacier. It is believed that many drumlins originate when glaciers advance over previously deposited drift and reshape the material.

Figure 12.25 Drumlin in upstate New York (Photo courtesy of Ward's Natural Science Establishment, Inc., Rochester, N.Y.)

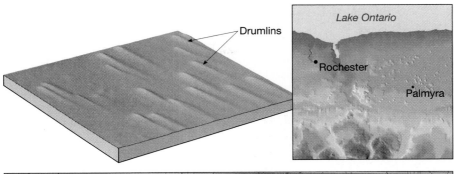

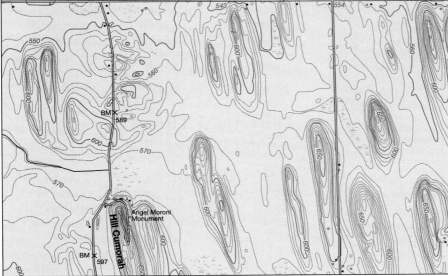

Figure 12.26 Portion of a drumlin field shown on the Palmyra, New York, 7.5-minute topographic map. North is at the top. The drumlins are steepest on the north side, indicating that the ice advanced from this direction.

Landforms Made of Stratified Drift

As the name implies, stratified drift is sorted according to the size and weight of the particles. Because ice is not capable of such sorting activity, these materials are not deposited directly by the glacier, as till is, but instead reflect the sorting action of glacial meltwater. Accumulations of stratified drift often consist largely of sand and gravel—that is, bed-load material—because the finer rock flour remains suspended and therefore is commonly carried far from the glacier by the meltwater streams. An indication that stratified drift consists primarily of sand and gravel is reflected in the fact that in many areas these deposits are actively mined as a source of aggregate for road work and other construction projects.

Outwash Plains and Valley Trains

At the same time that an end moraine is forming, water from the melting glacier cascades over the till, sweeping some of it out in front of the growing ridge of unsorted debris. Meltwater generally emerges from the ice in rapidly moving streams that are often choked with suspended material and carry a substantial bed load as well. As the water leaves the glacier, it moves onto the relatively flat surface beyond and rapidly loses velocity. As a consequence, much of its bed load is dropped and the meltwater begins weaving a complex pattern of braided channels (see Figure 10.15. p. 246). In this way a broad, ramplike surface composed of stratified drift is built adjacent to the downstream edge of most end moraines. When the feature is formed in association with an ice sheet, it is termed an **outwash plain**, and when largely confined to a mountain valley, it is usually referred to as a **valley train**.

Outwash plains and valley trains often are pockmarked with basins or depressions known as **kettles** (Figure 12.27). Kettles also occur in deposits of till. Kettles are formed when blocks of stagnant ice become wholly of partly buried in drift and eventually melt, leaving pits in the glacial sediment. Although most kettles do not exceed 2 kilometers in diameter, some with diameters exceeding 10 kilometers occur in Minnesota. Likewise, the typical depth of most kettles is less than 10 meters, although the

Figure 12.27 These ponds in Chippewa County, Wisconsin, occupy depressions called kettles. Kettles form when blocks of ice buried in drift melt creating pits. (Photo by Tom Bean)

vertical dimensions of some approach 50 meters. In many cases water eventually fills the depression and forms a pond or lake. One well-known example is Walden Pond near Concord, Massachusetts. It is here that Henry David Thoreau lived alone for two years in the 1840s and about which he wrote his famous book *Walden, or Life in the Woods.*

Ice-Contact Deposits

When the melting terminus of a glacier shrinks to a critical point, flow virtually stops and the ice becomes stagnant. Meltwater that flows over, within, and at the base of the motionless ice lays down deposits of stratified drift. Then, as the supporting ice melts away, the stratified sediment is left behind in the form of hills, terraces, and ridges. Such accumulations are collectively termed **ice-contact deposits** and are classified according to their shapes.

When the ice-contact stratified drift is in the form of a mound or steep-sided hill, it is called a **kame** (Figure 12.28). Some kames represent bodies of sediment deposited by meltwater in openings within or depressions on top of the ice. Others originate as deltas or fans built outward from the ice by meltwater streams. Later, when the stagnant ice melts away, these various accumulations of sediment collapse to form isolated, irregular mounds.

When glacial ice occupies a valley, **kame terraces** may be built along the sides of the valley. These features commonly are narrow masses of stratified drift laid down between the glacier and the side of the

valley by streams that drop debris along the margins of the shrinking ice mass.

A third type of ice-contact deposit is a long, narrow, sinuous ridge composed largely of sand and gravel. Some are more than 100 meters high with lengths in excess of 100 kilometers. The dimensions of many others are far less spectacular. Known as **eskers**, these ridges are deposited by meltwater rivers flowing within, on top of, and beneath a mass of motionless, stagnant glacial ice (Figure 12.29). Many sediment sizes are carried by the torrents of meltwater in the ice-banked channels, but only the coarser material can settle out of the turbulent stream.

Figure 12.30 depicts a hypothetical area during and following glaciation. It illustrates many of the landforms described in the preceding sections that may be found in regions affected by continental ice sheets.

The Glacial Theory and the Ice Age

In the preceding pages we mentioned the Ice Age, a time when ice sheets and alpine glaciers were far more extensive than they are today. As noted, there was a time when the most popular explanation for what we now know to be glacial deposits was that the materials had been drifted in by means of icebergs or perhaps simply swept across the landscape by a catastrophic flood. What convinced geologists that an extensive ice age was responsible for these deposits and many other glacial features?

In 1821, a Swiss engineer, Ignaz Venetz, presented a paper suggesting that glacial landscape features

Figure 12.28 Bodies of ice-contact stratified drift are classified according to their shape. When it is in the form of a steep-sided hill, the feature is called a kame. This is Dundee Kame in Wisconsin. (Photo by Richard P. Jacobs/JLM Visuals)

Figure 12.29 The retreat of the glacier reveals an esker, the sinuous ridge of sand and gravel in the center of the photograph. The esker consists of material deposited by meltwater in a channel flowing beneath the stagnant ice. Where the stream emptied into a small standing body of water, a delta was formed. (Photo by Bradford Washburn)

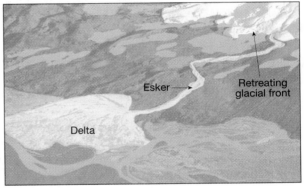

occurred at considerable distances from the existing glaciers in the Alps. This implied the glaciers had once been larger and occupied positions farther downvalley. Another Swiss scientist, Louis Agassiz, doubted the proposal of widespread glacial activity put forth by Venetz. He set out to prove that the idea was not valid. Ironically, his 1836 fieldwork in the Alps convinced him of the merits of his colleague's hypothesis. In fact, a year later Agassiz hypothesized a great ice age that had extensive and far-reaching effects—an idea that was to give Agassiz widespread fame (see Box 12.3).

The proof of the glacial theory proposed by Agassiz and others constitutes a classic example of applying the principle of uniformitarianism. Realizing that certain features are produced by no other known process but glacial action, they were able to begin reconstructing the extent of now-vanished ice sheets based on the presence of features and deposits found far beyond the margins of present-day glaciers. In this manner, the development and verification of the glacial theory continued during the nineteenth century, and through the efforts of many scientists, a knowledge of the nature and extent of former ice sheets became clear.

By the turn of the twentieth century, geologists had largely determined the areal extent of the Ice Age

glaciation. Further, during the course of their investigations, they had discovered that many glaciated regions had not one layer of drift but several. Moreover, close examination of these older deposits showed well-developed zones of chemical weathering and soil formations as well as the remains of plants that require warm temperatures. The evidence was clear; there had been not just one glacial advance but several, each separated by extended periods when climates were as warm or warmer than the present. The Ice Age had not simply been a time when the ice advanced over the land, lingered for a while, and then receded. Rather, the period was a very complex event, characterized by a number of advances and withdrawals of glacial ice.

By early in the twentieth century, a fourfold division of the Ice Age had been established for both North America and Europe. The divisions were based largely on studies of glacial deposits. In North

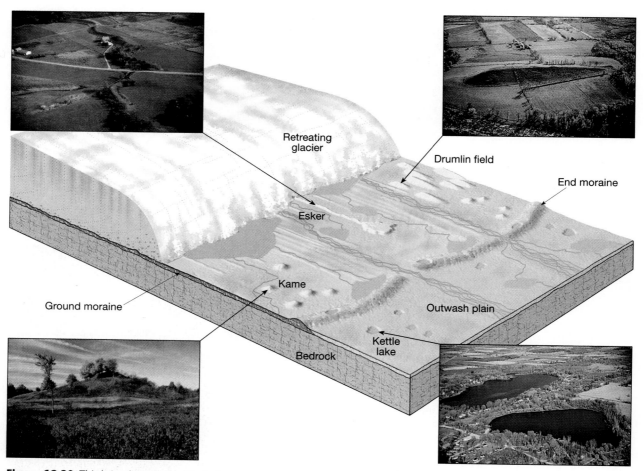

Figure 12.30 This hypothetical area illustrates many common depositional landforms. (Drumlin photo courtesy by Ward's Natural Science Establishment; Kame, Esker, and Kettle photos by Richard P. Jacobs/JLM Visuals)

America, each of the four major stages was named for the midwestern state where deposits of that stage were well exposed and/or were first studied. These are, in order of occurrence, the Nebraskan, Kansan, Illinoian, and Wisconsinan. These traditional divisions remained in place until relatively recently when it was learned that sediment cores from the ocean floor contain a much more complete record of climate change during the Ice Age.* Unlike the glacial record on land, which is punctuated by many unconformities, seafloor sediments provide an uninterrupted record of climatic cycles for this period. Studies of these seafloor sediments showed that glacial/interglacial cycles had occurred about every 100,000 years. About twenty such cycles of cooling and warming were identified for the span we call the Ice Age.

During the glacial age, ice left its imprint on almost 30 percent of Earth's land area, including about 10 million square kilometers of North America, 5 million square kilometers of Europe, and 4 million square kilometers of Siberia (Figure 12.31). The amount of glacial ice in the Northern Hemisphere was roughly twice that of the Southern Hemisphere. The primary reason is that the southern polar ice could not spread far beyond the margins of Antarctica. By contrast, North America and Eurasia provided great expanses of land for the spread of ice sheets.

Today we know that the Ice Age began between two million and three million years ago. This means that most of the major glacial stages occurred during a division of the geologic time scale called the **Pleistocene epoch**. Although the Pleistocene is commonly used as a synonym for the Ice Age, note that this epoch does not encompass all of the last glacial period. The Antarctic ice sheet, for example, probably formed at least 14 million years ago, and, in fact, might be much older.

---

*For more on this topic, see Box 6.2, "Seafloor Sediments and Climatic Change," p. 158.

Box 12.3

Louis Agassiz: Architect of the Ice Age[*]

Louis Agassiz, a Swiss scientist born in 1807, was instrumental in developing modern ideas about the Ice Age (Figure 12.C). Early in his scientific career he was professor of natural history at the University of Neuchâtel. His specialty was fossil fish, but his interests were broad, and through his studies he became aware of the work of Ignatz Venetz (1799–1859) and Johann de Charpentier (1786–1855) on the subject of glaciation. After a field trip into the Swiss Alps with Charpentier, Agassiz became convinced that glaciers had formed during a temporary period of worldwide climate change and had spread across much of Siberia and central Europe. He presented his hypothesis to the Helvetic Association in the 1837 Presidential Address, and to support his ideas he cited such evidence as ridges of drift along the Rhone River, movement of giant boulders, and scoured and scraped bedrock. Many scientists, including Agassiz' friend and benefactor, the noted scientist Alexander von Humboldt, disagreed with him.

During the next three years, Agassiz continued his research on glaciers in the Alps and, in 1840, published *Studies on Glaciers*. This book summarized his earlier work and also presented new information about the appearance and structure of glaciers, how they form, their internal temperatures, and how they carry loads of broken and pulverized rock.

Between 1840 and 1846, Agassiz and other scientists carried out additional studies on Switzerland's glaciers. Agassiz studied plant life in moraines as well as temperature variations and the movement of glaciers. The results of this work were published in *Glacial Systems* in 1846. Many scientists in the United States disagreed with Agassiz' ideas. In fact, the Association of American Geologists appointed a committee to study his theory of drift in 1842–1843. It concluded that Agassiz' work was unworthy of further consideration, an act that may have set American studies of this subject back a generation.

In 1846, Agassiz embarked on a lecture tour to Paris and the United States, sponsored by the Lowell Institute of Boston. He planned to return to Europe at the end of 1848, but the wars on the Continent and a lucrative offer from Harvard University changed his mind. He began his career at Harvard as a professor of natural history while continuing his research on glaciers in North America. Agassiz published *Principles of Zoology*, which went through sixteen editions during his lifetime.

Agassiz achieved the status of what today would be the equivalent of a rock star. His peers considered him to be an inspiration, and his lectures both on and off campus were standing-room only. A reporter was assigned to a series of his lectures so that they could be carried verbatim in the Boston newspapers. Agassiz established a natural history museum at Harvard, and his writings on the Ice Age led to a new interest in the geology of New England and other glaciated parts of North America.

Louis Agassiz died in 1873, leaving behind a rich legacy that changed the study of geology. His work on glaciers and the Ice Age provided a foundation upon which later scientists based their studies.

[*]Prepared by Nancy Lutgens

Figure 12.C Louis Agassiz (1807–1873) played a major role in the development of glacial theory. (Courtesy of Harvard University Archives)

Some Indirect Effects of Ice-Age Glaciers

In addition to the massive erosional and depositional work carried on by Pleistocene glaciers, the ice sheets had other effects, sometimes profound, on the landscape. For example, as the ice advanced and retreated, animals and plants were forced to migrate. This led to stresses that some organisms could not tolerate. Hence, a number of plants and animals became extinct. Furthermore, many present-day stream courses bear little resemblance to their preglacial routes. The Missouri River once flowed northward toward Hudson Bay, the Mississippi River followed a path through central Illinois, and the head of the Ohio River reached eastward only to Indiana. Other rivers that today carry only a trickle of water but nevertheless occupy broad channels are testimony to the fact that they once carried torrents of glacial meltwater.

In areas that were centers of ice accumulation, such as Scandinavia and the Canadian Shield, the land has been slowly rising over the past several thousand

Figure 12.31 Maximum extent of ice sheets in the Northern Hemisphere during the Ice Age.

years. Uplifting of almost 300 meters has occurred in the Hudson Bay region. This, too, is the result of the continental ice sheets. But how can glacial ice cause such vertical crustal movement? We now understand that the land is rising because the added weight of the 3-kilometer-thick mass of ice caused down-warping of Earth's crust. Following the removal of this immense load, the crust has been adjusting by gradually rebounding upward ever since (Figure 12.32).*

*For a more complete discussion of this concept, termed *isostatic adjustment*, see the section "Isostasy" in Chapter 20.

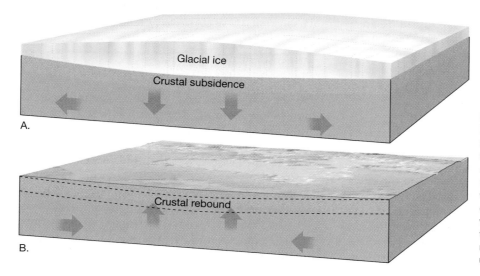

Figure 12.32 Simplified illustration showing crustal subsidence and rebound resulting from the addition and removal of continental ice sheets. **A.** In northern Canada and Scandinavia, where the greatest accumulation of glacial ice occurred, the added weight caused downwarping of the crust. **B.** Ever since the ice melted, there has been gradual uplift or rebound of the crust.

Certainly, one of the most interesting and perhaps dramatic effects of the Ice Age was the fall and rise of sea level that accompanied the advance and retreat of the glaciers. In Box 12.1 it was pointed out that sea level would rise by an estimated 60 or 70 meters if the water locked up the Antarctic Ice Sheet were to melt completely. Such an occurrence would flood many densely populated coastal areas.

Although the total volume of glacial ice today is great, exceeding 25 million cubic kilometers, during the Ice Age the volume of glacial ice amounted to about 70 million cubic kilometers, or 45 million cubic kilometers more than at present. Because we know that the snow from which glaciers are made ultimately comes from the evaporation of ocean water, the growth of ice sheets must have caused a worldwide drop in sea level (Figure 12.33). Indeed, estimates suggest that sea level was as much as 100 meters lower than today. Thus, land that is presently flooded by the oceans was dry. The Atlantic Coast of the United States lay more than 100 kilometers to the east of New York City; France and Britain were joined where the famous English Channel is today; Alaska and Siberia were connected across the Bering Strait; and Southeast Asia was tied by dry land to the islands of Indonesia.

While the formation and growth of ice sheets was an obvious response to significant changes in climate, the existence of the glaciers themselves triggered

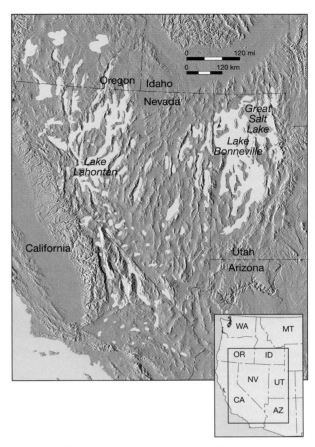

Figure 12.34 Pluvial lakes of the Western United States. (After R. F. Flint, *Glacial and Quaternary Geology*, New York: John Wiley & Sons)

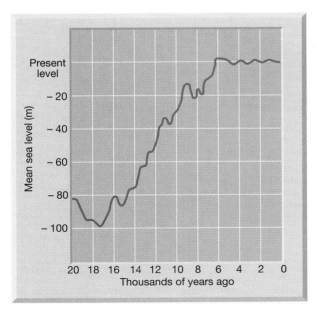

Figure 12.33 Changing sea level during the past 20,000 years. The lowest level shown on the graph represents the time about 18,000 years ago when the most recent ice advance was at a maximum.

important climatic changes in the regions beyond their margins. In arid and semiarid areas on all of the continents, temperatures were lower and thus evaporation rates were lower, but at the same time moderate precipitation totals were experienced. This cooler, wetter climate formed many **pluvial lakes** (from the Latin term *pluvia*, meaning *rain*). In North America, the greatest concentration of pluvial lakes occurred in the vast Basin and Range region of Nevada and Utah (Figure 12.34). By far the largest of the lakes in this region was Lake Bonneville. With maximum depths exceeding 300 meters and an area of 50,000 square kilometers, Lake Bonneville was nearly the same size as present-day Lake Michigan. As the ice sheets waned, the climate again grew more arid, and the lake levels lowered in response. Although most of the lakes completely disappeared, a few small remnants of Lake Bonneville remain, the Great Salt Lake being the largest and best known.

Causes of Glaciation

A great deal is known about glaciers and glaciation (see Box 12.4). Much has been learned about glacier formation and movement, the extent of glaciers past and present, and the features created by glaciers, both erosional and depositional. However, a widely accepted theory for the causes of glacial ages has not yet been established. Although more than 160 years have elapsed since Louis Agassiz proposed his theory of a great Ice Age, no complete agreement exists as to the causes of such events.

Although widespread glaciation has been rare in Earth's history, the Ice Age that encompassed the Pleistocene epoch is not the only glacial period for which a record exists. Earlier glaciations are indicated by deposits called **tillite**, a sedimentary rock formed when glacial till becomes lithified. Such deposits, found in strata of several different ages, usually contain striated rock fragments, and some overlie grooved and polished bedrock surfaces or are associated with sandstones and conglomerates that show features of outwash deposits. Two Precambrian glacial episodes have been identified in the geologic record, the first approximately two billion years ago and the second about 600 million years ago. Further, a well-documented record of an earlier glacial age is found in late Paleozoic rocks that are about 250 million years old and which exist on several landmasses.

Any theory that attempts to explain the causes of glacial ages must successfully answer two basic questions. (1) *What causes the onset of glacial conditions?* For continental ice sheets to have formed, average temperature must have been somewhat lower than at present and perhaps substantially lower than throughout much of geologic time. Thus, a successful theory would have to

---

 Box 12.4

Climate Change Recorded in Glacial Ice

Vertical cores taken from the Greenland and Antarctic ice sheets are important sources of data about climate change during and following the most recent cycle of glaciation. Scientists collect samples with a drilling rig, like a small version of an oil drill. A hollow shaft follows the drill head into the ice and an ice core is extracted. In this way, cores that sometimes exceed 2000 meters in length and may represent more than 200,000 years of climate history are acquired for study (Figure 12.D).

The ice provides a detailed record of changing air temperatures and snowfall. Air bubbles trapped in the ice record variations in atmospheric composition. Changes in carbon dioxide and methane are linked to fluctuating temperatures. The cores also include atmospheric fallout such as wind-blown dust, volcanic ash, pollen, and modern-day pollution.

Past temperatures are determined by *oxygen isotope analysis*. This technique is based on precise measurement of the ratio between two isotopes of oxygen: O^{16}, which is the most common, and the heavier O^{18}. More O^{18} is evaporated from the oceans when temperatures are high and less is evaporated when temperatures are low. Therefore, the heavier isotope is more abundant in the precipitation of warm eras and less

Figure 12.D Scientists at the National Ice Core Laboratory in Denver, Colorado, examine an ice core sample. Faint lines in the sample are annual dust layers deposited in summer months. Using these layers, the ice cores can be dated much like dating trees by studying their rings. (Photo by Ken Abbot/National Ice Core Laboratory)

abundant during colder periods. Using this principle, scientists are able to produce a record of past temperature changes. A portion of such a record is shown in Figure 12.E.

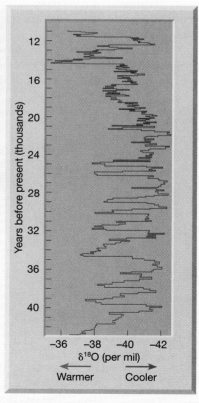

Figure 12.E Temperature variations as revealed by fluctuations in the O^{18}/O^{16} ratio in a portion of a Greenland ice core. An increase in O^{18} indicates an increase in air temperature. Levels of O^{18} decrease when cooler temperatures prevail.

account for the cooling that finally leads to glacial conditions. (2) *What caused the alternation of glacial and interglacial stages that have been documented for the Pleistocene epoch?* The first question deals with long-term trends in temperature on a scale of millions of years but this second question relates to much shorter-term changes.

Although the literature of science contains a vast array of hypotheses relating to the possible causes of glacial periods, we will discuss only a few major ideas to summarize current thought.

Plate Tectonics

Probably the most attractive proposal for explaining the fact that extensive glaciations have occurred only a few times in the geologic past comes from the theory of plate tectonics.* Because glaciers can form only on land, we know that landmasses must exist somewhere in the higher latitudes before an ice age can commence. Many scientists suggest that ice ages have occurred only when Earth's shifting crustal plates have carried the continents from tropical latitudes to more poleward positions.

---

*A brief overview of the theory appears in Chapter 1, and a more extensive discussion is presented in Chapter 19.

Glacial features in present-day Africa, Australia, South America, and India indicate that these regions, which are now tropical or subtropical, experienced an Ice Age near the end of the Paleozic era, about 250 million years ago. However, there is no evidence that ice sheets existed during this same period in what are today the higher latitudes of North America and Eurasia. For many years this puzzled scientists. Was the climate in these relatively tropical latitudes once like it is today in Greenland and Antarctica? Why did glaciers not form in North America and Eurasia? Until the plate tectonics theory was formulated, there had been no reasonable explanation.

Today, scientists understand that the areas containing these ancient glacial features were joined together as a single supercontinent located at latitudes far to the south of their present positions. Later, this landmass broke apart, and its pieces, each moving on a different plate, drifted toward their present locations (Figure 12.35). Now we know that during the geologic past, plate movements accounted for many dramatic climatic changes as landmasses shifted in relation to one another and moved to different latitudinal positions. Changes in oceanic circulation also must have occurred,

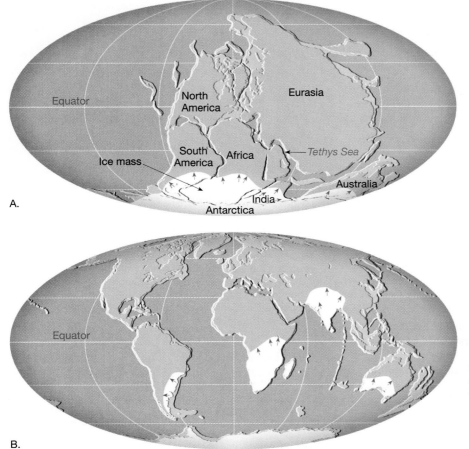

Figure 12.35 A. The supercontinent Pangaea showing the area covered by glacial ice 300 million years ago. **B.** The continents as they are today. The white areas indicate where evidence of the old ice sheets exists.

altering the transport of heat and moisture and consequently the climate as well. Because the rate of plate movement is very slow—a few centimeters annually—appreciable changes in the positions of the continents occur only over great spans of geologic time. Thus, climatic changes triggered by shifting plates are extremely gradual and happen on a scale of millions of years.

Variations in Earth's Orbit

Because climatic changes brought about by moving plates are extremely gradual, the plate tectonics theory cannot be used to explain the alternation between glacial and interglacial climates that occurred during the Pleistocene epoch. Therefore, we must look to some other triggering mechanism that might cause climatic change on a scale of thousands rather than millions of years. Many scientist today believe or strongly suspect that the climatic oscillations that characterized the Pleistocene may be linked to variations in Earth's orbit. This hypothesis was first developed and strongly advocated by the Yugoslavian scientist Milutin Milankovitch and is based on the premise that variations in incoming solar radiation are a principal factor in controlling Earth's climate.

Milankovitch formulated a comprehensive mathematical model based on the following elements (Figure 12.36):

1. Variations in the shape *(eccentricity)* of Earth's orbit about the Sun;

2. Changes in *obliquity*; that is, changes in the angle that the axis makes with the plane of Earth's orbit; and

3. The wobbling of Earth's axis, called *precession*.

Using these factors, Milankovitch calculated variations in the receipt of solar energy and the corresponding surface temperature of Earth back into time in an attempt to correlate these changes with the climatic fluctuations of the Pleistocene. In explaining climatic changes that result from these three variables, note that they cause little or no variation in the total solar energy reaching the ground. Instead, their impact is felt because they change the degree of contrast between the seasons. Somewhat milder winters in the middle to high latitudes means greater snowfall totals, whereas cooler summers would bring a reduction in snowmelt.

Among the studies that have added credibility to the astronomical theory of Milankovitch is one in which deep-sea sediments containing certain climatically sensitive microorganisms were analyzed to establish a chronology of temperature changes going back nearly

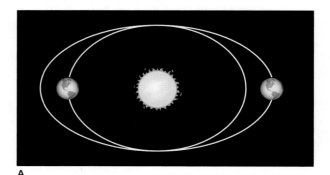

A.

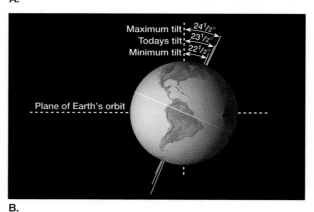

B.

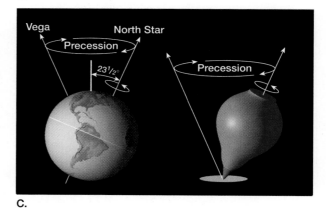

C.

Figure 12.36 Orbital variations. **A.** The shape of Earth's orbit changes during a cycle that spans about 100,000 years. It gradually changes from nearly circular to one that is more elliptical and then back again. This diagram greatly exaggerates the amount of change. **B.** Today the axis of rotation is titled about 23.5° to the plane of Earth's orbit. During a cycle of 41,000 years, this angle varies from 21.5° to 24.5°. **C.** Precession. Earth's axis wobbles like that of a spinning top. Consequently, the axis points to different spots in the sky during a cycle of about 26,000 years.

one-half million years.* This time scale of climatic change was then compared to astronomical calculations

---

*J. D. Hays, John Imbrie, and N. J. Shackelton, "Variations in the Earth's Orbit: Pacemaker of the Ice Ages," *Science* 194 (1976): 1121–32.

of eccentricity, obliquity, and precession to determine if a correlation did indeed exist. Although the study was very involved and mathematically complex, the conclusions were straightforward. The researchers found that major variations in climate over the past several hundred thousand years were closely associated with changes in the geometry of Earth's orbit; that is, cycles of climatic change were shown to correspond closely with the periods of obliquity, precession, and orbital eccentricity. More specifically, the authors stated: "It is concluded that changes in the earth's orbital geometry are the fundamental cause of the succession of Quaternary ice ages."*

---

*J. D. Hays et al., p. 1131. The term *Quaternary* refers to the period on the geologic time scale that encompasses the last 1.6 million years.

Let us briefly summarize the ideas that were just described. The theory of plate tectonics provides us with an explanation for the widely spaced and nonperiodic onset of glacial conditions at various times in the geologic past, whereas the theory proposed by Milankovitch and supported by the work of J. D. Hays and his colleagues furnishes an explanation for the alternating glacial and interglacial episodes of the Pleistocene.

In conclusion, we emphasize that the ideas just discussed do not represent the only possible explanations for glacial ages. Although interesting and attractive, these proposals are certainly not without critics; nor are they the only possibilities currently under study. Other factors may be, and probably are, involved.

Chapter Summary

- A *glacier* is a thick mass of ice originating on land as a result of the compaction and recrystallization of snow, and it shows evidence of past or present flow. Today, *valley* or *alpine glaciers* are found in mountain areas where they usually follow valleys that were originally occupied by streams. *Ice sheets* exist on a much larger scale, covering most of Greenland and Antarctica.

- Near the surface of a glacier, in the *zone of fracture*, ice is brittle. However, below about 50 meters, pressure is great, causing ice to *flow* like *plastic material*. A second important mechanism of glacial movement consists of the entire ice mass *slipping* along the ground.

- The average velocity of glacial movement is generally quite slow, but varies considerably from one glacier to another. The advance of some glaciers is characterized by periods of extremely rapid movements called *surges*.

- Glaciers form in areas where more snow falls in winter than melts during summer. Snow accumulation and ice formation occur in the *zone of accumulation*. Its outer limits are defined by the *snowline*. Beyond the snowline is the *zone of wastage*, where there is a net loss to the glacier. The *glacial budget* is the balance, or lack of balance, between accumulation at the upper end of the glacier, and loss, called *ablation*, at the lower end.

- Glaciers erode land by *plucking* (lifting pieces of bedrock out of place) and *abrasion* (grinding and scraping of a rock surface). Erosional features produced by valley glaciers include *glacial troughs*, *hanging valleys*, *pater noster lakes*, *fiords*, *cirques*, *arêtes*, *horns*, and *roches moutonnées*.

- Any sediment of glacial origin is called *drift*. The two distinct types of glacial drift are (1) *till*, which is material deposited directly by the ice; and (2) *stratified drift*, which is sediment laid down by meltwater from a glacier.

- The most widespread features created by glacial deposition are layers or ridges of till, called *moraines*. Associated with valley glaciers are *lateral moraines*, formed along the sides of the valley, and *medial moraines*, formed between two valley glaciers that have joined. *End moraines*, which mark the former position of the front of a glacier, and *ground moraines*, undulating layers of till deposited as the ice front retreats, are common to both valley glaciers and ice sheets. An *outwash plain* is often associated with the end moraine of an ice sheet. A *valley train* may form when the glacier is confined to a valley. Other depositional features include *drumlins* (streamlined asymmetrical hills composed of till), *eskers* (sinuous ridges composed largely of sand and gravel deposited by streams flowing in tunnels beneath the ice, near the terminus of a glacier), and *kames* (steep-sided hills consisting of sand and gravel).

- The *Ice Age*, which began about two million years ago, was a very complex period characterized by a number of advances and withdrawals of glacial ice. Most of the major glacial episodes occurred during a division of the geologic time scale called the *Pleistocene epoch*. Perhaps the most convincing evidence for the occurrence of several glacial advances during the Ice Age is the widespread existence of *multiple layers of drift* and an uninterrupted record of climate cycles preserved in *seafloor sediments*.

- In addition to massive erosional and depositional work, other effects of Ice Age glaciers included the *forced migration of organisms, changes in stream courses, adjustment of the crust* by rebounding after the removal of the immense load of ice, and *climate changes* caused by the existence of the glaciers

themselves. In the sea, the most far-reaching effect of the Ice Age was the *worldwide change* in *sea level* that accompanied each advance and retreat of the ice sheets.

- Any theory that attempts to explain the causes of glacial ages must answer two basic questions: (1) What causes the onset of glacial conditions? and (2) What caused the alternating glacial and interglacial stages that have been documented for the Pleistocene epoch? Two of the many hypotheses for the cause of glacial ages involve (1) plate tectonics and (2) variations in Earth's orbit.

Review Questions

1. What is a glacier? Under what circumstances does glacial ice form?

2. Describe how glaciers fit into the hydrologic cycle. What role do they play in the rock cycle?

3. Each statement below refers to a particular type of glacier. Name the type of glacier.

   (a) The term *continental* is often used to describe this type of glacier.

   (b) This type of glacier is also called an *alpine glacier*.

   (c) This is a stream of ice leading from the margin of an ice sheet through the mountains to the sea.

   (d) This is a glacier formed when one or more valley glaciers spreads out at the base of a steep mountain front.

   (e) Greenland is the only example in the Northern Hemisphere.

4. Where are glaciers found today? What percentage of Earth's land area do they cover? How does this compare to the area covered by glaciers during the Pleistocene?

5. Describe the two components of glacial flow. At what rates do glaciers move? In a valley glacier, does all of the ice move at the same rate? Explain.

6. Why do crevasses form in the upper portion of a glacier but not below 50 meters?

7. Under what circumstances will the front of a glacier advance? Retreat? Remain stationary?

8. Describe the processes of glacial erosion.

9. How does a glaciated mountain valley differ in appearance from a mountain valley that was not glaciated?

10. List and describe the erosional features you might expect to see in an area where valley glaciers exist or have recently existed.

11. What is glacial drift? What is the difference between till and stratified drift? What general effect do glacial deposits have on the landscape?

12. List the four basic moraine types. What do all moraines have in common? What is the significance of terminal and recessional moraines?

13. Why are medial moraines proof that valley glaciers must move?

14. How do kettles form?

15. What direction was the ice sheet moving that affected the area shown in Figure 12.26? Explain how you were able to determine this.

16. What are ice-contact deposits? Distinguish between kames and eskers.

17. The development of the glacial theory is a good example of applying the principle of uniformitarianism. Explain briefly.

18. During the Pleistocene epoch the amount of glacial ice in the Northern Hemisphere was about twice as great as in the Southern Hemisphere. Briefly explain why this was the case.

19. List three indirect effects of Ice Age glaciers.

20. How might plate tectonics help explain the cause of ice ages? Can plate tectonics explain the alternation between glacial and interglacial climates during the Pleistocene?

Key Terms

ablation (p. 299)
abrasion (p. 302)
alpine glacier (p. 294)
arête (p. 306)
basal slip (p. 298)
calving (p. 299)
cirque (p. 304)
col (p. 305)
crevasse (p. 298)
drumlin (p. 312)
end moraine (p. 308)
esker (p. 314)
fiord (p. 305)
firn (p. 297)
glacial budget (p. 299)
glacial drift (p. 307)

glacial erratic (p. 308)
glacial striations (p. 302)
glacial trough (p. 304)
glacier (p. 294)
ground moraine (p. 309)
hanging valley (p. 304)
horn (p. 306)
ice cap (p. 295)
ice-contact deposit
   (p. 314)
ice sheet (p. 295)
ice shelf (p. 295)
kame (p. 314)
kame terrace (p. 314)
kettle (p. 313)
lateral moraine (p. 309)

medial moraine (p. 311)
outlet glacier (p. 295)
outwash plain (p. 313)
pater noster lakes
   (p. 304)
piedmont glacier (p. 296)
plastic flow (p. 297)
Pleistocene epoch
   (p. 316)
plucking (p. 302)
pluvial lake (p. 319)
recessional moraine
   (p. 309)
roche moutonnée
   (p. 307)
rock flour (p. 302)

snowline (p. 299)
stratified drift (p. 307)
surge (p. 299)
tarn (p. 304)
terminal moraine
   (p. 309)
till (p. 307)
tillite (p. 320)
truncated spur (p. 304)
valley glacier (p. 294)
valley train (p. 313)
zone of accumulation
   (p. 299)
zone of fracture (p. 298)
zone of wastage (p. 299)

Web Resources

The *Earth* Home Page provides on-line resources for this chapter on the World Wide Web. You will find review exercises, specific updates for items in the chapter, suggested reading, and links to interesting related pathways on the Internet. Visit the *Earth* Home Page at **http://www.prenhall.com/tarbuck.**

CHAPTER 13

Deserts and Winds

Left Sand Mountain east of Fallon in Nevada's dry Great Basin Desert. — *Photo by Carr Clifton*

The word desert literally means deserted or unoccupied. For many dry regions this is a very appropriate description, although where water is available in deserts, plants and animals thrive. Nevertheless, the world's dry regions are probably the least familiar land areas on Earth outside of the polar realm.

Desert landscapes frequently appear stark. Their profiles are not softened by a carpet of soil and abundant plant life. Instead, barren rocky outcrops with steep, angular slopes are common. At some places the rocks are tinted orange and red. At others they are gray and brown and streaked with black. For many visitors, desert scenery exhibits a striking beauty; to others, the terrain seems bleak. No matter which feeling is elicited, it is clear that deserts are very different from the more humid places where most people live.

As you shall see, arid regions are not dominated by a single geologic process. Rather, the effects of tectonic forces, running water, and wind are all apparent. Because these processes combine in different ways from place to place, the appearance of desert landscapes varies a great deal as well (Figure 13.1).

Distribution and Causes of Dry Lands

The dry regions of the world encompass about 42 million square kilometers, a surprising 30 percent of Earth's land surface. No other climatic group covers so large a land area. Within these water-deficient regions, two climatic types are commonly recognized: **desert**, or arid, and **steppe**, or semiarid. The two share many features; their differences are primarily a matter of degree (see Box 13.1). The steppe is a marginal and more humid variant of the desert and is a transition zone that surrounds the desert and separates it from bordering humid climates. The world map showing the distribution of desert and steppe regions reveals that dry lands are concentrated in the subtropics and in the middle latitudes (Figure 13.2).

Low-Latitude Deserts

The heart of the low-latitude dry climates lies in the vicinities of the Tropics of Cancer and Capricorn. Figure 13.2 shows a virtually unbroken desert environment stretching for more than 9300 kilometers from the Atlantic coast of North Africa to the dry lands of northwestern India. In addition to this single great expanse, the Northern Hemisphere contains another much smaller area of tropical desert and steppe in northern Mexico and the southwestern United States.

In the Southern Hemisphere, dry climates dominate Australia. Almost 40 percent of the continent is desert, and much of the remainder is steppe. In addition, arid and semiarid areas occur in southern Africa and make a limited appearance in coastal Chile and Peru.

Figure 13.1 A scene in north-central New Mexico near Abiquiu. The appearance of desert landscapes varies a great deal from place to place. (Photo by Linda Waidhofer/Liaison International)

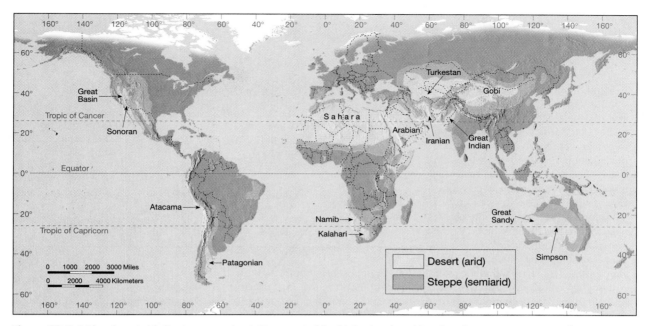

Figure 13.2 Arid and semiarid climates cover about 30 percent of Earth's land surface. No other climate group covers so large an area.

What causes these bands of low-latitude desert? The answer is the global distribution of air pressure and winds. Figure 13.3, an idealized diagram of Earth's general circulation, helps visualize the relationship. Heated air in the pressure belt known as the *equatorial low* rises to great heights (usually between 15 and 20 kilometers) and then spreads out. As the upper-level flow reaches 20° to 30° latitude, north or south, it sinks toward the surface. Air that rises through the atmosphere expands

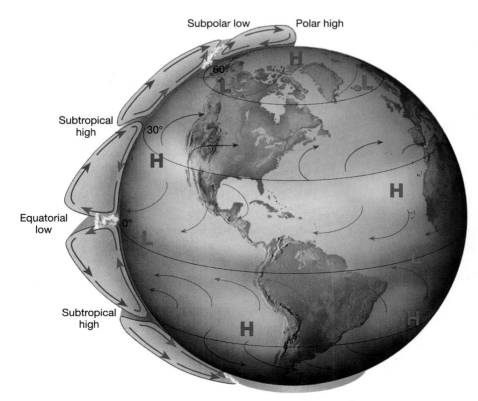

Figure 13.3 Idealized diagram of Earth's general circulation. The deserts and steppes that are centered in the latitude between 20° and 30° north and south coincide with the subtropical high-pressure belts. Here dry, subsiding air inhibits cloud formation and precipitation. By contrast, the pressure belt known as the equatorial low is associated with areas that are among the rainiest on Earth.

Box 13.1

What Is Meant by "Dry"?

Albuquerque, New Mexico, in the southwestern United States, receives an average of 20.7 centimeters (8.07 inches) of rainfall annually. As you might expect, because Albuquerque's precipitation total is modest, the station is classified as a desert when the commonly used Köppen climate classification is applied. The Russian city of Verkhoyansk is a remote station located near the Arctic Circle in Siberia. The yearly precipitation total there averages 15.5 centimeters (6.05 inches), about 5 centimeters less than Albuquerque's. Although Verkhoyansk receives less precipitation than Albuquerque, its classification is that of a humid climate. How can this occur?

We all recognize that deserts are dry places, but just what is meant by the term *dry*? That is, how much rain defines the boundary between humid and dry regions? Sometimes it is arbitrarily defined by a single rainfall figure, for example, 25 centimeters (10 inches) per year of precipitation. However, the concept of dryness is a relative one that refers to any situation in which a water deficiency exists. Hence, climatologists define *dry climate* as one in which yearly precipitation is not as great as the potential loss of water by evaporation. Dryness then not only is related to annual rainfall totals but is also a function of evaporation, which, in turn, is closely dependent upon temperature.

Table 13.A Average Annual Precipitation Defining the Boundary between Dry and Humid Climates

Average Annual Temp. (C°)	Winter Rainfall Maximum (centimeters)	Even Distribution (centimeters)	Summer Rainfall Maximum (centimeters)
5	10	24	38
10	20	34	48
15	30	44	58
20	40	54	68
25	50	64	78
30	60	74	88

As temperatures climb, potential evaporation also increases. Fifteen to twenty-five centimeters of precipitation may be sufficient to support coniferous forests in northern Scandinavia or Siberia, where evaporation into the cool, humid air is slight and a surplus of water remains in the soil. However, the same amount of rain falling on New Mexico or Iran supports only a sparse vegetative cover because evaporation into the hot, dry air is great. So clearly no specific amount of precipitation can serve as a universal boundary for dry climates.

To establish the boundary between dry and humid climates, the widely used Köppen classification uses formulas that involve three variables: (1) average annual precipitation, (2) average annual temperature, and (3) seasonal distribution of precipitation. The use of average annual temperature reflects its importance as an index of evaporation. The amount of rainfall defining the humid-dry boundary will be larger where mean annual temperatures are high and smaller where temperatures are low. The use of seasonal precipitation distribution as a variable is also related to this idea. If rain is concentrated in the warmest months, loss to evaporation is greater than if the precipitation is concentrated in the cooler months.

Table 13.A summarizes the precipitation amounts that divide dry and humid climates. Notice that a station with an annual mean of 20°C (68°F) and a summer rainfall maximum of 68 centimeters (26.5 inches) is classified as dry. If the rain falls primarily in winter, however, the station must receive only 40 centimeters (15.6 inches) or more to be considered humid. If the precipitation is more evenly distributed, the figure defining the humid–dry boundary is between the other two.

and cools, a process that leads to the development of clouds and precipitation. For this reason, the areas under the influence of the equatorial low are among the rainiest on Earth. Just the opposite is true for the regions in the vicinity of 30° north and south latitude, where high pressure predominates. Here, in the zones known as the *subtropical highs*, air is subsiding. When air sinks, it is compressed and warmed. Such conditions are just opposite of what is needed to produce clouds and precipitation. Consequently, these regions are known for their clear skies, sunshine, and ongoing drought.

Middle-Latitude Deserts

Unlike their low-latitude counterparts, middle-latitude deserts and steppes are not controlled by the subsiding air masses associated with high pressure. Instead, these dry lands exist principally because they are sheltered in the deep interiors of large landmasses. They are far removed from the ocean, which is the ultimate source of moisture for cloud formation and precipitation. One well-known example is the Gobi Desert of central Asia, shown on the map north of India.

The presence of high mountains across the paths of prevailing winds further separates these areas

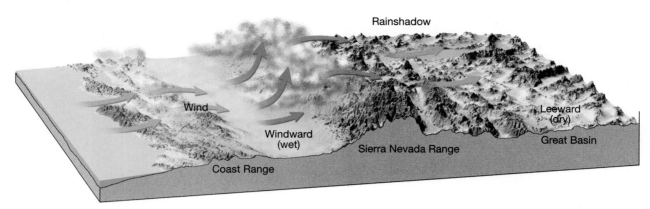

Figure 13.4 Many deserts in the middle latitudes are rainshadow deserts. As moving air meets a mountain barrier, it is forced to rise. Clouds and precipitation on the windward side often result. Air descending the leeward side is much drier. The mountains effectively cut the leeward side off from the sources of moisture, producing a rainshadow desert. The Great Basin desert is a rainshadow desert that covers nearly all of Nevada and portions of adjacent states.

from water-bearing, maritime air masses; the mountains force the air to lose much of its water. The mechanism is simple: As prevailing winds meet mountain barriers, the air is forced to ascend. When air rises it expands and cools, a process that can produce clouds and precipitation. The windward sides of mountains, therefore, often have high precipitation. By contrast, the leeward sides of mountains are usually much drier (Figure 13.4). This situation exists because air reaching the leeward side has lost much of its moisture and, if the air descends, it is compressed and warmed, making cloud formation even less likely. The dry region that results is often referred to as a **rainshadow desert**. Because many middle-latitude deserts occupy sites on the leeward sides of mountains, they can also be classified as rainshadow deserts. In North America, the Coast Ranges, Sierra Nevada, and Cascades are the foremost mountain barriers to moisture from the Pacific (Figure 13.4). In Asia, the great Himalayan chain prevents the summertime monsoon flow of moist Indian Ocean air from reaching the interior.

Because the Southern Hemisphere lacks extensive land areas in the middle latitudes, only a small area of desert and steppe occurs in this latitude range, existing primarily near the southern tip of South America in the rainshadow of the towering Andes.

The middle-latitude deserts provide an example of how tectonic processes affect climate. Rainshadow deserts exist by virtue of the mountains produced when plates collide. Without such mountain-building episodes, wetter climates would prevail where many dry regions exist today.

Geologic Processes in Arid Climates

The angular hills, the sheer canyon walls, and the desert surface of pebbles or sand contrast sharply with the rounded hills and curving slopes of more humid places. Indeed, to a visitor from a humid region, a desert landscape may seem to have been shaped by forces different from those operating in well-watered areas. However, although the contrasts may be striking, they do not reflect different processes. They merely disclose the differing effects of the same processes that operate under contrasting climatic conditions.

Weathering

In humid regions, relatively fine-textured soils support an almost continuous cover of vegetation that mantles the surface. Here the slopes and rock edges are rounded, reflecting the strong influence of chemical weathering in a humid climate. By contrast, much of the weathered debris in deserts consists of unaltered rock and mineral fragments—the result of mechanical weathering processes. In dry lands, rock weathering of any type is greatly reduced because of the lack of moisture and the scarcity of organic acids from decaying plants. However, chemical weathering is not completely lacking in deserts. Over long spans of time, clays and thin soils do form, and many iron-bearing silicate minerals oxidize, producing the rust-colored stain that tints some desert landscapes.

A.

B.

Figure 13.5 A. Most of the time, desert stream channels are dry. **B.** An ephemeral stream shortly after a heavy shower. Although such floods are short-lived, large amounts of erosion occurs. (Photos by E. J. Tarbuck)

The Role of Water

Permanent streams are normal in humid regions but practically all desert streambeds are dry most of the time (Figure 13.5A). Deserts have **ephemeral streams**, which means they carry water only in response to specific episodes of rainfall. A typical ephemeral stream might flow only a few days or perhaps just a few hours during the year. In some years, the channel might carry no water at all.

This fact is obvious even to the casual traveler who notices numerous bridges with no streams beneath them or numerous dips in the road where dry channels cross. However, when the rare heavy showers do come, so much rain falls in such a short time that all of it cannot soak in (Figure 13.6). Because desert vegetative cover is sparse, runoff is largely unhindered and consequently rapid, often creating flash floods along valley floors (Figure 13.5B). These floods are quite unlike floods in humid regions. A flood on a river like the Mississippi may take several days to reach its crest and then subside. But desert floods arrive suddenly and subside quickly. Because

Figure 13.6 Desert thunderstorm over Eagle Mesa in Arizona's Monument Valley. There are often many weeks, months, or occasionally even years separating periods of rain in the desert. When rains do occur, they are often heavy and of relatively short duration. Because the rainfall intensity is high, all of the water cannot soak in, and rapid runoff results. (Photo by Greg Gawlowski/Dembinsky Photo Assoc.)

much surface material in a desert is not anchored by vegetation, the amount of erosional work that occurs during a single short-lived rain event is impressive.

In the dry western United States, different names are used for ephemeral streams, including *wash* and *arroyo*. In other parts of the world, a dry desert stream may be a *wadi* (Arabia and North Africa), a *donga* (South America), or a *nullah* (India).

Humid regions are notable for their integrated drainage systems. But in arid regions, streams usually lack an extensive system of tributaries. In fact, a basic characteristic of desert streams is that they are small and die out before reaching the sea. Because the water table is usually far below the surface, few desert streams can draw upon it as streams do in humid regions (see Figure 11.4, p. 272). Without a steady supply of water, the combination of evaporation and infiltration soon depletes the stream.

The few permanent streams that do cross arid regions, such as the Colorado and Nile rivers, originate *outside* the desert, often in well-watered mountains. Here the water supply must be great to compensate for the losses occurring as the stream crosses the desert. For example, after the Nile leaves its headwaters in the lakes and mountains of central Africa, it traverses almost 3000 kilometers of the Sahara without a single tributary. By contrast, in humid regions, the discharge of a river grows as it flows downstream because tributaries and groundwater contribute additional water along the way.

It should be emphasized that *running water, although infrequent, nevertheless does most of the erosional work in deserts* (see Box 13.2). This is contrary to the common belief that wind is the most important erosional agent sculpturing desert landscapes. Although wind erosion is indeed more significant in dry areas than elsewhere, most desert landforms are carved by running water. As you will see shortly, the main role of wind is in the transportation and deposition of sediment, which creates and shapes the ridges and mounds we call dunes.

Basin and Range: The Evolution of a Desert Landscape

Because arid regions typically lack permanent streams, they are characterized as having **interior drainage**. This means that they have a discontinuous pattern of intermittent streams that do not flow out of the desert to the ocean. In the United States, the dry Basin and Range region provides an excellent example. The region includes southern Oregon, all of Nevada, western Utah,

southeastern California, southern Arizona, and southern New Mexico. The name Basin and Range is an apt description for this almost 800,000-square-kilometer region, because it is characterized by more than 200 relatively small mountain ranges that rise 900 to 1500 meters above the basins that separate them.

In this region, as in others like it around the world, erosion mostly occurs without reference to the ocean (ultimate base level) because the interior drainage never reaches the sea. Even where permanent streams flow to the ocean, few tributaries exist, and thus only a narrow strip of land adjacent to the stream has sea level as its ultimate level of land reduction.

The block models in Figure 13.7 depict how the landscape has evolved in the Basin and Range region. During and following the uplift of the mountains, running water begins carving the elevated mass and depositing large quantities of debris in the basin. During this early stage, relief is greatest, because as erosion lowers the mountains and sediment fills the basins, elevation differences gradually diminish.

When the occasional torrents of water produced by sporadic rains move down the mountain canyons, they are heavily loaded with sediment. Emerging from the confines of the canyon, the runoff spreads over the gentler slopes at the base of the mountains and quickly loses velocity. Consequently, most of its load is dumped within a short distance. The result is a cone of debris at the mouth of a canyon known as an **alluvial fan** (Figure 13.8). Because the coarsest material is dropped first, the head of the fan is steepest, having a slope of perhaps 10 to 15 degrees. Moving down the fan, the size of the sediment and the steepness of the slope decrease and merge imperceptibly with the basin floor. An examination of the fan's surface would likely reveal a braided channel pattern because of the water shifting its course as successive channels became choked with sediment. Over the years, a fan enlarges, eventually coalescing with fans from adjacent canyons to produce an apron of sediment called a **bajada** along the mountain front.

On the rare occasions of abundant rainfall, streams may flow across the bajada to the center of the basin, converting the basin floor into a shallow **playa lake**. Playa lakes are temporary features that last only a few days or at best a few weeks before evaporation and infiltration remove the water. The dry, flat lake bed that remains is called a **playa**. Playas are typically composed of fine silts and clays, and occasionally encrusted with salts precipitated during evaporation (see Figure 6.13, p. 154). These precipitated salts may be unusual. A case in point is the sodium borate (better known as borax) mined from ancient playa lake deposits in Death Valley, California.

Box 13.2

Common Misconceptions about Deserts

Deserts are hot, lifeless, sand-covered land-scapes shaped largely by the force of wind. The preceding statement summarizes the image of arid regions that many people hold, especially those living in more humid places. Is it an accurate view? The answer is no. Although there are clearly elements of reality in such an impression, it is a generalization that contains a number of misconceptions (Figure 13.A).

One common fallacy about deserts is that they are lifeless or almost lifeless. Although reduced in amount and different in character, plant and animal life are usually present. Desert plants may differ widely from one part of the world to another, but all have one characteristic in common: they have developed adaptations that make them highly tolerant of drought. Many have waxy leaves, stems, or branches or a thickened cuticle (outermost protective layer) to reduce water loss. Others have very small leaves or no leaves at all.

Also, the roots of some species often extend to great depths in order to tap the moisture found there, whereas others produce a shallow but widespread root system that enables them to absorb great amounts of moisture quickly from the infrequent desert downpours. Often the stems of these plants are thickened by a spongy tissue that can store enough water to sustain the plant until the next rainfall comes. Thus, although widely dispersed and providing little ground cover, plants of many kinds flourish in the desert.

A second widely held belief about the world's dry lands is that they are always hot. This fact seems to be reinforced by commonly quoted temperature statistics. The highest accepted temperature record for the United States as well as the entire Western Hemisphere is 57°C (134°F). This long-standing record was set at Death

Figure 13.A Snow covers the rocky ground in the Sonoran Desert near Tucson, Arizona. As this scene demonstrates, deserts are not necessarily always hot, lifeless, dune-covered expanses. (Photo by T. Wiewandt/DRK Photo)

Valley, California, on July 10, 1913. The nearly 59°C (137°F) reading in Azizia, Libya, in North Africa's Sahara Desert on September 13, 1922, is the world record. Despite these remarkably high figures, cold temperatures are also experienced in desert regions.

For example, the average daily minimum in January at Phoenix, Arizona, is 1.7°C (35°F), just barely above freezing. At Ulan Bator in Mongolia's Gobi Desert, the average *high* temperature on January days is only –19°C (–2°F)! Dry climates are found from the tropics poleward to the high middle latitudes. Consequently, although tropical deserts lack a cold season, deserts in the middle latitudes do experience seasonal temperature changes.

The last two commonly held misconceptions about the world's deserts are more geologic than climatic. One mistaken assumption is that they consist

of mile after mile of drifting sand. It is true that sand accumulations do exist in some areas and may be striking features, but they represent only a small percentage of the total desert area. For example, in the Sahara, the world's largest desert, accumulations of sand cover only one-tenth of its area. The sandiest of all deserts is the Arabian, one-third of which consists of sand.

The final mistaken assumption is the seemingly logical idea that wind is the most important agent of erosion in deserts. Although wind is relatively more significant in dry areas than anywhere else, most erosional landforms in deserts are created by running water. When rains come, they frequently take the form of thunderstorms. Because the heavy rain associated with these storms cannot all soak in, rapid runoff results. Without a thick vegetative cover to protect the ground, erosion is great.

With the ongoing erosion of the mountain mass and the accompanying sedimentation, the local relief continues to diminish. Eventually nearly the entire mountain mass is gone. Thus, by the late stages of erosion, the mountain areas are reduced to a few large bedrock knobs projecting above the surrounding sediment-filled basin. These isolated erosional remnants on a late stage desert landscape are called **inselbergs**, a German word meaning "island mountains" (see Box 13.3).

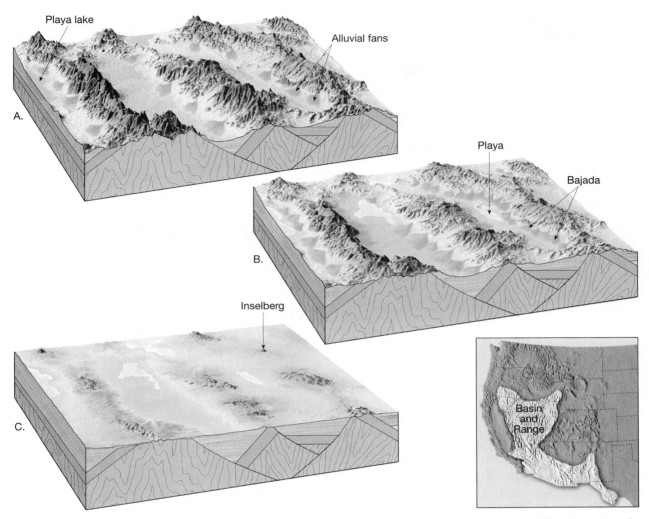

Figure 13.7 Stages of landscape evolution in a mountainous desert such as the Basin and Range region of the West. As erosion of the mountains and deposition in the basins continue, relief diminishes. **A.** Early stage. **B.** Middle stage. **C.** Late stage.

Each of the stages of landscape evolution in an arid climate depicted in Figure 13.7 can be observed in the Basin and Range region. Recently uplifted mountains in an early stage of erosion are found in southern Oregon and northern Nevada. Death Valley, California, and southern Nevada fit into the more advanced middle stage, whereas the late stage, with its inselbergs, can be seen in southern Arizona.

Transportation of Sediment by Wind

Moving air, like moving water, is turbulent and able to pick up loose debris and transport it to other locations. Just as in a stream, the velocity of wind increases with height above the surface. Also like a stream, wind transports fine particles in suspension while heavier ones are carried as bed load. However, the transport of sediment by wind differs from that of running water in two significant ways. First, wind's lower density compared to water renders it less capable of picking up and transporting coarse materials. Second, because wind is not confined to channels, it can spread sediment over large areas, as well as high into the atmosphere.

Bed Load

The **bed load** carried by wind consists of sand grains. Observations in the field and experiments using wind tunnels indicate that windblown sand moves by skipping and bouncing along the surface—a process termed **saltation**. The term is not a reference to salt, but instead derives from the Latin word meaning "to jump."

The movement of sand grains begins when wind reaches a velocity sufficient to overcome the inertia of the resting particles. At first, the sand rolls along the surface. When a moving sand grain strikes another grain, one or both of them may jump into

Figure 13.8 Aerial view of alluvial fans in Death Valley, California. The size of the fan depends on the size of the drainage basin. As the fans grow, they eventually coalesce to form a bajada. (Photo by Michael Collier)

the air. Once in the air, the grains are carried forward by the wind until gravity pulls them back toward the surface. When the sand hits the surface, it either bounces back into the air or dislodges other grains, which then jump upward. In this manner, a chain reaction is established, filling the air near the ground with saltating sand grains in a short period of time (Figure 13.9).

Bouncing sand grains never travel far from the surface. Even when winds are very strong, the height of the saltating sand seldom exceeds a meter and usually is no greater than one-half meter. Some sand grains are too large to be thrown into the air by impact from other particles. When this is the case, the energy provided by the impact of the smaller saltating grains drives the larger grains forward. Estimates indicate that between 20 and 25 percent of the sand transported in a sandstorm is moved in this way.

Suspended Load

Unlike sand, finer particles of dust can be swept high into the atmosphere by the wind. Because dust is often composed of rather flat particles that have large surface areas compared to their weight, it is relatively easy for turbulent air to counterbalance the pull of gravity and keep these fine particles airborne for hours or even days. Although both silt and clay can be carried in suspension, silt commonly makes up the bulk of the **suspended load** because the reduced level of chemical weathering in deserts provides only small amounts of clay.

Fine particles are easily carried by the wind, but they are not so easily picked up to begin with. The reason is that the wind velocity is practically zero within a very thin layer close to the ground. Thus, the wind cannot lift the sediment by itself. Instead, the dust must be ejected or spattered into the moving air by bouncing sand grains or other disturbances.

Figure 13.9 A cloud of saltating sand grains moving up the gentle slope of a dune. (Photo by Stephen Trimble)

Box 13.3

Australia's Ayers Rock

When travelers contemplating a trip to Australia consult brochures and other tourist literature, they are bound to see a photograph or read a description of Ayers Rock. As Figure 13.B illustrates, this well-known attraction is a massive feature that rises steeply from the surrounding plain. Located southwest of Alice Springs in the dry center of the continent, the roughly circular monolith is more than 350 meters (1200 feet) high, and its base is more than 9.5 kilometers (6 miles) in circumference. Its summit is flattened, its sides furrowed. The rock type is sandstone, and the hues of red and orange change with the light of day. In addition to being a striking geological attraction, Ayers Rock is of interest because it is a sacred place for the aboriginal tribes of the region.

Ayers Rock is a spectacular example of a feature known a an inselberg. *Inselberg* is a German word meaning "island mountain" and seems appropriate because these masses clearly resemble rocky islands standing above the surface of a broad sea. Similar features are scattered throughout many other arid and semiarid regions of the world. Ayers Rock is a special type of inselberg that consists of a very resistant rock mass exhibiting a rounded or domed form. Such masses are termed *bornhardts* for the nineteenth-century German explorer Wilhelm Bornhardt, who described similar features in parts of Africa.

Bornhardts form in regions where massive or resistant rock such as granite

Figure 13.B Ayers Rock rises conspicuously above the dry plains of central Australia. It is a type of inselberg known as a *bornhardt*. As erosion gradually lowers the surface, the less weathered massive rock remains standing high above the more jointed and more easily weathered rock that surrounds it. (Photo by Art Wolfe)

or sandstone is surrounded by rock that is more susceptible to weathering. The greater susceptibility of the adjacent rock is often the result of its being more highly jointed. Joints allow water and therefore weathering processes to penetrate to greater depths. When the adjacent, deeply weathered rock is stripped away by erosion, the far less weathered rock mass remains standing high. After a bornhardt forms, it tends to shed water. By contrast, the surrounding debris-covered plains absorb water and weather more rapidly. Therefore,

once formed, a bornhardt helps to perpetuate its existence by reinforcing the processes that created it. In fact, masses such as Ayers Rock can remain a part of the landscape for tens of millions of years.

Bornhardts are more common in the lower latitudes because the weathering that is responsible for their formation proceeds more rapidly in warmer climates. In regions that are now arid or semiarid, bornhardts may reflect times when the climate was wetter than it is today.

This idea is illustrated nicely by a dry, unpaved country road on a windy day. Left undisturbed, little dust is raised by the wind. However, as a car or truck moves over the road, the layer of silt is kicked up, creating a thick cloud of dust.

Although the suspended load is usually deposited relatively near its source, high winds are capable of carrying large quantities of dust great distances (Figure 13.10). In the 1930s, silt picked up in Kansas was transported to New England and beyond into the North Atlantic. Similarly, dust blown from the Sahara has been traced as far as the West Indies.

 Wind Erosion

Compared to running water and glaciers, wind is a relatively insignificant erosional agent. Recall that even in deserts, most erosion is performed by intermittent running water, not by the wind. Wind erosion is more effective in arid lands than in humid areas because in humid places moisture binds particles together and vegetation anchors the soil. For wind to be an effective erosional force, dryness and scanty vegetation are important prerequisites. When such circumstances exist, wind may pick up, transport, and

Figure 13.10 Dust blackens the sky on May 21, 1937, near Elkhart, Kansas. It was because of storms like this that portions of the Great Plains were called the "Dust Bowl" in the 1930s. (Photo reproduced from the collection of the Library of Congress)

deposit great quantities of fine sediment. During the 1930s, parts of the Great Plains experienced vast dust storms. The plowing under of the natural vegetative cover for farming, followed by severe drought, exposed the land to wind erosion, and led to the area being labeled the Dust Bowl.*

---

*For more information, see Box 5.3 in this text (p. 140), "Dust Bowl: Soil Erosion in the Great Plains."

Deflation, Blowouts, and Desert Pavement

One way that wind erodes is by **deflation**, the lifting and removal of loose material. Deflation sometimes is difficult to notice because the entire surface is being lowered at the same time, but it can be significant. In portions of the 1930s Dust Bowl, vast areas of land were lowered by as much as a meter in only a few years.

The most noticeable results of deflation in some places are shallow depressions appropriately called **blowouts** (Figure 13.11). In the Great Plains

Figure 13.11 Formation of a blowout. **A.** Area prior to deflation. **B.** Area after deflation has created a shallow depression. **C.** This 1.5-meter-high mound of soil that was anchored by vegetation shows the level of the land prior to the formation of the blowout. (Photo by E. J. Tarbuck)

C.

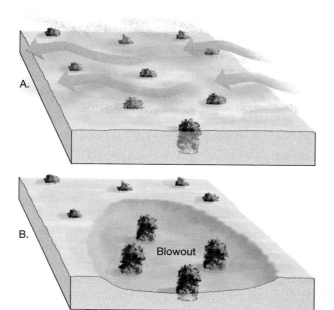

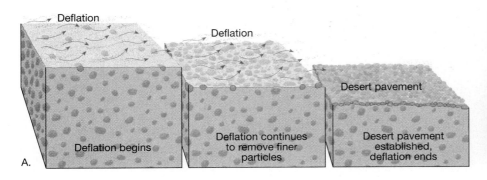

Figure 13.12 A. Formation of desert pavement. As these cross-sections illustrate, coarse particles gradually become concentrated into a tightly packed layer as deflation lowers the surface by removing sand and silt. **B.** If left undisturbed, desert pavement such as this in Arizona's Sonoran Desert will protect the surface from further deflation. (Photo by David Muench)

region, from Texas north to Montana, thousands of blowouts are visible on the landscape. They range from small dimples less than a meter deep and 3 meters wide to depressions that approach 50 meters in depth and several kilometers across. The factor that controls the depths of these basins (that is, acts as base level) is the local water table. When blowouts are lowered to the water table, damp ground and vegetation prevent further deflation.

In portions of many deserts the surface is a closely packed layer of coarse pebbles and cobbles too large to be moved by the wind. This stony veneer, called **desert pavement**, is created as deflation lowers the surface by removing sand and silt until eventually only a continuous cover of coarse particles remains (Figure 13.12). Once desert pavement becomes established, a process that can take hundreds of years, the surface is protected from further deflation if left undisturbed. However, because the layer is only one or two stones thick, disruption by vehicles or animals can dislodge the pavement and expose the fine-grained material below to deflation once again.

Abrasion

Like glaciers and streams, wind also erodes by **abrasion**. In dry regions as well as along some beaches, windblown sand cuts and polishes exposed rock surfaces. Abrasion sometimes creates interestingly shaped stones called **ventifacts** (Figure 13.13). The side of the stone exposed to the prevailing wind is abraded, leaving it polished, pitted, and with sharp edges. If the wind is not consistently from one direction, or if the pebble becomes reoriented, it may have several faceted surfaces.

Unfortunately, abrasion is often given credit for accomplishments beyond its capabilities. Such features as balanced rocks that stand high atop narrow

pedestals, and intricate detailing on tall pinnacles, are not the results of abrasion. Sand seldom travels more than a meter above the surface, so the wind's sand-blasting effect is obviously limited in vertical extent.

 Wind Deposits

Although wind is relatively unimportant in producing *erosional* landforms, significant *depositional* landforms are created by the wind in some regions. Accumulations of windblown sediment are particularly conspicuous in the world's dry lands and along many sandy coasts. Wind deposits are of two distinctive types: (1) mounds and ridges of sand from the wind's bed load, which we call dunes, and (2) extensive blankets of silt, called loess, that once were carried in suspension.

Sand Deposits

As is the case with running water, wind drops its load of sediment when velocity falls and the energy available for transport diminishes. Thus, sand begins to

Figure 13.13 Ventifacts are rocks that are polished and shaped by sandblasting. (Photo by Stephen Trimble)

Figure 13.14 Sand sliding down the steep slip face of a dune, in White Sands National Monument, New Mexico. (Photo by Michael Collier)

accumulate wherever an obstruction across the path of the wind slows its movement. Unlike many deposits of silt, which form blanket-like layers over large areas, winds commonly deposit sand in mounds or ridges called **dunes** (Figure 13.14).

As moving air encounters an object, such as a clump of vegetation or a rock, the wind sweeps around and over it, leaving a shadow of slower-moving air behind the obstacle as well as a smaller zone of quieter air just in front of the obstacle. Some of the saltating sand grains moving with the wind come to rest in these wind shadows. As the accumulation of sand continues, it becomes a more imposing barrier to the wind and thus a more efficient trap for even more sand. If there is a sufficient supply of sand and the wind blows steadily for a long enough time, the mound of sand grows into a dune.

Many dunes have an asymmetrical profile, with the leeward (sheltered) slope being steep and the windward slope more gently inclined (Figure 13.15). Sand moves up the gentler slope on the windward side by saltation. Just beyond the crest of the dune, where the wind velocity is reduced, the sand accumulates. As more sand collects, the slope steepens and eventually some of it slides under the pull of gravity (Figure 13.14). In this way the leeward slope of the dune, called the **slip face**, maintains an angle of about 34 degrees, the angle of repose for loose dry sand (recall from Chapter 9 that the angle of repose is the steepest angle at which loose material remains stable). Continued sand accumulation, coupled with periodic slides down the slip face, results in the slow migration of the dune in the direction of air movement.

For some areas, moving sand is troublesome. In portions of the Middle East, valuable oil rigs must be protected from encroaching dunes. In some cases, fences are built sufficiently upwind of the dunes to stop their migration. As sand continues to collect, however, the fences must be built higher. In Kuwait, protective fences extend for almost 10 kilometers around one important oil field. Migrating dunes can also pose a problem to the construction and maintenance of highways and railroads that cross sandy desert regions. For example, to keep a portion of Highway 95 near Winnemucca, Nevada, open to traffic, sand must be taken away about three times a year. Each time, between 1500 and 4000 cubic meters of sand are removed. Attempts at stabilizing the dunes by planting different varieties of grasses have been unsuccessful because the meager rainfall cannot support the plants.

As sand is deposited on the slip face, layers form that are inclined in the direction the wind is blowing. These sloping layers are called **cross beds** (Figure 13.15). When the dunes are eventually buried under other layers of sediment and become part of the sedimentary rock record, their asymmetrical shape is destroyed, but the cross beds remain as testimony to their origin. Nowhere is cross-bedding more prominent than in the sandstone walls of Zion Canyon in southern Utah (Figure 13.16).

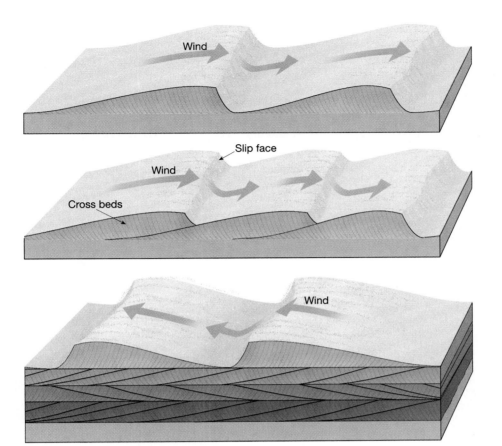

Figure 13.15 Dunes commonly have an asymmetrical shape. The steeper leeward side is called the slip face. Sand grains deposited on the slip face at the angle of repose create the cross-bedding of the dunes. A complex pattern develops in response to changes in prevailing winds.

Types of Sand Dunes

Dunes are not just random heaps of wind-blown sediment. Rather, they are accumulations that usually assume patterns that are surprisingly consistent (Figure 13.17). Addressing this point, a leading early investigator of dunes, the British engineer R. A. Bagnold, observed: "Instead of finding chaos and disorder, the observer never fails to be amazed at a simplicity of form, an exactitude of repetition, and a geometric order...." A broad assortment of dune forms exists, generally simplified to a few major types for discussion.

Of course, gradations exist among different forms as well as irregularly shaped dunes that do not fit easily into any category. Several factors influence the form and size that dunes ultimately assume. These include wind direction and velocity, availability of sand, and the amount of vegetation present. Six basic dune types are shown in Figure 13.17, with arrows indicating wind directions.

Barchan Dunes. Solitary sand dunes shaped like crescents and with their tips pointing downwind are called **barchan dunes** (Figures 13.17A and 13.18). These dunes form where supplies of sand are limited

Figure 13.16 Cross beds are an obvious characteristic of the Navajo Sandstone, here exposed in Zion National Park, Utah. When dunes are buried and become part of the sedimentary record, the cross-bedded structure is preserved. (Photo by Martin G. Miller)

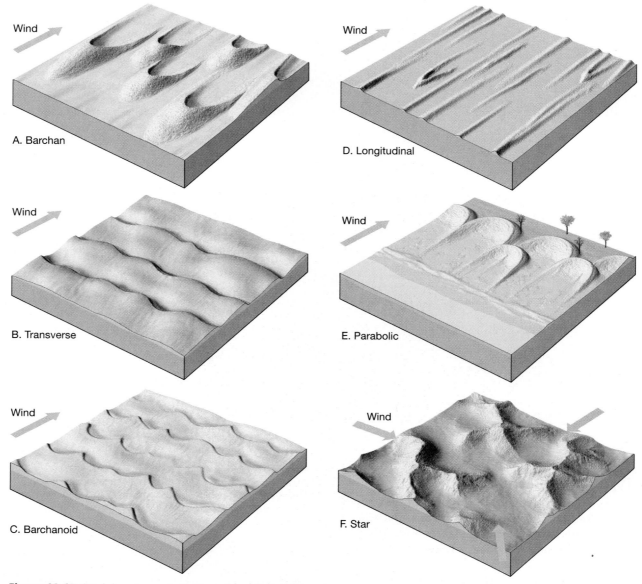

Figure 13.17 Sand dune types. **A.** Barchan dunes. **B.** Transverse dunes. **C.** Barchanoid dunes. **D.** Longitudinal dunes. **E.** Parabolic dunes. **F.** Star dunes.

and the surface is relatively flat, hard, and lacking vegetation. They migrate slowly with the wind at a rate of up to 15 meters per year. Their size is usually modest, with the largest barchans reaching heights of about 30 meters while the maximum spread between their horns approaches 300 meters. When the wind direction is nearly constant, the crescent form of these dunes is nearly symmetrical. However, when the wind direction is not perfectly fixed, one tip becomes larger than the other.

Transverse Dunes. In regions where the prevailing winds are steady, sand is plentiful, and vegetation is sparse or absent, the dunes form a series of long ridges

Figure 13.18 Aerial view of a solitary barchan. The gentle slope is on the side from which the wind is blowing. (Photo by John S. Shelton)

Box 13.4

Desertification: A Global Environmental Problem

The term *desertification* means the expansion of desertlike conditions into non-desert areas. Such a transformation can result from natural processes that act gradually over decades, centuries, and millennia. However, in recent years, desertification has come to mean the rapid alteration of land to desertlike conditions as the result of human activities.

Desertification commonly takes place on the margins of deserts. However, the advancement of desertlike conditions into areas that were previously useful for agriculture is not a uniform, clear-cut shifting of desert borders. Rather, degeneration into desert usually occurs as a patchy transformation of dry but habitable land into dry uninhabitable land. It results primarily from inappropriate land use and is aided and accelerated by drought. Desertification may be halted during wet years, only to advance rapidly during succeeding dry years.

Desertification begins when land near the desert's edge becomes used for growing crops or for grazing livestock. Either way, the natural vegetation is removed by plowing or grazing.

If crops are planted, and drought occurs, the unprotected soil is exposed to the forces of erosion. Gullying of slopes and accumulations of sediment in stream channels are visible signs on the landscape, as are the clouds of dust created as topsoil is removed by the wind.

Where livestock are raised, the land is also degraded. Although the modest natural vegetation on marginal lands can

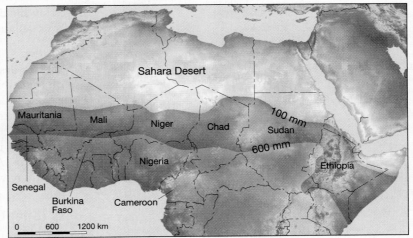

Figure 13.C Desertification is most serious in the southern margin of the Sahara in a region known as the Sahel. The lines defining the approximate boundaries of the Sahel represent average annual rainfall in millimeters.

maintain local wildlife, it cannot support the intensive grazing of large domesticated herds. Overgrazing reduces or eliminates plant cover. When the vegetative cover is destroyed beyond the minimum required to hold the soil against erosion, the destruction becomes irreversible.

Desertification first received worldwide attention when drought struck a region in Africa called the *Sahel* in the late 1960s (Figure 13.C). During that period, and others since, the people in this vast expanse south of the Sahara Desert have suffered malnutrition and death by starvation. Livestock herds have been decimated, and the loss of productive land has been great.

Hundreds of thousands of people have been forced to migrate. As agricultural lands shrink, people must rely on smaller areas for food production. This, in turn, stresses the environment and accelerates the desertification process.

Although human suffering from desertification is most serious in the Sahel, the problem is by no means confined to that region. Desertification exists in other parts of Africa and on every other continent except Antarctica. Each year millions of acres are lost beyond practical hope for reclamation. Recurrent droughts may seem to be the obvious reason for desertification, but the chief cause is stress placed by people on a tenuous environment with fragile soils.

that are separated by troughs and oriented at right angles to the prevailing wind. Because of this orientation, they are termed **transverse dunes** (Figure 13.17B). Typically, many coastal dunes are of this type. In addition, transverse dunes are common in many arid regions where the extensive surface of wavy sand is sometimes called a *sand sea*. In some parts of the Sahara and Arabian deserts, transverse dunes reach heights of 200 meters, are 1 to 3 kilometers across, and can extend for distances of 100 kilometers or more.

There is a relatively common dune form that is intermediate between isolated barchans and extensive waves of transverse dunes. Such dunes, called **barchanoid dunes**, form scalloped rows of sand oriented at right angles to the wind (Figure 13.17C). The rows resemble a series of barchans that have been positioned side by side. Visitors exploring the gypsum dunes at White Sands National Monument, New Mexico, will recognize this form (Figure 13.19).

downwind (Figure 13.17E). Parabolic dunes often form along coasts where there are strong onshore winds and abundant sand. If the sand's sparse vegetative cover is disturbed at some spot, deflation creates a blowout. Sand is then transported out of the depression and deposited as a curved rim, which grows higher as deflation enlarges the blowout.

Star Dunes. Confined largely to parts of the Sahara and Arabian deserts, **star dunes** are isolated hills of sand that exhibit a complex form (Figure 13.17F). Their name is derived from the fact that the bases of these dunes resemble multipointed stars. Usually three or four sharp-crested ridges diverge from a central high point that in some cases may approach a height of 90 meters. As their form suggests, star dunes develop where wind directions are variable.

Loess (Silt) Deposits

In some parts of the world the surface topography is mantled with deposits of windblown silt, called **loess**. Over periods of perhaps thousands of years, dust storms deposited this material. When loess is breached by streams or road cuts, it tends to maintain vertical cliffs and lacks any visible layers, as you can see in Figure 13.20.

The distribution of loess worldwide indicates that there are two primary sources for this sediment: deserts and glacial outwash deposits. The thickest and most extensive deposits of loess on Earth occur in western and northern China. They were blown here from the extensive desert basins of central Asia. Accumulations of 30 meters are common and thicknesses of more than 100 meters have been measured. It is this fine, buff-colored sediment that gives the Yellow River (Huang Ho) its name.

In the United States, deposits of loess are significant in many areas, including South Dakota, Nebraska, Iowa, Missouri, and Illinois as well as portions of the Columbia Plateau in the Pacific Northwest. The correlation between the distribution of loess and important farming regions in the Midwest and eastern Washington state is not just a coincidence, because soils derived from this wind-deposited sediment are among the most fertile in the world.

Unlike the deposits in China, which originated in deserts, the loess in the United States (and Europe) is an indirect product of glaciation. Its source is deposits of stratified drift. During the retreat of the ice sheets, many river valleys were choked with sediment deposited by meltwater. Strong westerly winds sweeping across the barren floodplains picked up the finer sediment and dropped it as a blanket on the eastern sides of the valleys. Such an origin is confirmed by

Figure 13.19 Barchanoid dunes represent a type that is intermediate between isolated barchans on the one hand and extensive transverse dunes on the other. The gypsum dunes at White Sands National Monument, New Mexico, are an example. (Photo by Michael Collier)

Longitudinal Dunes. **Longitudinal dunes** are long ridges of sand that form more or less parallel to the prevailing wind and where sand supplies are limited (Figure 13.17D). Apparently the prevailing wind direction must vary somewhat, but still remain in the same quadrant of the compass. Although the smaller types are only 3 or 4 meters high and several dozens of meters long, in some large deserts longitudinal dunes can reach great size. For example, in portions of North Africa, Arabia, and central Australia, these dunes may approach a height of 100 meters and extend for distances of more than 100 kilometers (62 miles).

Parabolic Dunes. Unlike the other dunes that have been described thus far, **parabolic dunes** form where vegetation partially covers the sand. The shape of these dunes resembles the shape of barchans except that their tips point into the wind rather than

Figure 13.20 This vertical loess bluff near the Mississippi River in southern Illinois is about 3 meters high. (Photo by James E. Patterson)

the fact that loess deposits are thickest and coarsest on the lee side of such major glacial drainage outlets as the Mississippi and Illinois rivers and rapidly thin with increasing distance from the valleys. Furthermore, the angular mechanically weathered particles composing the loess are essentially the same as the rock floor produced by the grinding action of glaciers.

Chapter Summary

- The *concept of dryness is relative*; it refers to any situation in which a water deficiency exists. Dry regions encompass about 30 percent of Earth's land surface. Two climatic types are commonly recognized: *desert*, which is arid, and *steppe* (a marginal and more humid variant of desert), which is semi-arid. *Low-latitude deserts* coincide with the zones of subtropical highs in lower latitudes. On the other hand, *middle-latitude deserts* exist principally because of their positions in the deep interiors of large landmasses far removed from the ocean.

- The same geologic processes that operate in humid regions also operate in deserts, but under contrasting climatic conditions. In dry lands *rock weathering of any type is greatly reduced* because of the lack of moisture and the scarcity of organic acids from decaying plants. Much of the weathered debris in deserts is the result of *mechanical weathering*. Practically all desert streams are dry most of the time and are said to be *ephemeral*. Stream courses in deserts are seldom well integrated and lack an extensive system of tributaries. Nevertheless, *running water is responsible for most of the erosional work in a desert*. Although wind erosion is more significant in dry areas than elsewhere, the main role of wind in a desert is in the transportation and deposition of sediment.

- Because arid regions typically lack permanent streams, they are characterized as having *interior drainage*. Many of the landscapes of the Basin and Range region of the western and southwestern United States are the result of streams eroding uplifted mountain blocks and depositing the sediment in interior basins. *Alluvial fans, playas*, and *playa lakes* are features often associated with these landscapes. In the late stages of erosion, the mountain areas are reduced to a few large bedrock knobs, called *inselbergs*, projecting above sediment-filled basins.

- The transport of sediment by wind differs from that by running water in two ways. First, wind has a low density compared to water; thus, it is not capable of picking up and transporting coarse materials. Second, because wind is not confined to channels, it can spread sediment over large areas. The *bed load* of wind consists of sand grains skipping and bouncing along the surface in a process termed *saltation*. Fine dust particles are capable of being carried by the wind great distances as *suspended load*.

- Compared to running water and glaciers, wind is a relatively insignificant erosional agent. *Deflation*, the lifting and removal of loose material, often produces shallow depressions called *blowouts*. In portions of many deserts the surface is a layer of coarse pebbles and gravels, called *desert pavement*, too large to be moved by the wind. Wind also erodes by *abrasion*, often creating interestingly shaped stones called *ventifacts*. Because sand seldom travels more than a meter above the surface, the effect of abrasion is obviously limited in vertical extent.

- Wind deposits are of two distinct types: (1) *mounds and ridges of sand*, called *dunes*, which are formed from sediment that is carried as part of the wind's bed load; and (2) extensive *blankets of silt*, called *loess*, that once were carried by wind in *suspension*. The profile of a dune shows an asymmetrical shape with the leeward (sheltered) slope being steep and the windward slope more gently inclined. The *types of sand dunes* include (1) *barchan dunes*; (2) *transverse dunes*; (3) *barchanoid dunes*; (4) *longitudinal dunes*; (5) *parabolic dunes*; and (6) *star dunes*. The thickest and most extensive deposits of loess occur in western and northern China. Unlike the deposits in China, which originated in deserts, the loess in the United States and Europe is an indirect product of glaciation.

Review Questions

1. How extensive are the desert and steppe regions of Earth?

2. What is the primary cause of subtropical deserts? Of middle-latitude deserts?

3. In which hemisphere (Northern or Southern) are middle-latitude deserts most common?

4. Why is the amount of precipitation that is used to determine whether a place has a dry climate or a humid climate a variable figure? (See Box 13.1, p. 329)

5. List four common misconceptions about deserts. (See Box 13.2, p. 334)

6. Why is rock weathering reduced in deserts?

7. As a permanent stream such as the Nile River crosses a desert, does discharge increase or decrease? How does this compare to a river in a humid region?

8. What is the most important erosional agent in deserts?

9. Why is sea level (ultimate base level) not a significant factor influencing erosion in desert regions?

10. Describe the features and characteristics associated with each of the stages in the evolution of a mountainous desert. Where in the United States can these stages be observed?

11. Describe the way in which wind transports sand. During very strong winds, how high above the surface can sand be carried?

12. Why is wind erosion relatively more important in arid regions than in humid areas?

13. What factor limits the depths of blowouts?

14. How do sand dunes migrate?

15. List three factors that influence the form and size of a sand dune.

16. Six major dune types are recognized. Indicate which type of dune is associated with each of the following statements.

    (a) Dunes whose tips point into the wind.

    (b) Long sand ridges oriented at right angles to the wind.

    (c) Dunes that often form along coasts where strong winds create a blowout.

    (d) Solitary dunes whose tips point downwind.

    (e) Long sand ridges that are oriented more or less parallel to the prevailing wind.

    (f) An isolated dune consisting of three or four sharp-crested ridges diverging from a central high point.

    (g) Scalloped rows of sand oriented at right angles to the wind.

17. Although sand dunes are the best-known wind deposits, accumulations of loess are very significant in some parts of the world. What is loess? Where are such deposits found? What are the origins of this sediment?

18. What term refers to the process by which desert-like conditions expand into areas that were previously productive? Is this strictly a natural process? (See Box 13.4, p. 343).

Key Terms

abrasion (p. 339)	cross beds (p. 340)	interior drainage (p. 333)	saltation (p. 335)
alluvial fan (p. 333)	deflation (p. 338)	loess (p. 344)	slip face (p. 340)
bajada (p. 333)	desert (p. 328)	longitudinal dune (p. 344)	star dune (p. 344)
barchan dune (p. 341)	desert pavement (p. 339)	parabolic dune (p. 344)	steppe (p. 328)
barchanoid dune (p. 343)	dune (p. 340)	playa (p. 333)	suspended load (p. 336)
bed load (p. 335)	ephemeral stream (p. 332)	playa lake (p. 333)	transverse dune (p. 343)
blowout (p. 338)	inselberg (p. 334)	rainshadow desert (p. 331)	ventifact (p. 339)

Web Resources

 The *Earth* Home Page provides on-line resources for this chapter on the World Wide Web. You will find review exercises, specific updates for items in the chapter, suggested reading, and links to interesting related pathways on the Internet. Visit the *Earth* Home Page at **http://www.prenhall.com/tarbuck.**

CHAPTER 14

Shorelines

Left Sea stacks and surf-rounded stones, Garrapata Beach, California. — *Photo by Carr Clifton*

A.

Figure 14.1 A. This satellite image includes the familiar outline of Cape Cod. Boston is in the upper left corner. The two large islands off the south shore of Cape Cod are Martha's Vineyard (left) and Nantucket (right). Although the work of waves constantly modifies this coastal landscape, shoreline processes are not responsible for creating it. Rather, the present size and shape of Cape Cod result from the positioning of moraines and other glacial materials deposited during the Pleistocene epoch. (Photo courtesy of Earth Satellite Corporation) **B. (opposite)** High-altitude image of the Point Reyes area north of San Francisco, California. The 5.5-kilometer-long south-facing cliffs at Point Reyes (lower left corner) are exposed to the full force of the waves from the Pacific Ocean. Nevertheless, this promontory retreats slowly because the bedrock from which it formed is very resistant. (Photo courtesy of USDA-ASCS)

The restless waters of the ocean are constantly in motion. Winds generate surface currents, the gravity of the Moon and Sun produce tides, and density differences create deep-ocean circulation. Further, waves carry the energy from storms to distant shores, where their impact erodes the land.

Nowhere is the restless nature of the ocean's water more noticeable than along the shore—the dynamic interface among air, land, and sea. Here we can observe the rhythmic rise and fall of tides and see waves constantly rolling in and breaking. Sometimes the waves are low and gentle. At other times, they pound the shore with awesome fury.

Although it may not be obvious, the shoreline is constantly being modified by waves. For example, along Cape Cod, Massachusetts, wave activity is eroding cliffs of poorly consolidated glacial sediment so aggressively that the cliffs are retreating inland up to 1 meter per year

(Figure 14.1A). By contrast, at Point Reyes, California, the far more durable bedrock cliffs are less susceptible to wave attack and therefore are retreating much more slowly (Figure 14.1B). Along both coasts, wave activity is moving sediment along the shore and building narrow sandbars that protrude into and across some bays.

The nature of present-day shorelines is not just the result of the relentless attack of the land by the sea. Indeed, the shore has a complex character that results from multiple geologic processes. For example, practically all coastal areas were affected by the worldwide rise in sea level that accompanied the melting of glaciers at the close of the Pleistocene epoch. As the sea encroached landward, the shoreline retreated, becoming superimposed upon existing landscapes that had resulted from such diverse processes as stream erosion, glaciation, volcanic activity, and the forces of mountain building.

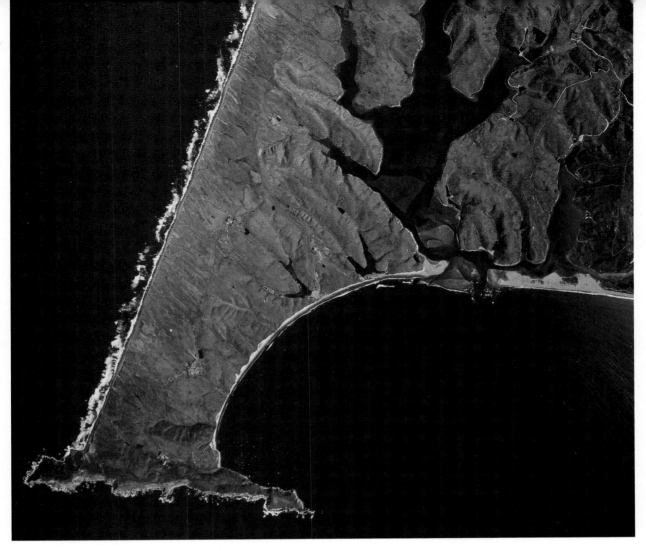

B.

Waves

Wind-generated waves provide most of the energy that shapes and modifies shorelines. Where the land and sea meet, waves that may have traveled unimpeded for hundreds or thousands of kilometers suddenly encounter a barrier that will not allow them to advance farther and must absorb their energy. Stated another way, the shore is the location where a practically irresistible force confronts an almost immovable object. The conflict that results is never-ending and sometimes dramatic.

Today the coastal zone is experiencing intensive human activity. Unfortunately, people often treat the shoreline as if it were a stable platform on which structures can safely be built. This attitude inevitably leads to conflicts between people and nature. As we shall see, many coastal landforms, especially beaches and barrier islands are relatively fragile, short-lived features that are inappropriate sites for development.

Characteristics of Waves

The undulations of the water surface, called waves, derive their energy and motion from the wind. If a breeze of less than 3 kilometers (2 miles) per hour starts to blow across still water, small wavelets appear almost instantly. When the breeze dies, the ripples disappear as suddenly as they formed. However, if the wind exceeds 3 kilometers per hour, more stable waves gradually form and advance with the wind.

All waves have the characteristics illustrated in Figure 14.2. The tops of the waves are *crests*, which are separated by *troughs*. The vertical distance between trough and crest is the **wave height.** The horizontal distance separating successive crests is the **wavelength.** The **wave period** is the time interval between the passage of two successive crests at a stationary point.

The height, length, and period that are eventually achieved by a wave depend on three factors: (1) the wind speed; (2) the length of time the wind has blown; and (3) the **fetch,** or distance that the wind has

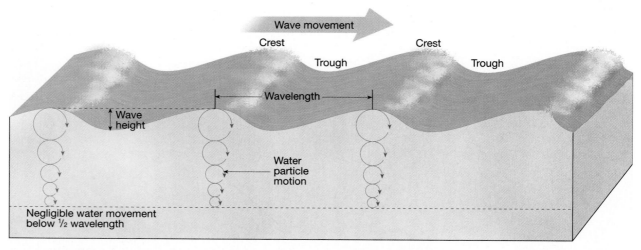

Figure 14.2 This diagram illustrates the basic parts of a wave as well as the movement of water particles with the passage of the wave. Negligible water movement occurs below a depth equal to one-half the wavelength (the level of the dashed line).

traveled across open water. As the quantity of energy transferred from the wind to the water increases, the height and steepness of the waves increase as well. Eventually a critical point is reached where waves grow so tall that they topple over, forming ocean breakers called *whitecaps*.

For a particular wind speed, there is a maximum fetch and duration of wind beyond which waves will no longer increase in size. When the maximum fetch and duration are reached for a given wind velocity, the waves are said to be "fully developed." The reason that waves can grow no further is that they are losing as much energy through the breaking of whitecaps as they are receiving from the wind.

When wind stops or changes direction or if waves leave the stormy area where they were created, they continue on without relation to local winds. The waves also undergo a gradual change to *swells*, which are lower and longer, and may carry a storm's energy to distant shores. Because many independent wave systems exist at the same time, the sea surface acquires a complex, irregular pattern. Hence, the sea waves we watch from the shore are often a mixture of swells from faraway storms and waves created by local winds.

Types of Waves

When observing waves, always remember that you are watching *energy* travel through a medium (water). If you make waves by tossing a pebble into a pond, or by splashing in a pool, or by blowing across the surface of an aquarium, you are imparting *energy* to the water, and the waves you see are just the visible evidence of the energy passing through.

In the open sea, it is the wave energy that moves forward, not the water itself. Each water particle moves in a circular path during the passage of a wave (Figure 14.2). As a wave passes, a water particle returns almost to its original position. The circular orbits followed by the water particles at the surface have a diameter equal to the wave height. When water is part of the wave crest, it moves in the same direction as the advancing wave form. In the trough, the water moves in the opposite direction. This is demonstrated by observing the behavior of a floating object as a wave passes. The toy boat in Figure 14.3 merely seems to bob up and down and sway slightly to and fro without advancing appreciably from its original position. (The wind does drag the water forward slightly, causing the surface circulation of the oceans.) Because of this, waves in the open sea are called **waves of oscillation.**

The energy contributed by the wind to the water is transmitted not only along the surface of the sea but also downward. However, beneath the surface, the circular motion rapidly diminishes until, at a depth equal to about one-half the wavelength, the movement of water particles becomes negligible. This is shown by the rapidly diminishing diameters of water-particle orbits in Figure 14.2.

As long as a wave is in deep water, it is unaffected by water depth (Figure 14.4, left). However, when a wave approaches the shore, the water becomes shallower and influences wave behavior. The wave begins to "feel bottom" at a water depth equal to about one-half its wavelength. Such depths interfere with water movement at the base of the wave and slow its advance (Figure 14.4, center). As a wave advances toward the shore, the slightly faster waves farther out to sea catch up, decreasing the wavelength. As the

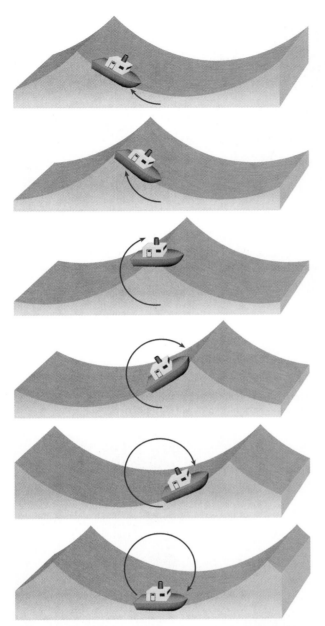

Figure 14.3 The movements of the toy boat show that the wave energy advances, but the water itself advances only slightly from its original position. In this sequence, the wave moves from left to right as the boat (and the water in which it is floating) rotates in a circular motion. The boat moves slightly to the left up the front of the approaching wave, then after reaching the crest, slides to the right down the back of the wave.

speed and length of the wave diminish, the wave steadily grows higher. Finally a critical point is reached when the steep wave front collapses, or *breaks* (Figure 14.4, right). What had been a wave of oscillation now becomes a **wave of translation** in which the water advances up the shore.

The turbulent water created by breaking waves is called **surf.** On the landward margin of the surf zone the turbulent sheet of water from collapsing breakers, called *swash*, moves up the slope of the beach. When the energy of the swash has been expended, the water flows back down the beach toward the surf zone as *backwash*.

 Wave Erosion

During calm weather wave action is minimal. However, just as streams do most of their work during floods, so too do waves accomplish most of their work during storms. The impact of high, storm-induced waves against the shore can be awesome in its violence (Figure 14.5). Each breaking wave may hurl thousands of tons of water against the land, sometimes causing the ground to literally tremble. The pressures exerted by Atlantic waves in wintertime, for example, average nearly 10,000 kilograms per square meter (more than 2000 pounds per square foot). The force during storms is even greater. During one such storm, a 1350-ton portion of a steel and concrete breakwater was ripped from the rest of the structure and moved to a useless position toward the shore at Wick Bay, Scotland. Five years later the 2600-ton unit that replaced the first met a similar fate.

There are many such stories that demonstrate the great force of breaking waves. It is no wonder that cracks and crevices are quickly opened in cliffs, seawalls, breakwaters, and anything else that is subjected to these enormous shocks. Water is forced into every opening, causing air in the cracks to become highly compressed by the thrust of crashing waves. When the wave subsides, the air expands rapidly, dislodging rock fragments and enlarging and extending fractures.

In addition to the erosion caused by wave impact and pressure, **abrasion,** the sawing and grinding action of the water armed with rock fragments, is also important. In fact, abrasion is probably more intense in the surf zone than in any other environment. Smooth, rounded stones and pebbles along the shore are obvious reminders of the relentless grinding action of rock against rock in the surf zone. Further, such fragments are used as "tools" by the waves as they cut horizontally into the land (Figure 14.6).

Along shorelines composed of unconsolidated material rather than hard rock, the rate of erosion by breaking waves can be extraordinary. In parts of Britain, where waves have the easy task of eroding glacial deposits of sand, gravel, and clay, the coast has been worn back 3 to 5 kilometers since Roman times (2000 years ago), sweeping away many villages and ancient landmarks.

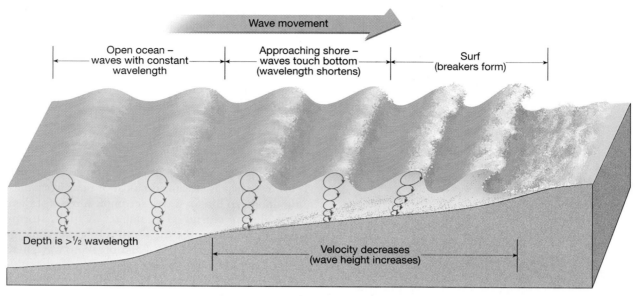

Wave movement

Open ocean – waves with constant wavelength | Approaching shore – waves touch bottom (wavelength shortens) | Surf (breakers form)

Depth is >½ wavelength

Velocity decreases (wave height increases)

Figure 14.4 Changes that occur when a wave moves onto shore.

Figure 14.5 When waves break against the shore, the force of the water can be powerful and the erosional work that is accomplished can be great. Marin headlands, Golden Gate National Recreation Area, California. (Photo by Galen Rowell)

Figure 14.6 Cliff undercut by wave erosion along the Oregon coast. (Photo by E. J. Tarbuck)

Wave Refraction and Longshore Transport

The bending of waves, called **wave refraction,** plays an important part in shoreline processes (Figure 14.7). It affects the distribution of energy along the shore and thus strongly influences where and to what degree erosion, sediment transport, and deposition will take place.

Waves seldom approach the shore straight on. Rather, most waves move toward the shore at an angle. However, when they reach the shallow water of a smoothly sloping bottom, they are bent and tend to become parallel to the shore. Such bending occurs because the part of the wave nearest the shore reaches shallow water and slows first, whereas the end that is still in deep water continues forward at its full speed. The net result is a wave front that may approach nearly parallel to the shore regardless of the original direction of the wave.

Because of refraction, wave impact is concentrated against the sides and ends of headlands that project into the water, whereas wave attack is weakened in bays. This differential wave attack along irregular coastlines is illustrated in Figure 14.8. As the waves reach the shallow water in front of the headland sooner than they do in adjacent bays, they are bent more nearly parallel to the protruding land and strike it from all three sides. By contrast, refraction in the bays causes waves to diverge and expend less energy. In these zones of weakened wave activity, sediments can accumulate and form sandy beaches. Over a long

Figure 14.7 Wave bending around the end of a beach at Stinson Beach, California. (Photo by James E. Patterson)

period, erosion of the headlands and deposition in the bays will straighten an irregular shoreline.

Although waves are refracted, most still reach the shore at some angle, however slight. Consequently, the uprush of water from each breaking wave (the swash) is oblique. However, the backwash is straight down the slope of the beach. The effect of this pattern of water movement is to transport sediment in a zigzag pattern along the beach (Figure 14.9). This movement is called **beach drift,** and it can transport sand and pebbles hundreds or even thousands of meters each day.

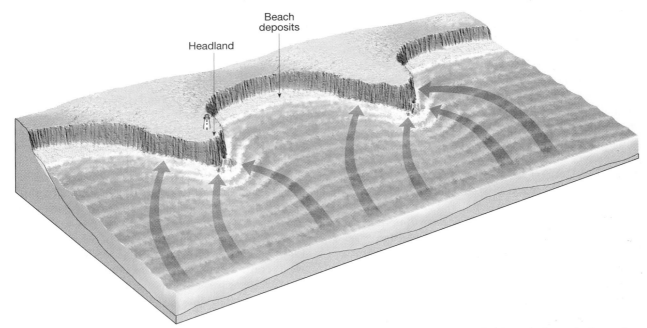

Figure 14.8 Because of wave refraction, the greatest erosional power is concentrated on the headlands. In the bays, the force of the waves is much weaker.

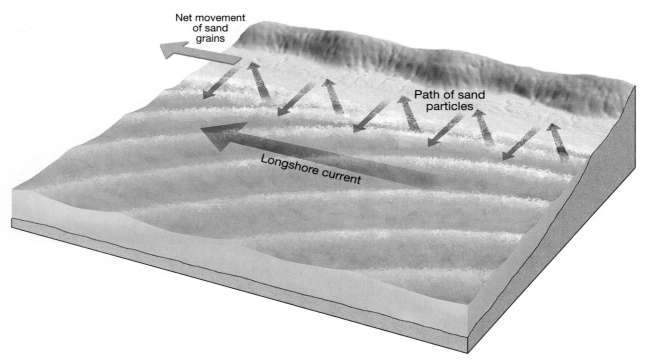

Net movement
of sand
grains

Path of sand
particles

Longshore current

Figure 14.9 Beach drift and longshore currents are created by obliquely breaking waves. Beach drift occurs as incoming waves carry sand obliquely up the beach, while the water from spent waves carries it directly down the slope of the beach. Similar movements occur offshore in the surf zone to create the longshore current. These processes transport large quantities of material along the beach and in the surf zone.

Oblique waves also produce currents within the surf zone that flow parallel to the shore and move substantially more sediment than beach drift. Because the water here is turbulent, these **longshore currents** easily move the fine suspended sand and roll larger sand and gravel along the bottom. When the sediment transported by longshore currents is added to the quantity moved by beach drift, the total amount can be very large. At Sandy Hook, New Jersey, for example, the quantity of sand transported along the shore over a 48-year period averaged almost 750,000 tons annually. For a 10-year period of Oxnard, California, more than 1.5 million tons of sediment moved along the shore each year.

There should be little wonder that beaches have been characterized as "rivers of sand." At any point along a beach there is likely to be more sediment that was derived elsewhere than material eroded from the land immediately behind it. It is also worth noting that much of the sediment composing beaches is not wave-eroded debris. Rather, in many areas sediment-laden rivers that discharge into the ocean are the major sources of material (see Box 14.1). Hence, if it were not for beach drift and longshore currents, many beaches would be nearly sandless.

Shoreline Features

Shoreline features vary depending on the rocks of the shore, currents, wave intensity, and whether the coast is stable, sinking, or rising. This section summarizes many common shoreline features.

Wave-Cut Cliffs and Platforms

Whether along the rugged and irregular New England coast or along the steep shorelines of the West Coast, the effects of wave erosion are often easy to see. **Wave-cut cliffs,** as the name implies, originate by the cutting action of the surf against the base of coastal land. As erosion progresses, rocks overhanging the notch at the base of the cliff crumble into the surf, and the cliff retreats (see Figure 14.6). A relatively flat, bench-like surface, the **wave-cut platform,** is left behind by the receding cliff (Figure 14.10). The platform broadens as wave attack continues. Some of the debris produced by the breaking waves remains along the water's edge as part of the beach, while the remainder is transported farther seaward.

Box 14.1

Louisiana's Vanishing Coastal Wetlands

Coastal wetlands form in sheltered environments that include swamps, tidal flats, coastal marshes, and bayous. They are rich in wildlife and provide nesting grounds and important stopovers for waterfowl and migratory birds, as well as spawning areas and valuable habitats for fish.

The delta of the Mississippi River in Louisiana contains about 40 percent of all coastal wetlands in the lower 48 states. Louisiana's wetlands are sheltered from the wave action of hurricanes and winter storms by low-lying barrier islands. Both the wetlands and the protecting islands have formed as a result of the shifting of the Mississippi River during the past 5000 to 6000 years.

The deltaic processes that control the movement of water and sediment have resulted in a system of complex drainage patterns, natural ridges and levees, and offshore barrier beaches—all of which restrict the advance and encroachment of salt water. The dependence of Louisiana's coastal wetlands on the Mississippi River and its distributaries as a direct source of sediment and fresh water leaves them vulnerable to changes in the river system. Moreover, the reliance on barrier islands for protection from storm waves leaves coastal wetlands vulnerable when these narrow offshore islands are eroded.

Today, the coastal wetlands of Louisiana are disappearing at an alarming rate. The U.S. Army Corps of Engineers estimates that about 90 square kilometers (35 square miles) of Louisiana's wetlands are lost *annually*. Fifty years from now the shoreline is predicted to be many kilometers from where it is today. As the shoreline retreats, southern Louisiana's wetlands are being submerged beneath the Gulf of Mexico. And if the predicted rise in sea level occurs, it will only add to the coastal land loss.

Why are Louisiana's wetlands shrinking? The answer is twofold: *natural change* and *human activity*. First, the Mississippi delta and its wetlands are continually changing naturally. As

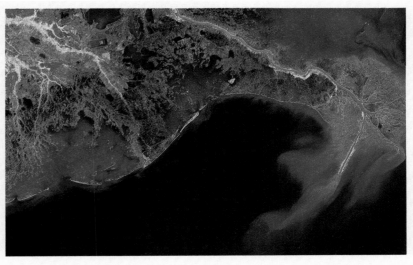

Figure 14.A Satellite image of the Mississippi delta. For the past 600 years or so, the main flow of the river has been along its present course, extending southeast from New Orleans. During that span, the delta advanced into the Gulf of Mexico at a rate of about 10 kilometers (6 miles) per century. (Photo courtesy of NASA)

sediment accumulates and builds the delta in one area, erosion and subsidence cause losses elsewhere (Figure 14.A). When the river shifts, the zones of delta growth and destruction also shift. Second, ever since people arrived, the rate at which the delta and its wetlands are destroyed has accelerated.

Before Europeans settled the delta, the Mississippi River regularly overflowed its banks in seasonal floods. The huge quantities of sediment that were deposited renewed the soil and kept the delta from sinking below sea level. However, with settlement came flood-control efforts and the desire to maintain and improve navigation on the river. Artificial levees were constructed to contain the rising river during flood stage. Over time, the levees were extended all the way to the mouth of the Mississippi to keep the channel open for navigation.

The effects have been straightforward. The levees prevent sediment and fresh water from being dispersed into the wetlands. Instead, the river is forced to carry its load to the deep waters at the mouth. Meanwhile, the processes

of compaction, subsidence, and wave erosion continue. Because not enough sediment is added to offset these forces, the size of the delta and the extent of its wetlands gradually shrink.

The problem has been aggravated by a decline in the sediment transported by the Mississippi, decreasing by approximately 50 percent over the past 100 years. A substantial portion of the reduction results from trapping of sediment in large reservoirs created by dams built on tributaries to the Mississippi, especially the Missouri and Arkansas rivers. The diversion of part of the Mississippi's flow to the Atchafalaya distributary also reduces the amount of sediment available to build and maintain the delta.

Understanding and modifying the impact of people is a necessary basis for any plan to reduce the loss of wetlands in the Mississippi delta. Although some projects are underway to divert more sediment and fresh water to the wetlands, plans to cope effectively with the problem have yet to be implemented.

Figure 14.10 Elevated wave-cut platform along the California coast near San Francisco. A new platform is being created at the base of the cliff. (Photo by John S. Shelton)

Arches, Stacks, Spits, and Bars

Headlands that extend into the sea are vigorously attacked by waves because of refraction. The surf erodes the rock selectively, wearing away the softer or more highly fractured rock at the fastest rate. At first, sea caves may form. When caves on opposite sides of a headland unite, a **sea arch** results. Eventually, the arch falls in, leaving an isolated remnant, or **sea stack,** on the wave-cut platform (Figure 14.11). In time, it too will be consumed by the action of the waves.

Where beach drift and longshore currents are active, several features related to the movement of sediments along the shore may develop. **Spits** are elongated ridges of sand that project from the land into the mouth of an adjacent bay (Figure 14.12). Often the end in the water hooks landward in response to wave-generated currents. The term **baymouth bar** is applied to a sand bar that completely crosses a bay, sealing it off from the open ocean (Figure 14.12). Such a feature tends to form across bays where currents are

Figure 14.11 Sea arch at the tip of Mexico's Baja Peninsula. (Photo by Mark A. Johnson/The Stock Market)

Figure 4.12 High-altitude image of a well-developed spit and baymouth bar along the coast of Martha's Vineyard, Massachusetts. Also notice the tidal delta in the lagoon adjacent to the inlet through the baymouth bar. (Photo courtesy of USDA-ASCS)

weak, allowing a spit to extend to the other side. A **tombolo,** a ridge of sand that connects an island to the mainland or to another island, forms in much the same manner as a spit.

Barrier Islands

The Atlantic and Gulf Coastal Plains are relatively flat and slope gently seaward. The shore zone is characterized by **barrier islands.** These low ridges of sand parallel the coast at distances from 3 to 30 kilometers offshore. From Cape Cod, Massachusetts, to Padre Island, Texas, nearly 300 barrier islands rim the coast (Figure 14.13).

Most barrier islands are from 1 to 5 kilometers wide and between 15 and 30 kilometers long. The highest features are sand dunes, which usually reach heights of 5 to 10 meters; in a few areas, unvegetated dunes are more than 30 meters high. The lagoons separating these narrow islands from the shore represent zones of relatively quiet water that allow small craft

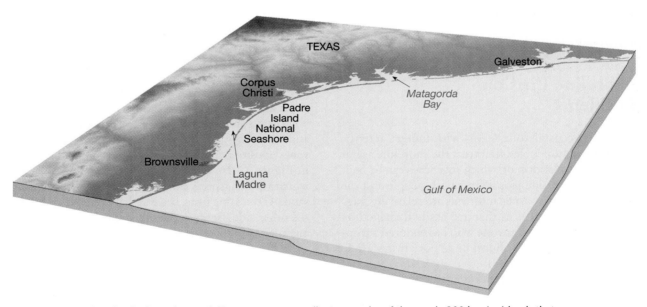

Figure 14.13 The islands along the south Texas coast are excellent examples of the nearly 300 barrier islands that rim the Atlantic and Gulf coasts.

traveling between New York and northern Florida to avoid the rough waters of the North Atlantic.

Barrier islands probably formed in several ways. Some originated as spits that were subsequently severed from the mainland by wave erosion or by the general rise in sea level following the last episode of glaciation. Others were created when turbulent waters in the line of breakers heaped up sand scoured from the bottom. Because these sand barriers rise above normal sea level, the piling of sand likely is the result of the work of storm waves at high tide. Finally, some barrier islands may be former sand dune ridges that originated along the shore during the last glacial period, when sea level was lower. When the ice sheets melted, sea level rose and flooded the area behind the beach-dune complex.

The Evolving Shore

A shoreline continually undergoes modification regardless of its initial configuration. At first most coastlines are irregular, although the degree of and reason for the irregularity may vary considerably from place to place. Along a coastline that is characterized by varied geology, the pounding surf may at first increase its irregularity because the waves will erode the weaker rocks more easily than the stronger ones. However, if a shoreline remains stable, marine erosion and deposition will eventually produce a straighter, more regular coast. Figure 14.14 illustrates the evolution of an initially irregular coast. As waves erode the headlands, creating cliffs and a wave-cut platform, sediment is carried along the shore. Some material is deposited in the bays, while other debris is formed into spits and baymouth bars. At the same time rivers fill the bays with sediment. Ultimately a generally straight, smooth coast results.

Shoreline Erosion Problems

Today the coastal zone teems with human activity. Unfortunately, people often treat the shoreline as if it were a stable platform on which structures can be built safely. This attitude jeopardizes both people and the shoreline. Many coastal landforms are relatively fragile, short-lived features that are easily damaged by development. And, as anyone who has endured a tropical storm knows, the shoreline is not always a safe place to live (see Box 14.2).

Compared with natural hazards such as earthquakes, volcanic eruptions, and landslides, shoreline erosion is often perceived to be a more continuous and predictable process that appears to cause relatively

modest damage to limited areas. In reality, the shoreline is a dynamic place that can change rapidly in response to natural forces. Exceptional storms are capable of eroding beaches and cliffs at rates that are far in excess of the long-term average. Such bursts of accelerated erosion can have a significant impact on the natural evolution of a coast; it can also have a profound impact on people who reside in the coastal zone (Figure 14.15). Erosion along our coasts causes significant property damage. Large sums are spent annually not only to repair damage, but also to prevent or control erosion. Already a problem at many sites, shoreline erosion is certain to become an increasingly serious problem as extensive coastal development continues.

Although the same processes cause change along every coast, not all coasts respond in the same way. Interactions among different processes and the relative importance of each process depend on local factors. The factors include: (1) the proximity of a coast to sediment-laden rivers; (2) the degree of tectonic activity; (3) the topography and composition of the land; (4) prevailing winds and weather patterns; and (5) the configuration of the coastline and nearshore areas.

During the past 100 years, growing affluence and increasing demands for recreation have brought unprecedented development to many coastal areas. As both the number and the value of buildings have increased, so too have efforts to protect property from storm waves. Also, controlling the natural migration of sand is an ongoing struggle in many coastal areas. Such interference can result in unwanted changes that are difficult and expensive to correct.

Jetties and Groins

From relatively early in America's history a principal goal in coastal areas was the development and maintenance of harbors. In many cases, this involved the construction of jetty systems. **Jetties** are usually built in pairs and extend into the ocean at the entrances to rivers and harbors. With the flow of water confined to a narrow zone, the ebb and flow caused by the rise and fall of the tides keep the sand in motion and prevent deposition in the channel. However, as illustrated in Figure 14.16, the jetty may act as a dam against which the longshore current and beach drift deposit sand. At the same time, wave activity removes sand on the other side. Because the other side is not receiving any new sand, there is soon no beach at all.

To maintain or widen beaches that are losing sand, groins are sometimes constructed. A **groin** is a barrier built at a right angle to the beach to trap sand that is moving parallel to the shore. The result is an irregular but wider beach. These structures often do their job so effectively that the longshore current

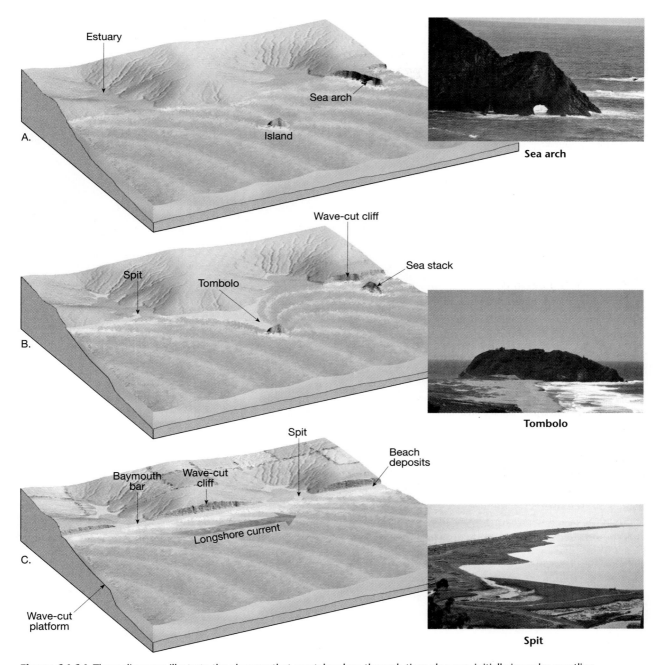

Figure 14.14 These diagrams illustrate the changes that can take place through time along an initially irregular coastline that remains relatively stable. The coastline shown in part **A** gradually evolves to **B** and then **C**. The diagrams also serve to illustrate many of the features described in the section on shoreline features. (Photos by E. J. Tarbuck)

beyond the groin becomes sand-starved. As a result, the current erodes sand from the beach on the leeward side of the groin. Such a situation is illustrated in Figure 14.17.

To offset this effect, property owners downcurrent from the structure may erect groins on their property. In this manner, the number of groins multiplies. An example of such proliferation is the shoreline of New Jersey, where hundreds of these structures have

been built. As it has been shown that groins often do not provide a satisfactory solution, they are no longer the preferred method of keeping beach erosion in check.

Breakwaters and Seawalls

In some coastal areas a **breakwater** may be constructed parallel to the shoreline. The purpose of such a structure is to protect boats from the force of large breaking waves by creating a quiet water zone near the

Figure 14.15 This destruction occurred when Hurricane Emily hit North Carolina's Outer Banks on August 31, 1993. (Photo courtesy of *Raleigh News and Observer*)

shore. However, when this is done, the reduced wave activity along the shore behind the structure may allow sand to accumulate. If this happens, the marina will eventually fill with sand while the downstream beach erodes and retreats. At Santa Monica, California, where the building of a breakwater created such a problem, the city had to install a dredge to remove sand from the protected quiet water zone and deposit it down the beach where longshore currents and beach drift could recirculate the sand (Figure 14.18).

As development has moved ever closer to the beach, seawalls are sometimes built to defend property from the force of breaking waves. **Seawalls** are simply massive barriers intended to prevent waves from reaching the areas behind the wall. Waves expend much of their energy as they move across an open beach. Seawalls cut this process short by reflecting the force of unspent waves seaward. As a consequence, the beach to the seaward side of the seawall experiences significant erosion and may, in some

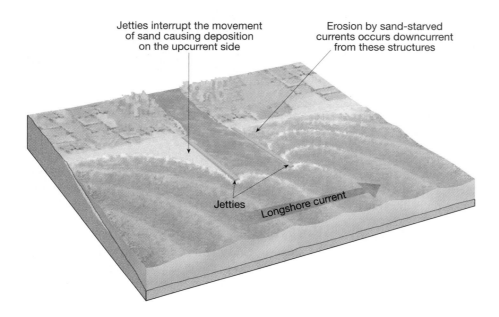

Jetties interrupt the movement of sand causing deposition on the upcurrent side

Erosion by sand-starved currents occurs downcurrent from these structures

Jetties

Longshore current

Figure 14.16 Jetties are built at the entrances to rivers and harbors and are intended to prevent deposition in the navigation channel. Jetties interrupt the movement of sand by beach drift and longshore currents. Beach erosion often results downcurrent from the site of the structure.

Figure 14.17 Groins along the New Jersey shore at Cape May. (Photo by John S. Shelton)

instances, be eliminated entirely. Once the width of the beach is reduced, the seawall is subjected to even greater pounding by the waves. Eventually this battering will cause the wall to fail, and a larger, more expensive wall must be built to take its place.

The wisdom of building temporary protective structures along shorelines is increasingly questioned. The feelings of many coastal scientists are expressed in the following excerpt from a position paper that grew out of a conference on America's Eroding Shoreline:

> It is now clear that halting the receding shoreline with protective structures benefits only a few and seriously degrades or destroys the natural beach and the value it holds for the majority. Protective structures divert the ocean's energy temporarily from private properties, but usually refocus that energy on the adjacent natural beaches. Many interrupt the natural sand flow in coastal currents, robbing many beaches of vital sand replacement.[*]

Beach Nourishment

Beach nourishment represents another approach to stabilizing shoreline sands. As the term implies, this practice simply involves the addition of large quantities of sand to the beach system. By building the beaches seaward, beach quality and storm protection

are both improved. Beach nourishment, however, is not a permanent solution to the problem of shrinking beaches. The same processes that removed the sand in the first place will eventually remove the replacement sand as well. In addition, beach nourishment is very expensive. When beach nourishment was used to renew 24 kilometers (15 miles) of Miami Beach, the cost was $64 million. Here the restoration must be redone every 10 to 12 years.

In some instances, beach nourishment can lead to unwanted environmental effects. For example, beach replenishment at Waikiki Beach, Hawaii, involved replacing coarse calcareous sand with softer, muddier calcareous sand. Destruction of the soft beach sand by breaking waves increased the water's turbidity and killed offshore coral reefs. At Miami Beach, increased turbidity also damaged local coral communities.

Beach nourishment appears to be an economically viable long-range solution to the beach preservation problem only in areas where there exists dense development, large supplies of sand, relatively low wave energy, and reconcilable environmental issues. Unfortunately, few areas possess all these attributes.

Abandonment and Relocation

So far two basic responses to shoreline erosion problems have been considered: (1) the building of structures such as seawalls to hold the shoreline in place and (2) the addition of sand to replenish eroding

Figure 14.18 Aerial view of a breakwater at Santa Monica, California. The breakwater appears as a faint line in the water behind which many boats are anchored. The construction of the breakwater disrupted longshore transport and caused the seaward growth of the beach. (Photo by John S. Shelton)

---

[*]"Strategy for Beach Preservation Proposed," *Geotimes* 30 (No. 12, December 1985): 15.

Box 14.2

Coastal Storms: A Conflict Between People and Nature

"The sea won this round."

These words were uttered by former President George Bush, November 3, 1991, as he surveyed the devastating damage to his vacation home in Kennebunkport, Maine, following a fierce Atlantic storm that ravaged a large portion of the East Coast for two days. The President vowed to rebuild the turn-of-the-century structure.

Conflicts such as the one experienced by Mr. Bush have always occurred along coasts. However, the increasing desirability and accessibility of coasts as places to live and work have increased the frequency and intensity of these conflicts over the past 50 years. The concentration of large numbers of people near the shoreline means that hurricanes and other large storms not only place millions at risk but can result in extraordinary property damage as well. The examples described here are Hurricanes Andrew and Hugo, two of the costliest natural disasters in U.S. history.

Hurricane destruction is divided into three categories: (1) wind damage, (2) storm surge, and (3) inland freshwater (rainwater) flooding. Storm surge is a dome of water that sweeps across the coast near the point where the eye makes landfall. In major storms, the surge may add 3 to 5 meters (10 to 16 feet) to normal tide heights.

Hurricane Andrew made landfall south of Key Biscayne, Florida, during the dark, early morning hours of August 24, 1992. It illustrates the

Figure 14.B Hurricane Andrew slammed into southern Florida on August 24, 1992, and then went on across the Gulf of Mexico to strike the Louisiana coast 2 days later. Most of the property damage caused by the storm was attributed to Andrew's fierce winds. (Photo by Allen Tannenbaum/SYGMA)

destructive effects of strong winds on a highly developed area. The storm's maximum sustained surface winds were about 230 kilometers (145 miles) per hour, with gusts exceeding 280 kilometers (175 miles) per hour.

It took Andrew just hours to cut its destructive path across Florida. Property damage from Andrew was primarily wind damage along a 40-kilometer-wide swath of destruction that was centered a few kilometers north of the town of Homestead (Figure 14.B). Fortunately, the highly developed coastline of Miami Beach was more than 27 kilometers (17 miles) from the eye of the storm and so did not experience sustained hurricane-

force winds. Two other aspects of the storm made it less damaging than it might otherwise have been. The area affected by the 5-meter (16-foot) storm surge was lightly developed and protected by mangrove forest, so surge damage was minimized. In addition, there was little rainwater flooding because the hurricane did not linger but advanced rapidly across the region.

The storm's destructive accomplishments were awesome. In one area, the wind carried 6-meter-long steel and concrete beams, with roofs still attached, more than 50 meters. Cars and boats were tossed about like toys, tens of thousands of homes and businesses were

beaches. However, a third option is also available. Many coastal scientists and planners are calling for a policy shift from defending and rebuilding beaches and coastal property in high hazard areas to removing storm-damaged buildings in those places and letting nature reclaim the beach. This approach is similar to that adopted by the federal government for river floodplains following the devastating 1993 Mississippi River floods in which vulnerable structures are abandoned and relocated on higher, safer ground.

Such proposals, of course, are controversial. People with significant nearshore investments shudder at the thought of not rebuilding and defending coastal developments from the erosional wrath of the sea. Others, however, argue that with sea level rising, the impact of coastal storms will only get worse in the decades to come (see Box 14.3). This group advocates that oft-damaged structures be abandoned or relocated to improve personal safety and to reduce costs. Such ideas will no doubt be the focus of much study and debate as states and communities evaluate and revise coastal land-use policies.

devastated, hundreds of acres of groves were uprooted, and Homestead Air Force Base was leveled. Government officials who toured the area said it looked like a "war zone."

After crossing southern Florida, Andrew went on to cross the Gulf of Mexico and lashed the Louisiana coastline on August 26. In all, Hurricane Andrew was responsible for at least 62 deaths and caused $20–30 billion in damages. It was the costliest natural disaster in U.S. history.

Hurricane Hugo was a strong storm that struck the South Carolina coast in September 1989. It illustrates the effects of a powerful storm surge. When the storm came ashore near Charleston, the storm surge washed over some barrier islands, demolishing seawalls, roads, and structures of all kinds (Figure 14.C). Where barrier islands were narrow and lacked dunes, they were breached by the storm-driven waters, and inlets were created. The damage Hugo caused along South Carolina's barrier islands and shore communities clearly shows the importance of good construction and the natural protection provided by a wide beach and high sand dunes.

Folly Beach, south of Charleston, was on the weaker (south) side of the hurricane, but its beaches were already severely narrowed by long-term erosion; homeowners had dumped boulders and concrete rubble on the beach for protection. The rocks and rubble proved useless during Hugo. The storm surge (over 3.5 meters at Folly Beach) overtopped the structure, caused major damage to beachfront houses, and totally swept away a popular seafood

Figure 14.C The damage caused by Hurricane Hugo in September 1989 to some of South Carolina's barrier islands underscores the need for knowledge of beach and storm processes in order to protect lives and property. (Photo by Jan Staller/The Stock Market)

restaurant. Isle of Palms and Sullivans Island, north of Charleston, suffered substantial damage, but the effects were lessened by a wide beach and dunes. Houses that survived were built to withstand high winds and flooding, in contrast to homes built to lower standards that were totally destroyed.

The total property loss resulting from Hugo was estimated to be $10 billion, with more than $7 billion of that in the continental United States. The loss of life included 28 on Caribbean islands and 21 in the continental United States.

Timely storm warnings helped to keep the loss of life from Andrew and Hugo at a low figure. However, the National Weather Service is concerned that the rapid population growth in coastal areas is setting the stage for a major hurricane disaster. Unfortunately, the rate of improvement in achieving forecast lead times is not keeping pace with the time requirements for evacuating the ever-increasing coastal population.

Contrasting the Atlantic and Pacific Coasts

The shoreline along the Pacific Coast of the United States is strikingly different from that characterizing the Atlantic and Gulf coast regions. Some of the differences are related to plate tectonics. The West Coast represents the leading edge of the North American plate, and because of this, it experiences active uplift and deformation. By contrast, the East Coast is a tectonically quiet region that is far from any active plate margin. Because of this basic geological difference,

the nature of shoreline erosion problems along America's opposite coasts is different.

Atlantic and Gulf Coasts. Much of the coastal development along the Atlantic and Gulf coasts has occurred on barrier islands. Typically, barrier islands, also termed *barrier beaches* or *coastal barriers*, consist of a wide beach that is backed by dunes and separated from the mainland by marshy lagoons. The broad expanses of sand and exposure to the ocean have made barrier islands exceedingly attractive sites for

Box 14.3

Is Global Warming Causing Sea Level to Rise?

The shifting dynamic nature of barrier islands and the ineffectiveness of most shoreline protection measures are now relatively well-established facts. Unfortunately, recent research, which indicates that sea level is rising, has compounded this already distressing situation. Studies indicate that sea level has risen between 10 and 15 centimeters over the past century. Furthermore, investigators project a significantly greater sea-level rise in the years to come—25 to 75 cm by the year 2100.* Although such a vertical change may seem modest, many coastal geologists believe that any given rise in sea level along the gently sloping Atlantic and Gulf coasts will cause from 10 to 1000 times as much horizontal shoreline retreat (Figure 14.D).

The idea that sea level will continue to rise in the coming decades is linked to the results of climatic studies that predict a global warming trend. Such predictions are based on the now well-established fact that the carbon dioxide (CO_2) content of the atmosphere has been rising at an accelerating rate for more than a century. The CO_2 is added primarily as a by-product of the combustion of ever-increasing quantities of fossil fuels.

If we assume that the use of fossil fuels will continue to rise at projected rates, current estimates indicate that the atmosphere's CO_2 content will grow by an additional 40 percent or more by some time in the second half of the twenty-first century.

The importance of CO_2 lies in the fact that it traps a portion of the radiation

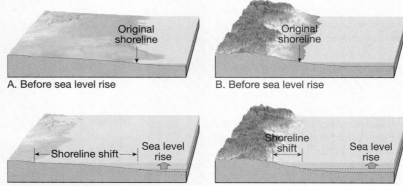

Figure 14.D The slope of a shoreline is critical to determining the degree to which sea-level changes will affect it. **A.** When the slope is gentle, small changes in sea level cause a substantial shift. **B.** The same sea-level rise along a steep coast results in only a small shoreline shift.

emitted by Earth and thereby keeps the air near the surface warmer than it would be without CO_2. Because CO_2 is an important heat-absorbing gas, an increase in the air's CO_2 content is believed to contribute to higher atmospheric temperatures. In addition, other trace gases generated by human activities also play a role.

How is a warmer atmosphere related to a global rise in sea level? First, higher temperatures can cause glacial ice to melt. About one-half of the 10- to 15-centimeter rise in sea level over the past century is attributed to the melting of small glaciers and ice caps. Second, a warmer atmosphere causes an increase in ocean volume through thermal expansion. That is, higher air temperatures raise the temperature of the upper layers of the ocean, which, in turn,

causes the water to expand and sea level to rise.

Because rising sea level is a gradual phenomenon, it may be overlooked by coastal residents as a significant contributor to shoreline erosion problems. Rather, the blame is assigned to other forces, especially storm activity. Although a given storm might be the immediate cause, the magnitude of its destruction may result from the relatively small sea level rise that allowed the storm's power to cross a much greater land area.

*J.D. Mahlman, "Uncertainties in Projections of Human-Caused Climate Warming," *Science* 278 (1997): 1416–1417.

development (Figure 14.19). Unfortunately, development has taken place more rapidly than our understanding of barrier island dynamics.

Because barrier islands face the open ocean, they receive the full force of major storms that strike the coast. When a storm occurs, the barriers absorb the energy of the waves primarily through the movement of sand. This process and the dilemma that results have been described as follows:

Waves may move sand from the beach to offshore areas or, conversely, into the dunes; they may erode the dunes, depositing sand onto the beach or carrying it out to sea; or they may carry sand from the beach and the dunes into the marshes behind the barrier, a process known as overwash. The common factor is movement. Just as a flexible reed may survive a wind that destroys an oak tree, so the barriers survive hurricanes

waves at bay than to admit that development was improperly placed to begin with.*

Pacific Coast. In contrast to the broad, gently sloping coastal plains of the East, much of the Pacific Coast is characterized by relatively narrow beaches that are backed by steep cliffs and mountain ranges. Recall that America's western margin is a more rugged and tectonically active region than the East. Because uplift continues, the apparent rise in sea level is not so readily apparent in the West. Nevertheless, like the shoreline erosion problems facing the East's barrier islands, West Coast difficulties also stem largely from the alteration of a natural system by people.

A major problem facing the Pacific shoreline, and especially portions of southern California, is a significant narrowing of many beaches. The bulk of the sand on many of these beaches is supplied by rivers that transport it from the mountains to the coast. Over the years this natural flow of material to the coast has been interrupted by dams built for irrigation and flood control. The reservoirs effectively trap the sand that would otherwise nourish the beach environment. When the beaches were wider, they served to protect the cliffs behind them from the force of storm waves. Now, however, the waves move across the narrowed beaches without losing much energy and cause more rapid erosion of the sea cliffs.

Although the retreat of the cliffs provides material to replace some of the sand impounded behind dams, it also endangers homes and roads built on the bluffs. In addition, development atop the cliffs aggravates the problem. Urbanization increases runoff which, if not carefully controlled, can result in serious bluff erosion. Watering lawns and gardens adds significant quantities of water to the slope. This water percolates downward toward the base of the cliff, where it may emerge in small seeps. This action reduces the slope's stability and facilitates mass wasting.

Shoreline erosion along the Pacific Coast varies considerably from one year to the next, largely because of the sporadic occurrence of storms. As a consequence, when the infrequent but serious episodes of erosion occur, the damage is often blamed on the unusual storms and not on coastal development or the sediment-trapping dams that may be great distances away. If, as predicted, sea level rises at an increasing rate in the years to come, increased shoreline erosion and sea-cliff retreat should be expected along many parts of the Pacific Coast.

Figure 14.19 This high-altitude infrared image shows a portion of an urbanized barrier island off the New Jersey coast. The problems and hazards associated with constructing buildings on barrier islands remain. It is just as unsafe to erect a building on shifting sand today as it was a century ago. In fact, considering the high population density on some islands, the potential for disaster is even greater. (Photo courtesy of USDA-ASCS)

and nor'easters not through unyielding strength but by giving before the storm.

This picture changes when a barrier is developed for homes or as a resort. Storm waves that previously rushed harmlessly through gaps between the dunes now encounter buildings and roadways. Moreover, since the dynamic nature of the barriers is readily perceived only during storms, homeowners tend to attribute damage to a particular storm, rather than to the basic mobility of coastal barriers. With their homes or investments at stake, local residents are more likely to seek to hold the sand in place and the

*Frank Lowenstein, "Beaches or Bedrooms—The Choice as Sea Level Rises," *Oceanus* 28 (No. 3, Fall 1985): 22.

Emergent and Submergent Coasts

The great variety of shorelines demonstrates their complexity. Indeed, to understand any particular coastal area, you must consider rock types, size and direction of waves, frequency of storms, tidal range, and submarine profile. Moreover, recent tectonic events and changes in sea level must also be taken into account. These many variables make shoreline classification difficult.

Many geologists classify coasts based on the changes that have occurred with respect to sea level. This commonly used classification divides coasts into two very general categories: emergent and submergent. **Emergent coasts** develop either because an area experiences uplift or as a result of a drop in sea level. Conversely, **submergent coasts** are created when sea level rises or the land adjacent to the sea subsides.

In some areas the coast is clearly emergent because rising land or a falling water level exposes wave-cut cliffs and platforms above sea level. Excellent examples include portions of coastal California where uplift has occurred in the recent geological past. The elevated wave-cut platform shown in Figure 14.10 illustrates this situation. In the case of the Palos Verdes Hills, south of Los Angeles, seven different terrace levels exist, indicating seven episodes of uplift. The ever-persistent sea is now cutting a new platform at the base of the cliff. If uplift follows, it too will become an elevated marine terrace.

Other examples of emergent coasts include regions that were once buried beneath great ice sheets. When glaciers were present, their weight depressed the crust; when the ice melted, the crust began gradually to spring back. Consequently, prehistoric shoreline features may now be found high above sea level. The Hudson Bay region of Canada is such an area, portions of which are still rising at a rate of more than a centimeter annually.

In contrast to the preceding examples, other coastal areas show definite signs of submergence. The shoreline of a coast that has been submerged in the relatively recent past is often highly irregular because the sea typically floods the lower reaches of river valleys. The ridges separating the valleys, however, remain above sea level and project into the sea as headlands. These drowned river mouths, which are called **estuaries**, characterize many coasts today. Along the Atlantic Coast, the Chesapeake and Delaware bays are examples of large estuaries created by submergence (Figure 14.20). The picturesque coast of Maine, particularly in the vicinity of Acadia National Park, is another excellent example of an area that was flooded

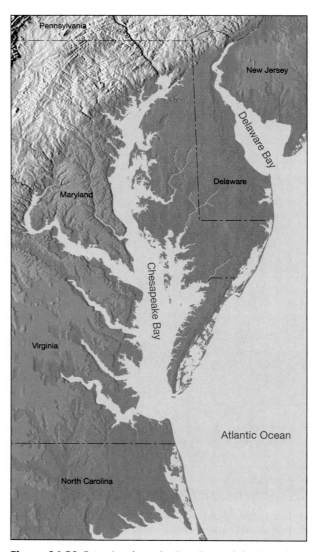

Figure 14.20 Estuaries along the East Coast of the United States. The lower portions of many river valleys were submerged by the rise in sea level that followed the end of the Ice Age. Chesapeake Bay and Delaware Bay are especially prominent examples.

by the post-glacial rise in sea level and transformed into a highly irregular coastline.

Keep in mind that most coasts have complicated geologic histories. With respect to sea level, many have at various times emerged and then submerged. Each time they may retain some of the features created during the previous situation.

Tides

Tides are daily changes in the elevation of the ocean surface. Their rhythmic rise and fall along coastlines have been known since antiquity. Other than waves, they are the easiest ocean movements to observe. An

Figure 14.21 High tide and low tide on Nova Scotia's Minas Basin in the Bay of Fundy. (Photos courtesy of Nova Scotia Department of Tourism)

exceptional example of extreme daily tides is shown in Figure 14.21.

Although known for centuries, tides were not explained satisfactorily until Sir Isaac Newton applied the law of gravitation to them. Newton showed that there is a mutual attractive force between two bodies, and that since oceans are free to move, they are deformed by this force. Hence, tides result from the gravitational attraction exerted upon Earth by the Moon, and to a lesser extent by the Sun.

Causes of Tides

It is easy to see how the Moon's gravitational force can cause the water to bulge on the side of Earth nearest the Moon. In addition, however, an equally large tidal bulge is produced on the side of Earth directly opposite the Moon (Figure 14.22).

Both tidal bulges are caused, as Newton discovered, by the pull of gravity. Gravity is inversely proportional to the square of the distance between two objects, meaning simply that it quickly weakens with distance. In this case, the two objects are the Moon and Earth. Because the force of gravity decreases with distance, the Moon's gravitational pull on Earth is slightly greater on the near side of Earth than on the far side. The result of this differential pulling is to stretch (elongate) the solid Earth very slightly. In contrast, the world ocean, which is mobile, is deformed quite dramatically by this effect to produce the two opposing tidal bulges.

Because the position of the Moon changes only moderately in a single day, the tidal bulges remain in place while Earth rotates "through" them. For this reason, if you stand on the seashore for 24 hours,

Earth will rotate you through alternating areas of deeper and shallower water. As you are carried into the regions of deeper water, the tide rises, and as you are carried away, the tide falls. Therefore, during one day you experience two high tides and two low tides.

Further, the tidal bulges migrate as the Moon revolves around Earth about every 29 days. As a result, the tides, like the time of moonrise, shift about 50 minutes later each day. After 29 days, the cycle is complete and a new one begins.

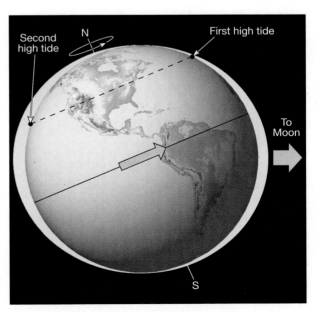

Figure 14.22 Tides on an Earth that is covered to a uniform depth with water. Depending on the Moon's position, tidal bulges may be inclined to the equator. In this situation, an observer will experience two unequal high tides.

There may be an inequality between the high tides during a given day. Depending on the position of the Moon, the tidal bulges may be inclined to the equator as in Figure 14.22. This figure illustrates that the first high tide experienced by an observer in the Northern Hemisphere is considerably lower than the high tide half a day later. On the other hand, a Southern Hemisphere observer would experience the opposite effect.

Spring and Neap Tides

The Sun also influences the tides. It is far larger than the Moon, but because it is so far away, the effect is considerably less than that of the Moon. In fact, the tide-generating potential of the Sun is slightly less than half that of the Moon.

Near the times of new and full moons, the Sun and Moon are aligned and their forces are added together (Figure 14.23A). Accordingly, the combined gravity of these two tide-producing bodies cause higher tidal bulges (high tides) and lower tidal troughs

(low tides). These are called the **spring tides.** Spring tides create the largest daily tidal range, that is, the largest variation between high and low tides. Conversely, at about the time of the first and third quarters of the Moon, the gravitational forces of the Moon and Sun act on Earth at right angles, and each partially offsets the influence of the other (Figure 14.23B). As a result, the daily tidal range is less. These are called **neap tides.**

So far, we have explained the basic causes and patterns of tides. But keep in mind these theoretical considerations cannot be used to predict either the height or the time of actual tides at a particular place. The shape of coastlines and the configuration of ocean basins greatly influence the tides. Consequently, tides at various places respond differently to the tide-producing forces. This being the case, the nature of the tide at any location can be determined most accurately by actual observation. The predictions in tidal tables and the tidal data on nautical charts are based on such observations.

Tidal Currents

Tidal current is the term used to describe the *horizontal* flow of water accompanying the rise and fall of the tide. These water movements induced by tidal forces can be important in some coastal areas. Tidal currents flow in one direction during a portion of the tidal cycle and reverse their flow during the remainder. Tidal currents that advance into the coastal zone as the tide rises are called **flood currents.** As the tide falls, seaward-moving water generates **ebb currents.** Periods of little or no current, called *slack water*, separate flood and ebb. The areas affected by these alternating tidal currents are **tidal flats.** Depending on the nature of the coastal zone, tidal flats vary from narrow strips seaward of the beach to extensive zones that may extend for several kilometers.

Although tidal currents are not important in the open sea, they can be rapid in bays, river estuaries, straits, and other narrow places. Off the coast of Brittany in France, for example, tidal currents that accompany a high tide of 12 meters (40 feet) may attain a speed of 20 kilometers (12 miles) per hour. While tidal currents are not generally major agents of erosion and sediment transport, notable exceptions occur where tides move through narrow inlets. Here they constantly scour the small entrances to many good harbors that would otherwise be blocked.

Sometimes deposits called **tidal deltas** are created by tidal currents (Figure 14.24). They may develop either as *flood deltas* landward of an inlet or as *ebb deltas* on the seaward side of an inlet. Because wave activity and longshore currents are reduced on the sheltered landward side, flood deltas are more

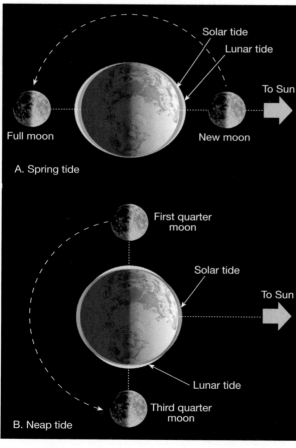

Figure 14.23 Relationship of the Moon to Earth during **A.** spring tides and **B.** neap tides.

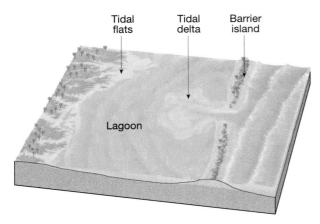

Figure 14.24 Because this tidal delta is forming in the relatively quiet waters on the landward side of a barrier island, it is termed a flood delta. As a rapidly moving tidal current emerges from the inlet, it slows and deposits sediment. The shapes of tidal deltas are variable.

common and more prominent (see Figure 14.12). They form after the tidal current moves rapidly through an inlet. As the current emerges from the narrow passage into more open waters, it slows and deposits its load of sediment.

Tides and Earth's Rotation

The tides by friction against the floor of the ocean basins, act as weak brakes that are steadily slowing Earth's rotation. The rate of slowing, however, is not great. Astronomers, who have precisely measured the length of the day over the past 300 years, have found that the day is increasing by 0.002 second per century. Although this may seem inconsequential, over millions of years this small effect will become very large. Eventually, billions of years into the future, rotation will cease and Earth will no longer have alternating days and nights.

If Earth's rotation is slowing, the length of each day must have been shorter and the number of days per year must have been greater in the geologic past. One method used to investigate this phenomenon involves the microscopic examination of shells of certain invertebrates. Clams and corals, as well as other organisms, grow a microscopically thin layer of new shell material each day. By studying the daily growth rings of some well-preserved fossil specimens, we can determine the number of days in a year. Studies using this ingenious technique indicate that early in the Cambrian period, about 570 million years ago, the length of the day was only 21 hours. Because the length of the year, which is determined by Earth's revolution about the Sun, does not change, the Cambrian year contained 424 twenty-one-hour days. By late Devonian time, about 365 million years ago, a year consisted of about 410 days, and as the Permian period opened, 286 million years ago, there were 390 days in a year.

Chapter Summary

- The three factors that influence the *height*, *wavelength*, and *period* of a wave are (1) *wind speed*, (2) *length of time the wind has blown*, and (3) *fetch*, the distance that the wind has traveled across the open water.

- The two types of wind-generated waves are (1) *waves of oscillation*, which are waves in the open sea in which the wave form advances as the water particles move in circular orbits, and (2) *waves of translation*, the turbulent advance of water formed near the shore as waves of oscillation collapse, or *break*, and form *surf*.

- Wave erosion is caused by *wave impact pressure* and *abrasion* (the sawing and grinding action of water armed with rock fragments). The bending of waves is called *wave refraction*. Owing to refraction, wave impact is concentrated against the sides and ends of headlands.

- Most waves reach the shore at an angle. The uprush (swash) and backwash of water from each breaking wave moves the sediment in a zigzag pattern along the beach. This movement, called *beach drift*, can transport sand hundreds or even thousands of meters each day. Oblique waves also produce *longshore currents* within the surf zone that flow parallel to the shore.

- Features produced by *shoreline erosion* include *wave-cut cliffs* (which originate from the cutting action of the surf against the base of coastal land), *wave-cut platforms* (relatively flat, benchlike surfaces left behind by receding cliffs), *sea arches* (formed when a headland is eroded and two caves from opposite sides unite), and *sea stacks* (formed when the roof of a sea arch collapses).

- Some of the depositional features formed when sediment is moved by beach drift and longshore currents are *spits* (elongated ridges of sand that project from the land into the mouth of an adjacent bay), *baymouth bars* (sand bars that completely cross a bay), and *tombolos* ridges of sand that connect an island to the mainland or to another island). Along the Atlantic and Gulf Coastal Plains, the shore zone is characterized by *barrier islands*, low ridges of sand that parallel the coast at distances from 3 to 30 kilometers offshore.

- Local factors that influence shoreline erosion are (1) the proximity of a coast to sediment-laden rivers, (2) the degree of tectonic activity, (3) the topography and composition of the land, (4) prevailing winds and weather patterns, and (5) the configuration of the coastline and nearshore areas.

- Three basic responses to shoreline erosion problems are (1) building *structures* such as *groins* (short walls built at a right angle to the shore to trap moving sand), *breakwaters* (structures built parallel to the shoreline to protect it from the force of large breaking waves), and *seawalls* (barriers constructed to prevent waves from reaching the area behind the wall) to hold the shoreline in place; (2) *beach nourishment*, which involves the addition of sand to replenish eroding beaches; and (3) *relocating* buildings away from the beach.

- Because of basic geological differences, the *nature of shoreline erosion problems along America's Pacific and Atlantic coasts is very different*. Much of the development along the Atlantic and Gulf coasts has occurred on barrier islands, which receive the full force of major storms. Much of the Pacific Coast is characterized by narrow beaches backed by steep cliffs and mountain ranges. A major problem facing the Pacific shoreline is a narrowing of beaches caused because the natural flow of materials to the coast has been interrupted by dams built for irrigation and flood control.

- One frequently used classification of coasts is based upon changes that have occurred with respect to sea level. *Emergent coasts*, often with wave-cut cliffs and wave-cut platforms above sea level, develop either because an area experiences uplift or as a result of a drop in sea level. Conversely, *submergent coasts*, with their drowned river mouths, called *estuaries*, are created when sea level rises or the land adjacent to the sea subsides.

- *Tides*, the daily rise and fall in the elevation of the ocean surface, are caused by the *gravitational attraction* of the Moon and, to a lesser extent, by the Sun. Near the times of new and full moons, the Sun and Moon are aligned, and their gravitational forces are added together to produce especially high and low tides. These are called the *spring tides*. Conversely, at about the times of the first and third quarters of the Moon, when the gravitational forces of the Moon and Sun are at right angles, the daily tidal range is less. These are called *neap tides*.

- *Tidal currents* are horizontal movements of water that accompany the rise and fall of tides. *Tidal flats* are the areas that are affected by the advancing and retreating tidal currents. When tidal currents slow after emerging from narrow inlets, they deposit sediment that may eventually create *tidal deltas*.

Review Questions

1. List three factors that determine the height, length, and period of a wave.
2. Describe the motion of a water particle as a wave passes (see Figure 14.3).
3. Explain what happens when a wave breaks.
4. Describe two ways in which waves cause erosion.
5. What is wave refraction? What is the effect of this process along irregular coastlines? (see Figure 14.8)
6. Why are beaches often called "rivers of sand"?
7. How has the construction of artificial levees and dams on the Mississippi River and its tributaries contributed to a shrinking of the Mississippi's delta and its extensive wetlands? (See Box 14.1)
8. Describe the formation of the following features: wave-cut cliff, wave-cut platform, sea stack, spit, baymouth bar, and tombolo.
9. List three possible ways in which barrier islands originate.
10. Hurricanes Andrew (1992) and Hugo (1989) were very costly natural disasters. How did each storm cause most of its property damage? (See Box 14.2)
11. For what purpose is a groin built? Why might the building of one groin lead to the building of others?
12. How might a seawall lead to increased beach erosion?
13. What are the drawbacks of beach nourishment?

14. What is the basis for the predictions that global air temperatures will rise? How can a warmer atmosphere lead to a rise in sea level? (See Box 14.3)

15. Relate the damming of rivers to the shrinking of beaches at many locations along the West Coast of the United States. Why do narrower beaches lead to accelerated sea cliff retreat?

16. What observable features would lead you to classify a coastal area as emergent?

17. Are estuaries associated with submergent or emergent coasts? Why?

18. Discuss the origin of ocean tides.

19. Explain why an observer can experience two unequal high tides during one day (see Figure 14.22).

20. How does the Sun influence tides?

21. Distinguish between flood current and ebb current.

22. How have tides affected Earth's rotation? How did geologists substantiate this idea?

Key Terms

abrasion (p. 353)
barrier island (p. 359)
baymouth bar (p. 358)
beach drift (p. 355)
beach nourishment (p. 363)
breakwater (p. 361)
ebb current (p. 370)
emergent coast (p. 368)
estuary (p. 368)

fetch (p. 351)
flood current (p. 370)
groin (p. 360)
jetty (p. 360)
longshore current (p. 356)
neap tide (p. 370)
sea arch (p. 358)
sea stack (p. 358)
seawall (p. 362)
spit (p. 358)

spring tide (p. 370)
submergent coast (p. 368)
surf (p. 353)
tidal current (p. 370)
tidal delta (p. 370)
tidal flats (p. 370)
tide (p. 368)
tombolo (p. 359)
wave-cut cliff (p. 356)
wave-cut platform (p. 356)

wave height (p. 351)
wavelength (p. 351)
wave of oscillation (p. 352)
wave of translation (p. 353)
wave period (p. 351)
wave refraction (p. 355)

Web Resources

The *Earth* Home Page provides on-line resources for this chapter on the World Wide Web. You will find review exercises, specific updates for items in the chapter, suggested reading, and links to interesting related pathways on the Internet. Visit the *Earth* Home Page at **http://www.prenhall.com/tarbuck.**

CHAPTER 15

Crustal Deformation

Deformation
  Stress and Strain
  Types of Deformation (Strain)

Mapping Geologic Structures
  Strike and Dip

Folds
  Types of Folds
  Domes and Basins

Faults
  Dip-Slip Faults
  Strike-Slip Faults

Joints

Left Deformation in the Canadian Rockies. Mount Kidd, Peter Lougheed Provincial Park, Alberta. — *Photo by Peter Frend/DRK Photo*

Figure 15.1 Uplifted and folded sedimentary strata at Stair Hole, near Lulworth, Dorset, England. These layers of Jurassic-age rock, originally deposited in horizontal beds, have been folded as a result of the collision between the African and European crustal plates. (Photo by Tom Bean)

Earth is a dynamic planet. In the preceding chapters, we learned that weathering, mass wasting, and erosion by water, wind, and ice continually sculpture the physical landscape. In addition, tectonic forces deform rocks in the crust. Evidence demonstrating the operation of enormous forces within Earth includes thousands of kilometers of rock layers that are bent, rumpled, overturned, and sometimes highly fractured (Figure 15.1). In the Canadian Rockies, for example, some rock units have been thrust over others in a nearly horizontal manner for hundreds of kilometers. On a smaller scale, crustal movements of a few meters occur along faults during large earthquakes. In addition, rifting (spreading) and extension of the crust produce elongated depressions and create ocean basins.

The results of tectonic activities are strikingly apparent in Earth's major mountain belts (see chapter-opening photo). Here, rocks containing fossils of marine organisms may be found thousands of meters above sea level, and massive rock units are intensely folded, as if they were made of putty.

Structural geologists study "how it got this way," for they can look only at the present state of deformation. By studying the orientation of folds and faults, they can often determine the original geologic setting and the nature and direction of the forces that produced these rock structures. The complex events that make up geologic history are thereby being unraveled.

In addition to their importance in interpreting the geologic past, rock structures are economically important. For example, most occurrences of oil and natural gas are associated with geologic structures that act to trap these fluids in valuable "reservoirs" of one type or another (see Chapter 21). Furthermore, rock fractures can be the site of hydrothermal mineralization, which is a major source of metallic ores. Moreover, the orientation and characteristics of rock structures must be considered when selecting sites for major construction projects such as bridges, hydroelectric dams, and nuclear power plants. In short, a working knowledge of these features is essential to our modern way of life.

In this chapter we will examine the forces that deform rock, as well as the structures that result. The basic geologic structures associated with deformation are folds, faults, joints, rock cleavage, and foliation. Because rock cleavage and foliation were examined in Chapter 7, this chapter will be devoted to the remaining rock structures and the tectonic forces that produce them.

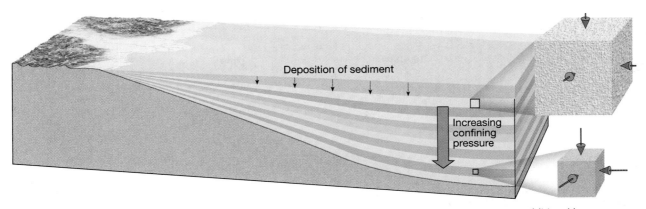

Figure 15.2 In a sedimentary basin, older layers at depth are subjected to increased confining pressures as additional layers are deposited. The higher pressures result in fluid expulsion and pore closure, causing a reduction in volume.

 Deformation

Deformation is a general term that refers to all changes in volume and/or shape of a rock body. Most crustal deformation occurs along plate margins. Recall from Chapter 1 that the lithosphere consists of large segments (plates) that move relative to one another. As the plates interact along their boundaries, tectonic forces deform the involved rock units.

Stress and Strain

To understand how tectonic forces cause rock to deform, we need to examine the concepts of *stress* and *strain*. **Stress** *is the amount of force acting on a rock unit to change its shape and/or volume*. As we saw in Chapter 7, stress may be applied uniformly in all directions (confining pressure) or nonuniformly in different directions (differential stress). By contrast, **strain** *is the change in the shape and/or volume of a rock unit caused by stress*. Figure 15.1 illustrates the strain (or deformation) exhibited by rock units near Dorset, England. When studying the deformation caused by stress as shown in Figure 15.1, geologists ask, "What do these structures tell us about the magnitude and direction of the forces operating at the time they formed?"

Among the forces that deform rock is **confining pressure**, which, like air pressure, is uniform in all directions. Confining pressure is caused by the load of the overlying rocks. For example, in basins where rapid deposition occurs, rock is buried deeper and deeper as additional layers of sediment accumulate. This creates tremendous compressive forces on the underlying rocks, causing a reduction in volume that makes the rocks more compact (Figure 15.2).

Confining pressure is also important because it affects the way rocks behave when deformed by differential forces. In near-surface environments, where the confining pressure and temperature are low, rocks are described as *brittle* because they will *fracture* when deformed. However, at great depths, where confining pressures are high, rocks become *ductile* and *flow* rather than fracture when a differential force is applied (Figure 15.3).

When stresses are applied unequally in different directions, they are termed **differential stresses**. Differential stresses that shorten a rock body are known as **compressional stresses** (Figure 15.4B). Conversely, when stresses are acting in opposing directions, they tend to elongate, or pull apart, the rock unit and are known as **tensional stresses** (Figure 15.4C).

Figure 15.3 Rocks exhibiting the results of ductile behavior. These rocks were deformed at great depth and were subsequently exposed at the bottom of the Grand Canyon. (Photo by E. J. Tarbuck)

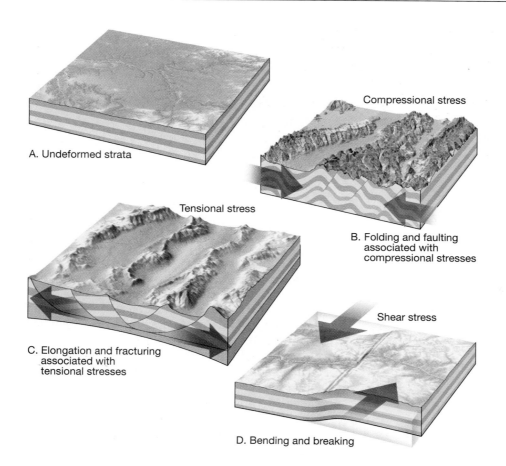

A. Undeformed strata

Compressional stress

B. Folding and faulting
associated with
compressional stresses

Tensional stress

C. Elongation and fracturing
associated with
tensional stresses

Shear stress

D. Bending and breaking

Figure 15.4 Simplified diagrams showing the deformation of **A.** flat-lying strata. **B.** Compressional stresses tend to shorten a rock body, often by folding. **C.** Tensional stresses act to elongate, or pull apart a rock unit. **D.** Shear stress acts to displace rocks by bending and breaking them.

In addition, differential stress can cause rock to **shear** (Figure 15.4D). Shearing is similar to the slippages that occur between individual playing cards when a deck is held between your hands and the top of the deck is moved relative to the bottom. In near-surface environments, shearing occurs when relatively brittle rock is broken into thin slabs that are forced to slide past one another. By contrast, rock located at great depths is at high temperatures and pressures and behaves in a ductile fashion during deformation. This accounts for its ability to flow and bend into intricate folds without fracturing when subjected to shearing.

Types of Deformation (Strain)

When rocks are subjected to stresses greater than their own strength, they begin to deform, usually by folding, flowing, or fracturing (Figure 15.5). It is easy to visualize how rocks break, for we normally think of them as being brittle. But how can large rock units be *bent* into intricate folds without being appreciably broken during the process? To answer this question, structural geologists performed laboratory experiments. Rocks were subjected to stresses under conditions that simulated those existing at various depths within the crust.

Although each rock type deforms somewhat differently, the general characteristics of rock deformation were determined from these experiments. Geologists discovered that when stress is applied, rocks first respond by deforming elastically. Changes resulting from **elastic deformation** are reversible; that is, like a rubber band, the rock will return to nearly its original size and shape when the stress is removed. As we shall see in the next chapter, the energy for most earthquakes comes from stored elastic energy that is released as rock snaps back to its original shape. Once the elastic limit is surpassed, rocks either deform plastically or they fracture. **Plastic deformation** results in permanent changes, that is, the size and shape of a rock unit are forever altered through folding and flowing.

Laboratory experiments confirmed the speculation that, at high temperatures and pressures simulating those deep in the crust, most rocks deform plastically once their elastic limit is surpassed (Figure 15.6). Rocks that deform plastically by folding and flowing are said to be **ductile**. By contrast, rocks tested under surface conditions also deform elastically, but once they exceed their elastic limit, most behave like a brittle solid and fracture. This type of deformation is called **brittle failure**.

Figure 15.5 Deformed sedimentary strata exposed in a road cut near Palmdale, California. In addition to the obvious folding, light-colored beds are offset along a fault located on the right side of the photograph. (Photo by E. J. Tarbuck)

In addition to environmental conditions, the mineral composition of rocks greatly influences how individual masses respond to deformation. For example, rocks composed of minerals that have strong internal molecular bonds tend to fail by brittle fracture, whereas those with weaker bonds are more susceptible to ductile flow. Rocks that are weak, and thus most likely to behave in a ductile manner when subjected to stress, include rock salt, gypsum, marble, and shale. By comparison, quartzite, granite, and gneiss are strong and brittle. In a near-surface environment, strong, brittle rocks will fail by fracture when subjected to stresses that exceed their strength.

One key factor that researchers are unable to duplicate in the laboratory is *geologic time*. We know that if stress is applied quickly, as with a hammer, rocks tend to fracture. On the other hand, these same materials may deform plastically if stress is applied over an extended period. For example, marble benches have been known to sag under their own weight over a

| Undeformed | Low confining pressure | Intermediate confining pressure | High confining pressure |

Figure 15.6 A marble cylinder deformed in the laboratory by applying thousands of pounds of load from above. Each sample was deformed in an environment that duplicated the confining pressure found at various depths. Notice that when the confining pressure was low, the sample deformed by brittle fracture, whereas when the confining pressure was high, the sample deformed plastically. (Photo courtesy of M.S. Paterson, Australian National University)

period of a hundred years or so. In nature, small forces applied over long time periods surely play an important role in the deformation of rock.

To summarize, three factors determine how rocks will behave when subjected to stresses that exceed their strength. First, the environment strongly influences how a rock will deform. In near-surface environments, where temperatures and pressures are low, most rocks exhibit brittle failure. However, in the high-temperature, high-pressure regimes that exist deep in the crust, the same rocks will deform by ductile flow. A second factor is the strength of the materials. Some rocks are very weak and are more likely to flow under conditions that would cause stronger rocks to fail by brittle fracture. Third, time plays a major role in rock deformation. Rocks that would fracture when stress is applied rapidly are known to flow when stress is applied gradually over a long time span.

Mapping Geologic Structures

The processes of deformation generate features at many different scales. At one extreme are Earth's major mountain systems. At the other extreme, highly localized stresses create minor fractures in bedrock. All of these phenomena, from the largest folds in the Alps to the smallest fractures in a slab of rock, are referred to as **rock structures**. Before beginning our discussion of rock structures, we need to become familiar with the way geologists describe and map them.

When conducting a study of a region, a geologist identifies and describes the dominant structures.

A structure often is so large that only a small portion is visible from any particular vantage point. In many situations, most of the bedrock is concealed by vegetation or by recent sedimentation. Consequently, the reconstruction must be done using data gathered from a limited number of *outcrops*, which are sites where bedrock is exposed at the surface (see Box 15.1). Despite such difficulties, a number of mapping techniques enable geologists to reconstruct the orientation and shape of the existing structures. In recent years, this work has been aided by advances in aerial photography and satellite imagery.

Geologic mapping is most easily accomplished where sedimentary strata are exposed. This is because sediments are usually deposited in horizontal layers. If the sedimentary rock layers are still horizontal, this tells geologists that the area is probably undisturbed structurally. But if strata are inclined, bent, or broken, this indicates that a period of deformation occurred following deposition.

Strike and Dip

Geologists use measurements called *strike* (trend) and *dip* (inclination) to help determine the orientation or attitude of a rock layer or fault surface (Figure 15.7). By knowing the strike and dip of rocks at the surface, geologists can predict the nature and structure of rock units and faults that are hidden beneath the surface, beyond their view.

Strike is the compass direction of the line produced by the intersection of an inclined rock layer or fault, with a horizontal plane (Figure 15.7). The strike, or compass bearing, is generally expressed as an angle

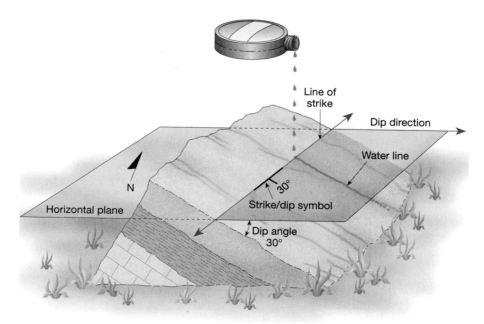

Figure 15.7 Strike and dip of a rock layer.

Box 15.1

Naming Local Rock Units

One of the primary goals of geology is to reconstruct Earth's long and complex history through the systematic study of rocks. In most areas, exposures of rocks are not continuous over great distances. Consequently, the study of rock layers must be done locally and then correlated with data from adjoining areas to produce a larger and more complete picture. The first step in the effort to unravel past geologic events is that of describing and mapping rock units exposed in local outcrops.

Describing anything as complex as a thick sequence of rocks requires subdividing the layers into units of manageable size. The most basic rock division is called a *formation*, which is simply a distinctive series of strata that originated through the same geologic processes. More precisely, a formation is a mappable rock unit that has definite boundaries (or contacts with other units) and certain obvious characteristics (rock type) by which it may be traced from place to place and distinguished from other rock units.

Figure 15.A shows several named formations that are exposed in the walls of the Grand Canyon. Just as these rock strata in the Grand Canyon were subdivided, geologists subdivide rock sequences throughout the world into formations.

Those who have had the opportunity to travel to some of the national parks in the West may already be familiar with the names of certain formations. Well-known formations include the Navajo Sandstone in Zion National Park, Redwall Limestone in the Grand Canyon, the Entrada Sandstone in Arches National Park, and the Wasatch Formation in Bryce Canyon National Park.

Figure 15.A Grand Canyon with a few of its rock units (formations) named. (Photo by E. J. Tarbuck)

Although formations can consist of igneous or metamorphic rocks, the vast majority of established formations are sedimentary rocks. A formation may be relatively thin and composed of a single rock type, for example, a 1-meter thick layer of limestone. At the other extreme, formations can be thousands of meters thick and consist of an interbedded sequence of rock types such as sandstones and shales. The most important condition to be met when establishing a formation is that *it constitutes a unit of rock produced by uniform or uniformly alternating conditions.*

In most regions of the world, the name of each formation consists of two parts, for example, the Oswego Sandstone and the Carmel Formation. The first part of the name is generally taken from a geologic structure or a locality where the formation is clearly and completely exposed. For instance, the expansive Morrison Formation is well exposed at Morrison, Colorado. As a result, this particular exposure is known as the *type locality*. Ideally, the second part of the name indicates the dominant rock type as exemplified by such names as the Dakota Sandstone, the Kaibab Limestone, and the Burgess Shale. When no single rock type dominates, the term *formation* is used, as, for example, the well-known Chinle Formation, exposed in Arizona's Petrified Forest National Park.

In summary, describing and naming formations is an important first step in the process of organizing and simplifying the study and analysis of Earth's history.

---

relative to north. For example, (N10°E) means the line of strike is 10 degrees to the east of north. The strike of the rock units illustrated in Figure 15.7 is approximately north 60°east (N60°E).

Dip is the angle of inclination of the surface of a rock unit or fault measured from a horizontal plane. Dip includes both an angle of inclination and a direction toward which the rock is inclined. In Figure 15.7 the dip angle of the rock layer is 30°. A good way to visualize dip is to imagine that water will always run down the rock surface parallel to the dip. The direction of dip will always be at a 90° angle to the strike.

In the field, geologists measure the strike (trend) and dip (inclination) of sedimentary rocks at as many

A. Map view

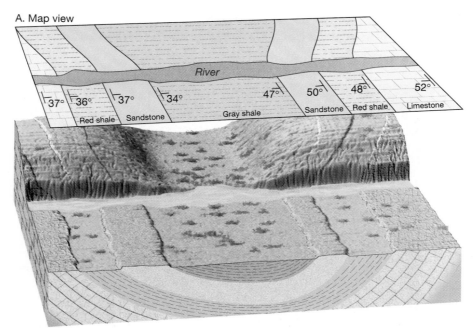

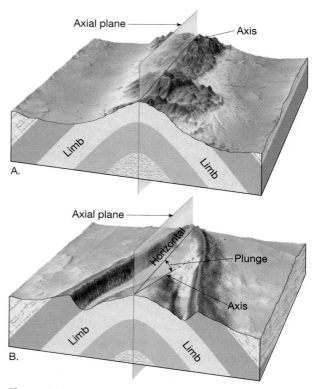

B. Block diagram

Figure 15.8 By establishing the strike and dip of outcropping sedimentary beds on a map **A.**, geologists can infer the orientation of the structure below ground **B.**

outcrops as practical. These data are then plotted on a topographic map or an aerial photograph along with a color-coded description of the rock. From the orientation of the strata, an inferred orientation and shape of the structure can be established, as shown in Figure 15.8. Using this information, the geologist can reconstruct the pre-erosional structures and can begin to interpret the region's geologic history.

 # Folds

During mountain building, flat-lying sedimentary and volcanic rocks are often bent into a series of wavelike undulations called **folds**. Folds in sedimentary strata are much like those that would form if you were to hold the ends of a sheet of paper and then push them together. In nature, folds come in a wide variety of sizes and configurations. Some folds are broad flexures in which rock units hundreds of meters thick have been slightly warped. Others are very tight, microscopic structures found in metamorphic rocks. Size differences notwithstanding, most folds are the result of compressional stresses that result in the shortening and thickening of the crust. Occasionally, folds are found singly, but most often they occur as a series of undulations.

To aid our understanding of folds and folding, we need to become familiar with the terminology used to name the parts of a fold. As shown in Figure 15.9, the two sides of a fold are called *limbs*. A line drawn along the points of maximum curvature of each layer is termed the *axis* of the fold. In some

folds, as Figure 15.9A illustrates, the axis is horizontal, or parallel to the surface. However, in more complex folding, the fold axis is often inclined at an angle known as the *plunge* (Figure 15.9B). Further,

Figure 15.9 Idealized sketches illustrating the features associated with symmetrical folds. The axis of the fold in **A** is horizontal, whereas the axis of the fold in **B** is plunging.

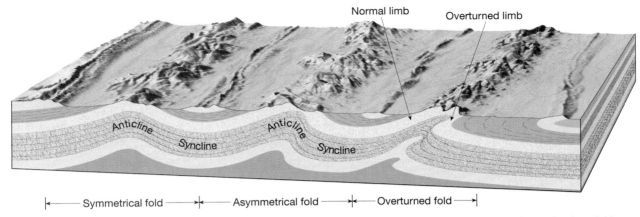

Figure 15.10 Block diagram of principal types of folded strata. The upfolded or arched structures are anticlines. The downfolds or troughs are synclines. Notice that the limb of an anticline is also the limb of the adjacent syncline.

the *axial plane* is an imaginary surface that divides a fold as symmetrically as possible.

Types of Folds

The two most common types of folds are called anticlines and synclines (Figure 15.10). An **anticline** is most commonly formed by the upfolding, or arching, of rock layers.* Figure 15.9 is an example of an anticline. Anticlines are sometimes spectacularly displayed where highways have been cut through deformed strata. Often found in association with anticlines are downfolds, or troughs, called **synclines**. Notice in

Figure 15.10 that the limb of an anticline is also a limb of the adjacent syncline.

Depending on their orientation, these basic folds are described as *symmetrical* when the limbs on either side of the axial plane diverge at the same angle and *asymmetrical* when they do not. An asymmetrical fold is said to be *overturned* if one limb is tilted beyond the vertical (Figure 15.10). An overturned fold can also "lie on its side" so that a plane extending through the axis of the fold actually would be horizontal. These *recumbent* folds are common in some mountainous regions (Figure 15.11). In the Alps, there is evidence

*By strict definition, an anticline is a structure in which the oldest strata are found in the center. This most typically occurs when strat are upfolded. Further, a syncline is strictly defined as a structure in which the youngest strata are found in the center. This occurs most commonly when strata are downfolded.

Figure 15.11 Recumbent fold in Precambrian rocks of the Umanak area, Greenland. (Photo by T.C.R. Pulvertaft, Geological Survey of Greenland)

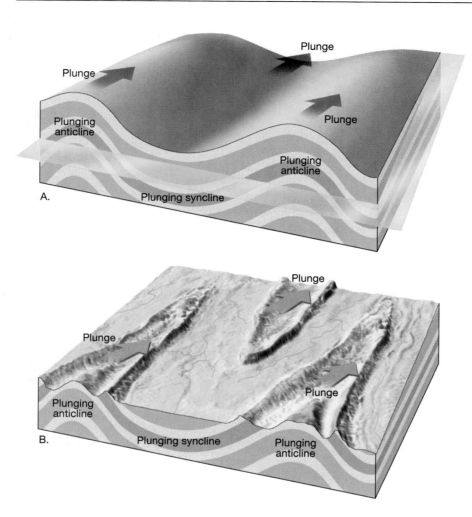

Figure 15.12 Plunging folds.
A. Idealized view of plunging folds in which a horizontal surface has been added. **B.** View of plunging folds as they might appear after extensive erosion. Notice that in a plunging anticline the outcrop pattern "points" in the direction of the plunge, while the opposite is true of plunging synclines.

that certain deformed strata have been shoved up to 50 kilometers over the adjacent rocks.

Folds do not continue forever; rather, their ends die out much like the wrinkles in cloth. Some folds *plunge* because the axis of the fold penetrates into the ground (Figure 15.12). As the figure shows, both anticlines and synclines can plunge. Figure 15.13 shows an example of a plunging anticline and the pattern produced when erosion removes the upper layers of the structure and exposes its interior. Note that the outcrop pattern of an anticline points in the direction it is plunging, whereas the opposite is true for a syncline. A good example of the kind of topography that results when erosional forces attack folded sedimentary strata is found in the Valley and Ridge Province of the Appalachians (see Figure 20.15).

It is important to understand that ridges are not necessarily associated with anticlines, nor are valleys related to synclines. Rather, ridges and valleys result because of differential weathering, and erosion. For example, in the Valley and Ridge Province, resistant sandstone beds remain as imposing ridges

Figure 15.13 Sheep Mountain, a doubly plunging anticline. Note that erosion has cut the flanking sedimentary beds into low ridges that make a "V" pointing in the direction of plunge. (Photo by John S. Shelton)

A.

Figure 15.14 Monocline. **A.** The San Rafael monocline, Utah. (Photo by Stephen Trimble) **B.** Monocline consisting of bent sedimentary beds that were deformed by faulting in the bedrock below.

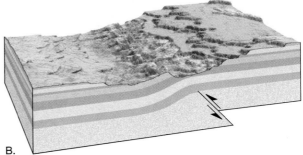

B.

separated by valleys cut into more easily eroded shale or limestone beds.

Although most folds are caused by compressional stresses that squeeze and crumble strata, some folds are a consequence of vertical displacement. **Monoclines**, broad flexures found on the Colorado Plateau and elsewhere, are such structures. Unlike anticlines and synclines, which have two limbs, monoclines have just one. These folds are thought to result from nearly vertical faulting in deep-lying basement rocks, as shown in Figure 15.14. Whereas the rigid basement complex responded to vertical stress by fracturing, the relatively flexible sedimentary strata above were deformed by folding.

Domes and Basins

Broad upwarps in basement rock may deform the overlying cover of sedimentary strata and generate large folds. When this upwarping produces a circular or elongated structure, the feature is called a **dome** (Figure 15.15A). Downwarped structures having a similar shape are termed **basins** (Figure 15.15B).

The Black Hills of western South Dakota is a large domed structure thought to be generated by upwarping. Here erosion has stripped away the highest portions of the unwarped sedimentary beds, exposing older igneous and metamorphic rocks in the center (Figure 15.16). Remnants of these once-continuous sedimentary layers are visible flanking the crystalline core of this mountain range. The more resistant strata are easy to identify because differential erosion has left them outcropping as prominent angular ridges called **hogbacks**. Because hogbacks can form whenever resistant strata are steeply inclined, they are also associated with other types of folds.

Domes can also be formed by the intrusion of magma (laccoliths) as exemplified by the Henry Mountains shown in Figure 3.21. In addition, the upward migration of salt formations can produce salt domes that are common in the Gulf of Mexico.

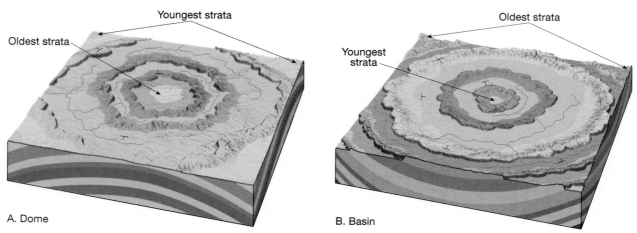

Figure 15.15 Gentle upwarping and downwarping of crustal rocks produce domes (**A**) and basins (**B**). Erosion of these structures results in an outcrop pattern that is roughly circular or elongated.

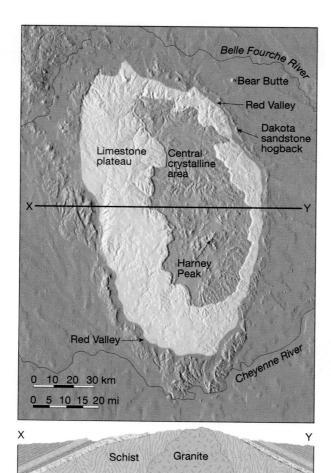

Figure 15.16 The Black Hills of South Dakota, a large domal structure with resistant igneous and metamorphic rocks exposed in the core. (After Arthur N. Strahler, *Introduction to Physical Geography*, 3rd ed., New York: John Wiley & Sons, 1973. Reprinted by permission.)

Several large basins exist in the United States (Figure 15.17). The basins of Michigan and Illinois have very gently sloping beds similar to saucers. These basins are thought to be the result of large accumulations of sediment, whose weight caused the crust to subside (see section on isostasy in Chapter 20). A few structural basins may have been the result of giant asteroid impacts.

Because large basins contain sedimentary beds sloping at such low angles, they are usually identified by the age of the rocks composing them. The youngest rocks are found near the center and the oldest rocks are at the flanks. This is just the opposite order of a domed structure, such as the Black Hills, where the oldest rocks form the core.

 Faults

Faults are fractures in the crust along which appreciable displacement has taken place. Occasionally, small faults can be recognized in road cuts where sedimentary beds have been offset a few meters, as shown in Figure 15.18. Faults of this scale usually occur as single discrete breaks. By contrast, large faults, like the San Andreas fault in California, have displacements of hundreds of kilometers and consist of many interconnecting fault surfaces. These *fault zones* can be several kilometers wide and are often easier to identify from high-altitude photographs than at ground level.

Sudden movements along faults are the cause of most earthquakes. However, many faults are inactive and thus are remnants of past deformation. Along active faults, rock is often broken and pulverized as crustal blocks on opposite sides of a fault grind past one another. The loosely coherent, clayey material that results from

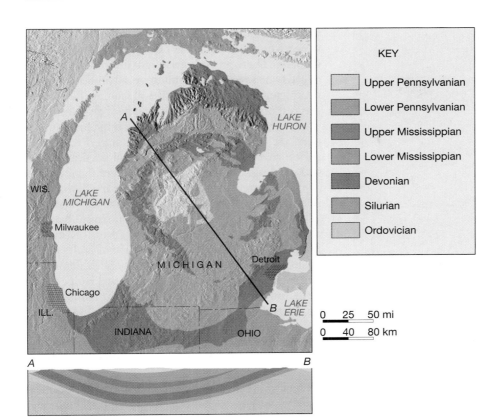

KEY

Upper Pennsylvanian

Lower Pennsylvanian

Upper Mississippian

Lower Mississippian

Devonian

Silurian

Ordovician

Figure 15.17 The bedrock geology of the Michigan Basin. Notice that the youngest rocks are centrally located, while the oldest beds flank this structure.

this activity is called *fault gouge*. On some fault surfaces, the rocks become highly polished and striated, or grooved, as the crustal blocks slide past one another. These polished and striated surfaces, called *slickensides*, provide geologists with evidence for the direction of the most recent displacement along the fault. Geologists classify faults by these relative movements, which can be predominantly horizontal, vertical, or oblique.

Dip-Slip Faults

Faults in which the movement is primarily parallel to the dip (or inclination) of the fault surface are called **dip-slip faults**. Vertical displacements along dip-slip faults may produce long, low cliffs called **fault scarps**. Fault scarps, such as the one shown in Figure 15.19, are produced by displacements that generate earthquakes. A strong earthquake in 1983 produced this 3-meter scarp.

Figure 15.18 Faulting caused the vertical displacement of these beds located near Kanab, Utah. Arrows show relative motion of rock units. (Photo by Tom Bean/DRK Photo)

Figure 15.19 A fault scarp in an alluvial fan is visible near the bottom of this photo. Death Valley, California. (Photo by Tom Bean)

It has become common practice to call the rock surface that is immediately above the fault the *hanging wall* and to call the rock surface below, the *footwall* (Figure 15.20). This nomenclature arose from prospectors and miners who excavated shafts and tunnels along fault zones because these are frequently sites of ore deposits. In these tunnels, the miners would walk on the rocks below the mineralized fault zone (the footwall) and hang their lanterns on the rocks above (the hanging wall).

Two major types of dip-slip faults are *normal faults* and *reverse faults*. In addition, when a reverse fault has an angle of dip (inclination) less than 45°, it is called a *thrust fault*. We will consider these types of dip-slip faults next.

Normal Faults. Dip-slip faults are classified as **normal faults** when the hanging wall block moves down relative to the footwall block (Figure 15.21). Most normal faults have steep dips of about 60°, which tend to flatten out with depth. However, some dip-slip faults have much lower dips, with some approaching horizontal. Because of the downward motion of the hanging wall, normal faults accommodate lengthening, or extension, of the crust.

Most normal faults are small, having displacements of only a meter or so, like the one shown in the road cut in Figure 15.18. Others extend for tens of kilometers where they may sinuously trace the boundary of a mountain front. In the western United States, large-scale normal faults like these are associated with structures called **fault-block mountains**.

Examples of fault-block mountains include the Teton Range of Wyoming and the Sierra Nevada of California. Both are faulted along their eastern flanks, which were uplifted as the blocks tilted downward to the west. These precipitous mountain fronts were produced over a period of 5 million to 10 million years by many irregularly spaced episodes of faulting. Each event was responsible for just a few meters of displacement.

Other excellent examples of fault-block mountains are found in the Basin and Range Province, a region that encompasses Nevada and portions of the surrounding states (Figure 15.22). Here the crust has been elongated and broken to create more than 200 relatively small mountain ranges. Averaging about 80 kilometers in length, the ranges rise 900 to 1500 meters above the adjacent down-faulted basins. Notice in Figure 15.22 that slopes of the normal faults in the Basin and Range Province decrease with depth and join together to form a nearly horizontal fault called a **detachment fault**. These detachment faults extend for several kilometers below the surface. Here they form a major boundary between the rocks below, which exhibit ductile deformation, and the rocks above, which demonstrate brittle deformation via faulting.

Normal faulting is also prevalent at spreading centers where plate divergence occurs. Here, a central block called a **graben** is bounded by normal faults and

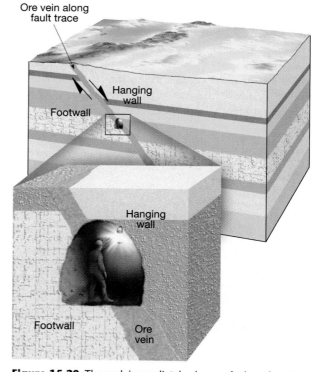

Figure 15.20 The rock immediately above a fault surface is the *hanging wall* and that below is called the *footwall*. These names came from miners who excavated ore along fault zones. The miners hung their lanterns on the rocks above the fault trace (hanging wall) and walked on the rocks below the fault trace (footwall).

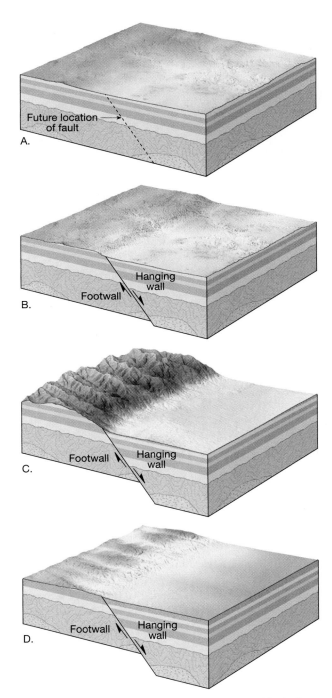

Figure 15.21 Block diagrams illustrating a normal fault. **A.** Rock strata prior to faulting. **B.** The relative movement of displaced blocks. Displacement may continue in a fault-block mountain range over millions of years and consist of many widely spaced episodes of faulting. **C.** How erosion might alter the upfaulted block. **D.** Eventually the period of deformation ends and erosion becomes the dominant geologic process.

drops as the plates separate (Figure 15.23). These grabens produce an elongated valley bounded by relatively uplifted structures called **horsts**. The Rift Valley of East Africa is made up of several large grabens, above which tilted horsts produce a linear

mountainous topography. This valley, nearly 6000 kilometers (3600 miles) long, contains the excavation sites of some of the earliest human fossils. Examples of inactive rift valleys include the Rhine Valley in Germany, and the Triassic-age grabens of the eastern United States. Even larger inactive normal fault systems are the rifted continental margins, such as along the east coasts of the Americas and the west coasts of Europe and Africa.

Fault motion provides geologists with a method of determining the nature of the forces at work within Earth. Normal faults indicate the existence of tensional stresses that pull the crust apart. This "pulling apart" can be accomplished either by uplifting that causes the surfaces to stretch and break or by opposing horizontal forces.

Reverse and Thrust Faults. **Reverse faults** and **thrust faults** are dip-slip faults in which the hanging wall block moves up relative to the footwall block (Figure 15.24). Recall that reverse faults have dips greater than 45° and thrust faults have dips less than 45°. Because the hanging wall block moves up and over the footwall block, reverse and thrust faults accommodate shortening of the crust.

Most high-angle reverse faults are small and accommodate local displacements in regions dominated by other types of faulting. Thrust faults, on the other hand, exist at all scales. Small thrust faults exhibit displacement on the order of millimeters to a few meters. Some large thrust faults have displacements on the order of tens to hundreds of kilometers.

Whereas normal faults occur in tensional environments, thrust faults result from strong compressional stresses. In these settings, crustal blocks are displaced *toward* one another, with the hanging wall being displaced upward relative to the footwall. Thrust faulting is most pronounced in subduction zones and other convergent boundaries where plates are colliding. Compressional forces generally produce folds as well as faults and result in a thickening and shortening of the material involved.

In mountainous regions such as the Alps, Northern Rockies, Himalayas, and Appalachians, thrust faults have displaced strata as far as 50 kilometers over adjacent rock units. The result of this large-scale movement is that older strata end up overlying younger rocks. The photo of Nevada's Keystone Overthrust in Figure 15.25 illustrates this phenomenon. Here, 500-million-year-old dark-colored limestone has been thrust on top of 150-million-year-old light-colored sandstone. The irregular line between the dark and light rocks marks the fault trace, which dips gently to the west (left).

A classic site of thrust faulting occurs in Glacier National Park (Figure 15.26). Here, mountain peaks that provide the park's majestic scenery have been carved

Figure 15.22 Normal faulting in the Basin and Range Province. Here, tensional stresses have elongated and fractured the crust into numerous blocks. Movement along these fractures has tilted the blocks producing parallel mountain ranges called fault-block mountains. (Photo by Michael Collier)

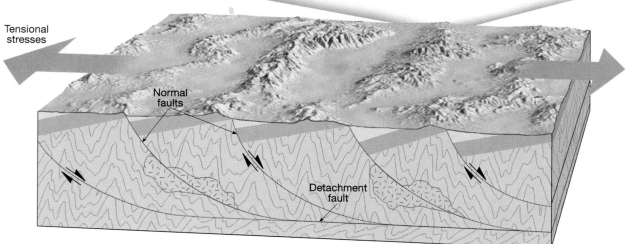

from Precambrian rocks that were displaced over much younger Cretaceous strata. At the eastern edge of Glacier National Park there is an outlying peak called Chief Mountain. This structure is an isolated remnant of the thrust sheet that was severed by the erosional forces of glacial ice and running water. An isolated block, such as Chief Mountain, is called a **klippe** (Figure 15.27).

Strike-Slip Faults

Faults in which the dominant displacement is horizontal, and parallel to the strike of the fault surface, are called **strike-slip faults**. Because of their large size and linear nature, many strike-slip faults produce a trace that is visible over a great distance. Rather than a single frac-

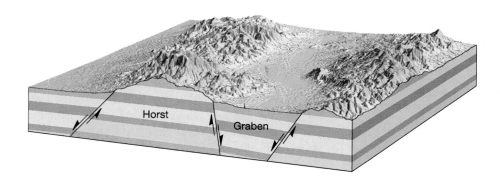

Figure 15.23 Diagrammatic sketch of downfaulted (graben) and upfaulted (horst) blocks.

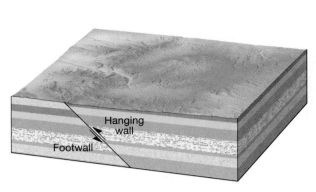

Figure 15.24 Block diagram showing the relative movement along a reverse fault.

ture along which movement takes place, large strike-slip faults consist of a zone of roughly parallel fractures. The zone may be up to several kilometers in width. The most recent movement, however, is often along a strand only a few meters wide which may offset features such as stream channels (Figure 15.28). Furthermore, crushed and broken rocks produced during faulting are more easily eroded, often producing linear valleys or troughs that mark the locations of strike-slip faults.

The earliest scientific records of strike-slip faulting were made following surface ruptures that produced

Figure 15.25 The Keystone Overthrust. Dark-colored Cambrian limestone has been thrust to the east over light-colored Jurassic sandstone. The irregular line between the dark- and light-colored rocks marks the fault trace, which dips gently to the west (left). (Photo by John S. Shelton)

large earthquakes. One of the most noteworthy of these was the great San Francisco earthquake of 1906. During this strong earthquake, structures such as fences that were built across the San Andreas fault were displaced as much as 4.7 meters (15 feet). Because the movement along the San Andreas causes the crustal block on the opposite side of the fault to move to the right as you

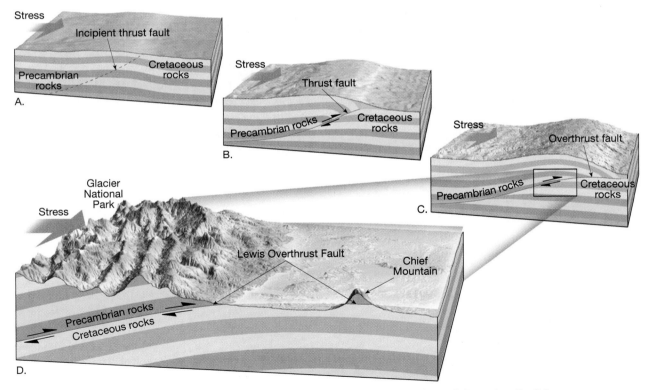

Figure 15.26 Idealized development of Lewis Overthrust fault. **A.** Geologic setting prior to deformation. **B., C.** Large-scale movement along a thrust fault displaced Precambrian rock over Cretaceous strata in the region of Glacier National Park. **D.** Erosion by glacial ice and running water sculptured the thrust sheet into a majestic landscape and isolated a remnant of the thrust sheet called Chief Mountain.

Figure 15.27 Chief Mountain, Glacier National Park, Montana is a klippe. (Photo by Carr Clifton)

continental plates that move horizontally with respect to each other. One of the best-known transform faults is California's San Andreas fault (see Box 15.2). This plate-bounding fault can be traced for about 950 kilometers (600 miles) from the Gulf of California to a point along the Pacific Coast north of San Francisco, where it heads out to sea. Ever since its formation about 29 million years ago, displacement along the San Andreas fault has exceeded 560 kilometers. This movement has accommodated the northward displacement of southwestern California and the Baja Peninsula of Mexico in relation to the remainder of North America. The nature of these important structures will be discussed in more detail in Chapter 19.

Strike-slip faults and dip-slip faults are on the opposite ends of a spectrum of fault structures. Faults that exhibit a combination of dip-slip and strike-slip movements are called **oblique-slip faults**.

 Joints

Among the most common rock structures are fractures called joints. Unlike faults, **joints** are fractures along which no appreciable displacement has occurred. Although some joints have a random orientation, most occur in roughly parallel groups (see Figure 5.7).

We have already considered two types of joints. Earlier we learned that *columnar joints* form when igneous rocks cool and develop shrinkage fractures that produce elongated, pillarlike columns (Figure 15.29). Also recall that sheeting produces a pattern of gently curved joints that develop more or less parallel to the surface of large exposed igneous bodies such as batholiths. Here the jointing results from the

face the fault, it is called a *right-lateral* strike-slip fault. The Great Glen fault in Scotland is a well-known example of a *left-lateral* strike-slip fault, which exhibits the opposite sense of displacement. The total displacement along the Great Glen fault is estimated to exceed 100 kilometers (60 miles). Also associated with this fault trace are numerous lakes, including Loch Ness, the home of the legendary monster.

Many major strike-slip faults cut through the lithosphere and accommodate motion between two large crustal plates. This special kind of strike-slip fault is called a **transform fault**. Numerous transform faults cut the oceanic lithosphere and link spreading oceanic ridges. Others accommodate displacement between

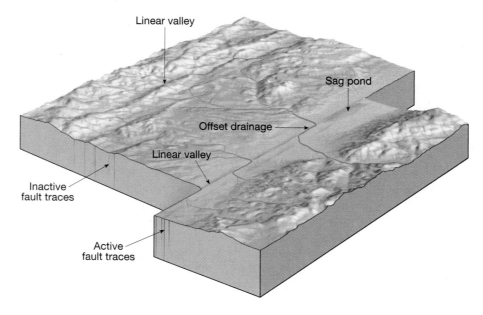

Figure 15.28 Block diagram illustrating the features associated with strike-slip faults. Note how the stream channels have been offset by fault movement. The faults in this diagram are right-lateral strike-slip faults. (Modified after R. L. Wesson and others)

Figure 15.29 Devil's Tower, Wyoming, exhibits columnar joints and the columns that result. (Photo by Bob Thomason/Tony Stone Images)

drainage pattern described in Chapter 10 is such a case.

Joints may also be significant from an economic standpoint. Some of the world's largest and most important mineral deposits were emplaced along joint systems. Hydrothermal solutions, which are basically mineralized fluids, can migrate into fractured host rocks and precipitate economically important amounts of copper, silver, gold, zinc, lead, and uranium.

Further, highly jointed rocks present a risk to the construction of engineering projects, including highways and dams. On June 5, 1976, fourteen lives were lost and nearly one billion dollars in property damage occurred when the Teton Dam in Idaho failed. This earthen dam was constructed of very erodible clays and silts and was situated on highly fractured volcanic rocks. Although attempts were made to fill the voids in the jointed rock, water gradually penetrated the subsurface fractures and undermined the dam's foundation. Eventually, the moving water cut a tunnel into the easily erodible clays and silts. Within minutes the dam failed, sending a 20-meter high wall of water down the Teton and Snake Rivers.

Figure 15.30 Chemical weathering is enhanced along joints in granitic rocks near the top of Lembert Dome, Yosemite National Park. (Photo by E. J. Tarbuck)

gradual expansion that occurs when erosion removes the overlying load.

In contrast to the situations just described, most joints are produced when rocks in the outermost crust are deformed. Here tensional and shearing stresses associated with crustal movements cause the rock to fail by brittle fracture. For example, when folding occurs, rocks situated at the axes of the folds are elongated and pulled apart to produce tensional joints. Extensive joint patterns can also develop in response to relatively subtle and often barely perceptible regional upwarping and downwarping of the crust. In many cases, the cause for jointing at a particular locale is not readily apparent.

Many rocks are broken by two or even three sets of intersecting joints that slice the rock into numerous regularly shaped blocks. These joint sets often exert a strong influence on other geologic processes. For example, chemical weathering tends to be concentrated along joints, and in many areas, groundwater movement and the resulting dissolution in soluble rocks is controlled by the joint pattern (Figure 15.30). Moreover, a system of joints can influence the direction that stream courses follow. The rectangular

Box 15.2

The San Andreas Fault System

The San Andreas, the best-known and largest fault system in North America, first attracted wide attention after the great 1906 San Francisco earthquake and fire. Following this devastating event, geologic studies demonstrated that a displacement of as much as 5 meters along the fault had been responsible for the earthquake. It is now known that this dramatic event is just one of many thousands of earthquakes that have resulted from repeated movements along the San Andreas throughout its 29-million-year history.

Where is the San Andreas fault system located? As shown in Figure 15.B, it trends in a northwesterly direction for nearly 1300 kilometers (780 miles) through much of western California. At its southern end, the San Andreas connects with a spreading center located in

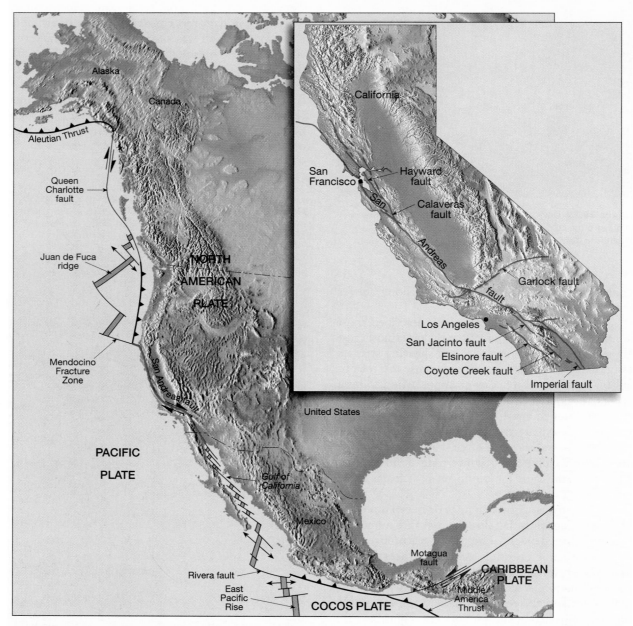

Figure 15.B Map showing the extent of the San Andreas fault system. Inset shows only a few of the many splinter faults that are part of this great fault system.

Table 15.A Major Earthquakes on the San Andreas Fault System

Date	Location	Magnitude	Remarks
1812	Wrightwood, CA	7	Church of San Juan Capistrano collapsed, killing 40 worshippers.
1812	Santa Barbara channel	7	Churches and other buildings wrecked in and around Santa Barbara.
1838	San Francisco peninsula	7	At one time thought to have been comparable to the great earthquake of 1906.
1857	Fort Tejon, CA	8.25	One of the greatest U.S. earthquakes. Occurred near Los Angeles, then a city of 4000.
1868	Hayward, CA	7	Rupture of the Hayward fault caused extensive damage in San Francisco Bay area.
1906	San Francisco, CA	8.25	The great San Francisco earthquake. As much as 80 percent of the damage caused by fire.
1940	Imperial Valley	7.1	Displacement on the newly discovered Imperial fault.
1952	Kern County	7.7	Rupture of the White Wolf fault. Largest earthquake in California since 1906. Sixty million dollars in damages and 12 people killed.
1971	San Fernando Valley	6.5	One-half billion dollars in damages and 58 lives claimed.
1989	Santa Cruz Mountains	7.1	Loma Prieta earthquake. Six billion dollars in damages, 62 lives lost, and 3757 people injured.
1994	Northridge (Los Angeles area)	6.9	Over 15 billion dollars in damages, 51 lives lost, and over 5000 injured.

the Gulf of California. In the north, the fault enters the Pacific Ocean at Point Arena, where it is thought to continue its northwesterly trend, eventually joining the Mendocino fracture zone. In the central section, the San Andreas is relatively simple and straight. However, at its two extremities, several branches spread from the main trace, so that in some areas the fault zone exceeds 100 kilometers (60 miles) in width.

Over much of its extent, a linear trough reveals the presence of the San Andreas fault. When the system is viewed from the air, linear scars, offset stream channels, and elongated ponds mark the trace in a striking manner. On the ground, however, surface expressions of the faults are much more difficult to detect. Some of the most distinctive landforms include long, straight escarpments, narrow ridges, and sag ponds formed by settling of blocks within the fault zone. Further, many stream channels characteristically bend sharply to the right where they cross the fault (Figure 15.C).

With the advent of the theory of plate tectonics, geologists began to realize the significance of this great fault system. The San Andreas fault is a transform boundary separating two crustal plates that move very slowly. The Pacific plate, located to the west, moves northwestward relative to the North American plate, causing earthquakes along the fault (Table 15.A).

The San Andreas is undoubtedly the most studied of any fault system in the world. Although many questions

remain unanswered, geologists have learned that each fault segment exhibits somewhat different behavior. Some portions of the San Andreas exhibit a slow creep with little noticeable seismic activity. Other segments regularly slip, producing small earthquakes, while still other segments seem to store elastic energy for hundreds of years and rupture in great earthquakes. This knowledge is useful when assigning earthquake hazard potential to a given segment of the fault zone.

Because of the great length and complexity of the San Andreas fault, it is more appropriately referred to as a "fault system." This major fault system consists primarily of the San Andreas fault and several major branches including the Hayward and Calaveras faults of central California and the San Jacinto and Elsinore faults of southern California (see Figure 15.B). These major segments, plus a vast number of smaller faults that include the Imperial fault, San Fernando fault, and the Santa Monica fault, collectively accommodate the relative motion between the North American and Pacific plates.

Blocks on opposite sides of the San Andreas fault move horizontally in opposite directions, such that if a person stood on one side of the fault, the block on the opposite side would appear to move to the right when slippage occurred. This type of displacement is known as *right-lateral strike-slip* by geologists (Figure 15.C).

Ever since the great San Francisco earthquake of 1906, when as much as 5

meters of displacement occurred, geologists have attempted to establish the cumulative displacement along this fault over its 29-million-year history. By matching rock units across the fault, geologists have determined that the total accumulated displacement from earthquakes and creep exceeds 560 kilometers (340 miles).

San Andreas Fault

Offset in stream

Figure 15.C Aerial view showing offset stream channel across the San Andreas fault on the Carrizo Plain west of Taft, California. (Photo by Michael Collier/DRK Photo)

Chapter Summary

- *Deformation* refers to changes in the volume and/or shape of a rock body and is most pronounced along plate margins. *Stress* is a measure of the amount of force that causes rocks to deform, whereas *strain* is the change (deformation) caused by stress. Stress that is uniform in all directions is called *confining pressure*, whereas *differential stresses* are applied unequally in different directions. Differential stresses that shorten a rock body are *compressional stresses*; those that elongate a rock unit are *tensional stresses*.

- Rocks deform differently depending on their chemical makeup, environment, and the rate at which stress is applied. Rocks first respond by deforming *elastically*, and will return to their original shape when the stress is removed. Once the elastic limit is surpassed, rocks either deform plastically or they fracture. *Plastic deformation* changes the shape of a rock unit through folding and flowing, and the rock is said to behave in a *ductile* manner. Plastic deformation occurs in a high temperature/high pressure environment. In a near-surface environment, when stress is applied rapidly, most rocks deform by *brittle failure*.

- The orientation of rock units or fault surfaces is established with measurements called strike and dip. *Strike* is the compass direction of a line produced by the intersection of an inclined rock layer or fault with a horizontal plane. *Dip* is the angle of inclination of the surface of a rock unit or fault measured from a horizontal plane.

- The most basic geologic structures associated with rock deformation are *folds* (flat-lying sedimentary and volcanic rocks bent into a series of wavelike undulations) and *faults*. The two most common types of folds are *anticlines*, formed by the upfolding, or arching, of rock layers, and *synclines*, which are downfolds. Most folds are the result of horizontal *compressional stresses*. Folds can be *symmetrical*, *asymmetrical*, or, if one limb has been tilted beyond the vertical, *overturned*. *Domes* (upwarped structures) and *basins* (downwarped structures) are circular or somewhat elongated folds formed by vertical displacements of strata.

- Faults are fractures in the crust along which appreciable displacement has occurred. Faults in which the movement is primarily vertical are called *dip-slip faults*. Dip-slip faults include both *normal* and *reverse faults*. Low-angle reverse faults are called *thrust faults*. Normal faults indicate *tensional stresses* that pull the crust apart. Along the spreading centers of plates, divergence can cause a central block called a *graben*, bounded by normal faults, to drop as the plates separate.

- Reverse and thrust faulting indicate that *compressional forces* are at work. Large *thrust faults* are found along subduction zones and other convergent boundaries where plates are colliding. In mountainous regions such as the Alps, Northern Rockies, Himalayas, and Appalachians, thrust faults have displaced strata as far as 50 kilometers over adjacent rock units.

- *Strike-slip faults* exhibit mainly horizontal displacement parallel to the strike of the fault surface. Large strike-slip faults, called *transform faults*, accommodate displacement between plate boundaries. Most transform faults cut the oceanic lithosphere and link spreading centers. The San Andreas fault cuts the continental lithosphere and accommodates the northward displacement of southwestern California.

- *Joints* are fractures along which no appreciable displacement has occurred. Joints generally occur in groups with roughly parallel orientations. Most joints are the result of brittle failure of rock units located in the outermost crust.

Review Questions

1. What is rock deformation?
2. Explain how confining pressure influences the way rocks deform.
3. In simple terms, what is the difference between brittle and ductile deformation?
4. Contrast compressional and tensional stresses.
5. How is elastic deformation different from plastic deformation?
6. List three factors that determine how rocks will behave when exposed to stresses that exceed their strength. Briefly explain the role of each.
7. What is an outcrop?

8. What two measurements are used to establish the orientation of deformed strata? Distinguish between them.

9. Distinguish between anticlines and synclines. Domes and basins. Anticlines and domes.

10. How is a monocline different from an anticline?

11. The Black Hills of South Dakota are a good example of what type of structural feature?

12. Contrast the movements that occur along normal and reverse faults. What type of stress is indicated by each fault?

13. Is the fault shown in Figure 15.18 a normal or a reverse fault?

14. Describe a horst and a graben. Explain how a graben valley forms and name one.

15. What type of faults are associated with fault-block mountains?

16. What type of fault is illustrated by Figure 15.25?

17. How are reverse faults different than thrust faults? In what way are they the same?

18. The San Andreas fault is an excellent example of a _____ fault.

19. With which of the three types of plate boundaries does normal faulting predominate? Reverse faulting? Strike-slip faulting?

20. How are joints different from faults?

Key Terms

anticline (p. 383)
basin (p. 385)
brittle failure (p. 378)
compressional stress (p. 377)
confining pressure (p. 377)
deformation (p. 377)
detachment fault (p. 388)
differential stress (p. 377)
dip (p. 381)

dip-slip fault (p. 388)
dome (p. 385)
ductile (p. 378)
elastic deformation (p. 378)
fault (p. 386)
fault-block mountain (p. 388)
fault scarp (p. 388)
fold (p. 382)
graben (p. 397)

hogback (p. 385)
horst (p. 389)
joint (p. 392)
klippe (p. 390)
monocline (p. 385)
normal fault (p. 388)
oblique-slip fault (p. 392)
plastic deformation (p. 378)
reverse fault (p. 389)
rock structure (p. 380)

shear (p. 378)
strain (p. 377)
stress (p. 377)
strike (p. 381)
strike-slip fault (p. 390)
syncline (p. 383)
tensional stress (p. 377)
thrust fault (p. 389)
transform fault (p. 392)

Web Resources

 The *Earth* Home Page provides on-line resources for this chapter on the World Wide Web. You will find review exercises, specific updates for items in the chapter, suggested reading, and links to interesting related pathways on the Internet. Visit the *Earth* Home Page at **http://www.prenhall.com/tarbuck.**

CHAPTER 16

Earthquakes

Left In January 1995, a strong earthquake toppled this elevated expressway near Kobe, Japan. — *Photo by Noboru Hashimoto/Sygma*

On October 17, 1989, at 5:04 P.M. Pacific Daylight Time, millions of television viewers around the world were settling in to watch the third game of the World Series. Instead, they saw their television sets go black as tremors hit San Francisco's Candlestick Park. Although the earthquake was centered in a remote section of the Santa Cruz Mountains, 100 kilometers to the south, major damage occurred in the Marina District of San Francisco.

The most tragic result of the violent shaking was the collapse of some double-decked sections of Interstate 880, also known as the Nimitz Freeway. The ground motions caused the upper deck to sway, shattering the concrete support columns along a mile-long section of the freeway. The upper deck then collapsed onto the lower roadway, flattening cars as if they were aluminum cans. This earthquake, named the Loma Prieta quake for its point of origin, claimed 67 lives.

In mid-January 1994, less than 5 years after the Loma Prieta earthquake devastated portions of the San Francisco Bay area, a major earthquake struck the Northridge area of Los Angeles. Although not the fabled "Big One," this moderate 6.6 to 6.9 magnitude earthquake left 51 dead, over 5000 injured, and tens of thousands of households without water and electricity. The damage exceeded $15 billion and was attributed to an unknown fault that ruptured 14 kilometers (9 miles) beneath Northridge (Figure 16.1).

The Northridge earthquake began at 4:31 A.M. and lasted roughly 40 seconds. During this brief period, the quake terrorized the entire Los Angeles area. In the three-story Northridge Meadows apartment complex, 16 people died when sections of the upper floors collapsed onto the first-floor units. Nearly 300 schools were seriously damaged and a dozen major roadways buckled. Among these were two of California's major arteries—the Golden State Freeway (Interstate 5), where an overpass collapsed completely and blocked the roadway, and the Santa Monica Freeway. Fortunately, these roadways had practically no traffic at this early morning hour.

In nearby Granada Hills, broken gas lines were set ablaze while the streets flooded from broken water mains. Seventy homes burned in the Sylmar area. A 64-car freight train derailed, including some cars carrying hazardous cargo. But it is remarkable that the destruction was not greater. Unquestionably, the

Figure 16.1 This parking deck in Northridge, California, collapsed during an earthquake in January 1994. (Photo by Spencer Grant/Liaison International)

Figure 16.2 San Francisco in flames after the 1906 earthquake. (Reproduced from the collection of the Library of Congress)

upgrading of structures to meet the requirements of building codes developed for this earthquake-prone area helped minimize what could have been a much greater human tragedy.

Across the Pacific from California lies Japan, no stranger to earthquakes, and among the most "quake-proofed" countries in the world. Yet at 5:46 A.M. on January 24, 1995, much of the "quake-proofing" proved futile as more than 5000 people perished in a 7.2-magnitude tremor centered near Kobe, the country's sixth-largest city (see chapter-opening photo).

Over 30,000 earthquakes that are strong enough to be felt occur worldwide annually. Fortunately, most are minor tremors and do very little damage. Generally, only about 75 significant earthquakes take place each year, and many of these occur in remote regions. However, occasionally a large earthquake occurs near a large population center. Under these conditions, an earthquake is among the most destructive natural forces on Earth (see Box 16.1).

The shaking of the ground, coupled with the liquefaction of some soils, wreaks havoc on buildings and other structures. In addition, when a quake occurs in a populated area, power and gas lines are often ruptured, causing numerous fires. In the famous 1906 San Francisco earthquake, much of the damage was caused by fires (Figure 16.2). They quickly became uncontrollable when broken water mains left firefighters with only trickles of water.

 ## What Is an Earthquake?

An **earthquake** is the vibration of Earth produced by the rapid release of energy. Most often earthquakes are caused by slippage along a fault in Earth's crust. The energy released radiates in all directions from its source, the **focus**, in the form of waves. These waves are analogous to those produced when a stone is dropped into a calm pond (Figure 16.3). Just as the impact of the stone sets water waves in motion, an earthquake generates seismic waves that radiate throughout Earth. Even though the energy dissipates rapidly with increasing distance from the focus, sensitive instruments located around the world record the event.

Earthquakes and Faults

The tremendous energy released by atomic explosions or by volcanic eruptions can produce an earthquake, but these events are relatively weak and infrequent. What mechanism produces a destructive earthquake? Ample evidence exists that Earth is not a static planet. We know that Earth's crust has been uplifted at times, because we have found numerous ancient wave-cut benches many meters above the level of the highest tides. Other regions

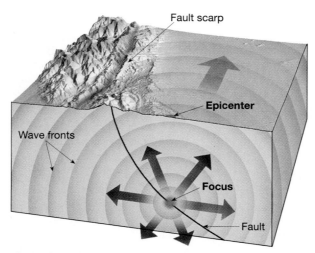

Figure 16.3 Earthquake focus and epicenter. The focus is the zone within Earth where the initial displacement occurs. The epicenter is the surface location directly above the focus.

exhibit evidence of extensive subsidence. In addition to these vertical displacements, offsets in fence lines, roads, and other structures indicate that horizontal movement is common (Figure 16.4). These movements are usually associated with large fractures in Earth's crust called **faults**.

Most of the motion along faults can be satisfactorily explained by the plate tectonics theory. This theory states that large slabs of Earth's crust are in continual slow motion. These mobile plates interact with neighboring plates, straining and deforming the rocks at their edges. In fact, it is along faults associated with plate boundaries that most earthquakes occur.

Furthermore, earthquakes are repetitive: as soon as one is over, the continuous motion of the plates resumes, adding strain to the rocks until they fail again.

Elastic Rebound

The actual mechanism of earthquake generation eluded geologists until H. F. Reid of Johns Hopkins University conducted a study following the great 1906 San Francisco earthquake. The earthquake was accompanied by horizontal surface displacements of several meters along the northern portion of the San Andreas fault. This 1300-kilometer (780-mile) fracture runs north-south through southern California. It is a large fault zone that separates two great sections of Earth's crust, the North American plate and the Pacific plate. Field investigations determined that, during this single earthquake, the Pacific plate lurched as much as 4.7 meters (15 feet) northward past the adjacent North American plate.

The mechanism for earthquake formation that Reid deduced from this information is illustrated in Figure 16.5. In part A of the figure, you see an existing fault, or break in the rock. In part B, tectonic forces ever so slowly deform the crustal rocks on both sides of the fault, as demonstrated by the bent features. Under these conditions, rocks are bending and storing elastic energy, much like a wooden stick does if bent. Eventually, the frictional resistance holding the rocks together is overcome. As slippage occurs at the weakest point (the focus), displacement will exert stress farther along the fault, where additional slippage will occur until most of the built-up strain is released (Figure 16.5C). This slippage allows the

Figure 16.4 This fence was offset 2.5 meters (8.5 feet) during the 1906 San Francisco earthquake. (Photo by G. K. Gilbert, U.S. Geological Survey)

Deformation of rocks

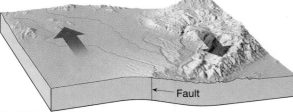

A. Original position

B. Buildup of strain

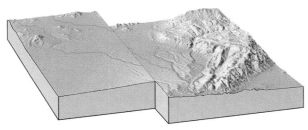

C. Slippage

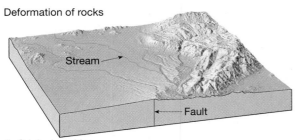

D. Strain released

Figure 16.5 Elastic rebound. As rock is deformed, it bends, storing elastic energy. Once strained beyond its breaking point, the rock cracks, releasing the stored-up energy in the form of earthquake waves.

deformed rock to "snap back." The vibrations we know as an earthquake occur as the rock elastically returns to its original shape. The "springing back" of the rock was termed **elastic rebound** by Reid, because the rock behaves elastically, much like a stretched rubber band does when it is released.

In summary, most earthquakes are produced by the rapid release of elastic energy stored in rock that has been subjected to great stress. Once the strength of the rock is exceeded, it suddenly ruptures, causing the vibrations of an earthquake. Earthquakes also occur along existing fault surfaces whenever the frictional forces on the fault surfaces are overcome.

The San Andreas is undoubtedly the most studied fault system in the world. Over the years, investigations have shown that displacement occurs along discrete segments that are 100 to 200 kilometers long. Further, each fault segment behaves somewhat differently from the others. Some portions of the San Andreas exhibit a slow, gradual displacement known as **fault creep**, which occurs relatively smoothly, and therefore with little noticeable seismic activity. Other segments regularly slip, producing small earthquakes.

Still other segments remain locked and store elastic energy for hundreds of years before rupturing in great earthquakes. The latter process is described as *stick-slip* motion, because the fault exhibits alternating periods of locked behavior followed by sudden slippage. It is estimated that great earthquakes should occur about every 50 to 200 years along those sections of the San Andreas fault that exhibit stick-slip motion. This knowledge is useful when assigning a potential earthquake risk to a given segment of the fault zone.

Not all movement along faults is horizontal. Vertical displacement, in which one side is lifted higher in relation to the other, is also common. Figure 16.6 shows a *fault scarp* (cliff) produced by such vertical movement. Further, many earthquakes occur at such great depths that no displacement is evident at the surface.

Figure 16.6 Fault scarp produced by vertical displacement during the 1964 Alaskan earthquake. (Courtesy of the U.S. Geological Survey)

Box 16.1

Damaging Earthquakes East of the Rockies

The majority of earthquakes occur near plate boundaries, as exemplified by California and Japan. However, areas distant from plate boundaries are not necessarily immune. A team of seismologists recently estimated that the probability of a damaging earthquake east of the Rocky Mountains during the next 30 years is roughly two-thirds as likely as an earthquake of comparable damage in California. Like all earthquake risk assessments, this prediction is based in part on the geographic distribution and average rate of earthquake occurrences in these regions.

At least six major earthquakes have occurred in the central and eastern United States since colonial times. Three of them, having estimated Richter magnitudes of 7.5, 7.3, and 7.8, were centered near the Mississippi River Valley in southeastern Missouri. Occurring over a 3-month period in December 1811, January 1812, and February 1812, these earthquakes and numerous smaller tremors destroyed the town of New Madrid, Missouri. They also triggered massive landslides, damaged a six-state area, altered the course of the Mississippi River, and enlarged Tennessee's Reelfoot Lake.

The distance over which these earthquakes were felt is truly remarkable. Chimneys were downed in Cincinnati and Richmond, and even Boston residents, 1770 kilometers (1100 miles) to the northeast, felt the tremor. Although destruction from the New Madrid earthquakes was slight compared to the Loma Prieta earthquake of 1989, the Midwest in the early 1800s was sparsely populated. Memphis, near the epicenter, had not yet been established, and St. Louis was a small frontier town. Other damaging earthquakes—Aurora, Illinois (1909) and Valentine, Texas (1931)—remind us that the central United States is vulnerable.

The greatest historical earthquake in the eastern states occurred in Charleston, South Carolina, in 1886. This 1-minute event caused 60 deaths, numerous injuries, and great economic

Figure 16.A Damage to Charleston, South Carolina caused by the August 31, 1886, earthquake. Damage ranged from toppled chimneys and broken plaster to total collapse. (Photo courtesy of U.S. Geological Survey)

loss within 200 kilometers (120 miles) of Charleston. Within 8 minutes, strong vibrations shook the upper floors of buildings in Chicago and St. Louis, causing people to rush outdoors. In Charleston alone over a hundred buildings were destroyed, and 90 percent of the remaining structures were damaged. It was difficult to find a chimney that was still standing (Figure 16.A).

New England and adjacent areas have experienced sizable shocks since colonial times including the 1683 quake in Plymouth and the 1755 quake in Cambridge, Massachusetts. Since records have been kept, New York State has experienced over 300 earthquakes large enough to be felt.

These eastern and central earthquakes occur far less frequently than do those in California. Yet the shocks east of the Rockies have generally produced structural damage over a larger area than tremors of similar magnitude in California. The reason is that the underlying bedrock in the central and

eastern United States is older and more rigid. As a result, seismic waves travel greater distances with less attenuation than in the western United States. For similar earthquakes, the region of maximum ground motion in the East may be up to 10 times larger than in the West. Consequently, the higher rate of earthquakes in the West is partly balanced by more widespread damage in the East.

Despite recent geologic history, Memphis, the largest population center in the area of the New Madrid earthquake, lacks adequate provision for earthquakes in its building code. Worse, Memphis rests on unconsolidated floodplain deposits, so its buildings are more susceptible to damage. A 1985 federal study concluded that a 7.6-magnitude earthquake in this area could cause an estimated 2500 deaths, collapse 3000 structures, cause $25 billion in damages, and displace a quarter of a million people in Memphis alone.

Foreshocks and Aftershocks

The intense vibrations of the 1906 San Francisco earthquake lasted about 40 seconds. Although most of the displacement along the fault occurred in this rather short period, additional movements along this and other nearby faults lasted for several days following the main quake. The adjustments that follow a major earthquake often generate smaller earthquakes called **aftershocks**. Although these aftershocks are usually much weaker than the main earthquake, they can sometimes destroy already badly weakened structures. This occurred, for example, during a 1988 earthquake in Armenia. A large aftershock of magnitude 5.8 collapsed many structures that had been weakened by the main tremor.

In addition, small earthquakes called **foreshocks** often precede a major earthquake by days or, in some cases, by as much as several years. Monitoring of these foreshocks has been used as a means of predicting forthcoming major earthquakes, with mixed success. We will consider the topic of earthquake prediction in a later section of this chapter.

Tectonic Forces and Earthquakes

It is important to understand that the tectonic forces creating the strain that was eventually released during the 1906 San Francisco earthquake are still active. Currently, laser beams are used to measure the relative motion between the opposite sides of this fault. These measurements reveal a displacement of 2 to 5 centimeters (1 to 2 inches) per year. Although this seems slow, it produces substantial movement over millions of years. To illustrate, in 30 million years, this rate of displacement would slide the western portion of California northward so that Los Angeles, on the Pacific plate, would be adjacent to San Francisco on the North American plate! More important in the short term, a displacement of just 2 centimeters per year produces 2 meters of offset every 100 years. Consequently, the 4 meters of displacement produced during the 1906 San Francisco earthquake should occur at least every 200 years along this segment of the fault zone. This fact lies behind California's concern for making buildings earthquake-resistant, in anticipation of the inevitable "big one."

 Seismology

The study of earthquake waves, **seismology**, dates back to attempts made by the Chinese almost 2000 years ago to determine the direction from which these waves originated. The seismic instrument used by the

Figure 16.7 Ancient Chinese seismograph. During an Earth tremor, the dragons located in the direction of the main vibrations would drop a ball into the mouths of the frogs below.

Chinese was a large hollow jar that probably contained a mass suspended from the top (Figure 16.7). This suspended mass (similar to a clock pendulum) was connected in some fashion to the jaws of several large dragon figurines that encircled the container. The jaws of each dragon held a metal ball. When earthquake waves reached the instrument, the relative motion between the suspended mass and the jar would dislodge some of the metal balls into the waiting mouths of frogs directly below.

The Chinese were probably aware that the first strong ground motion from an earthquake is directional, and when it is strong enough, all poorly supported items will topple over in the same direction. Apparently the Chinese used this fact plus the position of the dislodged balls to detect the direction to an earthquake's source. However, the complex motion of seismic waves makes it unlikely that the actual direction to an earthquake was determined with any regularity.

In principle at least, modern **seismographs**, instruments that record seismic waves, are not unlike the device used by the early Chinese. Seismographs have a mass freely suspended from a support that is attached to the ground (Figure 16.8). When the vibration from a distant earthquake reaches the instrument,

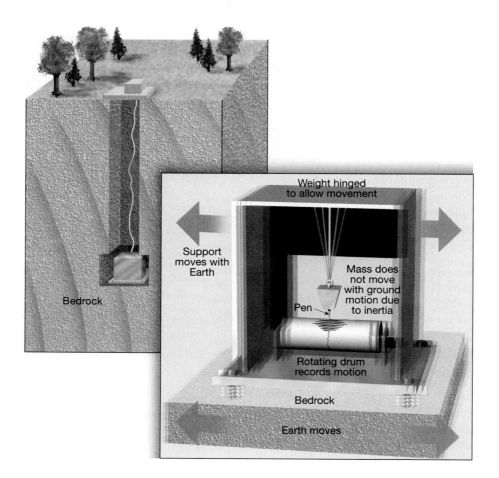

Figure 16.8 Principle of the seismograph. The inertia of the suspended mass tends to keep it motionless, while the recording drum, which is anchored to bedrock, vibrates in response to seismic waves. Thus, the stationary mass provides a reference point from which to measure the amount of displacement occurring as the seismic wave passes through the ground below.

the **inertia*** of the mass keeps it relatively stationary, while Earth and support move. The movement of Earth in relation to the stationary mass is recorded on a rotating drum or magnetic tape.

Earthquakes cause both vertical and horizontal ground motion; therefore, more than one type of seismograph is needed. The instrument shown in Figure 16.8 is designed so that the mass is permitted to swing from side-to-side and thus it detects horizontal ground motion. Usually two horizontal seismographs are employed, one oriented north-south and the other placed with an east-west orientation. Vertical ground motion can be detected if the mass is suspended from a spring, as shown in Figure 16.9.

To detect very weak earthquakes, or a great earthquake that occurred in another part of the world, seismic instruments are typically designed to magnify ground motion. Conversely, some instruments are designed to withstand the violent shaking that occurs very near the earthquake source.

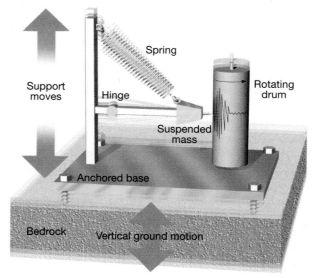

Figure 16.9 Seismograph designed to record vertical ground motion.

The records obtained from seismographs, called **seismograms**, provide a great deal of information concerning the behavior of seismic waves. Simply stated, seismic waves are elastic energy that radiates out in all directions from the focus. The propagation

---

*Inertia: Simply stated, objects at rest tend to stay at rest and objects in motion tend to remain in motion unless either is acted upon by an outside force. You probably have experienced this phenomenon when you tried to stop your automobile quickly and your body continued to move forward.

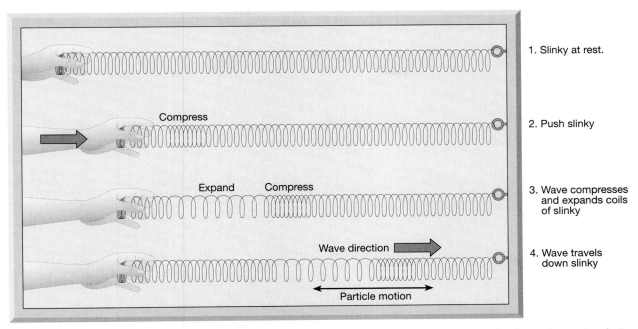

Figure 16.10 A toy slinky is used to illustrate the nature of P waves. Primary, or P waves, are compressional. As a P wave travels, it compresses and expands the material, causing it to vibrate back and forth in a direction parallel to the wave motion.

(transmission) of this energy can be compared to the shaking of gelatin in a bowl, which results as some is spooned out. Whereas the gelatin will have one mode of vibration, seismograms reveal that two main groups of seismic waves are generated by the slippage of a rock mass. One of these wave types travels along the outer part of Earth. These are called **surface waves**. Others travel through Earth's interior and are called **body waves**. Body waves are further divided into two types called **primary** or **P waves** and **secondary** or **S waves**.

Body waves are divided into P and S waves by their mode of travel through intervening materials. P waves are "push-pull" waves—they push (compress) and pull (expand) rocks in the direction the wave is traveling (Figure 16.10). Imagine holding someone by the shoulders and shaking that person. This push-pull movement is how P waves move through Earth. This wave motion is analogous to that generated by human vocal cords as they move air to create sound. Solids, liquids, and gases resist a change in volume when compressed and will elastically spring back once the force is removed. Therefore, P waves, which are compressional waves, can travel through all these materials.

On the other hand, S waves "shake" the particles at right angles to their direction of travel. This can be illustrated by fastening one end of a rope and shaking the other end, as shown in Figure 16.11. Unlike P waves, which temporarily change the *volume* of intervening material by alternately compressing and expanding it, S waves temporarily change the *shape* of

the material that transmits them. Because fluids (gases and liquids) do not respond elastically to changes in shape, they will not transmit S waves.

The motion of surface waves is somewhat more complex. As surface waves travel along the ground, they cause the ground and anything resting upon it to move, much like ocean swells toss a ship. In addition to their up-and-down motion, surface waves have a side-to-side motion similar to an S wave oriented in a horizontal plane. This latter motion is particularly damaging to the foundations of structures.

By observing a "typical" seismic record, as shown in Figure 16.12, you can see a major difference among these seismic waves: P waves arrive at the recording station first; then S waves; and then surface waves. This is a consequence of their speeds. To illustrate, the velocity of P waves through granite within the crust is about 6 kilometers per second. S waves under the same conditions travel at 3.6 kilometers per second. Differences in density and elastic properties of the rock greatly influence the velocities of these waves. Generally, in any solid material, P waves travel about 1.7 times faster than S waves, and surface waves can be expected to travel at 90 percent of the velocity of the S waves.

In addition to velocity differences, also notice in Figure 16.12 that the height, or more correctly, the amplitude, of these wave types varies. The S waves have a slightly greater amplitude than do the P waves, while the surface waves, which cause the greatest destruction, exhibit an even greater amplitude.

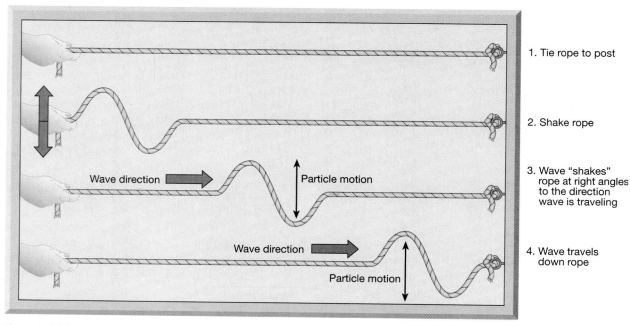

1. Tie rope to post

2. Shake rope

3. Wave "shakes" rope at right angles to the direction wave is traveling

4. Wave travels down rope

Wave direction

Particle motion

Wave direction

Particle motion

Figure 16.11 A rope is used to illustrate the nature of S waves. Secondary, or S waves, are shear waves which cause particles to vibrate at right angles to the direction of wave motion.

Because surface waves are confined to a narrow region near the surface and are not spread throughout Earth as P and S waves are, they retain their maximum amplitude longer. Surface waves also have longer periods (time interval between crests); therefore, they are often referred to as **long waves**, or **L waves**.

As we shall see, seismic waves are useful in determining the location and magnitude of earthquakes. In addition, seismic waves provide a tool for probing Earth's interior.

Locating the Source of an Earthquake

Recall that the *focus* is the place within Earth where earthquake waves originate. The **epicenter** is the location on the surface directly above the focus (see Figure 16.3).

The difference in velocities of P and S waves provides a method for locating the epicenter. The principle used is analogous to a race between two autos, one faster

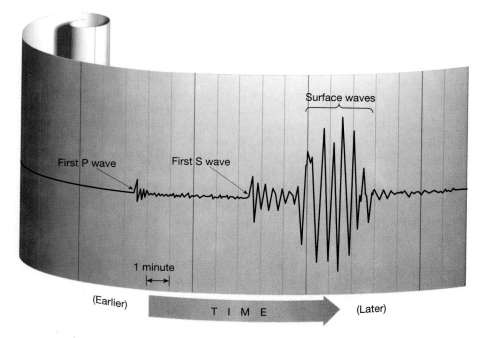

First P wave

First S wave

Surface waves

1 minute

(Earlier)

T I M E

(Later)

Figure 16.12 Typical seismogram. Note the time interval (about 5 minutes) between the arrival of the first P waves and the arrival of the first S waves.

than the other. The P wave always wins the race, arriving ahead of the S wave. But, the greater the length of the race, the greater will be the difference in the arrival times at the finish line (the seismic station). Therefore, the greater the interval measured on a seismogram between the arrival of the first P wave and the first S wave, the greater the distance to the earthquake source.

A system for locating earthquake epicenters was developed by using seismograms from earthquakes whose epicenters could be easily pinpointed from physical evidence. From these seismograms, travel-time graphs were constructed (Figure 16.13). The first travel-time graphs were greatly improved when seismograms became available from nuclear explosions, because the precise location and time of detonation were known.

Using the sample seismogram in Figure 16.12 and the travel-time curves in Figure 16.13, we can determine the distance separating the recording station from the earthquake in two steps: (1) determine the time interval between the arrival of the first P wave and the first S wave, and (2) find on the travel-time graph the equivalent time spread between the P and S wave curves. From this information, we can determine that this earthquake occurred 3800 kilometers (2350 miles) from the recording instrument.

Now we know the *distance*, but what *direction*? The epicenter could be in any direction from the seismic

station. As shown in Figure 16.14, the precise location can be found when the distance is known from three or more different seismic stations. On a globe, we draw a circle around each seismic station. Each circle represents the epicenter distance for each station. The point where the three circles intersect is the epicenter of the quake. This method is called *triangulation*.

The study of earthquakes was greatly bolstered during the 1960s through efforts to discriminate between underground nuclear explosions and natural earthquakes. The United States established a worldwide network of over 100 seismic stations coordinated through Golden, Colorado. The largest of these, located in Billings, Montana, consists of an array of 525 instruments grouped in 21 clusters covering a region 200 km in diameter. Using data from this array, seismologists employing high-speed computers are able to locate an epicenter by a trial-and-error technique.

 Earthquake Belts

About 95 percent of the energy released by earthquakes originates in a few relatively narrow zones that wind around the globe (Figure 16.15). The greatest energy is released along a path around the outer edge of the Pacific Ocean known as the *circum-Pacific belt*. Included in this zone are regions of great seismic activity such as Japan, the Philippines, Chile, and numerous volcanic island chains, as exemplified by the Aleutian Islands.

Another major concentration of strong seismic activity runs through the mountainous regions that flank the Mediterranean Sea and continues through Iran and on past the Himalayan complex. Figure 16.15 indicates that yet another continuous belt extends for thousands of kilometers through the world's oceans. This zone coincides with the oceanic ridge system, which is an area of frequent but low-intensity seismic activity.

The areas of the United States included in the circum-Pacific belt lie adjacent to California's San Andreas fault (Figure 16.16) and along the western coastal regions of Alaska, including the Aleutian Islands. In addition to these high-risk areas, other sections of the United States are regarded as regions where strong earthquakes are likely to occur.

One of these regions extends from southern Illinois southward along the Mississippi River (see Box 16.1). This region is the site of three strong shocks that devastated the town of New Madrid, Missouri, in 1811–1812. The drainage of the Mississippi was also altered by these quakes, and Tennessee's Reelfoot Lake was enlarged. Although the total amount of structural damage caused by the New Madrid earthquakes was slight, it should be remembered that in the early 1800s this was a sparsely populated frontier

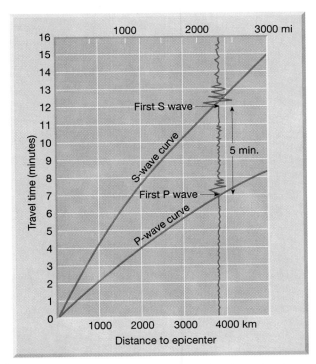

Figure 16.13 A travel-time graph is used to determine the distance to the epicenter. The difference in arrival times of the first P and S waves in the example is 5 minutes. Thus, the epicenter is roughly 3800 kilometers (2350 miles) away.

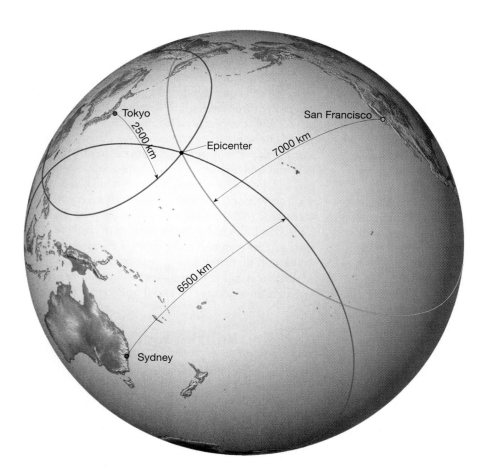

Figure 16.14 Earthquake epicenter is located using the distances obtained from three or more seismic stations.

region. A similar earthquake in this region today would be truly catastrophic.

Another strong earthquake centered away from the active circum-Pacific belt occurred August 31, 1886, in Charleston, South Carolina. The event, which spanned one minute, caused 60 deaths, numerous injuries, and great economic loss within a radius of 200 kilometers (120 miles) of Charleston. Within 8

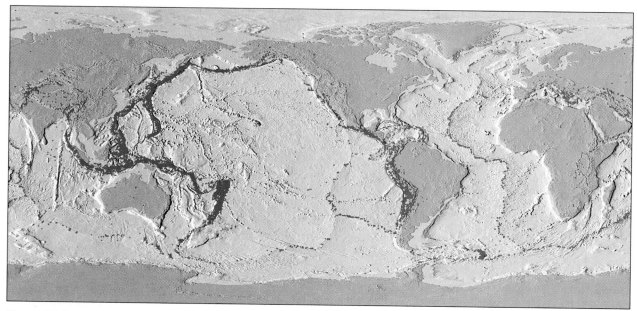

Figure 16.15 Distribution of the 14,229 earthquakes with magnitudes equal to or greater than 5 for the period 1980–1990. (Data from National Geophysical Data Center/NOAA)

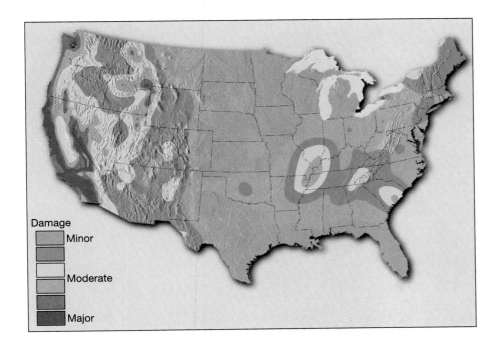

Figure 16.16 This U.S. Geological Survey shaking-hazard map is based on current information about the rate at which earthquakes occur in different areas and on how far strong shaking extends from quake sources. Colors show the levels of horizontal shaking that have a 1-in-10 chance of being exceeded in a 50-year period.

Damage
Minor
Moderate
Major

minutes, the effects of the vibrations were felt as far away as Chicago, Illinois, and St. Louis, Missouri, where strong vibrations on upper floors of buildings caused people to rush outdoors. According to reports made shortly after the earthquake, not a single building in Charleston escaped damage, and only a few buildings avoided serious damage.

Earthquake Depths

Evidence from seismic records reveals that earthquakes originate at depths ranging from 5 to nearly 700 kilometers. In a somewhat arbitrary fashion, earthquake foci have been classified by their depth of occurrence. Those with points of origin within 70 kilometers of the surface are referred to as *shallow*, while those generated between 70 and 300 km are considered *intermediate*, and those with a focus greater than 300 km are classified as *deep*. About 90 percent of all earthquakes occur at depths of less than 100 km, and nearly all very damaging earthquakes appear to originate at shallow depths.

For example, the 1906 San Francisco earthquake involved movement within the upper 15 km of Earth's crust, whereas the 1964 Alaskan earthquake had a focal depth of 33 km. Seismic data reveal that while shallow-focus earthquakes have been recorded with Richter magnitudes of 8.6, the strongest intermediate-depth quakes have had values below 7.5, and deep-focus earthquakes have not exceeded 6.9 in magnitude.

When earthquake data were plotted according to geographic location and depth, several interesting observations were noted. Rather than a random mixture of shallow and deep earthquakes, some very definite distribution patterns emerged. Earthquakes generated along the oceanic ridge system always have a shallow focus and none are very strong. Further, it was noted that almost all deep-focus earthquakes occurred in the circum-Pacific belt, particularly in regions situated landward of deep-ocean trenches.

In a study conducted in the southwestern Pacific near the Tonga trench, it was discovered that foci depths increased with distance from the trench, as shown in Figure 16.17. These seismic regions, called **Wadati-Benioff zones** after the two scientists who were the first to extensively study them, are oriented about 35 to nearly 90 degrees to the surface. Why should earthquakes be oriented along a narrow zone which plunges almost 700 kilometers into Earth's interior? We will consider this question further in Chapter 19.

Earthquake Intensity and Magnitude

Until a century ago, earthquake size and strength were described subjectively, making accurate classification of earthquake intensity difficult. Then, in 1902, Giuseppe Mercalli developed a fairly reliable intensity scale based on damage to various types of structures. The U.S. Coast and Geodetic Survey uses a modification of this scale today (Table 16.1).

The **Mercalli intensity scale** assesses the damage from a quake at a specific location. Please note that earthquake intensity depends not only on

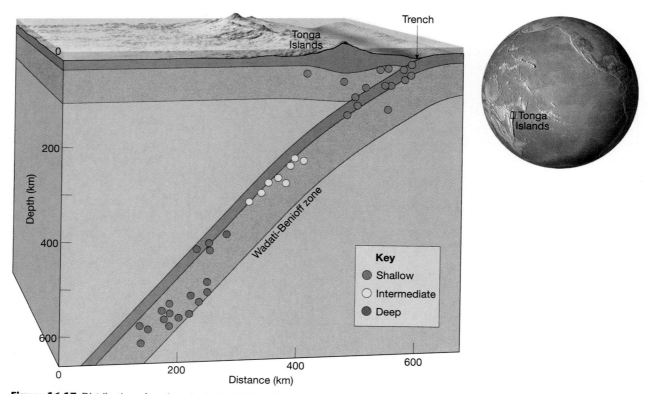

Figure 16.17 Distribution of earthquake foci in 1965 in the vicinity of the Tonga Islands. (Data from B. Isacks, J. Oliver, and L.R. Sykes)

the strength of the earthquake but also on other factors, such as distance from the epicenter, the nature of surface materials, and building design. A modest 6.9-Richter-magnitude earthquake in Armenia in 1988 was very destructive, mainly because of inferior building construction. A 1985 Mexico City quake was deadly because of the soft sediment upon which the city rests. Thus, the destruction wrought by earthquakes is very meaningful to people living there, but it is not a true measure of the earthquake's actual strength. Further, many earthquakes occur beneath the sea or at great depths in the crust and are not felt.

Today, earthquakes are ranked according to their **magnitude**, a measure of the amount of energy released during the event. Ideally, the magnitude of an earthquake can be determined from the amount of material that slides along the fault and the distance it is displaced. However, even in an ideal setting such as that of the 1906 San Francisco earthquake, where the fault trace is visible and displacement can be measured from physical evidence, this method can provide only a crude estimate of the forces involved. In most earthquakes, the fault does not penetrate the surface; therefore, the amount of displacement cannot be measured directly.

In 1935, Charles Richter of the California Institute of Technology attempted to rank the earthquakes of southern California into groups of large,

Table 16.1 Modified Mercalli Intensity Scale

I	Not felt except by a very few under especially favorable circumstances.
II	Felt only by a few persons at rest, especially on upper floors of buildings.
III	Felt quite noticeably indoors, especially on upper floors of buildings, but many people do not recognize it as an earthquake.
IV	During the day felt indoors by many, outdoors by few. Sensation like heavy truck striking building.
V	Felt by nearly everyone, many awakened. Disturbances of trees, poles, and other tall objects sometimes noticed.
VI	Felt by all; many frightened and run outdoors. Some heavy furniture moved; few instances of fallen plaster or damaged chimneys. Damage slight.
VII	Everybody runs outdoors. Damage negligible in buildings of good design and construction; slight to moderate in well-built ordinary structures; considerable in poorly built or badly designed structures.
VIII	Damage slight in specially designed structures; considerable in ordinary substantial buildings with partial collapse; great in poorly built structures. (Fall of chimneys, factory stacks, columns, monuments, walls.)
IX	Damage considerable in specially designed structures. Buildings shifted off foundations. Ground cracked conspicuously.
X	Some well-built wooden structures destroyed. Most masonry and frame structures destroyed. Ground badly cracked.
XI	Few, if any (masonry) structures remain standing. Bridges destroyed. Broad fissures in ground.
XII	Damage total. Waves seen on ground surfaces. Objects thrown upward into air.

medium, and small magnitude. The system he developed determines earthquake magnitudes from the deflections recorded on seismograms.

Today a refined **Richter scale** is used worldwide to describe earthquake magnitude. Richter magnitude is determined by measuring the amplitude of the largest wave recorded on the seismogram (Figure 16.18). For seismic stations worldwide to obtain the same magnitude for a given earthquake, adjustments must be made for the weakening of the seismic waves as they move from the focus, and for the sensitivity of the recording instrument. Richter established 100 kilometers as the standard distance and the Wood-Anderson instrument as the standard recording device.

The largest earthquakes ever recorded had Richter magnitudes of 8.9. These great shocks released approximately 10^{26} ergs of energy—roughly equivalent to the detonation of 1 billion tons of TNT. Apparently, earthquakes having a Richter magnitude greater than 9 do not occur. Conversely, earthquakes with a Richter magnitude of less than 2.0 are usually not felt by humans. With the advent of more sensitive instruments, tremors of a magnitude of <2 have been recorded. Table 16.2 shows how earthquake magnitudes and their effects are related.

Earthquakes vary enormously in strength, and great earthquakes produce traces having wave amplitudes that are thousands of times larger than those generated by weak tremors. To accommodate this wide variation, Richter could not use a linear scale, but instead used a *logarithmic scale* to express magnitude. On this scale a *tenfold* increase in wave amplitude corresponds to an increase of 1 on the magnitude scale. Thus, the amplitude of the largest surface wave for a 5-magnitude earthquake is 10 times greater than the wave amplitude produced by an earthquake having a magnitude of 4.

More important, each unit of Richter magnitude equates to roughly a *32-fold energy increase.* Thus, an earthquake with a magnitude of 6.5 release 32 times more energy than one with a magnitude of 5.5, and roughly 1000 times more energy than a 4.5-magnitude quake (Table 16.3).

A major earthquake with a magnitude of 8.5 releases millions of times more energy than the smallest earthquakes felt by humans. This should dispel the notion that a moderate earthquake decreases the chances for the occurrence of a major quake in the same region. Thousands of moderate tremors would be needed to release the vast amount of energy equal to one "great" earthquake.

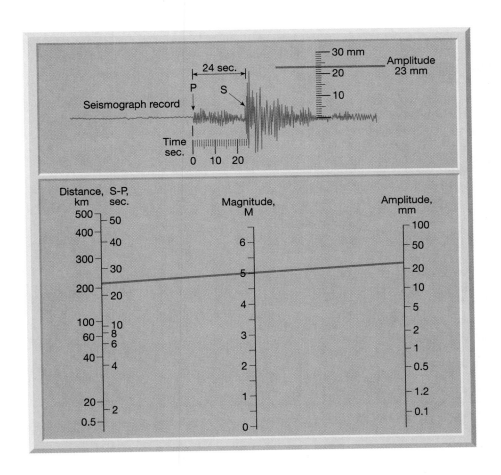

Figure 16.18 Illustration showing how the Richter magnitude of an earthquake can be determined graphically using a seismograph record from a Wood-Anderson instrument. (Data from California Institute of Technology)

Table 16.2 Earth Magnitudes and Expected World Incidence

Richter Magnitudes	Effects Near Epicenter	Estimated Number per Year
<2.0	Generally not felt, but recorded.	600,000
2.0–2.9	Potentially perceptible.	300,000
3.0–3.9	Felt by some.	49,000
4.0–4.9	Felt by most.	6200
5.0–5.9	Damaging shocks.	800
6.0–6.9	Destruction in populous regions.	266
7.0–7.9	Major earthquakes. Inflict serious damage.	18
≥8.0	Great earthquakes. Cause extensive destruction to communities near epicenter.	1.4

Researchers have shown that the Richter scale does not adequately differentiate those earthquakes that have very high magnitudes. Because all of the very strongest quakes have nearly equal wave amplitudes, the Richter scale becomes saturated at this level. Consequently, other methods of establishing the relative strengths of earthquakes were devised.

One method, called *moment magnitude*, analyzes very long-period seismic waves for this purpose. On this extended scale, the 1906 San Francisco earthquake with a surface wave magnitude of 8.3 would be demoted to 7.9, whereas the 1964 Alaskan earthquake with an 8.3–8.4 Richter magnitude would be increased to 9.2. Using moment magnitude, the strongest earthquake on record is the 1960 Chilean earthquake with a magnitude of 9.5.

Earthquake Destruction

The most violent earthquake to jar North America this century—the Good Friday Alaskan earthquake—occurred at 5:36 P.M. on March 27, 1964. Felt throughout that state, the earthquake had a magnitude of 8.3–8.4 on the Richter scale and reportedly lasted 3 to 4 minutes. This brief event left 131 people dead, thousands homeless, and the economy of the state badly disrupted. Had the schools and business districts been open, the toll surely would have been higher. Within 24 hours of the initial shock, 28 aftershocks were recorded, 10 of which exceeded a magnitude of 6 on the Richter scale. The location of the epicenter and the towns that were hardest hit by the quake are shown in Figure 16.19.

Many factors determine the degree of destruction that will accompany an earthquake. The most obvious is the magnitude of the earthquake and its proximity to a populated area. Fortunately, most earthquakes are small and occur in remote regions of Earth. However, about 20 major earthquakes are reported annually, one or two of which can be catastrophic.

During an earthquake, the region within 20 to 50 kilometers (12.5 to 30 miles) of the epicenter ordinarily will experience roughly the same degree of ground shaking, but beyond this limit the vibration deteriorates rapidly. Occasionally, during earthquakes

Table 16.3 Earthquake Magnitude and Energy Equivalence

Earthquake Magnitude	Energy Released* (Millions of Ergs)	Approximate Energy Equivalence
0	630,000	1 pound of explosives
1	20,000,000	
2	630,000,000	Energy of lightning bolt
3	20,000,000,000	
4	630,000,000,000	1000 pounds of explosives
5	20,000,000,000,000	
6	630,000,000,000,000	1946 Bikini atomic bomb test
		1994 Northridge Earthquake
7	20,000,000,000,000,000	1989 Loma Prieta Earthquake
8	630,000,000,000,000,000	1906 San Francisco Earthquake
		1980 Eruption of Mount St. Helens
9	20,000,000,000,000,000,000	1964 Alaskan Earthquake
		1960 Chilean Earthquake
10	630,000,000,000,000,000,000	Annual U.S. energy consumption

*For each unit increase in magnitude, the energy released increases about 31.6 times.
SOURCE: U.S. Geological Survey.

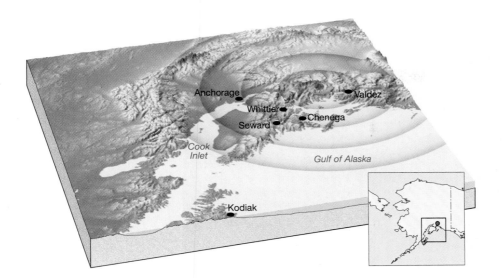

Figure 16.19 Region most affected by the Good Friday earthquake of 1964. Note the location of the epicenter (red dot). (After U.S. Geological Survey)

that occur in the stable continental interior, such as the New Madrid earthquake of 1811, the area of influence can be much larger. The epicenter of this earthquake was located directly south of Cairo, Illinois, and the vibrations were felt from the Gulf of Mexico to Canada, and from the Rockies to the Atlantic seaboard.

Destruction from Seismic Vibrations

The 1965 Alaskan earthquake provided geologists with new insights into the role of ground shaking as a destructive force. As the energy released by an earthquake travels along Earth's surface, it causes the ground to vibrate in a complex manner by moving up and down as well as from side to side. The amount of structural damage attributable to the vibrations depends on several factors, including (1) the intensity and (2) duration of the vibrations; (3) the nature of the material upon which the structure rests; and (4) the design of the structure.

All of the multistory structures in Anchorage were damaged by the vibrations. The more flexible wood frame residential buildings fared best. However, many homes were destroyed when the ground failed. A striking example of how construction variations affect earthquake damage is shown in Figure 16.20. You can see that the steel-frame building on the left withstood the vibrations, whereas the poorly designed J.C. Penney building was badly damaged. Engineers have learned that unreinforced masonry buildings are the most serious safety threat in earthquakes.

Most large structures in Anchorage were damaged, even though they were built according to the earthquake provisions of the Uniform Building Code. Perhaps some of that destruction can be attributed

to the unusually long duration of this earthquake. Most quakes consist of tremors that last less than a minute. For example, the 1994 Northridge earthquake was felt for about 40 seconds, and the strong vibrations of the 1989 Loma Prieta earthquake lasted less than 15 seconds. But the Alaska quake reverberated for 3 to 4 minutes.

Amplification of Seismic Waves. Although the region within 20 to 50 kilometers of the epicenter will experience about the same intensity of ground shaking, the destruction varies considerably within this area. This difference is mainly attributable to the nature of the ground on which the structures are built. Soft sediments, for example, generally amplify the vibrations more than solid bedrock. Thus, the buildings located in Anchorage, which were situated on unconsolidated sediments, experienced heavy structural damage. By contrast, most of the town of Whittier, although much nearer the epicenter, rests on a firm foundation of granite and hence suffered much less damage. However, Whittier was damaged by a seismic sea wave (described in the next section).

The 1985 Mexican earthquake gave seismologists and engineers a vivid reminder of what had been learned from the 1964 Alaskan earthquake. The Mexican coast, where the earthquake was centered, experienced unusually mild tremors despite the strength of the quake. As expected, the seismic waves became progressively weaker with increasing distance from the epicenter. However, in the central section of Mexico City, nearly 400 kilometers from the source, the vibrations intensified to five times that experienced in outlying districts. Much of this amplified ground motion can be attributed to soft sediments, remnants of an ancient lake bed, that underlie portions of the city (see Box 16.2).

Figure 16.20 Damage caused to the five-story J.C. Penney Co. building, Anchorage, Alaska. Very little structural damage was incurred by the adjacent building. (Courtesy of NOAA)

Liquefaction. In areas where unconsolidated materials are saturated with water, earthquake vibrations can generate a phenomenon known as **liquefaction**. Under these conditions, what had been a stable soil turns into a mobile fluid that is not capable of supporting buildings or other structures (Figure 16.21). As a result, underground objects such as storage tanks and sewer lines may literally float toward the surface of their newly liquefied environment. Buildings and other structures may settle and collapse. During the 1989 Loma Prieta earthquake, in San Francisco's Marina District, foundations failed and geysers of sand and water shot from the ground, indicating that liquefaction had occurred (Figure 16.22).

Figure 16.21 Effects of liquefaction. This tilted building rests on unconsolidated sediment that behaved like quicksand during the 1985 Mexican earthquake. (Photo by James L. Beck)

Figure 16.22 These "mud volcanoes" were produced by the Loma Prieta earthquake of 1989. They formed when geysers of sand and water shot from the ground, an indication that liquefaction occurred. (Photo by Richard Hilton, courtesy of Dennis Fox)

Seiches. The effects of great earthquakes may be felt thousands of kilometers from their source. Ground motion may generate *seiches*, the rhythmic sloshing of water in lakes, reservoirs, and enclosed basins such as the Gulf of Mexico. The 1964 Alaskan earthquake, for example, generated 2-meter waves off the coast of Texas, which damaged small craft while much smaller waves were noticed in swimming pools in both Texas and Louisiana.

Seiches can be particularly dangerous when they occur in reservoirs retained by earthen dams. These waves have been known to slosh over reservoir walls and weaken the structure, thereby endangering the lives of those downstream.

Tsunami

Most deaths associated with the 1964 Alaskan quake were caused by **seismic sea waves**, or **tsunami.***

---

*Seismic sea waves were given the name *tsunami* by the Japanese, who have suffered a great deal from them. The term *tsunami* is now used worldwide.

These destructive waves often are called "tidal waves" by the media. However, this name is inappropriate, for these waves are generated by earthquakes, not the tidal effect of the Moon or Sun.

Most tsunami result from vertical displacement of the ocean floor during an earthquake (Figure 16.23). Once created, a tsunami resembles the ripples formed when a pebble is dropped into a pond. In contrast to ripples, tsunami advance across the ocean at amazing speeds between 500 and 950 kilometers per hour. Despite this striking characteristic, a tsunami in the open ocean can pass undetected because its height is usually less than 1 meter and the distance between wave crests is great, ranging from 100 to 700 kilometers. However, upon entering shallower coastal waters, these destructive waves are slowed down and the water begins to pile up to heights that occasionally exceed 30 meters (Figure 16.23). As the crest of a tsunami approaches the shore, it appears as a rapid rise in sea level with a turbulent and chaotic surface. Tsunami can be very destructive (Figure 16.24).

Usually the first warning of an approaching tsunami is a relatively rapid withdrawal of water from beaches. Coastal residents have learned to heed this warning and move to higher ground, for about 5 to 30 minutes later, the retreat of water is followed by a surge capable of extending hundreds of meters inland. In a successive fashion, each surge is followed by rapid oceanward retreat of the water.

Tsunami are able to traverse large stretches of the ocean before their energy is totally dissipated. The tsunami generated by the 1960 Chilean earthquake, in addition to completely destroying villages along an 800-kilometer stretch of coastal South America, traveled 17,000 kilometers across the Pacific to Japan. Here, about 22 hours after the quake, considerable destruction was inflicted upon southern coastal villages of Honshu, the major island of Japan. For several days afterward, tidal gauges located in Hilo, Hawaii, were able to detect these diminishing waves as they bounced about the Pacific.

The tsunami generated in the 1964 Alaskan earthquake inflicted heavy damage to the communities in the vicinity of the Gulf of Alaska, completely destroying the town of Chenega. Kodiak was also heavily damaged and most of its fishing fleet destroyed when a seismic sea wave carried many vessels into the business district. The deaths of 107 persons have been attributed to this tsunami. By contrast, only 9 people died in Anchorage as a direct result of the vibrations.

Tsunami damage following the Alaskan earthquake extended along much of the west coast of North America, and despite a 1-hour warning, 12 people perished in Crescent City, California, where all of the deaths and most of the destruction were caused by the

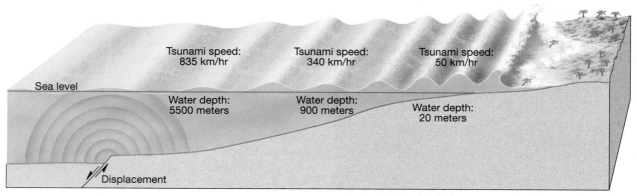

Figure 16.23 Schematic drawing of a tsunami generated by displacement of the ocean floor. The speed of a wave correlates with ocean depth. As shown, waves moving in deep water advance at speeds in excess of 800 kilometers per hour. Speed gradually slows to 50 kilometers per hour at depths of 20 meters. Decreasing depth slows the movement of the wave. As waves slow in shallow water, they grow in height until they topple and rush onto shore with tremendous force. The size and spacing of these swells are not to scale.

fifth wave. The first wave crested about 4 meters (13 feet) above low tide and was followed by three progressively smaller waves. Believing that the tsunami had ceased, people returned to the shore, only to be met by the fifth and most devastating wave, which, superimposed upon high tide, crested about 6 meters higher than the level of low tide.

In 1946, a large tsunami struck the Hawaiian Islands without warning. A wave more than 15 meters high left several coastal villages in shambles. This destruction motivated the United States Coast and Geodetic Survey to establish a tsunami warning system for the coastal areas of the Pacific. From seismic observatories throughout the region, warnings of large earthquakes are reported to the Tsunami Warning Center near Honolulu. Using tidal gauges, a determination is made as to whether a tsunami has been formed. Within an hour a warning is issued. Although tsunamis travel very rapidly, there is sufficient time to evacuate all but the region nearest the epicenter (Figure 16.25).

Figure 16.24 A man stands before a wall of water about to engulf him at Hilo, Hawaii, on April 1, 1946. This tsunami, which originated in the Aleutian Islands near Alaska, was still powerful enough when it hit Hawaii to rise 9 to 16 meters (30 to 55 feet). The *S.S. Brigham Victory*, from which this photograph was taken, managed to survive the onslaught, but 159 people in Hawaii, including the man seen here, were killed. (Photo courtesy of Water Resources Center Archives, University of California, Berkeley)

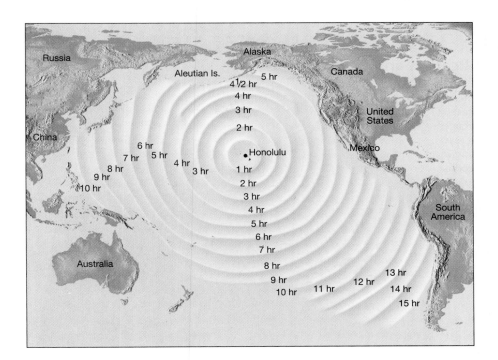

Figure 16.25 Tsunami travel times to Honolulu, Hawaii from locations throughout the Pacific. (From NOAA)

For example, a tsunami generated near the Aleutian Islands would take 5 hours to reach Hawaii, and one generated near the coast of Chile would travel 15 hours before reaching Hawaii. Fortunately, most earthquakes do not generate tsunami. On the average, only about 1.5 destructive tsunami are generated worldwide each year. Of these, only about one every 10 years is catastrophic.

Landslides and Ground Subsidence

In the 1964 Alaskan earthquake, the greatest damage to structures was from landslides and ground subsidence triggered by the vibrations. At Valdez and Seward, the violent shaking caused deltaic materials to experience liquefaction; the subsequent slumping carried both waterfronts away. Because of the threat of recurrence, the entire town of Valdez was relocated about 7 kilometers away on more stable ground. In Valdez, 31 people on a dock died when it slid into the sea.

Most of the damage in the city of Anchorage was also attributed to landslides. Many homes were destroyed in Turnagain Heights when a layer of clay lost its strength and over 200 acres of land slid toward the ocean (Figure 16.26). A portion of this spectacular landslide was left in its natural condition as a reminder of this destructive event. The site was appropriately named "Earthquake Park." Downtown Anchorage was also disrupted as sections of the main business district dropped by as much as 3 meters (10 feet).

Fire

The 1906 earthquake in San Francisco reminds us of the formidable threat of fire. The central city contained mostly large, older wooden structures and brick buildings. Although many of the unreinforced brick buildings were extensively damaged by vibrations, the greatest destruction was caused by fires, which started when gas and electrical lines were severed. The fires raged out of control for 3 days and devastated over 500 blocks of the city (see Figure 16.2). The problem was compounded by the initial ground shaking, which broke the city's water lines into hundreds of unconnected pieces.

The fire was finally contained when buildings were dynamited along a wide boulevard to provide a fire break, the same strategy used in fighting a forest fire. Although only a few deaths were attributed to the San Francisco fire, such is not always the case. A 1923 earthquake in Japan triggered an estimated 250 fires, which devastated the city of Yokohama and destroyed more than half the homes in Tokyo. Over 100,000 deaths were attributed to the fires, which were driven by unusually high winds.

Can Earthquakes Be Predicted?

The vibrations that shook Northridge, California, in 1994 inflicted 61 deaths and about $15 billion in damage (Figure 16.27). This was from a brief earthquake (about 40 seconds) of moderate rating (6.6–6.9

Box 16.2

Wave Amplification and Seismic Risks

Much of the damage and loss of life in the 1985 Mexico City earthquake occurred because downtown buildings were constructed on lake sediment that greatly amplified the ground motion. To understand why this happens, recall that as seismic waves pass through Earth, they cause the intervening material to vibrate much as a tuning fork when it is struck. Although most objects can be "forced" to vibrate over a wide range of frequencies, each has a natural period of vibration that is preferred. Different Earth materials, like different-length tuning forks, also have different natural periods of vibration.*

Ground-motion amplification results when the supporting material has a natural period of vibration (frequency), which matches that of the seismic waves. A common example of this phenomenon occurs when a parent pushes a child on a swing. When the parent periodically pushes the child in rhythm with the frequency of the swing, the child moves back and forth in a greater and greater arc (amplitude). By chance, the column of sediment beneath Mexico City had a natural period of vibration of about 2 seconds, matching that of the strongest seismic waves. Thus, when the seismic waves began shaking the soft sediments, a *resonance* developed, which greatly increased the amplitude of the vibrations. This amplification resulted in vibrations that exhibited 40 centimeters (1.3 feet) of back-and-forth ground motion every 2 seconds for nearly 2 minutes. Such movement was too intense for many poorly designed buildings in the city. In addition, intermediate-height structures (5 to 15 stories) sway back and forth with a period of about 2 seconds. Thus, resonance also developed between these buildings and the ground, with the result that most of the building failures occurred to structures in this height range (Figure 16.B).

Sediment-induced wave amplificaiton is also thought to have contributed significantly to the failure of the two-tiered Cypress section of Interstate 880 during the 1989 Loma Prieta earthquake (Figure 16.C). Studies

Figure 16.B During the 1985 Mexican earthquake, multistory buildings swayed back and forth as much as 1 meter. Many, including the hotel shown here, collapsed or were seriously damaged. (Photo by James L. Beck)

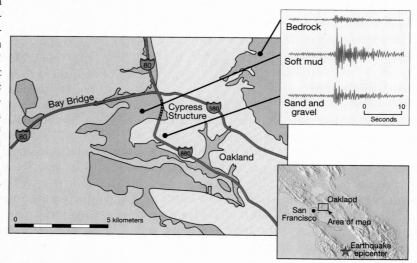

Figure 16.C The portion of the Cypress Freeway structure in Oakland, California, that stood on soft mud (dashed red line) collapsed during the 1989 Loma Prieta earthquake. Adjacent parts of the structure (solid red) that were built on firmer ground remained standing. Seismograms from an aftershock (upper right) show that shaking is greatly amplified in the soft mud as compared to the firmer materials.

conducted on the 1.4-kilometer section that did collapse showed that it was built on San Francisco Bay mud. Another section of this interstate that was damaged but did not collapse was constructed on firmer alluvial materials.

*To demonstrate the natural period of vibration of an object, hold a ruler over the edge of a desk so that most of it is not supported by the desk. Start it vibrating and notice the noise it makes. By changing the length of the unsupported portion of the ruler, the natural period of vibration will change accordingly.

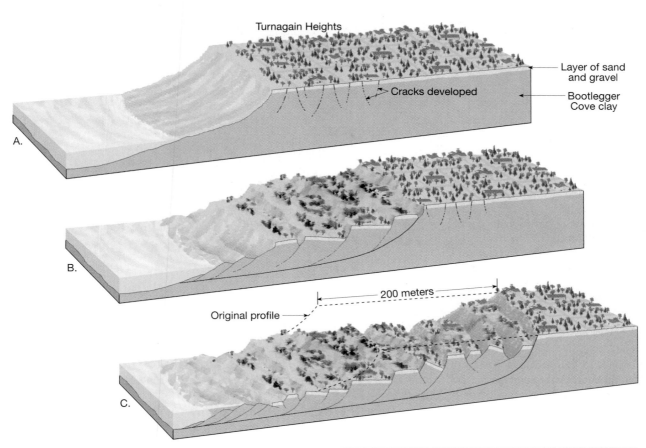

Figure 16.26 Turnagain Heights slide caused by the 1964 Alaskan earthquake. **A.** Vibrations from the earthquake caused cracks to appear near the edge of the bluff. **B.** Within seconds blocks of land began to slide toward the sea on a weak layer of clay. **C.** In less than 5 minutes, as much as 200 meters of the Turnagain Heights bluff area had been destroyed. **D.** Photo of a small portion of the Turnagain Heights slide. (Photo courtesy of U.S. Geological Survey)

D.

on the Richter scale). Seismologists warn that earthquakes of comparable or greater strength will occur along the San Andreas fault, which cuts a 1300-kilometer (800-mile) path through the state. Following the 1995 earthquake in Kobe, Japan, a U.S. Geological Survey physicist cautioned: "Kobe is almost a dress rehearsal for an earthquake on the Hayward fault (a fault parallel to the San Andreas, near San Francisco)." The obvious question is: Can earthquakes be predicted?

Short-Range Predictions

The goal of short-range earthquake prediction is to provide a warning of the location and magnitude of a large earthquake within a narrow time frame. Substantial efforts to achieve this objective are being put forth in Japan, the United States, China, and Russia—countries where earthquake risks are high (Table 16.4). This research has concentrated on monitoring possible *precursors*—phenomena that precede and thus provide a warning of a forthcoming earth-

Figure 16.27 Damage to Interstate 5 during the January 17, 1994, Northridge earthquake. (Photo by John Barr/Gamma Liaison)

quake. In California, for example, seismologists are measuring uplift, subsidence, and strain in the rocks near active faults. Some Japanese scientists are studying anomalous animal behavior that may precede a quake. Other researchers are monitoring changes in groundwater levels, while still others are trying to predict earthquakes based on changes in the electrical conductivity of rocks.

Among the most ambitious earthquake experiments is one being conducted along a segment of the

Table 16.4 Some Notable Earthquakes

Year	Location	Deaths (est.)	Magnitude	Comments
1290	Chihi (Hopei), Chile	100,000		
1556	Shensi, China	830,000		Possibly the greatest natural disaster.
1737	Calcutta, India	300,000		
1755	Lisbon, Portugal	70,000		Tsunami damage extensive.
*1811–1812	New Madrid, Missouri	Few		Three major earthquakes.
*1886	Charleston, South Carolina	60		Greatest historical earthquake in the eastern United States.
*1906	San Francisco, California	1500	8.1–8.2	Fires caused extensive damage.
1908	Messina, Italy	120,000		
1920	Kansu, China	180,000		
1923	Tokyo, Japan	143,000	7.9	Fire caused extensive destruction.
1960	Southern Chile	5700	8.5–8.6	Possibly the largest-magnitude earthquake ever recorded.
*1964	Alaska	131	8.3–8.4	
1970	Peru	66,000	7.8	Great rockslide.
*1971	San Fernando, California	65	6.5	Damage exceeded $1 billion.
1975	Liaoning Province, China	Few	7.5	First major earthquake to be predicted.
1976	Tangshan, China	240,000	7.6	Not predicted.
1985	Mexico City	9500	8.1	Major damage occurred 400 km from epicenter.
1988	Armenia	25,000	6.9	Poor construction practices contributed to destruction.
*1989	San Francisco Bay area	62	7.1	Damages exceeded $6 billion.
1990	Northwestern Iran	50,000	7.3	Landslides and poor construction practices caused great damage.
*1994	Northridge, California	61	6.7	Damages in excess of $15 billion.
1995	Kobe, Japan	5472	6.9	Damage estimated to exceed $100 billion.

*U.S. earthquakes.
SOURCE: U.S. National Oceanic and Atmospheric Administration.

San Andreas fault near the town of Parkfield in central California. Here, earthquakes of moderate intensity have occurred on a regular basis about once every 22 years since 1857. The most recent rupture was a 5.6-magnitude quake that occurred in 1966. With the next event already significantly "overdue," the U.S. Geological Survey has established an elaborate monitoring network. Included are creepmeters, tiltmeters, and bore-hole strain meters that are used to measure the accumulation and release of strain. Moreover, 70 seismographs of various designs have been installed to record foreshocks as well as the main event. Finally, a network of distance-measuring devices that employ lasers measures movement across the fault (Figure 16.28). The object is to identify ground movements that may precede a sizable rupture.

One claim of a successful short-range prediction was made by Chinese seismologists after the February 4, 1975, earthquake in Liaoning Province. According to reports, very few people were killed, although more than 1 million lived near the epicenter, because the earthquake was predicted and the population was evacuated. Recently, some Western seismologists have questioned this claim and suggest instead that an intense swarm of foreshocks, which began 24 hours before the main earthquake, may have caused many people to evacuate spontaneously. Further, an official Chinese government report issued 10 years later stated that 1328 people died and 16,980 injuries resulted from this earthquake.

One year after the Liaoning earthquake, at least 240,000 people died in the Tangshan, China, earthquake, which was not predicted. The Chinese have also issued false alarms. In a province near Hong Kong, people reportedly left their dwellings for over a month, but no earthquake followed. Clearly, whatever method the Chinese employ for short-range predictions, it is *not* reliable.

For a prediction scheme to warrant general acceptance, it must be both accurate and reliable. Thus, *it must have a small range of uncertainty as regards to location and timing, and it must produce few failures, or false alarms.* Can you imagine the debate that would precede an order to evacuate a large city in the United States, such as Los Angeles or San Francisco? The cost of evacuating millions of people, arranging for living accommodations, and providing for their lost work time and wages, would be staggering.

Currently, *no reliable method* for predicting earthquakes exists. In fact, except for a brief period of optimism during the 1970s, the leading seismologists of the past 100 years have generally concluded that short-range earthquake prediction is *not* feasible. To quote Charles Richter, developer of the well-known magnitude scale,

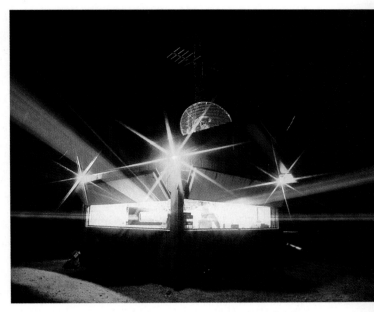

Figure 16.28 Lasers used to measure movement along the San Andreas fault. (Photo by John K. Nakata/U.S. Geological Survey)

"Prediction provides a happy hunting ground for amateurs, cranks, and outright publicity-seeking fakers." This statement was validated in 1990 when Iben Browning, a self-proclaimed expert, predicted that a major earthquake on the New Madrid fault would devastate an area around southeast Missouri on December 2 or 3. Many people in Missouri, Tennessee, and Illinois rushed out to buy earthquake insurance. Some schools and factories closed, while people as far away as northern Illinois stayed home rather than risk traveling to work. The designated date passed without even the slightest tremor.

Long-Range Forecasts

In contrast to short-range predictions, which aim to predict earthquakes within a time frame of hours or at most days, long-range forecasts give the probability of a certain magnitude earthquake occurring on a time scale of 30 to 100 years, or more. Stated another way, these forecasts give statistical estimates of the expected intensity of ground motion for a given area over a specified time frame. Although long-range forecasts may not be as informative as we might like, this data is important for updating the Uniform Building Code, which contains nationwide standards for designing earthquake-resistant structures.

Long-range forecasts are based on the premise that earthquakes are repetitive or cyclical, like the weather. In other words, as soon as one earthquake is over, the continuing motions of Earth's plates begin to build strain in the rocks again, until they fail once more. This has led seismologists to study historical records of earthquakes, for patterns, so their probability of recurrence might be established.

With this concept in mind, a group of seismologists plotted the distribution of rupture zones associated with great earthquakes that have occurred in the seismically active regions of the Pacific Basin. The maps revealed that individual rupture zones tended to occur adjacent to one another without appreciable overlap, thereby tracing out a plate boundary. Recall that most earthquakes are generated along plate boundaries by the relative motion of large crustal blocks. Because plates are in constant motion, the researchers predicted that over a span of one or two centuries, major earthquakes would occur along each segment of the Pacific plate boundary.

When the researchers studied historical records, they discovered that some zones had not produced a great earthquake in more than a century. These quiet zones, called **seismic gaps**, were identified as probable sites for major earthquakes in the next few decades (Figure 16.29). In the 25 years since the original studies were conducted, some of these gaps have ruptured. Included in this group is the zone that produced the earthquake which devastated portions of Mexico City in September 1985.

Another method of long-term forecasting, known as *paleoseismology*, has been implemented. One technique involves the study of layered deposits that were offset by prehistoric seismic disturbances. To date, the most complete investigation that employed this method focused on a segment of the San Andreas fault about 50 kilometers (30 miles) northeast of Los Angeles. Here the drainage of Pallet Creek has been repeatedly disturbed by successive ruptures along the fault zone. Ditches excavated across the creek bed have exposed

sediments that had apparently been displaced by nine large earthquakes over a period of 1400 years. From these data it was determined that a great earthquake occurs here an average of once every 140 to 150 years. The last major event occurred along this segment of the San Andreas fault in 1857. Thus, roughly 140 years have elapsed. If earthquakes are truly cyclic, a major event in southern California seems imminent. Such information led the U.S. Geological Survey to predict that there is a 50 percent probability that an earthquake of magnitude 8.3 will occur along the southern San Andreas fault within the next 30 years.

Using other paleoseismology techniques, researchers recently discovered strong evidence that very powerful earthquakes (magnitude of 8 or larger) have repeatedly struck the Pacific Northwest over the past several thousand years. The most recent event occurred about 300 years ago. As a result of these findings, public officials have taken steps to strengthen some of the region's existing dams, bridges, and water systems. Even the private sector responded. The U.S. Bancorp building in Portland, Oregon, was strengthened at a cost of $8 million and now exceeds the standards of the Uniform Building Code.

Another U.S. Geological Survey study gives the probability of a rupture occurring along various segments of the San Andreas fault for the 30 years between 1988 and 2018 (Figure 16.30). From this investigation, the Santa Cruz Mountains region was given a 30 percent probability of producing a 6.5-magnitude earthquake during this time period. In fact, it produced the Loma Prieta quake in 1989, of 7.1 magnitude.

The region along the San Andreas fault given the highest probability (90 percent) of generating a quake is the Parkfield section. This area has been called the "Old Faithful" of earthquake zones because activity

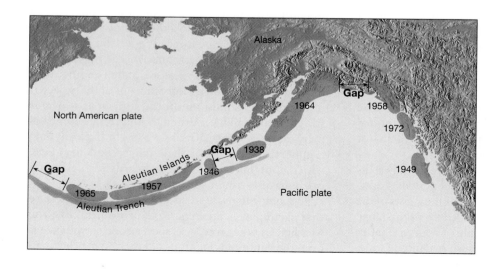

Figure 16.29 The distribution of rupture areas of large, shallow earthquakes from 1930 to 1979 along the southwestern coast of Alaska and the Aleutian Islands. The three seismic gaps denote the most likely locations for the next large earthquakes along this plate boundary. (After J.C. Savage et al., U.S. Geological Survey)

here has been very regular since record keeping began in 1857. Another section between Parkfield and the Santa Cruz Mountains is given a very low probability of generating an earthquake. This area has experienced very little seismic activity in historical times; rather, it exhibits a slow, continual movement known as *fault creep*. Such movement is beneficial because it prevents strain from building to high levels in the rocks.

In summary, it appears that the best prospects for making useful earthquake predictions involve forecasting magnitudes and locations on time scales of years, or perhaps even decades. These forecasts are important because they provide information used to develop the Uniform Building Code and assist in land-use planning.

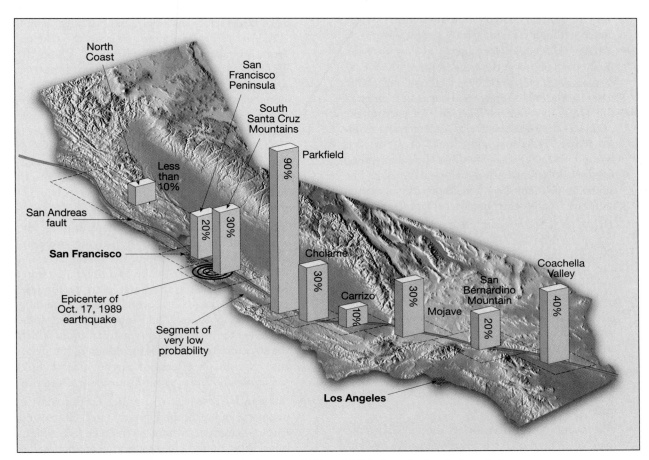

Figure 16.30 Probabilities of a major earthquake between 1988 and 2018 along the San Andreas fault.

Chapter Summary

- *Earthquakes* are vibrations of Earth produced by the rapid release of energy from rocks that rupture because they have been subjected to stresses beyond their limit. This energy, which takes the form of waves, radiates in all directions from the earthquake's source, called the *focus*. The movements that produce most earthquakes occur along large fractures, called *faults*, that are usually associated with plate boundaries.

- Along a fault, rocks store energy as they are bent. As slippage occurs at the weakest point (the focus), displacement will exert stress farther along a fault, where additional slippage will occur until most of the built-up strain is released. An earthquake occurs as the rock elastically returns to its original shape. The "springing back" of the rock is termed *elastic rebound*. Small earthquakes, called *foreshocks*, often precede a major earthquake. The adjustments that follow a major earthquake often generate smaller earthquakes called *aftershocks*.

- Two main types of *seismic waves* are generated during an earthquake: (1) *surface waves*, which travel along the outer layer of Earth; and (2) *body waves*, which travel through Earth's interior. Body waves are further divided into *primary*, or *P, waves*, which push (compress) and pull (expand) rocks in the direction the wave is traveling, and *secondary*, or *S, waves* which "shake" the particles in rock at right angles to their direction of travel. P waves can travel through solids, liquids, and gases. Fluids (gases and liquids) will not transmit S waves. In any solid material, P waves travel about 1.7 times faster than do S waves.

- The location on Earth's surface directly above the focus of an earthquake is the *epicenter*. An epicenter is determined using the difference in velocities of P and S waves. Using the difference in arrival times between P and S waves, the distance separating a recording station from the earthquake can be determined. When the distances are known from three or more seismic stations, the epicenter can be located using a method called *triangulation*.

- *A close correlation exists between earthquake epicenters and plate boundaries.* The principal earthquake epicenter zones are along the outer margin of the Pacific Ocean, known as the *circum-Pacific belt*, and through the world's oceans along the *oceanic ridge system.*

- Earthquake *intensity* depends not only on the strength of the earthquake but also on other factors, such as distance from the epicenter, the nature of surface materials, and building design. The *Mercalli intensity scale* assesses the damage from a quake at a specific location. Using the *Richter scale*, the *magnitude* (a measure of the total amount of energy released) of an earthquake is determined by measuring the *amplitude* (maximum displacement) of the largest seismic wave recorded. A logarithmic scale is used to express magnitude, in which a tenfold increase in recorded wave amplitude corresponds to an increase of 1 on the magnitude scale. Each unit of Richter magnitude equates to roughly a 32-fold energy increase.

- The most obvious factors determining the amount of destruction accompanying an earthquake are the magnitude of the earthquake and the proximity of the quake to a populated area. Structural damage attributable to earthquake vibrations depends on several factors, including (1) wave amplitudes, (2) the duration of the vibrations, (3) the nature of the material upon which the structure rests, and (4) the design of the structure. Secondary effects of earthquakes include *tsunamis*, landslides, ground subsidence, and fire.

- Substantial research to predict earthquakes is underway in Japan, the United States, China, and Russia—countries where earthquake risk is high. No reliable method of short-range prediction has yet been devised. Long-range forecasts are based on the premise that earthquakes are repetitive or cyclical. Seismologists study the history of earthquakes, for patterns, so their occurrences might be predicted. Long-range forecasts are important because they provide information used to develop the Uniform Building Code and assist in land-use planning.

Review Questions

1. What is an earthquake? Under what circumstances do earthquakes occur?

2. How are faults, foci, and epicenters related?

3. Who was first to explain the actual mechanism by which earthquakes are generated?

4. Explain what is meant by *elastic rebound*.

5. Faults that are experiencing no active creep may be considered "safe." Rebut or defend this statement.

6. Describe the principle of a seismograph.

7. List the major differences between P and S waves.

8. P waves move through solids, liquids, and gases, whereas S waves move only through solids. Explain.

9. Which type of seismic wave causes the greatest destruction to buildings?

10. Using Figure 16.13, determine the distance between an earthquake and a seismic station if the first S wave arrives 3 minutes after the first P wave.

11. Most strong earthquakes occur in a zone on the globe known as the _____.

12. Deep-focus earthquakes occur several hundred kilometers below what prominent features on the deep-ocean floor?

13. Distinguish between the Mercalli scale and the Richter scale.

14. For each increase of 1 on the Richter scale, wave amplitude increases _____ times.

15. An earthquake measuring 7 on the Richter scale releases about _____ times more energy than an earthquake with a magnitude of 6.

16. List four factors that affect the amount of destruction caused by seismic vibrations.

17. What factor contributed most to the extensive damage that occurred in the central portion of Mexico City during the 1985 earthquake?

18. The 1988 Armenian earthquake had a Richter magnitude of 6.9, far less than the great quakes in Alaska in 1964 and San Francisco in 1906. Nevertheless, the loss of life was far greater in the Armenian event. Why?

19. In addition to the destruction created directly by seismic vibrations, list three other types of destruction associated with earthquakes.

20. What is a tsunami? How is one generated?

21. Cite some reasons why an earthquake with a moderate magnitude might cause more extensive damage than a quake with a high magnitude.

22. Can earthquakes be predicted?

23. What is the value of long-range earthquake forecasts?

Key Terms

aftershock (p. 405)
body wave (p. 407)
earthquake (p. 401)
elastic rebound (p. 403)
epicenter (p. 408)
fault (p. 402)
fault creep (p. 403)
focus (p. 401)

foreshock (p. 405)
inertia (p. 406)
liquefaction (p. 416)
long (L) waves (p. 408)
magnitude (p. 412)
Mercalli intensity scale
    (p. 412)

primary (P) waves
    (p. 407)
Richter scale (p. 413)
secondary (S) waves
    (p. 407)
seismic gaps (p. 424)
seismic sea wave (p. 417)

seismogram (p. 406)
seismograph (p. 405)
seismology (p. 405)
surface wave (p. 407)
tsunami (p. 417)
Wadati–Benioff zones
    (p. 411)

Web Resources

 The *Earth* Home Page provides on-line resources for this chapter on the World Wide Web. You will find review exercises, specific updates for items in the chapter, suggested reading, and links to interesting related pathways on the Internet. Visit the *Earth* Home Page at **http://www.prenhall.com/tarbuck.**

CHAPTER 17

Earth's Interior

Left   An oil rig probes Earth's crust off the coast of
southern California — *Photo by Mark Lewis/Tony Stone
Worldwide*

arth's interior lies just below us: however, direct access to it remains very limited. Wells drilled into the crust in search of oil, gas, and other natural resources have generally been confined to the upper 7 kilometers (4 miles)—only a small fraction of Earth's 6370–kilometer radius. Even the Kola Well, a super-deep research well located in a remote northern outpost of Russia, has penetrated to a depth of only 13 kilometers. Although volcanic activity is considered a window into Earth's interior because materials are brought up from below, it allows only a glimpse of the outer 200 kilometers of our planet.

Fortunately, geologists have learned a great deal about Earth's composition and structure through computer modeling, by high-pressure laboratory experiments, and from samples of the solar system (meteorites) that collide with Earth. In addition, many clues to the physical conditions inside our planet have been acquired through the study of seismic waves generated by earthquakes and nuclear explosions. As seismic waves pass through Earth, they carry information to the surface about the materials through which they were transmitted. Hence, when carefully analyzed, seismic records provide an "X-ray" image of Earth's interior.

Probing Earth's Interior

Much of our knowledge of Earth's interior comes from the study of earthquake waves that penetrate Earth and emerge at some distant point. Simply stated, the technique involves accurately measuring the time required for P (*compressional*) and S (*shear*) waves to travel from an earthquake or nuclear explosion to a seismographic station. Because the time required for P and S waves to travel through Earth depends on the properties of the materials encountered, seismologists search for variations in travel times that cannot be accounted for simply by differences in the distances traveled. These variations correspond to changes in the properties of the materials encountered.

One major problem is that to obtain accurate travel times, seismologists must establish the exact location and time of an earthquake. This is often a difficult task because most earthquakes occur in remote areas. By contrast, the exact time and location of a nuclear test explosion is always known exactly. Despite the limitations of studying seismic waves generated by earthquakes, seismologists in the first half of the twentieth century were able to use them to detect the major layers of Earth. It was not until the early 1960s, when nuclear testing was in its heyday and networks consisting of hundreds of sensitive seismographs were deployed, that the finer structures of Earth's interior were established with certainty. (Testing of nuclear devices has been banned by international agreement for several years.)

The Nature of Seismic Waves

To examine Earth's composition and structure, we must first study some of the basic properties of wave transmission, or *propagation*. As stated in Chapter 16, seismic energy travels out from its source in all directions as waves. (For purposes of description, the common practice is to consider the paths taken by these waves as *rays*, or lines drawn perpendicular to the wave front, as shown in Figure 17.1) Significant characteristics of seismic waves include the following:

1. The velocity of seismic waves depends on the density and elasticity of the intervening material. Seismic waves travel most rapidly in rigid materials that elastically spring back to their original shapes when the stress caused by a seismic wave is removed. For instance, crystalline rock transmits seismic waves more rapidly than does a layer of unconsolidated mud.

2. Within a given layer the speed of seismic waves generally increases with depth because pressure increases and squeezes the rock into a more compact elastic material.

3. Compressional waves (P waves), which vibrate back and forth in the same plane as their direction of travel, are able to propagate through liquids as well as solids because, when compressed, these materials behave elastically; that is, they resist a change in volume and, like a rubber band, return to their original shape as a wave passes (Figure 17.2A).

4. Shear waves (S waves), which vibrate at right angles to their direction of travel, cannot propagate through liquids because, unlike solids, liquids have no shear strength (Figure 17.2B). That is, when

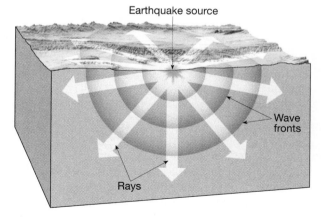

Figure 17.1 Seismic energy travels in all directions from an earthquake source (focus). The energy can be portrayed as expanding wave fronts or as rays drawn perpendicular to the wavefronts.

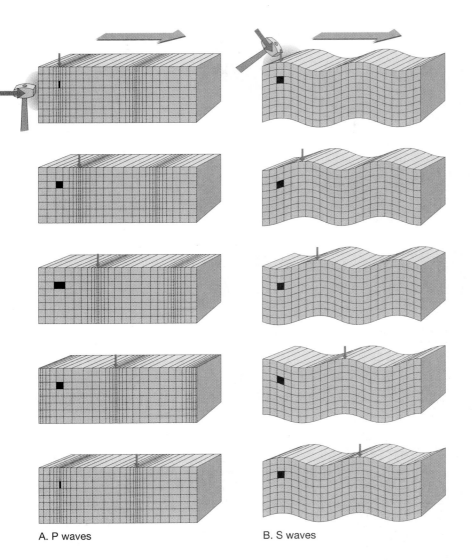

Figure 17.2 The transmission of P and S waves through a solid. **A.** The passage of P waves causes the intervening material to experience alternate compressions and expansions. **B.** The passage of S waves causes a change in shape without changing the volume of the material. Because liquids behave elastically when compressed (they spring back when the stress is removed), they will transmit P waves. However, since liquids do not resist changes shape, S waves cannot be transmitted through liquids. (After O. M. Phillips, *The Heart of the Earth,* San Francisco: Freeman, Cooper and Co., 1968)

A. P waves

B. S waves

liquids are subjected to forces that act to change their shapes, they simply flow.

5. In all materials, P waves travel faster than do S waves.

6. When seismic waves pass from one material to another, the path of the wave is refracted (bent).[*] In addition, some of the energy is reflected from the **discontinuity** (the boundary between the two dissimilar materials). This is similar to what happens to light when it passes from air into water.

Thus, depending on the nature of the layers through which they pass, seismic waves speed up or slow down, and may be refracted (bent), or reflected. These measurable changes in seismic wave motions enable seismologists to probe Earth's interior.

---
[*]Refraction occurs provided that the ray is not traveling perpendicular to the boundary.

Seismic Waves and Earth's Structure

If Earth were a perfectly homogeneous body, seismic waves would spread through it in all directions, as shown in Figure 17.3. Such seismic waves would travel in a straight line at a constant speed. However, this is not the case for Earth. It so happens that the seismic waves reaching seismographs located farther from an earthquake travel at faster average speeds than do those recorded at locations closer to the event. This general increase in speed with depth is a consequence of increased pressure, which enhances the elastic properties of deeply buried rock. As a result, the paths of seismic rays through Earth are refracted in the manner shown in Figure 17.4

As more sensitive seismographs were developed, it became apparent that in addition to gradual changes in seismic-wave velocities, rather abrupt

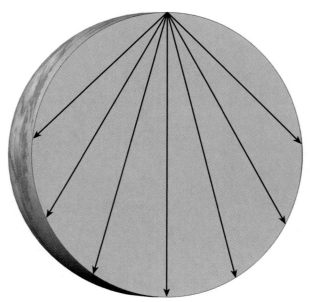

Figure 17.3 Seismic waves would travel through a hypothetical planet with uniform properties along straight-line paths and at constant velocities. Contrast with Figure 17.4.

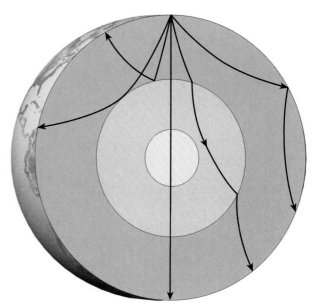

Figure 17.5 A few of many possible paths seismic rays take through Earth.

velocity changes also occur at particular depths. Because these discontinuities were detected world-wide, seismologists concluded that Earth must be composed of distinct shells having varying compositions and/or mechanical properties (Figure 17.5).

Compositional Layers

Compositional layering likely resulted from density sorting that took place during an early molten period in Earth's history. During this period the heavier ele-

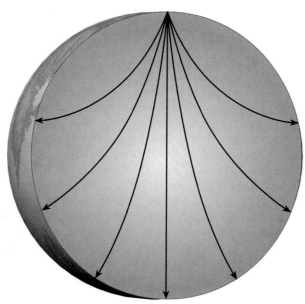

Figure 17.4 Wave paths through a planet where velocity increases with depth.

ments, principally iron and nickel, sank as the lighter rocky components floated upward. This segregation of material is still occurring, but at a much reduced rate. Because of this chemical differentiation, Earth's interior is not homogeneous. Rather, it consists of three major regions that have markedly different chemical compositions (Figure 17.6).

The principal compositional layers of Earth include (1) the **crust,** Earth's comparatively thin outer skin that ranges in thickness from 3 kilometers (2 miles) at the oceanic ridges to over 70 kilometers (40 miles) in some mountain belts such as the Andes and Himalayas; (2) the **mantle,** a solid rocky (silica-rich) shell that extends to a depth of about 2900 kilometers (1800 miles); and (3) the **core,** an iron-rich sphere having a radius of 3486 kilometers (2166 miles).

Mechanical Layers

Because both pressure and temperature greatly affect the mechanical behavior (strength) as well as the density of Earth materials, other structural divisions exist. For example the core, which is composed mostly of an iron-nickel-alloy, is divided into two regions that exhibit different mechanical behavior (Figure 17.6). The **outer core** is a *liquid* metallic layer 2270 kilometers (1410 miles) thick. This zone, which is capable of convective flow, surrounds the **inner core,** a *solid* sphere having a radius of 1216 kilometers (756 miles).

Other structural divisions have been discovered in which materials have undergone a phase change. Phase changes occur, for example, when rock melts, or when the atoms in minerals rearrange themselves into tighter crystalline structures in response to the

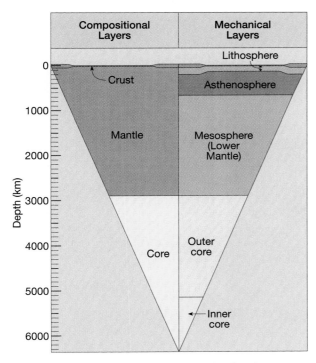

Figure 17.6 Comparison of the compositional and mechanical layers of Earth.

enormous pressures at great depths. The latter type of phase change occurs at depths between 400 and 660 kilometers. Later we will examine other divisions of Earth in which rocks having different chemical compositions behave as though they were a single coherent unit.

Discovering Earth's Major Boundaries

Over the past century, seismological data gathered from many seismographic stations have been compiled and analyzed. From this information, seismologists have developed a detailed image of Earth's interior (Figure 17.7). This model is continually being fine-tuned as more data become available and as new seismic techniques are employed. Furthermore, laboratory studies that experimentally determine the properties of various Earth materials under the extreme environments found deep in Earth add to this body of knowledge.

The Moho

In 1909, a pioneering Yugoslavian seismologist, Andrija Mohorovičić, presented the first convincing evidence for layering within Earth.[*] The boundary he

discovered separates crustal materials from rocks of different composition in the underlying mantle and was named the **Mohorovičić discontinuity** in his honor. For obvious reasons, the name for this boundary was quickly shortened to **Moho.**

By carefully examining the seismograms of shallow earthquakes, Mohorovičić found that seismographic stations located more than 200 kilometers from an earthquake obtained appreciably faster average travel velocities for P waves than did stations located nearer the quake (Figure 17.8). In particular, P waves that reached the closest stations first had velocities that averaged about 6 kilometers per second. By contrast, the seismic energy recorded at more distant stations traveled at speeds that approached 8 kilometers per second. This abrupt jump in velocity did not fit the general pattern that had been previously observed. From these data, Mohorovičić concluded that below 50 kilometers there exists a layer with properties markedly different from those of Earth's outer shell.

Figure 17.8 illustrates how Mohorovičić reached this important conclusion. Notice that the first wave to reach the seismograph located 100 kilometers from the epicenter traveled the shortest route directly through the crust. However, at the seismograph located 300 kilometers from the epicenter, the first P wave to arrive traveled through the mantle, a zone of higher velocity. Thus, although this wave traveled a greater distance, it reached the recording instrument sooner than did the rays taking the more direct route. This is because a large portion of its journey was through a region having a composition where seismic waves travel more rapidly. This principle is analogous to a driver taking a bypass route around a large city during rush hour. Although this alternate route is longer, it may be faster.

The Core–Mantle Boundary

A few years later, in 1914, the location of another major boundary was established by the German seismologist Beno Gutenberg. This discovery was based primarily on the observation that P waves diminish and eventually die out completely about 105 degrees from an earthquake (Figure 17.9). Then, about 140 degrees away, the P waves reappear, but about 2 minutes later than would be expected based on the distance traveled. This belt, where direct seismic waves are absent, is about 35 degrees wide and has been named the **P wave shadow zone**[+] (Figure 17.9).

Gutenberg, and others before him, realized that the P wave shadow zone could be explained if Earth

---

[*]The core–mantle boundary had been predicted by R. D. Oldham in 1906, but his arguments for a central core did not receive wide acceptance.

[+]As more sensitive instruments were developed, weak and delayed P waves that enter this zone via reflection were detected.

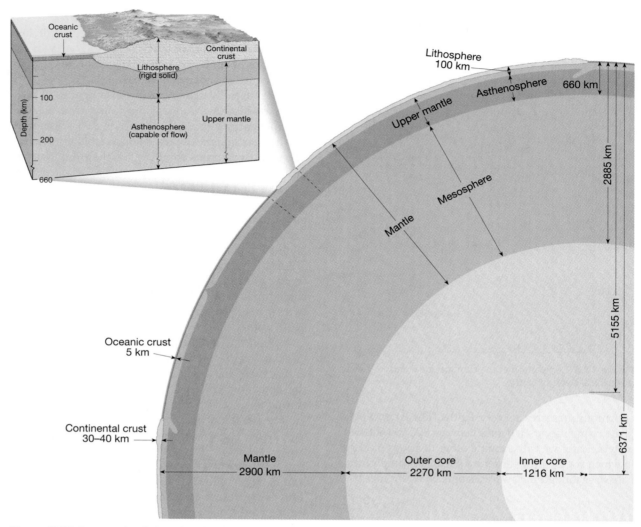

Figure 17.7 Cross-sectional view of Earth showing the internal structure.

contained a core that was composed of material unlike the overlying mantle. The core, which Gutenberg calculated to be located at a depth of 2900 kilometers, must somehow hinder the transmission of P waves similar to the way in which light rays are blocked by an object that casts a shadow. However, rather than actually stopping the P waves, the shadow zone is produced by bending (refracting) of the P waves, which enter the core as shown in Figure 17.9.

It was further determined that S waves do not travel through the core. This fact led geologists to conclude that at least a portion of this region is liquid (Figure 17.10). This conclusion was further supported by the observation that P-wave velocities suddenly decrease by about 40 percent as they enter the core. Because melting reduces the elasticity of rock, this evidence pointed to the existence of a liquid layer below the rocky mantle.

Discovery of the Inner Core

In 1936, the last major subdivision of Earth's interior was predicted by Inge Lehmann, a Danish seismologist (see Box 17.1). Lehmann discovered a new region of seismic reflection and refraction within the core. Hence, a core within a core was discovered. The size of the inner core was not precisely established until the early 1960s when underground nuclear tests were conducted in Nevada. Because the exact location and time of the explosions were known, echoes from seismic waves that bounced off the inner core provided a precise means of determining its size (Figure 17.11).

From these data, the inner core was found to have a radius of about 1216 kilometers. Furthermore, P waves passing through the inner core have appreciably faster average velocities than do those penetrating only the outer core. The apparent increase in

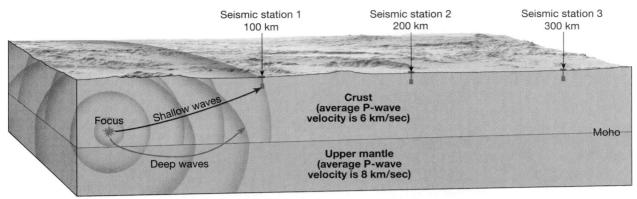

A. Time 1 – Slower shallow waves arrive at seismic station 1 first.

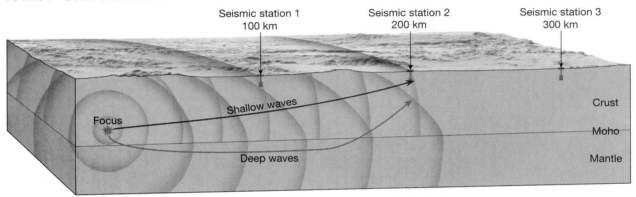

B. Time 2 – Slower shallow waves arrive at seismic station 2 first

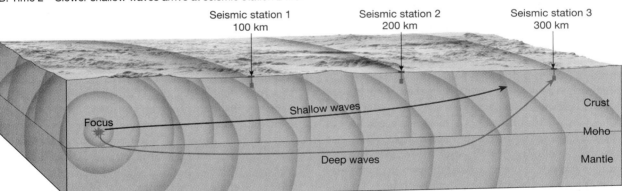

C. Time 3 – Faster deeper waves arrive at seismic station 3 first

Figure 17.8 Idealized paths of seismic waves traveling from an earthquake focus to three seismographic stations. In parts **A** and **B**, you can see that the two nearest recording stations receive the slower waves first because the waves traveled a shorter distance. However, as shown in part **C**, beyond 200 kilometers, the first waves received passed through the mantle, which is a zone of higher velocity.

the elasticity of the inner core is evidence that this innermost region is solid.

Over the past few decades, advances in seismology and rock mechanics have allowed for much refinement of the gross view of Earth's interior that has been presented to this point. Some of these refinements as well as other properties of the major divisions, including their densities and compositions, will be considered next.

The Crust

Earth's crust averages less than 20 kilometers thick, making it the thinnest of Earth's divisions (see Figure 17.7). Along this eggshell-thin layer great variations in thickness exist. Crustal rocks of the stable continental interiors are roughly 30 kilometers thick. However, in a few exceptionally prominent mountainous

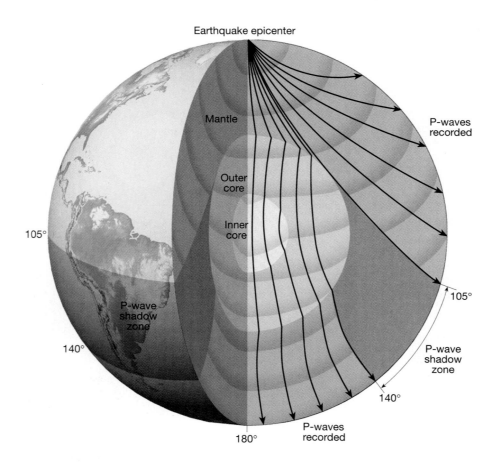

Earthquake epicenter

Mantle

Outer core

Inner core

P-waves recorded

105°

P-wave shadow zone

140°

105°

P-wave shadow zone

140°

180°

P-waves recorded

Figure 17.9 The abrupt change in physical properties at the mantle–core boundary causes the wave paths to bend sharply, resulting in a shadow zone for P waves between about 105 and 140 degrees.

regions, the crust obtains its greatest thickness, exceeding 70 kilometers. The oceanic crust is much thinner, ranging from 3 to 15 kilometers thick. Further, crustal rocks of the deep-ocean basins are compositionally different from those of the continents.

Continental rocks have an average density* of about 2.8 g/cm³ and some have been discovered that exceed 3.8 billion years in age. From both seismic studies and direct observations, the average composition of continental rocks is estimated to be comparable to the felsic igneous rock *granodiorite*. Like granodiorite, the continental crust is enriched in the elements potassium, sodium, and silicon. Although numerous granitic intrusions and chemically equivalent metamorphic rocks are very abundant, large outpourings of basaltic and andesitic rocks are also commonly found on the continents.

The rocks of the oceanic crust are younger (180 million years or less) and more dense (about 3.0 g/cm³) than continental rocks. The deep-ocean basins lie beneath 4 kilometers of seawater as well as hundreds of meters of sediment. Thus, until recently,

geologists had to rely on indirect evidence (such as slivers of what was thought to be oceanic crust that were thrust unto land) to estimate the composition of this inaccessible region. With the development of deep-sea drilling ships, the recovery of core samples from the ocean floor became possible. As predicted, the samples obtained were predominantly *basalt*. Recall that volcanic eruptions of basaltic lavas are known to have generated many islands, such as the Hawaiian chain, located within the deep-ocean basins.

The Mantle

Over 82 percent of Earth's volume is contained within the mantle, a nearly 2900-kilometer-thick shell of silicate rock extending from the base of the crust (Moho) to the liquid outer core. Our knowledge of the mantle's composition comes from experimental data and from the examination of material carried to the surface by volcanic activity. In particular, the rocks composing kimberlite pipes, in which diamonds are sometimes found, are often thought to have originated at depths approaching 200 kilometers, well within the mantle. Kimberlite deposits are composed of *peridotite*, a rock that contains iron and magnesium-rich silicate minerals, mainly olivine and pyroxene, plus lesser

---

*Liquid water has a density of 1 g/cm³; therefore, crustal rocks have a density nearly three times that of water.

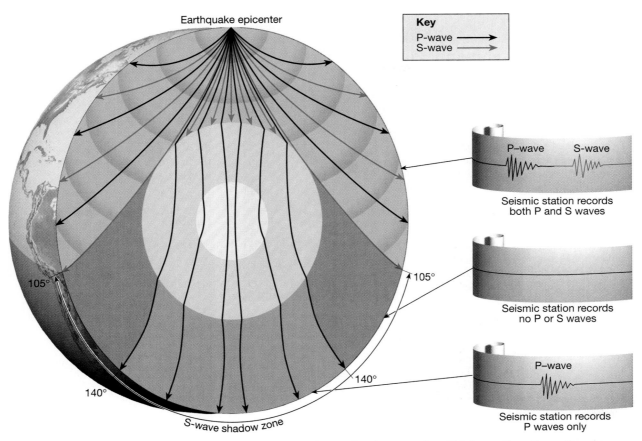

Figure 17.10 View of Earth's interior showing P and S wave paths. Any location more than 105 degrees from the earthquake epicenter will not receive direct S waves since the outer core will not transmit them. Although P waves are also absent beyond 105 degrees, they are recorded beyond 140 degrees, as shown in Figure 17.9.

amounts of garnet. Further, because S waves readily travel through the mantle, we know that it behaves as an elastic solid. Thus, the mantle is described as a solid rocky layer, the upper portion of which has the composition of the ultramafic rock peridotite.

The mantle is divided into the **mesosphere** or *lower mantle*, which extends from the core–mantle boundary to a depth of 660 kilometers, and the *upper mantle*, which continues to the base of the crust. In addition, other subdivisions have been identified. At the depth of about 400 kilometers a relatively abrupt increase in seismic velocity occurs (Figure 17.12). Whereas the crust–mantle boundary represents a compositional change, the zone of seismic velocity increase at the 400-kilometer level is the result of a *phase change*. (A phase change occurs when the crystalline structure of a mineral is altered in response to changes in temperature and/or pressure.) Laboratory studies show that the mineral olivine, $(Mg, Fe)_2SiO_4$, which is one of the main constituents in the rock peridotite, will collapse to a more compact, high-pressure mineral (spinel) at the pressures experienced at this depth. This change to a denser crystal form explains the increased seismic velocities observed.

Another boundary within the mantle has been detected from variations in seismic velocity at a depth of 660 kilometers (Figure 17.12). At this depth the mineral spinel is believed to undergo a transformation to the mineral perovskite $(Mg, Fe)SiO_3$. Because perovskite is thought to be pervasive in the lower mantle, it is perhaps Earth's most abundant mineral.

In the lowermost roughly 200 kilometers of the mantle, there exists an important region known as the **D″ layer** (see Figure 17.6). Recently, researchers reported that seismic waves traveling through some parts of the D″ layer experience a sharp decrease in P-wave velocities. So far, the best explanation for this phenomenon is that the lowermost layer of the mantle is partially molten in places.

If these zones of partially melted rock exist, they are very important because they would be capable of transporting heat from the core to the lower mantle much more efficiently than solid rock. A high rate of heat flow would, in turn, cause the solid mantle located above these partly molten zones to be warmed sufficiently to become buoyant and slowly rise toward the surface. Such rising plumes of super-heated rock may be the source of volcanic activity associated with

Box 17.1

Inge Lehmann: Pioneering Geophysicist*

Inge Lehmann was a pioneering scientist at a time when few women had careers in science and mathematics (Figure 17.A). Born in Denmark in 1888, Lehmann lived a long and productive life that included important contributions to our understanding of Earth's interior. She died in 1993 at the age of 105.

After undergraduate work at the University of Copenhagen and Cambridge University, Lehmann earned two master's degrees from the University of Copenhagen: one in mathematics in 1920 and another in geodesy in 1928. In later years, she studied in Germany, France, Belgium, and The Netherlands.

Inge Lehmann's career in seismology started in 1925 when she helped establish seismic networks in Denmark and Greenland. Three years later, in 1928, she was named the first chief of the seismology department at the Royal Danish Geodetic Institute, a position she held for 25 years. She recorded, analyzed, and cataloged seismograms from Denmark and Greenland, and published seismic bulletins.

It was a paper that she published in 1936 that established her place in the history of geophysics. Known simply as *P′ (P-prime)*, the paper identified a new region of seismic reflection and refraction in Earth's interior, now known as the *Lehmann discontinuity* (Figure 17.B).

Figure 17.A Inge Lehmann, 1888–1993. (Photo Courtesy of U.S. Geological Survey)

Because of her exacting scrutiny of seismic records, Inge Lehmann had discovered the boundary that divides Earth's core into inner and outer parts.

Many honors came to Lehmann in recognition of her extraordinary accomplishments. Among them were the Gold Medal from the Royal Danish Academy of Science in 1965, the Bowie Medal from the American Geophysical Union in 1971, and the Medal of the Seismological Society of America in 1977. In 1997, the Lehmann Medal was created by the American Geophysical

Union (AGU) to recognize outstanding research on the structure, composition, and dynamics of Earth's mantle and core. It was the first medal awarded by the AGU that is named for a woman and the first to be named for someone who worked outside the United States.

---

*This box was prepared by Nancy L. Lutgens

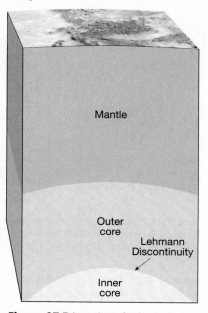

Figure 17.B Location of Lehmann Discontinuity.

---

hotspots, such as that experienced at Hawaii and Iceland. If these observations are accurate, some of the volcanic activity that we see at the surface is a manifestation processes occurring 2900 kilometers (1800 miles) below our feet.

The Lithosphere and Asthenosphere

Earth's outer layer, consisting of the uppermost mantle and crust, forms a relatively cool, rigid shell. Although this layer consists of materials with markedly different chemical compositions, it tends to act as a unit that behaves similarly to mechanical deformation. This outermost rigid unit of Earth is called the

lithosphere (*sphere of rock*). Averaging about 100 kilometers in thickness, the lithosphere may be 250 kilometers or more in thickness below the older portions (shields) of the continents (Figure 17.13). Within the ocean basins the lithosphere is only a few kilometers thick along the oceanic ridges and increases to perhaps 100 kilometers in regions of older and cooler crustal rocks.

Beneath the lithosphere (to a depth of about 660 kilometers) lies a soft, relatively weak layer located in the upper mantle known as the **asthenosphere** ("weak sphere"). The upper 150 kilometers or so of the asthenosphere has the temperature/pressure regime in which a small amount of melting takes place (perhaps 1 to 5 percent). This region of partial melting within the upper asthenosphere is known as the **low-velocity zone,**

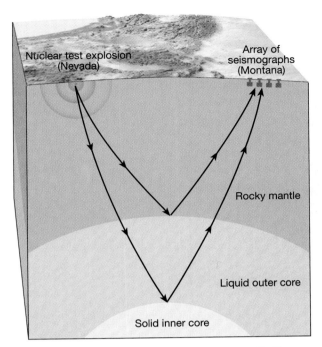

Figure 17.11 Travel times of seismic waves generated from nuclear test explosions were used to measure the depth of the inner core accurately. An array of seismographs located in Montana detected the "echoes" that bounced back from the boundary of the inner core.

because seismic waves show a marked decrease in velocity (Figure 17.13). Within this very weak zone, the lithosphere is effectively detached from the asthenosphere located below. The result is that the lithosphere is able to move independently of the asthenosphere, a topic we shall consider in Chapter 19.

It is important to emphasize that the strength of various Earth materials is a function of their composition, as well as the temperature and pressure of their environment. You should not get the idea that the entire lithosphere is brittle, like the rocks found at the surface. Rather, the rocks of the lithosphere get progressively weaker (more easily deformed) with increasing depth. At the depth of the upper asthenosphere (low-velocity zone) the rocks are very close to their melting temperature (some melting is thought to occur) so that they are easily deformed. Thus, the asthenosphere is weak because it is near its melting point, just as hot wax is weaker than cold wax. However, in the material located below this weak zone, increased pressure offsets the effects of increased temperature. Therefore, these materials gradually stiffen with depth, forming the more rigid lower mantle. Despite their greater strength, the materials of the lower mantle are still capable of very gradual flow.

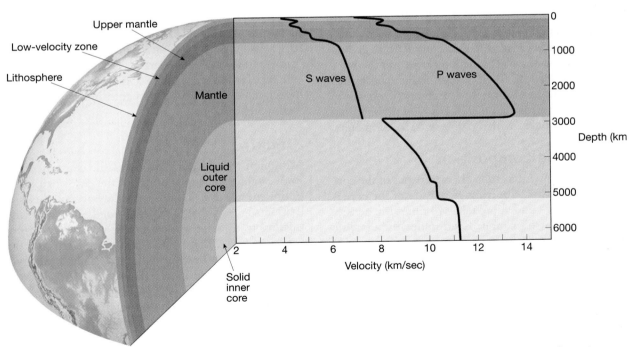

Figure 17.12 Variations in P ans S wave velocities with depth. Abrupt changes in average wave velocities delineate the major features of Earth's interior. At a depth of about 100 kilometers, the sharp decrease in wave velocity corresponds to the top of the low-velocity zone. Two other bends in the velocity curves occur in the upper mantle at depths of about 400 and 700 kilometers. These variations are thought to be caused by minerals that have undergone phase changes, rather than resulting from compositional differences. The abrupt decrease in P-wave velocity and the absence of S waves at 2885 kilometers marks the core–mantle boundary. The liquid outer core will not transmit S waves, and the propagation of P waves is slowed within this layer. As the P waves enter the solid inner core, their velocity once again increases. (Data from Bruce A. Bolt)

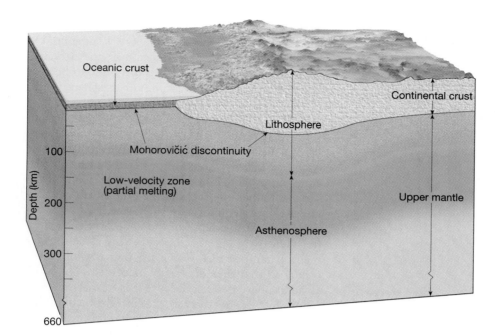

Figure 17.13 Respective locations of the asthenosphere and lithosphere.

The Core

Larger than the planet Mars, the core is Earth's dense central sphere with a radius of 3486 kilometers. Extending from the inner edge of the mantle to the center of Earth, the core constitutes about one-sixth of Earth's volume and nearly one-third of its total mass. Pressure at the center is millions of times greater than the air pressure at Earth's surface, and the temperatures can exceed 6700°C. As more precise seismic data became available, the core was found to consist of a liquid outer layer about 2270 kilometers thick, and a solid inner sphere with a radius of 1216 kilometers.

Density and Composition

One of the more interesting characteristics of the core is its great density. The average density of the core is about 11 g/cm³, and approaches 14 times the density of water at Earth's center. Even under the extreme pressures at these depths, the common silicate minerals found in the crust (with surface densities 2.6 to 3.5 g/cm³) would not be compacted enough to account for the density calculated for the core. Consequently, attempts were undertaken to determine the material that could account for this property.

Surprisingly enough, meteorites provided an important clue to Earth's internal composition. Because meteorites are part of the solar system, they are assumed to be representative samples of the material from which Earth originally accreted. Their composition ranges from metallic types, made primarily of iron and lesser amounts of nickel, to stony meteorites composed of rocky substances that closely resemble the rock peridotite. Because Earth's crust and mantle contain a much smaller percentage of iron than is found in the debris of the solar system, geologists concluded that the interior of Earth must be enriched in this heavy metal. Further, iron is by far the most abundant substance found in the solar system that possesses the seismic properties and density resembling that measured for the core. Current estimates suggest that the core is mostly iron, with 5 to 10 percent nickel and lesser amounts of lighter elements, including perhaps sulfur and oxygen.

Origin

Although the existence of a metallic central core is well established, explanations of the core's origin are more speculative. The most widely accepted explanation suggests that the core formed early in Earth's history from what was originally a relatively homogeneous body. During the period of accretion, the entire Earth was heated by energy released by the collisions of infalling material. Sometime late in this period of growth, Earth's internal temperature was sufficiently high to melt and mobilize the accumulated material. Blobs of heavy iron-rich materials collected and sank toward the center. Simultaneously, lighter substances may have floated upward to generate the crust. In a short time, geologically speaking, Earth took on a layered configuration, not significantly different from what we find today.

In its formative stage, the entire core was probably liquid. Further, this liquid iron alloy was in a state of vigorous mixing. However, as Earth began to cool, iron in the core began to crystallize and the inner core began to form. As the core continues to cool, the inner core should grow at the expense of the outer core.

Earth's Magnetic Field

Our picture of the core with its solid inner sphere surrounded by a mobile liquid shell is further supported by the existence of Earth's magnetic field. This field behaves as though a large bar magnet were situated deep within Earth. However, we know that the source of the magnetic field cannot be permanently magnetized material, because Earth's interior is too hot for any material to retain its magnetism. The most widely accepted explanation of Earth's magnetic field requires that the core be made of a material that conducts electricity, such as iron, and one that is mobile (see Box 17.2). Both of these conditions are met by the model of Earth's core that was established on the basis of seismological data.

One recently discovered consequence of Earth's magnetic field is that it affects the rotation of the solid inner core. Current estimates indicate that the inner core rotates in a west-to-east direction about 1 degree a year *faster* than the Earth's surface. Thus, the core makes one extra rotation about every 400 years. Further, the axis of rotation for the inner core is offset about 10 degrees from Earth's rotational poles.

Earth's Internal Heat Engine

As we discussed in Chapter 3, temperature gradually increases with an increase in depth at a rate known as the **geothermal gradient** (Figure 17.14). The geothermal gradient varies considerably from place to place. In the crust, temperatures increase rapidly, averaging between 20°C and 30°C per kilometer. However, the rate of increase is much less in the mantle and core. At a depth of 100 kilometers, the temperature is estimated to exceed 1200°C, whereas the temperature at the core–mantle boundary is calculated to be about 4500°C and it may exceed 6700°C at Earth's center (hotter than the surface of the Sun!).

Three major processes have contributed to Earth's internal heat: (1) heat emitted by radioactive decay of isotopes of uranium (U), thorium (Th), and potassium (K); (2) heat released as iron crystallized to form the solid inner core; and (3) heat released by colliding particles during the formation of our planet. Although the first two processes are still operating, their rate of heat generation is much less than in the geologic past. Today, our planet is radiating more of

its internal heat out to space than is being replaced by these mechanisms. Therefore, Earth is slowly, but continuously, cooling.

Heat Flow in the Crust

Heat flow in the crust occurs by the familiar process called **conduction.** Anyone who has attempted to pick up a metal spoon left in a hot pan is quick to realize that heat was conducted through the spoon. *Conduction*, which is the transfer of heat through matter by molecular activity, occurs at a relatively slow rate in crustal rocks. Thus, the crust tends to act as an insulator (cool on top and hot on the bottom), which helps account for the steep temperature gradient exhibited by the crust.

Certain regions of Earth's crust have much higher rates of heat flow than do others. In particular, along the axes of mid-ocean ridges where the crust is only a few kilometers thick, heat flow rates are relatively high. By contrast, a relatively low heat flow is observed in ancient shields (such as the Canadian and Baltic Shields). This may occur because these zones have a thick lithospheric root that effectively insulates the crust from the asthenospheric heat below. Other crustal regions exhibit a high heat flow, because of shallow igneous intrusions or because of higher-than-average concentrations of radioactive materials.

Mantle Convection

Any working model of the mantle must explain the temperature distribution calculated for this layer. Whereas a large increase in temperature with depth occurs within the crust, this same trend does not continue downward through the mantle. Rather, the temperature increase with depth in the mantle is much more gradual. This means that the mantle must have an effective method of transmitting heat from the core outward. Because rocks are relatively poor conductors of heat, most researchers conclude that some form of mass transport (convection) of rock must exist within the mantle. **Convection** is the transfer of heat by the mass movement or circulation in a substance. Consequently, the rock of the mantle must be capable of flow.

Convective flow in the mantle—in which warm, less dense rock rises, and cooler, denser material sinks—is the most important process operating in Earth's interior. This thermally driven flow is the force that propels the rigid lithospheric plates across the globe, and it ultimately generates Earth's mountain belts and worldwide earthquake and volcanic activity. Recall that plumes of superheated rock are thought to form near the core–mantle boundary and slowly rise toward the surface. These buoyant plumes would be the warm, upward

Box 17.2

Earth's Magnetic Field

Anyone who has used a compass to find direction knows that Earth's magnetic field has a north pole and a south pole. In many respects our planet's magnetic field resembles that produced by a simple bar magnet. Invisible lines of force pass through Earth and out into space while extending from one pole to the other (Figure 17.C). A compass needle, which is a small magnet free to move about, becomes aligned with these lines of force and points toward the magnetic poles. It should be noted that Earth's magnetic poles do not coincide exactly with the geographic poles. The north magnetic pole is located in northeastern Canada, near Hudson Bay, while the south magnetic pole is located near Antartica in the Indian Ocean south of Australia.

In the early 1960s, geophysicists learned that Earth's magnetic field periodically (every million years or so) reverses polarity; that is, the north magnetic pole becomes the south magnetic pole, and vice versa. The cause of these reversals is apparently linked to the fact that Earth's magnetic field experiences long-term fluctuations in intensity. Recent calculations indicate that the magnetic field has weakened by about 5 percent over the past century. If this trend continues for another 1500 years, Earth's magnetic field will become very weak or even nonexistent.

It has been suggested that the decline in magnetic intensity is related to changes in the convective flow in the core. In a similar manner, magnetic reversals may be triggered when something disturbs the main convection pattern of the fluid core. After a reversal takes place, the flow is reestablished and builds a magnetic field with opposite polarity.

Magnetic reversals are not unique to Earth. The Sun's magnetic field regularly reverses polarity, having an average

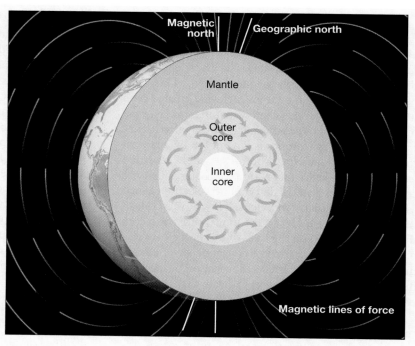

Figure 17.C Earth's magnetic field is thought to be generated by vigorous convection of molten iron alloy in the liquid outer core.

period of about 22 years. These solar reversals are closely tied to the well-known 11-year sunspot cycle.

When Earth's magnetic field was first described in 1600, it was thought to originate from permanently magnetized materials located deep within Earth. We have since learned that, except for the upper crust, the planet is much too hot for magnetic materials to retain their magnetism. Furthermore, permanently magnetized materials are not known to vary their intensity in a manner that would account for the observed long-term waxing and waning of Earth's magnetic field.

The details of how Earth's magnetic field is produced are still poorly understood. Nevertheless, most investigators agree that the gradual flow of molten iron in the outer core is an important

part of the process. The most widely accepted view proposes that the core behaves like a self-sustaining *dynamo*, a device that converts mechanical energy into magnetic energy. The driving forces of this system are Earth's rotation and the unequal distribution of heat in the interior, which propels the highly conductive molten iron in the outer core. As the iron churns in the outer core, it interacts with Earth's magnetic field. This interaction generates an electric current, just as moving a wire past a magnet creates a current in the wire. Once established, the electric current produces a magnetic field that reinforces Earth's magnetic field. As long as the flow within the molten iron outer core continues, electric currents will be produced and Earth's magnetic field will be sustained.

flowing arm in the convective mechanism at work in the mantle (see Box 17.3). Downwelling is thought to occur at convergent plate boundaries where cool, dense slabs of lithosphere are being subducted. Some studies predict

that this dense, cool material may eventually descend all the way to the core–mantle boundary.

If this convective mechanism exists, how does the rocky mantle transmit S waves, which can travel only

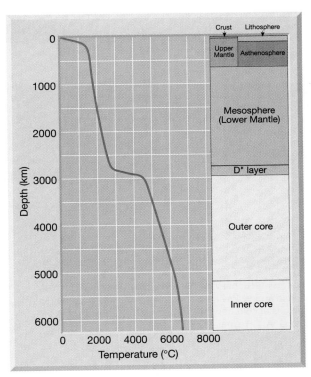

Figure 17.14 Estimated geothermal gradient for Earth. Temperatures in mantle and core are based on several assumptions and may vary ± 500°C. (Data from Kent C. Condie)

though solids, and at the same time flow like a fluid? This apparent contradiction can be resolved if the mantle behaves like a solid under certain conditions and like a fluid under other conditions. Geologists generally describe material of this type as exhibiting *plastic* behavior. When a material that exhibits plastic behavior encounters short-lived stresses, such as those produced by seismic waves, the material behaves like an elastic solid. However, in response to stresses applied over very long time periods, this same material will flow.

This behavior explains why S waves can penetrate the mantle, yet this rocky layer is able to flow. Plastic behavior is not restricted to mantle rocks. Manufactured substances such as Silly Putty and some taffy candies also exhibit this behavior. When struck with a hammer these materials shatter like a brittle solid. However, when slowly pulled apart they deform by flowing. From this analogy, do not get the idea that the mantle is composed of soft, putty-like material. Rather, it is composed of hot, solid rock, which under extreme confining pressures unknown on the surface of Earth, is able to flow.

Chapter Summary

- Much of our knowledge of Earth's interior comes from the study of earthquake waves that penetrate Earth and emerge at some distant point. In general, seismic waves travel faster in solid elastic materials and slower in weaker layers. Further, seismic energy is reflected and refracted (bent) at boundaries between compositionally or mechanically different materials. By carefully measuring the travel times of seismic waves, seismologists have been able to determine the major divisions of Earth's interior.

- The principal compositional layers of Earth include (1) the *crust*, Earth's comparatively thin outer skin that ranges in thickness from 3 kilometers (2 miles) at the oceanic ridges to over 70 kilometers (40 miles) in some mountainous belts such as the Andes and Himalayas; (2) the *mantle*, a solid rocky shell that extends to a depth of about

2900 kilometers (1800 miles); and (3) the *core*, an iron-rich sphere having a radius of 3486 kilometers (2166 miles).

- The crust, the rigid outermost layer of Earth, is divided into oceanic and continental crust. Oceanic crust ranges from 3 to 15 kilometers in thickness and is composed of basaltic igneous rocks. By contrast, the continental crust consists of a large variety of rock types having an average composition of felsic rock called granodiorite. The rocks of the oceanic crust are younger (180 million years or less) and more dense (about 3.0 g/cm³) than continental rocks. Continental rocks have an average density of about 2.7 g/cm³ and some have been discovered that exceed 3.8 billion years in age.

- Over 82 percent of Earth's volume is contained in the mantle, a rocky shell about 2900 kilometers

Box 17.3

Exploring Convective Flow in the Mantle

Recently, new technologies have become available that may significantly enhance our knowledge of convective flow in the mantle. One analytical tool, called *seismic tomography*, is similar in principle to CAT scanning (computer-aided tomography), which is used in medical diagnoses. Whereas CAT scanning uses X-rays to penetrate the human body, information on Earth's interior is obtained from seismic waves triggered by earthquakes. Like CAT scanning, seismic tomography uses computers to combine data from multiple sources to build a three-dimensional image of the object.

Recall that the velocities of seismic waves are strongly influenced by the properties of the transmitting materials. In tomographic studies, the information from many criss-crossing waves is combined to map regions of "slow" and "fast" seismic velocity. In general, regions of slow seismic velocity are associated with hot, upwelling rock, whereas regions of fast seismic velocity represent areas in which cool rock is descending.

Seismic tomography studies reveal that the flow in the mantle is far more complex than simple convection cells, in which hot material gradually rises and cold material sinks. It appears that upwelling is confined to a few large

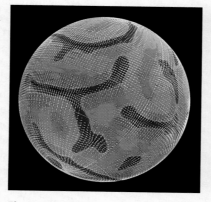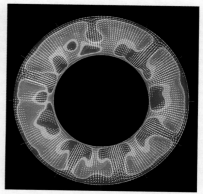

Figure 17.D Surface and cross-sectional view of numerically simulated thermal convection in the mantle. Red and yellow areas indicate hot upwelling currents, while blue areas depict regions of cold downwelling currents. (Courtesy of D. Bercovici, G. Schubert, and G. A. Glatzmaier)

cylindrical plumes. Furthermore, these studies show that areas of downwelling are located beneath convergent boundaries where plates are being subducted. This downward flow appears to extend to the lower mantle, but other interpretations of the data are possible.

Another innovative technique called *numerical modeling* has been employed to stimulate thermal convection in the mantle. Simply, this method uses high-speed computers to solve mathematical equations that describe the dynamics of mantlelike

fluids. Because of a number of uncertainties, including lack of precise knowledge of the viscosity of the mantle, various conditions are simulated. Results of these studies can be graphically represented, as shown in Figure 17.D. One study concludes that downwelling occurs in sheetlike structures, supporting seismic evidence that descending lithospheric slabs are an integral part of mantle circulation. Furthermore, large mantle plumes were found to be the main mechanism of upwelling in the mantle.

thick. The boundary between the crust and mantle represents a change in composition. Although the mantle behaves like a solid when transmitting earthquake waves, mantle rocks are able to flow at an infinitesimly slow rate. Some of the rocks in the lowermost mantle (D″ layer) are thought to be partially molten.

- The core is composed mostly of iron with lesser amounts of nickel and other elements. At the extreme pressure found in the core, this iron-rich material has an average density of about 11 g/cm³ and approaches 14 times the density of water at Earth's center. The inner and outer core are compositionally similar; however, the outer core is liquid and capable of flow. It is the circulation within

the core of our rotating planet that generates Earth's magnetic field.

- Earth's outer layer, including the uppermost mantle and crust, form a relatively cool, rigid shell known as the *lithosphere* (sphere of rock). Averaging about 100 kilometers in thickness, the lithosphere may be 250 kilometers or more in thickness below older portions (shields) of the continents. Within the ocean basins the lithosphere ranges from a few kilometers thick along the oceanic ridges to perhaps 100 kilometers in regions of older and cooler crustal rocks.

- Beneath the lithosphere (to a depth of about 660 kilometers) lies a soft, relatively weak layer located in the upper mantle known as the *asthenosphere* ("weak sphere"). The upper 150 kilometers or so of the asthenosphere has a temperature/pressure regime in which a small amount of melting takes place (perhaps 1 to 5 percent). Within this very weak zone, the lithosphere is effectively detached from the asthenosphere located below.

- Temperature gradually increases with depth in our planet's interior. Three processes contribute to Earth's internal heat: (1) heat emitted by radioactivity; (2) heat released as iron solidifies in the core; and (3) heat released by colliding particles during the formative years of our planet.

- Convective flow in the mantle is thought to consist of buoyant plumes of hot rock and downward flow of cool, dense slabs of lithosphere. This thermally generated convective flow is the driving force that propels lithospheric plates across the globe.

Review Questions

1. List the major characteristics of seismic waves.
2. What are the three compositional layers of Earth?
3. How does the boundary between the crust and mantle (Moho) differ from the boundaries that occur at depths of about 400 and 660 kilometers?
4. What evidence did Beno Gutenberg use for the existence of Earth's central core?
5. Suppose the shadow zone for P waves was located between 120 and 160 degrees, rather than between 105 and 140 degrees. What would this indicate about the size of the core?
6. Describe the method first used to measure accurately the size of the inner core.
7. Which of Earth's three compositional layers is the most voluminous?
8. What evidence is provided by seismology to indicate that the outer core is liquid?
9. Where is the D″ layer located and what role is it thought to play in transporting heat within Earth?
10. Describe the lithosphere. In what important way is it different from the asthenosphere?
11. Why are meteorites considered important clues to the composition of Earth's interior?
12. Describe the chemical (mineral) makeup of the four principal layers of Earth.
13. List three processes that have contributed to Earth's internal heat.
14. Describe the process of conduction.
15. Briefly explain how heat is transported through the mantle.

Key Terms

asthenosphere (p. 438)
core (p. 432)
crust (p. 432)
conduction (p. 441)
convection (p. 441)

D″ layer (p. 437)
discontinuity (p. 431)
geothermal gradient (p. 441)
inner core (p. 433)

lithosphere (p. 438)
low-velocity zone (p. 439)
mantle (p. 432)
mesosphere (p. 437)

Mohorovičić discontinuity, or Moho (p. 433)
outer core (p. 432)
P wave shadow zone (p. 434)

Web Resources

The *Earth* Home Page provides on-line resources for this chapter on the World Wide Web. You will find review exercises, specific updates for items in the chapter, suggested reading, and links to interesting related pathways on the Internet. Visit the *Earth* Home Page at **http://www.prenhall.com/tarbuck.**

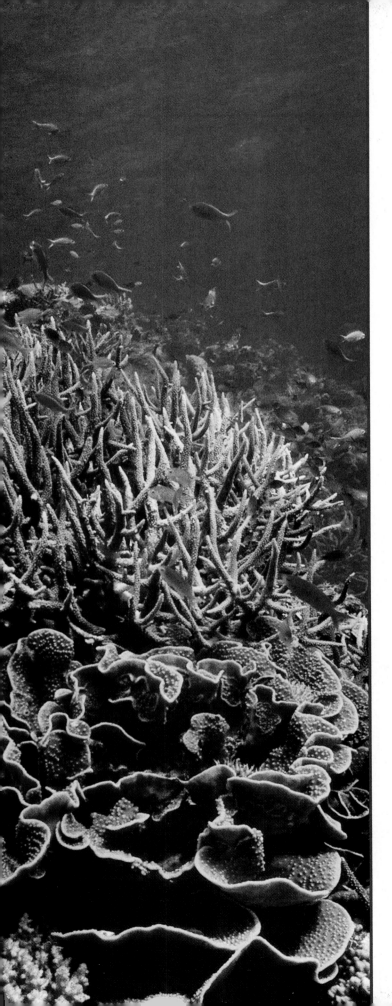

CHAPTER 18

The Ocean Floor and Seafloor Spreading

Left A coral reef in the tropical Pacific Ocean. — *Photo by Nancy Sefton/Photo Researchers, Inc.*

If all water were drained from the ocean basins, a great diversity of features would be seen including linear chains of volcanoes, deep canyons, rift valleys, and large submarine plateaus. The scenery would be nearly as varied as that on the continents.

An understanding of the ocean floor's varied topography did not unfold until the historic $3\frac{1}{2}$ year voyage of the H.M.S. *Challenger* (Figure 18.1). From December 1872 to May 1876, the *Challenger* expedition made the first, and still perhaps most comprehensive, study of the global ocean ever attempted by one agency. The 110,000-kilometer (68,000-mile) trip took the ship and its crew of scientists to every ocean except the Arctic. Throughout the voyage they sampled the depth of the water by laboriously lowering a weighted line overboard. Not many years later, the knowledge gained by the *Challenger* of the ocean's great depths and varied topography was further expanded with the laying of transatlantic cables. However, as long as ocean depth had to be measured with weighted line, our knowledge of the seafloor remained slight.

Mapping the Ocean Floor

In the 1920s a technological breakthrough occurred with the invention of electronic depth-sounding equipment (**echo sounder**). (The echo sounder is also referred to as *sonar*.) The echo sounder works by transmitting sound waves toward the ocean bottom

(Figure 18.2A). A delicate receiver intercepts the echo reflected from the bottom, and a clock precisely measures the time interval to fractions of a second. By knowing the velocity of sound waves in water (about 1500 meters per second) and the time required for the energy pulse to reach the ocean floor and return, depth can be established. The depths determined from continuous monitoring of these echoes are normally plotted so that a profile of the ocean floor is obtained. By laboriously combining profiles from several adjacent traverses, a chart of the seafloor is produced. Although much more complete and detailed than anything available before, these charts only show the largest topographic features of the ocean floor (see Figure 18.11).

In the last few decades, researchers have designed even more sophisticated echo sounders to map the ocean floor. (The major impetus for these developments comes from concerns for our national security.) In contrast to simple echo sounders, *multibeam sonar* employs an array of sound sources and listening devices. Thus, rather than obtaining the depth of a single point every few seconds, this technique obtains a profile of a narrow strip of seafloor (Figure 18.2B). These profiles are recorded every few seconds as the research vessel advances, building a continuous swath of relatively detailed coverage. (A ship will map a section of seafloor traveling in a back-and-forth pattern like that used when mowing a lawn.)

Figure 18.1 The H.M.S. *Challenger*. (From C.W. Thompson and Sir John Murray, *Report on the Scientific Results of the Voyage of the H.M.S. Challenger,* Vol. 1. Great Britain: Challenger Office, 1895, Plate 1)

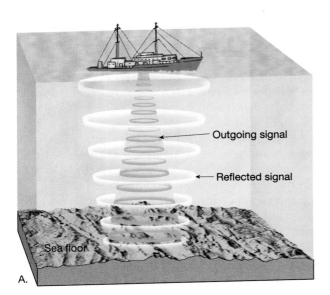
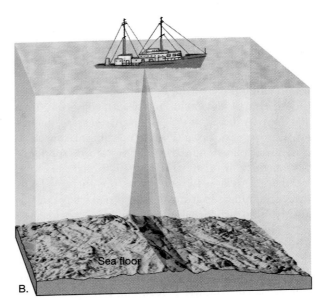

Figure 18.2 Echo sounders. **A.** An echo sounder determines the water depth by measuring the time interval required for an acoustic wave to travel from a ship to the seafloor and back. The speed of sound in water is 1500 m/sec. Therefore, depth = 1/2 (1500 m/sec × echo travel time). **B.** Modern multibeam sonar obtains a profile of a narrow swath of seafloor every few seconds.

Despite their greater efficiency and enhanced detail, research vessels equipped with multibeam sonar travel at a mere 10 to 20 kilometers per hour. It would take at least 100 vessels outfitted with this equipment hundreds of years to map the entire seafloor. By contrast, in 1991 and 1992, the *Magellan* spacecraft, traveling at 19,500 kilometers per hour, used radar to map more than 90 percent of the surface of Venus. Someday we hope to be able to view the ocean floor with the same detail we presently have for the Moon and some of the planets.

Oceanographers studying the topography of the ocean floor have delineated three major units: *continental margins*, *deep-ocean basins*, and *mid-ocean ridges*. The map in Figure 18.3 outlines these provinces for the North Atlantic, and the profile at the bottom illustrates the varied topography. Such profiles usually have their vertical dimension exaggerated many times—40 times in this case—to make topographic features more conspicuous. Because of this, the slopes shown in the seafloor profile appear to be *much* steeper than they actually are.

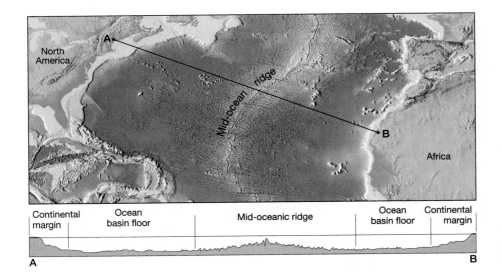

Figure 18.3 Major topographic divisions of the North Atlantic and a profile from New England to the coast of North Africa.

Continental Margins

Two main types of **continental margins** have been identified—passive and active. Passive margins are found along most of the coastal areas that surround the Atlantic Ocean, including the east coasts of North and South America, as well as the coastal areas of western Europe and Africa. Passive margins are *not* associated with plate boundaries and therefore experience very little volcanism and few earthquakes. Here, weathered materials eroded from the adjacent landmass accumulate to form a thick, broad wedge of relatively undisturbed sediments.

By contrast, active continental margins occur where oceanic lithosphere is being subducted beneath the edge of a continent. The result is a relatively narrow margin, consisting of highly deformed sediments that were scraped from the descending lithospheric slab. Active continental margins are common around the Pacific Rim, where they parallel deep oceanic trenches.

Passive Continental Margins

The features comprising a **passive continental margin** include the continental shelf, the continental slope, and the continental rise (Figure 18.4).

Continental Shelf. The **continental shelf** is a gently sloping submerged surface extending from the shoreline toward the deep-ocean basin. Because it is underlain by continental crust, it is clearly a flooded extension of the continents. The continental shelf varies greatly in width. Almost nonexistent along some continents, the shelf may extend seaward as far as 1500 kilometers (930 miles) along others. On the average, the continental shelf is about 80 kilometers (50 miles) wide and 130 meters (430 feet) deep at the seaward edge. The average inclination of the continental shelf is only about one-tenth of 1 degree, a drop of about 2 meters per kilometer (10 feet per mile). The slope is so slight that it would appear to an observer to be a horizontal surface.

Although continental shelves represent only 7.5 percent of the total ocean area, they have economic and political significance because they contain important mineral deposits, including large reservoirs of petroleum and natural gas, as well as huge sand and gravel deposits. The waters of the continental shelf also contain many important fishing grounds that are significant sources of food.

Although the continental shelf is relatively featureless, some areas are mantled by extensive glacial deposits and are thus quite rugged. The most profound features are long valleys running from the coastline into deeper waters. Many of these valleys are the seaward extensions of river valleys on the adjacent landmass. Such valleys appear to have been excavated during the Pleistocene epoch (Ice Age).

During this time great quantities of water were tied up in vast ice sheets on the continents. This caused sea level to drop by 100 meters or more, exposing large areas of the continental shelves (see Figure 12.A, p. 297). Because of this drop in sea level, rivers extended their courses, and land-dwelling plants and animals inhabited the newly exposed portions of the continents. Dredging off the east coast of North America has produced the remains of numerous land dwellers, including mammoths, mastodons, and

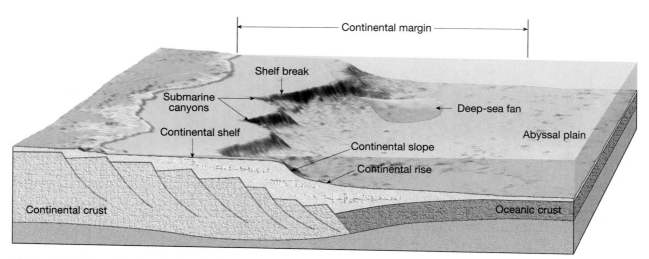

Figure 18.4 Schematic view showing the provinces of a passive continental margin. Note that the slopes shown for the continental shelf and continental slope are greatly exaggerated. The continental shelf has an average slope of one-tenth of 1 degree, while the continental slope has an average slope of about 5 degrees.

horses, adding to the evidence that portions of the continental shelves were once above sea level.

Most passive continental shelves, such as those along the east coast of the United States, consist of thick accumulations of shallow-water sediments. These sediments are frequently several kilometers thick and are interbedded with limestones that formed during earlier periods of coral reef building, a process that occurs only in shallow water. Such evidence led researchers to conclude that these thick accumulations of sediment are produced along a gradually subsiding continental margin.

Continental Slope. Marking the seaward edge of the continental shelf is the **continental slope**, a relatively steep structure (as compared with the shelf) that marks the boundary between continental crust and oceanic crust (Figure 18.4). Although the inclination of the continental slope varies greatly from place to place, it averages about 5 degrees and in places may exceed 25 degrees. Further, the continental slope is a relatively narrow feature, averaging only about 20 kilometers in width.

Continental Rise. In regions where trenches do not exist, the continental slope merges into a more gradual incline known as the **continental rise**. Here the slope drops to about one-third degree, or about 6 meters per kilometer. Whereas the width of the continental slope averages about 20 kilometers, the continental rise may extend for hundreds of kilometers into the deep-ocean basin.

The continental rise consists of a thick accumulation of sediment that moved downslope from the continental shelf to the deep-ocean floor. The sediments are delivered to the base of the continental slope by *turbidity currents* that follow submarine canyons (see Figure 18.6). (We will discuss these shortly.) When these muddy currents emerge from the mouth of a canyon onto the relatively flat ocean floor, they deposit sediment that forms a **deep-sea fan**. Deep-sea fans have the same basic shape as alluvial fans, which form at the foot of steep mountain slopes. As fans from adjacent submarine canyons grow, they coalesce to produce the continuous apron of sediment at the base of the continental slope that we call the continental rise.

Active Continental Margins

Along some coasts the continental slope descends abruptly into a deep-ocean trench located between the continent and ocean basin. Thus, the landward wall of the trench and the continental slope are essentially the same feature. In such locations, the continental shelf is very narrow, if it exists at all.

Active continental margins are located primarily around the Pacific Ocean in areas where the leading edge of a continent is overrunning oceanic lithosphere (Figure 18.5). Here sediments from the ocean floor and pieces of oceanic crust are scraped from the descending oceanic plate and plastered against the edge of the overriding continent. This chaotic accumulation of deformed sediment and scraps of oceanic crust is called an **accretionary wedge**. Prolonged subduction, along with the accretion of sediments on the landward side of the trench, can produce a large accumulation of sediments along a continental margin. For example, a large accretionary wedge is found along the northern coast of Japan's Honshu island.

Some subduction zones have little or no accumulation of sediments, indicating that ocean sediments are being carried into the mantle with the subducting plate. Here the continental margin is very narrow as the trench may lie a mere 50 kilometers off-shore.

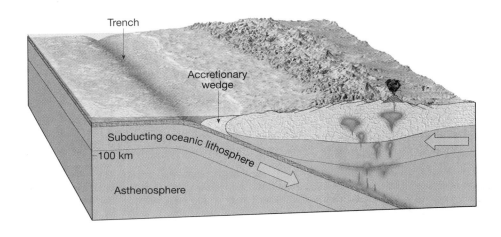

Figure 18.5 Active continental margin. Here sediments from the ocean floor are scraped from the descending plate and added to the continental crust as an accretionary wedge.

Submarine Canyons and Turbidity Currents

Deep, steep-sided valleys known as **submarine canyons** are cut into the continental slope and may extend across the entire continental rise to the deep-ocean basin (Figure 18.6). Although some of the canyons appear to be the seaward extensions of river valleys such as the Hudson valley and Amazon River, others are not directly associated with existing river systems. Furthermore, because these canyons extend to depths far below the lowest level of the sea during the Ice Age, we cannot attribute their formation to stream erosion. These features must be created by some process that operates far below the ocean surface.

Most available information favors the view that submarine canyons have been eroded, at least in part, by turbidity currents (Figure 18.6). **Turbidity currents** are downslope movements of dense, sediment-laden water. They are created when sand and mud on the continental shelf and slope are dislodged, perhaps by an earthquake, and are thrown into suspension. Because such mud-choked water is denser than normal seawater, it flows downslope, eroding and accumulating more sediment. The erosional work repeatedly carried on by these muddy torrents is thought to be the major force in the growth of most submarine canyons (see Box 18.1).

Narrow continental margins, such as the one located along the California coast, are dissected by numerous submarine canyons. Here, headward erosion has extended many of these canyons landward into shallow water where longshore currents are active. As a result, sediments carried to the coasts by rivers are transported along the shore until they reach a submarine canyon. This steady supply of sediment collects until it becomes unstable and moves as a massive landslide (turbidity current) to the deep ocean floor.

Turbidity currents eventually lose momentum and come to rest along the floor of the ocean basin (Figure 18.6). As these currents slow, the suspended sediments begin to settle out. First, the coarser, heavier sand is dropped, followed by successively finer deposits of silt and then clay. Consequently, these deposits, called **turbidites**, are characterized by a decrease in sediment grain size from bottom to top, a phenomenon known as **graded bedding** (Figure 18.6).

Although there is still more to be learned about the complex workings of turbidity currents, it has been well established that they are an important mechanism of sediment transport in the ocean. By the action of turbidity currents, submarine canyons are created and sediments are carried to the deep-ocean floor.

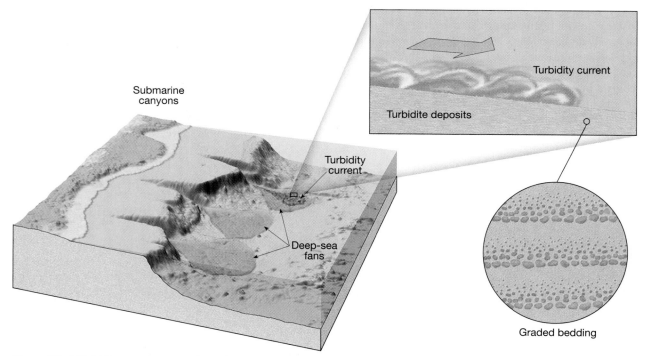

Figure 18.6 Turbidity currents move downslope, eroding the continental margin to enlarge submarine canyons. These sediment-laden density currents eventually lose momentum and deposit their loads of sediment as deep-sea fans. Beds deposited by these currents are called turbidites. Each event produces a single bed characterized by a decrease in sediment size from bottom to top, a feature known as a graded bed.

Evidence for the Existence of Turbidity Currents

For many years the existence of turbidity currents in the ocean was a matter of considerable debate among marine geologists. In the 1950s two lines of evidence helped establish turbidity currents as important mechanisms of submarine erosion and sediment transportation.

The first important evidence came from records of a rather severe earthquake that took place off the coast of Newfoundland in 1929 and resulted in the breakage of 13 transatlantic telephone and telegraph cables. At the time, it was presumed that the tremor had caused the multiple breaks. However, when the data were analyzed, it appeared that this was not the case.

After plotting the locations of the breaks on a map, researchers saw that all the breaks had occurred along the steep continental slope and the gentler continental rise. Because the time of each break was known from information provided by automatic recorders, a pattern of what had happened could be deduced. The breaks high up on the continental slope took place first, almost concurrently with the earthquake. The other breaks happened in succession, the last occurring 13 hours later, some 720 kilometers (450 miles) from the source of the quake (Figure 18.A).

The breaks downslope had obviously taken place too long after the tremor to have been caused by the shock of the earthquake. The existence of a turbidity current, triggered by the quake, thus appeared as a plausible alternative. As the avalanche of sediment-choked water raced downslope, it snapped the cables in its path. Investigators calculated that the current reached speeds approaching 80 kilometers (50 miles) per hour on the continental slope and about 24 kilometers (15 miles) per hour on the more gently sloping continental rise.

A second compelling line of evidence relating turbidity currents to submarine erosion and transportation of sediment came from the examination of deep-sea sediment samples. These cores show that extensive graded beds of sand, silt, and clay exist in the quiet waters of the deep ocean. Some samples also include fragments of plants and animals that live only in the shallower waters of the continental shelves. No mechanism other than turbidity currents has been identified to satisfactorily explain the existence of these deposits.

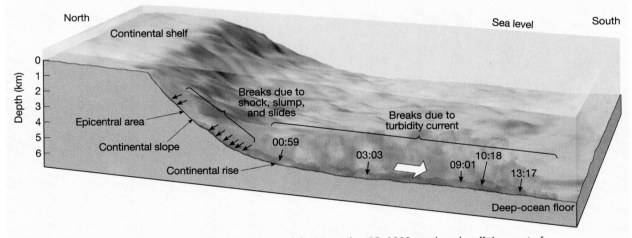

Figure 18.A Profile of the seafloor showing the events of the November 18, 1929, earthquake off the coast of Newfoundland. The arrows point to cable breaks; the numbers show times of breaks in hours and minutes after the earthquake. Vertical scale is greatly exaggerated. (After B. C. Heezen and M. Ewing, "Turbidity Currents and Submarine Slump and the 1929 Grand Banks Earthquake," *American Journal of Science* 250:867)

Features of the Deep-Ocean Basin

Between the continental margin and the oceanic ridge system lies the **deep-ocean basin**. The size of this region—almost 30 percent of Earth's surface—is roughly comparable to the percentage of the surface that projects above sea level as land. Here we find remarkably flat regions known as abyssal plains, broad volcanic peaks called seamounts, and deep-ocean trenches, which are extremely deep linear depressions in the ocean floor.

Deep-Ocean Trenches

Deep-ocean **trenches** are long, relatively narrow features that form the deepest parts of the ocean. Most trenches are located in the Pacific Ocean where some approach or exceed 10,000 meters in depth, and at least a portion of one, the Challenger Deep in the Mariana trench, is more than 11,000 meters below sea level (Figure 18.7). Table 18.1 presents dimensional characteristics of some of the larger trenches.

Although deep-ocean trenches represent only a very small portion of the area of the ocean floor, they are nevertheless significant geologic features. Trenches are the sites where moving lithospheric plates plunge back into the mantle. In addition to the earthquakes created as one plate descends beneath another, volcanic activity is also associated with these regions. Thus, trenches are often paralleled by volcanic island arcs. Furthermore, volcanic mountains, such as those making up portions of the Andes, are located parallel to trenches that lie adjacent to continental margins. The release of volatiles

(water) from a descending plate triggers melting in the wedge of asthenosphere above it. This buoyant material slowly migrates upward and gives rise to volcanic activity at the surface.

Abyssal Plains

Abyssal plains are incredibly flat features; in fact, these regions are likely the most level places on Earth. The abyssal plain found off the coast of Argentina, for example, has less than 3 meters of relief over a distance exceeding 1300 kilometers. The monotonous topography of abyssal plains will occasionally be interrupted by the protruding summit of a partially buried volcanic structure.

Using seismic profilers, instruments whose signals penetrate far below the ocean floor, researchers

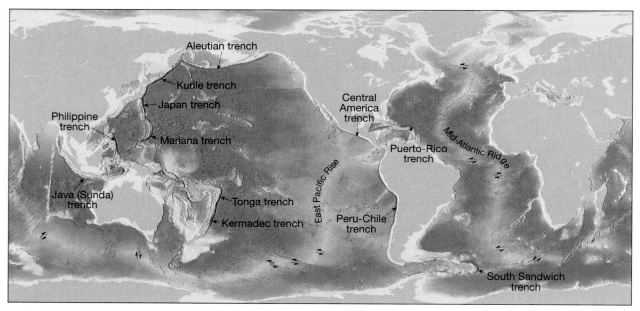

Figure 18.7 Distribution of the world's major oceanic trenches.

Table 18.1 Dimensions of Some Deep-Ocean Trenches

Trench	Depth (kilometers)	Average Width (kilometers)	Length (kilometers)
Aleutian	7.7	50	3700
Japan	8.4	100	800
Java	7.5	80	4500
Kurile–Kamchatka	10.5	120	2200
Mariana	11.0	70	2550
Central America	6.7	40	2800
Peru–Chile	8.1	100	5900
Philippine	10.5	60	1400
Puerto Rico	8.4	120	1550
South Sandwich	8.4	90	1450
Tonga	10.8	55	1400

have determined that abyssal plains owe their relatively featureless topography to thick accumulations of sediment that have buried an otherwise rugged ocean floor. The nature of the sediment indicates that these plains consist primarily of sediments transported far out to sea by turbidity currents.

Abyssal plains are found in all the oceans. However, because the Atlantic Ocean has fewer trenches to act as traps for the sediments carried down the continental slope, it has more extensive abyssal plains than does the Pacific.

Seamounts

Dotting the ocean floors are isolated volcanic peaks called **seamounts**. These features may rise hundreds of meters above the surrounding topography. Although these broad conical peaks have been discovered in all the oceans, the greatest number have been identified in the Pacific.

Many of these undersea volcanoes form near oceanic ridges, regions of seafloor spreading. If a volcano grows rapidly, it may emerge as an island. Examples of volcanic islands in the Atlantic include the Azores, Ascension, Tristan da Cunha, and St. Helena.

While they exist as islands, some of these volcanoes are eroded to near sea level by running water and wave action. Over a span of millions of years the islands gradually sink as the moving plate slowly carries them from the oceanic ridge area. These submerged, flat-topped seamounts are called **guyots**. In other instances, guyots may be remnants of eroded volcanic islands that were formed away from the ridge crest, possibly by hot-spot activity. Here subsidence occurs after the volcanic activity ceases and the seafloor cools and contracts.

Coral Reefs and Atolls

Coral reefs are among the most picturesque features found in the ocean. They are constructed primarily from the calcareous (calcite-rich) skeletal remains and secretions of corals and certain algae. The term *coral reef* is somewhat misleading in that it makes no mention of the skeletons of many small animals and plants found inside the branching framework built by the corals; nor does it reveal that secretions of algae help bind the entire structure together.

Coral reefs are confined largely to the warm, clear waters of the Pacific and Indian oceans, although a few occur elsewhere. Reef-building corals grow best in waters with an average annual temperature of about 24°C (75°F). They can survive neither sudden temperature changes nor prolonged exposure

to temperatures below 18°C (64°F). In addition, these reef-builders require clear sunlit water. Consequently, the limiting depth of active reef growth is only about 45 meters. Clear blue waters such as those in the Bahamas support active reef building.

In 1831, the naturalist Charles Darwin set out aboard the British ship H.M.S. *Beagle* on its famous 5-year expedition that circumnavigated the globe. One outcome of Darwin's studies was the development of an hypothesis on the formation of coral islands, called **atolls**. As Figure 18.8 illustrates, atolls consist of a nearly continuous ring of coral reef surrounding a central lagoon. Darwin's hypothesis explained what seemed to be a paradox; that is, how can corals, which require warm, shallow, sunlit water no deeper than a few dozen meters to live, create structures that reach thousands of meters to the floor of the ocean? Commenting on this in his book *The Voyage of the Beagle* Darwin observed:

> …from the fact of the reef-building corals not living at great depths, it is absolutely certain that throughout these vast areas, wherever there is now an atoll, a foundation must have originally existed within a depth of from 20 to 30 fathoms from the surface.*

The essence of Darwin's hypothesis was that coral reefs form on the flanks of sinking volcanic islands. As the island slowly sinks, the corals continue to build the reef complex upward (Figure 18.9).

As Darwin further noted:

> For as mountain after mountain, and island after island slowly sank beneath the water, fresh bases would be successively afforded for the growth of the corals.

Thus, atolls, like guyots, are thought to owe their existence to the gradual sinking of oceanic crust. In succeeding years there were numerous challenges to Darwin's proposal. These arguments were not completely put to rest until after World War II when the United States made extensive studies of two atolls (Eniwetok and Bikini), which became sites for testing atomic bombs. Drilling operations at these atolls revealed that volcanic rock did indeed underlie the thick coral reef structure. This finding was a striking confirmation of Darwin's explanation.

---

*One fathom equals 1.8 meters (6 feet), the approximate distance from fingertip to fingertip of a person with outstretched arms.

A.

B.

Figure 18.8 **A.** An aerial view of Tetiaroa Atoll in the Pacific. The light blue waters of the relatively shallow lagoon contrast with the dark blue color of the deep ocean surrounding the atoll. (Photo by Douglas Peebles Photography) **B.** View from space of a group of atolls in the Pacific Ocean. (Courtesy of NASA)

Seafloor Sediments

Except for a few areas, such as near the crests of mid-ocean ridges, the ocean floor is mantled with sediment. Part of this material has been deposited by turbidity currents, and the rest has slowly settled to the bottom from above. The thickness of this carpet of debris varies greatly. In some trenches, which act as traps for sediments originating on the continental margin, accumulations may exceed 9 kilometers. In general, however, sediment accumulations are considerably less. In the Pacific Ocean, uncompacted sediment measures about 600 meters or less, while on the floor of the Atlantic, the thickness varies from 500 to 1000 meters.

Although deposits of sand-sized particles are found on the deep-ocean floor, mud is the most common sediment covering this region. Muds also predominate on the continental shelf and slopes; however, the sediments in these areas are coarser overall because of greater quantities of sand.

Sampling has revealed that sand deposits are most prevalent as beach deposits along the shore. In some cases, though, coarse sediment, which we normally expect to be deposited near the shore, occurs in irregular patches at greater depths near the seaward limits of the continental shelves. Some sand may have been deposited by strong localized currents capable of moving coarse sediment far from shore, but much of it appears to be the result of sand deposition on ancient beaches. Such beaches formed during the Ice Age, when sea level was much lower than today.

Types of Seafloor Sediments

Seafloor sediments can be classified according to their origin into three broad categories: (1) **terrigenous** ("derived from the land"); (2) **biogenous** ("derived from organisms"); and (3) **hydrogenous** ("derived from water"). Although each category is discussed separately, remember that all seafloor sediments are mixtures. No body of sediment comes from a single source.

Terrigenous Sediments. Terrigenous sediment consists primarily of mineral grains that were weathered from continental rocks and transported to the ocean. The sand-sized particles settle near shore. However, because the very smallest particles take years to settle to the ocean floor, they may be carried for thousands of kilometers by ocean currents. As a consequence, virtually every part of the ocean receives some terrigenous sediment. The rate at which this sediment accumulates on the deep-ocean floor, though, is very slow. From 5000 to 50,000 years are necessary for a 1-centimeter layer to form. Conversely, on the continental margins near the mouths of large rivers,

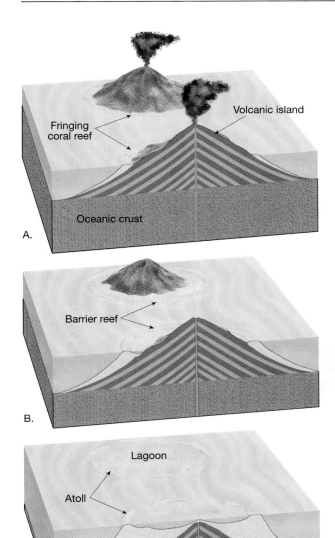

A.

B.

C.

Figure 18.9 Formation of a coral atoll due to the gradual sinking of oceanic crust and upward growth of the coral reef. As the volcanic island sinks, the fringing reef (**A**) gradually becomes a barrier reef (**B**). Eventually, the volcano is completely submerged and an atoll remains (**C**).

terrigenous sediment accumulates rapidly. In the Gulf of Mexico, for example, the sediment has reached a depth of many kilometers.

Because fine particles remain suspended in the water for a very long time, ample opportunity exists for chemical reactions to occur. Because of this, the colors of the deep-sea sediments are often red or brown. This results when iron in the particle or in the water reacts with dissolved oxygen in the water and produces a coating of iron oxide (rust).

Biogenous Sediments. Biogenous sediment consists of shells and skeletons of marine animals and plants (Figure 18.10). This debris is produced mostly by microscopic organisms living in the sunlit waters near the ocean surface. Their remains constantly "rain" down on the seafloor.

The most common biogenous sediments are known as *calcareous* ($CaCO_3$) *oozes*, and, as their name implies, they have the consistency of thick mud. These sediments are produced by organisms that inhabit warm surface waters. When calcareous hard parts slowly sink through a cool layer of water, they begin to dissolve. This results because cold seawater is rich in carbon dioxide and is thus more acidic than is warm

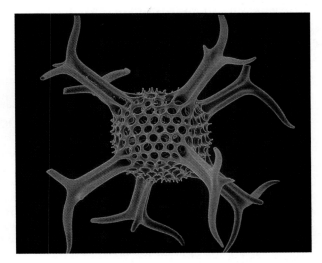

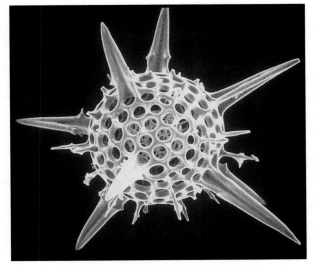

Figure 18.10 Microscopic radiolarian hard parts are examples of biogenous sediments. These photomicrographs have been enlarged hundreds of times. (Photos by Manfred Kage/Peter Arnold, Inc.)

water. In seawater deeper than about 4500 meters (15,000 feet), calcareous shells will completely dissolve before they reach the bottom. Consequently, calcareous ooze does not accumulate in the deep-ocean basins.

Other biogenous sediments include *siliceous* (SiO_2) *oozes* and phosphate-rich materials. The former are composed primarily of opaline skeletons of diatoms (single-celled algae) (Figure 18.10) and radiolarians (single-celled animals), whereas the latter are derived from the bones, teeth, and scales of fish and other marine organisms.

Hydrogenous Sediments. Hydrogenous sediment consists of minerals that crystallize directly from seawater through various chemical reactions. For example, some limestones are formed when calcium carbonate precipitates directly from the water; however, most limestone is composed of biogenous sediment.

One of the principal examples of hydrogenous sediment, and one of the most important sediments on the ocean floor in terms of economic potential, are **manganese nodules**. These rounded blackish lumps are composed of a complex mixture of minerals that form very slowly on the floors of the ocean basins (see Box 18.2).

Mid-Ocean Ridges

Our knowledge of **mid-ocean ridges** comes from soundings taken of the ocean floor, core samples obtained from deep-sea drilling, visual inspection using deep-diving submersibles, and even firsthand inspection of slices of ocean floor that have been shoved up onto dry land. The ocean ridge system is characterized by an elevated position, extensive faulting, and

---

Box 18.2

Resources from the Deep-Ocean Floor

A rapidly growing world population, coupled with the desire of people everywhere to have a higher standard of living, is increasing the demand for mineral resources. Today, oil, natural gas, titanium, tin, gold, and diamonds are being recovered not only on the continents but also from shallow offshore areas.

In the future, the deep-ocean basins may also become sites of mineral production. Exploration already has discovered large deposits of manganese nodules and rich accumulations of metallic sulfides, similar to ores mined on land.

Manganese nodules occur in many parts of the deep-ocean basins. Despite technological and political problems, mining of these deposits may one day be possible. At least one company has attempted to file a claim on manganese nodules covering a vast area of the Pacific Ocean floor.

Manganese nodules are rounded, dark lumps of mixed composition (Figure 18.B). About 30 percent of a typical nodule is manganese dioxide (MnO_2). Roughly 20 percent of each nodule is iron oxide (Fe_2O_3). The manganese and iron are important, but the real interest lies in more valuable

metals that may be enriched in the nodules: copper, nickel, and cobalt.

Nodules are widely distributed, but not all regions are equally promising for development. Good locations have abundant nodules (more than 5 kilograms per square meter) that contain the economically optimum mix of copper, nickel, and cobalt. Sites meeting these criteria are relatively limited. Also, the logistics of extracting nodules from deep-ocean basins must be worked out, in addition to political problems of ownership and access.

Sulfide deposits are a second potential source of deep-ocean mineral resources. The richest submarine metallic sulfide deposits yet known have been discovered in a series of deep basins along the axis of the Red Sea. The largest deposit ranks with the largest deposits on land. It may contain 100 million metric tons of potential ore with 29 percent iron, over 3 percent zinc, 1 percent copper, and substantial gold and silver.

Located above these deposits are pools of metallic brines that have salinities two to three times normal and temperatures that exceed 36°C (97°F). Hydrothermal convection of these reactive brines is believed to extract

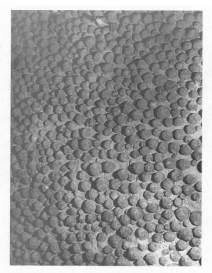

Figure 18.B Manganese nodules on the floor of the Pacific at a depth of more than 5000 meters. (Photo courtesy of Lawrence Sullivan, Lamont-Doherty Earth Observatory)

metallic ions from the underlying crustal rocks and to deposit them as metallic sulfides on the seafloor. Despite their depth of 2000 meters (6500 feet) or more, mining is foreseeable as world demand drives prices upward.

numerous volcanic structures that have developed on the newly formed crust (Figure 18.11).

The interconnected ocean ridge system is the longest topographic feature on Earth's surface, exceeding 70,000 kilometers (43,000 miles) in length. Representing 20 percent of Earth's surface, the ocean ridge system winds through all major oceans in a manner similar to the seam on a baseball (Figure 18.11). The term *ridge* may be misleading as these features are not narrow but have widths of from 3000 to 4000 kilometers and, in places, may occupy as much as one-half the total area of the ocean floor. An examination of Figure 18.11 shows that the ridge system is broken into segments that are offset by large transform faults. Further, along the axis of some segments are deep down-faulted structures called **rift valleys**.

Although ocean ridges stand 2 to 3 kilometers above the adjacent deep-ocean basins, they are much different from mountains found on the continents. Rather than consisting of thick sequences of folded and faulted sedimentary rocks, ocean ridges consist of layer upon layer of basaltic rocks that have been faulted and uplifted.

The flanks of most ridges are topographically relatively featureless and rise very gradually (slope less than 1 degree) toward the ridge crest. Approaching the ridge crest, the topography begins to exhibit greater local relief as volcanic structures and faulted valleys become more prominent. The most rugged topography is found on those ridges that exhibit large rift valleys.

Partly because of its accessibility to both American and European scientists, the Mid-Atlantic Ridge has been studied more thoroughly than other ridge systems (Figure 18.11). The Mid-Atlantic Ridge is a broad, submerged structure standing 2500 to 3000 meters above the adjacent ocean basin floor. In a few places, such as Iceland, the ridge has actually grown above sea level. Throughout most of its length, however, this divergent plate boundary lies 2500 meters below sea level. Another prominent feature of the Mid-Atlantic Ridge is its deep linear rift valley extending along the ridge axis. In places this rift valley is deeper than the Grand Canyon of the Colorado River and two or three times as wide. The name *rift valley* has been applied to this feature because it is so strikingly similar to continental rift valleys such as the East African Rift Valley.

Seafloor Spreading

The concept of seafloor spreading was formulated in the early 1960s by Harry Hess of Princeton University. Later geologists were able to verify Hess's contention that seafloor spreading occurs along relatively narrow zones, called **rift zones**, located at the crests of ocean ridges (see Box 18.3). As plates move apart, magma wells up into the newly created fractures and generates new slivers of oceanic lithosphere (Figure 18.12). This apparently unending process generates new lithosphere that moves from the ridge crest in a conveyor-belt fashion.

Spreading Rates and Ridge Topography

Active rift zones are characterized by frequent but generally weak earthquakes and a rate of heat flow that is greater than most other crustal segments. Here vertical displacement of large slabs of oceanic crust caused by faulting and the growth of volcanic piles contribute to the characteristically rugged topography of the oceanic ridge system. Further, the rocks along the ridge axis appear very fresh and are nearly devoid of sediment. Away from the ridge axis the topography becomes more subdued, and the thickness of the sediments and the depth of the water increase. Gradually the ridge system grades into the flat, sediment-laden abyssal plains of the deep-ocean basin.

When various segments of the oceanic ridge system were studied in detail, numerous differences came to light. For example, the East Pacific Rise has a relatively fast rate of spreading that averages about 6 centimeters per year and reaches a maximum of 10 centimeters per year along a section of the ridge located near Easter Island (Figure 18.13). By contrast, the spreading rate in the North Atlantic is much less, averaging about 2 centimeters per year. Apparently the rate of spreading strongly influences the appearance of the ridge system. The slow spreading along the Mid-Atlantic Ridge contributes to its rugged topography and large central rift valley. By contrast, the rapid spreading of the East Pacific Rise is thought to account for its more subdued topography and the lack of a rift valley. Despite these differences, all ridge systems generate new seafloor in a similar manner.

Because newly formed sections of the ocean floor are warm, they are also rather buoyant. This buoyancy causes large blocks to shear from the seafloor and be elevated. Because the rate of spreading along the Mid-Atlantic Ridge is relatively slow, the upward displacement of oceanic slabs is more pronounced there than along faster-spreading centers such as the East Pacific Rise. Thus, along the Mid-Atlantic Ridge, uplifted sections form nearly vertical walls that border the central rift zone. As seafloor spreading continues, the earlier-formed blocks are wedged away from the ridge axis and replaced by more recently formed, and therefore warmer and

Figure 18.11 The topography of Earth's solid surface is shown on these two pages.

more buoyant, segments of ocean crust. This process contributes both to the height of the Mid-Atlantic Ridge and to its rugged topography.

The primary reason for the elevated position of a ridge system is the fact that newly created oceanic crust is hot, and therefore it occupies more volume than do cooler rocks of the deep-ocean basin. As the young lithosphere travels away from the spreading center, it gradually cools and contracts. This thermal contraction accounts in part for the greater ocean depths that exist away from the ridge. It takes almost 100 million years before cooling and contraction cease completely. By this time, rock that was once part of an elevated ocean ridge system is located in

the deep-ocean basin, where it is mantled by thick accumulations of sediment.

During seafloor spreading new material is added about equally to the two diverging plates; hence, we would expect new ocean floor to grow symmetrically on each side of a centrally located ridge crest. Indeed, the ridge systems of the Atlantic and Indian oceans are located near the middle of these water bodies and as a consequence are named mid-ocean ridges. However, the East Pacific Rise is situated far from the center of the Pacific Ocean. Despite uniform spreading along the East Pacific Rise, much of the Pacific Basin that once lay east of this spreading center has been overridden by the westward migration of the American plate.

Box 18.3

A Close-up View of the Ocean Floor

Much has been (and continues to be) learned about the floor of the ocean from echo sounders and other remote sensing equipment, as well as from drilling and sampling from surface ships. However, oceanographers in the 1970s became aware that direct manned observation was essential to bring about a clearer understanding of many deep-sea phenomena. What was needed was a firsthand view of the previously unseen world below.

Today, the names and accomplishments of deep-diving, manned submersibles such as the *Alvin* are common knowledge among oceanographers (Figure 18.C). Manned submersibles are now extending the coverage provided by traditional oceanographic research vessels by allowing scientists to investigate the fine-scale features that previously eluded detection.

One of the pioneering research projects that used deep-diving submersibles was a cooperative venture called Project FAMOUS (French-American Mid-Ocean Undersea Study). In 1974, after 3 years of preliminary surveying and study by surface ships, two French submersibles and one American vessel made a total of 44 dives to the floor of the Atlantic. The primary purpose of the project was to examine the structure of the rift valley in the Mid-Atlantic Ridge. The data collected by the three vessels proved invaluable and led to more realistic explanations of how the spreading process works in creating new ocean floor.

Later, dives were made by the *Alvin* to a spreading center at 20°N latitude on the East Pacific Rise near the mouth of the Gulf of California. Here, in addition to gathering large quantities of basic data, the scientists aboard the *Alvin* discovered the existence of spectacular, geyserlike hot springs. They witnessed 2- to 5-meter-high chimney-like structures spewing dark, mineral-rich, hot (350°C–400°C) water (Figure 18.D). As the heated solutions hit the surrounding 2°C seawater, sulfides of copper, iron, and zinc precipitated immediately, forming mounds of minerals around the steaming vents.

In addition to viewing firsthand the formation of massive sulfide deposits, the scientists aboard the *Alvin* found communities of exotic, bottom-dwelling animals living near cooler (20°C) hot springs. The discovery of an animal community thriving more than 3 kilometers below the surface where no light can reach was totally unexpected. An analysis of the sulfur-rich vent water, as well as the stomach contents of some animals, revealed that the base of the food chain was sulfur-oxidizing bacteria.

Dives such as the ones briefly highlighted here have demonstrated the value of deep-diving, manned submersibles in detecting and studying the fine-scale features of the ocean floor. These vessels now occupy a permanent and important place as tools in oceanographic research.

Figure 18.C The deep-diving submersible *Alvin* is 7.6 meters long, weighs 16 tons, has a cruising speed of 1 knot, and can reach depths as great as 4000 meters. A pilot and two scientific observers are aboard during a normal 6- to 10-hour dive. (Photo courtesy of Woods Hole Oceanographic Institution)

Figure 18.D A black smoker spewing hot, mineral-rich water along the East Pacific Rise. As heated solutions meet cold seawater, sulfides of copper, iron, and zinc precipitate immediately, forming mounds of minerals around these vents. (Photo by Dudley Foster, Woods Hole Oceanographic Institution)

Figure 18.12 A photograph taken from the *Alvin* during project FAMOUS shows lava extrusions in the rift valley of the Mid-Atlantic Ridge. Large toothpastelike extrusions such as this were common features. A mechanical arm is sampling an adjacent blisterlike extrusion. (Photo courtesy of Woods Hole Oceanographic Institution)

Figure 18.13 The relative age of oceanic crust beneath deep-sea deposits. Notice that the youngest rocks (bright red areas) are found along the oceanic ridge crests and the oldest oceanic crust (brown areas) is located adjacent to the continents and subduction zones in the western Pacific. When you observe the Atlantic Basin, a symmetrical pattern centered on the Mid-Atlantic Ridge crest becomes apparent. This pattern verifies the fact that seafloor spreading generates new oceanic crust equally on both sides of a spreading center. Further, compare the widths of the yellow stripes in the Pacific Basin with those in the South Atlantic. Because these stripes were produced during the same time period, this comparison verifies that the rate of seafloor spreading was faster in the Pacific than in the South Atlantic. (After *The Bedrock Geology of the World*, by R. L. Larson et al. Copyright © by W. H. Freeman)

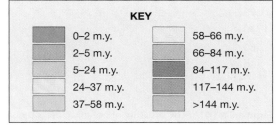

KEY

0–2 m.y.	58–66 m.y.
2–5 m.y.	66–84 m.y.
5–24 m.y.	84–117 m.y.
24–37 m.y.	117–144 m.y.
37–58 m.y.	>144 m.y.

Structure of the Oceanic Crust

Although most oceanic crust forms out of view, far below sea level, geologists have been able to examine the structure of the ocean floor firsthand. In such locations as Newfoundland, Cyprus, and California, slivers of oceanic crust have been thrust high above sea level. From these outcrops, researchers conclude that the ocean floor consists of three distinct layers (Figure 18.14A). The upper layer is composed mainly of basaltic **pillow lavas**. The middle layer is made up of numerous interconnected dikes called **sheeted dikes**. Finally, the lower layer is made up of gabbro, the coarse-grained equivalent of basalt, that crystallized at depth. This sequence of rocks is called an **ophiolite complex** (Figure 18.14A). From studies of various ophiolite complexes and related data, geologists have pieced together a scenario for the formation of the ocean floor.

The magma that migrates upward to create new ocean floor originates from partially melted peridotite in the asthenosphere. In the region of the rift zone, this magma source may lie no more than 35 kilometers below the seafloor. Being molten and less dense than the surrounding solid rock, the magma gradually moves upward where it enters large reservoirs located only a few kilometers below the ridge crest (Figure 18.14B). As the ocean floor is pulled or pushed apart, numerous fractures develop in the crust, permitting this molten rock to migrate to the surface. (Eventually magma in these vertical fractures will crystallize, generating the zone of sheeted dikes.)

During each eruptive phase, the initial lava flows are quite fluid and spread over the rift zone in broad, thin sheets. As new flows are added to the ocean floor, each is cut by fractures that allow additional lava to migrate upward and form overlying layers. Later in each eruptive cycle, as the magma in the shallow reservoir cools and thickens, shorter flows with a more characteristic pillow form occur. Recall that pillow lava has the appearance of large, elongate sand bags stacked one atop another (Figure 18.15). Depending on the rate of flow, the thick pillow lavas may build into volcano-sized mounds. These mounds will eventually be cut off from their supply of magma and be carried away from the ridge crest by seafloor spreading.

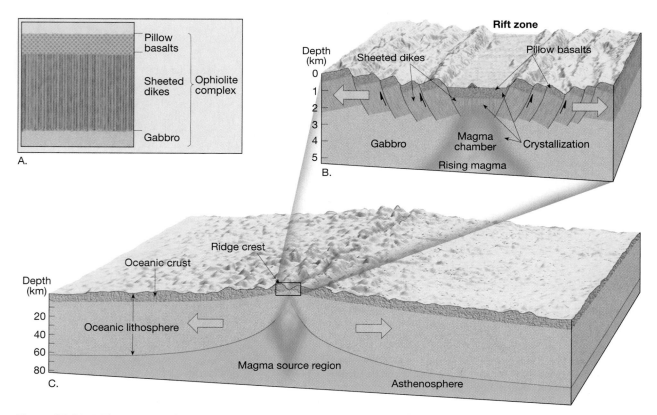

Figure 18.14 **A.** The structure of oceanic crust is thought to be equivalent to the ophiolite complexes that have been discovered elevated above sea level in such places as California and Newfoundland. **B.** The formation of the three units of an ophiolite complex in the rift zone of an oceanic ridge. **C.** Diagram illustrating the site where new ocean crust is generated.

Figure 18.15 Ancient pillow lava at Trinity Bay, Newfoundland. (Photo courtesy of the Geological Survey of Canada, photo no. 152581)

The magma that remains in the subterranean chamber will crystallize at depth to generate thick units of coarse-grained gabbro. This lowest rock unit forms as crystallization takes place along the walls and floor of the magma chamber. In this manner the processes at work along a ridge system generate the entire sequence of rocks found in an ophiolite complex.

Chapter Summary

- Ocean depths are determined using an *echo sounder*, a device carried by a ship that bounces sound off the ocean floor. The time it takes for the sound waves to make the round trip to the bottom and back to the ship is directly related to the depth. Continuous data from the echoes are plotted to produce a profile of the ocean floor. Although much of the ocean floor has been mapped using echo sounders, only the large topographic features are shown.

- Oceanographers studying the topography of the ocean basins have delineated three major units: *continental margins*, *deep ocean basins*, and *mid-ocean ridges*.

- The zones that collectively make up a *passive continental margin* include the *continental shelf* (a gently sloping, submerged surface extending from the shoreline toward the deep-ocean basin); *continental slope* (the true edge of the continent, which has a steep slope that leads from the continental shelf into deep water); and in regions where trenches do not exist, the steep continental slope merges into a more gradual incline known as the *continental rise*. The continental rise consists of sediments that have moved downslope from the continental shelf to the deep-ocean floor.

- *Active continental margins* are located primarily around the Pacific Ocean in areas where the leading edge of a continent is overrunning oceanic lithosphere. Here sediment scraped from the descending oceanic plate is plastered against the continent to form a collection of sediments called an *accretionary wedge*. An active continental margin generally has a narrow continental shelf, which grades into a deep-ocean trench.

- *Submarine canyons* are deep, steep-sided valleys that originate on the continental slope and may extend to depths of 3 kilometers. Some of these canyons appear to be the seaward extensions of river valleys. However, most information seems to favor the view that many submarine canyons are excavated by *turbidity currents* (downslope movements of dense, sediment-laden water). *Turbidites*, sediments deposited by turbidity currents, are characterized by a decrease in sediment grain size from bottom to top, a phenomenon known as *graded bedding*.

- The deep ocean basin lies between the continental margin and the mid-ocean ridge system. Its features include *deep-ocean trenches* (long, narrow depressions that are the deepest parts of the ocean, and where moving crustal plates descend back into the mantle); *abyssal plains* (among the most level places on Earth, consisting of thick accumulations of sediments that were deposited atop the low, rough portions of the ocean floor by turbidity currents); and *seamounts* (isolated, steep-sided volcanic peaks on the ocean floor that originate near oceanic ridges or in association with volcanic hot spots).

- *Coral reefs*, which are confined largely to the warm, sunlit waters of the Pacific and Indian oceans, are constructed over thousands of years primarily from the accumulation of skeletal remains and secretions of corals and certain algae. A coral island, called an *atoll*, consists of a continuous or broken ring of coral reef surrounding a central lagoon. Atolls form from corals that grow on the flanks of sinking volcanic islands, where the corals continue to build the reef complex upward as the island slowly sinks.

- *There are three broad categories of seafloor sediments.* Terrigenous sediment consists primarily of mineral grains that were weathered from continental rocks and transported to the ocean. *Biogenous sediment* consists of shells and skeletons of marine animals

and plants. *Hydrogenous sediment* includes minerals that crystallize directly from seawater through various chemical reactions.

- *Mid-ocean ridges*, the sites of seafloor spreading, are found in all major oceans and represent more than 20 percent of Earth's surface. These broad features are certainly the most prominent features in the oceans, for they form an almost continuous mountain range. Ridges are characterized by an *elevated position, extensive faulting,* and *volcanic structures* that have developed on newly formed oceanic crust. Most of the geologic activity associated with ridges occurs along a narrow region on the ridge crest, called the *rift zone*, where magma from the asthenosphere moves upward to create new slivers of oceanic crust.

- New oceanic crust is formed in a continuous manner by the process of seafloor spreading. The upper crust is composed of *pillow lavas* of basaltic composition. Underlying this layer are numerous interconnected dikes (*sheeted dikes*) that are connected to a layer of gabbro. This entire sequence of rock is called an *ophiolite complex*.

Review Questions

1. Assuming that the average speed of sound waves in water is 1500 meters per second, determine the water depth if the signal sent out by an echo sounder requires 6 seconds to strike bottom and return to the recorder (see Figure 18.2).

2. List the three major features that comprise the continental margin. Which of these features is considered a flooded extension of the continent? Which has the steepest slope?

3. How does the continental margin along the west coast of South America differ from the continental margin along the east coast of North America?

4. Defend or rebut the following statements: "Most of the submarine canyons found on the continental slope and rise were formed during the Ice Age when rivers extended their valleys seaward."

5. What are turbidites? What is meant by the term *graded bedding*?

6. Discuss the evidence that helped confirm the existence of turbidity currents in the ocean. (See Box 18.1)

7. Why are abyssal plains more extensive on the floor of the Atlantic than on the floor of the Pacific?

8. What is an atoll? Describe Darwin's proposal on the origin of atolls. Was it ever confirmed?

9. Distinguish among the three basic types of seafloor sediment.

10. If you were to examine recently deposited biogenous sediment taken from a depth in excess of 4500 meters (15,000 feet), would it more likely be rich in calcareous materials or siliceous materials? Explain.

11. How are mid-ocean ridges related to seafloor spreading?

12. What is the primary reason for the elevated position of the oceanic ridge system?

Key Terms

abyssal plain (p. 454)
accretionary wedge
  (p. 451)
active continental margin
  (p. 451)
atoll (p. 455)
biogenous sediment
  (p. 456)
continental margin
  (p. 450)

continental rise (p. 451)
continental shelf (p. 450)
continental slope (p. 451)
deep-ocean basin (p. 453)
deep-sea fan (p. 451)
echo sounder (p. 448)
graded bedding (p. 452)
guyot (p. 455)
hydrogenous sediment
  (p. 456)

manganese nodule
  (p. 458)
mid-ocean ridge (p. 458)
ophiolite complex
  (p. 464)
passive continental mar-
  gin (p. 450)
pillow lavas (p. 464)
rift valleys (p. 459)
rift zone (p. 459)

seamount (p. 455)
sheeted dike (p. 464)
submarine canyon
  (p. 452)
terrigenous sediment
  (p. 456)
trench (p. 453)
turbidite (p. 452)
turbidity current (p. 452)

Web Resources

The *Earth* Home Page provides on-line resources for this chapter on the World Wide Web. You will find review exercises, specific updates for items in the chapter, suggested reading, and links to interesting related pathways on the Internet. Visit the *Earth* Home Page at **http://www.prenhall.com/tarbuck.**

CHAPTER 19

Plate Tectonics

Left Composite satellite image of a part of North Africa and the Arabian Peninsula. — *Image by Worldsat International, Inc.*

Early in this century geologic thought about the age of the ocean basins was dominated by a belief in their antiquity. Moreover, most geologists accepted the geographic permanency of the oceans and continents. Mountains were thought to result from the contractions of Earth caused by gradual cooling from a once-molten state. As the interior cooled and contracted, Earth's solid outer skin was deformed by folding to fit the shrinking planet. Mountains were therefore regarded as analogous to the wrinkles on a dried-out piece of fruit. This model of Earth's tectonic processes, however inadequate, was firmly entrenched in the geologic thought of the time.[*]

Since the 1960s, vast amounts of new data have dramatically changed our understanding of the nature and workings of our planet. Earth scientists now realize that the continents gradually migrate across the globe. Where landmasses split apart, new ocean basins are created between the diverging blocks. Meanwhile, older portions of the seafloor are being carried back into the mantle in regions where trenches occur in the deep ocean floor. Because of this movement, blocks of continental material eventually collide and form Earth's great mountain ranges (Figure 19.1). In short, a revolutionary new model of Earth's tectonic processes has emerged.

This profound reversal of scientific opinion has been appropriately described as a scientific revolution. Like other scientific revolutions, considerable time elapsed between the idea's inception and its general acceptance. The revolution began early in the twentieth century as a relatively straightforward proposal that the continents drift about the face of Earth. After many years of heated debate, the idea of drifting continents was rejected by the vast majority of Earth scientists as being improbable.

The concept of a mobile Earth was particularly distasteful to North American geologists, perhaps because much of the supporting evidence for movement had been gathered from the southern continents, with which most North American geologists were essentially unfamiliar. However, during the 1950s and 1960s, new evidence began to rekindle interest in this nearly abandoned proposal. By 1968, these new developments led to the unfolding of a far more encompassing theory that incorporated aspects of continental drift and seafloor spreading—a theory known as **plate tectonics**.

---

[*]*Tectonics* refers to the deformation of Earth's crust and results in the formation of structural features such as mountains.

Figure 19.1 Converging lithospheric plates were responsible for creating the Andes. Shown is Mount Huascaran, Peru. (Photo by Peter Feibert/Liaison International)

In this chapter we examine the events that led to this dramatic reversal of scientific opinion in an attempt to provide some insight into how science works. We will briefly trace the developments that took place from the inception of the concept of continental drift through the general acceptance of the theory of plate tectonics. Evidence gathered to support the concept of a mobile Earth will also be provided.

Continental Drift: An Idea Before Its Time

The idea that continents, particularly South America and Africa, fit together like pieces of a jigsaw puzzle originated with the development of reasonably accurate world maps. However, little significance was given this idea until 1915, when Alfred Wegener, a German meteorologist and geophysicist, published *The Origin of Continents and Oceans.** In this book, Wegener set forth the basic outline of his radical hypothesis of **continental drift**.

Wegener suggested that a single *supercontinent* called **Pangaea** (meaning "all land") once existed (Figure 19.2). He further hypothesized that about 200 million years ago this supercontinent began breaking into smaller continents, which then "drifted" to their present positions.

---

*Wegener's ideas were actually preceded by those of an American geologist, F. B. Taylor, who in 1910 published a paper on continental drift. Taylor's paper provided little corroborating evidence for continental drift, which may have been the reason that it had a relatively small impact on the scientific community.

Wegener and others who advocated this hypothesis collected substantial evidence to support these claims. The fit of South America and Africa, fossil evidence, rock structures, and ancient climates all seemed to support the idea that these now separate landmasses were once joined. Let us examine their evidence.

Fit of the Continents

Like a few others before him, Wegener first suspected that the continents might once have been joined when he noticed the remarkable similarity between the coastlines on opposite sides of the South Atlantic (Figure 19.3). However, his use of present-day shorelines to make a fit of the continents was challenged immediately by other Earth scientists. These opponents correctly argued that shorelines are continually modified by erosional and depositonal processes. Even if continental displacement had taken place, a good fit today would be unlikely. Furthermore, abundant fossil evidence indicates that most of the world's land areas have experienced periods of either uplift or subsidence in the recent geologic past. This would have markedly altered the position of the global coastlines. Wegener appeared to be aware of these problems, and, in fact, his original jigsaw fit of the continents was only very crude.

A much better approximation of the true outer boundary of the continents is the continental shelf. Today, the seaward edge of the continental shelf lies submerged, a few hundred meters below sea level. In the early 1960s, Sir Edward Bullard and two associates produced a map that attempted to fit the edges of the continental shelves of South America and Africa at a

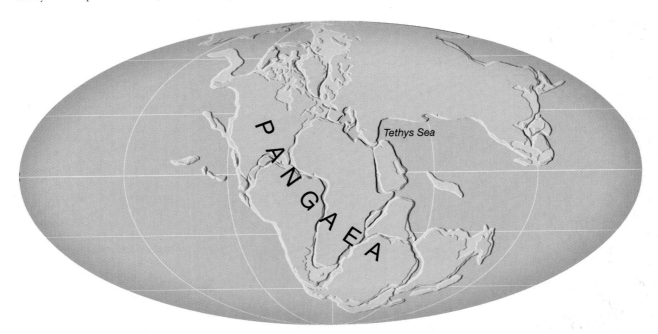

Figure 19.2 Reconstruction of Pangaea as it is thought to have appeared 200 million years ago.

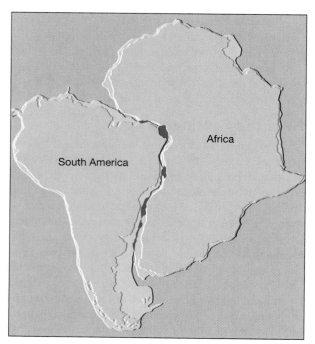

Figure 19.3 This shows the best fit of South America and Africa along the continental slope at a depth of 500 fathoms (about 900 meters). The areas where continental blocks overlap appear in brown. (After A. G. Smith, "Continental Drift." In *Understanding the Earth*, edited by I. G. Gass. Courtesy of Artemis Press)

depth of 900 meters. The remarkable fit that was obtained is shown in Figure 19.3. Although the continents overlap in a few places, these are regions where streams have deposited large quantities of sediment, thus enlarging the continental shelves. The overall fit was even better than the supporters of continental drift suspected it would be.

Fossil Evidence

Although Wegener was intrigued by the remarkable similarities of the shorelines on opposite sides of the Atlantic, he at first thought the idea of a mobile Earth improbable. Not until he came across an article citing fossil evidence for the existence of a land bridge connecting South America and Africa did he begin to take his own idea seriously. Through a search of the literature, Wegener learned that most paleontologists (scientists who study the fossilized remains of plants and animals) were in agreement that some type of land connection was needed to explain the existence of identical fossils on widely separated landmasses. This requirement was particularly true for late Paleozoic and early Mesozoic life-forms.

Mesosaurus. To add credibility to his argument for the existence of the supercontinent Pangaea, Wegener cited documented cases of several fossil organisms that had been found on different landmasses but which could not have crossed the vast oceans presently separating the continents. The classic example is *Mesosaurus*, a presumably aquatic, snaggle-toothed reptile whose fossil remains are limited to eastern South America and southern Africa (Figure 19.4). If *Mesosaurus* had been able to swim well enough to cross the vast South Atlantic Ocean, its remains should be more widely distributed. As this was not the case, Wegener argued that South America and Africa must have been joined somehow.

How did scientists during Wegener's era explain the discovery of identical fossil organisms in places separated by thousands of kilometers of open ocean? The idea of land bridges was the most widely accepted solution to the problem of migration (Figure 19.5).

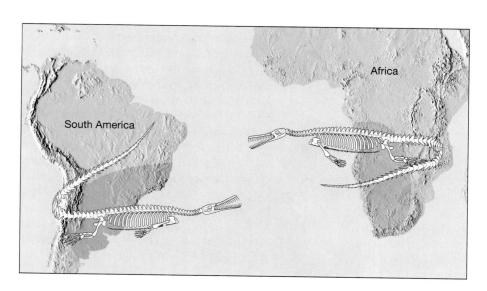

Figure 19.4 Fossils of *Mesosaurus* have been found on both sides of the South Atlantic and nowhere else in the world. Fossil remains of this and other organisms on the continents of Africa and South America appear to link these landmasses during the late Paleozoic and early Mesozoic eras.

RAFTING

ISTHMIAN LINKS

ISLAND STEPPING STONES

CONTINENTAL DRIFT

Figure 19.5 These sketches by John Holden illustrate various explanations for the occurrence of similar species on landmasses that are presently separated by vast oceans. (Reprinted with permission of John Holden)

We know, for example, that during the recent Ice Age, the lowering of sea level allowed animals to cross the narrow Bering Strait between Asia and North America. Was it possible that land bridges once connected Africa and South America? We are now quite certain that land bridges of this magnitude did not exist. If they had, their remnants should still lie below sea level, but they are nowhere to be found.

Glossopteris. Wegener also cited the distribution of the fossil fern *Glossopteris* as evidence for the existence of Pangaea. This plant, identified by its large seeds that could not be blown very far, was known to be widely dispersed among Africa, Australia, India, and South America during the late Paleozoic era. Later, fossil remains of *Glossopteris* were discovered in Antarctica as well. Wegener also learned that these seed ferns and associated flora grew only in a subpolar climate. Therefore, he concluded that the landmasses must have been joined, because they are presently located in climatic zones that are far too diverse to support these cold-loving plants. For Wegener, fossils were convincing evidence that a supercontinent had existed.

Present-Day Organisms. In his book, Wegener also cited the distribution of present-day organisms as evidence to support the concept of drifting continents. For example, modern organisms with similar ancestries clearly had to evolve in isolation during the last few tens of millions of years. Most obvious of these are the Australian marsupials (such as kangaroos), which have a direct fossil link to the marsupial opossums found in the Americas.

Rock Type and Structural Similarities

Anyone who has worked a picture puzzle knows that, in addition to the pieces fitting together, the picture must be continuous as well. The "picture" that must match in the "Continental Drift Puzzle" is one of rock types and mountain belts found on the continents. If the continents were once together, the rocks found in a particular region on one continent should closely match in age and type with those found in adjacent positions on the matching continent.

Such evidence exists in the form of mountain belts that terminate at one coastline, only to reappear on a landmass across the ocean. For instance, the mountain belt that includes the Appalachians trends northeastward through the eastern United States and disappears off the coast of Newfoundland (Figure 19.6A). Mountains of comparable age and structure are found in the British Isles and Scandinavia. When these landmasses are reassembled, as in Figure 19.6B, the mountain chains form a nearly continuous belt.

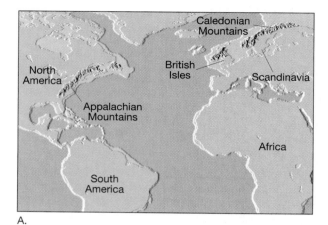

A.

B.

Figure 19.6 Matching mountain ranges across the North Atlantic. **A.** The Appalachian Mountains trend along the eastern flank of North America and disappear off the coast of Newfoundland. Mountains of comparable age and structure are found in the British Isles and Scandinavia. **B.** When these landmasses are placed in their predrift locations, these ancient mountain chains form a nearly continuous belt. These folded mountain belts formed roughly 300 million years ago as the landmasses collided during the formation of the supercontinent of Pangaea.

Wegener was very satisfied that the similarities in rock structure on both sides of the Atlantic linked these landmasses. In his own words, "It is just as if we were to refit the torn pieces of a newspaper by matching their edges and then check whether the lines of print run smoothly across. If they do, there is nothing left but to conclude that the pieces were in fact joined in this way."*

---

*Alfred Wegener, *The Origin of Continents and Oceans*. Translated from the 4th revised German edition of 1929 by J. Birman (Lond Methuen, 1966).

Paleoclimatic Evidence

Because Alfred Wegener was a meteorologist by profession, he was keenly interested in obtaining paleoclimatic (ancient climate) data in support of continental drift. His efforts were rewarded when he found evidence for apparently dramatic global climatic changes. In particular, he learned of ancient glacial deposits which indicated that near the end of the Paleozoic era (between 220 million and 300 million years ago), ice sheets covered extensive areas of the Southern Hemisphere. Layers of glacial till of the same age were found in southern Africa and South America, as well as in India and Australia. Below these beds of glacial debris lay striated and grooved bedrock. In some locations the striations and grooves indicated that the ice had moved from what is now the sea onto land (Figure 19.7A). Much of the land area containing evidence of this late Paleozoic glaciation presently lies within 30 degrees of the equator in a subtropical or tropical climate.

Could Earth have gone through a period of sufficient cooling to have generated extensive continental ice sheets in what is presently a tropical region? Wegener rejected this explanation because during that same time period, large tropical swamps existed in the Northern Hemisphere. These swamps, with their lush vegetation, eventually became the major coal fields of the eastern United States, Europe, and Siberia.

Fossils from these coal fields indicate that the tree ferns that produced the coal deposits had large fronds, which are indicative of a tropical setting. Furthermore, unlike trees in colder climates, these trees lacked growth rings, a characteristic of tropical plants caused by minimal seasonal fluctuations in temperature.

Wegener believed that a more plausible explanation for the late Paleozoic glaciation was provided if the landmasses are fitted together as a supercontinent with South Africa centered over the South Pole (Figure 19.7B). This would account for the conditions necessary to generate extensive expanses of glacial ice over much of the Southern Hemisphere. At the same time, this geography would place today's northern landmasses nearer the equator and account for their vast coal deposits. Wegener was so convinced that his explanation was correct that he wrote, "This evidence is so compelling that by comparison all other criteria must take a back seat."

How does a glacier develop in hot, arid Australia? How do land animals migrate across wide expanses of open water? As compelling as this evidence may have been, 50 years passed before most of the scientific community would accept it and the logical conclusions to which it led.

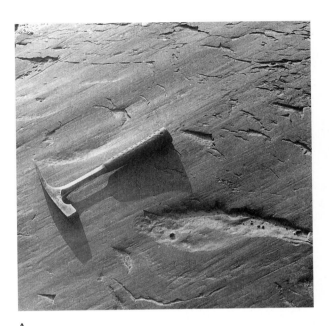

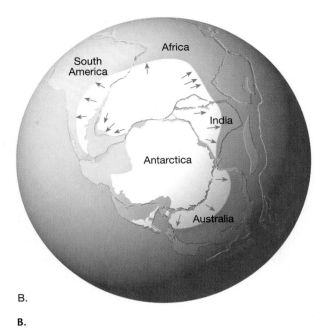

Figure 19.7 A. Glacial striations in the bedrock of Hallet Cove, South Australia, indicate direction of ice movement. (Photo by W. B. Hamilton, U.S. Geological Survey) **B.** Direction of ice movement in the southern supercontinent called Gondwanaland according to those who developed the continental drift hypothesis.

The Great Debate

Wegener's proposal did not attract much open criticism until 1924, when his book was translated into English, French, Spanish, and Russian. From that point until his death in 1930, his drift hypothesis encountered a great deal of hostile criticism. To quote the respected American geologist R. T. Chamberlain: "Wegener's hypothesis in general is of the foot-loose type, in that it takes considerable liberty with our globe, and is less bound by restrictions or tied down by awkward, ugly facts than most of its rival theories. Its appeal seems to lie in the fact that it plays a game in which there are few restrictive rules and no sharply drawn code of conduct."

W. B. Scott, former president of the American Philosophical Society, expressed the prevalent American view of continental drift in fewer words when he described the hypothesis, as "utter damned rot!"

Objections to the Continental Drift Hypothesis

One of the main objections to Wegener's hypothesis stemmed from his inability to provide a mechanism capable of moving the continents across the globe. Wegener proposed that the tidal influence of the Moon was strong enough to give the continents a westward motion. However, the prominent physicist Harold Jeffreys quickly countered with the argument that tidal friction of the magnitude needed to displace the continents would bring Earth's rotation to a halt in a matter of a few years.

Wegener also suggested that the larger and sturdier continents broke through the oceanic crust, much like ice breakers cut through ice. However, no evidence existed to suggest that the ocean floor was weak enough to permit passage of the continents without themselves being appreciably deformed in the process.

By 1929, criticisms of Wegener's ideas were pouring in from all areas of the scientific community. Despite these affronts, Wegener wrote the fourth and final edition of his book, maintaining his basic hypothesis and adding supporting evidence.

In 1930, Wegener made his third and final trip to the Greenland Ice Sheet (Figure 19.8). Although the primary focus of this expedition was to study the harsh winter weather on the ice-covered island, Wegener planned to test his continental drift hypothesis as well. Wegener felt that by precisely establishing the locations of specific points and then measuring their changes over a period of years, he could demonstrate the westward drift of Greenland with respect to Europe (see Box 19.1). In November 1930, while returning from Eismitte (an experimental station located in the center of Greenland),

Figure 19.8 Alfred Wegener shown waiting out the 1912–1913 Arctic winter during an expedition to Greenland, where he made a 1200-kilometer traverse across the widest part of the island's ice sheet. (Photo courtesy of Bildarchiv Preussischer Kulturbesitz, Berlin)

Wegener perished along with his companion. His intriguing idea, however, did not die with him.

Continental Drift and the Scientific Method

Why was Wegener not able to overturn the established scientific view of his day? Although his hypothesis was correct in principle, it contained many incorrect details. For example, the continents do not break through the ocean floor, and tidal energy is not the driving mechanism for continental motion. In order for any scientific viewpoint to gain wide acceptance, supporting evidence from all realms of science must be found.

This same idea was stated very well by Wegener himself in response to his critics, when he said, "Scientists still do not appear to understand sufficiently that all earth sciences must contribute evidence toward unveiling the state of our planet in earlier times, and the truth of the matter can only be reached by combining all this evidence." Wegener's great contribution to our understanding of Earth notwithstanding, not *all* of the evidence supported the continental drift hypothesis as he had formulated it. Therefore, Wegener himself answered the very question he must have asked many times: "Why do they reject my proposal?"

Although most of Wegener's contemporaries opposed his views, even to the point of open ridicule,

a few considered his ideas plausible. Among the most notable of this latter group were the eminent South African geologist Alexander du Toit and the well-known Scottish geologist Arthur Holmes. In 1937, du Toit published *Our Wandering Continents*, in which he eliminated some of Wegener's weaker points and added a wealth of new evidence in support of this revolutionary idea. Arthur Holmes proposed a plausible driving mechanism for continental drift. In Holmes's book *Physical Geology*, he suggested that convection currents operating within the mantle were responsible for propelling the continents across the globe.

For these few geologists who continued the search, the exciting concept of continents in motion held their interest. Others viewed continental drift as a solution to previously unexplainable observations.

Continental Drift and Paleomagnetism

Very little new light was shed on the continental drift hypothesis between the time of Wegener's death in 1930 and the early 1950s. Little was known about the land beneath the sea, which represents over 70 percent of Earth's surface. This relatively unexplored area held the key to unraveling many secrets of our planet.

Perhaps the initial impetus for the renewed interest in continental drift came from rock magnetism, a

Box 19.1

A New Test for Plate Tectonics

Until the late 1980s, evidence supporting the theory of plate tectonics was indirectly acquired from the study of geologic phenomena such as volcanoes, earthquakes, and seafloor sediments. With satellite technology, it now is possible to test the theory directly. Scientists have confirmed the fact that Earth's plates shift in relation to one another in the way that the plate tectonics theory predicts.

The new evidence comes from two techniques that allow unprecedented accuracy of distance measurement between widely separated points. Called *Satellite Laser Ranging* and *Very Long Baseline Interferometry*, these methods can detect the motion of any one site with respect to another at a level of better than 1 centimeter per year. Thus, for the first time, scientists can directly measure the relative motions of Earth's plates. Further, because these techniques are quite different, scientists use them to cross-check one against the other by comparing measurements made for the same sites.

The *Satellite Laser Ranging System* (SLR) employs ground-based stations that bounce laser pulses off satellites whose orbital positions are well-established. Precise timing of the round-trip travel of these pulses allows scientists to calculate the precise locations of the ground-based stations. By monitoring these stations over time, researchers can establish the relative motions of the sites.

The *Very Long Baseline Interferometry System* (VLBI) uses large radio telescopes to record signals from very distant quasars (quasi-stellar objects) (Figure 19.A). Quasars lie billions of light-years from Earth, so they act as stationary reference points. The millisecond differences in the arrival time of the same signal at different earthbound observational sites provide a means of establishing the precise distance between receivers.

Confirming data from these two techniques leave little doubt that real plate motion has been detected. Calculations show that Hawaii is moving in a northwesterly direction and approaching Japan at 8.3 centimeters per year. A site located in Maryland is retreating from one in England at a rate of about 1.7 centimeters per year. This rate is close to the 2.2 centimeters per year of seafloor spreading that was established from paleomagnetic evidence.

Figure 19.A Radio telescopes like these located at Socorro, New Mexico, are used to determine accurately the distance between two distant sites. Data collected by repeated measurements have detected relative plate motions of 1 to 12 centimeters per year between various sites worldwide. (Photo by Geoff Chester)

comparatively new field of study. Early workers studying rock magnetism set out to investigate ancient changes in Earth's magnetic field in hopes of better understanding the nature of the present-day magnetic field. Anyone who has used a compass to find direction knows that the magnetic field has a north pole and a south pole. These magnetic poles align closely, but not exactly, with the geographic poles. (The geographic poles are simply the top and bottom of the spinning sphere we live on, the points through which passes the imaginary axis of rotation.)

Paleomagnetism

In many respects the magnetic field is like that produced by a simple bar magnet. Invisible lines of force pass through Earth and extend from one pole to the other (Figure 19.9). A compass needle, itself a small magnet free to move about, becomes aligned with these lines of force and thus points toward the magnetic poles.

The technique used to study ancient magnetic fields relies on the fact that certain rocks contain minerals that serve as "fossil compasses." These iron-rich minerals, such as magnetite, are abundant in lava flows of basaltic composition. When heated above a temperature known as the **Curie point**, these magnetic minerals lose their magnetism. However, when these iron-rich grains cool below their Curie point (about 580°C), they become magnetized in the direction parallel to the existing magnetic lines of force. Once the minerals solidify, the magnetism they possess will remain "frozen" in this position. In this regard, they behave much like a compass needle; they "point" toward the existing magnetic poles at the time of their cooling. Then, if the rock is moved, or if the magnetic pole changes position, the rock magnetism will, in most instances, retain its original alignment. Rocks that formed thousands or millions of years ago and "recorded" the location of the magnetic poles at the

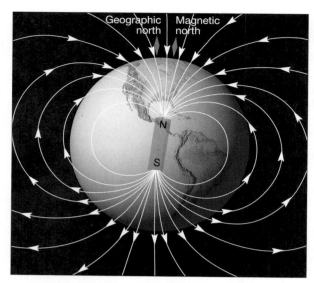

Figure 19.9 Earth's magnetic field consists of lines of force much like those a giant bar magnet would produce if placed at the center of Earth.

time of their formation are said to possess fossil magnetism, or **paleomagnetism**.

Another important aspect of rock magnetism is that the magnetized minerals not only indicate the direction to the poles (like a compass), but they also provide a means of determining the latitude of their origin. To envision how latitude can be established from paleomagnetism, imagine a compass needle mounted in a vertical plane rather than horizontally like an ordinary compass. As shown in Figure 19.10, when this modified compass (*dip needle*) is situated over the north magnetic pole, it aligns with the magnetic lines of force and points straight down. However, as this dip needle is moved closer to the equator, the angle of inclination is reduced until the needle aligns with the horizontal lines of force at the equator. Thus, from the dip needle's angle of inclination, one can determine the latitude. In a similar manner, the orientation of the paleomagnetism in rocks indicates the latitude of the rock at the time it became magnetized.

Polar Wandering

A study conducted in Europe by S. K. Runcorn and his associates during the 1950s led to an unexpected discovery. The magnetic alignment in the iron-rich minerals in lava flows of different ages was found to vary widely. A plot of the apparent position of the magnetic north pole revealed that, during the past 500 million years, the position of the pole had gradually wandered from a location near Hawaii northward through eastern Siberia and finally to its present location (Figure 19.11). This was strong evidence that

either the magnetic poles had migrated through time, an idea known as *polar wandering*, or that the lava flows moved—in other words, the continents had drifted.

Although the magnetic poles are known to move, studies of the magnetic field indicate that the average positions of the magnetic poles correspond closely to the positions of the geographic poles. This is consistent with our knowledge of Earth's magnetic field, which is generated in part by the rotation of Earth about its axis. If the geographic poles do not wander appreciably, which we believe is true, neither can the magnetic poles. Therefore, a more acceptable explanation for the apparent polar wandering was provided by the continental drift hypothesis. If the magnetic poles remain stationary, their *apparent movement* is produced by the drifting of the continents.

The latter idea was further supported by comparing the latitude of Europe as determined from rock magnetism with evidence from paleoclimatic studies. In particular, during the period when coal-producing swamps covered much of Europe, paleomagnetic evidence places Europe near the equator—a fact consistent with the tropical environment indicated by these coal deposits.

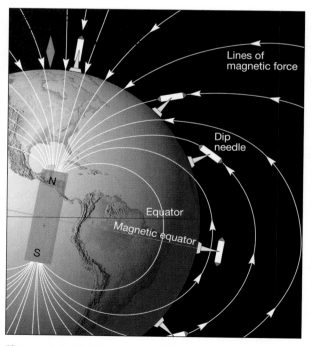

Figure 19.10 Earth's magnetic field causes a dip needle (compass oriented in a vertical plane) to align with the lines of magnetic force. The dip angle decreases uniformly from 90 degrees at the magnetic poles to 0 degrees at the magnetic equator. Consequently, the distance to the magnetic poles can be determined from the dip angle.

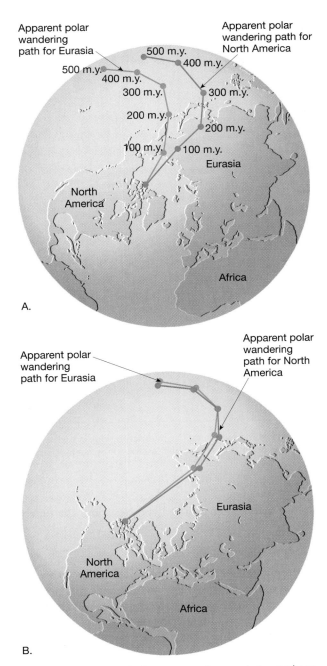

Figure 19.11 Simplified apparent polar-wandering paths as established from North American and Eurasian paleomagnetic data. **A.** The more westerly path determined from North American data was caused by the westward movement of North America by about 24 degrees from Eurasia. **B.** The positions of the wandering paths when the landmasses are reassembled.

Further evidence for continental drift came a few years later when a polar-wandering curve was constructed for North America and Europe (Figure 19.11A). To nearly everyone's surprise, the curves for North America and Europe had similar paths, except

that they were separated by about 24° longitude. When these rocks solidified, could there have been two magnetic north poles that migrated parallel to each other? This is very unlikely. The differences in these migration paths, however, can be reconciled if the two presently separated continents are placed next to one another, as we now believe they were prior to opening of the Atlantic Ocean (Figure 19.11B). Then, the two polar-wandering curves nearly overlap each other.

Although these new data rekindled interest in continental drift, they by no means caused a major swing in opinion. For one thing, the techniques used in extracting paleomagnetic data were relatively new and untested. Furthermore, rock magnetism tends to weaken with time. Despite these problems and other conflicting evidence, some researchers were convinced that continental drift had indeed occurred. A new era had begun.

A Scientific Revolution Begins

During the 1950s and 1960s, great technological strides permitted extensive mapping of the ocean floor. From this work came the discovery of a global oceanic ridge system. Furthermore, the Mid-Atlantic Ridge was found to parallel the continental margins on both sides of the Atlantic (see Figure 18.11, p. 460). Also of importance was the discovery of a central rift valley extending the length of the Mid-Atlantic Ridge, an indication that great tensional forces were at work. In addition, high heat flow and volcanism were found to characterize the oceanic ridge system.

In other parts of the ocean additional discoveries were being made. Earthquake studies conducted in the vicinity of the deep-ocean trenches demonstrated that tectonic activity was occurring at great depths beneath the ocean. Flat-topped seamounts hundreds of meters below sea level showed signs of formerly being islands that somehow became submerged. Of equal importance, dredging of the oceanic crust did not bring up rock that was older than 160 million years. Could the ocean floor actually be a geologically young feature?

Seafloor Spreading

In the early 1960s, these newly discovered facts were put together by Harry Hess of Princeton University into a hypothesis that was later named **seafloor spreading**. Hess, who was a modest individual, presented his paper as an "essay in geopoetry." Unlike its forerunner, continental drift, which essentially neglects the ocean basins, seafloor spreading is centered on the activity beyond our direct view.

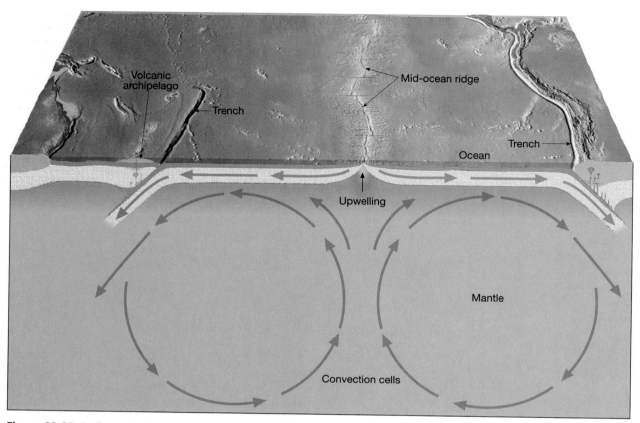

Figure 19.12 Seafloor spreading. Harry Hess proposed that upwelling of mantle material along the mid-ocean ridge system created new seafloor. The convective motion of mantle material carries the seafloor in a conveyor-belt fashion to the deep-ocean trenches, where the seafloor descends into the mantle.

In Hess's now-classic paper, he proposed that the ocean ridges are located above zones of upwelling in the mantle (Figure 19.12). As rising material from the mantle spreads laterally, seafloor is carried in a conveyor-belt fashion away from the ridge crest. Here, tensional forces fracture the crust and provide pathways for magma to intrude and generate new slivers of oceanic crust. Thus, as the seafloor moves away from the ridge crest, newly formed crust replaces it. Hess further proposed that the deep-ocean trenches, such as the Peru–Chile trench, are sites where ocean crust is drawn back into the planet. Here, according to Hess, the older portions of the seafloor are gradually consumed as they descend into the mantle. As one researcher summarized, "No wonder the ocean floor was young—it was constantly being renewed!"

With the seafloor-spreading hypothesis in place, Harry Hess had initiated another phase of this scientific revolution. The conclusive evidence to support his ideas came a few years later from the work of a young English graduate student, Fred Vine, and his supervisor, D. H. Matthews. The greatness of Vine's and Matthews's work was that they were able to connect two ideas previously thought to be unrelated:

Hess's seafloor-spreading hypothesis and the newly discovered geomagnetic reversals.

Geomagnetic Reversals

About the time that Hess formulated seafloor spreading, geophysicists had begun to accept the fact the Earth's magnetic field periodically reverses polarity; that is, the north magnetic pole becomes the south magnetic pole, and vice versa (see Box 17.2, p. 431). A rock solidifying during one of the periods of reverse polarity will be magnetized with the polarity opposite that of rocks being formed today. When rocks exhibit the same magnetism as the present magnetic field, they are said to possess **normal polarity**, whereas rocks exhibiting the opposite magnetism are said to have **reverse polarity**.

Evidence for magnetic reversals was obtained from lavas and sediments from around the world. Once the concept of magnetic reversals was confirmed, researchers set out to establish a time scale for polarity reversals. There are many areas where volcanic activity has occurred sporadically for periods of millions of years. The task was to measure the polarity of paleo-magnetism in numerous lava flows of various ages

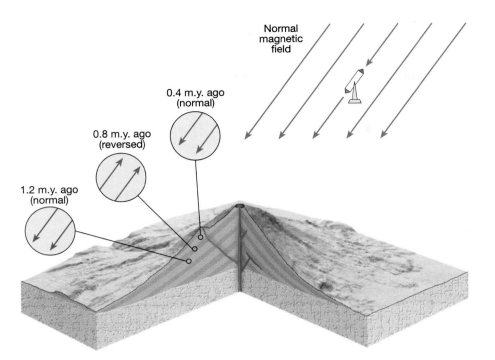

Figure 19.13 Schematic illustration of paleomagnetism preserved in lava flows of various ages. Data such as these from various locales were used to establish the time scale of polarity reversals shown in Figure 19.14.

(Figure 19.13). These data, which were collected from around the globe, were used to determine the dates when the polarity of Earth's magnetic field changed. Figure 19.14 shows the time scale of polarity reversals established for the last few million years.

A significant relationship was uncovered between magnetic reversals and the seafloor-spreading hypothesis. Researchers discovered alternating stripes of high- and low-intensity magnetism that trend roughly parallel to the ridge crest. This was accomplished using very sensitive instruments called **magnetometers**, which were towed by research vessels across segments of the ocean floor as shown in Figure 19.15.

This relatively simple pattern of magnetic variation defied explanation until 1963, when Fred Vine and D. H. Matthews tied the discovery of the high- and low-intensity stripes to Hess's concept of seafloor spreading.* Vine and Matthews suggested that the stripes of high-intensity magnetism are regions where the paleomagnetism of the ocean crust exhibits normal polarity. Consequently, these rocks *enhance* (reinforce) the existing magnetic field. Conversely, the low-intensity stripes are regions where the ocean crust is polarized in the reverse direction and, therefore, *weaken* the existing magnetic field. But how do parallel stripes of normally and reversely magnetized rock become distributed across the ocean floor?

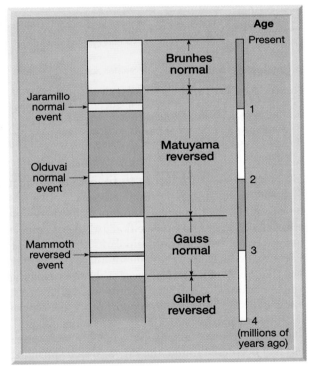

Figure 19.14 Time scale of Earth's magnetic field in the recent past. This time scale was developed by establishing the magnetic polarity for lava flows of known age. (Data from Allen Cox and G. B. Dalrymple)

Vine and Matthews reasoned that as magma intrudes and solidifies along the ridges, its magnetic components take on the polarity of the existing magnetic field. During the past 700,000 years, normally magnetized crust formed along the global ridge system.

---

*This idea was also advanced a few months earlier by L. W. Morely, but his paper was rejected for publication because of its highly speculative nature.

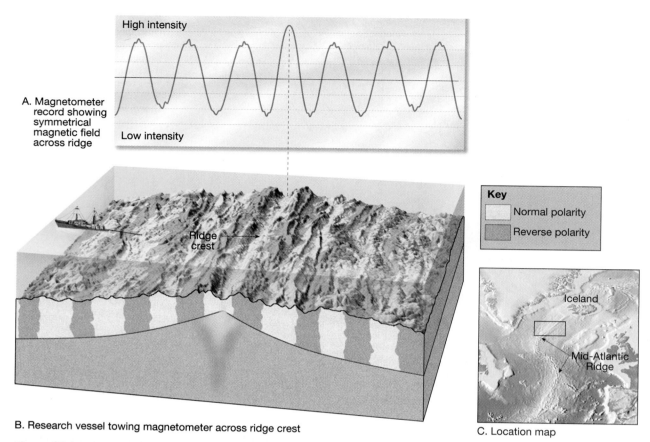

A. Magnetometer record showing symmetrical magnetic field across ridge

High intensity

Low intensity

Ridge crest

Key

☐ Normal polarity
■ Reverse polarity

Iceland

Mid-Atlantic Ridge

B. Research vessel towing magnetometer across ridge crest

C. Location map

Figure 19.15 The ocean floor as a magnetic tape recorder. **A.** Schematic representation of magnetic intensities recorded as a magnetometer is towed across a segment of the Mid-Atlantic Ridge. **B.** Notice the symmetrical stripes of low- and high-intensity magnetism that parallel the ridge crest. Vine and Matthews suggested that the stripes of high-intensity magnetism occur where normally magnetized oceanic basalts enhance the existing magnetic field. Conversely, the low-intensity stripes are regions where the crust is polarized in the reverse direction, which weakens the existing magnetic field.

However, as shown in Figure 19.14, ocean crust that formed 1.5 million years ago exhibits reverse polarity. As new rock is added in equal amounts to both trailing edges of the spreading oceanic floor, we should expect the pattern of stripes (size and polarity) found on one side of an ocean ridge to be a mirror image of the other side (Figure 19.16). The discovery of the re-alternating stripes of normal and reverse polarity was the strongest evidence so far presented in support of the concept of seafloor spreading.

Once the dates of the most recent magnetic reversals had been established, the rate at which spreading occurs at the various ridges could be determined accurately. In the Pacific Ocean, for example, the magnetic stripes are much wider for corresponding time intervals than those of the Atlantic Ocean. We can conclude that a faster spreading rate exists for the spreading center of the Pacific as compared with that of the Atlantic.

There is now general agreement that paleomagnetism was the most convincing evidence set forth to support the concepts of continental drift and seafloor spreading. By 1968, geologists began reversing their stand on this issue in a manner not unlike a magnetic reversal. The tide of scientific opinion had indeed switched in favor of a mobile Earth.

Plate Tectonics: A Modern Version of an Old Idea

By 1968, the concepts of continental drift and seafloor spreading were united into a much more encompassing theory known as **plate tectonics**. Plate tectonics can be defined as the composite of a variety of ideas that explain the observed motion of Earth's lithosphere through the mechanisms of subduction and seafloor spreading, which, in turn, generate Earth's major features, including continents and ocean basins. The implications of plate tectonics are so far-reaching that this theory has become the basis for viewing most geologic processes.

According to the plate tectonics model, the uppermost mantle, along with the overlying crust,

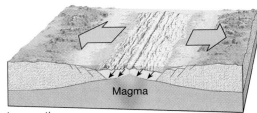

A. Period of normal magnetism

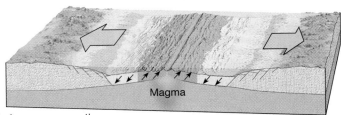

B. Period of reverse magnetism

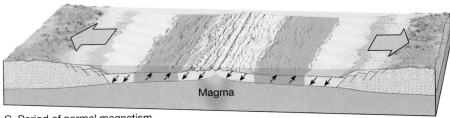

C. Period of normal magnetism

Figure 19.16 As new basalt is added to the ocean floor at mid-ocean ridges, it is magnetized according to Earth's existing magnetic field. Hence, it behaves much like a tape recorder as it records each reversal of the planet's magnetic field.

behave as a strong, rigid layer, known as the **lithosphere**. This outermost shell overlies a weaker region in the mantle known as the **asthenosphere**. Further, the lithosphere is broken into numerous segments, called **plates**, that are in motion and are continually changing in shape and size. As shown in Figure 19.17, seven major plates are recognized. They are the North American, South American, Pacific, African, Eurasian, Australian, and Antarctic plates. The largest is the Pacific plate, which is located mostly within the ocean. Notice from Figure 19.17 that several of the large plates include an entire continent plus a large area of seafloor (for example, the South American plate). This is a major departure from Wegener's continental drift hypothesis, which proposed that the continents moved through the ocean floor, not with it. Note also that none of the plates are defined entirely by the margins of a continent.

Intermediate-sized plates include the Caribbean, Nazca, Philippine, Arabian, Cocos, and Scotia plates. In addition, there are over a dozen smaller plates that have been identified but are not shown in Figure 19.17.

We know that lithospheric plates move at very slow, but continuous, rates of a few centimeters a year. This movement is ultimately driven by the unequal distribution of heat within Earth. The titanic grinding movements of Earth's lithospheric plates generate earthquakes, create volcanoes, and deform large masses of rock into mountains.

Plate Boundaries

Plates move as coherent units relative to all other plates. Although the interiors of plates may deform, all major interactions among individual plates (and therefore most deformation) occur along their *boundaries*. In fact, the first attempts to outline plate boundaries were made using locations of earthquakes. Later work showed that plates are bounded by three distinct types of boundaries, which are differentiated by the type of movement they exhibit. These boundaries are depicted in Figure 19.18 and are briefly described here:

1. **Divergent boundaries**—where plates move apart, resulting in upwelling of material from the mantle to create new seafloor (Figure 9.18A).

2. **Convergent boundaries**—where plates move together, resulting in the subduction (consumption) of oceanic lithosphere into the mantle (Figure 19.18B).

3. **Transform fault boundaries**—where plates grind past each other without the production or destruction of lithosphere (Figure 19.18C).

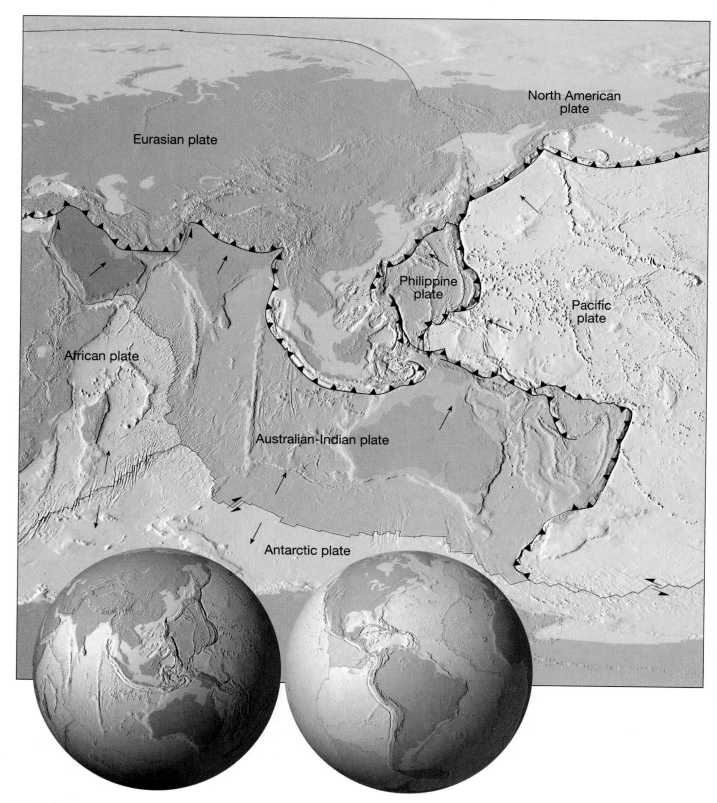

Figure 19.17 A mosaic of rigid plates constitutes Earth's outer shell. (After W. B. Hamilton, U.S. Geological Survey)

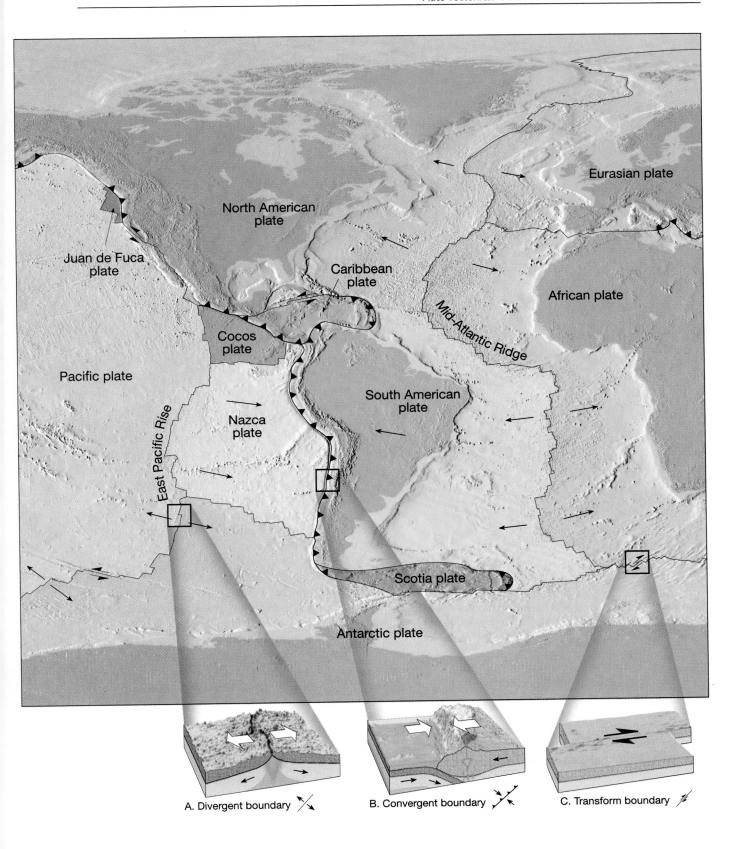

North American plate

Juan de Fuca plate

Caribbean plate

Cocos plate

Pacific plate

East Pacific Rise

Nazca plate

South American plate

Mid-Atlantic Ridge

African plate

Eurasian plate

Scotia plate

Antarctic plate

A. Divergent boundary

B. Convergent boundary

C. Transform boundary

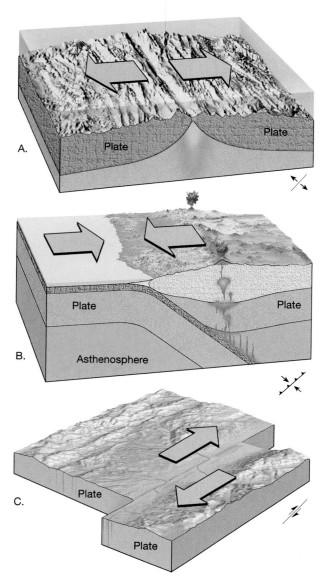

Figure 19.18 Schematic representation of plate boundaries showing the relative motion of plates. **A.** Divergent boundary. **B.** Convergent boundary. **C.** Transform fault boundary.

Each plate is bounded by a combination of these zones, as can be seen in Figure 19.17. For example, the Nazca plate has a divergent zone on the west, a convergent boundary on the east, and numerous transform faults, which offset segments of the divergent boundary. Although the total surface of Earth does not change, individual plates may diminish or grow in area depending on the distribution of convergent and divergent boundaries. For example, the Antarctic and African plates are almost entirely bounded by spreading centers and hence are growing larger. By contrast, the Pacific plate is being consumed into the mantle along its northern and western flanks and is therefore diminishing in size.

Furthermore, new plate boundaries can be created in response to changes in the forces acting on these rigid slabs. For example, a relatively new divergent boundary is located in Africa, in a region known as the East African Rift Valleys. If spreading continues there, the African plate will split into two plates separated by a new ocean basin. At other locations, plates carrying continental crust are presently moving toward each other. Eventually, these continents may collide and be sutured together. Thus, the boundary that once separated two plates disappears as the plates become one. The result of such a continental collision is a majestic mountain range such as the Himalayas.

In the following sections we will briefly summarize the nature of the three types of plate boundaries.

Divergent Boundaries

Most divergent boundaries, where plate spreading occurs, are situated along the crests of oceanic ridges (Figure 19.19). Here, as the plates move away from the ridge axis, the fractures created are immediately filled with molten rock that wells up from the hot asthenosphere below. This hot material cools to hard rock, producing new slivers of seafloor. In a continuous manner, plate spreading and upwelling of magma add new oceanic crust (lithosphere) between the diverging plates.

As noted earlier, this mechanism has created the floor of the Atlantic Ocean during the past 160 million years and is appropriately called *seafloor spreading*. A typical rate of seafloor spreading is 5 centimeters (2 inches) per year, although it varies from 2 to 20 centimeters per year. This extremely slow rate of lithosphere production is nevertheless rapid enough so that all of Earth's ocean basins could have been generated within the last 200 million years. In fact, none of the ocean floor that has been dated exceeds 180 million years in age.

Along divergent boundaries where molten rock emerges, the ocean floor is elevated. Worldwide, this ridge extends for over 70,000 kilometers through all major ocean basins. You can see parts of the mid-ocean ridge system in Figure 19.17. As new lithosphere is formed along the oceanic ridge it is slowly, yet continually, displaced away from the zone of upwelling found along the ridge axis. Thus, it begins to cool and contract, thereby increasing in density. This partially accounts for the greater depth of the older and cooler oceanic crust found in the deep ocean basins. In addition, the mantle rocks located below the oceanic crust cool and strengthen with increased distance from the ridge axis, thereby adding to the plate thickness. Stated another way, both the thickness and the density of oceanic lithosphere are age dependent. The older (cooler) it is, the greater its thickness and density.

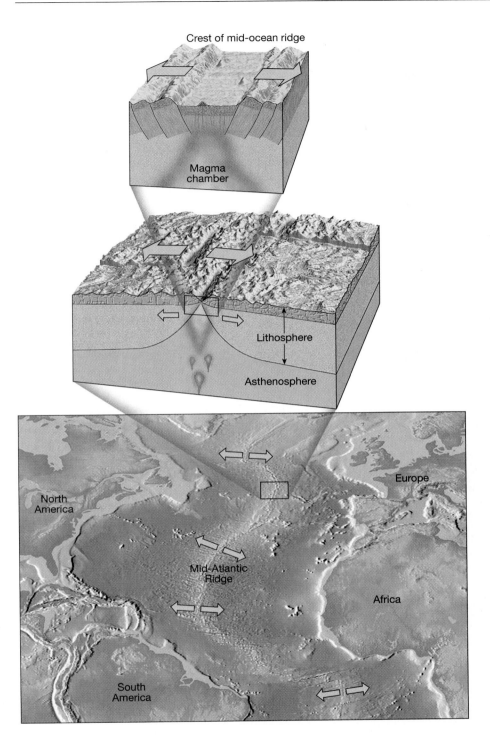

Figure 19.19 Most divergent plate boundaries are situated along the crests of oceanic ridges.

Not all spreading centers are found in the middle of large oceans. The Red Sea is the site of a recently formed divergent boundary. Here the Arabian Peninsula separated from Africa and began to move toward the northeast. Consequently, the Red Sea is providing oceanographers with a view of how the Atlantic Ocean may have looked in its infancy. Another narrow, linear sea produced by seafloor spreading in the recent geologic past is the Gulf of California, which separates the Baja Peninsula from the rest of Mexico.

When a spreading center develops within a continent, the landmass may split into two or more smaller segments as Alfred Wegener had proposed for the breakup of Pangaea. The fragmentation of a continent is thought to be associated with the upward movement of hot rock from the mantle. The effect of this activity is doming in the crust directly above the hot rising plume. This uplifting produces extensional forces that stretch and thin the crust, as shown in Figure 19.20A. Extension of the crust is accompanied by alternate episodes of faulting and volcanism. Adjacent to the spreading axis, faulted crustal blocks create downfaulted valleys called **rifts** or **rift valleys** (Figure 19.20B). As the spreading continues, the rift valley will lengthen and deepen, eventually extending out into the ocean. At this point the valley will become a narrow linear sea with an outlet to the ocean, similar to the Red Sea today (Figure 19.20C). The zone of rifting will remain the site of igneous activity, continually generating new seafloor in an ever-expanding ocean basin (Figure 19.20D).

The East African rift valleys represent the initial stage in the breakup of a continent as just described (Figure 19.21). The extensive volcanic activity that accompanies continental rifting is exemplified by large volcanic mountains such as Kilimanjaro and Mount Kenya. If the rift valleys in Africa remain active, East Africa will eventually separate from the mainland in much the same way the Arabian Peninsula split just a few million years ago.

Not all rift valleys develop into full-fledged spreading centers. Running through the central United States is an aborted rift zone extending from Lake Superior to Kansas. This once-active rift valley is filled with rock that was extruded onto the crust more than a billion years ago. The Rio Grande rift in New Mexico is a relatively young structure that may

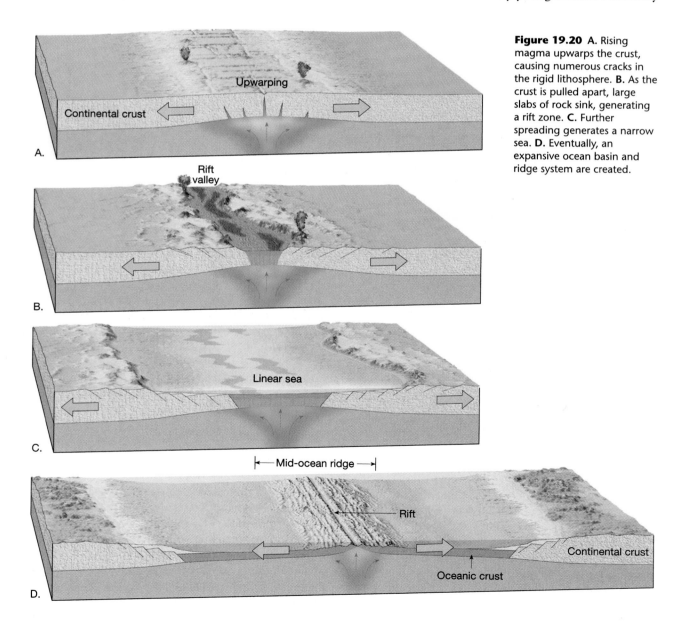

Figure 19.20 **A.** Rising magma upwarps the crust, causing numerous cracks in the rigid lithosphere. **B.** As the crust is pulled apart, large slabs of rock sink, generating a rift zone. **C.** Further spreading generates a narrow sea. **D.** Eventually, an expansive ocean basin and ridge system are created.

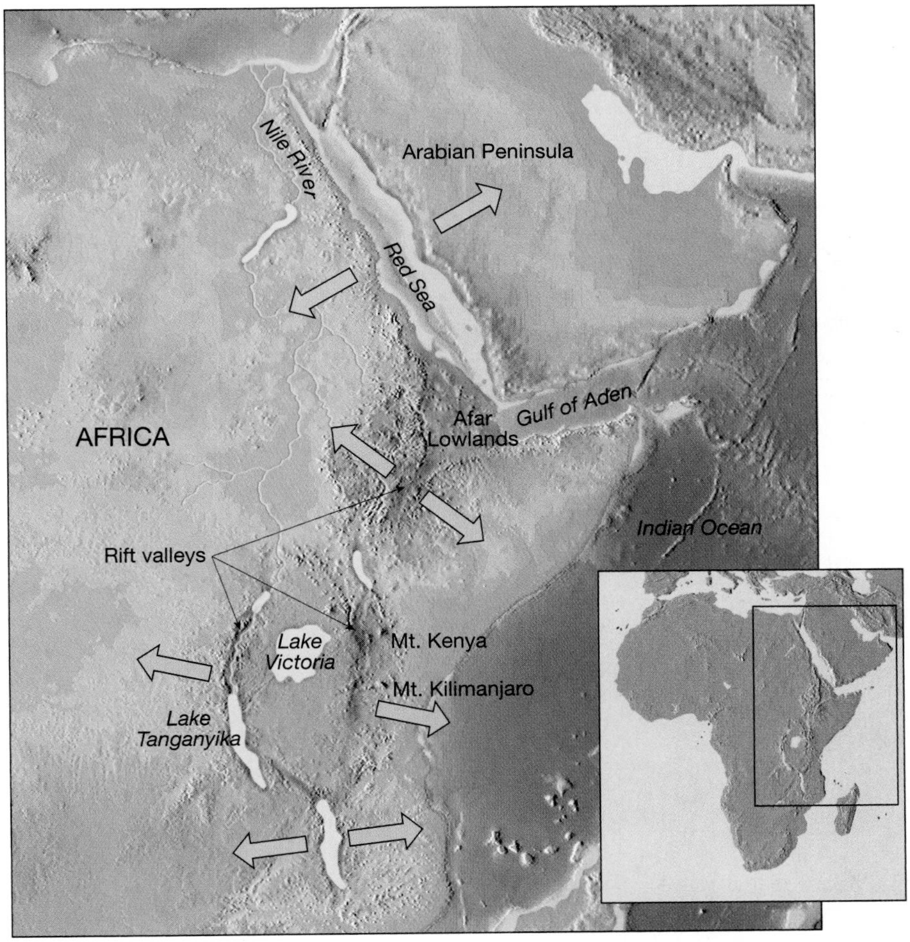

Figure 19.21 East African rift valleys and associated features.

or may not continue to grow. Why one rift valley develops into a full-fledged oceanic spreading center while others are abandoned is not yet known.

Convergent Boundaries

At spreading centers, new oceanic lithosphere is continually being generated. However, because the total surface area of Earth remains constant, lithosphere must also be consumed. Zones of plate convergence are the sites where oceanic lithosphere is subducted and absorbed into the mantle. When two plates converge, the leading edge of one is bent downward, allowing it to descend beneath the other. The region where an oceanic plate descends into the asthenosphere is called a **subduction zone**. As the oceanic plate slides beneath the overriding plate, the oceanic plate bends, thereby producing a **deep-ocean trench**, like the Peru–Chile trench (Figure 19.22A). Trenches formed in this manner may be thousands of kilometers long, 8 to 12 kilometers deep, and about 100 kilometers wide (Figure 19.23).

The average angle at which oceanic lithosphere descends into the asthenosphere is about 45 degrees. However, depending on its buoyancy, a plate may descend at an angle as small as a few degrees or it may plunge vertically (90 degrees) into the mantle. When a spreading axis is located near a subduction zone, the lithosphere is young and, therefore, warm and buoyant. Therefore, the angle of descent is small. This is the situation along parts of the Peru–Chile trench. Low dip angles are usually associated with a strong coupling between the descending slab and the overriding plate. Consequently, these regions experience great earthquakes. By contrast, some subduction zones, such as the Mariana trench, have steep angles of descent and few strong earthquakes.

Although all convergent zones have the same basic characteristics, they are highly variable features. Each is controlled by the type of crustal material involved and the tectonic setting. Convergent boundaries can form between two oceanic plates, one oceanic and one continental plate, or two continental plates. All three situations are illustrated in Figure 19.22.

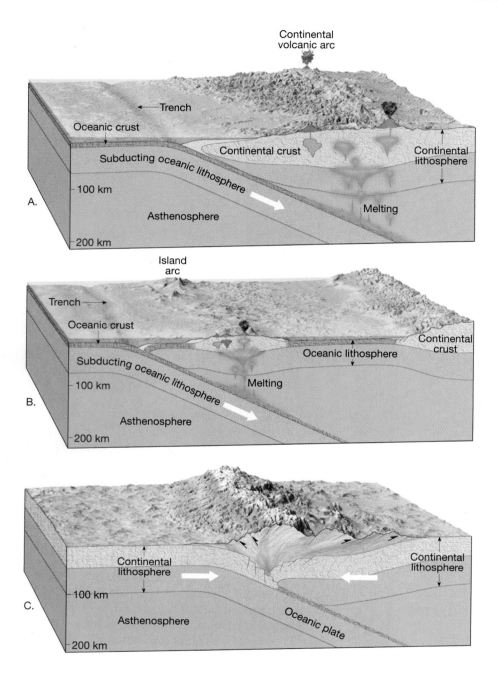

Figure 19.22 Zones of plate convergence.
A. Oceanic–continental.
B. Oceanic–oceanic.
C. Continental–continental.

Oceanic–Continental Convergence

Whenever the leading edge of a plate capped with continental crust converges with a plate capped with oceanic crust, the plate with the less dense continental material remains "floating," while the denser oceanic slab sinks into the asthenosphere (Figure 19.22A). As the oceanic slab descends, some of the sediments carried on the subducting plate, as well as pieces of oceanic crust, are scraped off and plastered against the edge of the overriding continental block. This chaotic accumulation of deformed sediment and scraps of oceanic crust is called an **accretionary wedge**. Studies conducted in the coastal regions of western Mexico, where the Cocos plate is being subducted, indicate that at least half of the sediment carried on the descending plate can be removed in this manner. Therefore, this process contributes to the already substantial accumulation of sediment deposited along continental margins.

When a descending plate reaches a depth of about 100 to 150 kilometers, heat drives water and other volatile components from the subducted sediments into the overlying mantle. These substances act as a flux does at a foundry, inducing partial melting of mantle rocks at reduced temperatures. The partial melting of mantle rock generates magmas having a basaltic or, possibly, andesitic composition. The newly formed magma, being less dense than the rocks of the mantle, will buoyantly rise. Often the magma will pond (accumulate) beneath the overlying continental crust where it may melt some of the silica-enriched crustal rocks. Eventually some of this silica-rich magma may migrate to the surface, where it can give rise to volcanic eruptions, some of which are explosive.

The Andean arc that runs along the western flank of South America is the product of magma generated as the Nazca plate descends beneath the continent (see Figure 19.17). In the central section of the southern Andes the subduction angle is very shallow, which probably accounts for the lack of volcanism in that area. As the South American plate moves westward, it overruns the Nazca plate. The result is a seaward migration of the Peru–Chile trench and a reduction in the size of the Nazca plate.

Mountains such as the Andes, which are produced in part by volcanic activity associated with the subduction of oceanic lithosphere, are called **continental volcanic arcs**. Another active continental volcanic arc is located in the western United States. The Cascade Range of Washington, Oregon, and California consists of several well-known volcanic mountains, including Mount Rainier, Mount Shasta, and Mount St. Helens (see Figure 4.33, p.114). As the continuing activity of Mount St. Helens testifies, the Cascade Range is still active. The magma here arises from melting triggered by the subduction of a small remaining segment of the Farallon plate, of which the Juan de Fuca plate is the largest northern segment.

A remnant of a formerly extensive continental volcanic arc is California's Sierra Nevada, in which Yosemite National Park is located. The Sierra Nevada is much older than the Cascade Range, and it has been inactive for several million years as evidenced by the absence of volcanic cones. Here erosion has stripped away most of the obvious traces of volcanic activity and left exposed the large, crystallized magma chambers that once fed lofty volcanoes.

Oceanic–Oceanic Convergence

When two oceanic slabs converge, one descends beneath the other, initiating volcanic activity in a manner similar to that which occurs at an oceanic–continental convergent boundary. In this case, however, the volcanoes form on the ocean floor rather than on a continent (Figure 19.22B). If this activity is sustained, it will eventually build volcanic structures that emerge as islands. The volcanic islands are spaced about 80 kilometers apart and are built upon submerged ridges a few hundred kilometers wide. This newly formed land consisting of an arc-shaped chain of small volcanic islands is called a **volcanic island arc**. The Aleutian, Mariana, and Tonga islands are examples of volcanic island arcs. Island arcs such as these are generally located 200 to 300 kilometers from the trench axis (see Box 19.2). Located adjacent to the island arcs just mentioned are the Aleutian trench, the Mariana trench, and the Tonga trench (see Figure 19.23).

Most volcanic island arcs are located in the western Pacific. Some, such as the New Hebrides and Mariana arcs, have a small accretionary wedge or none at all. Either very little sediment is deposited in these trenches or most of the sediment is carried into the mantle by the subducting plate. At these sites the subducting Pacific crust is relatively old and dense and therefore will readily sink into the mantle. This is thought to account for the steep angle of descent (often approaching 90 degrees) common in the trenches of the western Pacific. Many of these subduction zones lack the large earthquakes that are associated with some other convergent zones, such as the Peru–Chile trench.

Only two volcanic island arcs are located in the Atlantic—the Lesser Antilles arc adjacent to the Caribbean Sea, and the Sandwich Islands in the South Atlantic. The Lesser Antilles are a product of the subduction of the Atlantic beneath the Caribbean plate. Located within this arc is the island of Martinique where Mount Pelee erupted in 1902 destroying the town of St. Pierre and killing an estimated 28,000 people, and the island of Montserrat, were volcanic activity has occurred very recently (see Box 4.4, p.104).

In a few places, volcanic arcs are built upon both oceanic and continental crust. For example, the western section of the Aleutian arc consists of numerous islands built on oceanic crust, whereas the volcanoes at the eastern end of the chain are located on the Alaska Peninsula. Further, some volcanic island arcs are built on fragments of continental crust that have been rifted from the mainland. This type of volcanic island arc is exemplified by the Philippines and by Japan.

Continental–Continental Convergence

When two plates carrying continental crust converge, neither plate will subduct beneath the other because of the low density and thus the buoyant nature of continental rocks. The result is a collision between the two continental blocks (Figure 19.22C). Such a collision occurred when the subcontinent of India "rammed" into Asia and produced the Himalayas—the most

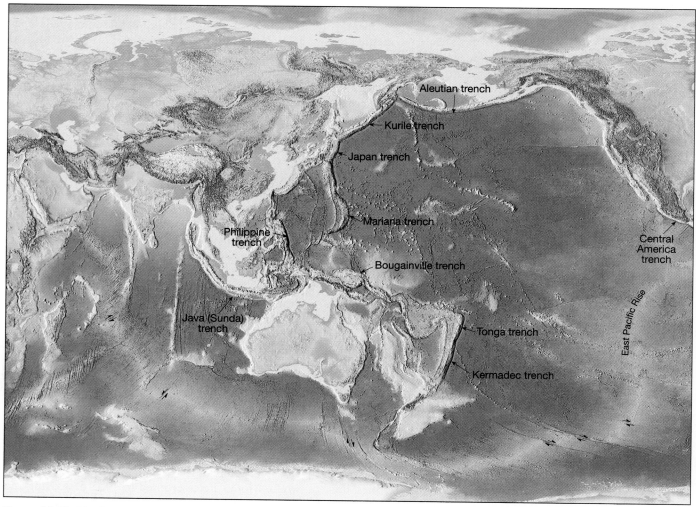

Figure 19.23 Distribution of the world's oceanic trenches, ridge system, fracture zones, and transform faults. Where transform faults offset ridge segments, they permit the ridge to change direction (curve) as can be seen in the Atlantic Ocean.

spectacular mountain range on Earth (Figure 19.24). During this collision, the continental crust buckled, fractured, and was generally shortened and thickened. In addition to the Himalayas, several other major mountain systems, including the Alps, Appalachians, and Urals, formed during continental collisions.

Prior to a continental collision, the landmasses involved are separated by an ocean basin. As the continental blocks converge, the intervening seafloor is subducted beneath one of the plates. Subduction initiates partial melting in the overlying mantle rocks, which, in turn, results in the growth of a volcanic arc. Depending on the location of the subduction zone, the volcanic arc could develop on either of the converging landmasses, or if the subduction zone developed several hundred kilometers seaward from the coast, a volcanic island arc would form. Eventually, as the intervening seafloor is

consumed, these continental masses collide. This folds and deforms the accumulation of sediments along the continental margin as if they were placed in a gigantic vise. The result is the formation of a new mountain range composed of deformed and metamorphosed sedimentary rocks, fragments of the volcanic arc, and possibly slivers of oceanic crust.

After continents collide, the subducted oceanic plate may separate from the continental block and continue its downward movement. However, because of its buoyancy, continental lithosphere cannot be carried very far into the mantle. In the case of the Himalayas, the leading edge of the Indian plate was forced partially under Asia, generating an unusually great thickness of continental lithosphere. This accumulation accounts, in part, for the high elevation of the Himalayas and helps explain the elevated Tibetan Plateau to the north.

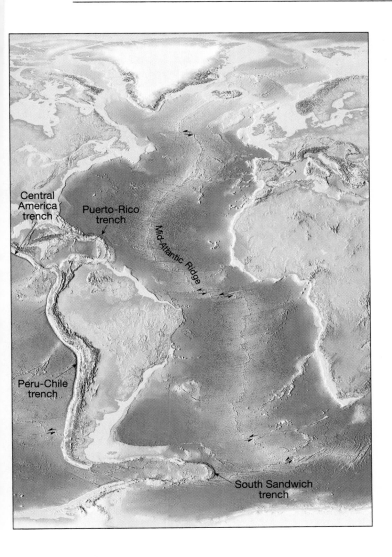

Central America trench

Puerto-Rico trench

Mid-Atlantic Ridge

Peru-Chile trench

South Sandwich trench

Transform Fault Boundaries

The third type of plate boundary is the transform fault, which is characterized by strike-slip faulting where plates grind past one another without the production or destruction of lithosphere. Transform faults were first identified where they join offset segments of an ocean ridge (see Figure 19.23). At first it was erroneously assumed that the ridge system originally formed a long and continuous chain that was later offset by horizontal displacement along these large faults. However, the displacement along these faults was found to be in the exact opposite direction required to produce the offset ridge segments.

The true nature of transform faults was discovered in 1965 by J. Tuzo Wilson of the University of Toronto. Wilson suggested that these large faults connect the global active belts (convergent boundaries,

divergent boundaries, and other transform faults) into a continuous network that divides Earth's outer shell into several rigid plates. Thus, Wilson became the first to suggest that Earth was made of individual plates, while at the same time identifying the faults along which relative motion between the plates is made possible.

Most transform faults join two segments of a mid-ocean ridge (see Figure 19.23). Here, they are part of prominent linear breaks in the oceanic crust known as **fracture zones**, which include both the transform faults and their inactive extentions into the plate interior. These fracture zones are present approximately every 100 kilometers along the trend of a ridge axis. As shown in Figure 19.25, active transform faults lie only *between* the two offset ridge segments. Here seafloor produced at one ridge axis moves in the opposite direction as seafloor produced at an opposing ridge segment. Thus, between the ridge segments these adjacent slabs of oceanic crust are grinding past each other along a transform fault. Beyond the ridge crests are the inactive zones, where the fractures produced by strike-slip faulting are preserved as linear topographic scars. These fracture zones tend to curve such that small segments roughly parallel the direction of plate motion at the time of their formation.

In another role, transform faults provide the means by which the oceanic crust created at ridge crests can be transported to a site of destruction, the deep-ocean trenches. Figure 19.26 illustrates this situation. Notice that the Juan de Fuca plate moves in a southeasterly direction, eventually being subducted under the west coast of the United States. The southern end of this relatively small plate is bounded by the Mendocino transform fault. This transform fault boundary connects the Juan de Fuca ridge to the Cascade subduction zone (Figure 19.26). Therefore, it facilitates the movement of the crustal material created at the ridge crest to its destination beneath the North American continent (Figure 19.26). Another example of a *ridge-trench* transform fault is found southeast of the tip of South America. Here transform faults on the north and south margins of the Scotia plate connect the trench to a short spreading axis (see Figure 19.17).

Although most transform faults are located within the ocean basins, a few, including California's famous San Andreas fault, cut through continental crust. Notice in Figure 19.26 that the San Andreas fault connects a spreading center located in the Gulf of California to the Cascade subduction zone and the Mendocino transform fault located along the northwest coast of the United States. Along the San Andreas fault, the Pacific plate is moving toward the northwest, past the North American plate. If this movement continues, that part of California west of

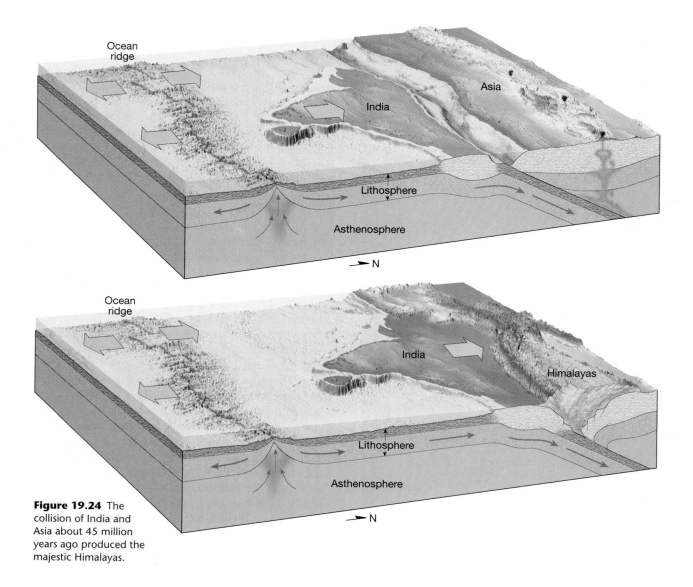

Ocean ridge

Asia

India

Lithosphere

Asthenosphere

N

Ocean ridge

India

Himalayas

Lithosphere

Asthenosphere

N

Figure 19.24 The collision of India and Asia about 45 million years ago produced the majestic Himalayas.

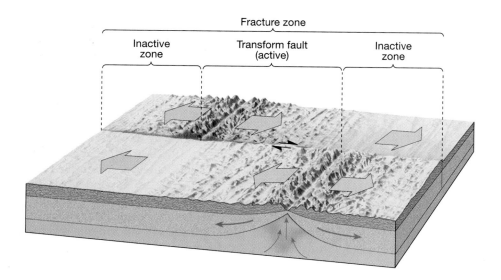

Fracture zone

Inactive zone

Transform fault (active)

Inactive zone

Figure 19.25 Diagram illustrating a transform fault boundary offsetting segments of a divergent boundary (oceanic ridge).

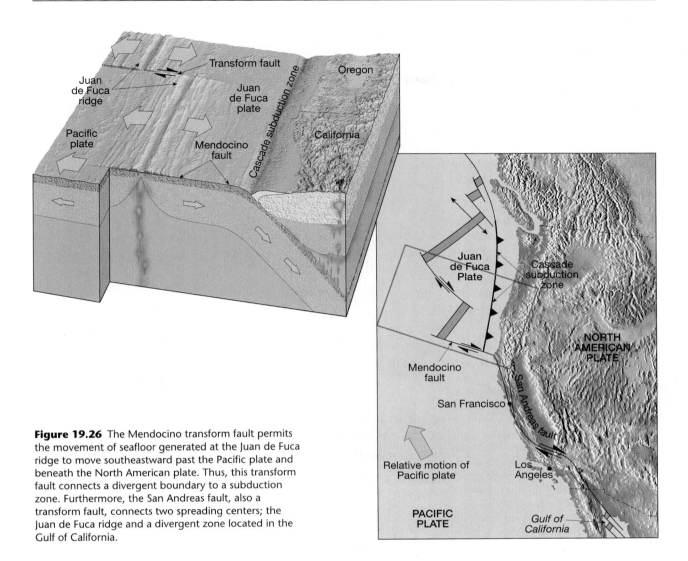

Figure 19.26 The Mendocino transform fault permits the movement of seafloor generated at the Juan de Fuca ridge to move southeastward past the Pacific plate and beneath the North American plate. Thus, this transform fault connects a divergent boundary to a subduction zone. Furthermore, the San Andreas fault, also a transform fault, connects two spreading centers; the Juan de Fuca ridge and a divergent zone located in the Gulf of California.

the fault zone, including the Baja Peninsula, will become an island off the west coast of the United States and Canada. It could eventually reach Alaska. However, a more immediate concern is the earthquake activity triggered by movements along this fault system.

Testing the Plate Tectonics Model

With the development of plate tectonics, researchers from all Earth sciences began testing this model of how Earth works. Some of the evidence supporting continental drift and seafloor spreading has already been presented. In addition,

some of the evidence that was instrumental in solidifying the support for this new idea follows. Note that much of this evidence was not new; rather, it was new interpretations of already existing data that swayed the tide of opinion.

Moreover, some of the data were compiled to refute rather than support global tectonics. As one researcher phrased it, "My observations are not compatible with seafloor spreading, and I shall prepare a critical demonstration that this is so and thus demolish this nutty idea and we can all get back to work." However, he, like many other scientists, found that his results were indeed compatible with this new theory. The revolution had produced a new model from which to view all tectonic processes operating on Earth.

Box 19.2

Susan DeBari: A Career in Geology

I discovered geology the summer that I worked doing trail maintenance in the North Cascades mountains of Washington State. I had just finished my freshman year in college and had never before studied Earth Science. But a co-worker (now my best friend) began to describe the geological features of the mountains that we were hiking in—the classic cone shape of Mt. Baker volcano, the U-shaped glacial valleys, the advance of active glaciers, and other wonders. I was hooked and went back to college that fall with a geology passion that hasn't abated. As an undergraduate, I worked as a field assistant to a graduate student and did a senior thesis project on rocks from the Aleutian island arc. From that initial spark, island arcs have remained my top research interest, on through Ph.D. research at Stanford University, postdoctoral work at the University of Hawaii, and as a faculty member at San Jose State University and Western Washington University. Especially the deep crust of arcs, the material that lies close to the Mohorovičić discontinuity (fondly known as the Moho). What kinds of processes are occurring down there at the base of the crust in island arcs? What is the source of magmas that make their way to the surface (the mantle? the deep crust itself?). How do these magmas interact with the crust as they make their way upwards? What do these early magmas look like chemically? Are they very different from what is erupted at the surface?

Obviously, geologists cannot go down to the base of the crust (typically 20 to 40 kilometers beneath Earth's surface). So what they do is play a bit of a detective game. They must use rocks that are *now exposed at the surface* that were originally formed in the deep crust of an island arc. The rocks must have been brought to the surface rapidly along fault zones to preserve their original features. Thus, I can walk on rocks of the deep crust without really leaving the Earth's surface! There are a few places around the world where these rare rocks are

Figure 19.B Susan DeBari photographed with the Japanese submersible, *Shinkai 6500*, which she used to collect rock samples from the Izu Bonin trench.

exposed. Some of the places that I have worked include the Chugach Mountains of Alaska, the Sierras Pampeanas of Argentina, the Karakorum range in Pakistan, Vancouver Island's west coast, and the North Cascades of Washington. Fieldwork has involved hiking most commonly, but also extensive use of mules and trucks.

I also went looking for exposed pieces of the deep crust of island arcs in a less obvious place, in one of the deepest oceanic trenches of the world, the Izu Bonin trench (Figure 19.B). Here I dove into the ocean in a submersible called the *Shinkai 6500* (pictured to my left in the background). The *Shinkai 6500* is a Japanese submersible that has the capability to dive to 6500 meters below the surface of the ocean (approximately 4 miles). My plan was to take rock samples from the wall of the trench at its deepest levels

using the submersible's mechanical arm. Because preliminary data suggested that vast amounts of rock were exposed for several kilometers in a vertical sense, this could be a great way to sample the deep arc basement. I dove in the submersible three times, reaching a maximum depth of 6497 meters. Each dive lasted 9 hours, spent in a space no bigger than the front seat of a Honda, shared with two of the Japanese pilots that controlled the submersible's movements. It was an exhilarating experience!

I am now on the faculty at Western Washington University where I continue to do research on the deep roots of volcanic arcs, and get students involved as well. I am also involved in Science Education training for K-12 teachers, hoping to get young people motivated to ask questions about the fascinating world that surrounds them!

Plate Tectonics and Earthquakes

By 1968, the basic outline of global tectonics was firmly established. In that year, three seismologists published papers demonstrating how successfully the new plate tectonics model accounted for the global distribution of earthquakes (Figure 19.27). In particular, these scientists were able to account for the close association between deep-focus earthquakes and subduction zones. Furthermore, the absence of deep-focus earthquakes along the oceanic ridge system was also shown to be consistent with the new theory.

Earthquakes Along Subduction Zones. Note the close association between plate boundaries and earthquakes by comparing the distribution of earthquakes shown in Figure 19.27 with the map of plate boundaries in Figure 19.17 (p. 484). In trench regions, where dense slabs of lithosphere plunge into the mantle, this association is especially striking. When the depths of earthquake foci and their locations within the trench systems are plotted, an interesting pattern emerges. Figure 19.28, which shows the distribution of earthquakes in the vicinity of the Japan trench, is an example. Here most shallow-focus earthquakes occur within, or adjacent to, the trench, whereas intermediate- and deep-focus earthquakes occur toward the mainland. A similar distribution pattern exists along the western margin of South America where the Nazca plate is being subducted beneath the South American continent.

In the plate tectonics model, deep-ocean trenches form where dense slabs of oceanic lithosphere plunge into the mantle (Figure 19.28). Shallow-focus earthquakes are produced as the descending plate interacts with the overriding lithosphere. As the slab descends farther into the asthenosphere, deeper-focus earthquakes are generated. Because the earthquakes occur within the rigid subducting plate rather than in the "plastic" mantle, they provide a method for tracking the plate's descent. Recall from Chapter 16 that the zones of inclined seismic activity that extend from the trench into the mantle are called *Wadati-Benioff zones* after two seismologists who conducted extensive studies on the distribution of earthquake foci. Very few earthquakes have been recorded below 690 kilometers, possibly because the slab has been heated sufficiently to lose its rigidity.

Intermediate- and Deep-focus Earthquakes. The cause of intermediate- and deep-focus earthquakes occurring between 70 and 690 kilometers has been a long-standing problem in geology. At depths below about 70 kilometers, rocks are expected to deform by ductile flow rather than by brittle fracture or frictional sliding. With the development of the plate tectonics theory, researchers were provided a new avenue to explore this old problem. Experimental evidence

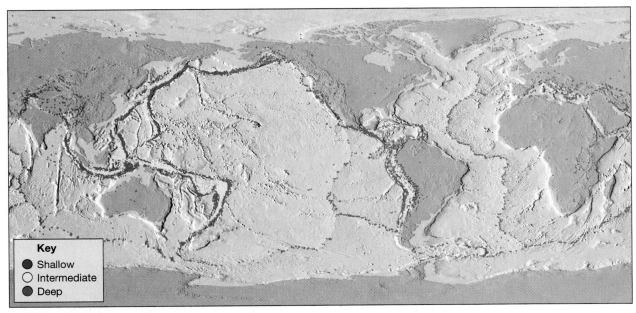

Figure 19.27 Distribution of shallow-, intermediate-, and deep-focus earthquakes. Note that deep-focus earthquakes only occur in association with convergent plate boundaries and subduction zones. (Data from NOAA)

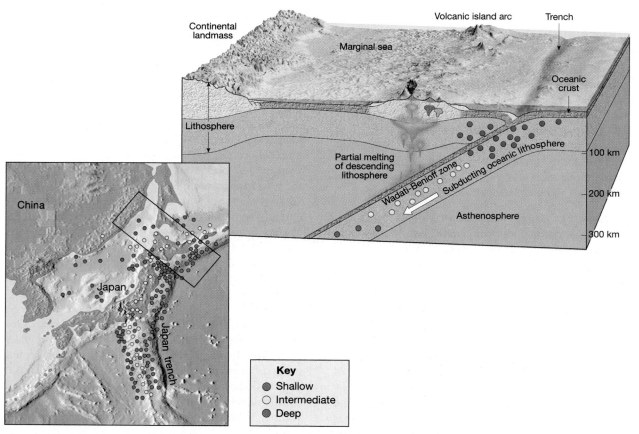

Figure 19.28 Distribution of earthquake foci in the vicinity of the Japan trench. Note that intermediate- and deep-focus earthquakes occur only within the sinking slab of oceanic lithosphere. (Data from NOAA)

indicates a cold descending slab may warm up so slowly that it exhibits brittle fracture to depths as great as 300 kilometers. Thus, intermediate-focus earthquakes could be generated in a manner similar to shallow quakes—that is, through the release of elastic energy along a fault surface. But what causes deep-focus earthquakes?

The mechanisms responsible for earthquakes that occur between 300 and 690 kilometers remain elusive. One hypothesis that has gained substantial support is based on evidence that during subduction, increasing pressures cause some minerals, such as olivine, to go through a phase change. A phase change produces a mineral having a more compact crystalline structure. It has been suggested that when this transformation occurs in the coldest plate interiors, it causes a type of high-pressure faulting, which in turn produces a deep-focus earthquake.

To test this and other proposals, researchers are duplicating mantle conditions by using specially designed anvils that can apply enormous pressures to minute rock samples between two diamonds. The samples are heated by shining a laser through one of the diamonds, and sensors are used to detect acoustical events thought to be analogous to earthquakes. Although these studies are preliminary, they do indicate that at high pressures and temperatures, minerals can fail through slippage in the manner predicted.

Although the exact cause of deep-focus earthquakes is still debated, their close association with subduction zones is well documented. Because subduction zones are the only regions where cold crustal rocks are forced to great depths, these should be the only sites of deep-focus earthquakes. Indeed, the absence of deep-focus earthquakes along divergent and transform boundaries supports the theory of plate tectonics.

Evidence from Ocean Drilling

Some of the most convincing evidence confirming seafloor spreading has come from drilling directly into ocean-floor sediment. From 1968 until 1983, the source of these important data was the Deep Sea Drilling Project, an international program sponsored by several major oceanographic institutions and the National Science Foundation. The primary goal was to gather firsthand information about the age and processes that formed the ocean basins.

To accomplish this, a new drilling ship, the *Glomar Challenger*, was built. This ship represented a significant technological breakthrough; it was capable of lowering drill pipe thousands of meters to the ocean floor and then drilling hundreds of meters into the sediments and underlying basaltic crust.

Operations began in August 1968, in the South Atlantic. At several sites, holes were drilled through the entire thickness of sediments to the basaltic rock below. An important objective was to gather samples of sediment from just above the igneous crust as a means of dating the seafloor at each site.* Because sedimentation begins immediately after the oceanic crust forms, remains of microorganisms found in the oldest sediments—those resting directly on the basalt—can be used to date the ocean floor at that site.

When the oldest sediment from each drill site was plotted against its distance from the ridge crest, the plot demonstrated that the age of the sediment increased with increasing distance from the ridge. This finding agreed with the seafloor-spreading hypothesis, which predicted that the youngest oceanic crust would be found at the ridge crest, and that the oldest oceanic crust would be at the continental margins.

Further, the rate of seafloor spreading determined from the ages of sediments was identical to the rate previously estimated from magnetic evidence. Subsequent drilling in the Pacific Ocean verified these findings. These excellent correlations provided a striking confirmation of seafloor spreading.

The data from the Deep Sea Drilling Project also reinforced the idea that the ocean basins are geologically youthful because no sediment with an age in excess of 160 million years was found. By comparison, continental crust that exceeds four billion years in age has been dated.

The thickness of ocean-floor sediments provided additional verification of seafloor spreading. Drill cores from the *Glomar Challenger* revealed that sediments are almost entirely absent on the ridge crest and that the sediment thickens with increasing distance from the ridge. Because the ridge crest is younger than the areas farther away from it, this pattern of sediment distribution should be expected if the seafloor spreading hypothesis is correct.

Furthermore, measurements in the open ocean have shown that sediment accumulates at a rate of about 1 centimeter per 1000 years. Therefore, if the ocean

floor were an ancient feature, sediments would be many kilometers thick. However, data from hundreds of drilling sites indicate that the greatest thickness of sediment in the deep-ocean basins is only a few hundred meters, equivalent to intervals of only a few tens of millions of years. Thus, here is additional data confirming the fact that the ocean floor is a young geologic feature.

During its 15 years of operation, the *Glomar Challenger* logged more than 600,000 kilometers on 96 voyages across every ocean. The drilling of 1092 holes yielded more than 96 kilometers (60 miles) of invaluable core samples. Although the Deep Sea Drilling Project ended and the *Glomar Challenger* was retired, the important work of sampling the floors of the world's ocean basins continues.

The Ocean Drilling Project has succeeded the Deep Sea Drilling Project and, like its predecessor, it is a major international program. The more technologically advanced drilling ship, the *JOIDES Resolution*, now continues the work of the *Glomar Challenger*.† The *JOIDES Resolution* can drill in water depths as great as 8100 meters (27,000 feet) and contains onboard laboratories equipped with the largest and most varied array of seagoing scientific research equipment in the world (Figure 19.29).

Hot Spots

Mapping of seamounts in the Pacific Ocean revealed a chain of volcanic structures extending from the Hawaiian Islands to Midway Island and continuing northward toward the Aleutian trench (Figure 19.30). Radiometric dating in this chain showed that the volcanoes increase in age with increasing distance from Hawaii. Hawaii, the youngest volcano in the chain, rose from the seafloor less than a million years ago, whereas Midway Island is 27 million years old, and Suiko Seamount, near the Aleutian trench, is 65 million years old (Figure 19.30).

Researchers have proposed that a rising plume of mantle material is located beneath the island of Hawaii. Partial melting of this hot rock as it enters the low-pressure environment near the surface generates a volcanic area known as a **hot spot**. Presumably, as the Pacific plate moved over this hot spot, successive volcanic structures were built. The age of each volcano indicates the time when it was situated over the relatively stationary mantle plume.

This pattern is shown in Figure 19.30. Kauai is the oldest of the large islands in the Hawaiian chain. Five million years ago, when it was positioned over the

---

*Radiometric dates of the ocean crust itself are unreliable because of the alteration of basalt by seawater.

†JOIDES stands for Joint Oceanographic Institutions for Deep Earth Sampling.

Figure 19.29 The *JOIDES Resolution*, the drilling ship of the Ocean Drilling Program. This modern drilling ship has replaced the *Glomar Challenger* in the important work of sampling the floors of the world's oceans. (Photo courtesy of Ocean Drilling Program)

hot spot, Kauai was the only Hawaiian Island (Figure 19.30). Visible evidence of the age of Kauai can be seen by examining its extinct volcanoes, which have been eroded into jagged peaks and vast canyons. By contrast, the south slopes of the comparatively young island of Hawaii consist of fresh lava flows, and two of Hawaii's volcanoes, Mauna Loa and Kilauea, remain active.

Recent evidence indicates that a new volcanic pile, named Loihi Seamount, is forming on the ocean floor about 35 kilometers off the southeast coast of Hawaii. Geologically speaking, it should not be long before another tropical island will be added to the Hawaiian chain (see Box 4.3, p. 96).

Two island groups parallel the Hawaiian Island–Emperor Seamount chain. One chain consists of the Tuamotu and Line islands, and the other includes the Austral, Gilbert, and Marshall islands. In each case, the most recent volcanic activity has occurred at the southeastern end of the chain, and the islands get progressively older to the northwest. Thus, like the Hawaiian Island–Emperor Seamount chain, these volcanic structures apparently formed by the same motion of the Pacific plate over fixed mantle plumes. Not only does this evidence support the fact that the plates do indeed move relative to Earth's interior, but also the hot spot "tracks" trace the direction of the plate motion. Notice in Figure 19.30, for example, that the Hawaiian Island–Emperor Seamount chain bends. This particular bend in the track occurred about 40 million years ago when the motion of the Pacific plate changed from nearly due north to a northwesterly path. Similarly, hot spots found on the floor of the Atlantic have increased our understanding of the migration of landmasses following the breakup of Pangaea.

Although the existence of mantle plumes is well documented, their exact role in plate tectonics is a topic of ongoing research. Some researchers suggest that mantle plumes originate deep in the mantle, perhaps at the core–mantle boundary. Most evidence indicates that hot spots remain relatively stationary. Of the 50 to 120 hot spots that exist, about a dozen or so are located near divergent plate boundaries. A hot spot beneath Iceland is probably responsible for the unusually large accumulation of volcanic rocks found in that section of the Mid-Atlantic Ridge. Another hot

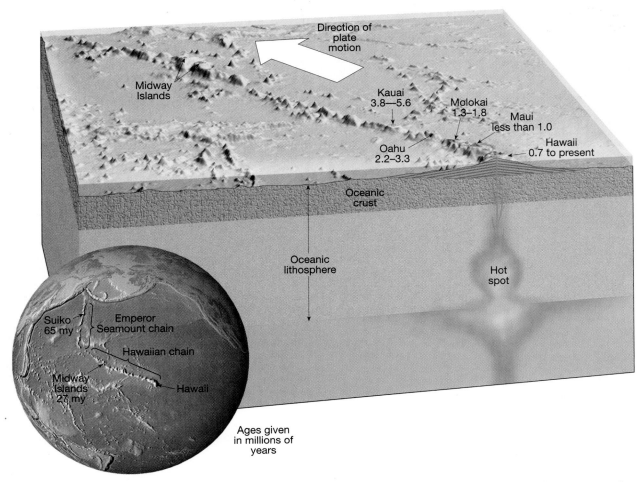

Figure 19.30 The chain of islands and seamounts that extends from Hawaii to the Aleutian trench results from the movement of the Pacific plate over an apparently stationary hot spot. Radiometric dating of the Hawaiian Islands shows that the volcanic activity decreases in age toward the island of Hawaii.

spot is probably located beneath Yellowstone National Park and is likely responsible for the large outpourings of lava and volcanic ash covering that area.

Pangaea: Before and After

A great deal of evidence has been gathered to support the fact that Alfred Wegener's supercontinent of Pangaea began to break apart about 200 million years ago. An important consequence of this continental rifting was the creation of a "new" ocean basin, the Atlantic. The breakup of Pangaea and the formation of the Atlantic Ocean Basin apparently occurred over a span of nearly 160 million years. The last phase, the separation of Greenland and Eurasia, began only about 50 million years ago.

Although continental rifting is well documented, one remaining question is: What causes a continent to break apart? Researchers have suggested that when a thick segment of continental lithosphere remains stationary for an extended period, the conditions are right for continental rifting. Because continental crust is a poor conductor of heat, it acts like a thick blanket retarding the outward flow of heat from the mantle. Thus, a buildup of heat causes the supercontinent to bulge upward. This activity generates extensional forces that thin and eventually break the lithosphere apart. Upwelling of hot material between the rifted continental fragments generates a new ocean floor.

Breakup of Pangaea

The fragmentation of Pangaea began about 200 million years ago. Figure 19.31 illustrates the breakup and subsequent paths taken by the landmasses. As we can readily see in Figure 19.31B, two major rifts initiated the breakup. The rift zone between North America and Africa generated numerous outpourings of Jurassic-age basalts. These basalts are

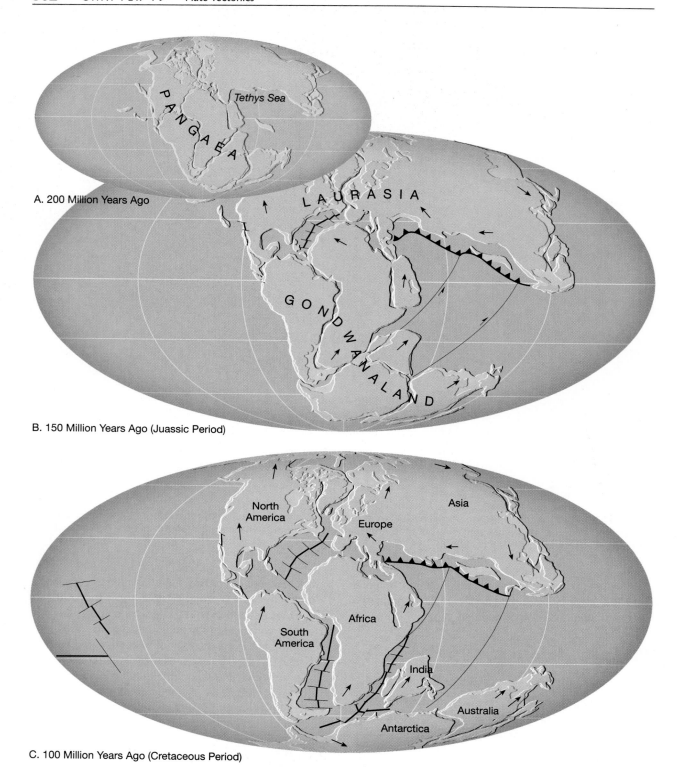

A. 200 Million Years Ago

B. 150 Million Years Ago (Juassic Period)

C. 100 Million Years Ago (Cretaceous Period)

Figure 19.31 Several views of the breakup of Pangaea over a period of 200 million years.

presently visible along the eastern seaboard of the United States. Radiometric dating of this material indicates that rifting began between 200 million and 165 million years ago. This time span can be used as the birth date of this section of the North Atlantic. The rift that formed in the southern landmass of Gondwanaland developed a Y-shaped fracture, which sent India on a northward journey and

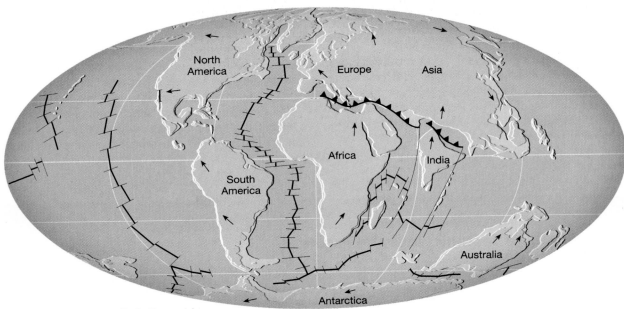

D. 50 Million Years Ago (Early Cenozoic)

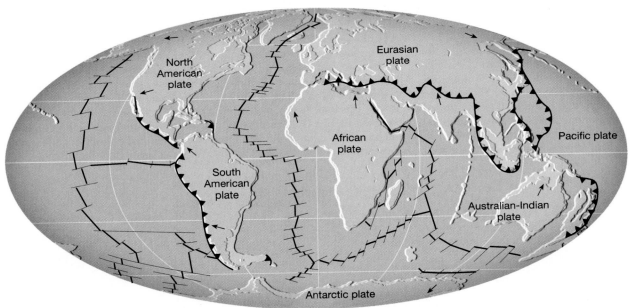

E. Present

simultaneously separated South America–Africa from Australia–Antarctica.

Figure 19.31C illustrates the position of the continents 135 million years ago, about the time Africa and South America began splitting apart to form the South Atlantic. India can be seen halfway into its journey to Asia, while the southern portion of the North Atlantic has widened greatly. By the beginning of the Cenozoic, about 65 million years ago, Madagascar had separated from Africa, and the South Atlantic had emerged as a full-fledged ocean (Figure 19.31D). At

this juncture, India had drifted over a hot spot that generated numerous fluid basalt flows across a region in western India now called the Deccan Plateau. These lava flows are very similar to those that make up the Columbia Plateau in the Pacific Northwest.

A modern map (Figure 19.31E) shows India in contact with Asia, an event that began about 45 million years ago and created the highest mountains on Earth, the Himalayas, along with the Tibetan Highlands. It is interesting to note that the average height of Tibet is 5000 meters, higher than any spot in the contiguous

United States. India's continued northward migration causes the numerous and often destructive earthquakes that plague that part of the world.

By comparing Figures 19.31D and 19.31E, we can see that the separation of Greenland from Eurasia was a recent event in geologic history. Also notice the recent formation of the Baja Peninsula along with the Gulf of California. This event began less than 10 million years ago (see Box 19.3).

Before Pangaea

Prior to the formation of Pangaea, the landmasses had probably gone through several episodes of fragmentation similar to what we see happening today. Also like today, these ancient continents moved away from each other only to collide again at some other location. The oldest well-documented supercontinent, known as *Rodinia*, formed a little over a billion years ago. Rodinia appears to have included most of the continents, but in

Box 19.3

Plate Tectonics into the Future

Two geologists, Robert Dietz and John Holden, extrapolated present-day plate movements into the future. Figure 19.C illustrates where they envision Earth's landmasses will be 50 million years from now if present plate movements persist for this time span.

In North America we see that the Baja Peninsula and the portion of southern California that lies west of the San Andreas fault will have slid past the North American plate. If this northward migration takes place, Los Angeles and San Francisco will pass each other in about 10 million years, and in about 60

million years, Los Angeles will begin to descend into the Aleutian trench.

Significant changes are seen in Africa, where a new sea emerges as East Africa parts company with the mainland. In addition, Africa will have moved slowly into Europe, perhaps initiating the next major mountain-building stage on our dynamic planet. Meanwhile, the Arabian Peninsula continues to diverge from Africa, allowing the Red Sea to widen and closing the Persian Gulf.

In other parts of the world, Australia is now astride the equator and, along

with New Guinea, is on a collision course with Asia. Meanwhile, North and South America are beginning to separate, while the Atlantic and Indian oceans continue to grow at the expense of the Pacific Ocean.

These projections into the future, although interesting, must be viewed with caution because many assumptions must be correct for these events to unfold as just described. Nevertheless, changes in the shapes and positions of continents that are equally profound will undoubtedly occur for millions of years to come.

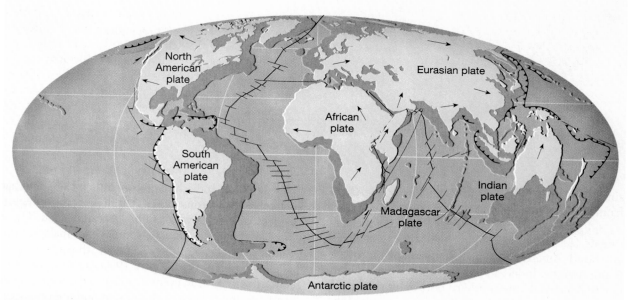

Figure 19.C The world as it may look 50 million years from now. (From "The Breakup of Pangaea," Robert S. Dietz and John C. Holden. Copyright 1970 by Scientific American, Inc. All rights reserved.)

a configuration quite different from Pangaea. Roughly 725 million years ago, Rodinia began to fragment. It was these rifted continental fragments that combined to form Pangaea during the period between 500 million and 250 million years ago. Evidence for the formation of Pangaea include Russia's Ural Mountains and the Appalachian Mountains, which flank the eastern coast of North America.

About 500 million years ago the northern continent (known as *Laurasia*) was fragmented into three major sections—North America, northern Europe (southern Europe was part of Africa), and Siberia—with each section separated by a sizable ocean. The southern continent of *Gondwana* probably was intact and lay near the South Pole. The first collision occurred as North America and Europe closed the pre-North Atlantic. This activity resulted in the formation of the northern Appalachians. Parts of the floor of the former ocean can be seen today high above sea level in Nova Scotia. The sliver of eastern Canada and the United States that lies seaward of the zone of this collision is truly a gift from Europe. It is also thought that before North America and Europe collided, part of Scotland, Ireland, and Norway were attached to the North American plate. While North America and Europe were joining, Siberia was closing the gap between itself and Europe, which lay to the west. This closing culminated about 300 million to 350 million years ago in the formation of the Ural Mountains. The consolidation of these landmasses completed the northern continent of Laurasia.

During the next 50 million years the northern and southern landmasses converged, producing the supercontinent of Pangaea. At this time (about 300 million to 250 million years ago), Africa and North America collided to produce the southern Appalachians.

The Driving Mechanism

The plate tectonics theory *describes* plate motion and the *effects* of this motion. Therefore, acceptance of this model does not rely on a knowledge of the force or forces moving the plates. This is fortunate, because none of the driving mechanisms yet proposed can account for *all* major facets of plate motion. Nevertheless, it is clear that the unequal distribution of heat within Earth is the underlying driving force for plate movement.

Convection Currents

One of the first models to explain the movements of plates was proposed by the eminent English geologist Arthur Holmes as a possible driving mechanism for continental drift. Adapted to plate tectonics, this hypothesis suggests that large convection currents in the mantle—in which warm, less dense rock rises and cooler, denser material sinks—drive plate motion (Figure 19.32A). According to this proposal, the warm, less dense material of the lower mantle rises very slowly in the regions of oceanic ridges. As the material spreads laterally, it drags the lithosphere along like packages on a conveyor belt. Eventually, the material cools and begins to sink back into the lower mantle, where it is reheated.

Partly because of its simplicity, this proposal once had wide appeal. However, researchers employing modern research techniques have learned that the flow of material in the mantle is far more complex than such simple convection cells. Further, it is now clear that lithospheric plates are not passengers carried along by convective flow, but rather they are part of that circulation.

Slab-Push and Slab-Pull

Many other mechanisms that may contribute to plate motion have been suggested. One relies on the fact that, as a newly formed slab of oceanic crust moves away from the ridge crest, it gradually cools and becomes denser. Eventually, the cold oceanic slab becomes denser than the underlying asthenosphere and begins to sink. When this occurs, the cold sinking slab pulls the trailing lithosphere along. This so-called **slab-pull** is thought to be an important mechanism for transporting cold material back into the mantle (Figure 19.32B).

Another mechanism proposes that the elevated position of an oceanic ridge could cause the lithosphere to slide under the influence of gravity. However, some ridge systems are subdued, which would reduce the effectiveness of the **slab-push** mechanism. Nevertheless, the slab-push phenomenon appears to be active in some ridge systems but is probably much less effective than slab-pull (Figure 19.32B).

Rising Plumes and Descending Slabs

One version of the thermal convection model suggests that hot, buoyant plumes of rock are the upward flowing arm in the convective mechanism at work in the mantle (Figure 19.32C). These hot plumes are presumed to extend upward from the vicinity of the mantle–core boundary. Upon reaching the lithosphere, they spread laterally and facilitate plate motion away from the zone of upwelling. These mantle plumes reveal themselves as long-lived volcanic areas (hot spots) in such places as Iceland. A dozen or so hot spots

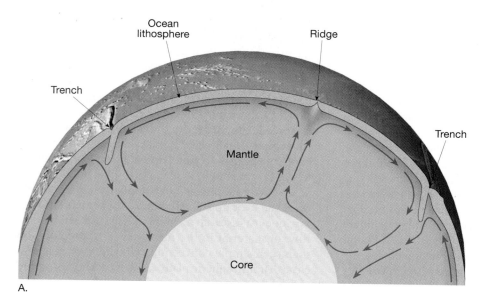

A.

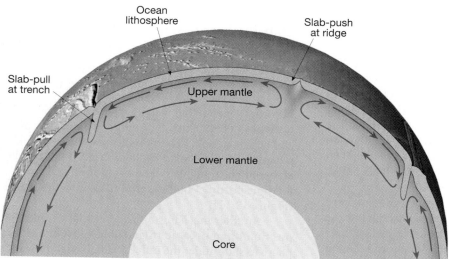

B.

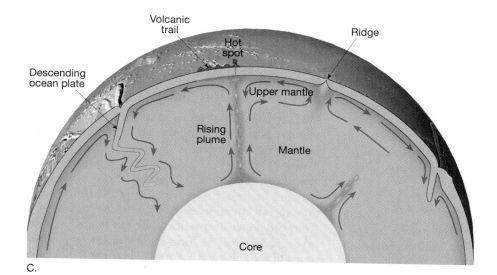

C.

Figure 19.32 Proposed models of the driving force for plate tectonics. **A.** Large convection cells in the mantle may carry the lithosphere in a conveyor-belt fashion. **B.** Slab-pull results because the subducting slab is more dense than the underlying material. Slab-push is a form of gravity sliding caused by the elevated position of lithosphere at a ridge crest. **C.** The hot plume model suggests that all upward convection is confined to a few narrow plumes, while the downward limbs of these convection cells are the cold, dense subducting oceanic plates.

have been identified along ridge systems where they may contribute to plate divergence. Recall, however, that many hot spots, including the one which generated the Hawaiian Islands, are not located in ridge areas.

In another version of the hot plume model, all upward convection is confined to a few large cylindrical structures. Embedded in these few large zones of upwelling are most of Earth's hot spots. The downward limbs of these convection cells are the cold, dense, subducting lithospheric plates. Moreover, advocates of this view suggest that subducting slabs may descend all the way to the core–mantle boundary.

Although there is still much to be learned about the mechanisms that cause plates to move, some facts are clear. The unequal distribution of heat in Earth generates some type of thermal convection in the mantle that ultimately drives plate motion. Furthermore, the descending lithospheric plates are active components of downwelling, and they serve to transport cold material into the mantle. Whether the upwelling is mainly in the form of narrow rising plumes or large cylindrical plumes of various shapes is yet to be determined.

Chapter Summary

- In the early 1900s *Alfred Wegener* set forth the *continental drift* hypothesis. One of its major tenets was that a supercontinent called *Pangaea* began breaking apart into smaller continents about 200 million years ago. The smaller continental fragments then "drifted" to their present positions. To support the claim that the now-separate continents were once joined, Wegener and others used the *fit of South America and Africa, fossil evidence, rock types and structures*, and *ancient climates*. One of the main objections to the continental drift hypothesis was its inability to provide an acceptable mechanism for the movement of continents.

- From the study of *paleomagnetism* researchers learned that the continents had wandered as Wegener proposed. In 1962, Harry Hess formulated the idea of *seafloor spreading*, which states that new seafloor is continually being generated at mid-oceanic ridges and old, dense seafloor is being consumed at the deep ocean trenches. Support for seafloor spreading followed, with the discovery of alternating stripes of high- and low-intensity magnetism that parallel the ridge crests.

- By 1968, continental drift and seafloor spreading were united into a far more encompassing theory known as *plate tectonics*. According to plate tectonics, Earth's rigid outer layer (*lithosphere*) overlies a weaker region called the *asthenosphere*. Further, the lithosphere is broken into seven large and numerous smaller segments, called *plates*, that are in motion and continually changing in shape and size. Plates move as relatively coherent units and are deformed mainly along their boundaries.

- *Divergent plate boundaries* occur where plates move apart, resulting in upwelling of material from the mantle to create new seafloor. Most divergent boundaries occur along the axis of the oceanic ridge system and are associated with seafloor spreading, which occurs at rates of 2 to 20 centimeters per year. New divergent boundaries may form within a continent (for example, the East African Rift Valleys) where they may fragment a landmass and develop a new ocean basin.

- *Convergent plate boundaries* occur where plates move together, resulting in the subduction (consumption) of oceanic lithosphere into the mantle along a deep oceanic trench. Convergence between an oceanic and continental block results in subduction of the oceanic slab and the formation of a *continental volcanic arc* such as the Andes of South America. Oceanic–oceanic convergence results in an arc-shaped chain of volcanic islands called a *volcanic island arc*. When two plates carrying continental crust converge, both plates are too buoyant to be subducted. The result is a "collision" resulting in the formation of a mountain belt such as the Himalayas.

- *Transform fault boundaries* occur where plates grind past each other without the production or destruction of lithosphere. Most transform faults join two segments of a mid-oceanic ridge. Others connect spreading centers to subduction zones and thus facilitate the transport of oceanic crust created at a ridge crest to its site of destruction, at a deep ocean trench. Still others, like the San Andreas fault, cut through continental crust.

- The theory of plate tectonics is supported by (1) the global distribution of *earthquakes* and their close association with plate boundaries; (2) the ages and thickness of *sediments* from the floors of the deep-ocean basins; and (3) the existence of island chains that formed over *hot spots* and provide a frame of reference for tracing the direction of plate motion.

- The gross details of the migrations of individual continents over the past billion years have been reconstructed. *Pangaea began breaking apart about 200 million years ago.* North America separated from Africa between 200 million and 165 million years ago. Prior to the formation of Pangaea the landmasses had gone through several episodes of fragmentation similar to what we see happening today.

- Several models for the driving mechanism of plates have been proposed. One model, the *convection current hypothesis*, involves various *convection cells* within the mantle that carry the overlying plates like packages on a conveyor belt. The *slab-pull hypothesis* proposes that when cold, dense oceanic material is subducted it pulls the trailing lithosphere along. *Slab-push* may occur when gravity sets the elevated slabs astride the ridge crest in motion. Another model suggests that relatively narrow *hot plumes* of rock within the mantle contribute to plate motion. No single driving mechanism can account for all major facets of plate motion.

Review Questions

1. What first led scientists such as Alfred Wegener to suspect that the continents were once joined?

2. What was Pangaea?

3. List the evidence that Wegener and his followers gathered to support the continental drift hypothesis.

4. Early in this century, what was the prevailing view of how land animals migrated across vast expanses of ocean?

5. Briefly explain why the recent acceptance of plate tectonics has been described as a scientific "revolution."

6. How does evidence for a late Paleozoic glaciation in the Southern Hemisphere support the continental drift hypothesis?

7. Explain how paleomagnetism can be used to establish the latitude of a specific place at some distant time.

8. What is meant by seafloor spreading? Who is credited with formulating the concept of seafloor spreading?

9. Describe how Fred Vine and D. H. Matthews related the seafloor-spreading hypothesis to magnetic reversals.

10. On what basis were plate boundaries first established?

11. Where is lithosphere being formed? Consumed? Why must the production and destruction of the lithosphere be going on at about the same rate?

12. Why is the oceanic portion of a lithospheric plate subducted while the continental portion is not?

13. In what ways may the origin of the Japanese Islands be considered similar to the formation of the Andes Mountains? How do they differ?

14. Differentiate between transform faults and the two other types of plate boundaries.

15. Some people predict that California will sink into the ocean. Is this idea consistent with the concept of plate tectonics?

16. Applying the idea that hot spots remain fixed, in what direction was the Pacific plate moving while the Emperor Seamounts were being produced? (See Figure 19.30, p. 501.) While the Hawaiian Islands were being produced?

17. With what type of plate boundary are the following places or features associated (be as specific as possible): Himalayas, Aleutian Islands, Red Sea, Andes Mountains, San Andreas fault, Iceland, Japan, Mount St. Helens?

Key Terms

accretionary wedge (p. 490)
asthenosphere (p. 483)
continental drift (p. 471)
continental volcanic arc (p. 491)
convergent boundary (p. 483)
Curie point (p. 477)

deep-ocean trench (p. 489)
divergent boundary (p. 483)
fracture zone (p. 493)
hot spot (p. 499)
lithosphere (p. 483)
magnetometer (p. 481)
normal polarity (p. 480)

paleomagnetism (p. 478)
Pangaea (p. 471)
plate (p. 483)
plate tectonics (p. 482)
reverse polarity (p. 480)
rift, or rift valley (p. 488)
seafloor spreading (p. 479)
slab-pull (p. 505)

slab-push (p. 505)
subduction zone (p. 489)
transform fault boundary (p. 483)
volcanic island arc (p. 491)

Web Resources

The *Earth* Home Page provides on-line resources for this chapter on the World Wide Web. You will find review exercises, specific updates for items in the chapter, suggested reading, and links to interesting related pathways on the Internet. Visit the *Earth* Home Page at **http://www.prenhall.com/tarbuck.**

CHAPTER 20

Mountain Building and the Evolution of Continents

Left Lone Pine Peak in California's Sierra Nevada. —
Photo by Carr Clifton

ountains are often spectacular features that rise several hundred meters or more above the surrounding terrain. Some occur as single isolated masses; the volcanic cone Kilimanjaro, for example, stands almost 6000 meters (20,000 feet) above sea level, overlooking the expansive grasslands of East Africa. Other peaks are parts of extensive mountain belts, such as the American Cordillera, which runs continuously from the tip of South America through Alaska, including the Rockies and Andes Mountains. Chains such as the Himalayas consist of youthful, towering peaks that are still rising, whereas others, including the Appalachian Mountains in the eastern United States, are much older and have been eroded much lower than their original lofty heights (Figure 20.1).

All major mountain belts show evidence of enormous horizontal forces that have folded, faulted, and generally deformed large sections of Earth's crust. Although the processes of folding and faulting have contributed to the majestic appearance of mountains, much of the credit for their beauty must be given to weathering and erosion by running water and glacial ice, which sculpture these uplifted masses in an unending effort to lower them to sea level.

Geologists have come to realize that, in addition to providing spectacular scenery, mountains play a significant role in the evolution of continental crust. In particular, most geologists agree that the continents have gradually grown larger by the addition of linear mountainous terrains to their flanks. Examples of this growth include North America's Appalachian Mountains as well as the Andes of South America. This hypothesis predicts that nearly all continental areas once stood as mountains and were subsequently lowered to their present elevations by erosion. We will return to this idea later in the chapter, but first let us consider the nature of mountain belts and the forces that elevate them above adjacent lowlands.

Mountain Belts

Like other people, geologists have been inspired more by Earth's mountains than by any other landform. Through extensive scientific exploration over the last 150 years, much has been learned about the internal processes that generate these often spectacular terrains. The name for the processes that collectively produce a mountain belt is **orogenesis**, from the Greek *oros* ("mountain") and *genesis* ("to come into being"). The rocks comprising mountains provide striking visual evidence of the enormous

Figure 20.1 The Valley of Ten Peaks, Banff National Park, Alberta, Canada. (Photo by Carr Clifton/Minden Pictures)

compressional forces that have deformed large sections of Earth's crust and subsequently elevated them to their present positions. Although folding is often the most conspicuous sign of these forces, thrust faulting, metamorphism, and igneous activity are always present in varying degrees.

When geologists speak of mountain-building processes, they are generally referring to the major mountain belts shown in Figure 20.2. Included in this group are the Alps, Urals, Himalayas, Appalachians, and the American Cordillera. Mountain belts are found on every continent, extending as linear chains for hundreds or even thousands of kilometers. Typically, mountain belts consist of numerous *mountain ranges* that show evidence of being formed during the same mountain-building episode.

This chapter deals with the most conspicuous result of crustal deformation—Earth's major mountain belts. Yet it is worth noting that some regions exhibit mountainous topography that is produced without appreciable crustal deformation. For example, plateaus, which are areas of high-standing rocks that are essentially horizontal, can be deeply dissected by erosional forces into rugged, mountainlike landscapes. Although these highlands resemble mountains topographically, they lack the structures associated with orogenesis. The opposite situation also exists. For instance, the Piedmont section of the eastern Appalachians exhibits topography that is nearly as subdued as that seen in the Great Plains. Yet, because this region is composed of deformed metamorphic rocks, it is clearly part of the Appalachian Mountains.

Isostasy and Crustal Uplift

Two major questions have been paramount to the understanding of the processes that build mountains. First, what is the origin of the enormous horizontal forces that deform crustal rocks during orogenesis? Second, what keeps mountains elevated above the surrounding lowlands?

Evidence for Crustal Uplift

The fossilized shells of marine organisms are often found at high elevations in mountains, an indication that the rocks composing the peaks were once below sea level. This is convincing evidence that some drastic changes have occurred between the time these animals died and the time when their fossilized remains were discovered.

Evidence for crustal uplift such as this is common in the geologic record and is even present in the historical record. For example, Figure 20.3 shows three columns remaining from a Roman ruins. The columns have clam borings to a height of about 6 meters (20 feet), indicating that the land upon which this structure was built submerged and was later partially uplifted. These elevated clam borings might also be explained by a recent change in sea level; however, a similar change is not recorded at any other location for that same period. Further evidence for crustal uplift can be found along the coastline of the western United States. When a coastal area remains undisturbed for an extended period, wave action cuts a gently sloping platform. In parts of

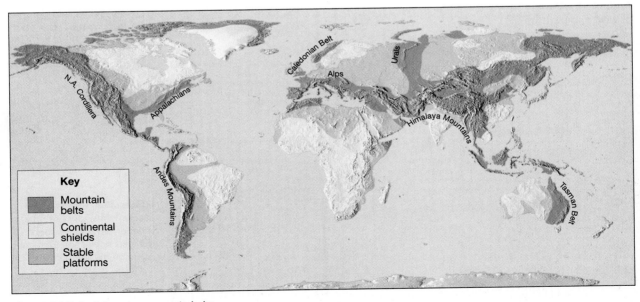

Figure 20.2 Earth's major mountain belts.

Clam borings

Figure 20.3 Remaining columns of the ancient Roman temple of Serapis, Pozzuoli, Italy, in 1836. Clam borings 6 meters above sea level indicate former submergence. (Charles Lyell, *Principles of Geology*, 10th ed., 1867)

California, ancient wave-cut platforms can now be found as terraces, hundreds of meters above sea level (Figure 20.4). Each terrace represents a period when that area was at sea level. Such evidence of crustal uplift is easy to find; unfortunately, the reasons for uplift are not always as easy to determine.

Do Mountains Have Roots?

One of the major advances in determining the structure of mountains occurred in the 1840s when Sir George Everest (after whom Mount Everest is named) conducted the first topographical survey in India. During this survey the distance between the towns of Kalianpur and Kaliana, located south of the Himalayan range, was measured using two different methods. One method employed the conventional surveying technique of triangulation, and the other method determined the distance astronomically. Although the two techniques should have given similar results, the astronomical calculations placed these towns nearly 150 meters closer to each other than did the triangulation survey.

The discrepancy was attributed to the gravitational attraction exerted by the massive Himalayas on the plumb bob used for leveling the astronomical instrument. (A plumb bob is a metal weight suspended by a cord, used to determine a vertical orientation.) It was suggested that the deflection of the plumb bob would be greater at Kaliana than at Kalianpur because it is closer to the mountains (Figure 20.5).

A few years later, J. H. Pratt estimated the mass of the Himalayas and calculated the error that should have been caused by the gravitational influence of the mountains. To his surprise, Pratt discovered that the mountains should have produced an error three times larger than was actually observed. Simply stated, the mountains were not "pulling their weight." It was as if they had a hollow central core.

A hypothesis to explain the apparent "missing" mass was developed by George Airy. Airy suggested that Earth's lighter crustal rocks float on the denser, more plastic mantle. Further, he correctly argued that the crust must be thicker under mountains than

Figure 20.4 Former wave-cut platforms now exist as a series of elevated terraces on the west side of San Clemente Island off the southern California coast. Once at sea level, the highest terraces have risen about 400 meters above it. (Photo by John S. Shelton)

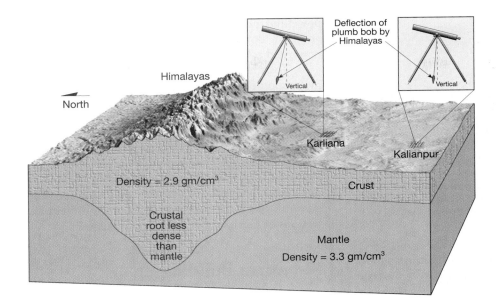

beneath the adjacent lowlands. In other words, mountainous terrains are supported by light crustal material that extends as "roots" into the denser mantle (Figure 20.6). This phenomenon is exhibited by icebergs, which are buoyed up by the weight of the displaced water. If the Himalayas do have roots of light crustal rocks that extend far beneath them, then these mountains would exert the smaller gravitational attraction that Pratt had calculated. Hence, Airy's model explained why the plumb bob was deflected much less than expected.

Seismological and gravitational studies have confirmed the existence of crustal roots under some mountain ranges. The thickness of continental crust is normally about 35 kilometers, but crustal thicknesses exceeding 70 kilometers have been determined for some mountain belts. Elevated regions that are supported primarily by the buoyant nature of crustal roots are the Andes Mountains, the Sierra Nevada, and the Tibetan Plateau.

Isostasy

We know that the force of gravity must play an important role in determining the elevation of the land. The concept of a floating crust in gravitational balance, as Airy proposed, is called **isostasy**.

Perhaps the easiest way to grasp the concept of isostasy is to envision a series of wooden blocks of different heights floating in water (Figure 20.7). Note that the thicker wooden blocks float higher in water than the thinner blocks. In a similar manner, mountain belts stand high above the surface and have "roots" that extend deeper into the supporting material below.

Now visualize what would happen if another small block of wood were placed atop one of the blocks in Figure 20.7. The combined block would sink until a new isostatic balance was reached. However, the top of the combined block would actually be higher than before and the bottom would be lower. This process of establishing a new level of equilibrium is called **isostatic adjustment**.

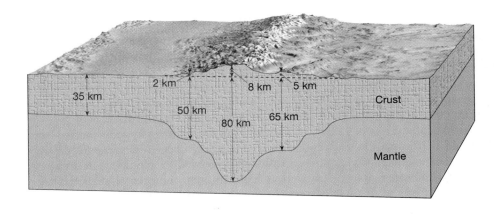

Figure 20.6 Mountains are supported by the buoyancy of light crustal roots floating in the denser mantle (like icebergs floating in seawater). (Modified after Peter Molnar)

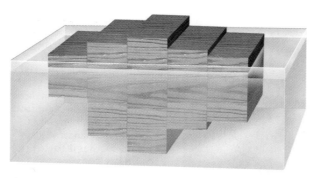

Figure 20.7 This drawing illustrates how wooden blocks of different thicknesses float in water. In a similar manner, thick sections of crustal material float higher than do thinner crustal slabs.

If the concept of isostasy is correct, we should expect that when weight is added to the crust, the crust will respond by subsiding, and when that weight is removed, the crust will rebound. (Visualize what happens to a ship as cargo is being loaded and unloaded.) Evidence for crustal subsidence and crustal rebound exists. A classic example is provided by Ice Age glaciers. When continental ice sheets occupied portions of North America during the Pleistocene epoch, the added weight of the 3-kilometer-thick mass of ice caused downwarping of Earth's crust, by hundreds of meters. In the 8,000 to 10,000 years since the last ice sheet melted, uplifting of as much as 330 meters has occurred in Canada's Hudson Bay region, where the thickest ice had accumulated (see Figure 12.32, p. 318).

As this example illustrates, isostatic adjustment can account for considerable crustal movement. We can now understand why, as erosion lowers the summits of mountains, the crust will rise in response to the reduced load. However, each episode of isostatic uplift is somewhat less than the elevation loss due to erosion. The processes of uplifting and erosion will continue until the mountain block reaches "normal" crustal thickness (Figure 20.8). When this occurs, the mountains will be eroded to near sea level, and the deeply buried portions of the mountains will be exposed at the surface. In addition, as the mountains are worn down, the weight of the eroded sediment deposited on the adjacent continental margin will cause the margin to subside.

Although the major mountain belts seem to be supported by isostatic buoyancy, another important mechanism also contributes to the elevated position of some landmasses. With the development of the plate tectonics theory, it became clear that some form of convective circulation is occurring in the mantle. Although the exact nature of this movement is still not fully understood, it is known that some regions, such as Yellowstone, are underlain by zones of hot,

upwelling mantle plumes. In these regions, the buoyancy of hot rising magma accounts for broad upwarped zones in the overlying lithosphere.

To summarize, we have learned that Earth's major mountain belts consist of unusually thick sections of deformed crustal material that has been elevated above the surrounding terrain. The support for these massive features comes from the buoyancy of deep crustal roots. As erosion lowers the peaks, isostatic adjustment gradually raises the mountains in response. Eventually, the deepest portions of the mountains are brought to the shallower depths of the surrounding crust. The question still to be answered is, How do these thick sections of Earth's crust come into existence?

Mountain Building

Mountain building has occurred during the recent geologic past in several locations around the world. These young mountainous belts include the American Cordillera (Figure 20.9), which runs along the western margin of the Americas from Cape Horn to Alaska; the Alpine-Himalaya chain, which extends from the Mediterranean through Iran to northern India and into Indochina; and the mountainous terrains of the western Pacific, which include volcanic island arcs such as Japan, the Philippines, and Sumatra (see Figure 20.2, p. 513). Most of these young mountain belts have come into existence within the last 100 million years. Some, including the Himalayas, began their growth as recently as 50 million years ago (see Box 20.1).

In addition to these young mountain belts, several chains of Paleozoic- and Precambrian-age mountains exist on Earth as well. Although these older structures are deeply eroded and topographically less prominent, they clearly possess the same structural features found in younger mountains. Typical of this older group are the Appalachians in the eastern United States and the Urals in Russia.

Mountain Structures

Although major mountain belts differ from one another in particular details, all possess the same basic structures. Generally mountain belts consist of roughly parallel ridges of folded and faulted sedimentary and volcanic rocks, portions of which have been strongly metamorphosed and intruded by younger igneous bodies (Figure 20.10).

In most cases the sedimentary rocks formed from enormous accumulations of deep-water marine deposits that occasionally exceeded 15 kilometers in thickness, as well as from thinner continental shelf deposits. Moreover, most of these deformed sedimentary rocks

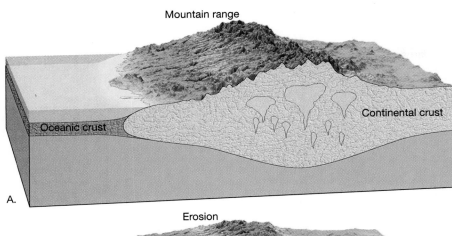

Mountain range

Mountain range

Oceanic crust

Continental crust

A.

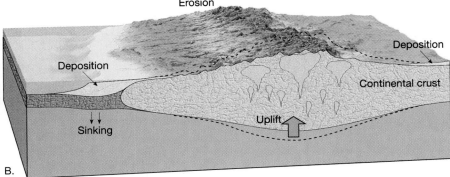

Erosion

Deposition

Deposition

Continental crust

Sinking

Uplift

B.

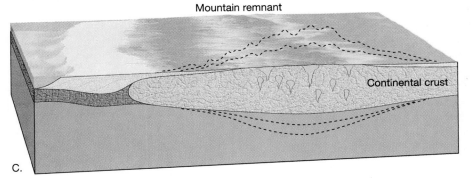

Mountain remnant

Continental crust

C.

Figure 20.8 This sequence illustrates how the combined effect of erosion and isostatic adjustment results in a thinning of the crust in mountainous regions. **A.** When mountains are young, the continental crust is thickest. **B.** As erosion lowers the mountains, the crust rises in response to the reduced load. **C.** Erosion and uplift continue until the mountains reach "normal" crustal thickness.

are older than the mountain-building event. These facts indicate that an extensive period of quiescent deposition along a continental margin was followed by a dramatic episode of deformation.

Careful study of mountainous terrains has revealed that the period of mountain building is generally quite long, in some cases exceeding 100 million years. In addition, reconstruction of these events has shown that deformation generally progressed from the margin of the continent toward the interior, so that the deep-water sediments were the first to be deformed. These sediments, which consist of poorly sorted sandstones, volcanic debris, and shales, have been intensely folded, faulted, and strongly metamorphosed as if squeezed by a gigantic vise whose moving jaw migrated from the sea toward the land. In many mountain belts a period of volcanism, with the emplacement of granitic intrusions, accompanies this episode of deformation.

Next to be deformed are the shallow-water strata of the continental shelf, composed of relatively clean sandstones, limestones, and shales. These beds are usually deformed by folding and thrust faulting, which causes large slices of rocks to slide up and over younger layers (see Chapter 15).

For an extended period after a mountain-building episode has ended, the area experiences regional uplift. Generally, little additional deformation is associated with this activity. As the deformed strata are elevated to great heights, the processes of erosion accelerate, carving the deformed strata into a dissected mountainous landscape.

Figure 20.9 Map of major mountainous landforms of the western United States. (After Thelin and Pike, USGS)

Over the years, several hypotheses have been put forward regarding the formation of Earth's major mountain belts. One early proposal suggested that mountains are simply wrinkles in Earth's crust, produced as the planet cooled from its original semi-molten state. As Earth lost heat, it contracted and

Figure 20.10 Highly deformed sedimentary strata in the Rocky Mountains of British Columbia. These sedimentary rocks are continental shelf deposits that were displaced toward the interior of Canada by low-angle thrust faults. (Photo by John Montagne)

shrank. In response to this process the crust was deformed much as the peel of an orange wrinkles as the fruit dries out. However, neither this nor any other early hypothesis was able to withstand careful scrutiny.

With the development of the plate tectonics theory, a different model for understanding orogenesis became available. According to the theory of plate tectonics, mountain building occurs at convergent plate boundaries. Here colliding plates provide the horizontal compressional stress to fold, fault, and metamorphose the thick accumulations of sediments that are deposited along the flanks of landmasses. In addition, partial melting of mantle rock provides a source of magma that intrudes and further deforms these deposits.

Mountain Building at Convergent Boundaries

To unravel the events that produce mountains, studies examine ancient mountain structures as well as sites where orogenesis is currently active. Of particular interest are subduction zones, where lithospheric plates are converging. At most modern-day subduction zones, volcanic arcs are forming. This situation is typified by Alaska's Aleutian Islands (Figure 20.11) and by the Andean arc of western South America.

Aleutian-type subduction zones occur where two oceanic plates converge (Figure 20.12). By contrast, *Andean-type subduction zones* are situated where oceanic crust is being thrust beneath a continental mass. Events that generate an Aleutian-type volcanic arc follow an evolutionary path different from those that produce an Andean-type volcanic arc. Further, although the development of a volcanic arc can result

in the formation of mountainous topography, this activity is viewed as just one phase in the development of a major mountain belt.

Andean-Type Convergent Zones

Mountain building along continental margins involves the convergence of an oceanic plate and a plate whose leading edge contains continental crust. Exemplified by the Andes Mountains, this *Andean type* of convergence generates structures resembling those of a developing volcanic island arc (Figure 20.13).

Passive Margins. The first stage in the development of an Andean-type mountain belt occurs prior to the formation of the subduction zone. During this period the continental margin is a **passive margin**; that is, it is not a plate boundary but a part of the same plate as the adjoining oceanic crust. The East Coast of the United States provides a present-day example of a passive continental margin. Here, as at other passive continental margins surrounding the Atlantic, deposition of sediment on the continental shelf is producing a thick wedge of shallow-water sandstones, limestones, and shales (Figure 20.13A). Beyond the continental shelf, turbidity currents are depositing sediments on the continental slope and rise (see Chapter 18).

Active Continental Margins. At some point the continental margin becomes active. A subduction zone forms and the deformation process begins (Figure 20.13B). A good place to examine an active continental margin is the west coast of South America. Here the Nazca plate is being subducted beneath the South American plate along the Peru–Chile trench (see Figure 19.17, p.484). This subduction zone probably

Box 20.1

Fault-Block Mountains

Most mountain belts, including the Alps, Himalayas, and Appalachians, form in compressional environments, as evidenced by the predominance of large thrust faults and folded strata. However, some mountain building is associated with tensional stresses. The mountains that form in such circumstances, termed *fault-block mountains*, are bounded on at least one side by high- to moderate-angle normal faults. Most fault-block mountains form in response to broad uplifting, which causes elongation and faulting. Such a situation is exemplified by the fault blocks that rise above the rift valleys of East Africa. Other fault-block mountains, such as Wyoming's Grand Tetons, develop where an isolated block is tilted as one side of the range is lifted high above adjacent undeformed valleys (Figure 20.A).

The Basin and Range province, a region that encompasses Nevada and portions of Utah, New Mexico, Arizona, and California, contains excellent examples of fault-block mountains (see Figure 20.9). Here, the brittle upper crust has literally been broken into hundreds of pieces, giving rise to

Figure 20.A The Grand Tetons, Wyoming, an example of fault-block mountains. (Photo by Carr Clifton)

formed in conjunction with the breakup of the supercontinent Pangaea. As South America separated from Africa and migrated westward, the oceanic crust adjacent to the west coast of South America was bent and thrust under the continent.

In an idealized Andean-type subduction, convergence of the continental block and the subducting oceanic plate leads to deformation and metamorphism of the continental margin. Once the oceanic plate descends to about 100 kilometers, partial melting of mantle rock above the subducting slab generates magma that migrates upward, intruding and further deforming these strata (Figure 20.13B). During the development of this continental volcanic arc, sediment derived from the land as well as that scraped from the subducting plate is plastered against the landward side of the trench. This chaotic accumulation of sedimentary and metamorphic rocks with occasional scraps of ocean crust is called an **accretionary wedge** (Figure 20.13B). Prolonged subduction can build an accretionary wedge that is large enough to stand above sea level (Figure 20.13C).

Andean-type mountain belts are composed of two roughly parallel zones. The landward segment is the continental volcanic arc, made up of volcanoes and large intrusive bodies intermixed with high-temperature metamorphic rocks. The seaward segment is the accretionary wedge. It consists of folded, faulted, and metamorphosed sediments and volcanic debris (Figure 20.13C).

nearly parallel mountain ranges, averaging about 80 kilometers in length, which rise precipitously above the adjacent sediment-laden basins.

A debate about the mechanisms that created the Basin and Range has gone on for 30 years. It is clear that important geological events preceded the period of extension and faulting. During this earlier time span it is thought that an oceanic plate (Farallon plate) was subducted eastward nearly horizontally under western North America (Figure 20.B, top). This caused compressional stresses in the overlying continental crust, which became tectonically active as far inland as the Rocky Mountains in Colorado.

Then, about 50 million years ago, the once compressional environment became extensional. Some geologists suggest that rifting in the Basin and Range was triggered when the oceanic slab that had been thrust (subducted) nearly horizontally under the continent began to sink (Figure 20.B, bottom). Sinking of the oceanic slab produced an upwelling of hot material from the asthenosphere. The buoyancy of the warm material caused upwarping and rifting that stretched the overlying crust by 200 to 300 kilometers. The lower crust is ductile and easily stretched. The upper crust, on the other hand, is brittle and deformed by faulting (see Figure 15.22, p. 390). The extensional faulting rifted the brittle upper crust, causing individual blocks to slump. The

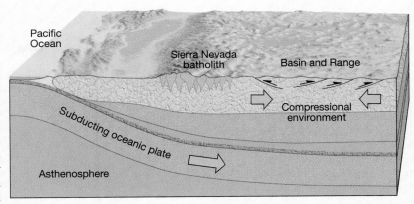

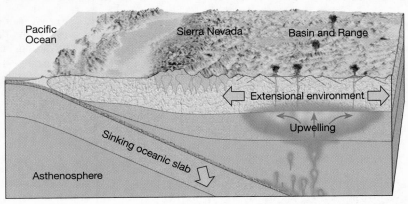

Figure 20.B One proposal for the formation of the Basin and Range Province. Top: Nearly horizontal subduction of an oceanic plate produced compressional stresses that generally thickened the crust in the Basin and Range. Bottom: Sinking of this oceanic slab allowed for the upwelling of hot material from the asthenosphere. The buoyancy of the warm material caused upwarping and tensional fracturing in the crust above. This event was associated with volcanism and east-west extension of the crust by 200 to 300 kilometers.

high parts of the tilted blocks make up the mountain ranges, while their low parts form the basins, now partially filled with sediment.

Sierra Nevada and Coast Ranges. One of the best examples of an inactive Andean-type orogenic belt is found in the western United States. It includes the Sierra Nevada and the Coast Ranges in California (see Figure 20.9, p. 518). These parallel mountain belts were produced by the subduction of a portion of the Pacific Basin under the western edge of the North American plate. The Sierra Nevada batholith is a remnant of a portion of the continental volcanic arc that was produced by several surges of magma over tens of millions of years. Subsequent uplifting and erosion have removed most of the evidence of past volcanic activity and exposed a core of crystalline metamorphic and igneous rocks.

In the trench region, sediments scraped from the subducting plate, plus those provided by the eroding continental volcanic arc, were intensely folded and faulted into an accretionary wedge. This chaotic mixture of rocks presently constitutes the Franciscan Formation of California's Coast Ranges. Uplifting of the Coast Ranges took place only recently, as evidenced by the young unconsolidated sediments that still mantle portions of these highlands.

Continental Collisions

So far, we have discussed the formation of mountain belts where the leading edge of just one of the two lithospheric plates was capped by continental crust.

Figure 20.11 Three of many volcanic islands that comprise the Aleutian arc. This narrow band of volcanism results from the subduction of the Pacific plate. In the distance is the Great Sitkin volcano (1772 meters), which the Aleuts call the "Great Emptier of Bowels," because of its frequent activity. (Photo by Bruce D. Marsh)

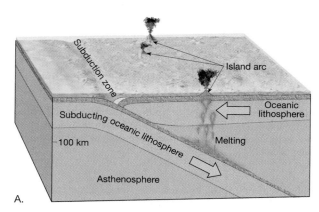

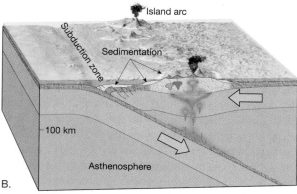

Figure 20.12 The development of a volcanic island arc by the convergence of two oceanic plates. These Aleutian-type subduction zones produce a mountainous topography that is similar to the volcanic arc that comprises the Aleutian islands of Alaska.

However, it is possible that both plates may be carrying continental crust. Because continental crust is apparently too buoyant to undergo any appreciable subduction, a collision between the continental fragments results (Figure 20.14). We will now look at two examples of this: an example of youthful mountains, the Himalayas, and an example of mountains that formed long ago and have been substantially lowered by erosion, the Appalachians.

The Himalayas. An example of such a collision began about 45 million years ago when India collided with the Eurasian plate. Prior to this event, India, which was once part of Antarctica, had split from that continent and over the course of millions of years slowly moved a few thousand kilometers due north (Figure 20.14A). The result of the collision was the formation of the spectacular Himalaya Mountains and the Tibetan Plateau (Figure 20.14B).

Although most of the oceanic crust that separated these landmasses prior to the collision was subducted, some was caught up in the squeeze, along with sediment that lay offshore. Today these sedimentary rocks and slivers of oceanic crust are elevated high above sea level. Geologists believe that, following such a collision, the subducted oceanic plate decouples from the rigid continental plate and continues its descent into the mantle.

The spreading center that propelled India northward is still active; hence, India continues to be

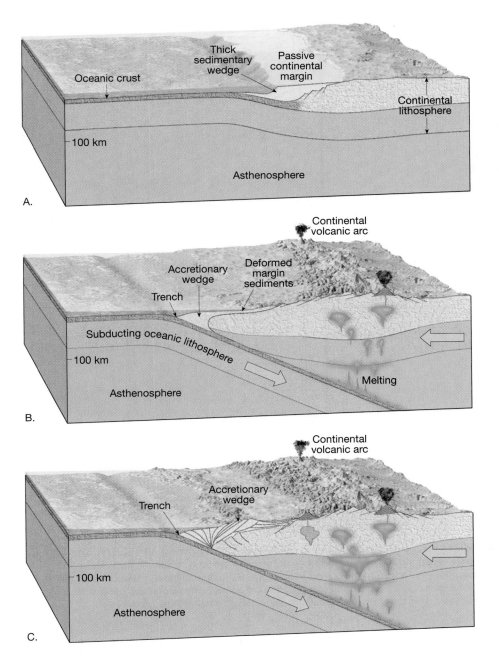

A.

B.

C.

Figure 20.13 Orogenesis along an Andean-type subduction zone. **A.** Passive continental margin with extensive wedge of sediments. **B.** Plate convergence generates a subduction zone, and partial melting produces a developing volcanic arc. **C.** Continued convergence and igneous activity further deform and thicken the crust, elevating the mountain belt, while the accretionary wedge grows.

thrust into Asia at an estimated rate of a few centimeters each year. However, numerous earthquakes recorded off the southern coast of India indicate that a new subduction zone may be in the making. If formed, it would provide a subduction site for the floor of the Indian Ocean, which is continually being produced at a spreading center located to the southwest. Should this occur, India's northward journey, relative to Asia, would come to an end and the growth of the Himalayas would cease.

A similar but much older collision is believed to have taken place when the European continent collided with the Asian continent to produce the Ural Mountains, which extend in a north-south direction through Russia. Prior to the discovery of plate tectonics, geologists had difficulty explaining the existence of mountain ranges, such as the Urals, that are located deep within continental interiors. How could thousands of meters of marine sediment be deposited and then become highly deformed while situated in the middle of a large stable landmass?

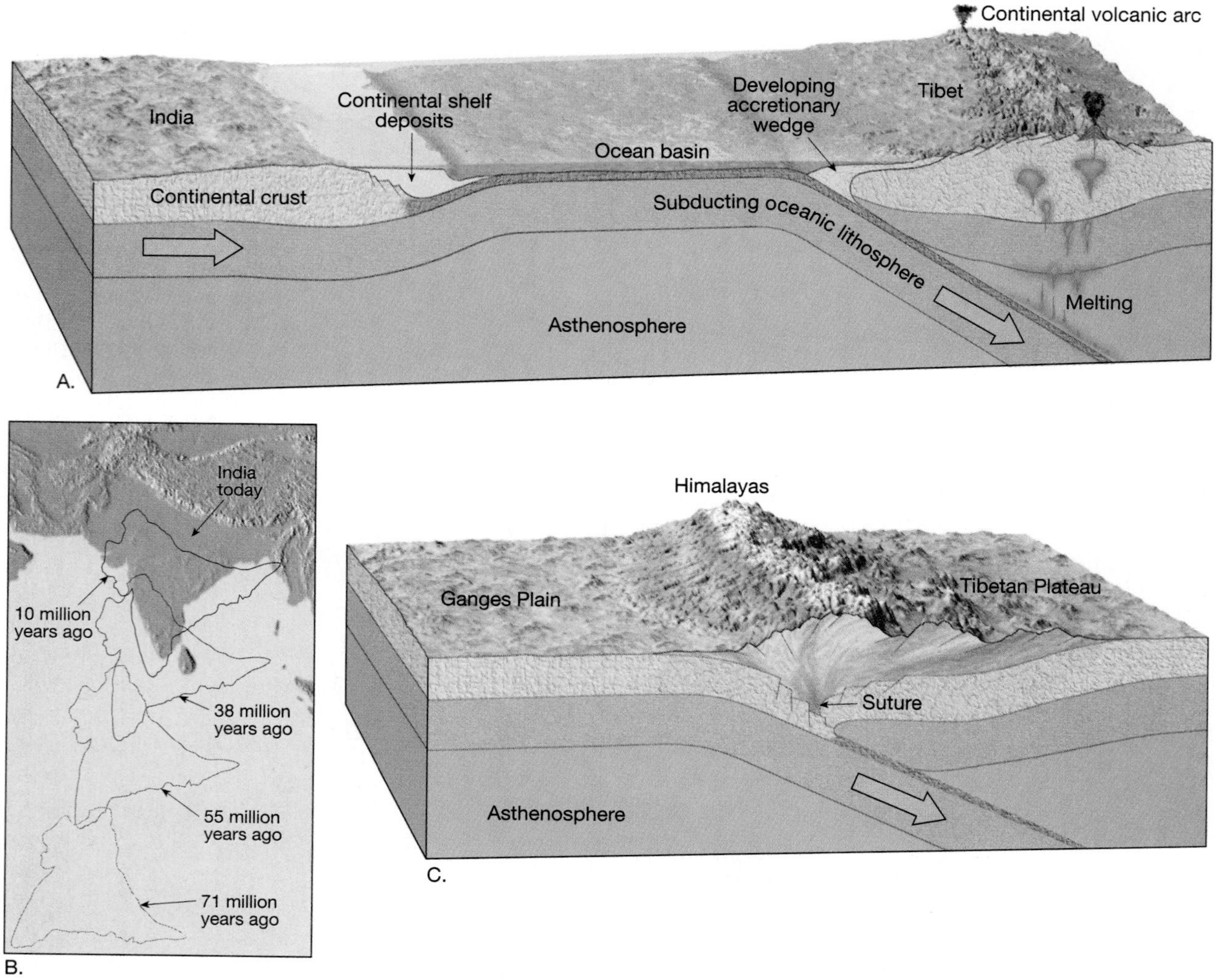

Figure 20.14 Simplified diagrams showing the northward migration and collision of India with the Eurasian plate. **A.** Converging plates generated a subduction zone, while partial melting of the subducting oceanic slab produced a continental volcanic arc. Sediments scraped from the subducting plate were added to the accretionary wedge. **B.** Position of India in relation to Eurasia at various times. **C.** Eventually the two landmasses collided, deforming and elevating the accretionary wedge and continental shelf deposits. In addition, slices of the Indian crust were thrust up onto the Indian plate. (Modified after Peter Molnar)

The Appalachians. The Appalachian Mountains provide great scenic beauty near the eastern margin of North America from Alabama to Newfoundland. In addition, mountains that formed contemporaneously with the Appalachians are found in the British Isles, Scandinavia, Western Europe, and Greenland. The orogeny that generated this extensive mountain system lasted nearly 300 million years and intensely metamorphosed and deformed the rocks of the Appalachian central core.

The Appalachians resulted from a collision among North America, Europe, and northern Africa. Although they have since separated, these landmasses were juxtaposed as part of the supercontinent Pangaea less than 200 million years ago. Detailed studies in the southern Appalachians indicate that the formation of this mountain belt was more complex than once thought. Rather than forming during a single continental collision, the Appalachians resulted from several distinct episodes of mountain building.

The final orogeny occurred about 250 million to 300 million years ago when Africa and Europe collided with North America. At some locations, the total landward displacement may have exceeded 250 kilometers (155 miles). This displacement further deformed the shallow-water sediments that had flanked North America. Today these folded and faulted sandstones, limestones, and shales make up the essentially unmetamorphosed rocks of the Valley and Ridge Province (Figure 20.15).

In summary, the orogenesis of a complex mountain chain, as typified by the Himalayas and Appalachians, is thought to occur as follows:

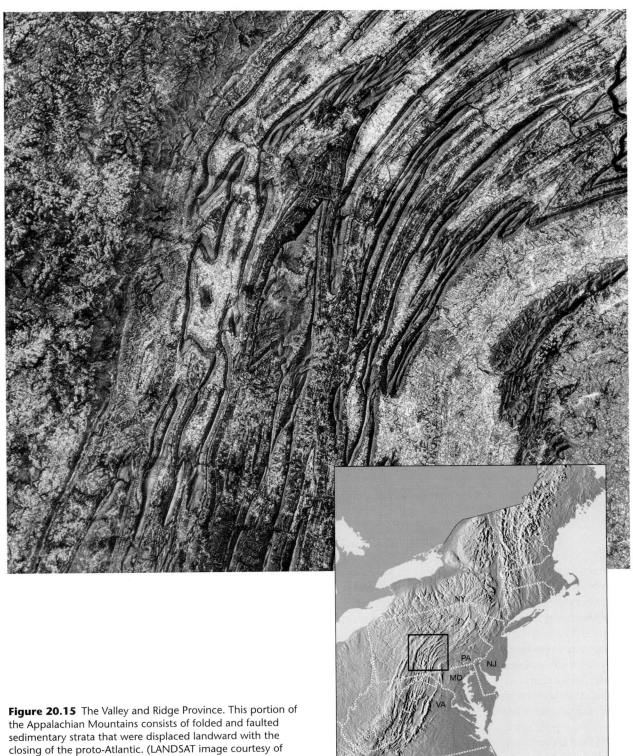

Figure 20.15 The Valley and Ridge Province. This portion of the Appalachian Mountains consists of folded and faulted sedimentary strata that were displaced landward with the closing of the proto-Atlantic. (LANDSAT image courtesy of Phillips Petroleum Company, Exploration Projects Section)

1. After the breakup of a continental landmass, a thick wedge of sediments is deposited along the passive continental margins, thereby increasing the size of the newly formed continent.

2. For reasons not yet understood, the ocean basin begins to close and the continents begin to converge.

3. Plate convergence results in the subduction of the intervening oceanic slab and initiates an extended period of igneous activity. This activity results in the formation of a continental volcanic arc with associated intrusions (see Figure 20.14A).

4. Eventually, the continental blocks collide. This event severely deforms and metamorphoses the entrapped sediments (see Figure 20.14C). Continental convergence causes these deformed materials, and occasionally slabs of crustal material, to be displaced up onto the plates along thrust faults. This activity shortens and thickens the crustal rocks, producing an elevated mountain belt.

5. Finally, a change in the plate boundary ends the growth of the mountains. Only at this point do the processes of erosion become the dominant forces in altering the landscape. Prolonged erosion coupled with isostatic adjustments eventually reduce this mountainous landscape to the average thickness of the continents.

This sequence of events is thought to have been duplicated many times during Earth's long history. However, the rate of deformation and geologic and climatic settings varied in each instance. Thus, the formation of each mountain chain must be regarded as a unique event (see Box 20.2).

Orogenesis and Continental Accretion

Plate tectonics theory originally suggested two mechanisms for orogenesis: (1) Continental collisions were proposed to explain the formation of such mountainous terrains as the Alps, Himalayas, and Urals; and (2) subduction of oceanic lithosphere was thought to be the underlying tectonic process for many circum-Pacific mountain chains, as typified by the Andes. Subsequent investigations, however, indicate a third mechanism of orogenesis: Smaller crustal fragments collide and merge with continental margins, and through this process of collision and accretion, many of the mountainous regions rimming the Pacific have been generated (Figure 20.16).

What is the nature of these small crustal fragments, and where do they come from? Research suggests that,

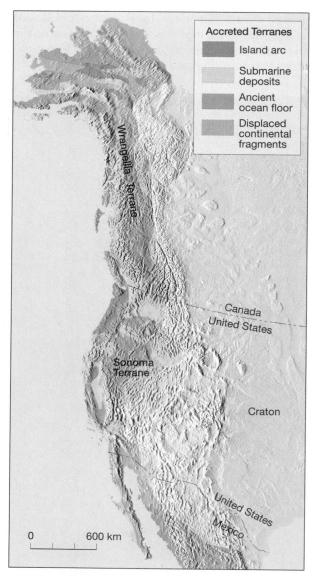

Figure 20.16 Map showing terranes that have been added to western North America during the past 200 million years. Data from paleomagnetic and fossil evidence indicate that some of these terranes originated thousands of kilometers to the south of their present location. (Redrawn after D. R. Hutchinson and others)

prior to their accretion to a continental block, some of the fragments may have been microcontinents similar to the present-day island of Madagascar located east of Africa. Many others were island arcs like Japan, the Philippines, and the Aleutian Islands. Still others may have been below sea level; their analogs today may be submerged crustal fragments rising high above the floor of the western Pacific (see Figure 19.11, p. 479). Over a hundred of these "oceanic plateaus" are known to exist. It is believed that these plateaus originated as submerged

Figure 20.17 Uplifted strata of the Endicott Mountains, Brooks Range, Gates of Arctic National Park and Preserve, Alaska. Presumably these beds were deformed when they were accreted to Alaska within the last 200 million years. (Photo by Jeff Gnass/The Stock Market)

continental fragments, extinct volcanic island arcs, or submerged volcanic plateaus with hot-spot activity.

Accretion of Foreign Terranes. The widely accepted view today is that, as oceanic plates move, they carry the embedded oceanic plateaus or microcontinents to a subduction zone. Here the upper portions of these thickened zones are peeled from the descending plate and thrust in relatively thin sheets upon the adjacent continental block. This newly added material increases the width of the continent. The material may later be overridden and displaced farther inland by the addition of other fragments.

Geologists refer to these accreted crustal blocks as terranes. Simply, the term **terrane** designates any crustal fragment whose geologic history is distinct from that of the adjoining terranes. (Do not confuse the term *terrane* with the word *terrain*, which indicates the topography or lay of the land.) Terranes come in varied shapes and sizes; some are no larger than volcanic islands. Others, such as the one making up the entire Indian subcontinent, are very large.

Accretion and Orogenesis. The idea that orogenesis occurs in association with the accretion of crustal fragments to a continental mass arose principally from studies conducted in the northern portion of the North American Cordillera (Figure 20.16). Here it was determined that some mountainous areas, principally those in the orogenic belts of Alaska and British Columbia, contain fossil and magnetic evidence indicating that these strata once lay nearer the equator.

It is now assumed that many other terranes found in the North American Cordillera were once scattered throughout the eastern Pacific, much as we find island arcs and oceanic plateaus distributed in the western Pacific today. Since the breakup of Pangaea 200 million years ago, North America has been migrating westward overriding the Pacific Basin (Figure 20.17). Apparently, this activity resulted in the piecemeal addition of fragments to the entire Pacific margin of the continent, from the Baja Peninsula to northern Alaska. In a like manner, many modern microcontinents will eventually be accreted to active continental margins, producing new orogenic belts.

Box 20.2

The Rocky Mountains

The portion of the Rocky Mountains that extends from southern Montana to New Mexico was produced by a period of deformation known as the *Laramide Orogeny*. This event, which created some of the most picturesque scenery in the United States, peaked about 60 million years ago (Figure 20.C). The mountain ranges generated during the Laramide Orogeny include the Front Range of Colorado, the Sangre de Cristo of New Mexico and Colorado, and the Bighorns of Wyoming.

These mountains are structurally much different from the northern Rockies, which include the Canadian Rockies and those portions of the Rockies found in Idaho, western Wyoming, and western Montana. Whereas the latter ranges are folded mountains, composed of thick sequences of sedimentary rocks that were deformed by folding and low-angle thrust faulting, the middle and southern Rockies are upwarped mountains pushed almost vertically upward as part of broad upwarping of the crust.

Figure 20.C The spectacular Maroon Bells are part of the Colorado Rockies. (Photo by Peter Saloutos/The Stock Market)

In general, the middle and southern Rockies consist of ancient basement rocks overlain by relatively thin layers of younger strata. Since the time of their formation, much of the sedimentary cover has been eroded from the highest portions of the uplifted blocks, exposing their igneous and metamorphic cores.

The Origin and Evolution of Continental Crust

Earlier in this chapter, we learned that the theory of plate tectonics provides a model from which to examine the formation of Earth's major mountainous belts. But what roles have plate tectonics and mountain building played in the events that led to the origin and evolution of the continents? At this time no single answer to this question has met with overwhelming acceptance.

The lack of agreement among geologists can in part be attributed to the complex nature and antiquity of most continental material, which makes deciphering its history very difficult. Whereas oceanic crust has a relatively simple layered structure and a rather uniform composition, the much older continental crust consists of a collage of highly deformed and metamorphosed igneous and sedimentary rocks. Moreover, in many places, such as the Plains states, the basement rocks are mantled by thick accumulations of much younger sedimentary rocks—a fact that inhibits detailed study. Nevertheless, during the last few decades, great strides have been made in unraveling the secrets held by the rocks that make up the stable continental interiors.

Early Evolution of the Continents

At one extreme is a proposal that most, if not all, continental crust formed early in Earth's history. According to this view, the continental crust originated during a primeval molten stage that coincided with the segregation of material that produced Earth's core and mantle (see Chapter 22). During this period of chemical segregation, the less dense silica-rich constituents of the mantle rose to the surface to produce a scum of continental-type rocks. The more dense silicate minerals, enriched in iron and magnesium, remained in the mantle, while metallic iron sank to form the core.

Shortly after this segregation, a mechanism that may have resembled plate tectonics continually reworked and recycled the continental crust. Through

Examples include a number of granitic outcrops that project as steep summits, such as Pikes Peak and Longs Peak in Colorado's Front Range. In many areas, remnants of the sedimentary strata that once covered this region are visible as prominent angular ridges called *hogbacks* flanking the crystalline cores of the mountain ranges (Figure 20.D).

Although the Rockies have been extensively studied for over a century, there is still a good deal of debate regarding the mechanisms that led to uplift. According to the plate tectonics model, most mountain belts are produced along continental margins in association with convergent plate boundaries. Here, crustal buckling generates thick sequences of folded, faulted, and metamorphosed strata that are often intruded by massive igneous bodies.

However, this model cannot be easily adapted to the middle and southern Rocky Mountains, which consist of elongated, uplifted blocks of Precambrian basement rocks separated by sediment-filled basis.

One hypothesis for the formation of the middle and southern Rocky Mountains proposes that about 65 million years ago a subducting plate, called the Farallon plate, remained nearly horizontal as it pushed eastward under North America as far inland as the Black Hills of South Dakota (see Figure 20.B, p. 521). As the subducted slab scraped beneath North America, compressional stresses shortened and thickened the rocks in the lower crust. Furthermore, this event locally uplifted blocks of ancient basement rocks along reverse faults to produce the Rockies and intermountain basins. Thus, the Laramide Orogeny may be associated with a special type of convergent plate boundary. Whereas plate subduction along a typical convergent plate boundary occurs at an angle of about 45 degrees, the middle and southern Rocky Mountains may have been produced by the nearly horizontal subduction of an oceanic plate.

Other researchers suggest that the near horizontal subduction of the Farallon plate may have altered the underside of the continents, but not enough to uplift the Rockies. Rather, once the Farallon plate finally sank, hot underlying mantle rocks ascended to replace it. Thus, it was hot, buoyant mantle rocks that provided the force to raise the Rockies. Upwelling in the mantle would also explain the elevated positions of the adjacent Colorado Plateau and the Basin and Range Province (see Box 20.1).

Figure 20.D Hogback ridges in the Rocky Mountains of Colorado. Shown is a view looking south along the east flank of the Front Range. These upturned sedimentary rocks are remnants of strata that once covered the Precambrian igneous and metamorphic core of the mountains to the west (right). (Photo by Tom Till/DRK Photo)

such activity, the primitive continental crust was deformed, metamorphosed, and even remelted. However, because of their buoyancy, these silica-rich rocks either were never consumed into the mantle or, if they were subducted, they melted and returned as magma! Thus, the essence of this hypothesis is that the total amount of continental crust has not changed appreciably since its origin; only the distribution and shape of the landmasses have been modified by tectonic activity.

Gradual Evolution of the Continents

An opposing view contends that the continents have grown larger through geologic time by the gradual accretion of material derived from the upper mantle. A main tenet of this hypothesis is that the primitive crust was of an oceanic type and the continents were small or possibly nonexistent. Then, through the chemical differentiation of mantle material, the continents slowly grew.

This view proposes that the formation of continental material takes place in multiple stages as shown in Figure 20.18. The first step occurs in the upper mantle directly beneath the oceanic ridges. Here partial melting of the rock peridotite yields basaltic magma which rises to form oceanic crust. The rocks of the ocean floor are higher in silica, potassium, and sodium and lower in iron and magnesium than the rocks of the upper mantle from which they were derived.

As new ocean floor is generated at the ridge crests, older oceanic crust is being destroyed at the oceanic trenches. As an oceanic plate is thrust into the mantle, heat and pressure drives water from the subducting crustal rocks. These volatiles migrate into the wedge of hot mantle that lie above and trigger melting. Once enough molten rock forms, it will buoyantly rise toward the surface. During its ascent, the magma usually undergoes further chemical differentiation. This gives rise to comparatively silica-rich magma, which is emplaced in a volcanic arc.

According to this view, the earliest continental rocks came into existence at a few isolated island arcs. Once formed, these island arcs coalesced to form

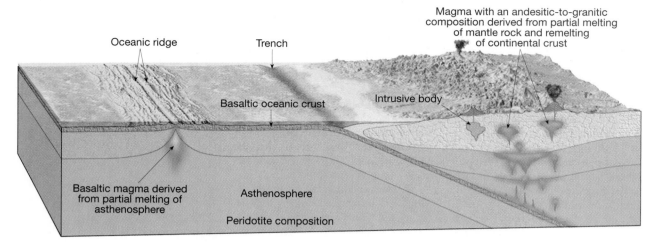

Figure 20.18 The multistage process for transforming material from the asthenosphere into continental crust. Once continental crust is generated, its low density apparently keeps it afloat indefinitely.

larger continental masses, while deforming the volcanic and sedimentary rocks that were deposited in the intervening oceans. Eventually this process generated masses of continental crust having the size and thickness of modern continents.

Evidence supporting this view of continental growth comes from research in regions of plate subduction, such as Japan and the western flanks of the Americas. Equally important, however, has been the research conducted in the stable interiors of the continents, particularly in the shield areas (see Figure 20.2, p. 513). All continents have these vast, flat-lying expanses of highly deformed and metamorphosed igneous and sedimentary rocks. The most common shield areas consist of immense bodies of granite and granodiorite, which have been strongly metamorphosed into gneisses and later intruded by younger igneous bodies.

In addition to these granite-gneiss terranes are the greenstone belts, so named because of the green tinge common to volcanic rocks of basaltic composition that have undergone low-grade regional metamorphism. The greenstone belts also contain sedimentary rocks that were strongly deformed and metamorphosed during what appears to be a mountain-building episode.

Recent indications are that the rocks of the shield areas are mineralogically and structurally similar to the rocks found at active continental margins where oceanic crust is being consumed. More specifically, the granite-gneiss terranes are chemically similar to the intrusive igneous bodies, such as the Sierra Nevada batholith, which extend along the western margin of the Americas. The ancient rock masses and these

younger Mesozoic batholiths are not identical. However, the differences can be explained by the fact that the rocks of the granite-gneiss terranes located in shield areas formed at great depth and were exposed only after long periods of uplift and erosion.

On the other hand, the exposed portions of the younger Mesozoic batholiths formed much nearer the surface and thus lack the signs of deep-seated metamorphic alteration. Further, the rocks of the greenstone belts appear to have a chemical composition similar to that of the lavas and sediments found near modern volcanic arcs, such as the islands of Japan. The rocks of the greenstone belts are thought to have been deformed along with slivers of the ocean floor and firmly plastered to the continental margin during closings of ancient ocean basins. These rocks would therefore be equivalent to the rocks of an accretionary wedge, such as the ones found in the Coast Range in California.

Radiometric dating of rocks from shield areas, including those in Minnesota and Greenland, has revealed that the oldest terranes formed some 3.8 billion years ago. This date represents one of the earliest periods of mountain building. At that time, possibly only 10 percent of the present continental crust existed. The next major period of continental evolution may have taken place between 3 billion and 2.5 billion years ago as indicated by radiometric dates of similar terranes found in the shield areas of Canada, Africa, and western Australia. This period of mountain building generated the Superior and Churchill Cratons shown in Figure 20.19. The locations of these ancient continental blocks during their formation are not known. However, about 1.9 billion years ago these

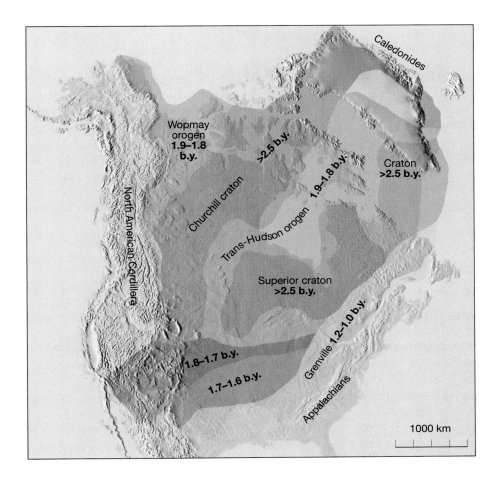

Figure 20.19 The map shows the major Precambrian mountain belts (orogens) and cores of ancient continental blocks (cratons) and their ages in billions of years (b.y.). It appears that North America was assembled from crustal blocks that were joined by processes very similar to modern plate tectonics. These ancient collisions produced mountainous belts that include remnant volcanic island arcs which were trapped by the colliding continental fragments.

two continental fragments collided to produce the Trans-Hudson orogen (Figure 20.19). This mountain-building episode was not restricted to North America, because ancient deformed strata of similar age are found on other continents.

It is not known with certainty how many periods of mountain building have occurred since the formation of Earth. The last major period evidently coincided with the closing of the proto-Atlantic and other ancient ocean basins during the formation of the supercontinent Pangaea.

If the continents do in fact grow by accretion of material to their flanks, then the continents have grown larger at the expense of oceanic crust. Like the other proposal, this view assumes a buoyant and non-subductable continental crust. Although continental crust remains afloat, continents have occasionally fragmented and been carried along in a conveyor-belt fashion until they collided with other landmasses. Presently, Australia, which separated from Antarctica, is being rafted northward and will probably join Asia

in much the same manner as India did about 45 million years ago. Thus, according to this view, fragmentation and the formation of new crustal rocks that accompanied the reshuffling of these fragments are responsible for the present volume, structure, and configuration of continents.

A word of caution is in order. The views set forth in this section to explain the origin and evolution of the continents are speculative. Plate tectonics appears to be the major force in crustal evolution over the last two billion years. However, during the early history of Earth, the heat released by the decay of uranium, thorium, and potassium may have been more than twice as great as it is today. Was plate tectonics active early in Earth's history, only at a different rate, or were there much different processes in operation? Was the primitive crust composed primarily of continental rocks, or was it of the oceanic type? These are some of the questions that remain areas of active research.

Chapter Summary

- The name for the processes that collectively produce a mountain system is *orogenesis*. Earth's less dense crust floats on top of the denser and deformable rocks in the mantle, much like wooden blocks floating in water. This concept of a floating crust in gravitational balance is called *isostasy*. Most mountains are located where the crust has been shortened and thickened. Therefore, mountains have deep crustal roots that support them. As erosion lowers the peaks, *isostatic adjustment* gradually raises the mountains in response. The processes of uplifting and erosion will continue until the mountain block reaches "normal" crustal thickness. Mountains can also rise where hot rising magma upwarps the overlying crust.

- Most mountains consist of roughly parallel ridges of folded and faulted sedimentary and volcanic rocks, portions of which have been strongly metamorphosed and intruded by younger igneous bodies.

- Major mountain systems form along *convergent plate boundaries*. *Andean-type mountain building* along continental margins involves the convergence of an oceanic plate and a plate whose leading edge contains continental crust. At some point in the formation of Andean-type mountains a *subduction zone* forms along with a *continental volcanic arc*. Sediment from the land, as well as material scraped from the subducting plate, becomes plastered against the landward side of the trench, forming an *accretionary*

wedge. One of the best examples of an inactive Andean-type mountain belt is found in the western United States, and includes the Sierra Nevada and the Coast Range in California.

- *Continental collisions*, in which both plates are carrying continental crust, have resulted in the formation of the Himalaya Mountains and the Tibetan Plateau. Continental collisions also formed many other mountain belts, including the Alps, Urals, and Appalachians.

- Recent investigations indicate that *accretion*, a third mechanism of orogenesis, takes place where *small crustal fragments collide and accrete to continental margins* along plate boundaries. Many of the mountainous regions rimming the Pacific have formed in this manner. The accreted crustal blocks are referred to as *terranes*. The mountainous topography of Alaska and British Columbia formed as the result of the accretion of terranes to northwestern North America.

- Geologists are trying to determine what role plate tectonics and mountain building play in the origin and evolution of continents. At one extreme is the view that most continental crust was formed early in Earth's history and has simply been reworked by the processes of plate tectonics. An opposing view contends that the continents have gradually grown larger by accretion of material derived from the mantle.

Review Questions

1. List three examples of evidence that support the crustal uplift concept.

2. What evidence initially led geologists to conclude that mountains have deep crustal roots?

3. What happens to a floating object when weight is added? Subtracted? How does this principle apply to changes in the elevation of mountains? What term is applied to the adjustment that causes crustal uplift of this type?

4. List one line of evidence in support of the idea that the lithosphere tries to remain in isostatic balance.

5. What do we call the site where sediments are deposited along the margin of the continent where they have a good chance of being squeezed into a mountain range?

6. Which type of plate boundary is most directly associated with mountain building?

7. What is an accretionary wedge? Briefly explain its formation.

8. What is a passive margin? Give an example. Give an example of an active continental margin.

9. In what way are the Sierra Nevada and the western Andes similar?

10. Suppose a sliver of oceanic crust was discovered 1000 kilometers into the interior of a continent. Would this support or refute the theory of plate tectonics? Why?

11. How does the plate tectonics theory help explain the existence of fossil marine life in rocks atop the Ural Mountains?

12. In your own words, briefly enumerate the steps involved in the formation of a major mountain system according to the plate tectonics model.

13. Define the term *terrane*. How is it different from the term *terrain*?

14. On the basis of current knowledge, describe the major difference between the evolution of the Appalachian Mountains and the North American Cordillera.

15. Briefly describe the formation of fault-block mountains. (See Box 20.1, p. 520)

16. Compare the forces of deformation associated with fault-block mountains to those of most other major mountain belts.

17. Contrast the opposing views on the origin of the continental crust.

Key Terms

accretionary wedge (p. 520)

isostasy (p. 515)

isostatic adjustment (p. 515)

orogenesis (p. 512)

passive margin (p. 519)

terrane (p. 527)

Web Resources

The *Earth* Home Page provides on-line resources for this chapter on the World Wide Web. You will find review exercises, specific updates for items in the chapter, suggested reading, and links to interesting related pathways on the Internet. Visit the *Earth* Home Page at **http://www.prenhall.com/tarbuck.**

CHAPTER 21

Energy and Mineral Resources

Left Aerial view of the giant Bingham Canyon copper mine near Salt Lake City, Utah. — *Photo by Michael Collier*

Figure 21.1 Manhattan from the Empire State Building. As this scene reminds us, mineral and energy resources are the basis of modern civilization. (Photo by Kim Heacox/DRK Photo)

Materials that we extract from Earth are the basis of modern civilization (Figure 21.1). Mineral and energy resources from the crust are the raw materials from which the products used by society are made. Like most people who live in highly industrialized nations, you may not realize the quantity of resources needed to maintain your present standard of living. Figure 21.2 shows the annual per capita consumption of several important metallic and nonmetallic mineral resources for the United States. This is each person's prorated share of the materials required by industry to provide the vast array of homes, cars, electronics, cosmetics, packaging, and so on that modern society demands. Figures for other highly industrialized countries such as Canada, Australia, and several nations in Western Europe are comparable.

The number of different mineral resources required by modern industries is large. Although some countries, including the United States, have substantial deposits of many important minerals, no nation is entirely self-sufficient. This reflects the fact that important deposits are limited in number and localized in occurrence. All countries must rely on international trade to fulfill at least some of their needs.

Renewable and Nonrenewable Resources

Resources are commonly divided into two broad categories—renewable and nonrenewable. **Renewable resources** can be replenished over relatively short time spans such as months, years, or decades. Common examples are plants and animals for food, natural fibers for clothing, and trees for lumber and paper. Energy from flowing water, wind, and the Sun are also considered renewable.

By contrast, **nonrenewable resources** continue to be formed in Earth, but the processes that create them are so slow that significant deposits take millions of years to accumulate. For human purposes, Earth contains fixed quantities of these substances. When the present supplies are mined or pumped from the ground, there will be no more. Examples are fuels (coal, oil, natural gas) and many important metals (iron, copper, uranium, gold). Some of these nonrenewable resources, such as aluminum, can be used over and over again; others, such as oil, cannot be recycled.

Occasionally, some resources can be placed in either category depending on how they are used. Groundwater is one such example. Where it is pumped from the ground at a rate that can be replenished, groundwater can be classified as a renewable resource. However, in places where groundwater is withdrawn faster than it is replenished, the water table drops steadily. In this case, the groundwater is being "mined" just like other nonrenewable resources.*

Figure 21.3 shows that the population of our planet is growing dramatically. Although the number did not reach one billion until the beginning of the

nineteenth century, just 130 years were needed for the population to double to two billion. Between 1930 and 1975 the figure doubled again to four billion. The year 2005 coincides with nearly seven billion people on our planet. Clearly, as population grows, the demand for resources expands as well. However, the rate of mineral and energy resource usage has climbed faster than population growth. This results from an increasing standard of living. In the United States, only 6 percent of the world's population uses approximately 30 percent of the world's annual production of mineral and energy resources!

How long can our remaining resources sustain the rising standard of living in today's industrialized countries and still provide for the growing needs of developing regions? How much environmental deterioration are we willing to accept in pursuit of resources? Can alternatives be found? If we are to cope

---

*The problem of declining water table levels is discussed in Chapter 11.

Figure 21.2 The annual per capita consumption of nonmetallic and metallic mineral resources for the United States is nearly 10,000 kilograms (11 tons)! About 94 percent of the materials used are nonmetallic. (After U.S. Bureau of Mines)

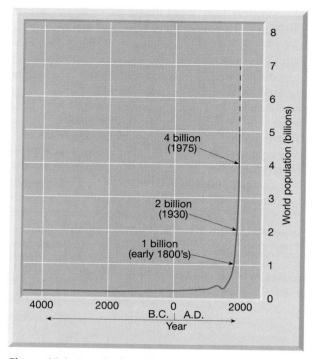

Figure 21.3 Growth of world population. It took until 1800 for the number to reach 1 billion. By 2005, nearly 7 billion people will inhabit the planet. The demand for basic resources is growing faster than the rate of population increase. (Data from the Population Reference Bureau)

with an increasing per capita demand and a growing world population, we must understand our resources and their limits.

Energy Resources

Coal, petroleum, and natural gas are the primary fuels of our modern industrial economy. Nearly 90 percent of the energy consumed in the United States today comes from these basic fossil fuels. Although major shortages of oil and gas may not occur for many years, proven reserves are declining. Despite new exploration, even in very remote regions and severe environments, new sources of oil are not keeping pace with consumption.

Unless large, new petroleum reserves are discovered (which is possible, but not likely), a greater share of our future needs will have to come from coal and from alternative energy sources such as nuclear, geothermal, solar, wind, tidal, and hydroelectric power. Two fossil fuel alternatives, tar sands and oil shale, are sometimes mentioned as promising new sources of liquid fuels. In the following sections, we will briefly examine the fuels that have traditionally supplied our energy needs as well as sources that will provide an increasing share of our future requirements.

Coal

Along with oil and natural gas, coal is commonly called a **fossil fuel**. Such a designation is appropriate because each time we burn coal we are using energy from the Sun that was stored by plants many millions of years ago. We are indeed burning a "fossil."

Coal has been an important fuel for centuries. In the nineteenth and early twentieth centuries, cheap and plentiful coal powered the industrial revolution. By 1900, coal was providing 90 percent of the energy used in the United States. Although still important, coal currently accounts for about 20 percent of the nation's energy needs (Figure 21.4).

Until the 1950s, coal was an important domestic heating fuel as well as a power source for industry. However, its direct use in the home has been largely replaced by oil, natural gas, and electricity. These fuels are preferred because they are more readily available (delivered via pipes, tanks, or wiring) and cleaner to use.

Nevertheless, coal remains the major fuel used in power plants to generate electricity, and it is therefore indirectly an important source of energy for our homes. More than 70 percent of present-day coal usage is for the generation of electricity. As oil reserves gradually diminish in the years to come, the use of coal will probably increase. Expanded coal production is possible because the world has enormous reserves and the technology to mine coal efficiently. In the United States, coal fields are widespread and contain supplies that should last for hundreds of years (Figure 21.5).

Although coal is plentiful, its recovery and use present a number of problems. Surface mining can turn the countryside into a scarred wasteland if careful (and costly) reclamation is not carried out to restore the land. (Today all U.S. surface mines must

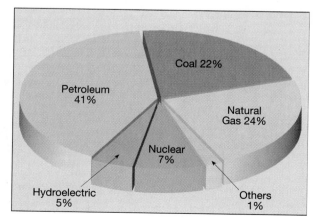

Figure 21.4 Consumption of energy in the United States, 1997. (Source: U.S. Department of Energy)

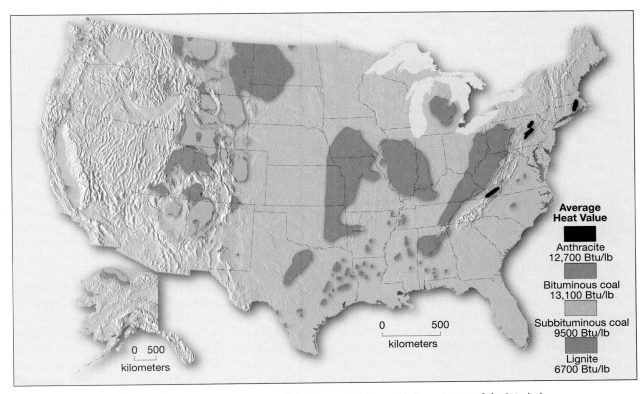

Average Heat Value

Anthracite
12,700 Btu/lb

Bituminous coal
13,100 Btu/lb

Subbituminous coal
9500 Btu/lb

Lignite
6700 Btu/lb

0   500
kilometers

0   500
kilometers

Figure 21.5 Coal fields of the United States. (Courtesy of the Bureau of Mines, U.S. Department of the Interior)

reclaim the land.) Although underground mining does not scar the landscape to the same degree, it has been costly in terms of human life and health.

Moreover, underground mining long ago ceased to be a pick-and-shovel operation, and is today a highly mechanized and computerized process (Figure 21.6). Strong federal safety regulations have made U.S. mining quite safe. However, the hazards of collapsing roofs, gas explosions, and working with heavy equipment remain.

Figure 21.6 Modern underground coal mining is highly mechanized and relatively safe. (Photo by Melvin Grubb)

Air pollution is a major problem associated with the burning of coal. Much coal contains significant quantities of sulfur. Despite efforts to remove sulfur before the coal is burned, some remains; when the coal is burned, the sulfur is converted into noxious sulfur oxide gases. Through a series of complex chemical reactions in the atmosphere, the sulfur oxides are converted to sulfuric acid, which then falls to Earth's surface as rain or snow. This acid precipitation can have adverse ecological effects over widespread areas (see Box 5.1, p. 126).

As none of the problems just mentioned are likely to present the increased use of this important and abundant fuel, stronger efforts must be made to correct the problems associated with the mining and use of coal.

Oil and Natural Gas

Petroleum and natural gas are found in similar environments and typically occur together. Both consist of various hydrocarbon compounds (compounds consisting of hydrogen and carbon) mixed together. They may also contain small quantities of other elements such as sulfur, nitrogen, and oxygen. Like coal, petroleum and natural gas are biological products derived from the remains of organisms. However, the environments in which they form are very different, as are the organisms. Coal is formed mostly from plant material that accumulated in a swampy environment above sea level. Oil and gas are derived from the remains of both plants and animals having a marine origin (see Box 21.1).

Petroleum Formation

Petroleum formation is complex and not completely understood. Nonetheless, we know that it begins with the accumulation of sediment in ocean areas that are rich in plant and animal remains. These accumulations must occur where biological activity is high, such as in nearshore areas. However, most marine environments are oxygen-rich, which leads to the decay of organic remains before they can be buried by other sediments. Therefore, accumulations of oil and gas are not as widespread as are the marine environments that support abundant biological activity. This limiting factor notwithstanding, large quantities of organic matter are buried and protected from oxidation in many offshore sedimentary basins. With increasing burial over millions of years, chemical reactions gradually transform some of the original organic matter into the liquid and gaseous hydrocarbons we call petroleum and natural gas.

Unlike the organic matter from which they formed, the newly created petroleum and natural gas are mobile. These fluids are gradually squeezed from the compacting, mud-rich layers where thy originate into adjacent permeable beds such as sandstone, where openings between sediment grains are larger. Because this occurs under water, the rock layers containing the oil and gas are saturated with water. But oil and gas are less dense than water, so they migrate upward through the water-filled pore spaces of the enclosing rocks. Unless something halts this upward migration, the fluids will eventually reach the surface, at which point the volatile components will evaporate.

Oil Traps

Sometimes the upward migration is halted. A geologic environment that allows for economically significant amounts of oil and gas to accumulate underground is termed an **oil trap**. Several geologic structures may act as oil traps, but all have two basic conditions in common: a porous, permeable **reservoir rock** that will yield petroleum and natural gas in sufficient quantities to make drilling worthwhile; and a **cap rock**, such as shale, that is virtually impermeable to oil and gas. The cap rock halts the upwardly mobile oil and gas and keeps the oil and gas from escaping at the surface.

Figure 21.7 illustrates some common oil and natural gas traps. One of the simplest traps is an *anticline*, an uparched series of sedimentary strata (Figure 21.7A). As the strata are bent, the rising oil and gas collect at the apex of the fold. Because of its lower density, the natural gas collects above the oil. Both rest upon the denser water that saturates the reservoir rock. One of the world's largest oil fields, El Nala in Saudi Arabia, is the result of an anticlinal trap, as is the famous Teapot Dome in Wyoming.

Fault traps form when strata are displaced in such a manner as to bring a dipping reservoir rock into position opposite an impermeable bed as shown in Figure 21.7B. In this case the upward migration of the oil and gas is halted where it encounters the fault.

In the Gulf coastal plain region of the United States, important accumulations of oil occur in association with *salt domes*. Such areas have thick accumulations of sedimentary strata, including layers of rock salt. Salt occurring at great depths has been forced to rise in columns by the pressure of overlying beds. These rising salt columns gradually deform the overlying strata. Because oil and gas migrate to the highest level possible, they accumulate in the upturned sandstone beds adjacent to the salt column (Figure 21.7C).

Yet another important geologic circumstance that may lead to significant accumulations of oil and gas is termed a *stratigraphic trap*. These oil-bearing structures result primarily from the original pattern of sedimentation rather than structural deformation. The stratigraphic trap illustrated in Figure 21.7D exists because a sloping bed of sandstone thins to the point of disappearance.

Box 21.1

Career Essay

Susan M. Landon, Independent Petroleum Geologist*

My career as a petroleum geologist has been rewarding—and fun (Figure 21.A)! My interest in geology began collecting rocks as a young girl. I enjoyed science in high school and majored in geology at a small liberal arts college. I went on to a state university where I studied environmental controls on coral growth rates on Caribbean reefs. When I learned that oil companies were hiring geologists, I interviewed with several and was offered a job with a major oil company. After fifteen years with that company, I decided to resign to become an independent. I no longer have the support and resources a large company provides and the risk I personally assume is much greater. The rewards, on the other hand, can also be much greater. Personal satisfaction, the potential for significant income, and the pleasure of solving a mystery are my rewards.

As an independent petroleum geologist, I use geological principles to identify areas that may contain economically viable accumulations of oil or natural gas. My job is similar to solving a mystery. Earth provides clues to the location of an oil or natural gas reservoir and my job is to find those clues and develop a geologically logical interpretation. My job also includes finding an investor, usually an oil company, willing to fund additional data acquisition, leasing the mineral rights from the land owner, and drilling one or more wells. Although less than one out of ten wells discovers a new oil or natural gas field, my job is to provide geologically

sound opportunities (good places to drill) to the petroleum industry.

Oil and natural gas accumulations are not randomly distributed. They are controlled by several geological requirements. One necessity is a source for hydrocarbons. Oil and natural gas are generated from organic matter (plants and animals) that is deposited with sand, silt, and mud. If enough organic matter is present in the rock and it is subjected to temperatures high enough to cause the necessary chemical reactions, then the organic matter is converted to petroleum. The oil or natural gas must then be able to migrate from the source rock into a porous and permeable reservoir rock. The reservoir rock must then form a trap.

Information collected from outcrops and wells previously drilled in an area provide some of the geological clues. Potential source rocks and reservoir rocks can be identified and mapped. Working with geophysicists, we can image and interpret the geometry of rock layers in the subsurface using gravity and magnetic intensity measurements and seismic techniques. Once a specific location is chosen, the geologist is responsible for getting as much geological information from the well as possible. A well may cost from tens of thousands to millions of dollars depending on the depth and location. If the well is not successful at locating an economically viable oil or natural gas field, the geological information still provides new clues for the continuing search.

Geologists and geophysicists work with engineers, after a field is discovered, to maximize the efficiency of producing the oil or natural gas. Sophisticated

Figure 21.A Susan Landon is an independent petroleum geologist and president of the American Geological Institute (AGI).

technology allows the team to image the reservoir in 3D, helping them predict variations as a result of faulting or changes in rock type. This job, whether you are looking for an undiscovered field or describing a newly found field, is like putting together a jigsaw puzzle without benefit of the picture on the box lid.

*In addition to being a highly successful petroleum geologist, Susan Landon is President of the American Geological Institute (AGI). The AGI is a nonprofit federation of 31 geoscientific and professional associations that represent more than 100,000 geoscientists. Among its many functions, the AGI strives to increase public awareness of the vital role the geosciences play in humanity's use of resources and interaction with the environment.

When the lid created by the cap rock is punctured by drilling, the oil and natural gas, which are under pressure, migrate from the pore spaces of the reservoir rock to the drill hole. On rare occasions the fluid pressure may force oil up the drill hole to the surface in extreme cases causing a "gusher" or oil fountain at the surface. Usually, however, a pump is required to lift the oil out.

A drill hole is not the only means by which oil and gas can escape from a trap. Traps can be broken by natural forces. For example, Earth movements may create fractures that allow the hydrocarbon fluids to escape. Surface erosion may breach a trap with similar results. The older the rock strata, the greater the chance that deformation or erosion has affected a trap. Indeed, not all ages of rock yield oil and gas in the

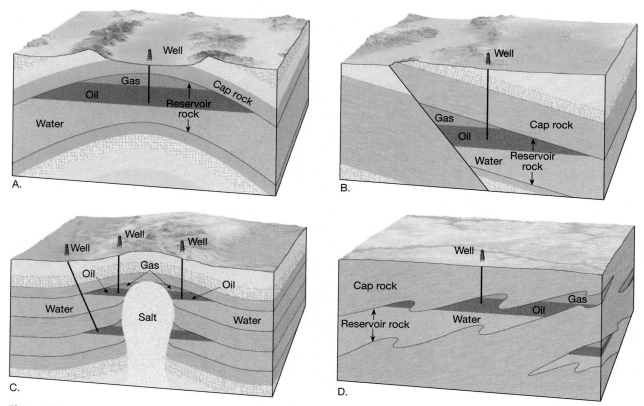

Figure 21.7 Common oil traps. **A.** Anticline. **B.** Fault trap. **C.** Salt dome. **D.** Stratigraphic (pinch-out) trap.

same proportions. The greatest production comes from the youngest rocks, those of Cenozoic age. Older Mesozoic rocks produce considerably less, followed by even smaller yields from the still older Paleozoic strata. There is virtually no oil produced from the most ancient rocks, those of Precambrian age.

Some Environmental Effects of Burning Fossil Fuels

Humanity faces a broad array of human-caused environmental problems. Among the most serious are the impacts on the atmosphere that occur because of fossil fuel combustion. Urban air pollution, acid rain, and global (greenhouse) warming are all closely linked to the use of these basic energy resources. (The acid rain problem is discussed in Box 5.1, p. 126.)

Urban Air Pollution

Air pollutants are airborne particles and gases that occur in concentrations that endanger the health and well-being of organisms or disrupt the orderly functioning of the environment.

To people living in cities, air pollution is a serious matter. The city has been accurately described as a giant chemical reactor that can produce a remarkable

variety of undesirable products. Figure 21.8 shows the major primary pollutants and the sources that produce them. *Primary pollutants* are emitted directly from identifiable sources. They pollute the air immediately upon being emitted.

The significance of the transportation category is obvious. Fuel combustion for transportation accounts for nearly half of our pollution (by weight). The hundreds of millions of cars and trucks on the roads are the greatest contributors in this category. Figure 21.8 also shows that the second major source of primary pollutants is fuel combustion by stationary sources such as electrical-generating plants.

When chemical reactions take place among primary pollutants, *secondary pollutants* are formed. The noxious mixture of gases and particles that make up urban smog is an important example that is created when volatile organic compounds and nitrogen oxides from vehicle exhausts react in the presence of sunlight.

Carbon Dioxide and Global Warming

Warming of the lower atmosphere is global in scale. Unlike acid rain and urban air pollution, this issue is not associated with any of the primary pollutants in Figure 21.8. Rather, the connection between global

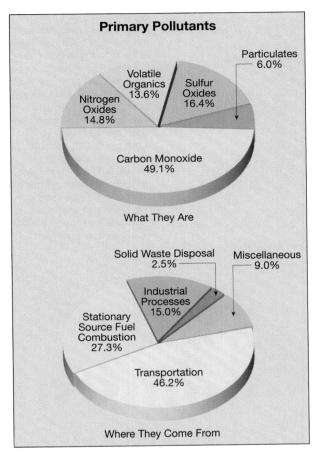

Primary Pollutants

Volatile Organics 13.6%

Sulfur Oxides 16.4%

Particulates 6.0%

Nitrogen Oxides 14.8%

Carbon Monoxide 49.1%

What They Are

Solid Waste Disposal 2.5%

Miscellaneous 9.0%

Industrial Processes 15.0%

Stationary Source Fuel Combustion 27.3%

Transportation 46.2%

Where They Come From

Figure 21.8 Major primary pollutants and their sources. Percentages are calculated on the basis of weight. (Data from the U.S. Environmental Protection Agency)

warming and burning fossil fuels relates to a basic product of combustion—carbon dioxide.

Carbon dioxide (CO_2) is a gas that is found naturally in the atmosphere and that is being augmented by fuel combustion. Although CO_2 represents only about 0.036 percent of clean, dry air, it is nevertheless meteorologically significant. The importance of carbon dioxide lies in the fact that it is transparent to incoming, short-wavelength solar radiation, but it is not transparent to some of the longer-wavelength, outgoing terrestrial radiation. A portion of the energy leaving the ground is absorbed by carbon dioxide and subsequently reemitted, part of it toward the surface, thereby keeping the air near the ground warmer than it would be without carbon dioxide. Thus, carbon dioxide is one of the gases responsible for warming the lower atmosphere. The process is called the *greenhouse effect* (Figure 21.9). Because carbon dioxide is an important heat absorber, any change in the air's carbon dioxide content could alter temperatures in the lower atmosphere.

Although the proportion of carbon dioxide in the air is relatively uniform at any given time, its percentage has been rising steadily for more than a century (Figure 21.10). Much of this rise is the result of burning ever-increasing quantities of fossil fuels.* From the mid-nineteenth century until 1998, there has been an increase of more than 25 percent in the carbon dioxide content of the air.

---

*Although the use of fossil fuels is the primary means by which humans add CO_2 to the atmosphere, the clearing of forests, especially in the tropics, also contributes substantially. Carbon dioxide is released as vegetation is burned or decays.

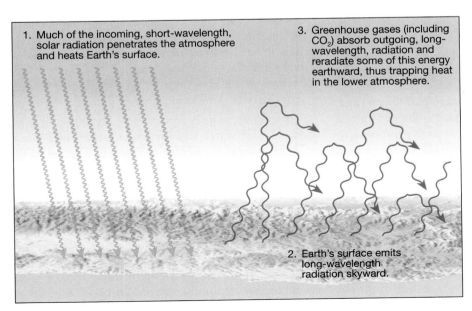

1. Much of the incoming, short-wavelength, solar radiation penetrates the atmosphere and heats Earth's surface.

3. Greenhouse gases (including CO_2) absorb outgoing, long-wavelength, radiation and reradiate some of this energy earthward, thus trapping heat in the lower atmosphere.

2. Earth's surface emits long-wavelength radiation skyward.

Figure 21.9 The heating of the atmosphere. Most of the solar radiation that is not reflected back to space passes through the atmosphere and is absorbed at Earth's surface. Earth's surface, in turn, emits longer wavelength radiation. A portion of this energy is absorbed by certain gases in the atmosphere. Carbon dioxide is one of these gases. The process is called the *greenhouse effect.*

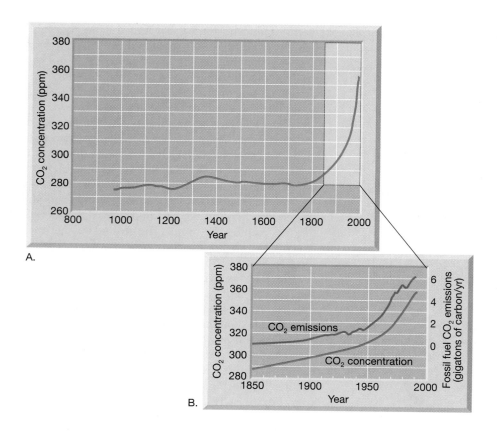

Figure 21.10 A. Carbon dioxide (CO_2) concentrations over the past 1000 years. Most of the record is based on data obtained from Antarctic ice cores. Bubbles of air trapped in the glacial ice provide samples of past atmospheres. The record since 1958 comes from direct measurements of atmospheric CO_2 taken at Mauna Loa Observatory, Hawaii. B. The rapid increase in CO_2 concentration since the onset of industrialization in the late 1700s is clear and has followed closely the rise in CO_2 emissions from fossil fuels.

If we assume that the use of fossil fuels will continue to increase at projected rates, current estimates indicate that the atmosphere's present carbon dioxide content of almost 360 parts per million (ppm) will approach 400 ppm by the year 2010 and will reach 600 ppm by sometime in the second half of the twenty-first century. With such an increase in carbon dioxide, the greenhouse effect would become much more dramatic and measurable than in the past. The most realistic models predict an increase in the mean global surface temperature of between 1.5°C and 4.5°C. A change of this magnitude would be unprecedented in human history. Such an increase would come close to equaling the warming that has taken place since the peak of the most recent glacial stage 18,000 years ago, but would occur *much more rapidly*. The effects of a rapid temperature change are a matter of great concern and much uncertainty (see Box 21.2). Consequently, global warming is and will continue to be a major scientific and political issue for years to come.

Coal, oil, and natural gas are vital energy sources that power the modern world. However, the benefits that these basic and relatively low cost fuels provide do not come without environmental costs. Among the consequences are three serious atmospheric impacts: urban air pollution, acid rain, and global warming. People are clearly altering the composition of the air.

Not only is this influence felt locally and regionally, but it extends worldwide and to many kilometers above Earth's surface.

Tar Sands and Oil Shale—Petroleum for the Future?

In the years to come world oil supplies will dwindle. When this occurs, a lower grade of hydrocarbon may have to be substituted. Are fuels derived from tar sands and oil shales good candidates?

Tar Sands

Tar sands are usually mixtures of clay and sand combined with water and varying amounts of a black, highly viscous tar known as *bitumen*. The use of the term *sand* can be misleading, because not all deposits are associated with sands and sandstones. Some occur in other materials, including shales and limestones. The oil in these deposits is very similar to heavy crude oils pumped from wells. The major difference between conventional oil reservoirs and tar sand deposits is in the viscosity (thickness) of the oil they contain. In tar sands, the oil is much more viscous and cannot be simply pumped out.

Box 21.2

Some Possible Consequences of a Global Warming

What consequences can be expected if the carbon dioxide content of the atmosphere reaches a level that is twice what it was early in the twentieth century? Because the climate system is so complex, predicting the distribution of particular regional changes is still very speculative. It is not yet possible to pinpoint specifics such as where or when it will be drier or wetter. Plausible scenarios can only be given for larger scales of space and time. This discussion examines some potential consequences. As computers become more powerful and data improve, scientists will gradually develop models that can provide more specific and reliable results.

The magnitude of the temperature increase will not be the same everywhere. The temperature rise will probably be smallest in the tropics and increase toward the poles. Furthermore, the models indicate that some regions will experience a significant increase in precipitation and runoff, whereas others will experience a decrease in runoff either because of reduced precipitation or because of increased evaporation rates brought about by higher temperatures. Such changes could have a profound impact on the distribution of the world's water resources and hence affect the productivity of agricultural regions that depend on rivers for irrigation water.

A 2°C warming and 10 percent precipitation decrease in the region drained by the Colorado River, for example, could diminish the river's flow by an estimated 50 percent or more. Because the present flow of the river is barely enough to meet current demands for irrigated agriculture, the negative effect would be serious (Figure 21.B). Many other rivers form the basis for extensive systems of irrigated agriculture, and the projected reduction of their flow could have equally grave consequences. In contrast, large precipitation increases in

Figure 21.B It is not yet possible to specify the magnitude and location of particular climate changes that may result from greenhouse warming. Many consequences are possible. Decreased rainfall and increased evaporation rates could diminish the flow of certain rivers and force the abandonment of some presently productive irrigated farmland. (Photo by E. J. Tarbuck)

other areas would increase the flow of some rivers and bring more frequent destructive floods.

The effects of increased atmospheric carbon dioxide on nonirrigated crops that depend on rain for moisture are complex and difficult to estimate. Some places will no doubt experience productivity losses due to decreases in rainfall or increases in evaporation rates. Still, these losses may be offset by gains elsewhere. Increased temperatures in the high latitudes could lengthen the growing season, for instance. This factor, in turn, could allow the expansion of agriculture into areas that are presently not suited to crop production.

Another impact of a human-induced global warming is a probable rise in sea level. This effect is examined in Chapter 14, Box 14.2 (p. 364).

Atmospheric scientists expect that weather patterns will change as a result of the projected global warming. Potential weather changes include:

1. A greater intensity of tropical storms because of warmer ocean temperatures.

2. Shifts in the paths of large-scale cyclonic storms, which, in turn, would affect the distribution of precipitation and the occurrence of severe storms.

3. Increases in the frequency and intensity of heat waves and droughts.

In concluding this discussion, it should be stated again that the impact on global climate of an increase in the atmosphere's content of CO_2 is obscured by many unknowns and uncertainties. The changes that occur will probably take the form of gradual environmental shifts that will be imperceptible to most people from year to year. Nevertheless, although the changes may seem gradual, the effects will clearly have powerful economic, social, and political consequences.

Substantial tar sand deposits occur in several locations around the world, but it is the Canadian province of Alberta that has the largest tar sand deposits (Figure 21.11).

Currently, tar sands are mined at the surface, in a manner similar to the strip mining of coal. The excavated material is then heated with pressurized steam until the bitumen softens and rises. Once collected,

Figure 21.11 In North America, the largest tar sand deposits occur in the Canadian province of Alberta. Known as the Athabasca Tar Sands, these deposits cover thousands of square kilometers and contain an estimated reserve of 35 billion barrels of oil that might be recovered.

the oily material is treated to remove impurities and then hydrogen is added. This last step upgrades the material to a synthetic crude, which can then be refined. Extracting and refining tar sand requires a great deal of energy—nearly half as much as the end product yields! Never the less, tar sands from Alberta's vast deposits are the source of about 15 percent of Canada's oil production.

Obtaining oil from tar sand has significant environmental drawbacks. Substantial land disturbance is associated with mining huge quantities of rock and sediment. Moreover, large quantities of water are required for processing, and when processing is completed, contaminated water and sediment accumulate in toxic disposal ponds.

Only about 10 percent of Alberta's tar sands can be economically recovered by surface mining. Obtaining the more deeply buried oil will require recovery in place, that is, without mining. This will likely involve injecting hot fluids to reduce viscosity and then pumping the material to the surface. If such techniques are put into practice, the hope is that the land surface will only be disturbed slightly and that environmental impacts will be reduced.

Oil Shale

Oil shale contains enormous amounts of untapped oil. Worldwide, the U.S. Geological Survey estimates that there are more than 3000 billion barrels of oil contained in shales that would yield more than 38 liters (10 gallons) of oil per ton of shale. But this figure is misleading because fewer than 200 billion barrels are known to be recoverable with present technology. Still, estimated U.S. resources are about 14 times as great as those of conventionally recoverable oil, and estimates will probably increase as more geological information is gathered.

Roughly half of the world supply is in the Green River Formation in Colorado, Utah, and Wyoming (Figure 21.12). Within this region the oil shales are part of sedimentary layers that accumulated at the bottoms of two vast, shallow lakes during the Eocene epoch (57 million to 36 million years ago).

Oil shale has been suggested as a partial solution to dwindling fuel supplies. However, the heat energy in oil shale is about one-eighth that in crude oil, owing to the large proportion of mineral matter in the shales.

This mineral matter adds cost to mining, processing, and waste disposal. Producing oil from oil shale has the same problems as producing oil from tar sands. Surface mining causes widespread land disturbance and presents significant waste-disposal problems. Moreover, processing requires large amounts of water, something that is in short supply in the semi-arid region occupied by the Green River Formation.

At present, oil is plentiful and relatively inexpensive on world markets. Therefore, with current technologies, most oil shales are not worth mining. Oil shale research and development efforts have been nearly abandoned by industry. Nevertheless, the U.S. Geological Survey suggests that the large amount of oil that potentially can be extracted from oil shales in the United States probably ensures its eventual inclusion in the national energy mix.

Alternate Energy Sources

An examination of Figure 21.4 (p. 538) clearly shows that we live in the era of fossil fuels. Nearly 90 percent of the world's energy needs are derived from these nonrenewable resources. Present estimates indicate that the amount of recoverable fossil fuels may equal 10 trillion barrels of oil, which is enough to last 170 years at the present rate of consumption. Of course, as world population soars, the rate of consumption will climb. Thus, reserves will eventually be in short supply. In the meantime, the environmental impact of

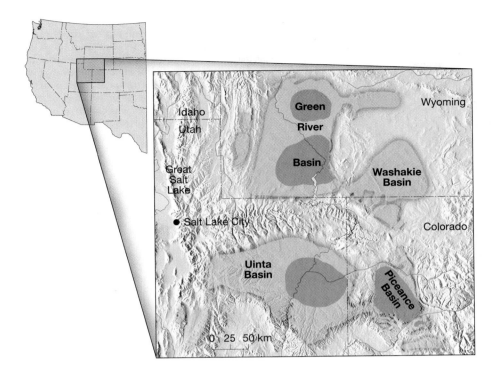

Figure 21.12 Distribution of oil shale in the Green River Formation of Colorado, Utah, and Wyoming. The areas shaded with the darker color represent the richest deposits. Government and industry have invested large sums to make these oil shales an economic resource, but costs have always been higher than the price of oil. However, as prices of competing fuels rise, these vast deposits will become economically more attractive. (After D. C. Duncan and V. E. Swanson, U.S. Geological Survey Circular 523, 1965)

burning huge quantities of fossil fuels will undoubtedly have an adverse effect on the environment.

How can a growing demand for energy be met without radically affecting the planet we inhabit? Although no clear answer has yet been formulated, the need to rely more heavily on alternate energy sources must be considered. In this section we will examine several possible sources, including nuclear, solar, wind, hydroelectric, geothermal, and tidal energy.

Nuclear Energy

Roughly 7 percent of the energy demand of the United States is being met by nuclear power plants. The fuel for these facilities comes from radioactive materials that release energy by the process of **nuclear fission**. Fission is accomplished by bombarding the nuclei of heavy atoms, commonly uranium-235, with neutrons. This causes the uranium nuclei to split into smaller nuclei and to emit neutrons and heat energy. The ejected neutrons, in turn, bombard the nuclei of adjacent uranium atoms, producing a *chain reaction*. If the supply of fissionable material is sufficient and if the reaction is allowed to proceed in an uncontrolled manner, an enormous amount of energy would be released in the form of an atomic explosion.

In a nuclear power plant, the fission reaction is controlled by moving neutron-absorbing rods into or out of the nuclear reactor. The result is a controlled nuclear chain reaction that releases great amounts of heat. The energy produced is transported from the reactor and used to drive steam turbines that turn electrical generators, just as occurs in most conventional power plants.

Uranium. Uranium-235 is the only naturally occurring isotope that is readily fissionable and is therefore the primary fuel used in nuclear power plants.[*] Although large quantities of uranium ore have been discovered, most contain less than 0.05 percent uranium. Of this small amount, 99.3 percent is the nonfissionable isotope uranium-238 and just 0.7 percent consists of the fissionable isotope uranium-235. Because most nuclear reactors operate with fuels that are at least 3 percent uranium-235, the two isotopes must be separated in order to concentrate the fissionable uranium-235. The process of separating the uranium isotopes is difficult and substantially increases the cost of nuclear power.

Although uranium is a rare element in Earth's crust, it does occur in enriched deposits. Some of the most important occurrences are associated with what are believed to be ancient placer deposits in stream beds. For example, in Witwatersrand, South Africa, grains of uranium ore (as well as rich gold deposits) were concentrated by virtue of their high density in rocks made largely of quartz pebbles. In the United States the richest uranium deposits are found in Jurassic

---

[*]Thorium, although not capable by itself of sustaining a chain reaction, can be used with uranium-235 as a nuclear fuel.

Figure 21.13 Drablo Canyon nuclear power plant near San Luis Obispo, California. Reactors are in the dome-shaped buildings and cooling water is escaping to the ocean. The siting of this facility was controversial because of its close proximity to faults capable of producing potentially damaging earthquakes. (Photo by Comstock)

and Triassic sandstones in the Colorado Plateau and in younger rocks in Wyoming. Most of these deposits have formed through the precipitation of uranium compounds from groundwater. Here precipitation of uranium occurs as a result of a chemical reaction with organic matter, as evidenced by the concentration of uranium in fossil logs and organic-rich, black shales.

Obstacles to Development. At one time nuclear power was heralded as the clean, cheap source of energy to replace fossil fuels. However, several obstacles have emerged to hinder the development of nuclear power as a major energy source. Not the least of these is the skyrocketing costs of building nuclear facilities. More important, perhaps, is the concern over the possibility of a serious accident at one of the nearly 200 nuclear plants in existence worldwide (Figure 21.13). The 1979 accident at Three Mile Island near Harrisburg, Pennsylvania, helped bring this point home. Here, a malfunction led the plant operators to believe there was too much water in the primary system instead of too little. This confusion allowed the reactor core to lie uncovered for several hours. Although there was little danger to the public, substantial damage to the reactor resulted.

Unfortunately, the 1986 accident at Chernobyl in the former Soviet Union was far more serious. In this incident, the reactor ran out of control and two small explosions lifted the roof of the structure, allowing pieces of uranium to be thrown over the immediate area. During the 10 days that it took to quench the fire that ensued, high levels of radioactive material were carried by the atmosphere and deposited as far away as Norway. In addition to the 18 people who died within 6 weeks of

the accident, many thousands more face an increased risk of death from cancers associated with the fallout.

It should be emphasized that the concentrations of fissionable uranium-235 and the design of reactors are such that nuclear power plants cannot explode like an atomic bomb. The dangers arise from the possible escape of radioactive debris during a meltdown of the core or other malfunction. In addition, hazards such as the disposal of nuclear waste and the relationship that exists between nuclear energy programs and the proliferation of nuclear weapons must be considered as we evaluate the pros and cons of employing nuclear power.

Solar Energy

The term *solar energy* generally refers to the direct use of the Sun's rays to supply energy for the needs of people. The simplest, and perhaps most widely used, *passive solar collectors* are south-facing windows. As sunlight passes through the glass, its energy is absorbed by objects in the room. These objects, in turn, radiate heat that warms the air. In the United States we often use south-facing windows, along with better-insulated and more airtight construction, to reduce heating costs substantially.

More elaborate systems used for home heating involve an *active solar collector*. These roof-mounted devices are usually large, blackened boxes that are covered with glass. The heat they collect can be transferred to where it is needed by circulating air or fluids through piping. Solar collectors are also used successfully to heat water for domestic and commercial needs. For example, solar collectors provide hot water for more than 80 percent of Israel's homes.

Figure 21.14 Solar One, a solar installation used to generate electricity in the Mojave Desert near Barstow, California. (Photo by Thomas Braise/The Stock Market)

Although solar energy is free, the necessary equipment and its installation are not. The initial costs of setting up a system, including a supplemental heating unit for times when solar energy is diminished (cloudy days and winter) or unavailable (nighttime), can be substantial. Nevertheless, over the long term, solar energy is economical in many parts of the United States and will become even more cost effective as the price of other fuels increases.

Research is currently under way to improve the technologies for concentrating sunlight. One method being examined uses mirrors that track the Sun and keep its rays focused on a receiving tower. A facility, with 2000 mirrors, has been constructed near Barstow, California (Figure 21.14). Solar energy focused on the lower tower heats water in pressurized panels to over 500°C. The superheated water is then transferred to turbines, which turn electrical generators.

Another type of collector uses photovoltaic (solar) cells that convert the sun's energy directly into electricity. A large experimental facility that uses photovoltaic cells is located near Hesperia, California, and supplies electricity to customers of Southern California Edison.

Recently, small rooftop photovoltaic systems have begun appearing in rural households of some Third World countries, including the Dominican Republic, Sri Lanka, and Zimbabwe. These units are about the size of an open briefcase and use a battery to store electricity that is generated during the daylight hours. In the tropics, these small photovoltaic systems are capable of running a television or radio, plus a few light bulbs, for 3 to 4 hours. Although much cheaper than building conventional electric generators, these units are still too expensive for poor families. Consequently, an estimated two billion people in developing countries still lack electricity.

Wind Energy

Approximately 0.25 percent (one-quarter of 1 percent) of the solar energy that reaches the lower atmosphere is transformed into wind. Although it is just a minuscule percentage, the absolute amount of energy is enormous. According to one estimate, if the winds of North and South Dakota could be harnessed, they would provide 80 percent of the electrical energy used in the United States.

Wind has been used for centuries as an almost free and nonpolluting source of energy. Sailing ships and wind-powered grist mills represent two of the early ways that this renewable resource was harnessed. Further, as rural America was settled, farmers relied on wind power to pump water and later to generate electricity. However, as centrally generated electricity was extended to rural areas in the 1930s, the reliance on wind power dwindled.

Following the "energy crisis" that was precipitated by an oil embargo in the 1970s, interest in wind power as well as other alternative forms of energy increased dramatically. In 1980, the federal government initiated a program to develop wind-power systems. Projects sponsored by the U.S. Department of Energy involved setting up experimental wind farms at sites known to have strong persistent winds (Figure 21.15). By the late 1990s, more than 17,000 wind turbines in California were integrated into the state's electrical grid. Their output represented about 1.3 percent of California's electrical demand.

Although the future for wind power is promising, it is not without difficulties. Technical problems must be overcome in building large, efficient turbines. In addition, noise pollution and the cost of large tracts of land in populated areas present significant obstacles to development.

Hydroelectric Power

Falling water has been an energy source for centuries. Through most of history, the mechanical energy produced by water wheels was used to power mills and other machinery. Today the power generated by falling water is used to drive turbines that produce electricity, hence the term **hydroelectric power**. In the United States, hydroelectric power plants contribute about 5 percent of the country's demand. Most of this energy is produced at large dams, which allow for a controlled flow of water (Figure 21.16). The water impounded in a reservoir is a form of stored energy that can be released at any time to produce electricity.

Although water power is considered a renewable resource, the dams built to provide hydroelectricity have finite lifetimes. All rivers carry suspended sediment that is deposited behind the dam as soon as it is built. Eventually the sediment will completely fill the reservoir. This takes 50 to 300 years depending on the quantity of suspended material transported by the river. An example is Egypt's huge Aswan High Dam, which was completed in the 1960s. It is estimated that half of the reservoir will be filled with sediment from the Nile River by the year 2025.

The availability of appropriate sites is an important limiting factor in the development of large-scale

Figure 21.15 This wind farm near Palm Springs, California, consists of several hundred wind turbines. (Photo by Michael Collier)

Figure 21.16 Lake Powell is the reservoir that was created when Glen Canyon Dam was built across the Colorado River. As water in the reservoir is released, it drives turbines and produces electricity. Eventually the reservoir will be filled with sediment deposited by the Colorado River. (Photo by Michael Collier)

hydroelectric power plants. A good site provides a significant height for the water to fall and a high rate of flow. Hydroelectric dams exist in many parts of the United States, with the greatest concentrations occurring in the Southeast and the Pacific Northwest. Most of the best U.S. sites have already been developed, limiting the future expansion of hydroelectric power. The total power produced by hydroelectric sources might still increase, but the relative share provided by this source may decline because other alternate energy sources may increase at a faster rate.

In recent years a different type of hydroelectric power production has come into use. Called a *pumped water storage system*, it is actually a type of energy management. During times when demand for electricity is low, unneeded power produced by nonhydroelectric sources is used to pump water from a lower reservoir to a storage area at a higher elevation. Then, when demand for electricity is great, the water stored in the higher reservoir is available to drive turbines and produce electricity to supplement the power supply.

Geothermal Energy

We harness **geothermal energy** by tapping natural underground reservoirs of steam and hot water. These occur where subsurface temperatures are high owing to relatively recent volcanic activity. We put geothermal energy to use in two ways: the steam and hot water are used for heating and to generate electricity.

Iceland is a large volcanic island with current volcanic activity. In Iceland's capital, Reykjavik, steam and hot water are pumped into buildings throughout the city for space heating. They also warm greenhouses, where fruits and vegetables are grown all year (Figure 21.17). In the United States, localities in several western states use hot water from geothermal sources for space heating.

As for generating electricity geothermally, the Italians were first to do so in 1904, so the idea is not new. By the mid-1990s, the U.S. Geological Survey reported that nearly 200 separate power units in 17 countries were operating, with a combined capacity of almost 4800 megawatts (million watts). The leaders are the United States, the Philippines, Indonesia, Mexico, Italy, and New Zealand.

The first commercial geothermal power plant in the United States was built in 1960 at The Geysers, north of San Francisco (Figure 21.18). By 1986, development at this location had grown to almost 1800 megawatts, enough to satisfy the needs of San Francisco and neighboring Oakland. However, this peak soon passed and the production of electricity began to decline. Nevertheless, it remains the world's premier geothermal field, and it continues to provide about 1000 megawatts of electrical power with little

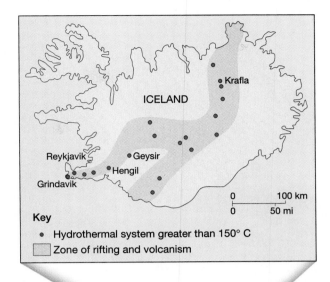

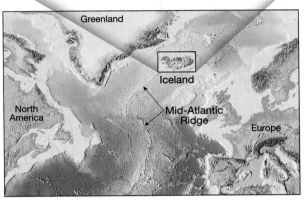

Figure 21.17 Iceland straddles the Mid-Atlantic Ridge. This divergent plate boundary is the site of numerous active volcanoes and geothermal systems. Because the entire country consists of geologically young volcanic rocks, warm water can be encountered in holes drilled almost anywhere. More than 45 percent of Iceland's energy comes from geothermal sources.

environmental impact. In addition to The Geysers, geothermal development is occurring elsewhere in the western United States, including Nevada, Utah, and the Imperial Valley in southern California.

What geologic factors favor a geothermal reservoir of commercial value?

1. *A potent source of heat,* such as a large magma chamber deep enough to ensure adequate pressure and slow cooling, yet not so deep that the natural water circulation is inhabited. Such magma chambers are most likely in regions of recent volcanic activity.

2. *Large and porous reservoirs with channels connected to the heat source,* near which water can circulate and then be stored in the reservoir.

3. *A cap of low permeability rocks* that inhibits the flow of water and heat to the surface. A deep,

Figure 21.18 The Geysers, near the city of Santa Rosa in northern California, is the world's largest electricity-generating geothermal development. Most of the steam wells are about 3000 meters deep. (Photo courtesy of Pacific Gas and Electric)

well-insulated reservoir contains much more stored energy than a similar but uninsulated reservoir.

We must recognize that geothermal power is not inexhaustible. When hot fluids are pumped from volcanically heated reservoirs, water cannot be replaced and then heated sufficiently to recharge the reservoir. Experience shows that steam and hot water from individual wells usually lasts no more than 10 to 15 years, so more wells must be drilled to maintain power production. Eventually the field is depleted.

As with other alternative methods of power production, geothermal sources are not expected to provide a high percentage of the world's growing energy needs. Nevertheless, in regions where its potential can be developed, its use will no doubt continue to grow.

Tidal Power

Several methods of generating electrical energy from the oceans have been proposed, but the ocean's energy potential remains largely untapped. The development of tidal power is the principal example of energy production from the ocean.

Tides have been used as a source of power for centuries. Beginning in the twelfth century, water wheels driven by the tides were used to power gristmills and sawmills. During the seventeenth and eighteenth centuries, much of Boston's flour was produced at a tidal mill. Today, far greater energy demands must be satisfied, and more sophisticated ways of using the force created by the perpetual rise and fall of the ocean must be employed.

Tidal power is harnessed by constructing a dam across the mouth of a bay or an estuary in a coastal area having a large tidal range (Figure 21.19). The narrow opening between the bay and the open ocean magnifies the variations in water level that occur as the tides rise and fall. The strong in-and-out flow that results at such a site is then used to drive turbines and electrical generators.

Tidal energy utilization is exemplified by the tidal power plant at the mouth of the Rance River in France. By far the largest yet constructed, this plant went into operation in 1966 and produces enough power to satisfy the needs of Brittany and also contribute to the demands of other regions. Much smaller experimental facilities have been built near Murmansk in Russia and near Taliang in China, and on the Bay of Fundy in the Canadian province of Nova Scotia (Figure 21.20).

Along most of the world's coasts it is not possible to harness tidal energy. If the tidal range is less than 8 meters (25 feet) or if narrow, enclosed bays are absent, tidal power development is uneconomical. For this reason, the tides will never provide a very high portion of our ever-increasing electrical energy requirements. Nevertheless, the development of tidal power may be worth pursuing at feasible sites because electricity produced by the tides consumes no exhaustible fuels and creates no noxious wastes.

Mineral Resources

Earth's crust is the source of a wide variety of useful and essential substances. In fact, practically every manufactured product contains substances derived from minerals. Table 21.1 and Table 2.4 (p. 53) list some important examples.

Mineral resources are the endowment of useful minerals ultimately available commercially. Resources

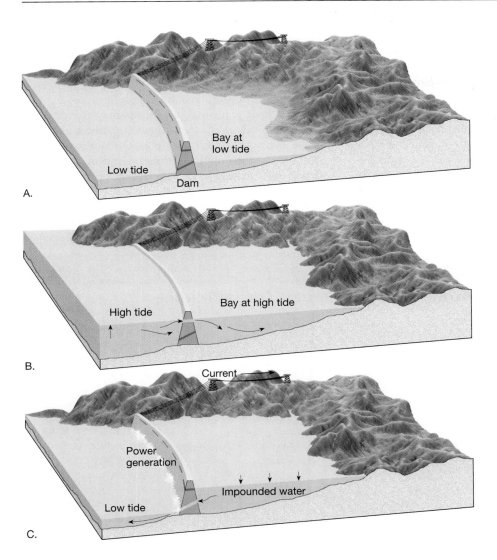

include already identified deposits from which minerals can be extracted profitably, called **reserves**, as well as known deposits that are not yet economically or technologically recoverable. Deposits inferred to exist, but not yet discovered, are also considered as mineral resources. In addition, the term **ore** is used to denote those useful metallic minerals that can be mined at a profit. In common usage, the term *ore* is also applied to some nonmetallic minerals such as fluorite and sulfur. However, materials used for such purposes as building stone, road aggregate, abrasives, ceramics, and fertilizers are not usually called ores; rather they are classified as industrial rocks and minerals.

Recall that more than 98 percent of the crust is composed of only eight elements. Except for oxygen and silicon, all other elements make up a relatively small fraction of common crustal rocks (see Table 2.3, p. 44). Indeed, the natural concentrations of many elements are exceedingly small. A deposit containing the average percentage of a valuable element

Figure 21.20 The first tidal power generating station in North America opened in 1984 at Annapolis Royal, Nova Scotia. This plant on Canada's Bay of Fundy is being used to assess the merits and drawbacks of building a larger facility. (Photo by Stephen J. Krasemann)

is worthless if the cost of extracting it exceeds the value of the material recovered. To be considered of value, an element must be concentrated above the level of its average crustal abundance. Generally, the lower the crustal abundance, the greater the concentration must be.

For example, copper makes up about 0.0135 percent of the crust. However, for a material to be considered a copper ore, it must contain a concentration that is about 50 times this amount. Aluminum, in contrast, represents 8.13 percent of the crust and must be concentrated to only about four times its average crustal percentage before it can be extracted profitably.

It is important to realize that a deposit may become profitable to extract or lose its profitability because of economic changes. If demand for a metal increases and prices rise, the status of a previously unprofitable deposit changes, and it becomes an ore. The status of unprofitable deposits may also change if a technological advance allows the useful element to be extracted at a lower cost than before. This occurred at the copper mining operation located at Bingham Canyon, Utah, the largest open-pit mine on Earth (see Box 21.3). Mining was halted here in 1985 because outmoded equipment had driven the cost of extracting the copper beyond the selling price.

Box 21.3

Bingham Canyon, Utah: The Largest Open-Pit Mine

In the photo, a mountain once stood where there is now a huge pit (Figure 21.C). This is the world's largest open-pit mine, Bingham Canyon copper mine, about 40 kilometers (25 miles) southwest of Salt Lake City, Utah. The rim is nearly 4 kilometers (2.5 miles) across and covers almost 8 square kilometers (3 square miles). Its depth is 900 meters (3000 feet). If a steel tower were erected at the bottom, it would have to be five times taller than the Eiffel Tower to reach the top of the pit!

It began in the late 1800s as an underground mine for veins of silver and lead. Later, copper was discovered. Similar deposits occur at several sites in the American Southwest and in a belt from southern Alaska to northern Chile.

As in other places in this belt, the ore at Bingham Canyon is disseminated throughout *porphyritic* igneous rocks; hence, the name *porphyry copper deposits*. The deposit formed after magma was intruded to shallow depths. Following this, shattering created extensive fractures that were penetrated by hydrothermal solutions from which the ore minerals precipitated.

Although the percentage of copper in the rock is small, the total volume of copper is huge. Ever since open-pit operations started in 1906, some 5 billion tons of material have been removed, yielding more than 12 million tons of copper. Significant amounts of gold, silver, and molybdenum have also been recovered.

Figure 21.C Aerial view of Bingham Canyon copper mine near Salt Lake City, Utah. This huge open-pit mine is about 4 kilometers across and 900 meters deep. Although the amount of copper in the rock is less than 1 percent, the huge volumes of material removed and processed each day (about 200,000 tons) yield significant quantities of metal. (Photo by Michael Collier)

The ore body is far from exhausted today. Over the next 25 years, plans call for an additional 3 billion tons of material to be removed and processed. This largest of artificial excavations has generated most of Utah's mineral production for more than 80 years and has been called the "richest hole on Earth."

Like many older mines, the Bingham pit was unregulated during most of its history. Development occurred prior to the present-day awareness of the environmental impacts of mining and prior to effective environmental legislation. Today, problems of groundwater and surface water contamination, air pollution, solid and hazardous wastes, and land reclamation are receiving long overdue attention at Bingham Canyon.

The owners responded by replacing an antiquated 1000-car railroad with conveyor belts and pipelines for transporting the ore and waste. These devices achieved a cost reduction of nearly 30 percent and returned this mining operation to profitability.

Over the years, geologists have been keenly interested in learning how natural processes produce localized concentrations of essential metallic minerals. One well-established fact is that occurrences of valuable mineral resources are closely related to the rock cycle. That is, the mechanisms that generate igneous, sedimentary, and metamorphic rocks, including the processes of weathering and erosion, play a major role in producing concentrated accumulations of useful elements. Moreover, with the development of the plate tectonics theory, geologists added yet another tool for understanding the processes by which one rock is transformed into another.

Mineral Resources and Igneous Processes

Some of the most important accumulations of metals, such as gold, silver, copper, mercury, lead, platinum, and nickel, are produced by igneous processes (Table 21.1). These mineral resources, like most others, result from processes that concentrate desirable materials to the extent that extraction is economically feasible.

Magmatic Segregation

The igneous processes that generate some of these metal deposits are quite straightforward. For example, as a large magma body cools, the heavy minerals that crystallize early tend to settle to the lower portion of the magma chamber. This type of magmatic segregation is particularly active in large basaltic magmas

Table 21.1 Occurrence of Metallic Minerals

Metal	Principal Ores	Geological Occurrences
Aluminum	Bauxite	Residual product of weathering
Chromium	Chromite	Magmatic segregation
Copper	Chalcopyrite Bornite Chalcocite	Hydrothermal deposits; contact metamorphism; enrichment by weathering processes
Gold	Native gold	Hydrothermal deposits; placers
Iron	Hematite Magnetite Limonite	Banded sedimentary formations; magmatic segregation
Lead	Galena	Hydrothermal deposits
Magnesium	Magnesite Dolomite	Hydrothermal deposits
Manganese	Pyrolusite	Residual product of weathering
Mercury	Cinnabar	Hydrothermal deposits
Molybdenum	Molybdenite	Hydrothermal deposits
Nickel	Pentlandite	Magmatic segregation
Platinum	Native platinum	Magmatic segregation; placers
Silver	Native silver Argentite	Hydrothermal deposits; enrichment by weathering processes
Tin	Cassiterite	Hydrothermal deposits; placers
Titanium	Ilmenite Rutile	Magmatic segregation; placers
Tungsten	Wolframite Scheelite	Pematites; contact metamorphic deposits; placers
Uranium	Uraninite (pitchblende)	Pegmatites; sedimentary deposits
Zinc	Sphalerite	Hydrothermal deposits

where chromite (ore of chromium), magnetite, and platinum are occasionally generated. Layers of chromite, interbedded with other heavy minerals, are mined from such deposits in the Stillwater Complex of Montana. Another example is the Bushveld Complex in South Africa, which contains over 70 percent of the world's known reserves of platinum.

Magmatic segregation is also important in the late stages of the magmatic process. This is particularly true of granitic magmas in which the residual melt can become enriched in rare elements and heavy metals. Further, because water and other volatile substances do not crystallize along with the bulk of the magma body, these fluids make up a high percentage of the melt during the final phase of solidification. Crystallization in a fluid-rich environment, where ion migration is enhanced, results in the formation of crystals several centimeters, or even a few meters, in length. The resulting rocks, called **pegmatites**, are composed of these unusually large crystals (see Box 3.1, p. 65).

Most pegmatites are granitic in composition and consist of unusually large crystals of quartz, feldspar, and muscovite. Feldspar is used in the production of ceramics and muscovite is used for electrical insulation and glitter. Further, pegmatites often contain some of the least abundant elements. Thus, in addition to the common silicates, some pegmatites include semiprecious gems such as beryl, topaz, and tourmaline. Moreover, minerals containing the elements lithium, cesium, uranium, and the rare earths* are occasionally found. Most pegmatites are located within large igneous masses or as dikes or veins that cut into the host rock surrounding the magma chamber (Figure 21.21).

Not all late-stage magmas produce pegmatites; nor do all have a granitic composition. Rather, some magmas become enriched in iron or occasionally copper. For example, at Kirava, Sweden, magma composed of over 60 percent magnetite solidified to produce one of the largest iron deposits in the world.

Diamonds

Another economically important mineral with an igneous origin is diamond. Although best known as gems, diamonds are used extensively as abrasives.

---

*The rare earths are a group of 15 elements (atomic numbers 57 through 71) that possess similar properties. They are useful catalys in petroleum refining and are used to improve color retention in television picture tubes.

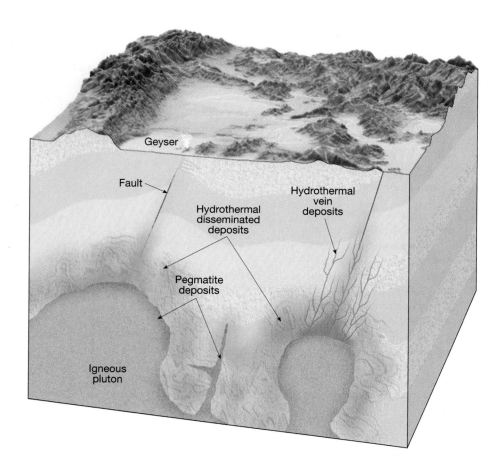

Figure 21.21 Illustration of the relationship between a parent igneous body and the associated pegmatite and hydrothermal deposits.

Figure 21.22 Native copper from northern Michigan's Keweenaw Peninsula is an excellent example of a hydrothermal deposit. At one time this area was an important source of copper, but it is now largely depleted. (Photo by E. J. Tarbuck)

Diamonds originate at depths of nearly 200 kilometers, where the confining pressure is great enough to generate this high-pressure form of carbon. Once crystallized, the diamonds are carried upward through pipe-shaped conduits that increase in diameter toward the surface. In diamond-bearing pipes, nearly the entire pipe contains diamond crystals that are disseminated throughout an ultramafic rock called *kimberlite*. The most productive kimberlite pipes are those found in South Africa. The only equivalent source of diamonds in the United States is located near Murfreesboro, Arkansas, but this deposit is exhausted and serves today merely as a tourist attraction.

Hydrothermal Solutions

Among the best-known and most important ore deposits are those generated from **hydrothermal** (hot-water) **solutions**. Included in this group are the gold deposits of the Homestake mine in South Dakota; the lead, zinc, and silver ores near Coeur d'Alene, Idaho; the silver deposits of the Comstock Lode in Nevada; and the copper ores of the Keweenaw Peninsula in Michigan (Figure 21.22).

The majority of hydrothermal deposits originate from hot, metal-rich fluids that are remnants of late-stage magmatic processes. During solidification, liquids plus various metallic ions accumulate near the top of the magma chamber. Because of their mobility, these ion-rich solutions can migrate great distances through the surrounding rock before they are eventually deposited, usually as sulfides of various metals (Figure 21.21). Some of this fluid moves along openings such as fractures or bedding planes, where it cools and precipitates the metallic ions to produce **vein deposits** (Figure 21.23). Many of the most productive deposits of gold, silver, and mercury occur as hydrothermal vein deposits.

Another important type of accumulation generated by hydrothermal activity is called a **disseminated deposit**. Rather than being concentrated in narrow veins and dikes, these ores are distributed as minute masses throughout the entire rock mass. Much of the world's copper is extracted from disseminated deposits, including those at Chuquicamata, Chile, and the huge Bingham Canyon copper mine in Utah (see Box 21.2). Because these accumulations contain only 0.4 to 0.8 percent copper, between 125 and 250 kilograms of ore must be mined for every kilogram of metal recovered. The environmental

Figure 21.23 Gneiss laced with quartz veins at Diablo Lake Overlook, North Cascades National Park, Washington.

Figure 21.24 Some hydrothermal deposits are formed by the circulation of heated groundwater in regions where magma is near the surface as in the Yellowstone National Park area. (Photo by Craig J. Brown/Liaison International)

impact of these large excavations, including the problem of waste disposal, is significant.

Some hydrothermal deposits have been generated by the circulation of ordinary groundwater in regions where magma was emplaced near the surface. The Yellowstone National Park area is a modern example of such a situation (Figure 21.24). When groundwater invades a zone of recent igneous activity, its temperature rises, greatly enhancing its ability to dissolve minerals. These migrating hot waters remove metallic ions from intrusive igneous rocks and carry them upward, where they may be deposited as an ore body. Depending on the conditions, the resulting accumulations may occur as vein deposits, as disseminated deposits or, where hydrothermal solutions reach the surface in the form of hot springs or geysers, as surface deposits.

With the development of the plate tectonics theory it became clear that some hydrothermal deposits originated along ancient oceanic ridges. A well-known example is found on the island of Cyprus, where copper has been mined for more than 4000 years. Apparently, these deposits represent ores that formed on the seafloor at an ancient oceanic spreading center.

Since the mid-1970s, active hot springs and metal-rich sulfide deposits have been detected at several sites, including study areas along the East Pacific Rise and the Juan de Fuca Ridge. The deposits are forming where heated seawater, rich in dissolved metals and sulfur, gushes from the seafloor as particle-filled clouds called *black smokers*. As shown in Figure 21.25, seawater infiltrates the hot oceanic crust along the flanks of the ridge. As the water moves through the newly formed material, it is heated and chemically interacts with the basalt,

extracting and transporting sulfur, iron, copper, and other metals. Near the ridge axis, the hot, metal-rich fluid rises along faults. Upon reaching the seafloor, the spewing liquid mixes with the cold seawater, and the sulfides precipitate to form massive sulfide deposits.

Mineral Resources and Metamorphic Processes

The role of metamorphism in producing mineral deposits is frequently tied to igneous processes. For example, many of the most important metamorphic ore deposits are produced by contact metamorphism. Here the host rock is recrystallized and chemically altered by heat, pressure, and hydrothermal solutions emanating from an intruding igneous body. The extent to which the host rock is altered depends on the nature of the intruding igneous mass as well as the nature of the host rock.

Some resistant materials, such as quartz sandstone, may show very little alteration, whereas others, including limestone, might exhibit the effects of metamorphism for several kilometers from the igneous pluton. As hot, ion-rich fluids move through limestone, chemical reactions take place, which produce useful minerals such as garnet and corundum. Further, these reactions release carbon dioxide, which greatly facilitates the outward migration of metallic ions. Thus, extensive aureoles of metal-rich deposits commonly surround igneous plutons that have invaded limestone strata.

The most common metallic minerals associated with contact metamorphism are sphalerite (zinc), galena (lead), chalcopyrite (copper), magnetite (iron),

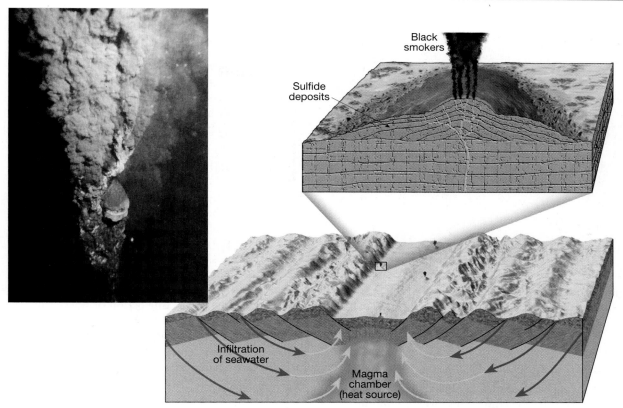

Figure 21.25 Massive sulfide deposits can result from the circulation of seawater through the oceanic crust along active spreading centers. As seawater infiltrates the hot basaltic crust, it leaches sulfur, iron, copper, and other metals. The hot, enriched fluid returns to the seafloor near the ridge axis along faults and fractures. Some metal sulfides may be precipitated in these channels as the rising fluid begins to cool. When the hot liquid emerges from the seafloor and mixes with cold seawater, the sulfides precipitate to form massive deposits. Photo shows a close-up view of a black smoker spewing hot, mineral-rich seawater along the East Pacific Rise. (Photo by Dudley Foster, Woods Hole Oceanographic Institution)

and bornite (copper). The hydrothermal ore deposits may be disseminated throughout the altered zone or exist as concentrated masses that are located either next to the intrusive body or at the periphery of the metamorphic zone.

Regional metamorphism can also generate useful mineral deposits. Recall that at convergent plate boundaries the oceanic crust, along with sediments that have accumulated at the continental margins, are carried to great depths. In these high-temperature, high-pressure environments the mineralogy and texture of the subducted materials are altered, producing deposits of nonmetallic minerals such as talc and graphite.

Weathering and Ore Deposits

Weathering creates many important mineral deposits by concentrating minor amounts of metals that are scattered through unweathered rock into economically valuable concentrations. Such a transformation is often termed **secondary enrichment** and takes place in one of two ways. In one situation chemical weathering coupled with downward-percolating water removes undesirable materials from decomposing rock, leaving the desirable elements enriched in the upper zones of the soil. The second way is basically the reverse of the first. That is, the desirable elements that are found in low concentrations near the surface are removed and carried to lower zones, where they are redeposited and become more concentrated.

Bauxite

The formation of *bauxite*, the principal ore of aluminum, is one important example of an ore created as a result of enrichment by weathering processes (Figure 21.26). Although aluminum is the third most abundant element in Earth's crust, economically valuable concentrations of this important metal are not common, because most aluminum is tied up in silicate minerals from which it is extremely difficult to extract.

Bauxite forms in rainy tropical climates in association with laterites. (In fact, bauxite is sometimes referred to as aluminum laterite.) When aluminum-rich source rocks are subjected to the intense and prolonged

Figure21.26 Bauxite is the ore of aluminum and forms as a result of weathering processes under tropical conditions. Its color varies from red or brown to nearly white. (Photo by E. J. Tarbuck)

chemical weathering of the tropics, most of the common elements, including calcium, sodium, and silicon, are removed by leaching. Because aluminum is extremely insoluble, it becomes concentrated in the soil as bauxite, a hydrated aluminum oxide. Thus, the formation of bauxite depends both on climatic conditions in which chemical weathering and leaching are pronounced and on the presence of an aluminum-rich source rock. Important deposits of nickel and cobalt are also found in laterite soils that develop from igneous rocks rich in ferromagnesian silicate minerals.

Other Deposits

Many copper and silver deposits result when weathering processes concentrate metals that are deposited through a low-grade primary ore. Usually such enrichment occurs in deposits containing pyrite (FeS_2), the most common and widespread sulfide mineral. Pyrite is important because when it chemically weathers, sulfuric acid forms, which enables percolating waters to dissolve the ore metals. Once dissolved, the metals gradually migrate downward through the primary ore body until they are precipitated. Deposition takes place because of changes that occur in the chemistry of the solution when it reaches the groundwater zone (the zone beneath the surface where all pore spaces are filled with water). In this manner the small percentage of dispersed metal can be removed from a large volume of rock and redeposited as a higher-grade ore in a smaller volume of rock.

This enrichment process is responsible for the economic success of many copper deposits, including one located in Miami, Arizona. Here the ore was upgraded from less than 1 percent copper in the primary deposit to as much as 5 percent copper in some localized zones of enrichment. When pyrite weathers (oxidizes) near the surface, residues of iron oxide remain. The presence of these rusty-colored caps at the surface indicates the possibility of an enriched ore below, and this represents a visible sign for prospectors.

Placer Deposits

Sorting typically results in like-sized grains being deposited together. However, sorting according to the specific gravity of particles also occurs. This latter type of sorting is responsible for the creation of **placers**, which are deposits formed when heavy minerals are mechanically concentrated by currents. Placers associated with streams are among the most common and best known, but the sorting action of waves can also create placers along the shore. Placer deposits usually involve minerals that are not only heavy but also tough and chemically resistant enough to withstand destruction by weathering processes and transporting currents. Placers form because heavy minerals settle quickly from a current, whereas less dense particles remain suspended and are carried onward. Common sites of accumulation include point bars on the insides of meanders as well as cracks, depressions, and other irregularities on stream beds.

Many economically important placer deposits exist, with accumulations of gold the best known. Indeed, it was the placer deposits discovered in 1848 that led to the famous California gold rush. Years later, similar deposits created a gold rush to Alaska as well (Figure 21.27). Panning for gold by washing sand and gravel from a flat pan to concentrate the fine "dust" at the bottom was a common method used by early prospectors to recover the precious metal, and it is a process similar to that which created the placer in the first place.

In addition to gold, other heavy and durable minerals form placers. These include platinum, diamonds, and tin. The Ural Mountains contain placers rich in platinum, and placers are important sources of diamonds in southern Africa. Significant portions of the world's supply of cassiterite, the principal ore of tin, have come from placer deposits in Malaysia and Indonesia. Cassiterite is often widely disseminated through granitic igneous rocks. In this state, the mineral is not sufficiently concentrated to be extracted profitably. However, as the enclosing rock dissolves and disintegrates, the heavy and durable cassiterite grains are set free. Eventually the freed particles are washed to a stream where they are deposited in placers that are significantly more concentrated than the original deposit. Similar circumstances and events are common to many minerals mined from placers.

Figure 21.27 **A.** It was placer deposits that led to the 1848 California gold rush. Here, a prospector in 1850 swirls his gold pan, separating sand and mud from flecks of gold. (Photo courtesy of Seaver Center for Western History Research, Los Angeles County Museum of Natural History) **B.** Modern gold mining of a placer deposit near Nome, Alaska. The dredge scoops up thawed gravel. (Photo by Fred Bruemmer)

A.

B.

In some cases, if the source rock for a placer deposit can be located, it too may become an important ore body. By following placer deposits upstream, one can sometimes locate the original deposit. This is how the gold-bearing veins of the Mother Lode in California's Sierra Nevada batholith were found, as well as the famous Kimberly diamond mines of South Africa. The placers were discovered first, their source at a later time.

Nonmetallic Mineral Resources

Earth materials that are not used as fuels or processed for the metals they contain are referred to as **nonmetallic mineral resources**. Realize that use of the word "mineral" is very broad in this economic context, and is quite different from the geologist's strict definition of mineral found in Chapter 2. Nonmetallic mineral resources are extracted and processed either for the nonmetallic elements they contain or for the physical and chemical properties they possess.

People often do not realize the importance of nonmetallic minerals because they see only the products that resulted from their use and not the minerals themselves. That is, many nonmetallics are used up in the process of creating other products. Examples include the fluorite and limestone that are part of the steelmaking process, the abrasives required to make a piece of machinery, and the fertilizers needed to grow a food crop (Table 21.2).

The quantities of nonmetallic minerals used each year are enormous. A glance at Figure 21.2 (p. 537) reminds us that the per capita consumption of nonfuel resources in the United States totals more than 11 metric tons, of which about 94 percent are nonmetallics. Nonmetallic mineral resources are commonly divided into two broad groups—*building materials* and *industrial minerals*. Because some substances have many different uses, they are found in both categories. Limestone, perhaps the most versatile and widely used rock of all, is the best example. As a building material, it is used not only as crushed rock and building stone, but also in the making of cement. Moreover, as an industrial mineral, limestone is an ingredient in the manufacture of steel and is used in agriculture to neutralize soils.

Building Materials

Natural aggregate consists of crushed stone, sand, and gravel. From the standpoint of quantity and value, aggregate is a very important building material. The United States produces nearly two billion tons of aggregate per year, which represents about one-half

Table 21.2 Occurrences and Uses of Nonmetallic Minerals

Mineral	Uses	Geological Occurrences
Apatite	Phosphorus fertilizers	Sedimentary deposits
Asbestos (chrysotile)	Incombustible fibers	Metamorphic alteration
Calcite	Aggregate; steelmaking; soil conditioning; chemicals; cement; building stone	Sedimentary deposits
Clay minerals (kaolinite)	Ceramics; china	Residual product of weathering
Corundum	Gemstones; abrasives	Metamorphic deposits
Diamond	Gemstones; abrasives	Kimberlite pipes; placers
Fluorite	Steelmaking; aluminum refining; glass; chemicals	Hydrothermal deposits
Garnet	Abrasives; gemstones	Metamorphic deposits
Graphite	Pencil lead; lubricant; refractories	Metamorphic deposits
Gypsum	Plaster of Paris	Evaporite deposits
Halite	Table salt; chemicals; ice control	Evaporite deposits; salt domes
Muscovite	Insulator in electrical applications	Pegmatites
Quartz	Primary ingredient in glass	Igneous intrusions; sedimentary deposits
Sulfur	Chemicals; fertilizer manufacture	Sedimentary deposits; hydrothermal deposits
Sylvite	Potassium fertilizers	Evaporite deposits
Talc	Powder used in paints, cosmetics, etc.	Metamorphic deposits

of the entire nonenergy mining volume in the country. It is produced commercially in every state and is used in nearly all building construction and in most public works projects (Figure 21.28).

Besides aggregate, other important building materials include gypsum for plaster and wallboard, clay for tile and bricks, and cement, which is made from limestone and shale. Cement and aggregate go into the making of concrete, a material that is essential to practically all construction. Aggregate gives concrete its strength and volume, and cement binds the mixture into a rock-hard substance. Just two kilometers of four-lane highway require more than 85 metric tons of aggregate. On a smaller scale, 90 tons of aggregate are needed just to build an average six-room house.

Because most building materials are widely distributed and present in almost unlimited quantities, they have little intrinsic value. Their economic worth comes only after the materials are removed from the ground and processed. As their per-ton value compared with metals and industrial minerals is low, mining and quarrying operations are usually undertaken to satisfy local needs. Except for special types of cut stone used for buildings and monuments, transportation costs greatly limit the distances that most building materials can be moved.

Industrial Minerals

Many nonmetallic resources are classified as industrial minerals. In some instances these materials are important because they are sources of specific chemical elements or compounds. Such minerals are used in the manufacture of chemicals and the production of fertilizers. In other cases their importance is related to the physical properties they exhibit. Examples include minerals such as corundum and garnet, which are used as abrasives. Although supplies are generally plentiful, most industrial minerals are not nearly as abundant as building materials. Moreover, deposits are far more restricted in distribution and extent. As a result, many of these nonmetallic resources must be transported considerable distances, which, of course, adds to their cost. Unlike most building materials, which need a minimum of processing before they are ready to use, many industrial minerals require considerable processing to extract the desired substance at the proper degree of purity for its ultimate use.

Fertilizers. The growth in world population toward seven billion by the year 2005 requires that the production of basic food crops continue to expand. Therefore fertilizers, primarily nitrate, phosphate, and potassium compounds, are extremely important to agriculture. The synthetic nitrate industry, which

Figure 21.28 Crushed stone and sand and gravel are primarily used for aggregate in the construction industry, especially in cement concrete for residential and commercial buildings, bridges, and airports, and as cement concrete or bituminous concrete (asphalt) for highway construction. A large percentage is used without a binder as road base, for road surfacing, and as railroad ballast. (Photo by Robert Ginn/PhotoEdit)

derives nitrogen from the atmosphere, is the source of practically all the world's nitrogen fertilizers. The primary source of phosphorus and potassium, however, remains Earth's crust. The mineral apatite is the primary source of phosphate. In the United States most production comes from marine sedimentary deposits in Florida and North Carolina. Although potassium is an abundant element in many minerals, the primary commercial sources are evaporite deposits containing the mineral sylvite. In the United States, deposits near Carlsbad, New Mexico, have been especially important.

Sulfur. Because of its diverse uses, sulfur is an important nonmetallic resource. In fact, the quantity of sulfur used is considered one index of a country's level of industrialization. More than 80 percent is used to produce sulfuric acid. Although its principal use is in the manufacture of phosphate fertilizer, sulfuric acid has a large number of other applications as well. Sources include deposits of native sulfur associated with salt domes and volcanic areas, as well as common

iron sulfides such as pyrite. In recent years an increasingly important source has been the sulfur removed from coal, oil, and natural gas in order to make these fuels less polluting.

Salt. Common salt, known by the mineral name *halite*, is another important and versatile resource. It is among the more prominent nonmetallic minerals used as a raw material in the chemical industry. In addition, large quantities are used to "soften" water and to keep streets and highways free of ice. Of course most people are aware that it is also a basic nutrient and a part of many food products.

Salt is a common evaporite, and thick deposits are exploited using conventional underground mining techniques (Figure 21.29). Subsurface deposits are also tapped using brine wells in which a pipe is introduced into a salt deposit and water is pumped down the pipe. The salt dissolved by the water is brought to the surface through a second pipe. In addition, seawater continues to serve as a source of salt as it has for centuries. The salt is harvested after the Sun evaporates the water. Box 6.1 (p. 153) looks at this process more closely.

Figure 21.29 Common salt is an important and versatile nonmetallic resource. The thick deposit at Grand Saline, Texas, is being exploited using conventional underground mining techniques. (Photo by Morton Salt Company, provided by John S. Shelton)

Chapter Summary

- *Renewable resources* can be replenished over relatively short time spans. Examples include natural fibers for clothing and trees for lumber. *Nonrenewable resources* form so slowly that from a human standpoint, Earth contains fixed supplies. Examples include fuels such as oil and coal, and metals such as copper and gold. A rapidly growing world population and the desire for an improved living standard cause nonrenewable resources to become depleted at an increasing rate.

- *Coal, petroleum,* and *natural gas,* the *fossil fuels* of our modern economy, are all associated with sedimentary rocks. Coal originates from large quantities of plant remains that accumulate in an oxygen-deficient environment, such as a swamp. More than 70 percent of present-day coal usage is for the generation of electricity. Air pollution produced by the sulfur oxide gases that form from burning most types of coal is a significant environmental problem.

- Oil and natural gas, which commonly occur together in the pore spaces of some sedimentary rocks, consist of various *hydrocarbon compounds* (compounds made of hydrogen and carbon) mixed together. Petroleum formation is associated with the accumulation of sediment in ocean areas that are rich in plant and animal remains that become buried and isolated in an oxygen-deficient environment. As the mobile petroleum and natural gas form, they migrate and accumulate in adjacent permeable beds such as sandstone. If the upward migration is halted by an impermeable rock layer, referred to as a *cap rock,* a geologic environment that allows for economically significant amounts of oil and gas to accumulate underground, called an *oil trap,* develops. The two basic conditions common to all oil traps are (1) a porous, permeable *reservoir rock* that will yield petroleum and/or natural gas in sufficient quantities, and (2) a cap rock.

- Environmental problems associated with burning fossil fuels include urban air pollution and global warming. The *primary pollutants* emitted by sources such as motor vehicles can react in the atmosphere to produce the *secondary pollutants* that make up urban smog. Combustion of fossil fuels is one of the ways that humans are increasing the atmosphere's carbon dioxide content. Greater quantities of this heat-absorbing gas could lead to global warming.

- When conventional petroleum resources are no longer adequate, fuels derived from *tar sands* and *oil shale* may become substitutes. Presently, tar sands from the province of Alberta are the source of about 15 percent of Canada's oil production. Oil from oil shale is presently uneconomical to produce. Oil production from both tar sands and oil shale has significant environmental drawbacks.

- Nearly 90 percent of our energy is derived from fossil fuels. In the United States the most important alternative energy source is *nuclear energy,* which provides about 7 percent of our demand. *Hydroelectric power* ranks next, providing about 5 percent. Other alternative energy sources are locally important but collectively provide about 1 percent of the U.S. energy demand. These include *solar power, geothermal energy, wind energy,* and *tidal power.*

- *Mineral resources* are the endowment of useful minerals ultimately available commercially. Resources include already identified deposits from which minerals can be extracted profitably, called *reserves,* as well as known deposits that are not yet economically or technologically recoverable. Deposits inferred to exist, but not yet discovered, are also considered as mineral resources. The term *ore* is used to denote those useful metallic minerals that can be mined for a profit, as well as some nonmetallic minerals, such as fluorite and sulfur, that contain useful substances.

- Some of the most important accumulations of metals, such as gold, silver, lead, and copper, are produced by igneous processes. The best-known and most important ore deposits are generated from *hydrothermal* (hot-water) *solutions.* Hydrothermal deposits originate from hot, metal-rich fluids that are remnants of late-stage magmatic processes. These ion-rich solutions move along fractures or bedding planes, cool, and precipitate the metallic ions to produce *vein deposits.* In a *disseminated deposit* (e.g., much of the world's copper deposits) the ores from hydrothermal solutions are distributed as minute masses throughout the entire rock mass.

- Many of the most important metamorphic ore deposits are produced by contact metamorphism. Extensive aureoles of metal-rich deposits commonly surround igneous bodies where ions have invaded limestone strata. The most common metallic minerals associated with contact metamorphism are sphalerite (zinc), galena (lead), chalcopyrite (copper), magnetite (iron), and bornite (copper). Of equal economic importance are the metamorphic rocks themselves. In many regions, slate, marble, and quartzite are quarried for a variety of construction purposes.

- Weathering creates ore deposits by concentrating minor amounts of metals into economically valuable deposits. The process, often called *secondary enrichment*, is accomplished by either (1) removing undesirable materials and leaving the desired elements enriched in the upper zones of the soil, or (2) removing and carrying the desirable elements to lower zones where they are redeposited and become more concentrated. *Bauxite*, the principal ore of aluminum, is one important ore created as a result of enrichment by weathering processes. In addition, many copper and silver deposits result when weathering processes concentrate metals that were formerly dispersed through low-grade primary ore.

- Earth materials that are not used as fuels or processed for the metals they contain are referred to as *nonmetallic resources*. Many are sediments or sedimentary rocks. The two broad groups of nonmetallic resources are *building materials* and *industrial minerals*. Limestone, perhaps the most versatile and widely used rock of all, is found in both groups.

Review Questions

1. Contrast renewable and nonrenewable resources. Give one or more examples of each.

2. What is the estimated world population for the year 2005? How does this compare to the figures for 1930 and 1975? Is demand for resources growing as rapidly as world population?

3. More than 70 percent of present-day coal usage is for what purpose?

4. Describe two impacts on the atmospheric environment of burning fossil fuels.

5. What is an oil trap? List two conditions common to all oil traps.

6. List two drawbacks associated with the processing of tar sands recovered by surface mining.

7. The United States has huge oil shale deposits, but does not produce shale oil commercially. Explain.

8. What is the main fuel for nuclear fission reactors?

9. List two obstacles that have hindered the development of nuclear power as a major energy source.

10. Briefly describe two methods by which solar energy might be used to produce electricity.

11. Explain why dams built to provide hydroelectricity do not last indefinitely.

12. Is geothermal power considered an inexhaustible energy source? Explain.

13. What advantages does tidal power production offer? Is it likely that tides will ever provide a significant proportion of the world's electrical energy requirements?

14. Contrast *resource* and *reserve*.

15. What might cause a mineral deposit that had not been considered an ore to be reclassified as an ore?

16. List two general types of hydrothermal deposits.

17. Metamorphic ore deposits are often related to igneous processes. Provide an example.

18. Name the primary ore of aluminum and describe its formation.

19. A rusty-colored zone of iron oxide at the surface may indicate the presence of a copper deposit at depth. Briefly explain.

20. Briefly describe the way in which minerals accumulate in placers. List four minerals that are mined from such deposits.

21. Which is greater, the per capita consumption of metallic or nonmetallic mineral resources?

22. Nonmetallic resources are commonly divided into two broad groups. List the two groups and some examples of materials that belong to each. Which group is most widely distributed?

Key Terms

cap rock (p. 540)
disseminated deposit (p. 557)
fossil fuel (p. 538)
geothermal energy (p. 551)
hydroelectric power (p. 550)

hydrothermal solution (p. 557)
mineral resource (p. 552)
nonmetallic mineral resource (p. 561)
nonrenewable resource (p. 536)

nuclear fission (p. 547)
oil trap (p. 540)
ore (p. 553)
pegmatite (p. 556)
placer (p. 560)
renewable resource (p. 536)

reserve (p. 553)
reservoir rock (p. 540)
secondary enrichment (p. 559)
vein deposit (p. 557)

Web Resources

The *Earth* Home Page provides on-line resources for this chapter on the World Wide Web. You will find review exercises, specific updates for items in the chapter, suggested reading, and links to interesting related pathways on the Internet. Visit the *Earth* Home Page at **http://www.prenhall.com/tarbuck**.

CHAPTER 22

Planetary Geology

Left Montage of Saturn and some of its satellites. —
NASA Image/Photo Researchers, Inc.

The Sun is the hub of a huge rotating system consisting of nine planets, their satellites, and numerous small but interesting bodies, including asteroids, comets, and meteoroids. An estimated 99.85 percent of the mass of our solar system is contained within the Sun, while the planets collectively make up most of the remaining 0.15 percent. The planets, in order from the Sun, are Mercury, Venus, Earth, Mars, Jupiter, Saturn, Uranus, Neptune, and Pluto (Figure 22.1).

Under the control of the Sun's gravitational force, each planet maintains an elliptical orbit and all of them travel in the same direction. The nearest planet to the sun, Mercury, has the fastest orbital motion, 48 kilometers per second, and the shortest period of revolution, 88 days. By contrast, the most distant planet, Pluto, has an orbital speed of 5 kilometers per second and requires 248 years to complete one revolution.

Imagine a planet's orbit drawn on a flat sheet of paper. The paper represents the planet's *orbital plane*. The orbital planes of all nine planets lie within 3 degrees of the plane of the Sun's equator, except for those of Mercury and Pluto, which are inclined 7 and 17 degrees, respectively.

When people first came to recognize that the planets are "worlds" much like Earth, a great deal of interest was generated. A primary concern has always been the possibility of intelligent life existing elsewhere in the universe. This expectation has not as yet come to pass. Nevertheless, as all of the planets most probably formed from the same primordial cloud of dust and gases, they should provide valuable information concerning Earth's history. Recent space explorations have been organized with this goal in mind. To date, Mercury, Venus, Mars, Jupiter, Saturn, Uranus, Neptune, and the Moon have been explored by space probes (Figure 22.2).

The Planets: An Overview

Careful examination of Table 22.1 shows that the planets fall quite nicely into two groups: the **terrestrial** (Earth-like) **planets** (Mercury, Venus, Earth, and Mars) and the **Jovian** (Jupiter-like) **planets** (Jupiter, Saturn, Uranus, and Neptune). Pluto is not included in either category, because its great distance from Earth and its small size make this planet's true nature a mystery.

The most obvious difference between the terrestrial and the Jovian planets is their size. The largest terrestrial planet (Earth) has a diameter only one-quarter as great as the diameter of the smallest Jovian planet (Neptune), and its mass is only one-seventeenth as great. Hence, the Jovian planets are often called *giants*. Also, because of their relative locations, the four Jovian planets are referred to as the *outer planets*, while the terrestrial planets are called the *inner planets*. As we shall see, there appears to be a correlation between the positions of these planets and their sizes.

Other dimensions along which the two groups markedly differ include density, composition, and rate of rotation. The densities of the terrestrial planets average about 5 times the density of water, whereas the Jovian planets have densities that average only 1.5 times that of water. One of the outer planets, Saturn, has a density only 0.7 that of water, which means that Saturn would float in water. Variations in the compositions of the planets are largely responsible for the density differences.

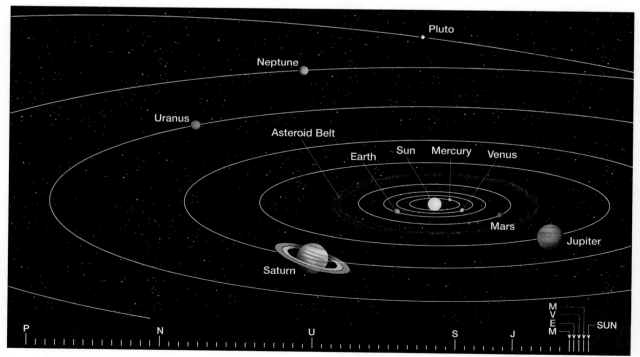

Figure 22.1 Orbits of the planets (not to scale).

Figure 22.2 Painting of *Voyager 2* as it might have appeared when it encountered Uranus on January 24, 1986. (By Don Davis, courtesy of NASA)

The substances that make up both groups of planets are divided into three groups—*gases, rocks,* and *ices*—based on their melting points.

1. The gases, hydrogen and helium, are those with melting points near absolute zero (0 Kelvin or –273°C), the lowest possible temperature.

2. The rocks are principally silicate minerals and metallic iron, which have melting points exceeding 700°C.

3. The ices have intermediate melting points (for example, H_2O has a melting point of 0°C) and include ammonia (NH_3), methane (CH_4), carbon dioxide (CO_2), and water (H_2O).

The terrestrial planets are mostly rocks: dense rocky and metallic material, with minor amounts of gases. The Jovian planets, on the other hand, contain a large percentage of gases (hydrogen and helium), with varying amounts of ices (mostly water, ammonia, and methane). This accounts for their low densities. (The outer planets may contain as much rocky and metallic material as the terrestrial planets, but this material would be concentrated in their central cores.)

The Jovian planets have very thick atmospheres consisting of varying amounts of hydrogen, helium, methane, and ammonia. By comparison, the terrestrial planets have meager atmospheres at best. A planet's ability to retain an atmosphere depends on its temperature and mass. Simply stated, a gas molecule can "evaporate" from a planet if it reaches a speed known as the **escape velocity**. For Earth, this velocity is 11 kilometers (7 miles) per second. Any material, including a rocket, must reach this speed before it can leave Earth and go into space.

The Jovian planets, because of their greater surface gravities, have higher escape velocities (21–60 km/sec) than the terrestrial planets. Consequently, it is more difficult for gases to "evaporate" from them. Also, because the molecular motion of a gas is temperature-dependent, at the low temperatures of the Jovian planets even the lightest gases are unlikely to acquire the speed needed to escape.

On the other hand, a comparatively warm body with a small surface gravity, like our Moon, is unable to hold even the heaviest gas and thus lacks an atmosphere. The slightly larger terrestrial planets of Earth, Venus, and Mars retain some heavy gases like carbon

Table 22.1 Planetary Data

		Mean Distance from Sun					Orbital Velocity	
Planet	Symbol	AU*	Millions of Miles	Millions of Kilometers	Period of Revolution	Inclination of Orbit	mi/s	km/s
Mercury	☿	0.39	36	58	88^d	7°00′	29.5	47.5
Venus	♀	0.72	67	108	225^d	3°24′	21.8	35.0
Earth	⊕	1.00	93	150	365.25^d	0°00″	18.5	29.8
Mars	♂	1.52	142	228	687^d	1°51′	14.9	24.1
Jupiter	♃	5.20	483	778	12^{yr}	1°18′	8.1	13.1
Saturn	♄	9.54	886	1427	29.5^{yr}	2°29′	6.0	9.6
Uranus	⛢	19.18	1783	2870	84^{yr}	0°46′	4.2	6.8
Neptune	♆	30.06	2794	4497	165^{yr}	1°46′	3.3	5.3
Pluto	♇	39.44	3666	5900	248^{yr}	17°12′	2.9	4.7

		Diameter						
Planet	Period of Rotation	Miles	Kilometers	Relative Mass (Earth = 1)	Average Density (g/cm^3)	Polar Flattening (%)	Eccentricity	Number of Known Satellites
Mercury	59^d	3015	4878	0.06	5.4	0.0	0.206	0
Venus	244^d	7526	12,104	0.82	5.2	0.0	0.007	0
Earth	$23^h56^m04^s$	7920	12,756	1.00	5.5	0.3	0.017	1
Mars	$24^h37^m23^s$	4216	6794	0.11	3.9	0.5	0.093	2
Jupiter	9^h50^m	88,700	143,884	317.87	1.3	6.7	0.048	16
Saturn	10^h14^m	75,000	120,536	95.14	0.7	10.4	0.056	21
Uranus	17^h14^m	29,000	51,118	14.56	1.2	2.3	0.047	15
Neptune	16^h03^m	28,900	50,530	17.21	1.7	1.8	0.009	8
Pluto	6.4^d	~1500	2445	0.002	1.8	0.0	0.250	1

*AU = astronomical unit, Earth's mean distance from the Sun.

dioxide, but even their atmospheres make up only an infinitesimally small portion of their total mass.

It is hypothesized that the primordial cloud of dust and gas from which all the planets are thought to have condensed had a composition somewhat similar to that of Jupiter. However, unlike Jupiter, the terrestrial planets today are nearly void of light gases and ices. Were the terrestrial planets once much larger? Did they contain these materials but lose them because of their relative closeness to the Sun? In the following section we will consider the evolutionary histories of these two diverse groups of planets in an attempt to answer these questions.

Origin and Evolution of the Planets

The orderly nature of our solar system leads most astronomers to conclude that the planets formed at essentially the same time and from the same primordial material as the Sun. This material formed a vast cloud of dust and gases called a *nebula*. The **nebular hypothesis** suggests that all bodies of the solar system formed from an enormous nebular cloud consisting of approximately 80 percent hydrogen, 15 percent helium, and a few percent of all the other heavier elements known to exist (Figure 22.3). The heavier substances in this frigid cloud of dust and gases consisted mostly of elements such as silicon, aluminum, iron, and calcium—the substances of today's common rocky materials. Also prevalent were other familiar elements, including oxygen, carbon, and nitrogen.

About five billion years ago, and for reasons not yet fully understood, this huge cloud of minute rocky fragments and gases began to contract under its own gravitational influence (see Figure 1.10, p. 10). The contracting clump of material apparently had some rotational motion. As this slowly rotating cloud gravitationally contracted, it rotated faster and faster for the same reason ice skaters do when they draw their arms toward their bodies. This rotation caused the nebular cloud to assume a disk shape. Within this rotating disk, relatively small contractions, like eddies in a stream, formed the nuclei from which the planets would eventually develop. However, the greatest concentration of material was gravitationally pulled toward the center, forming the *protosun*.

As this gaseous cloud collapsed, the temperature of the central mass continued to increase. Nebular material near the protosun reached temperatures of several thousand degrees and was completely vaporized.

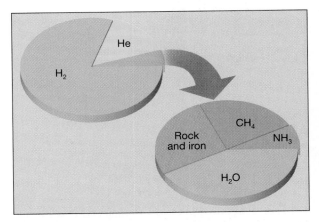

Figure 22.3 Composition of the primordial cloud of dust and gases from which the solar system is thought to have evolved. (Data from W.B. Hubbard)

However, at distances beyond the orbit of Mars, the temperatures probably remained very low, both then and now. Here, at –200°C, the dust fragments were most likely covered with a thick layer of ices made of water, carbon dioxide, ammonia, and methane. The disk-shaped cloud also contained appreciable amounts of the lighter gases, hydrogen and helium, which had not been consumed by the protosun.

In a relatively short time after the protosun formed, the temperature in the inner portion of the nebula dropped significantly. This temperature decrease caused those substances with high melting points to condense into sand-sized particles. Materials such as iron and nickel solidified first. Next to condense were the elements of which the rock-forming minerals are composed—silicon, calcium, iron, and so forth. As these fragments collided, they joined into larger asteroid-sized objects, which, in a few tens of millions of years, accreted into the four inner planets we call Mercury, Venus, Earth, and Mars.

As more and more of the nebular debris was swept up by these *protoplanets*, the inner solar system began to clear, allowing solar radiation to pass through to heat the planets' surfaces. Because of their relatively high temperatures and weak gravitational fields, the inner planets were unable to accumulate much of the lighter components of the nebular cloud. These lighter components—hydrogen, ammonia, methane, and water—were eventually whisked from the inner solar system by the solar winds.

Shortly after the four terrestrial planets formed, the decay of radioactive isotopes within them, plus the heat from the colliding particles, produced at least some melting of the planets' interiors. Melting, in turn, allowed the heavier elements, principally iron and nickel, to sink toward the center, while the lighter silicate minerals floated upward.

During this period of *chemical differentiation*, gaseous materials escaped from the planets' interiors, much like what happens during a volcanic event on Earth. The hottest and second-smallest planet, Mercury, was unable to retain even the heaviest of these gases. Mars, on the other hand, being 40 percent larger and considerably cooler than Mercury, retained a thin layer of carbon dioxide and some water in the form of ice. The largest of the terrestrial planets, Venus and Earth, have surface gravitations strong enough to retain a substantial amount of the heavier gases. However, when compared to the four Jovian planets, the atmospheres of these two terrestrial planets must be looked upon as meager at best.

At the same time that the terrestrial planets were forming, the larger Jovian planets, along with their extensive satellite systems, were also developing. However, because of the frigid temperatures existing far from the Sun, the fragments from which these planets formed contained a high percentage of ices— water, carbon dioxide, ammonia, and methane. Perhaps by random chance, two of the outer planets, Jupiter and Saturn, grew many times larger (by mass) than Uranus and Neptune. (For comparison, Jupiter is 318 and Saturn is 95 times more massive than Earth. However, Uranus and Neptune have masses about 15 and 17 times greater, respectively, than Earth.)

When Jupiter and Saturn reached a certain size, estimated to be about 10 Earth masses, their surface gravitation was sufficient to attract and hold even the lightest materials—hydrogen and helium. It is thought that these gases gravitationally collapsed onto these large protoplanets as they swept through their region of the solar system. Thus, much of their size is attributable to the large envelope of light elements, which exists as a dense liquid below a thick hydrogen-rich atmosphere. Jupiter and Saturn therefore consist of a central core of ices and rock, and a much larger outer envelope containing mostly hydrogen and helium (Figure 22.4A).

By contrast, the smaller Jovian planets, Uranus and Neptune, grew more slowly and contain proportionately much smaller amounts of hydrogen and helium. Nevertheless, hydrogen, methane, and ammonia are still the major constituents of their dense atmospheres. Perhaps a thin outer ocean of hydrogen exists on these planets as well. Thus, Uranus and Neptune are proposed to have a small rocky-iron core and a large mantle of water, ammonia, and methane surrounded by a thin ocean of liquid hydrogen (Figure 22.4B). Consequently, these planets structurally resemble Jupiter and Saturn, but without their large hydrogen-helium envelopes.

In many respects, the development of the outer planets with their large satellite systems roughly parallels

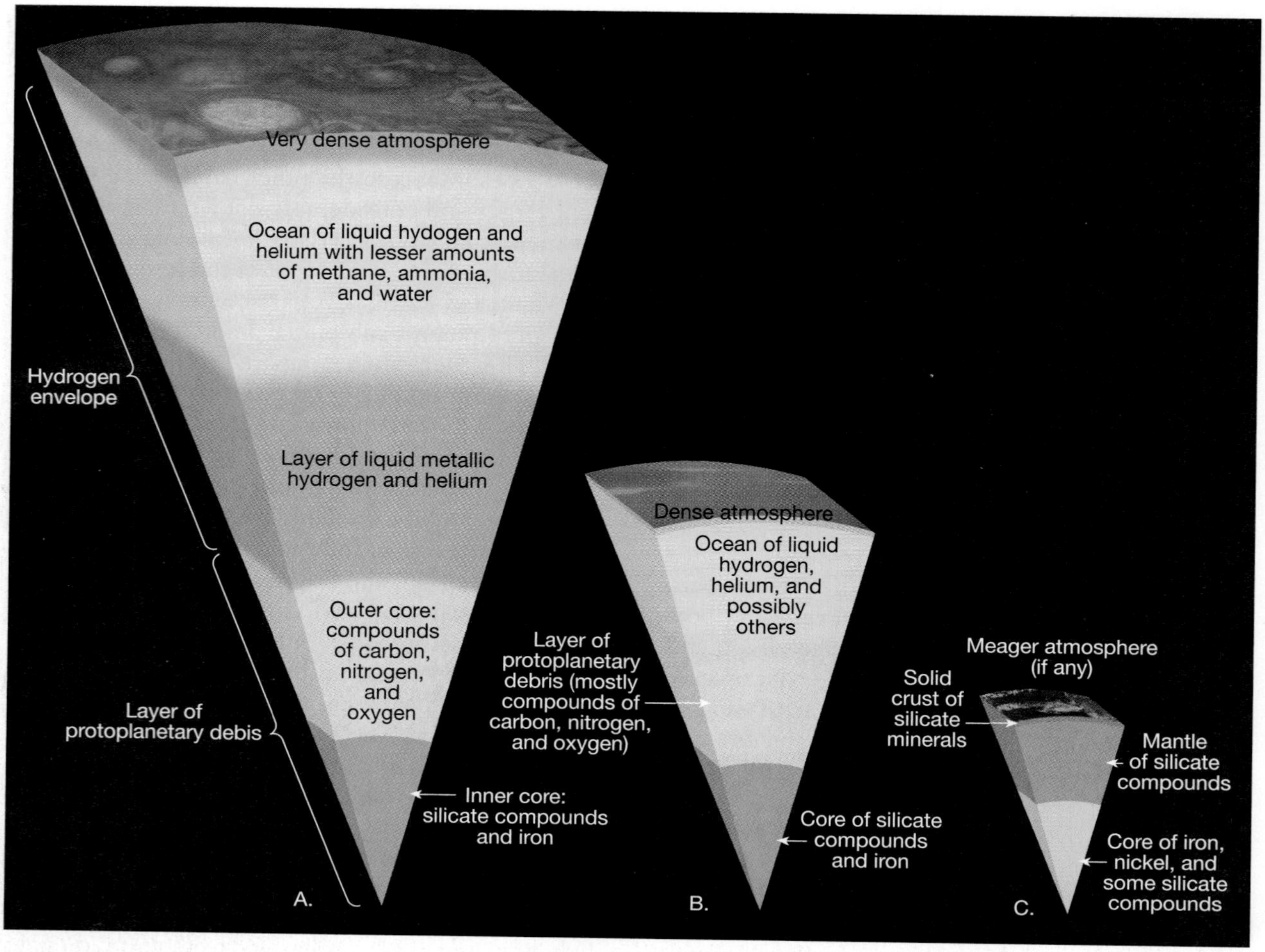

Figure 22.4 Idealized models for the internal structure of the Jovian and terrestrial planets. **A.** Jupiter and Saturn. **B.** Uranus and Neptune. **C.** Terrestrial planets. (Data from W.B. Hubbard et al.)

the events that formed the solar system as a whole. Like their parent planets, the satellites of the outer planets are composed primarily of icy materials with lesser amounts of rocky substances. However, because of their small size, they could not retain appreciable amounts of hydrogen and helium.

In the remainder of this chapter, we will consider each planet in more detail, as well as some minor members of the solar system. First, however, a discussion of our Moon, Earth's companion in space, is appropriate.

Earth's Moon

Earth now has hundreds of satellites, but only one natural satellite, the Moon, accompanies us on our annual journey around the Sun. Although other planets have moons, our planet-satellite system is unique in the solar system, because Earth's moon is unusually large compared to its parent planet. The diameter of the Moon is 3475 kilometers (2150 miles), about one-fourth of the Earth's 12,751 kilometers.

From calculations of the Moon's mass, its density is 3.3 times that of water. This density is comparable to that of *crustal* rocks on Earth but is considerably less than Earth's average density, which is 5.5 times that of water. Geologists have suggested that this difference can be accounted for if the Moon's iron core is small. The gravitational attraction at the lunar surface is one-sixth of that experienced on Earth's surface (a 100-pound person on Earth weighs only 17 pounds on the Moon). This difference allows an astronaut to carry a "heavy" life-support system with relative ease. If not burdened with such a load, an astronaut could jump six times higher than on Earth.

The Lunar Surface

When Galileo first pointed his telescope toward the Moon, he saw two different types of terrain—dark lowlands and bright, cratered highlands (Figure 22.5). Because the dark regions resembled seas on Earth, they were later named **maria**, Latin for "sea" (singular **mare**). This name is unfortunate, because the

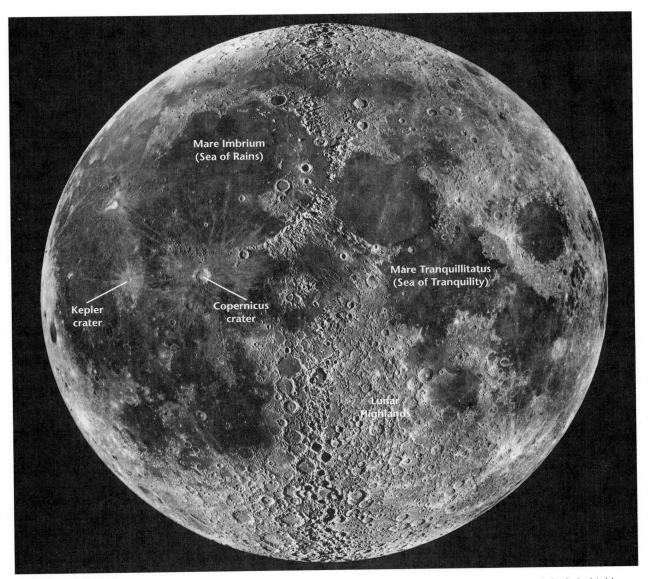

Figure 22.5 Telescopic view from Earth of the lunar surface. The major features are the dark "seas" (maria) and the light highly cratered highlands. (Courtesy of Lick Observatory)

Moon's surface is totally devoid of water. Figure 22.6 shows typical features of the lunar surface.

Today we know that the Moon has no atmosphere or water. Therefore, the weathering and erosion that continually modify Earth's surface are virtually lacking on the Moon. In addition, tectonic forces are not active on the Moon, so earthquakes and volcanic eruptions no longer occur. However, because the Moon is unprotected by an atmosphere, a different kind of erosion occurs: tiny particles from space (micrometeorites) continually bombard its surface and ever-so-gradually smooth the landscape. Moon rocks become slightly rounded on top if they are long exposed at the lunar surface. Nevertheless, it is unlikely that the Moon has changed appreciably in the last three billion years, except for a few craters created by large meteorites.

Craters. The most obvious features of the lunar surface are craters. They are so profuse that craters-within-craters are the rule! The larger ones in the lower portion of Figure 22.5 are about 250 kilometers (150 miles) in diameter, roughly the width of Indiana. Most craters were produced by the impact of rapidly moving debris (meteoroids), a phenomenon that was considerably more common in the early history of the solar system than it is today.

By contrast, Earth has only about a dozen easily recognized impact craters. This difference can be

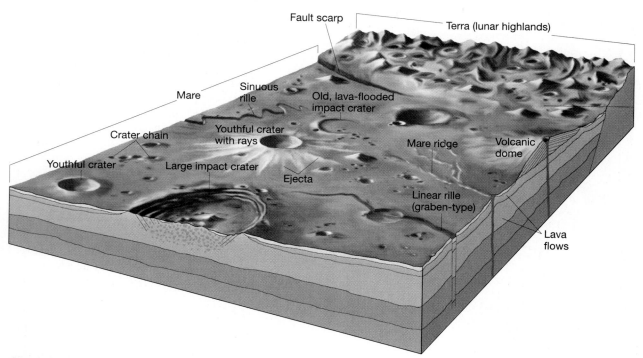

Figure 22.6 Block diagram illustrating major topographic features on the lunar surface.

attributed to Earth's atmosphere. Friction with the air burns up small debris before it reaches the ground. In addition, evidence for most of the craters that formed in Earth's history has been obliterated by erosion or tectonic processes.

The formation of an impact crater is illustrated in Figure 22.7. Upon impact, the high-speed meteoroid compresses the material it strikes, then almost instantaneously the compressed rock rebounds, ejecting material from the crater. This process is analogous to the splash that occurs when a rock is dropped into water, and it often results in the formation of a central peak, as seen in the large crater in Figure 22.8. Most of the ejected material (*ejecta*) lands near the crater, building a rim around it. Heat generated by the impacts is sufficient to melt some of the impacted rock. Astronauts have brought back samples of glass beads produced in this manner, as well as rock formed when angular fragments and dust were welded together by the impact.

A meteoroid only 3 meters (10 feet) in diameter can blast out a 150-meter (500-foot) wide crater. A few of the large craters such as Kepler and Copernicus, shown in Figure 22.5, formed from the impact of bodies 1 kilometer or more in diameter. These two large craters are thought to be relatively young because of the bright *rays* ("splash" marks) that radiate outward for hundreds of kilometers.

Highlands. Densely pockmarked highland areas make up most of the lunar surface. In fact, all of the "back" side of the Moon is characterized by such topography. (Only astronauts have seen the "back" side, because the Moon rotates on its axis once with each revolution around Earth, always keeping the same side facing Earth.) Within the highland regions are mountain ranges. The highest lunar peaks reach elevations approaching 8 kilometers, only 1 kilometer lower than Mount Everest.

Maria. The "seas" of basaltic lava originated when asteroids punctured the lunar surface, letting basaltic magma "bleed" out (Figure 22.9). Apparently the craters were flooded with layer upon layer of very fluid basaltic lava somewhat resembling the Colombia Plateau in the northwestern United States. The lava flows are often over 30 meters (100 feet) thick, and the total thickness of the material that fills the maria must approach thousands of meters.

Regolith. All lunar terrains are mantled with a layer of gray, unconsolidated debris derived from a few billion years of meteoric bombardment (Figure 22.10). This soil-like layer, properly called **lunar regolith**, is composed of igneous rocks, breccia, glass beads, and fine *lunar dust*. In the maria that have been explored by *Apollo* astronauts, the lunar regolith is apparently just over 3 meters (10 feet) thick.

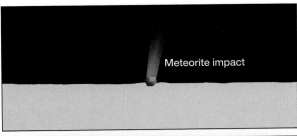

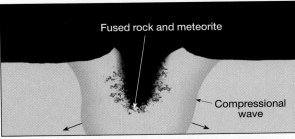

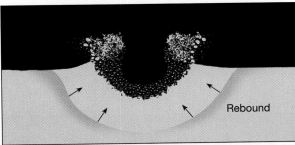

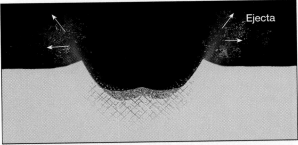

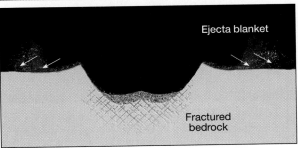

Lunar History

Although the Moon is our nearest planetary neighbor and astronauts have sampled its surface, much is still unknown about its origin. Until recently, most scientists argued that the Moon and Earth formed together. That is, the Moon and Earth consolidated from minute rock fragments and gases that orbited the protosun and accreted into planetary-sized bodies.

A newer hypothesis supported by many scientists suggests that a giant asteroid collided with Earth to produce the Moon. The explosion caused by the impact of a Mars-sized body upon a semimolten Earth would have ejected huge quantities of mantle rock from the primordial Earth. A portion of this ejecta would have remained in orbit around Earth, gradually accumulating to form the Moon. This giant-impact hypothesis is plausible, but raises many questions.

Despite the Moon's uncertain origin, planetary geologists have worked out basic details of its later history. One of their methods is to observe variations in crater density (quantity per unit area). The greater the crater density, the longer the topographic feature must have existed. From such evidence, scientists concluded that the Moon evolved in three phases: the original crust (highlands), maria basins, and rayed craters.

During its early history the Moon was continually impacted as it swept up debris. This continuous bombardment, and perhaps radioactive decay, generated enough heat to melt the Moon's outer shell, and quite possibly the interior as well. Remnants of this original crust occupy the densely cratered highlands, which have been estimated to be as much as 4.5 billion years old—about the same age as Earth.

The second major event in the Moon's evolution was the formation of maria basins (see Figure 22.9). Radiometric dating of the maria basalts puts their age between 3.2 billion and 3.8 billion years, roughly a billion years younger than the initial crust. In places, the lava flows overlap the highlands, another testimonial to the younger age of the maria deposits.

The last prominent features to form were the rayed craters, as exemplified by Copernicus (see Figure 22.5). Material ejected from, these "young" depressions is clearly seen blanketing the surface of the maria and many older rayless craters. Even a relatively young crater like Copernicus must be millions

Figure 22.7 Formation of an impact crater. The energy of the rapidly moving meteoroid is transformed into heat energy and compressional waves. The rebound of the compressed rock causes debris to be ejected from the crater, and the heat melts some material, producing glass beads. Small secondary craters are formed by the material "splashed" from the impact crater. (After E. M. Shoemaker)

Figure 22.8 The 20-kilometer-wide lunar crater Euler in the southwestern part of Mare Imbrium. Clearly visible are the bright rays, central peak, secondary craters, and the large accumulation of ejecta near the crater rim. (Courtesy of NASA)

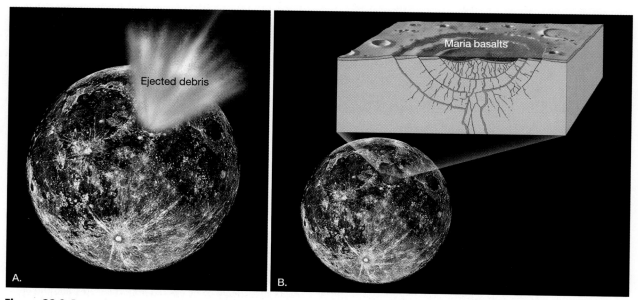

Figure 22.9 Formation of lunar maria. **A.** Impact of an asteroid-sized mass produced a huge crater hundreds of kilometers in diameter and disturbed the lunar crust far beyond the crater. **B.** Filling of the impact area with fluid basalts, perhaps derived from partial melting deep within the lunar mantle.

Figure 22.10 Astronaut Harrison Schmitt sampling the lunar surface. Notice the footprints (insert) in the lunar "soil." (Courtesy of NASA)

of years old. Had it formed on Earth, erosional forces would have long since obliterated it.

If photos of the Moon taken several hundreds of millions of years ago were available, they would reveal that the Moon changed little in the intervening years. By all measures, the Moon is a dead body wandering through space and time.

The Planets: A Brief Tour

Mercury: The Innermost Planet

Mercury, the innermost and second-smallest planet, is hardly larger than Earth's moon and is smaller than three other moons in the solar system. Like the Moon, it absorbs most of the sunlight that strikes it, reflecting only 6 percent into space. This is characteristic of terrestrial bodies that have no atmosphere. (Earth reflects about 30 percent of the light that strikes it, most of it from clouds.)

Mercury's close proximity to the Sun makes viewing from Earthbound telescopes difficult because of the glare. The first good glimpse of this planet came in 1974 when *Mariner 10* passed within 800 kilometers (500 miles) of its surface (Table 22.2). Its striking resemblance to the Moon was immediately evident from the high-resolution images that were sent back (Figure 22.11).

Mercury has cratered highlands, much like the Moon, and vast smooth terrains that resemble maria. However, unlike the Moon, Mercury is a very dense planet, which implies that it contains an iron core, perhaps larger than Earth's. Also, Mercury has very long scarps that cut across the plains and craters alike. These scarps may have resulted from crustal shortening as the planet cooled and shrank.

Mercury revolves quickly but rotates slowly. One full day-night cycle on Earth takes 24 hours but on Mercury it requires 179 Earth-days. Thus, a night on Mercury lasts for about 3 months and is followed by 3 months of daylight. Nighttime temperatures drop as low as –173°C (–280°F) and noontime temperatures exceed 427°C (800°F), hot enough to melt tin and lead. Mercury has the greatest extremes of any planet. The odds of life as we know it existing on Mercury are nil.

Venus: The Veiled Planet

Venus, second only to the Moon in brilliance in the night sky, is named for the goddess of love and beauty. It orbits the Sun in a nearly perfect circle once every 255 Earth-days. Venus is similar to Earth in size, density, mass, and location in the solar system. Thus, it has been referred to as "Earth's twin." Because of these similarities, it is hoped that a detailed study of Venus will provide geologists with a better understanding of Earth's evolutionary history.

Table 22.2 Most Significant Space Probes

Mariner 2	1962	Fly-by of Venus (first to any planet)
Mariner 4, 6, 7	1965	Fly-by missions to Mars
	1969	
	1969	
Mariner 9	1971	Mars orbiter
Apollo 8	1968	Astronauts circled the Moon and returned to Earth
Apollo 11	1969	First astronaut landed on the Moon
Apollo 17	1972	Last of six manned *Apollo* missions to the Moon
Mariner 10	1974	Orbited the Sun, passing Mercury several times; fly-by of Venus
Pioneer 10, 11	1973	First close-up views of Jupiter
	1974	
Venera 8, 9, 10	1972	Soviet landers on Venus (each survived about one hour)
	1975	
Venera 13, 14	1982	First color images of Venus
Viking 1, 2	1976	Mars orbiters and landers
Voyager 1	1979	Fly-by of Jupiter
	1980	Fly-by of Saturn
Voyager 2	1979	Fly-by of Jupiter
	1981	Fly-by of Saturn
	1986	Fly-by of Uranus
	1989	Fly-by of Neptune
Giotto	1986	First photo of comet's nucleus
Magellen	1990	Radar imaging orbiter of Venus
Mars Observer	1993	Mars orbiter
Galileo	1990	Fly-by of Earth, Venus
	1992	Fly-by of Earth, Moon
	1995	Jupiter orbiter
Mars Pathfinder	1997	Mars lander (Sojourner vehicle sampled surface)

Figure 22.11 Photomosaic of Mercury. This view of Mercury is remarkably similar to the "far side" of the Moon. (Courtesy of NASA)

Venus is shrouded in thick clouds impenetrable to visible light. Nevertheless, radar mapping by unmanned spacecraft and Earthbound instruments has revealed a varied topography with features somewhat between those of Earth and Mars. Simply, radar pulses in the microwave range are sent toward the Venusian surface, and the heights of plateaus and mountains are measured by timing the return of the radar echo. These data have confirmed that basaltic volcanism and tectonic deformation are the dominant processes operating on Venus. Further, based on the low density of impact craters, volcanism and tectonic deformation must have been very active during the recent geologic past (Figure 22.12).

Over 80 percent of the Venusian surface consists of subdued plains that are mantled by volcanic flows. Some lava channels extend hundreds of kilometers; one meanders 6800 kilometers across the planet. Thousands of volcanic structures have been identified, mostly small shield volcanoes, although over 1500 volcanoes greater than 20 kilometers across have been mapped. One is Sapas Mons, 400 kilometers (250 miles) across and 1.5 kilometers (0.9 mile) high. Many flows from this volcano erupted from its flanks rather than the summit, in the manner of Hawaiian shield volcanoes. Other volcanic structures discovered on Venus are circular, pancake-shaped domes about 25 kilometers (15 miles) in diameter and nearly 1 kilometer high (Figure 22.13). These domes

Figure 22.12 Computer-generated image of Venus. On the horizon is Maat Mons, a large volcano. Below it is a volcanic cone, Sapas Mons, from which light-colored lava flows extend hundreds of kilometers. (Courtesy of NASA)

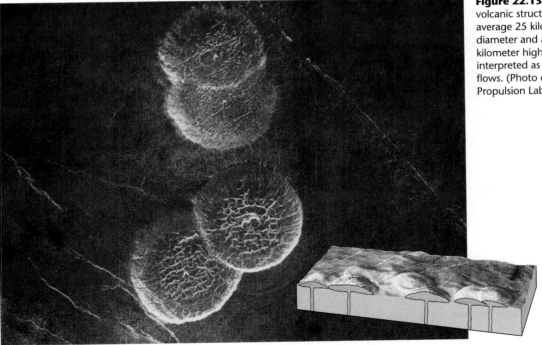

Figure 22.13 These domelike volcanic structures on Venus average 25 kilometers in diameter and are less than 1 kilometer high. They are interpreted as very thick lava flows. (Photo courtesy of Jet Propulsion Laboratory)

are thought to result from outpouring of very viscous lava, much like volcanic domes on Earth.

Only 8 percent of the Venusian surface consists of highlands that may be likened to continental areas on Earth. Tectonic activity on Venus seems to be driven by upwelling and downwelling of material in the planet's interior. Although mantle convection still operates on Venus, the processes of plate tectonics, which recycle rigid lithosphere, do not appear to have contributed to the present Venusian topography.

Before the advent of space vehicles, Venus was considered to be a potentially hospitable site for living organisms. However, evidence from *Mariner* fly-by space probes and Russian *Venera* landers indicates differently. The surface of Venus reaches temperatures of 475°C (900°F), and the Venusian atmosphere is 97 percent carbon dioxide. Only scant water vapor and nitrogen have been detected. The Venusian atmosphere contains an opaque cloud deck about 25 kilometers thick, which begins approximately 70 kilometers from the surface. Although the unmanned *Venera 8* survived less than an hour on the Venusian surface, it determined that the atmospheric pressure on that planet is 90 times that on Earth's surface. This hostile environment makes it unlikely that life as we know it exists on Venus.

Mars: The Red Planet

Mars has evoked greater interest than any other planet, for both scientists and nonscientists (see Box 22.1). When we imagine intelligent life on other worlds, "little green Martians" may come to mind. Interest in Mars stems mainly from this planet's accessibility to observation. All other planets within telescopic range have their surfaces hidden by clouds, except for Mercury, whose nearness to the Sun makes viewing difficult. Through the telescope, Mars appears as a reddish ball interrupted by some permanent dark

Box 22.1

Pathfinder: The First Geologist on Mars

On July 4, 1997, the *Mars Pathfinder* bounced onto the rock-littered surface of Mars and deployed its wheeled companion, Sojourner. For the next 3 months the lander sent back to Earth three gigabits of data, including 16,000 images and 20 chemical analyses. The landing site was a vast rolling landscape carved by ancient floods. The flood-deposit locale was selected in hope that a variety of rock types would be available for the rover Sojourner to examine.

Sojourner carried an alpha photon X-ray spectrometer (APXS) used to determine the chemical composition of rocks and Martian "soil" (regolith) at the landing site (Figure 22.A). In addition, the rover was able to take close-up images of the rocks. From these images, researchers concluded that the rocks were igneous. However, one hard, white, flat object named Scooby Doo was originally thought to be sedimentary rock, but the APXS data suggest its chemistry is like that of the soil found at the site. Thus, Scooby Doo is probably a well-cemented soil.

During its first week on Mars, Sojourner's APXS obtained data for a patch of windblown soil and a medium-sized rock, known affectionately as Barnacle Bill. Preliminary evaluation of the APXS data on Barnacle Bill shows that it contains over 60 percent silica. If these data are confirmed, it could indicate that Mars contains the volcanic rock andesite. Researchers had expected that most volcanic rocks on Mars would be

Figure 22.A *Pathfinder's* rover Sojourner obtaining data on the chemical composition of a Martian rock. (Photo courtesy of NASA)

basalt, which is lower in silica (less than 50 percent). On Earth, andesites are associated with tectonically active regions where oceanic crust is subducted into the mantle. Examples include the volcanoes of South America's Andes Mountains and the Cascades of North America.

Sojourner analyzed eight rocks and seven soils. The results thus far are only preliminary. Because these rocks are covered with a reddish dust that is high in sulfur, some controversy has arisen as to the exact composition of these Martian rocks. Some researchers

believe they are all of the same composition. The differences in measurements, they claim, are the result of varying thicknesses of dust.

In January 2002, NASA plans to land another rover on Mars. It will be bigger than Sojourner and will explore what is thought to be an ancient sedimentary environment for up to a year. This rover will be capable of drilling a 5-centimeter bore into rocks and extracting a core for possible later return to Earth.

regions that change intensity during the Martian year. The most prominent telescopic features of Mars are its brilliant white polar caps, resembling Earth's.

The Martian Atmosphere. The Martian atmosphere is only 1 percent as dense as Earth's and it is primarily carbon dioxide with tiny amounts of water vapor. Data from Mars probes confirm that the polar caps of Mars are made of water ice, covered by a thin layer of frozen carbon dioxide. As winter nears in either hemisphere, we see the equatorward growth of that hemisphere's ice cap as temperatures drop to −125°C (−193°F) and additional carbon dioxide is deposited.

Although the atmosphere of Mars is very thin, extensive dust storms occur and may cause the color changes observed from Earth-based telescopes. Hurricane-force winds up to 270 kilometers (170 miles) per hour can persist for weeks. Images from *Viking 1* and *Viking 2* revealed a Martian landscape remarkably similar to a rocky desert on Earth (Figure 22.14), with abundant sand dunes and impact craters partially filled with dust.

Mars' Dramatic Surface. *Mariner 9*, the first artificial satellite to orbit another planet, reached Mars in 1971 amid a raging dust storm. When the dust cleared, images of Mars' northern hemisphere revealed numerous large volcanoes. The biggest, Mons Olympus, is the size of Ohio and 23 kilometers (75,000 feet) high. This gigantic volcano and others resemble Hawaiian shield volcanoes on Earth (Figure 22.15). Their extreme size is thought to result from the absence of plate movements on Mars. Therefore, rather than a chain of smaller volcanoes forming as we find in Hawaii, large single cones developed.

Less-abundant impact craters indicate that at least some of the volcanic topography formed more recently, following the early period of heavy bombardment. Nevertheless, most Martian surface features are old by Earth standards. The highly cratered Martian southern hemisphere is probably similar in age to comparable lunar highlands (3.5–4.5 billion years old). Even the relatively fresh-appearing volcanic features of the northern hemisphere may be older than a billion years. This fact plus the absence of "Marsquake" recordings by *Viking* seismographs point to a tectonically dead planet.

Another surprising find made by *Mariner 9* was the existence of several canyons that dwarf even Earth's Grand Canyon of the Colorado River. One of the largest, Valles Marineris, is thought to have formed by slippage of material along huge faults in the crustal layer. In this respect, it would be comparable to the rift valleys of Africa (Figure 22.16).

Water on Mars? Not all Martian valleys have a tectonic origin. Many have tributaries in a pattern similar to that of stream valleys on Earth. In addition, *Viking* orbiter images have revealed unmistakable ancient islands in what is now a dry stream bed. When these streamlike channels were first discovered, some observers speculated that a thick water-laden atmosphere capable of generating torrential downpours once existed on Mars. If so, what happened to this water? The present Martian atmosphere contains only traces. Moreover, the environment of Mars is far too harsh to let water exist as a liquid. Despite these difficulties, the work of flowing water still remains the most acceptable explanation for many Martian channels.

Figure 22.14 This picture of the Martian landscape by the *Viking 1* lander shows a dune field with features remarkably similar to many seen in the deserts of Earth. The dune crests indicate that recent wind storms were capable of moving sand over the dunes in the direction from lower right to upper left. The large boulder at the left is about 10 meters from the spacecraft and measures 1 by 3 meters. (Courtesy of NASA)

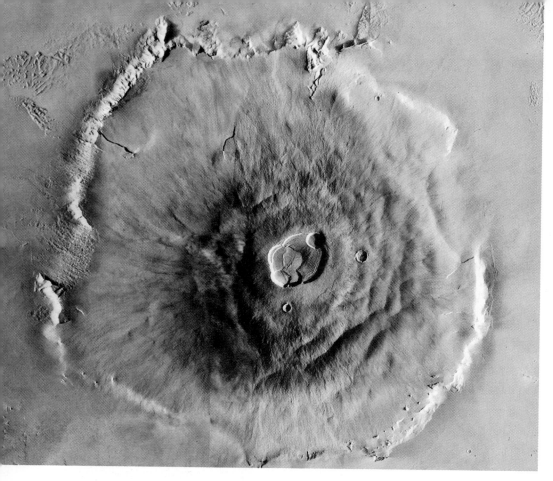

Figure 22.15 Image of Mons Olympus, an inactive shield volcano on Mars that covers an area about the size of the state of Ohio. (Courtesy of the U.S. Geological Survey)

Figure 22.16 This image shows the entire Valles Marineris canyon system, over 5000 kilometers long and up to 8 kilometers deep. The dark red spots on the left edge of the image are huge volcanoes, each about 25 kilometers high. (Courtesy of U.S. Geological Survey)

Many planetary geologists do not accept the premise that Mars once had an active water cycle similar to Earth's. Rather, they believe that many of the large streamlike valleys were created by the collapse of surface material caused by the slow melting of subsurface ice. If this is the case, these large valleys would be more akin to features formed by mass wasting processes on Earth.

Martian Satellites. Tiny Phobos and Deimos, the two satellites of Mars, were not discovered until 1977 because they are only 24 and 15 kilometers in diameter. Phobos is nearer to its parent than any other natural satellite in the solar system—only 5500 kilometers—and requires just 7 hours and 39 minutes for one revolution. *Mariner 9* revealed that both satellites are irregularly shaped and have numerous impact craters, much like their parent (Figure 22.17).

Undoubtedly, these moons are asteroids captured by Mars. A most interesting coincidence in astronomy and literature is the close resemblance between Phobos and Deimos and two fictional satellites of Mars described by Jonathan Swift in *Gulliver's Travels*, written about 150 years before these satellites were actually discovered.

Jupiter: Lord of the Heavens

Jupiter, truly a giant among planets, has a mass 2.5 times greater than the combined mass of all the remaining planets, satellites, and asteroids (see Box 22.2). In fact, had Jupiter been about 10 times larger, it would have evolved into a small star. Despite its great size, however, it is only 1/800 as massive as the Sun.

Figure 22.18 Artist's view of Jupiter with Great Red Spot visible in its southern hemisphere. Earth is shown for scale.

Jupiter also rotates more rapidly than any other planet, completing one rotation in slightly less than 10 Earth-hours. The effect of this fast spin is to make the equatorial region bulge and to make the polar dimension flatten (see "Polar Flattening" column in Table 22.1).

When viewed through a telescope or binoculars, Jupiter appears to be covered with alternating bands of multicolored clouds aligned parallel to its equator (Figure 22.18). The most striking feature is the *Great Red Spot* in the southern hemisphere (Figure 22.18). The Great Red Spot has been a prominent feature since it was first discovered more than three centuries ago. When *Voyager 2* swept by Jupiter in 1979, it was the size of two Earth-sized circles placed side by side. On occasion, it has grown even larger.

Images from *Pioneer 11* as it moved within 42,000 kilometers of Jupiter's cloud tops in 1974 indicated that the Great Red Spot is a counterclockwise-rotating (cyclonic) storm. It is caught between two jetstream-like bands of atmosphere flowing in opposite directions. This huge hurricane-like storm rotates once every 12 Earth-days. Although several smaller

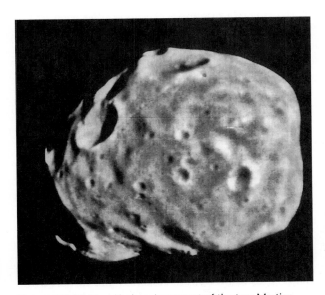

Figure 22.17 Tiny Phobos, innermost of the two Martian satellites. (Courtesy of NASA)

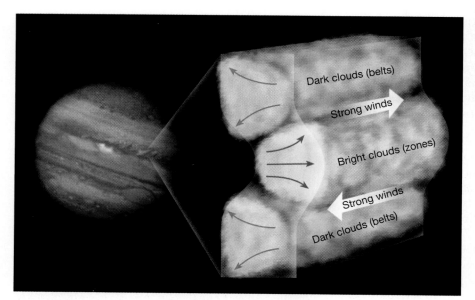

Figure 22.19 The structure of Jupiter's atmosphere. The areas of light clouds (*zones*) are regions where gases are ascending and cooling. Sinking dominates the flow in the darker cloud layers (*belts*). This convective circulation, along with the rapid rotation of the planet, generates the high-speed winds observed between the belts and zones.

Dark clouds (belts)

Strong winds

Bright clouds (zones)

Strong winds

Dark clouds (belts)

storms have been observed in other regions of Jupiter's atmosphere, none of these have survived for more than a few days.

Structure of Jupiter. Jupiter's hydrogen-helium atmosphere also has methane, ammonia, water, and sulfur compounds as minor constituents. The wind systems generate the light- and dark-colored bands that encircle this giant (Figure 22.19). Unlike the winds on Earth, which are driven by solar energy, Jupiter itself gives off nearly twice as much heat as it receives from the Sun. Thus, it is the *interior* heat from Jupiter that produces huge convection currents in the atmosphere.

Atmospheric pressure at the top of the clouds is equal to sea-level pressure on Earth. Because of Jupiter's immense gravity, the pressure increases rapidly toward its surface. At 1000 kilometers below the cloudtops, the pressure is great enough to compress hydrogen gas into a liquid. Consequently, Jupiter's surface is thought to be a gigantic ocean of liquid hydrogen. Less than halfway into Jupiter's interior, pressures of extreme magnitude cause the liquid hydrogen to turn into *liquid metallic* hydrogen. Jupiter is also believed to contain as much rocky and metallic material as is found in the terrestrial planets, probably in a central core.

Jupiter's Moons. Jupiter's satellite system, consisting of 16 moons discovered so far, resembles a miniature solar system. The four largest satellites, discovered by Galileo, travel in nearly circular orbits around the parent with periods of from 2 to 17 Earth-days (Figure 22.20). The two largest Galilean satellites, Callisto and Ganymede, surpass Mercury in size, whereas the two smaller ones, Europa and Io, are about the size of Earth's moon. These Galilean moons can be observed with binoculars or a small telescope and are interesting in their own right. Because their orbits are along Jupiter's equatorial plane, and because they all have the same orbital direction, these moons most probably formed from "leftover" debris in much the same way as the planets did.

By contrast, Jupiter's four outermost satellites are very small (20 kilometers in diameter), revolve in a direction that is opposite the other moons, and have orbits that are steeply inclined to the Jovian equator. These satellites appear to be asteroids that passed near enough to be captured gravitationally by Jupiter.

Images from *Voyagers 1* and *2* in 1979 revealed, to the surprise of almost everyone, that each of the four Galilean satellites has a character all its own, as shown in Figure 22.20. The innermost of the Galilean moons, Io, is the only volcanically active body discovered in our solar system other than Earth and Neptune's moon Triton (Figure 22.21). To date, eight active sulfurous volcanic centers have been discovered. Umbrella-shaped plumes have been seen rising from the surface of Io to heights approaching 200 kilometers (Figure 22.21).

The heat source for Io's volcanic activity is tidal energy generated by a relentless "tug of war" between Jupiter and the Galilean satellites. Because Io is gravitationally locked to Jupiter, the same side always faces the giant planet, like Earth's moon. The gravitational power of Jupiter and the other nearby satellites pulls and pushes on Io's tidal bulge as its slightly eccentric orbit takes it alternately closer to and farther away from Jupiter. This gravitational flexing of Io is transformed into heat energy.

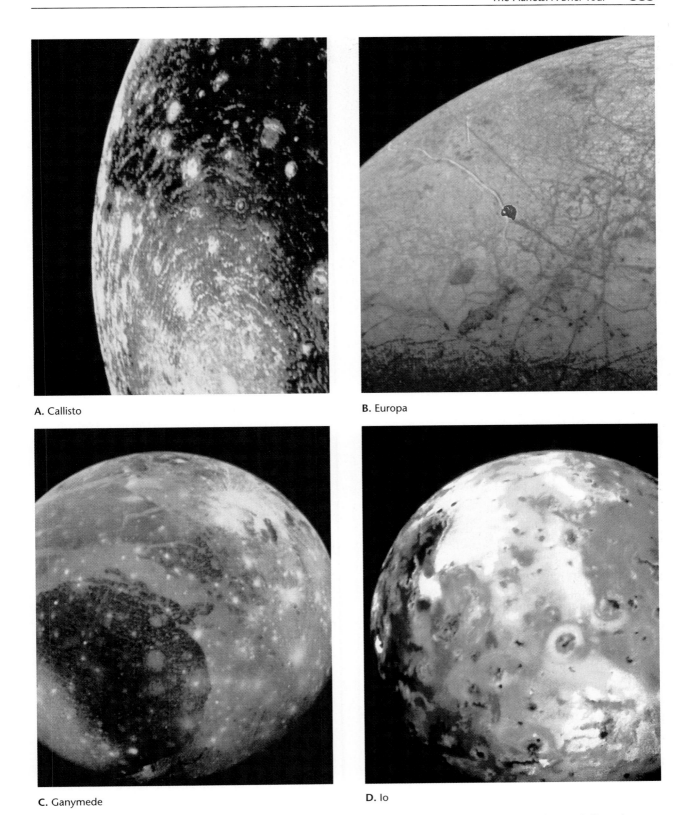

A. Callisto

B. Europa

C. Ganymede

D. Io

Figure 22.20 Jupiter's four largest moons called the Galilean moons because they were discovered by Galileo. **A.** Callisto, the outermost of the Galilean satellites, is densely cratered, much like Earth's moon. **B.** Europa, smallest of the Galilean moons, has an icy surface that is criss-crossed by many linear features. **C.** Ganymede, the largest Jovian satellite, contains cratered areas, smooth regions, and areas covered by numerous parallel grooves. **D.** The innermost moon, Io, is one of only three volcanically active bodies in the solar system. (Courtesy of NASA)

Figure 22.21 A volcanic eruption on Io. This plume of volcanic gases and debris is rising over 100 kilometers (60 miles) above Io's surface. (Courtesy of NASA)

One of the most interesting discoveries made by *Voyager 1* is Jupiter's thin ring system, believed to be different from that encircling Saturn. Rather than particles held in planetary-type orbits, the particles making up Jupiter's rings appear to be temporarily entrapped by the planet's intense magnetic field. The ring material may be sulfur particles from Io's volcanoes.

Saturn: The Elegant Planet

Requiring 29.46 Earth-years to make one revolution, Saturn is almost twice as far from the Sun as Jupiter, yet its atmosphere, composition, and internal structure appear to be remarkably similar to Jupiter's.

Saturn's Rings. The most prominent feature of Saturn is its system of rings (Figure 22.22), discovered by Galileo in 1610. Because he could not resolve them with his primitive telescope, they appeared to him as two smaller bodies adjacent to the planet. Their ring nature was revealed 50 years later by the Dutch astronomer Christian Huygens. Until the recent discovery that Jupiter, Uranus, and Neptune also have very faint ring systems, this phenomenon was thought to be unique to Saturn.

When viewed from Earth, Saturn's rings appear as distinct bands.

Our view of Saturn slowly changes as both planets proceed along their orbits, continually shifting relative positions. This changes our angle of view of Saturn's rings. Once every 15 years, we see them edge-on, and they appear as an extremely fine line.

Saturn Close-Up. In 1980 and 1981, fly-by missions of the nuclear-powered *Voyagers 1* and *2* space vehicles came within 100,000 kilometers of Saturn. More information was gained in a few days than had been acquired since Galileo first viewed this elegant planet telescopically (Figure 22.23):

1. Saturn's atmosphere is very dynamic, with winds roaring at up to 1500 kilometers (930 miles) per hour.

2. Large cyclonic "storms" similar to Jupiter's Great Red Spot, although much smaller, occur in Saturn's atmosphere.

3. Eleven additional moons were discovered.

4. The icy rings of Saturn were discovered to be more complex than expected. Each of the seven rings is made of numerous ringlets. The ringlets in one ring are intertwined in a braidlike configuration.

5. Satellite images reveal the thickness of the ring system to be no more than a few hundred meters. We easily see the thin rings from more than a billion kilometers distance because they are highly reflective.

No image obtained so far can resolve the fine structures of the rings, but they undoubtedly are composed of small particles (moonlets) that orbit the planet much like any other satellite. Radar observations indicate particles no larger than 10 meters, and the more abundant particles are perhaps as small as 10 centimeters (4 inches).

Beyond the outermost bright ring (A ring), some moonlets have accreted to form very small satellites having diameters on the order of 100 kilometers (60 miles). Five of these asteroid-sized moons have been discovered orbiting within the faint outer rings, and others probably exist. Planetary geologists are very interested in the gravitational interaction in Saturn's ring system. They hope this will reveal how material from the primordial cloud of dust and gases condensed to produce the planets.

Saturn's Moons. The Saturnian satellite system consists of at least 21 bodies (Figure 22.23). The largest, Titan, is bigger than Mercury and is the second-largest satellite in the solar system (after

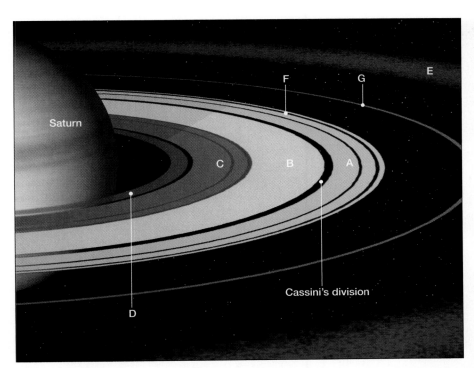

Figure 22.22 A view of the dramatic ring system of Saturn.

Box 22.2

Comet Shoemaker-Levy Impacts Jupiter

In July 1994, Comet Shoemaker-Levy impacted Jupiter with the force equal to six million megatons of energy. (A megaton is the equivalent of a million tons of the explosive, TNT). Clearly, this was the most dramatic event in the solar system ever observed by people. Concern that a similar event might occur on Earth led NASA to establish the Near-Earth Object Search Committee to detect asteroids or comets with trajectories that might cross Earth's orbit.

It is possible that a few small fragments of a comet may have actually impacted Earth in 1908. That year, a strong explosion flattened more than 1000 square kilometers of a remote Siberian forest.

Comet Shoemaker-Levy was discovered at California's Mount Palomar Observatory barely a year before it impacted Jupiter. Careful observation showed that it consisted of two dozen fragments. Researchers concluded that the comet had broken up during an earlier pass by Jupiter. As the larger fragments penetrated Jupiter's outer atmosphere, they produced brilliant impact flashes and debris plumes soaring

Figure 22.B Dark blemishes on Jupiter produced by the impact of fragments of Comet Shoemaker-Levy in July 1994. (Courtesy of NASA)

thousands of kilometers above the planet. The result of these fiery impacts were dark zones in Jupiter's atmosphere that exceeded Earth in size. The largest of these dark blemishes lasted for months

(Figure 22.B). Investigators learned more about the dynamics of Jupiter's atmosphere by observing how these blemishes dispersed.

Figure 22.23 Montage of the Saturnian satellite system. The moon Dione is in foreground; Tethys and Mimas are at lower right; Enceladus and Rhea are off ring's left; and Titan is upper right. (Photo courtesy of NASA)

Jupiter's Ganymede). It is the only satellite in the solar system, known to have a substantial atmosphere. Because of its dense gaseous cover, the atmospheric pressure at Titan's surface is about 1.5 times that at Earth's surface.

Data from *Voyager 1* reveal Titan's atmosphere to be roughly 80 percent nitrogen and perhaps 6 percent methane. This planet-sized moon appears to have polar ice caps that show seasonal variations in size. Its surface, if unfrozen, would be an ocean of liquid nitrogen.

Uranus and Neptune: The Twins

Earth and Venus have similar traits, but Uranus and Neptune are nearly twins. Only 1 percent different in diameter, both appear a pale greenish-blue, attributable to the methane in their atmospheres. Their structure and composition appear to be similar. Neptune,

however, is colder, because it is half again as distant from the Sun's warmth as is Uranus.

Uranus, the Sideways Planet. A unique feature of Uranus is that it rotates "on its side" (see Figure 22.2). Its axis of rotation, instead of being generally perpendicular to the plane of its orbit, like the other planets, lies only 8 degrees from the plane of its orbit. Its rotational motion, therefore, has the appearance of rolling, rather than the toplike spinning of the other planets. Because the axis of Uranus is inclined almost 90 degrees, the Sun is nearly overhead at one of its poles once each revolution, and then half a revolution later, it is overhead at the other pole.

A surprise discovery in 1977 revealed that Uranus has rings, much like those encircling Jupiter. This find occurred as Uranus passed in front of a distant star and blocked its view, a process called *occultation* (the word *occult* means "hidden"). Observers saw the star "wink" briefly five times (meaning five rings) before the primary occultation and again five times afterward. Later studies indicate that Uranus has at least nine distinct belts of debris orbiting its equatorial region.

Spectacular views from *Voyager 2* of the five largest moons of Uranus show quite varied terrains. Some have long, deep canyons and linear scars, whereas others possess large, smooth areas on otherwise crater-riddled surfaces. The Jet Propulsion Laboratory described Miranda, the innermost of the five largest moons, as having a greater variety of landforms than any body yet examined in the solar system.

Neptune. Even when the most powerful telescope is focused on Neptune, it appears as a bluish fuzzy disk. Until *Voyager 2*'s 1989 encounter, astronomers knew very little about this planet. However, *Voyager 2*'s 12-year, nearly three-billion-mile journey provided investigators with a great deal of new information about Neptune and its satellites.

Neptune has a dynamic atmosphere, much like those of Jupiter and Saturn (Figure 22.24). Winds exceeding 1000 kilometers per hour (600 miles per hour) encircle the planet making it one of the windiest places in the solar system. It also has an Earth-sized blemish called the *Great Dark Spot*, which is reminiscent of Jupiter's Great Red Spot, and is assumed to be a large rotating storm.

Perhaps most surprising are white, cirruslike clouds that occupy a layer about 50 kilometers above the main cloud deck, probably frozen methane. Six new satellites were discovered in the *Voyager* images, bringing Neptune's family to eight. All of the newly discovered moons orbit the planet in a direction opposite that of the two larger satellites. *Voyager* images also revealed a ring system around Neptune.

Triton, Neptune's largest moon, is a most interesting object (Figure 22.25). Its diameter is nearly that of Earth's moon. Triton is the only large moon in the solar system that orbits its parent in the direction opposite to the direction in which all the planets travel. This indicates that Triton formed independently of Neptune and was gravitationally captured.

Triton also has the lowest surface temperature of any body in the solar system, –200°C (–391°F). Its very thin atmosphere is mostly nitrogen with a little methane. The surface of Triton apparently is largely water ice, covered with layers of solid nitrogen and methane.

Pluto: Planet X

Pluto lies on the fringe of the solar system, almost 40 times farther from the Sun than Earth. It is 10,000 times too small to be visible to the unaided eye. Because of its great distance and slow orbital speed, it takes Pluto 248 Earth-years to orbit the Sun. Since its discovery in 1930, it has completed about one-fourth of a revolution. Pluto's orbit is noticeably elongated (highly eccentric), causing it to occasionally travel inside the orbit of Neptune, where it currently resides.

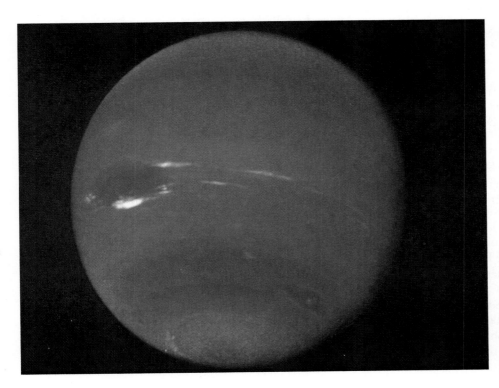

Figure 22.24 This image of Neptune shows the Great Dark Spot (left center). Also visible are bright cirruslike clouds that travel at high speed around the planet. A second oval spot is at 54° south latitude on the east limb of the planet. (Courtesy of the Jet Propulsion Laboratory)

Figure 22.25 A photomosaic of Neptune's largest moon, Triton. The large south polar cap is on the right half of the image. Seasonal ice (probably nitrogen) covers the region. Because spring in Triton's southern hemisphere extends from 1960 to the year 2000, some of the polar cap has evaporated. (Courtesy of the Jet Propulsion Laboratory)

Figure 22.26 Pluto and its moon Charon. The Hubble Space Telescope produced the first image (upper right) that resolved these two icy worlds into separate objects. The image in the upper left is the best ground-based photo produced to date. (Courtesy of NASA)

There is no chance that Pluto and Neptune will ever collide, because their orbits are inclined to each other and do not actually cross (see Figure 22.1).

In 1978, the moon Charon was discovered orbiting Pluto. Because of its close proximity to the planet, the best ground-based images of Charon show it only as an elongated bulge (Figure 22.26, left). In 1990, the Hubble Space Telescope produced an image that clearly resolves the separation between these two icy worlds (Figure 22.26, right). Charon orbits Pluto once every 6.4 Earth-days at a distance 20 times closer to Pluto than our Moon is to Earth.

The discovery of Charon greatly altered earlier estimates of Pluto's size. Current data indicate that Pluto has a diameter of 2445 kilometers, about one-fifth the size of Earth, making it the smallest planet in the solar system. Charon is about 1300 kilometers across, exceptionally large in proportion to its parent.

The average temperature of Pluto is estimated at −210°C, cold enough to solidify most gases that might be present. Thus, Pluto might best be described as a dirty iceball of frozen gases with lesser amounts of rocky substances.

A recent proposal suggests that Pluto was a satellite of Neptune that was displaced from its original orbit by a collision with something. The discovery of a satellite around Pluto is considered evidence that this event broke the Neptunian satellite into two pieces and sent them into an elongated orbit around the Sun.

Minor Members of the Solar System

Asteroids

Asteroids are smaller bodies that have been likened to "flying mountains." The largest, Ceres, is about 1000 kilometers (600 miles) in diameter, but most of the 50,000 that have been observed are only about a kilometer across. The smallest asteroids are assumed to be no larger than grains of sand. Most lie between the orbits of Mars and Jupiter and have periods of 3 to 6 years (Figure 22.27). Some have very eccentric orbits and travel very near the Sun, and a few larger ones regularly pass close to Earth and its moon. Many of the most recent impact craters on the Moon and Earth were probably caused by collisions with asteroids. Inevitably, future Earth–asteroid collisions will occur—see Box 22.3.

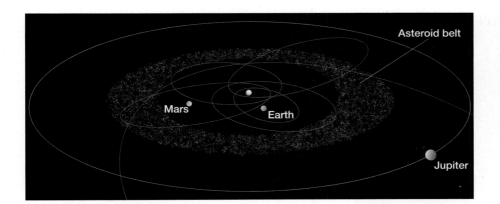

Figure 22.27 The orbits of most asteroids lie between Mars and Jupiter. Also shown are the orbits of a few known near-Earth asteroids. Perhaps a thousand or more asteroids have near-Earth orbits. Luckily, only a few dozen are thought to be larger than 1 kilometer in diameter.

Box 22.3

Is Earth on a Collision Course?

The solar system is cluttered with meteoroids, asteroids, active comets, and extinct comets. These fragments travel at great speeds and can strike Earth with the explosive force of a powerful nuclear weapon.

In the last few decades, it has become increasingly clear that comets and asteroids have collided with Earth far more frequently than was previously known. The evidence is giant impact structures called *astroblemes*. More than 100 have been identified (Figure 22.C). Many were once misunderstood to result from some volcanic process. Although most astroblemes are so old that they no longer resemble impact craters, evidence of their intense impact remains (Figure 22.D). One notable exception is a very fresh-looking crater near Winslow, Arizona, known as Meteor Crater (see Figure 22.31).

Evidence is mounting that about 65 million years ago a large asteroid about 10 kilometers (6 miles) in diameter collided with Earth. This impact may have caused the extinction of the dinosaurs, as well as nearly 50 percent of all plant and animal species.

More recently, a spectacular explosion has been attributed to the collision of our planet with a comet or asteroid. In 1908, in a remote region of Siberia, a "fire-ball" that appeared more brilliant than the Sun exploded with a violent force. The shock waves rattled windows and caused sounds that were heard up to 1000 kilometers away. The "Tunguska event," as it is called, scorched, delimbed, and flattened trees to 30 kilometers from the epicenter. But expeditions to the area

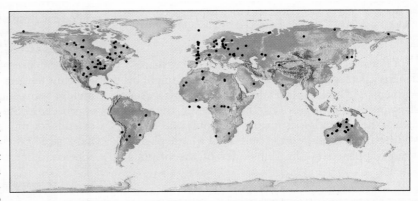

Figure 22.C World map of major impact structures (astroblemes). Others are being identified every year. (Data from Griffith Observatory)

found no evidence of an impact crater, nor any fragments. Evidently, the explosion, which equaled at least a 10-megaton nuclear bomb, occurred a few kilometers above Earth's surface. Most likely it was the demise of a comet or perhaps a stony asteroid. Why it exploded prior to impact is uncertain.

The dangers of living with these small but deadly objects from space again came to public attention in 1989 when an asteroid nearly a kilometer across shot past Earth. It was a near miss, about twice the distance to the Moon. Traveling at 70,000 kilometers (44,000 miles) per hour, it could have produced a crater 10 kilometers (6 miles) in diameter and perhaps 2 kilometers (1.2 miles) deep. As an observer noted, "Sooner or later it will be back." As it was, it crossed our orbit just 6 hours ahead of Earth. Statistics show that collisions of this tremendous magnitude should take place every few hundred million years

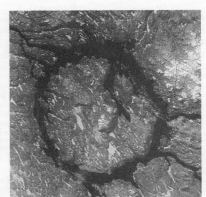

Figure 22.D Manicouagan, Quebec, is a 200-million-year-old eroded impact crater. The lake outlines the crater remnant, which is 70 kilometers (42 miles) across. Fractures related to this event extend outward for an additional 30 kilometers. (Courtesy of U.S. Geological Survey)

and could have drastic consequences for life on Earth.

Figure 22.28 Image of asteroid 951 (Gaspra) obtained by the Jupiter-bound *Galileo* spacecraft. Like other asteroids, Gaspra is probably a collision-produced fragment of a larger body. (Courtesy of NASA)

Because many asteroids have irregular shapes, planetary geologists first speculated that they might be fragments of a broken planet that once orbited between Mars and Jupiter (Figure 22.28). However, the total mass of the asteroids is estimated to be only one-thousandth that of Earth, which itself is not a large planet. What happened to the remainder of the original planet? Others have hypothesized that several larger bodies once coexisted in close proximity and that their collisions produced numerous smaller ones. The existence of several "families" of asteroids has been used to support this explanation. However, no conclusive evidence has been found for either hypothesis.

Comets

Comets are among the most interesting and unpredictable bodies in the solar system. They have been compared to dirty snowballs, because they are made of frozen gases (water, ammonia, methane, carbon dioxide, and carbon monoxide) that hold together small pieces of rocky and metallic materials. Many comets travel very elongated orbits that carry them beyond Pluto.

When first observed, a comet appears very small; but as it approaches the Sun, solar energy begins to vaporize the frozen gases, producing a glowing head called the **coma** (Figure 22.29). The size of the coma varies greatly from one comet to another. Extremely rare ones exceed the size of the Sun, but most approximate the size of Jupiter. Within the coma, a small glowing nucleus with a diameter of only a few kilometers can sometimes be detected. As comets approach the Sun, some, but not all, develop a tail that extends for millions of kilometers. Despite the enormous size of their tails and comas, the mass of comets appears to be insignificant.

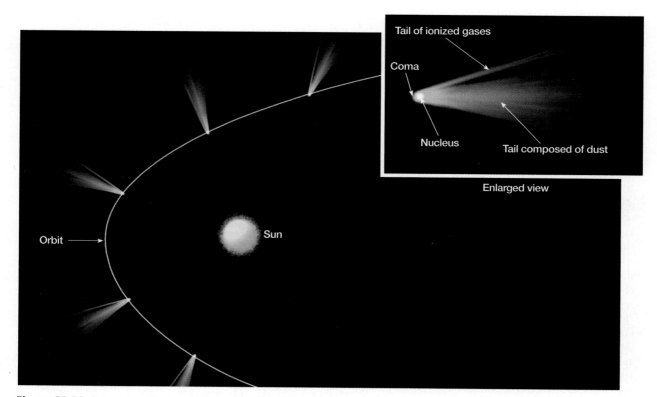

Figure 22.29 Orientation of a comet's tail as it orbits the Sun.

A comet's tail is not a *trailing* tail, but a plume composed of dust and ionized gases repelled by solar forces. This explains why the tail always points away from the Sun (Figure 22.29). The solar forces are *radiation pressure*, which pushes dust particles away from the coma, and the *solar wind*, which moves the ionized gases, particularly carbon monoxide (Figure 22.29). Sometimes a single tail composed of both dust and gases is produced, but often two tails are observed.

As a comet moves away from the Sun, the gases forming the coma recondense, the tail disappears, and the comet returns to "cold storage." Material that was blown from the coma to form the tail is lost from the comet forever. Consequently, most comets cannot survive more than a few hundred close orbits of the Sun. Once all the gases are expelled, the remaining material—a swarm of unconnected metallic and stony particles—continues the orbit without a coma or a tail.

Little is known about the origin of comets. Millions are believed to orbit the Sun beyond Pluto, and it is proposed that the gravitational effect of nearby stars sends some of them into the highly eccentric orbits that carry them toward the center of our solar system. Here, the gravitational pull of the larger planets, particularly Jupiter, alters a comet's orbit and accelerates its period of revolution.

The most famous short-period comet is Halley's comet. Its orbital period averages 76 years, and every one of its 29 appearances since 240 B.C. has been recorded by Chinese astronomers. This record is a testimonial to their dedication as astronomical observers and to the endurance of their culture. When seen in 1910, Halley's comet had developed a tail nearly 1.6 million kilometers (1 million miles) long and was visible during the daylight hours.

In 1986, the unspectacular showing of Halley's comet was a disappointment to many people in the Northern Hemisphere. Yet it was during this most recent visit to the inner solar system that a great deal of new information was learned about this most famous of comets. The new data were gathered by space probes sent to rendezvous with the comet. Most notably, the European probe *Giotto* approached to within 600 kilometers of the comet's nucleus and obtained the first images of this elusive structure.

We now know that the nucleus is potato-shaped, 16 kilometers by 8 kilometers. The surface is irregular and full of craterlike pits. Gases and dust that vaporize from the nucleus to form the coma and tail appear to gush from its surface as bright jets or streams. Only about 10 percent of the comet's total surface was emitting these jets at the time of the rendezvous. The remaining surface area of the comet appeared to be covered with a dark layer that may consist of organic material.

In 1997, the comet Hale-Bopp made for spectacular viewing around the globe. As comets go, the nucleus of Hale-Bopp was unusually large, about 40 kilometers (25 miles) in diameter. As shown in Figure 22.30, two tails nearly 200 million miles long extended from this comet. The bluish gas-tail is composed of positively charged ions, and it points almost directly away from the Sun. The yellowish tail is composed of dust and other rocky debris. Because the rocky material is more massive than the ionized gases, it is less affected by the solar wind and follows a different trajectory away from the comet.

Figure 22.30 Comet Hale-Bopp. The two tails seen in the photograph are about 10 million to 15 million miles long. (A Peoria Astronomical Society Photograph by Eric Clifton and Greg Neaveill)

Meteoroids

Nearly everyone has seen a **meteor**, popularly (but inaccurately) called a "shooting star." This streak of light lasts from an eyeblink to a few seconds and occurs when a small solid particle, a **meteoroid**, enters Earth's atmosphere from interplanetary space. The friction between the meteoroid and the air heats both and produces the light we see. Although a rare meteoroid is as large as an asteroid, most are the size of sand grains and weigh less than 1/100 gram. Consequently, they vaporize before reaching Earth's surface. Some, called *micrometeorites*, are so tiny that their rate of fall becomes too slow to cause them to burn up, so they drift down as "space dust." Each day, the number of meteoroids that enter Earth's atmosphere must reach into the thousands. After sunset, a half dozen or more are bright enough to be seen with the naked eye each hour from anywhere on Earth.

Occasionally, meteor sightings increase dramatically to 60 or more per hour. These displays, called **meteor showers**, result when Earth encounters a swarm of meteoroids traveling in the same direction and at nearly the same speed as Earth. The close association of these swarms to the orbits of some short-term comets strongly suggests that they represent material lost by these comets (Table 22.3). Some swarms not associated with orbits of known comets are probably the remains of the nucleus of a long-defunct comet. The notable Perseid meteor shower that occurs each year around August 12 is probably material from the Comet 1862 III, which has a period of 110 years.

Meteoroids associated with comets are small and not known to reach the ground. Most meteoroids large enough to survive the heated fall are thought to originate among the asteroids, where chance collisions modify their orbits and send them toward Earth.

Earth's gravitational force does the rest (see Box 22.3).

The remains of meteoroids, when found on Earth, are referred to as **meteorites**. A few very large meteoroids have blasted out craters on Earth's surface that strongly resemble those on the lunar surface. The most famous is Meteor Crater in Arizona (Figure 22.31). This huge cavity is about 1.2 kilometers (0.75 miles) across, 170 meters (560 feet) deep, and has an upturned rim that rises 50 meters (165 feet) above the surrounding countryside. Over 30 tons of iron fragments have been found in the immediate area, but attempts to locate a main body have been unsuccessful. Based on erosion, the impact likely occurred within the last 20,000 years.

Prior to Moon rocks brought back by lunar explorers, meteorites were the only extraterrestrial materials that could be directly examined (Figure 22.32). Meteorites are classified by their composition: (1) **irons**—mostly iron with 5 to 20 percent nickel, (2) **stony**—silicate minerals with inclusions of other minerals, and (3) **stony-irons**—mixtures. Although stony meteorites are probably more common, people find mostly irons. This is understandable, for irons withstand the impact better, weather more slowly, and are much easier for a lay person to distinguish from terrestrial rocks. Iron meteorites are probably fragments of once-molten cores of large asteroids or small planets.

One rare kind of meteorite, called a *carbonaceous chondrite*, was found to contain simple amino acids and other organic compounds, which are the basic building blocks of life. This discovery confirms similar findings in observational astronomy, which indicate that numerous organic compounds exist in the frigid realm of outer space.

If meteorites represent the makeup of Earth-like planets, as some planetary geologists suggest, then Earth must contain a much larger percentage of iron

Table 22.3 Major Meteor Showers

Shower	Approximate Dates	Associated Comet
Quadrantids	January 4–6	—
Lyrids	April 20–23	Comet 1861 I
Eta Aquarids	May 3–5	Halley's comet
Delta Aquarids	July 30	—
Perseids	August 12	Comet 1862 III
Draconids	October 7–10	Comet Giacobini-Zinner
Orionids	October 20	Halley's comet
Taurids	November 3–13	Comet Encke
Andromedids	November 14	Comet Biela
Leonids	November 18	Comet 1866 I
Germinids	December 4–16	—

Figure 22.31 Meteor Crater, about 32 kilometers (20 miles) west of Winslow, Arizona. (Photo by Michael Collier)

than is indicated by surface rocks. This is one reason why geologists suggest that Earth's core may be mostly iron and nickel. In addition, meteorite dating indicates that our solar system's age certainly exceeds 4.5 billion years. This "old age" has been confirmed by data from lunar samples.

Figure 22.32 Iron meteorite found near Meteor Crater, Arizona. (Courtesy of Meteor Crater, Northern Arizona, USA)

Chapter Summary

- The planets can be arranged into two groups: the *terrestrial* (Earth-like) *planets* (Mercury, Venus, Earth, and Mars) and the *Jovian* (Jupiter-like) *planets* (Jupiter, Saturn, Uranus, and Neptune). Pluto is not included in either group. When compared to the Jovian planets, the terrestrial planets are smaller, more dense, contain proportionally more rocky material, and lesser amounts of gases (hydrogen and helium) and ices (water, ammonia, methane, and carbon dioxide).

- The *nebular hypothesis* describes the formation of the solar system. The planets and Sun began forming about five billion years ago from a large cloud of dust and gases called a *nebula*. As the nebular cloud contracted, it began to rotate and flatten into a disk. Material that was gravitationally pulled toward the center became the *protosun*. Within the rotating disk, small centers formed, called *protoplanets*, sweeping up more and more of the nebular debris. Because of their high temperatures and weak gravitational fields, the inner planets (Mercury, Venus, Earth, and Mars) were unable to accumulate much of the lighter components (hydrogen, ammonia, methane, and water) of the nebula. However, because of the very cold temperatures existing far from the Sun, the fragments from which the Jovian planets formed contained a high percentage of ices—water, carbon dioxide, ammonia, and methane.

- The lunar surface exhibits several types of features. Most *craters* were produced by the impact of rapidly moving debris (meteoroids). Bright, densely cratered highlands make up much of the lunar surface. Dark, fairly smooth lowlands are called *maria*. Maria basins are enormous impact craters that have been flooded with layer upon layer of very fluid basaltic lava. All lunar terrains are mantled with a soil-like layer of gray, unconsolidated debris, called *lunar regolith*, which has been derived from a few billion years of meteoric bombardment.

- Closest to the Sun, *Mercury* is small, dense, has no atmosphere, and exhibits the greatest temperature extremes of any planet. *Venus*, the brightest planet in the sky, has a thick, dense, carbon dioxide atmosphere, a surface of relatively subdued plains and inactive volcanoes, a surface atmospheric pressure 90 times that of Earth's, and surface temperatures of 475°C. *Mars*, the red planet, has a carbon dioxide atmosphere only 1 percent as dense as Earth's, extensive winds and dust storms, numerous inactive volcanoes, many large canyons, and several valleys of debatable origin resembling drainage patterns similar to stream valleys on Earth. *Jupiter*, the largest planet, rotates rapidly, has a banded appearance, a Great Red Spot that varies in size, a ring system, and at least 16 moons (one of which, Io, is volcanically active). *Saturn*, best known for its rings, also has a dynamic atmosphere with winds up to 1500 kilometers (930 miles) per hour and "storms" similar to Jupiter's Great Red Spot. *Uranus* and *Neptune* are often called the twins because of similar structure and composition. Uranus is unique in rotating "on its side." Neptune has white, cirruslike clouds above its main cloud deck and an Earth-sized Great Dark Spot, assumed to be a large rotating storm similar to Jupiter's Great Red Spot. *Pluto*, a small frozen world with one moon, may have once been a satellite of Neptune. Pluto's elongated orbit causes it to occasionally travel inside the orbit of Neptune, but with no chance of collision.

- The minor members of the solar system include *asteroids, comets,* and *meteoroids*. No conclusive evidence has been found to explain the origin of the asteroids. Comets are made of ices with small pieces of rocky and metallic material. Many travel in very elongated orbits that carry them beyond Pluto. Meteoroids, small solid particles that travel through interplanetary space, become *meteors* when they enter Earth's atmosphere and vaporize with a flash of light. *Meteor showers* occur when Earth encounters a swarm of meteoroids, probably fragments of a comet. *Meteorites* are the remains of meteoroids. The *three types of meteorites* are iron, stony, and stony-iron.

Review Questions

1. By what criteria are the planets placed into either the Jovian or terrestrial group?

2. What are the three types of materials that make up the planets? How are they different? How does their distribution account for the density differences between the terrestrial and Jovian planetary groups?

3. Explain why different planets have different atmospheres.

4. How is crater density used in the relative dating of lunar features?

5. Briefly outline the history of the Moon.

6. How are the maria of the Moon similar to the Columbia Plateau?

7. Why has Mars been the planet most studied telescopically?

8. What surface features does Mars have that are also common on Earth?

9. Although Mars has valleys that appear to be products of stream erosion, what fact makes it unlikely that Mars has had a water cycle like that on Earth?

10. The two "moons" of Mars were once suggested to be artificial. What characteristics do they have that would cause such speculation?

11. What is the nature of Jupiter's Great Red Spot?

12. Why are the Galilean satellites of Jupiter so named?

13. What is distinctive about Jupiter's satellite Io?

14. Why are the four outer satellites of Jupiter thought to have been captured?

15. What evidence indicates that Saturn's rings are composed of individual moonlets rather than consisting of solid disks?

16. Explain why a large satellite cannot exist closer to Saturn than the outer edge of its ring system.

17. What is unique about Saturn's satellite Titan?

18. What three bodies in the solar system exhibit volcanic-like activity?

19. What do you think would happen if Earth passed through the tail of a comet?

20. Describe the origin of comets according to the most widely accepted hypothesis.

21. Compare meteoroid, meteor, and meteorite.

22. Why are meteorite craters more common on the Moon than on Earth, even though the Moon is a much smaller target?

23. It has been estimated that Haley's comet has a mass of 100 billion tons. Further, this comet is estimated to lose 100 million tons of material during the few months that its orbit brings it close to the Sun. With an orbital period of 76 years, what is the maximum remaining life span of Haley's comet?

Key Terms

asteroid (p. 590)
coma (p. 592)
comet (p. 592)
escape velocity (p. 569)
iron meteorite (p. 594)

Jovian planet (p. 568)
lunar regolith (p. 574)
maria (p. 572)
meteor (p. 593)
meteorite (p. 594)

meteoroid (p. 593)
meteor shower (p. 594)
nebular hypothesis
    (p. 570)

stony-iron meteorite
    (p. 594)
stony meteorite (p. 594)
terrestrial planet
    (p. 568)

Web Resources

The *Earth* Home Page provides on-line resources for this chapter on the World Wide Web. You will find review exercises, specific updates for items in the chapter, suggested reading, and links to interesting related pathways on the Internet. Visit the *Earth* Home Page at **http://www.prenhall.com/tarbuck.**

APPENDIX A

Metric and English Units Compared

Units

1 kilometer (km)	=	1000 meters (m)
1 meter (m)	=	100 centimeters (cm)
1 centimeter (cm)	=	0.39 inch (in.)
1 mile (mi)	=	5280 feet (ft)
1 foot (ft)	=	12 inches (in.)
1 inch (in.)	=	2.54 centimeters (cm)
1 square mile (mi²)	=	640 acres (a)
1 kilogram (kg)	=	1000 grams (g)
1 pound (lb)	=	16 ounces (oz)
1 fathom	=	6 feet (ft)

Conversions

When you want
to convert:	multiply by:	to find:

Length

inches	2.54	centimeters
centimeters	0.39	inches
feet	0.30	meters
meters	3.28	feet
yards	0.91	meters
meters	1.09	yards
miles	1.61	kilometers
kilometers	0.62	miles

Area

square inches	6.45	square centimeters
square centimeters	0.15	square inches
square feet	0.09	square meters
square meters	10.76	square feet
square miles	2.59	square kilometers
square kilometers	0.39	square miles

Volume

cubic inches	16.38	cubic centimeters
cubic centimeters	0.06	cubic inches
cubic feet	0.028	cubic meters
cubic meters	35.3	cubic feet
cubic miles	4.17	cubic kilometers
cubic kilometers	0.24	cubic miles
liters	1.06	quarts
liters	0.26	gallons
gallons	3.78	liters

Masses and Weights

ounces	28.35	grams
grams	0.035	ounces
pounds	0.45	kilograms
kilograms	2.205	pounds

Temperature

When you want to convert degrees Fahrenheit (°F) to degrees Celsius (°C), subtract 32 degrees and divide by 1.8.

When you want to convert degrees Celsius (°C) to degrees Fahrenheit (°F), multiply by 1.8 and add 32 degrees.

When you want to convert degrees Celsius (°C) to Kelvins (K), delete the degree symbol and add 273. When you want to convert Kelvins (K) to degrees Celsius (°C), add the degree symbol and subtract 273.

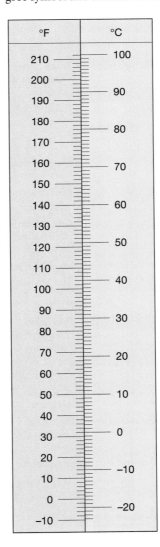

Figure A.1 A comparison of Fahrenheit and Celsius temperature scales.

599

APPENDIX B

Common Minerals of Earth's Crust

Mineral or Group Name	Composition	Cleavage/ Fracture	Color	Hardness	Other Properties/ Comments
Albite See *Plagioclase feldspar*					
Amphibole (common member: hornblende)	Complex family of hydrous Ca, Na, Mg, Fe, Al silicates	Two at 60 and 120 degrees	Deep green to black	5–6	Forms elongated crystals. Commonly found in igneous and metamorphic rocks.
Anorthite See *Plagioclase feldspar*					
Augite See *Pyroxene*					
Bauxite	Mixture of weathered clay minerals	Irregular fracture	Varied, reddish brown common	Variable	Earthy luster, commonly contains small spheres. Ore of aluminum.
Biotite	$K (Mg, Fe)_3(AlSi_3O_{10})(OH)_2$	Perfect cleavage in one direction	Black to dark brown	2–2.5	Splits into thin, flexible sheets. Common mica found in igneous and metamorphic rocks.
Bornite	Cu_5FeS_4	Uneven fracture	Brownish bronze on a fresh surface	3	Tarnishes to a variegated purple blue; hence, called peacock ore. High specific gravity (5). Ore of copper.
Calcite	$CaCO_3$ Calcium carbonate	Three perfect cleavages at 75 degrees	White or colorless	3	Common in sedimentary rocks. When transparent exhibits double refraction. Reacts with weak acid.
Chalcedony	SiO_2 Silicon dioxide	Conchoidal fracture	White when pure. Often multicolored	5–6.5	Microcrystalline form of quartz. Multicolored. Called agates when banded. Opal is an amorphous variety.
Chalcopyrite	$CuFeS_2$	Irregular fracture	Brass yellow	3.5–4	Usually massive. Specific gravity 4–4.5. Ore of copper.
Chlorite	$(Mg, Fe)_5(Al, Fe)_2Si_3 O_{10}(OH)_8$	One direction of cleavage	Light to dark green	2–2.5	Occurs as mass of flaky scales. Common in metamorphic rocks.
Cinnabar	HgS	One direction, but not generally observed	Scarlet red	2.5	Occurs in masses mixed with other materials. Often dull earthy luster. Important ore of mercury.
Clay minerals (common member, kaolinite)	Complex group of hydrous aluminum silicates	Irregular fracture	Buff to brownish gray	1–2.5	Found in earthy masses as a main constituent of soil. Also abundant in shales and other sedimentary rocks.
Corundum	Al_2O_3	Two good cleavages with striations	Variable; red, blue, yellow and green	9	Important gemstone. Red variety called ruby; blue variety is sapphire. Also used as an abrasive.
Dolomite	$CaMg(CO_3)_2$	Three good cleavages at 75 degrees	Variable; white when pure	3.5–4	Similar to calcite, but will effervesce with acid only when powdered. Common in sedimentary rocks.
Epidote	Complex Ca, Fe and Al silicate	One good cleavage, one poor	Yellow-green to dark green	6–7	Commonly occurs as small elongated crystals in metamorphic rocks.
Feldspar See *Orthoclase feldspar* and *Plagioclase feldspar*					

Mineral or Group Name	Composition	Cleavage/ Fracture	Color	Hardness	Other Properties/ Comments
Fluorite	CaF_2	Perfect cleavage in 4 directions	Colorless; violet, green, or yellow	4	Commonly found with ores of metals.
Galena	PbS	Three cleavages at right angles	Silver gray	2.5	Shiny metallic mineral with high specific gravity (7.6). Ore of lead.
Garnet	Complex family of silicate minerals containing Ca, Mg, Fe, Mn, Al, Ti, Cr	Uneven to conchoidal fracture	Various colors; commonly deep red to brown	6.5–7.5	Forms 12- or 24-sided crystals commonly found in metamorphic rocks.
Graphite	C	One direction of cleavage	Steel gray	1–2	Occurs in scaly, foliated masses. Used as a lubricant. Greasy feel.
Gypsum	$CaSo_42H_2O$	Cleavage good in one direction poor in two others	Colorless to white	2	Occurs as tabular crystals, or fibrous or finely crystalline masses. Common in sedimentary layers. Used for plaster.
Halite	NaCl	Three cleavages at right angles	Colorless to white	2.5	Common table salt. Occurs as granular masses. Common sedimentary mineral.
Hematite	Fe_2O_3	Uneven fracture	Reddish brown to steel gray	5.5–6.5	Occurs as earthy masses. High specific gravity (4.8–5.5). Important ore of iron.
Hornblende	See *Amphibole*				
Kaolinite	See *Clay minerals*				
Kyanite	Al_2SiO_5	One good direction of cleavage	White to light blue	5–7	Forms long, bladed or tabular crystals. Common mineral in metamorphic rocks.
Labradorite	See *Plagioclase feldspar*				
Limonite (goethite)	Mixture of hydrous iron oxides	Uneven fracture	Yellowish to brown	1–5.5	Earthy masses. Forms from the alteration of other iron-rich minerals. Gives rock surfaces and soils a yellow color.
Magnetite	Fe_3O_4	Uneven fracture	Black	5.5–6.5	Submetallic to metallic luster. Magnetic. High specific gravity (5). Generally occurs in granular masses. Ore of iron.
Malachite	$Cu_2CO_3(OH)_2$	Uneven fracture	Bright green	3.5–4	Effervesces in acid. Ore of copper.
Mica	See *Biotite* and *Muscovite*				
Muscovite	$KAl_3Si_3O_{10}(OH)_2$	Perfect cleavage in one direction	Colorless to light gray	2–2.5	Splits into thin elastic sheets. Transparent in thin sheets. Common in all rock types.
Olivine	$(Mg, Fe)_2SiO_4$	Conchoidal fracture	Olive to dark green	6.5–7	Occurs as granular masses or grains in dark-colored igneous rocks.
Orthoclase feldspar (K feldspar)	$KAlSi_3O_8$	Two cleavages at nearly right angles	White to gray. Frequently salmon pink	6	Forms elongated crystals in igneous rocks. Also, commonly found in sedimentary and metamorphic rocks.
Plagioclase feldspar	$NaAlSi_3O_8$ (albite) $CaAl_2Si_2O_8$ (anorthite)	Two cleavages at nearly right angles	White to gray	6	Forms elongated crystals in igneous rocks. Also commonly found in sedimentary and metamorphic rocks. Striations on some cleavage planes.

Mineral or Group Name	Composition	Cleavage/ Fracture	Color	Hardness	Other Properties/ Comments
Pyrite	FeS_2	Uneven fracture	Brass yellow	6–6.5	Occurs as granular masses or well-formed cubic crystals. High specific gravity (4.8–5.2). Often called "fool's gold."
Pyroxene (common member: augite)	Complex family of Mg, Fe, Ca, Na, and Al silicates	Good cleavage in two directions at nearly right angles	Green to black	5–6	Occurs as individual grains in igneous and metamorphic rocks.
Quartz	SiO_2	Conchoidal fracture	Colorless when pure	7	Common in all rock types. Often lightly colored, including gray, pink, yellow, and violet.
Serpentine	$Mg_3Si_2O_5(OH)_4$	Uneven fracture	Light to dark green	2.5–5	Fibrous variety is asbestos. Occurs most often in metamorphic rocks.
Sillimanite	Al_2SiO_5	One direction of cleavage	White to gray	6–7	High-grade metamorphic mineral.
Sphalerite	ZnS	Six directions of cleavage	Yellow to brown	3.5–4	Moderate specific gravity (4.1–4.3). Smell of sulfur when powdered. Ore of zinc.
Staurolite	$FeAl_4(SiO_4)_2(OH)_2$	Cleavage not prominent	Brown to reddish brown	7	Elongated crystals, occasionally twinned to form a cross-shaped crystal. Commonly found in metamorphic rocks.
Sulfur	S	Irregular fracture	Yellow	1.5–2.5	Bright yellow mineral most often associated with sedimentary deposits, in coal, and near volcanoes.
Talc	$Mg_3(Si_4O_{10})(OH)_2$	Good cleavage in one direction	White to light green	1–1.5	Soapy feel. Found in foliated masses consisting of thin flakes or scales. Most often associated with metamorphic rocks.
Wollastonite	$CaSiO_3$	Two perfect cleavages	Colorless to white	4.5–5	Forms fibrous or bladed crystals. Common in contact metamorphic rocks.

Topographic Maps

A map is a representation on a flat surface of all or a part of Earth's surface drawn to a specific scale. Maps are often the most effective means for showing the locations of both natural and human features, their sizes, and their relationships to one another. Like photographs, maps readily display information that would be impractical to express in words.

While most maps show only the two horizontal dimensions, geologists, as well as other map users, often require that the third dimension, elevation, be shown on maps. Maps that show the shape of the land are called **topographic maps**. Although various techniques may be used to depict elevations, the most accurate method involves the use of contour lines.

Contour Lines

A **contour line** is a line on a map representing a corresponding imaginary line on the ground that has the same elevation above sea level along its entire length. While many map symbols are pictographs, resembling the objects they represent, a contour line is an abstraction that has no counterpart in nature. It is, however, an accurate and effective device for representing the third dimension on paper.

Some useful facts and rules concerning contour lines are listed as follows. This information should be studied in conjunction with Figure C.1.

1. Contour lines bend upstream or upvalley. The contours form Vs that point upstream, and in the upstream direction the successive contours represent higher elevations. For example, if you were standing on a stream bank and wished to get to the point at the same elevation directly opposite you on the other bank, without stepping up or down, you would need to walk upstream along the contour at that elevation to where it crosses the stream bed, cross the stream, and then walk back downstream along the same contour.

2. Contours near the upper parts of hills form closures. The top of a hill is higher than the highest closed contour.

3. Hollows (depressions) without outlets are shown by closed, hatched contours. Hatched contours are contours with short lines on the inside pointing downslope.

4. Contours are widely spaced on gentle slopes.

5. Contours are closely spaced on steep slopes.

6. Evenly spaced contours indicate a uniform slope.

7. Contours usually do not cross or intersect each other, except in the rare case of an overhanging cliff.

8. All contours eventually close, either on a map or beyond its margins.

9. A single high contour never occurs between two lower ones, and vice versa. In other words, a change in slope direction is always determined by the repetition of the same elevation either as two different contours of the same value or as the same contour crossed twice.

10. Spot elevations between contours are given at many places, such as road intersections, hill summits, and lake surfaces. Spot elevations differ from control elevation stations, such as bench marks, in not being permanently established by permanent markers.

Relief

Relief refers to the difference in elevation between any two points. *Maximum relief* refers to the difference in elevation between the highest and lowest points in the area being considered. Relief determines the **contour interval**, which is the difference in elevation between succeeding contour lines that is used on topographic maps. Where relief is low, a small contour interval, such as 10 or 20 feet, may be used. In flat areas, such as wide river valleys or broad, flat uplands, a contour interval of 5 feet is often used. In rugged mountainous terrain, where relief is many hundreds of feet, contour intervals as large as 50 or 100 feet are used.

Scale

Map **scale** expresses the relationship between distance or area on the map to the true distance or area on Earth's surface. This is generally expressed as a ratio or fraction, such as 1:24,000 or 1/24,000. The numerator, usually 1, represents map distance, and the denominator, a large number, represents ground

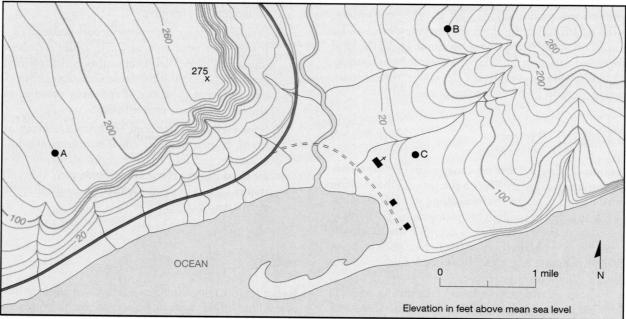

Elevation in feet above mean sea level

Figure C.1 Perspective view of an area and a contour map of the same area. These illustrations show how features are depicted on a topographic map. The upper illustration is a perspective view of a river valley and the adjoining hills. The river flows into a bay, which is partly enclosed by a hooked sandbar. On either side of the valley are terraces through which streams have cut gullies. The hill on the right has a smoothly eroded form and gradual slopes, whereas the one on the left rises abruptly in a sharp precipice, from which it slopes gently, and forms an inclined plateau traversed by a few shallow gullies. A road provides access to a church and the two houses situated across the river from a highway that follows the seacoast and curves up the river valley. The lower illustration shows the same features represented by symbols on a topographic map. The contour interval (vertical distance between adjacent contours) is 20 feet. (After U.S. Geological Survey)

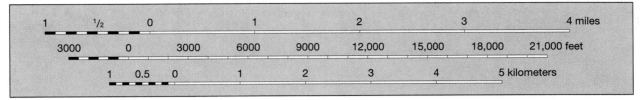

Figure C.2 Graphic scale.

distance. Thus, 1:24,000 means that a distance of 1 unit on the map represents a distance of 24,000 such units on the surface of Earth. It does not matter what the units are.

Often, the graphic or bar scale is more useful than the fractional scale, because it is easier to use for measuring distances between points. The graphic scale (Figure C.2) consists of a bar divided into equal segments, which represent equal distances on the map. One segment on the left side of the bar is usually divided into smaller units to permit more accurate estimates of fractional units.

Topographic maps, which are also referred to as *quadrangles*, are generally classified according to publication scale. Each series is intended to fulfill a specific type of map need. To select a map with the proper scale for a particular use, remember that large-scale maps show more detail and small-scale maps show less detail. The sizes and scales of topographic maps published by the U.S. Geological Survey are shown in Table C.1.

Color and Symbol

Each color and symbol used on U.S. Geological Survey topographic maps has significance. Common topographic map symbols are shown in Figure C.3. The meaning of each color is as follows:

Blue— water features

Black— works of humans, such as homes, schools, churches, roads, and so forth

Brown— contour lines

Green— woodlands, orchards, and so forth

Red— urban areas, important roads, public land subdivision lines

Table C.1 National Topographic Maps

Series	Scale	1 inch Represents	Standard Quadrangle Size (latitude-longitude)	Quadrangle Area (square miles)	Paper Size E-W N-S Width Length (inches)
7 1/2 -minute	1:24,000	2000 feet	7 1/2 ' × 7 1/2 '	49–70	22 × 27
Puerto Rico 7 1/2 -minute	1:20,000	about 1667 feet	7 1/2 ' × 7 1/2 '	71	29 1/2 × 32 1/2
15-minute	1:62,500	nearly 1 mile	15' × 15'	197–282	17 × 21
Alaska 1:63,360	1:63,360	1 mile	15' × 20'–36'	207–281	18 × 21
U.S. 1:250,000	1:250,000	nearly 4 miles	1° × 2°	4580–8669	34 × 22
U.S. 1:1,000,000	1:1,000,000	nearly 16 miles	4° × 6°	73,734–102,759	27 × 27

Source: U.S. Geological Survey

TOPOGRAPHIC MAP SYMBOLS

VARIATIONS WILL BE FOUND ON OLDER MAPS

Primary highway, hard surface	Boundaries: National
Secondary highway, hard surface	State
Light-duty road, hard or improved surface	County, parish, municipio
Unimproved road	Civil township, precinct, town, barrio
Road under construction, alinement known	Incorporated city, village, town, hamlet
Proposed road	Reservation, National or State
Dual highway, dividing strip 25 feet or less	Small park, cemetery, airport, etc.
Dual highway, dividing strip exceeding 25 feet	Land grant
Trail	Township or range line, United States land survey
	Township or range line, approximate location
Railroad: single track and multiple track	Section line, United States land survey
Railroads in juxtaposition	Section line, approximate location
Narrow gage: single track and multiple track	Township line, not United States land survey
Railroad in street and carline	Section line, not United States land survey
Bridge: road and railroad	Found corner: section and closing
Drawbridge: road and railroad	Boundary monument: land grant and other
Footbridge	Fence or field line
Tunnel: road and railroad	
Overpass and underpass	Index contour ... Intermediate contour
Small masonry or concrete dam	Supplementary contour ... Depression contours
Dam with lock	Fill ... Cut
Dam with road	Levee ... Levee with road
Canal with lock	Mine dump ... Wash
	Tailings ... Tailings pond
Buildings (dwelling, place of employment, etc.)	Shifting sand or dunes ... Intricate surface
School, church, and cemetery	Sand area ... Gravel beach
Buildings (barn, warehouse, etc.)	
Power transmission line with located metal tower	Perennial streams ... Intermittent streams
Telephone line, pipeline, etc. (labeled as to type)	Elevated aqueduct ... Aqueduct tunnel
Wells other than water (labeled as to type)	Water well and spring ... Glacier
Tanks: oil, water, etc. (labeled only if water)	Small rapids ... Small falls
Located or landmark object; windmill	Large rapids ... Large falls
Open pit, mine, or quarry; prospect	Intermittent lake ... Dry lake bed
Shaft and tunnel entrance	Foreshore flat ... Rock or coral reef
	Sounding, depth curve ... Piling or dolphin
Horizontal and vertical control station:	Exposed wreck ... Sunken wreck
Tablet, spirit level elevation BM△5653	Rock, bare or awash; dangerous to navigation
Other recoverable mark, spirit level elevation △5455	
Horizontal control station: tablet, vertical angle elevation VABM△95/9	Marsh (swamp) ... Submerged marsh
Any recoverable mark, vertical angle or checked elevation △3775	Wooded marsh ... Mangrove
Vertical control station: tablet, spirit level elevation BM×957	Woods or brushwood ... Orchard
Other recoverable mark, spirit level elevation ×954	Vineyard ... Scrub
Spot elevation ×7369 ×7369	Land subject to controlled inundation ... Urban area
Water elevation 670 670	

Figure C.3 U.S. Geological Survey topographic map symbols. (Variations may be found on older maps.)

APPENDIX D

Landforms of the Conterminous United States

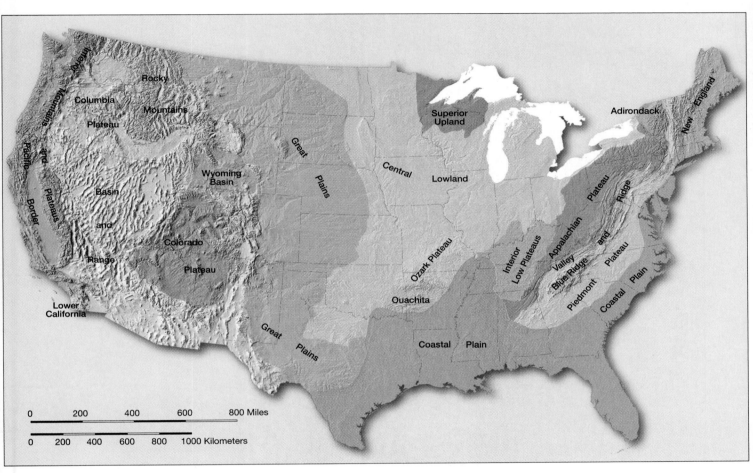

Figure D.1 Outline map showing major physiographic provinces of the United States.

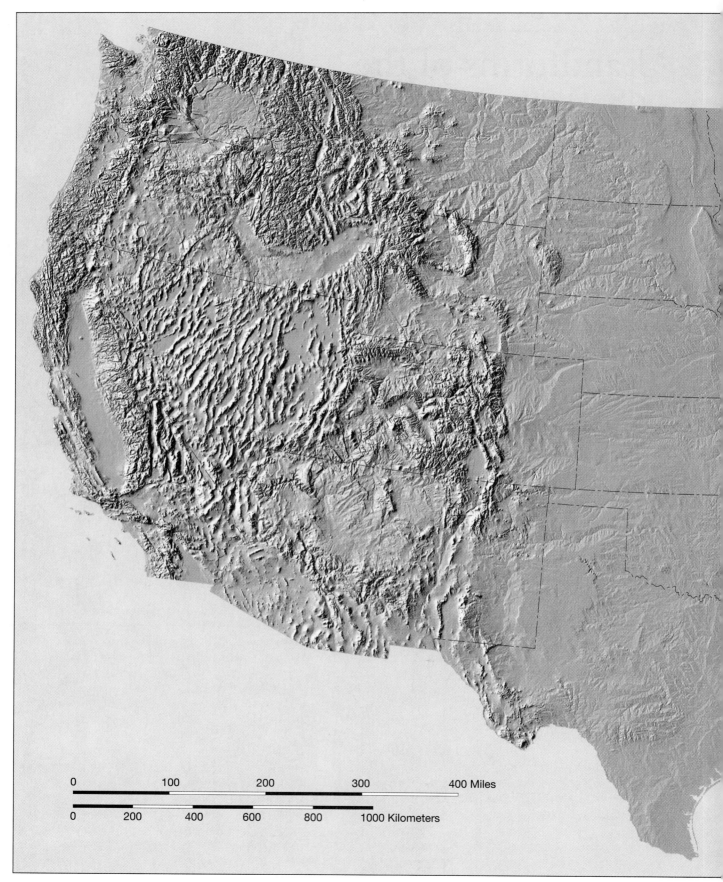

0 100 200 300 400 Miles

0 200 400 600 800 1000 Kilometers

Figure D.2 Digital shaded relief landform map of the United States (Data provided by the U.S. Geological Survey)

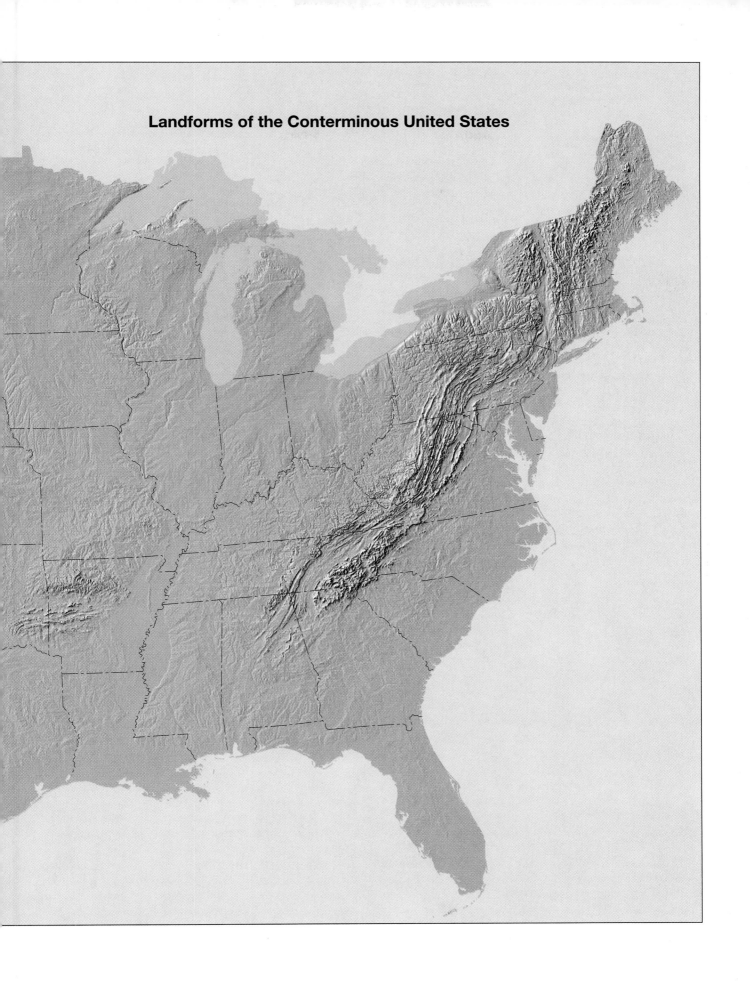

Landforms of the Conterminous United States

Glossary

Aa A type of lava flow that has a jagged, blocky surface.

Ablation A general term for the loss of ice and snow from a glacier.

Abrasion The grinding and scraping of a rock surface by the friction and impact of rock particles carried by water, wind, and ice.

Absolute dating Determination of the number of years since the occurrence of a given geologic event.

Abyssal plain Very level area of the deep-ocean floor, usually lying at the foot of the continental rise.

Accretionary wedge A large wedge-shaped mass of sediment that accumulates in subduction zones. Here sediment is scraped from the subducting oceanic plate and accreted to the overriding crustal block.

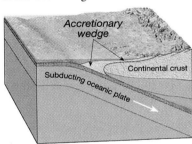

Active layer The zone above the permafrost that thaws in summer and refreezes in winter.

Aftershock A smaller earthquake that follows the main earthquake.

Alluvial fan A fan-shaped deposit of sediment formed when a stream's slope is abruptly reduced.

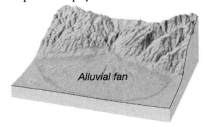

Alluvium Unconsolidated sediment deposited by a stream.

Alpine glacier A glacier confined to a mountain valley, which in most instances had previously been a stream valley.

Angle of repose The steepest angle at which loose material remains stationary without sliding downslope.

Angular unconformity An unconformity in which the older strata dip at an angle different from that of the younger beds.

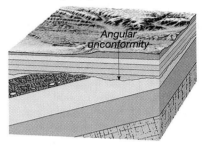

Antecedent stream A stream that continued to downcut and maintain its original course as an area along its course was uplifted by faulting or folding.

Anthracite A hard, metamorphic form of coal that burns cleanly and hot.

Anticline A fold in sedimentary strata that resembles an arch.

Aphanitic A texture of igneous rocks in which the crystals are too small for individual minerals to be distinguished with the unaided eye.

Aquiclude An impermeable bed that hinders or prevents groundwater movement.

Aquifer Rock or sediment through which groundwater moves easily.

Archeon eon The second eon of Precambrian time. The eon following the Hadean and preceding the Proterozoic. It extends between 3.8 and 2.5 billion years.

Arête A narrow, knifelike ridge separating two adjacent glaciated valleys.

Arkose A feldspar-rich sandstone.

Artesian well A well in which the water rises above the level where it was initially encountered.

Assimilation In igneous activity, the process of incorporating country rock into a magma body.

Asteroid One of thousands of small planetlike bodies, ranging in size from a few hundred kilometers to less than one kilometer across. Most asteroids' orbits lie between those of Mars and Jupiter.

Asthenosphere A subdivision of the mantle situated below the lithosphere. This zone of weak material exists below a depth of about 100 kilometers and in some regions extends as deep as 700 kilometers. The rock within this zone is easily deformed.

Astronomical theory A theory of climatic change first developed by the Yugoslavian astronomer Milankovitch. It is based on changes in the shape of Earth's orbit, variations in the obliquity of Earth's axis, and the wobbling of Earth's axis.

Atmosphere The gaseous portion of a planet, the planet's envelope of air. One of the traditional subdivisions of Earth's physical environment.

Atoll A continuous or broken ring of coral reef surrounding a central lagoon.

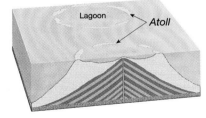

Atom The smallest particle that exists as an element.

Atomic mass unit A mass unit equal to exactly one-twelfth the mass of a carbon-12 atom.

Atomic number The number of protons in the nucleus of an atom.

Atomic weight The average of the atomic masses of isotopes for a given element.

Aureole A zone or halo of contact metamorphism found in the country rock surrounding an igneous intrusion.

611

Backswamp A poorly drained area on a floodplain resulting when natural levees are present.

Bajada An apron of sediment along a mountain front created by the coalescence of alluvial fans.

Barchan dune A solitary sand dune shaped like a crescent with its tips pointing downwind.

Wind

Barchan dune

Barchanoid dune Dunes forming scalloped rows of sand oriented at right angles to the wind. This form is intermediate between isolated barchans and extensive waves of transverse dunes.

Barrier island A low, elongate ridge of sand that parallels the coast.

Barrier island

Basal slip A mechanism of glacial movement in which the ice mass slides over the surface below.

Basalt A fine-grained igneous rock of mafic composition.

Base level The level below which a stream cannot erode.

Basin A circular downfolded structure.

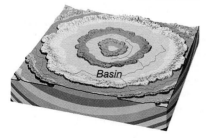

Basin

Batholith A large mass of igneous rock that formed when magma was emplaced at depth, crystallized, and was subsequently exposed by erosion.

Baymouth bar A sandbar that completely crosses a bay, sealing it off from the main body of water.

Baymouth bar

Beach drift The transport of sediment in a zigzag pattern along a beach caused by the uprush of water from obliquely breaking waves.

Beach nourishment Process in which large quantities of sand are added to the beach system to offset losses caused by wave erosion. Building beaches seaward improves beach quality and storm protection.

Bed See *strata*.

Bedding plane A nearly flat surface separating two beds of sedimentary rock. Each bedding plane marks the end of one deposit and the beginning of another having different characteristics.

Bed load Sediment moved along the bottom of a stream by moving water, or particles moved along the ground surface by wind.

Belt of soil moisture A zone in which water is held as a film on the surface of soil particles and may be used by plants or withdrawn by evaporation. The uppermost subdivision of the zone of aeration.

Benioff zone The zone of inclined seismic activity that extends from a trench downward into the asthenosphere.

Biochemical Describing a type of chemical sediment that forms when material dissolved in water is precipitated by water-dwelling organisms. Shells are common examples.

Biogenous sediment Seafloor sediments consisting of material of marine-organic origin.

Biosphere The totality of life-forms on Earth. The parts of the lithosphere, hydrosphere, and atmosphere in which living organisms can be found.

Bituminous coal The most common form of coal, often called soft, black coal.

Blowout (deflation hollow) A depression excavated by wind in easily eroded materials.

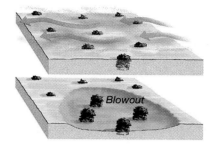

Blowout

Body wave A seismic wave that travels through Earth's interior.

Bottomset bed A layer of fine sediment deposited beyond the advancing edge of a delta and then buried by continued delta growth.

Bowen's reaction series A concept proposed by N. L. Bowen that illustrates the relationships between magma and the minerals crystallizing from it during the formation of igneous rocks.

Braided stream A stream consisting of numerous intertwining channels.

Breakwater A structure protecting a nearshore area from breaking waves.

Breccia A sedimentary rock composed of angular fragments that were lithified.

Caldera A large depression typically caused by collapse or ejection of the summit area of a volcano.

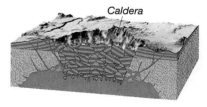

Caldera

Caliche A hard layer, rich in calcium carbonate, that forms beneath the *B* horizon in soils of arid regions.

Calving Wastage of a glacier that occurs when large pieces of ice break into the water.

Capacity The total amount of sediment a stream is able to transport.

Capillary fringe A relatively narrow zone at the base of the zone of aeration. Here water rises from the water table in tiny threadlike openings between grains of soil or sediment.

Cap rock A necessary part of an oil trap. The cap rock is impermeable and hence keeps upwardly mobile oil and gas from escaping at the surface.

Cassini gap A wide gap in the ring system of Saturn between the *A* ring and the *B* ring.

Catastrophism The concept that Earth was shaped by catastrophic events of a short-term nature.

Cavern A naturally formed underground chamber or series of chambers most commonly produced by solution activity in limestone.

Cementation One way in which sedimentary rocks are lithified. As material precipitates from water that percolates through the sediment, open spaces are filled and particles are joined into a solid mass.

Cenozoic era A time span on the geologic time scale beginning about 65 million years ago following the Mesozoic era.

Chemical sedimentary rock Sedimentary rock consisting of material that was precipitated from water by either inorganic or organic means.

Chemical weathering The processes by which the internal structure of a mineral is altered by the removal and/or addition of elements.

Cinder cone A rather small volcano built primarily of pyroclastics ejected from a single vent.

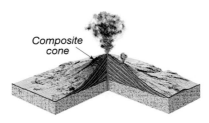
Cinder cone

Cirque An amphitheater-shaped basin at the head of a glaciated valley produced by frost wedging and plucking.

Clastic A sedimentary rock texture consisting of broken fragments of pre-existing rock.

Cleavage The tendency of a mineral to break along planes of weak bonding.

Col A pass between mountain valleys where the headwalls of two cirques intersect.

Color A phenomenon of light by which otherwise identical objects may be differentiated.

Column A feature found in caves that is formed when a stalactite and stalagmite join.

Columnar joints A pattern of cracks that forms during cooling of molten rock to generate columns.

Coma The fuzzy, gaseous component of a comet's head.

Comet A small body that generally revolves about the sun in an elongated orbit.

Compaction A type of lithification in which the weight of overlying material compresses more deeply buried sediment. It is most important in the fine-grained sedimentary rocks such as shale.

Competence A measure of the largest particle a stream can transport; a factor dependent on velocity.

Composite cone A volcano composed of both lava flows and pyroclastic material.

Composite cone

Compound A substance formed by the chemical combination of two or more elements in definite proportions and usually having properties different from those of its constituent elements.

Compressional stress A stress that pushes together material on either side of a real or imaginary plane.

Compressional stress

Concordant A term used to describe intrusive igneous masses that form parallel to the bedding of the surrounding rock.

Cone of depression A cone-shaped depression in the water table immediately surrounding a well.

Cone of depression

Confining pressure An equal, all-sided pressure.

Conformable layers Rock layers that were deposited without interruption.

Conglomerate A sedimentary rock composed of rounded gravel-sized particles.

Contact metamorphism Changes in rock caused by the heat from a nearby magma body.

Continental drift A hypothesis, credited largely to Alfred Wegener, that suggested all present continents once existed as a single supercontinent. Further, beginning about 200 million years ago, the supercontinent began breaking into smaller continents which then "drifted" to their present positions.

Continental margin That portion of the sea floor adjacent to the continents. It may include the continental shelf, continental slope, and continental rise.

Continental rise The gently sloping surface at the base of the continental slope.

Continental shelf The gently sloping submerged portion of the continental margin extending from the shoreline to the continental slope.

Continental slope The steep gradient that leads to the deep-ocean floor and marks the seaward edge of the continental shelf.

Convergent boundary A boundary in which two plates move together, causing one of the slabs of lithosphere to be consumed into the mantle as it descends beneath an overriding plate.

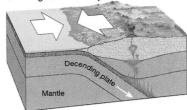
Convergent boundary

Decending plate

Mantle

Correlation Establishing the equivalence of rocks of similar age in different areas.

Covalent bond A chemical bond produced by the sharing of electrons.

Crater The depression at the summit of a volcano, or that which is produced by a meteorite impact.

Creep The slow downhill movement of soil and regolith.

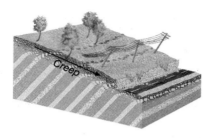

Crevasse A deep crack in the brittle surface of a glacier.

Cross-bedding Structure in which relatively thin layers are inclined at an angle to the main bedding. Formed by currents of wind or water.

Cross-cutting A principle of relative dating. A rock or fault is younger than any rock (or fault) through which it cuts.

Crust The very thin outermost layer of Earth.

Crystal An orderly arrangement of atoms.

Crystal form The external appearance of a mineral as determined by its internal arrangement of atoms.

Crystal settling During the crystallization of magma, the earlier-formed minerals are denser than the liquid portion and settle to the bottom of the magma chamber.

Crystallization The formation and growth of a crystalline solid from a liquid or gas.

Curie point The temperature above which a material loses its magnetization.

Cut bank The area of active erosion on the outside of a meander.

Cutoff A short channel segment created when a river erodes through the narrow neck of land between meanders.

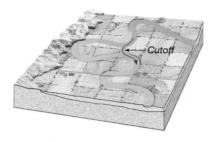

Darcy's law When permeability is uniform, the velocity of groundwater increases as the slope of the water table increases. It is expressed by the formula: $V = K(\frac{h}{l})$, where V is velocity, h the head, l the length of flow, and K the coefficient of permeability.

Daughter product An isotope resulting from radioactive decay.

Debris slide A slide involving downslope movement of relatively dry, unconsolidated regolith and rock debris. The mass does not exhibit backward rotation as in a slump but slides or rolls forward.

Deep-ocean basin The portion of sea floor that lies between the continental margin and the oceanic ridge system. This region comprises almost 30 percent of Earth's surface.

Deep-ocean trench A narrow, elongated depression of the sea floor.

Deflation The lifting and removal of loose material by wind.

Deformation General term for the processes of folding, faulting, shearing, compression, or extension of rocks as the result of various natural forces.

Delta An accumulation of sediment formed where a stream enters a lake or an ocean.

Dendritic pattern A stream system that resembles the pattern of a branching tree.

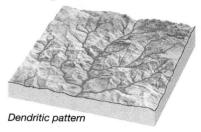

Dendritic pattern

Density The weight per unit volume of a particular material.

Desalination The removal of salts and other chemicals from seawater.

Desert One of the two types of dry climate; the driest of the dry climates.

Desert pavement A layer of coarse pebbles and gravel created when wind removed the finer material.

Detrital sedimentary rocks Rocks that form from the accumulation of materials that originate and are transported as solid particles derived from both mechanical and chemical weathering.

Differential weathering The variation in the rate and degree of weathering caused by such factors as mineral makeup, degree of jointing, and climate.

Dike A tabular-shaped intrusive igneous feature that cuts through the surrounding rock.

Dip The angle at which a rock layer or fault is inclined from the horizontal. The direction of dip is at a right angle to the strike.

Dip-slip fault A fault in which the movement is parallel to the dip of the fault.

Discharge The quantity of water in a stream that passes a given point in a period of time.

Disconformity A type of unconformity in which the beds above and below are parallel.

Discontinuity A sudden change with depth in one or more of the physical properties of the material making up Earth's interior. The boundary between two dissimilar materials in Earth's interior as determined by the behavior of seismic waves.

Discordant A term used to describe plutons that cut across existing rock structures, such as bedding planes.

Disseminated deposit Any economic mineral deposit in which the desired mineral occurs as scattered particles in the rock but in sufficient quantity to make the deposit an ore.

Dissolved load That portion of a stream's load carried in solution.

Distributary A section of a stream that leaves the main flow.

Diurnal tide A tide characterized by a single high and low water height each tidal day.

Divergent boundary A boundary in which two plates move apart, resulting in upwelling of material from the mantle to create new sea floor.

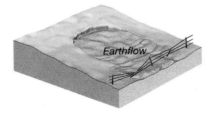

Divergent boundary

Divide An imaginary line that separates the drainage of two streams; often found along a ridge.

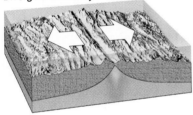

Divide →

Dome A roughly circular upfolded structure.

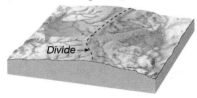

Dome

Drainage basin The land area that contributes water to a stream.

Drawdown The difference in height between the bottom of a cone of depression and the original height of the water table.

Drift The general term for any glacial deposit.

Drumlin A streamlined symmetrical hill composed of glacial till. The steep side of the hill faces the direction from which the ice advanced.

Dry climate A climate in which yearly precipitation is less than the potential loss of water by evaporation.

Dune A hill or ridge of wind-deposited sand.

Earthflow The downslope movement of water-saturated, clay-rich sediment. Most characteristic of humid regions.

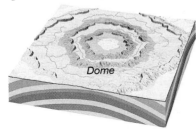

Earthflow

Earthquake Vibration of Earth produced by the rapid release of energy.

Ebb current The movement of tidal current away from the shore.

Echo sounder An instrument used to determine the depth of water by measuring the time interval between emission of a sound signal and the return of its echo from the bottom.

Effluent stream A stream channel that intersects the water table. Consequently, groundwater feeds into the stream.

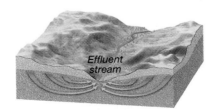

Effluent stream

Elastic deformation Nonpermanent deformation in which rock returns to its original shape when the stress is released.

Elastic rebound The sudden release of stored strain in rocks that results in movement along a fault.

Electron A negatively charged subatomic particle that has a negligible mass and is found outside an atom's nucleus.

Element A substance that cannot be decomposed into simpler substances by ordinary chemical or physical means.

Eluviation The washing out of fine soil components from the *A* horizon by downward-percolating water.

Emergent coast A coast where land formerly below sea level has been exposed by crustal uplift or a drop in sea level or both.

End moraine A ridge of till marking a former position of the front of a glacier.

Energy-level shell The region occupied by electrons with a specific energy level.

Entrenched meander A meander cut into bedrock when uplifting rejuvenated a meandering stream.

Eon The largest time unit on the geologic time scale, next in order of magnitude above era.

Ephemeral stream A stream that is usually dry because it carries water only in response to specific episodes of rainfall. Most desert streams are of this type.

Epicenter The location on Earth's surface that lies directly above the focus of an earthquake.

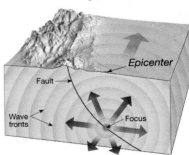

Epicenter
Fault →
Wave fronts →
Focus

Epoch A unit of the geologic time scale that is a subdivision of a period.

Era A major division on the geologic time scale; eras are divided into shorter units called periods.

Erosion The incorporation and transportation of material by a mobile agent, such as water, wind, or ice.

Esker Sinuous ridge composed largely of sand and gravel deposited by a stream flowing in a tunnel beneath a glacier near its terminus.

Estuary A funnel-shaped inlet of the sea that formed when a rise in sea level or subsidence of land caused the mouth of a river to be flooded.

Evaporite A sedimentary rock formed of material deposited from solution by evaporation of the water.

Evapotranspiration The combined effect of evaporation and transpiration.

Exfoliation dome Large, dome-shaped structure, usually composed of granite, formed by sheeting.

Exotic stream A permanent stream that traverses a desert and has its source in well-watered areas outside the desert.

Extrusive Igneous activity that occurs at Earth's surface.

Facies A portion of a rock unit that possesses a distinctive set of characteristics that distinguishes it from other parts of the same unit.

Fall A type of movement common to mass-wasting processes that refers to the free falling of detached individual pieces of any size.

Fault A break in a rock mass along which movement has occurred.

Fault-block mountain A mountain formed by the displacement of rock along a fault.

Fault scarp A cliff created by movement along a fault. It represents the exposed surface of the fault prior to modification by weathering and erosion.

Faunal succession Fossil organisms succeed one another in a definite and determinable order, and any time period can be recognized by its fossil content.

Fetch The distance that the wind has traveled across the open water.

Fiord A steep-sided inlet of the sea formed when a glacial trough was partially submerged.

Firn Granular recrystallized snow. A transitional stage between snow and glacial ice.

Fissility The property of splitting easily into thin layers along closely spaced, parallel surfaces, such as bedding planes in shale.

Fission (nuclear) The splitting of a heavy nucleus into two or more lighter nuclei caused by the collision with a neutron. During this process a large amount of energy is released.

Fissure eruption An eruption in which lava is extruded from narrow fractures or cracks in the crust.

Flood basalts Flows of basaltic lava that issue from numerous cracks or fissures and commonly cover extensive areas to thicknesses of hundreds of meters.

Floodplain The flat, low-lying portion of a stream valley subject to periodic inundation.

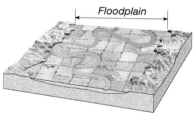

Flood current The tidal current associated with the increase in the height of the tide.

Flow A type of movement common to mass-wasting processes in which water-saturated material moves downslope as a viscous fluid.

Flowing artesian well An artesian well in which water flows freely at Earth's surface because the pressure surface is above ground level.

Fluorescence The absorption of ultraviolet light, which is re-emitted as visible light.

Focus (earthquake) The zone within Earth where rock displacement produces an earthquake.

Fold A bent layer or series of layers that were originally horizontal and subsequently deformed.

Foliated A texture of metamorphic rocks that gives the rock a layered appearance.

Foliation A term for a linear arrangement of textural features often exhibited by metamorphic rocks.

Foreset bed An inclined bed deposited along the front of a delta.

Foreshocks Small earthquakes that often precede a major earthquake.

Fossil The remains or traces of organisms preserved from the geologic past.

Fossil fuel General term for any hydrocarbon that may be used as a fuel, including coal, oil, natural gas, bitumen from tar sands, and shale oil.

Fracture Any break or rupture in rock along which no appreciable movement has taken place.

Frost wedging The mechanical breakup of rock caused by the expansion of freezing water in cracks and crevices.

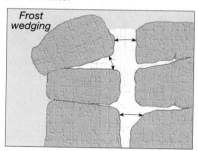

Fumarole A vent in a volcanic area from which fumes or gases escape.

Geology The science that examines Earth, its form and composition, and the changes that it has undergone and is undergoing.

Geothermal energy Natural steam used for power generation.

Geothermal gradient The gradual increase in temperature with depth in the crust. The average is 30°C per kilometer in the upper crust.

Geyser A fountain of hot water ejected periodically from the ground.

Glacial erratic An ice-transported boulder that was not derived from the bedrock near its present site.

Glacial striations Scratches and grooves on bedrock caused by glacial abrasion.

Glacial trough A mountain valley that has been widened, deepened, and straightened by a glacier.

Glacier A thick mass of ice originating on land from the compaction and recrystallization of snow that shows evidence of past or present flow.

Glass (volcanic) Natural glass produced when molten lava cools too rapidly to permit recrystallization. Volcanic glass is a solid composed of unordered atoms.

Glassy A term used to describe the texture of certain igneous rocks, such as obsidian, that contain no crystals.

Gondwanaland The southern portion of Pangaea consisting of South America, Africa, Australia, India, and Antarctica.

Graben A valley formed by the downward displacement of a fault-bounded block.

Graben

Graded bed A sediment layer characterized by a decrease in sediment size from bottom to top.

Graded stream A stream that has the correct channel characteristics to maintain exactly the velocity required to transport the material supplied to it.

Gradient The slope of a stream; generally expressed as the vertical drop over a fixed distance.

Greenhouse effect Carbon dioxide and water vapor in a planet's atmosphere absorb and re-radiate infrared wavelengths, effectively trapping solar energy and raising the temperature.

Groin A short wall built at a right angle to the seashore to trap moving sand.

Groins

Groundmass The matrix of smaller crystals within an igneous rock that has porphyritic texture.

Ground moraine An undulating layer of till deposited as the ice front retreats.

Groundwater Water in the zone of saturation.

Guyot A submerged flat-topped seamount.

Hadean eon The first eon on the geologic time scale. The eon ending 3.8 billion years ago that preceded the Archean eon.

Half-life The time required for one-half of the atoms of a radioactive substance to decay.

Hanging valley A tributary valley that enters a glacial trough at a considerable height above the floor of the trough.

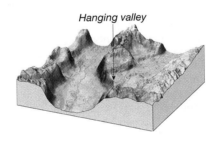

Hanging valley

Hardness A mineral's resistance to scratching and abrasion.

Head The vertical distance between the recharge and discharge points of a water table. Also the source area or beginning of a valley.

Headward erosion The extension upslope of the head of a valley due to erosion.

Historical geology A major division of geology that deals with the origin of Earth and its development through time. Usually involves the study of fossils and their sequence in rock beds.

Hogback A narrow, sharp-crested ridge formed by the upturned edge of a steeply dipping bed of resistant rock.

Horizon A layer in a soil profile.

Horn A pyramid-like peak formed by glacial action in three or more cirques surrounding a mountain summit.

Horst An elongate, uplifted block of crust bounded by faults.

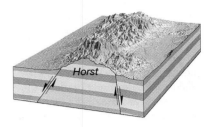

Horst

Hot spot A concentration of heat in the mantle capable of producing magma which, in turn, extrudes onto Earth's surface. The intraplate volcanism that produced the Hawaiian Islands is one example.

Hot spring A spring in which the water is 6–9°C (10–15°F) warmer than the mean annual air temperature of its locality.

Humus Organic matter in soil produced by the decomposition of plants and animals.

Hydraulic gradient The slope of the water table. Expressed as h/l, where h is the head and l the length of flow.

Hydroelectric power Electricity generated by falling water that is used to drive turbines.

Hydrogenous sediment Seafloor sediment consisting of minerals that crystallize from seawater. An important example is manganese nodules.

Hydrologic cycle The unending circulation of Earth's water supply. The cycle is powered by energy from the sun and is characterized by continuous exchanges of water among the oceans, the atmosphere, and the continents.

Hydrolysis A chemical weathering process in which minerals are altered by chemically reacting with water and acids.

Hydrosphere The water portion of our planet; one of the traditional subdivisions of Earth's physical environment.

Hydrothermal solution The hot, watery solution that escapes from a mass of magma during the latter stages of crystallization. Such solutions may alter the surrounding country rock and are frequently the source of significant ore deposits.

Hypothesis A tentative explanation that is then tested to determine if it is valid.

Ice cap A mass of glacial ice covering a high upland or plateau and spreading out radially.

Ice-contact deposit An accumulation of stratified drift deposited in contact with a supporting mass of ice.

Ice sheet A very large, thick mass of glacial ice flowing outward in all directions from one or more accumulation centers.

Ice shelf Forming where glacial ice flows into bays, it is a large, relatively flat mass of floating ice that extends seaward from the coast but remains attached to the land along one or more sides.

Igneous rock Rock formed from the crystallization of magma.

Immature soil A soil lacking horizons.

Inclusion A piece of one rock unit contained within another. Inclusions are used in relative dating. The rock mass adjacent to the one containing the inclusion must have been there first in order to provide the fragment.

Index fossil A fossil that is associated with a particular span of geologic time.

Index mineral A mineral that is a good indicator of the metamorphic environment in which it formed. Used to distinguish different zones of regional metamorphism.

Inertia Objects at rest tend to remain at rest and objects in motion tend to stay in motion unless either is acted upon by an outside force.

Infiltration The movement of surface water into rock or soil through cracks and pore spaces.

Infiltration capacity The maximum rate at which soil can absorb water.

Influent stream A stream channel that is above the water table level. Water seeps downward from the channel to the zone of saturation to produce an upward bulge in the water table.

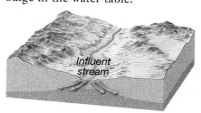
Influent stream

Inner core The solid innermost layer of Earth, about 1216 kilometers (754 miles) in radius.

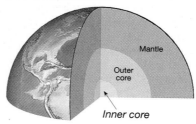
Mantle
Outer core
Inner core

Inselberg An isolated mountain remnant characteristic of the late stage of erosion in a mountainous arid region.

Intensity (earthquake) An indication of the destructive effects of an earthquake at a particular place. Intensity is affected by such factors as distance to the epicenter and the nature of the surface materials.

Interior drainage A discontinuous pattern of intermittent streams that do not flow to the ocean.

Intrusive rock Igneous rock that formed below Earth's surface.

Ion An atom or molecule that possesses an electrical charge.

Ionic bond A chemical bond between two oppositely charged ions formed by the transfer of valence electrons from one atom to the other.

Irons One of the three main categories of meteorites. This group is composed largely of iron with varying amounts of nickel (5–20 percent). Most meteorite finds are irons.

Island arc A chain of volcanic islands generally located a few hundred kilometers from a trench where active subduction of one oceanic slab beneath another is occurring.

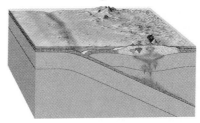

Isostasy The concept that Earth's crust is "floating" in gravitational balance upon the material of the mantle.

Isostatic adjustment Compensation of the lithosphere when weight is added or removed. When weight is added, the lithosphere will respond by subsiding, and when weight is removed there will be uplift.

Isotopes Varieties of the same element that have different mass numbers; their nuclei contain the same number of protons but different numbers of neutrons.

Jetties A pair of structures extending into the ocean at the entrance to a harbor or river that are built for the purpose of protecting against storm waves and sediment deposition.

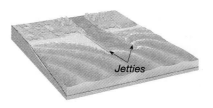
Jetties

Joint A fracture in rock along which there has been no movement.

Jovian planet One of the Jupiter-like planets; Jupiter, Saturn, Uranus, and Neptune. These planets have relatively low densities.

Kame A steep-sided hill composed of sand and gravel originating when sediment collected in openings in stagnant glacial ice.

Kame terrace A narrow, terrace-like mass of stratified drift deposited between a glacier and an adjacent valley wall.

Karst A topography consisting of numerous depressions called sinkholes.

Kettle holes Depressions created when blocks of ice become lodged in glacial deposits and subsequently melt.

Klippe A remnant or outlier of a thrust sheet that was isolated by erosion.

Laccolith A massive igneous body intruded between preexisting strata.

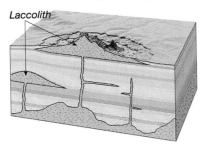

Laccolith

Lag time The amount of time between a rainstorm and the occurrence of flooding.

Lahar Mudflows on the slopes of volcanoes that result when unstable layers of ash and debris become saturated and flow downslope, usually following stream channels.

Laminar flow The movement of water particles in straight-line paths that are parallel to the channel. The water particles move downstream without mixing.

Lateral moraine A ridge of till along the sides of a valley glacier composed primarily of debris that fell to the glacier from the valley walls.

Laterite A red, highly leached soil type found in the tropics that is rich in oxides of iron and aluminum.

Laurasia The northern portion of Pangaea consisting of North America and Eurasia.

Lava Magma that reaches Earth's surface.

Lava dome A bulbous mass associated with an old-age volcano, produced when thick lava is slowly squeezed from the vent. Lava domes may act as plugs to deflect subsequent gaseous eruptions.

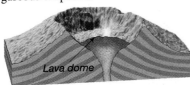

Lava dome

Law A formal statement of the regular manner in which a natural phenomenon occurs under given conditions; e.g., the "law of superposition."

Law of superposition In any undeformed sequence of sedimentary rocks, each bed is older than the one above and younger than the one below.

Leaching The depletion of soluble materials from the upper soil by downward-percolating water.

Liquefaction The transformation of a stable soil into a fluid that is often unable to support buildings or other structures.

Lithification The process, generally cementation and/or compaction, of converting sediments to solid rock.

Lithosphere The rigid outer layer of Earth, including the crust and upper mantle.

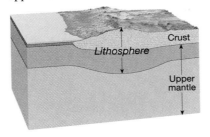

Local base level See *Temporary base level.*

Loess Deposits of windblown silt, lacking visible layers, generally buff-colored, and capable of maintaining a nearly vertical cliff.

Longitudinal dunes Long ridges of sand oriented parallel to the prevailing wind; these dunes form where sand supplies are limited.

Longitudinal dunes

Longitudinal profile A cross-section of a stream channel along its descending course from the head to the mouth.

Longshore current A nearshore current that flows parallel to the shore.

Long (L) waves These earthquake-generated waves travel along the outer layer of Earth and are responsible for most of the surface damage. L waves have longer periods than other seismic waves.

Low-velocity zone A subdivision of the mantle located between 100 and 250 kilometers and discernible by a marked decrease in the velocity of seismic waves. This zone does not encircle Earth.

Lunar breccia A lunar rock formed when angular fragments and dust are welded together by the heat generated by the impact of a meteoroid.

Lunar regolith A thin, gray layer on the surface of the moon, consisting of loosely compacted, fragmented material believed to have been formed by repeated meteoritic impacts.

Luster The appearance or quality of light reflected from the surface of a mineral.

Magma A body of molten rock found at depth, including any dissolved gases and crystals.

Magma mixing The process of altering the composition of a magma through the mixing of material from another magma body.

Magmatic differentiation The process of generating more than one rock type from a single magma.

Magnetometer A sensitive instrument used to measure the intensity of Earth's magnetic field at various points.

Magnitude (earthquake) The total amount of energy released during an earthquake.

Manganese nodules A type of hydrogenous sediment scattered on the ocean floor, consisting mainly of manganese and iron and usually containing small amounts of copper, nickel, and cobalt.

Mantle The 2885-kilometer (1789-mile) thick layer of Earth located below the crust.

Maria The smooth areas on our moon's surface that were incorrectly thought to be seas.

Massive An igneous pluton that is not tabular in shape.

Mass number The sum of the number of neutrons and protons in the nucleus of an atom.

Mass wasting The downslope movement of rock, regolith, and soil under the direct influence of gravity.

Meander A looplike bend in the course of a stream.

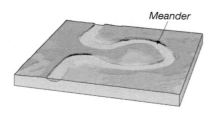

Meander

Meander scar A floodplain feature created when an oxbow lake becomes filled with sediment.

Mechanical weathering The physical disintegration of rock, resulting in smaller fragments.

Medial moraine A ridge of till formed when lateral moraines from two coalescing alpine glaciers join.

Melt The liquid portion of magma excluding the solid crystals.

Mercalli intensity scale A 12-point scale originally developed to evaluate earthquake intensity based on the amount of damage to various structures.

Mesozoic era A time span on the geologic time scale between the Paleozoic and Cenozoic eras—from about 225 to 65 million years ago.

Metallic bond A chemical bond present in all metals that may be characterized as an extreme type of electron sharing in which the electrons move freely from atom to atom.

Metamorphic rock Rock formed by the alteration of preexisting rock deep within Earth (but still in the solid state) by heat, pressure, and/or chemically active fluids.

Metamorphism The changes in mineral composition and texture of a rock subjected to high temperatures and pressures within Earth.

Meteorite Any portion of a meteoroid that survives its traverse through Earth's atmosphere and strikes the surface.

Meteoroid Any small solid particle that has an orbit in the solar system.

Meteor shower Numerous meteoroids traveling in the same direction and at nearly the same speed. They are thought to be material lost by comets.

Micrometeorite A very small meteorite that does not create sufficient friction to burn up in the atmosphere, but slowly drifts down to Earth.

Mid-ocean ridge A continuous mountainous ridge on the floor of all the major ocean basins and varying in width from 500 to 5000 kilometers (300 to 3000 miles). The rifts at the crests of these ridges represent divergent plate boundaries.

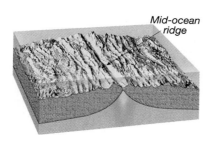

Mid-ocean ridge

Migmatite A rock exhibiting both igneous and metamorphic rock characteristics. Such rocks may form when light-colored silicate minerals melt and then crystallize, while the dark silicate minerals remain solid.

Mineral A naturally occurring, inorganic crystalline material with a unique chemical structure.

Mineral resource All discovered and undiscovered deposits of a useful mineral that can be extracted now or at some time in the future.

Mohorovičić discontinuity (Moho) The boundary separating the crust and the mantle, discernible by an increase in seismic velocity.

Mohs scale A series of ten minerals used as a standard in determining hardness.

Monocline A one-limbed flexure in strata. The strata are usually flat lying or very gently dipping on both sides of the monocline.

Mouth The point downstream where a river empties into another stream or water body.

Mud crack A feature in some sedimentary rocks that forms when wet mud dries out, shrinks, and cracks.

Mudflow The flowage of debris containing a large amount of water; most characteristic of canyons and gullies in dry, mountainous regions.

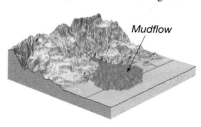

Mudflow

Natural levees The elevated landforms composed of alluvium that parallel some streams and act to confine their waters, except during floodstage.

Neap tide The lowest tidal range, occurring near the times of the first and third quarters of the moon.

Nebular hypothesis A model for the origin of the solar system that supposes a rotating nebula of dust and gases that contracted to form the sun and planets.

Neutron A subatomic particle found in the nucleus of an atom. The neutron is electrically neutral with a mass approximately equal to that of a proton.

Nonclastic A term for the texture of sedimentary rocks in which the minerals form a pattern of interlocking crystals.

Nonconformity An unconformity in which older metamorphic or intrusive igneous rocks are overlain by younger sedimentary strata.

Nonflowing artesian well An artesian well in which water does not rise to the surface because the pressure surface is below ground level.

Nonfoliated Metamorphic rocks that do not exhibit foliation.

Nonmetallic mineral resource Mineral resource that is not a fuel or processed for the metals it contains.

Nonrenewable resource Resource that forms or accumulates over such long time spans that it must be considered as fixed in total quantity.

Normal fault A fault in which the rock above the fault plane has moved down relative to the rock below.

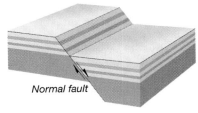

Normal fault

Normal polarity A magnetic field the same as that which presently exists.

Nuclear fission The splitting of atomic nuclei into smaller nuclei, causing neutrons to be emitted and heat energy to be released.

Nucleus The small, heavy core of an atom that contains all of its positive charge and most of its mass.

Nuée ardente Incandescent volcanic debris buoyed up by hot gases that moves downslope in an avalanche fashion.

Oblique-slip fault A fault having both vertical and horizontal movement.

Occultation The disappearance of light resulting when one object passes behind an apparently larger one. For example, the passage of Uranus in front of a distant star.

Oceanic ridge system See *Mid-ocean ridge.*

Octet rule Atoms combine in order that each may have the electron arrangement of a noble gas; that is, the outer energy level contains eight neutrons.

Oil trap A geologic structure that allows for significant amounts of oil and gas to accumulate.

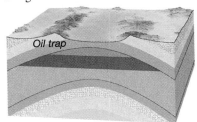

Oil trap

Ophiolite complex The sequence of rocks that make up the oceanic crust. The three-layer sequence includes an upper layer of pillow basalts, a middle zone of sheeted dikes, and a lower layer of gabbro.

Ore Usually a useful metallic mineral that can be mined at a profit. The term is also applied to certain nonmetallic minerals such as fluorite and sulfur.

Original horizontality Layers of sediment are generally deposited in a horizontal or nearly horizontal position.

Orogenesis The processes that collectively result in the formation of mountains.

Outer core A layer beneath the mantle about 2270 kilometers (1410 miles) thick which has the properties of a liquid.

Outlet glacier A tongue of ice normally flowing rapidly outward from an ice cap or ice sheet, usually through mountainous terrain to the sea.

Outwash plain A relatively flat, gently sloping plain consisting of materials deposited by meltwater streams in front of the margin of an ice sheet.

Oxbow lake A curved lake produced when a stream cuts off a meander.

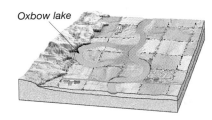

Oxbow lake

Oxidation The removal of one or more electrons from an atom or ion. So named because elements commonly combine with oxygen.

Pahoehoe A lava flow with a smooth-to-ropy surface.

Paleomagnetism The natural remnant magnetism in rock bodies. The permanent magnetization acquired by rock which can be used to determine the location of the magnetic poles and the latitude of the rock at the time it became magnetized.

Paleontology The systematic study of fossils and the history of life on Earth.

Paleozoic era A time span on the geologic time scale between the Precambrian and Mesozoic eras—from about 600 million to 225 million years ago.

Pangaea The proposed supercontinent which 200 million years ago began to break apart and form the present landmasses.

Parabolic dune A sand dune similar in shape to a barchan dune except that its tips point into the wind. These dunes often form along coasts that have strong onshore winds, abundant sand, and vegetation that partly covers the sand.

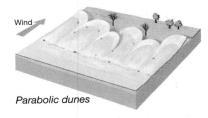

Parabolic dunes

Parasitic cone A volcanic cone that forms on the flank of a larger volcano.

Parent material The material upon which a soil develops.

Partial melting The process by which most igneous rocks melt. Since individual minerals have different melting points, most igneous rocks melt over a temperature range of a few hundred degrees. If the liquid is squeezed out after some melting has occurred, a melt with a higher silica content results.

Passive margin An inactive continental margin that is characterized by a thick accumulation of undeformed sediments and sedimentary rocks.

Pater noster lakes A chain of small lakes in a glacial trough that occupies basins created by glacial erosion.

Pedalfer Soil of humid regions characterized by the accumulation of iron oxides and aluminum-rich clays in the *B* horizon.

Pediment A sloping bedrock surface fringing a mountain base in an arid region, formed when erosion causes the mountain in front to retreat.

Pedocal Soil associated with drier regions and characterized by an accumulation of calcium carbonate in the upper horizons.

Pegmatite A very coarse-grained igneous rock (typically granite) commonly found as a dike associated with a large mass of plutonic rock that has smaller crystals. Crystallization in a water-rich environment is believed to be responsible for the very large crystals.

Peneplain In the idealized cycle of landscape evolution in a humid region, an undulating plain near base level associated with old age.

Perched water table A localized zone of saturation above the main water table created by an impermeable layer (aquiclude).

Peridotite An igneous rock of ultramafic composition thought to be abundant in the upper mantle.

Period A basic unit of the geologic time scale that is a subdivision of an era. Periods may be divided into smaller units called epochs.

Permafrost Any permanently frozen subsoil. Usually found in the subarctic and arctic regions.

Permeability A measure of a material's ability to transmit water.

Phaneritic An igneous rock texture in which the crystals are roughly equal in size and large enough so the individual minerals can be identified with the unaided eye.

Phanerozoic eon That part of geologic time represented by rocks containing abundant fossil evidence. The eon extending from the end of the Proterozoic eon (570 million years ago) to the present.

Phenocryst Conspicuously large crystal embedded in a matrix of finer-grained crystals.

Physical geology A major division of geology that examines the materials of Earth and seeks to understand the processes and forces acting beneath and upon Earth's surface.

Piedmont glacier A glacier that forms when one or more alpine glaciers emerge from the confining walls of mountain valleys and spread out to create a broad sheet in the lowlands at the base of the mountains.

Pillow lava Basaltic lava that solidifies in an underwater environment and develops a structure that resembles a pile of pillows.

Pipe A vertical conduit through which magmatic materials have passed.

Placer Deposit formed when heavy minerals are mechanically concentrated by currents, most commonly streams and waves. Placers are sources of gold, tin, platinum, diamonds, and other valuable minerals.

Plastic deformation Permanent deformation that results in a change in size and shape through folding or flowing.

Plastic flow A type of glacial movement that occurs within the glacier, below a depth of approximately 50 meters, in which the ice is not fractured.

Plate One of numerous rigid sections of the lithosphere that moves as a unit over the material of the asthenosphere.

Plate tectonics The theory which proposes that the earth's outer shell consists of individual plates which interact in various ways and thereby produce earthquakes, volcanoes, mountains, and the crust itself.

Playa The flat central area of an undrained desert basin.

Playa lake A temporary lake in a playa.

Playfair's law A well-known and oft-quoted statement by John Playfair which states that a valley is the result of the work of the stream that flows in it.

Pleistocene epoch An epoch of the Quaternary period beginning about 2.5 million years ago and ending about 10,000 years ago. Best known as a time of extensive continental glaciation.

Plucking (quarrying) The process by which pieces of bedrock are lifted out of place by a glacier.

Pluton A structure that results from the emplacement and crystallization of magma beneath the surface of Earth.

Pluvial lake A lake formed during a period of increased rainfall. For example, this occurred in many non-glaciated areas during periods of ice advance elsewhere.

Point bar A crescent-shaped accumulation of sand and gravel deposited on the inside of a meander.

Polymorphs Two or more minerals having the same chemical composition but different crystalline structures. Exemplified by the diamond and graphite forms of carbon.

Porosity The volume of open spaces in rock or soil.

Porphyritic An igneous rock texture characterized by two distinctively different crystal sizes. The larger crystals are called phenocrysts, whereas the matrix of smaller crystals is termed the groundmass.

Porphyry An igneous rock with a porphyritic texture.

Pothole A depression formed in a stream channel by the abrasive action of the water's sediment load.

Precambrian All geologic time prior to the Paleozoic era.

Primary (P) wave A type of seismic wave that involves alternating compression and expansion of the material through which it passes.

Principle of faunal succession Fossil organisms succeed one another in a definite and determinable order, and any time period can be recognized by its fossil content.

Principle of original horizontality Layers of sediment are generally deposited in a horizontal or nearly horizontal position.

Proterozoic eon The eon following the Archean and preceding the Phanerozoic. It extends between 2500 and 570 million years ago.

Proton A positively charged subatomic particle found in the nucleus of an atom.

P wave The fastest earthquake wave, which travels by compression and expansion of the medium.

Pyroclastic An igneous rock texture resulting from the consolidation of individual rock fragments that are ejected during a violent eruption.

Pyroclastic flow A highly heated mixture, largely of ash and pumice fragments, traveling down the flanks of a volcano or along the surface of the ground.

Pyroclastic material The volcanic rock ejected during an eruption. Pyroclastics include ash, bombs, and blocks.

Radial drainage A system of streams running in all directions away from a central elevated structure, such as a volcano.

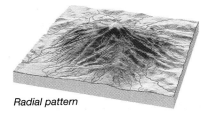

Radial pattern

Radioactivity The spontaneous decay of certain unstable atomic nuclei.

Radiocarbon (carbon-14) The radioactive isotope of carbon, which is produced continuously in the atmosphere and used in dating events as far back as 75,000 years.

Radiometric dating The procedure of calculating the absolute ages of rocks and minerals that contain certain radioactive isotopes.

Rainshadow desert A dry area on the lee side of a mountain range. Many middle-latitude deserts are of this type.

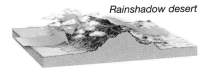

Rainshadow desert

Rapids A part of a stream channel in which the water suddenly begins flowing more swiftly and turbulently because of an abrupt steepening of the gradient.

Rays Bright streaks that appear to radiate from certain craters on the lunar surface. The rays consist of fine debris ejected from the primary crater.

Recessional moraine An end moraine formed as the ice front stagnated during glacial retreat.

Rectangular pattern A drainage pattern characterized by numerous right angle bends that develops on jointed or fractured bedrock.

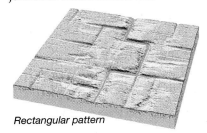

Rectangular pattern

Refraction A change in direction of waves as they enter shallow water. The portion of the wave in shallow water is slowed, which causes the wave to bend and align with the underwater contours.

Regional metamorphism Metamorphism associated with large-scale mountain building.

Regolith The layer of rock and mineral fragments that nearly everywhere covers Earth's land surface.

Rejuvenation A change in relation to base level, often caused by regional uplift, that causes the forces of erosion to intensify.

Relative dating Rocks and structures are placed in their proper sequence or order. Only the chronological order of events is determined.

Renewable resource A resource that is virtually inexhaustible or that can be replenished over relatively short time spans.

Reserve Already identified deposits from which minerals can be extracted profitably.

Reverse fault

Reservoir rock The porous, permeable portion of an oil trap that yields oil and gas.

Residual soil Soil developed directly from the weathering of the bedrock below.

Reverse fault A fault in which the material above the fault plane moves up in relation to the material below.

Reverse polarity A magnetic field opposite to that which presently exists.

Richter scale A scale of earthquake magnitude based on the motion of a seismograph.

Rift A region of Earth's crust along which divergence is taking place.

Rift zone See *Rift*.

Rills Tiny channels that develop as unconfined flow begins producing threads of current.

Ripple marks Small waves of sand that develop on the surface of a sediment layer by the action of moving water or air.

Roche moutonnée An asymmetrical knob of bedrock formed when glacial abrasion smoothes the gentle slope facing the advancing ice sheet and plucking steepens the opposite side as the ice overrides the knob.

Rock A consolidated mixture of minerals.

Rock avalanche The very rapid downslope movement of rock and debris. These rapid movements may be aided by a layer of air trapped beneath the debris, and they have been known to reach speeds in excess of 200 kilometers per hour.

Rock cleavage The tendency of rocks to split along parallel, closely spaced surfaces. These surfaces are often highly inclined to the bedding planes in the rock.

Rock cycle A model that illustrates the origin of the three basic rock types and the interrelatedness of Earth materials and processes.

Rock flour Ground-up rock produced by the grinding effect of a glacier.

Rockslide The rapid slide of a mass of rock downslope along planes of weakness.

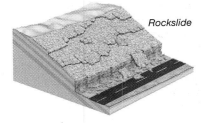

Rockslide

Runoff Water that flows over the land rather than infiltrating into the ground.

Salinity The proportion of dissolved salts to pure water, usually expressed in parts per thousand (0/000).

Saltation Transportation of sediment through a series of leaps or bounces.

Salt flat A white crust on the ground produced when water evaporates and leaves its dissolved materials behind.

Schistosity A type of foliation characteristic of coarser-grained metamorphic rocks. Such rocks have a parallel arrangement of platy minerals such as the micas.

Scoria Hardened lava that has retained the vesicles produced by the escaping gases.

Sea arch An arch formed by wave erosion when caves on opposite sides of a headland unite.

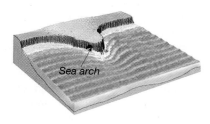

Sea arch

Seafloor spreading The hypothesis first proposed in the 1960s by Harry Hess which suggested that new oceanic crust is produced at the crests of mid-ocean ridges, which are the sites of divergence.

Seamount An isolated volcanic peak that rises at least 1000 meters (3300 feet) above the deep-ocean floor.

Sea stack An isolated mass of rock standing just offshore, produced by wave erosion of a headland.

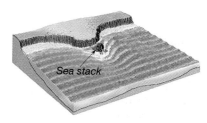

Sea stack

Seawall A barrier constructed to prevent waves from reaching the area behind the wall. Its purpose is to defend property from the force of breaking waves.

Secondary enrichment The concentration of minor amounts of metals that are scattered through unweathered rock into economically valuable concentrations by weathering processes.

Secondary (S) wave A seismic wave that involves oscillation perpendicular to the direction of propagation.

Sediment Unconsolidated particles created by the weathering and erosion of rock, by chemical precipitation from solution in water, or from the secretions of organisms, and transported by water, wind, or glaciers.

Sedimentary rock Rock formed from the weathered products of pre-existing rocks that have been transported, deposited, and lithified.

Seismic sea wave A rapidly moving ocean wave generated by earthquake activity, which is capable of inflicting heavy damage in coastal regions.

Seismogram The record made by a seismograph.

Seismograph An instrument that records earthquake waves.

Seismology The study of earthquakes and seismic waves.

Settling velocity The speed at which a particle falls through a still fluid. The size, shape, and specific gravity of particles influence settling velocity.

Shadow zone The zone between 105 and 140 degrees distance from an earthquake epicenter which direct waves do not penetrate because of refraction by Earth's core.

Shear Stress that causes two adjacent parts of a body to slide past one another.

Sheeted dikes A large group of nearly parallel dikes.

Sheet flow Runoff moving in unconfined thin sheets.

Sheeting A mechanical weathering process characterized by the splitting off of slablike sheets of rock.

Shelf break The point at which a rapid steepening of the gradient occurs, marking the outer edge of the continental shelf and the beginning of the continental slope.

Shield A large, relatively flat expanse of ancient metamorphic rock within the stable continental interior.

Shield volcano A broad, gently sloping volcano built from fluid basaltic lavas.

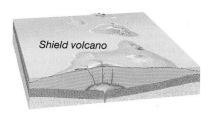

Shield volcano

Silicate Any one of numerous minerals that have the silicon-oxygen tetrahedron as their basic structure.

Silicon-oxygen tetrahedron A structure composed of four oxygen atoms surrounding a silicon atom that constitutes the basic building block of silicate minerals.

Sill A tabular igneous body that was intruded parallel to the layering of pre-existing rock.

Sinkhole A depression produced in a region where soluble rock has been removed by groundwater.

Sinkhole

Slaty cleavage The type of foliation characteristic of slates in which there is a parallel arrangement of fine-grained metamorphic minerals.

Slide A movement common to mass-wasting processes in which the material moving downslope remains fairly coherent and moves along a well-defined surface.

Slip face The steep, leeward surface of a sand dune which maintains a slope of about 34 degrees.

Slump The downward slipping of a mass of rock or unconsolidated material moving as a unit along a curved surface.

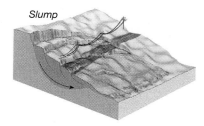

Slump

Snowfield An area where snow persists throughout the year.

Snowline The lower limit of perennial snow.

Soil A combination of mineral and organic matter, water, and air; that portion of the regolith that supports plant growth.

Soil horizon A layer of soil that has identifiable characteristics produced by chemical weathering and other soil-forming processes.

Soil profile A vertical section through a soil showing its succession of horizons and the underlying parent material.

Solifluction Slow, downslope flow of water-saturated materials common to permafrost areas.

Solum The *O*, *A*, and *B* horizons in a soil profile. Living roots and other plant and animal life are largely confined to this zone.

Solution The change of matter from the solid or gaseous state into the liquid state by its combination with a liquid.

Sorting The degree of similarity in particle size in sediment or sedimentary rock.

Specific gravity The ratio of a substance's weight to the weight of an equal volume of water.

Speleothem A collective term for the dripstone features found in caverns.

Spheroidal weathering Any weathering process that tends to produce a spherical shape from an initially blocky shape.

Spit An elongate ridge of sand that projects from the land into the mouth of an adjacent bay.

Spring A flow of groundwater that emerges naturally at the ground surface.

Spring tide The highest tidal range. Occurs near the times of the new and full moons.

Stalactite The iciclelike structure that hangs from the ceiling of a cavern.

Stalagmite The columnlike form that grows upward from the floor of a cavern.

Star dune An isolated hill of sand that exhibits a complex form and develops where wind directions are variable.

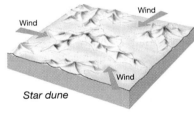

Star dune

Steppe One of the two types of dry climate. A marginal and more humid variant of the desert that separates it from bordering humid climates.

Stock A pluton similar to but smaller than a batholith.

Stony-irons One of the three main categories of meteorites. This group, as the name implies, is a mixture of iron and silicate minerals.

Stony meteorite One of the three main categories of meteorites. Such meteorites are composed largely of silicate minerals with inclusions of other minerals.

Strata Parallel layers of sedimentary rock.

Stratified drift Sediments deposited by glacial meltwater.

Stratovolcano See *Composite cone*.

Streak The color of a mineral in powdered form.

Stream A general term to denote the flow of water within any natural channel. Thus, a small creek and a large river are both streams.

Stream piracy The diversion of the drainage of one stream resulting from the headward erosion of another stream.

Stress The force per unit area acting on any surface within a solid. Also known as *directed pressure*.

Striations (glacial) Scratches or grooves in a bedrock surface caused by the grinding action of a glacier and its load of sediment.

Strike The compass direction of the line of intersection created by a dipping bed or fault and a horizontal surface. Strike is always perpendicular to the direction of dip.

Strike-slip fault A fault along which the movement is horizontal.

Subduction The process by which oceanic lithosphere plunges into the mantle along a convergent zone.

Subduction zone A long, narrow zone where one lithospheric plate descends beneath another.

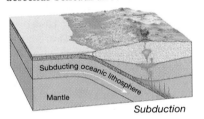

Subduction

Submarine canyon A seaward extension of a valley that was cut on the continental shelf during a time when sea level was lower, or a canyon carved into the outer continental shelf, slope, and rise by turbidity currents.

Submergent coast A coast whose form is largely the result of the partial drowning of a former land surface due to a rise of sea level or subsidence of the crust, or both.

Subsoil A term applied to the *B* horizon of a soil profile.

Superposed stream A stream that cuts through a ridge lying across its path. The stream established its course on uniform layers at a higher level without regard to underlying structures and subsequently downcut.

Superposition, law of In any undeformed sequence of sedimentary rocks, each bed is older than the one above and younger than the one below.

Surf A collective term for breakers; also the wave activity in the area between the shoreline and the outer limit of breakers.

Surface waves Seismic waves that travel along the outer layer of the earth.

Surge A period of rapid glacial advance. Surges are typically sporadic and short-lived.

Suspended load The fine sediment carried within the body of flowing water or air.

S wave An earthquake wave, slower than a P wave, that travels only in solids.

Swells Wind-generated waves that have moved into an area of weaker winds or calm.

Syncline A linear downfold in sedimentary strata; the opposite of anticline.

Syncline

Tabular Describing a feature such as an igneous pluton having two dimensions that are much longer than the third.

Talus An accumulation of rock debris at the base of a cliff.

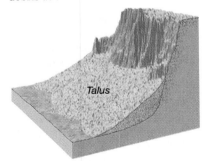

Talus

Tarn A small lake in a cirque.

Tectonics The study of the large-scale processes that collectively deform Earth's crust.

Temporary (local) base level The level of a lake, resistant rock layer, or any other base level that stands above sea level.

Tensional stress The type of stress that tends to pull a body apart.

Terminal moraine The end moraine marking the farthest advance of a glacier.

Terrace A flat, benchlike structure produced by a stream, which was left elevated as the stream cut downward.

Terrane A crustal block bounded by faults, whose geologic history is distinct from the histories of adjoining crustal blocks.

Terrestrial planet One of the Earthlike planets: Mercury, Venus, Earth, and Mars. These planets have similar densities.

Terrigenous sediment Seafloor sediments derived from terrestrial weathering and erosion.

Texture The size, shape, and distribution of the particles that collectively constitute a rock.

Theory A well-tested and widely accepted view that explains certain observable facts.

Thrust fault A low-angle reverse fault.

Thrust fault

Tidal current The alternating horizontal movement of water associated with the rise and fall of the tide.

Tidal delta A deltalike feature created when a rapidly moving tidal current emerges from a narrow inlet and slows, depositing its load of sediment.

Tidal flat A marshy or muddy area that is alternately covered and uncovered by the rise and fall of the tide.

Tide Periodic change in the elevation of the ocean surface.

Till Unsorted sediment deposited directly by a glacier.

Tillite A rock formed when glacial till is lithified.

Tombolo A ridge of sand that connects an island to the mainland or to another island.

Tombolo

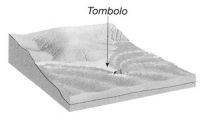

Topset bed An essentially horizontal sedimentary layer deposited on top of a delta during floodstage.

Transform boundary A boundary in which two plates slide past one another without creating or destroying lithosphere.

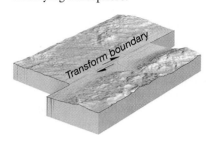

Transpiration The release of water vapor to the atmosphere by plants.

Transported soil Soils that form on unconsolidated deposits.

Transverse dunes A series of long ridges oriented at right angles to the prevailing wind; these dunes form where vegetation is sparse and sand is very plentiful.

Travertine A form of limestone ($CaCO_3$) that is deposited by hot springs or as a cave deposit.

Trellis drainage A system of streams in which nearly parallel tributaries occupy valleys cut in folded strata.

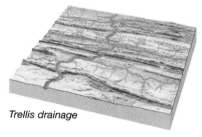

Trellis drainage

Trench An elongate depression in the sea floor produced by bending of oceanic crust during subduction.

Truncated spurs Triangular-shaped cliffs produced when spurs of land that extend into a valley are removed by the great erosional force of a valley glacier.

Tsunami The Japanese word for a seismic sea wave.

Turbidite Turbidity current deposit characterized by graded bedding.

Turbidity current A downslope movement of dense, sediment-laden water created when sand and mud on the continental shelf and slope are dislodged and thrown into suspension.

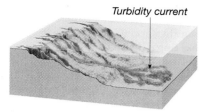

Turbidity current

Turbulent flow The movement of water in an erratic fashion often characterized by swirling, whirlpool-like eddies. Most streamflow is of this type.

Ultimate base level Sea level; the lowest level to which stream erosion could lower the land.

Unconformity A surface that represents a break in the rock record, caused by erosion and nondeposition.

Uniformitarianism The concept that the processes that have shaped Earth in the geologic past are essentially the same as those operating today.

Valence electron The electrons involved in the bonding process; the electrons occupying the highest principal energy level of an atom.

Valley glacier See *Alpine glacier*.

Valley train A relatively narrow body of stratified drift deposited on a valley floor by meltwater streams that issue from the terminus of an alpine glacier.

Vein deposit A mineral filling a fracture or fault in a host rock. Such deposits have a sheetlike, or tabular, form.

Vent A pipelike conduit that connects a magma chamber to a volcanic crater.

Ventifact A cobble or pebble polished and shaped by the sandblasting effect of wind.

Vesicles Spherical or elongated openings on the outer portion of a lava flow that were created by escaping gases.

Vesicular A term applied to igneous rocks that contain small cavities called vesicles, which are formed when gases escape from lava.

Viscosity A measure of a fluid's resistance to flow.

Volcanic Pertaining to the activities, structures, or rock types of a volcano.

Volcanic arc   Mountains formed in part by igneous activity associated with the subduction of oceanic lithosphere beneath a continent. Examples include the Andes and the Cascades.

Volcanic bomb   A streamlined pyroclastic fragment ejected from a volcano while still semimolten.

Volcanic neck   An isolated, steep-sided, erosional remnant consisting of lava that once occupied the vent of a volcano.

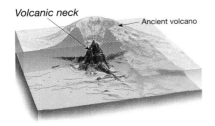

Volcanic neck
Ancient volcano

Volcano   A mountain formed from lava and/or pyroclastics.

Waterfall   A precipitous drop in a stream channel that causes water to fall to a lower level.

Water gap   A pass through a ridge or mountain in which a stream flows.

Water table   The upper level of the saturated zone of groundwater.

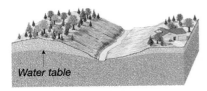

Water table

Wave-cut cliff   A seaward-facing cliff along a steep shoreline formed by wave erosion at its base and mass wasting.

Wave-cut platform   A bench or shelf along a shore at sea level, cut by wave erosion.

Wave height   The vertical distance between the trough and crest of a wave.

Wave length   The horizontal distance separating successive crests or troughs.

Wave of oscillation   A water wave in which the wave form advances as the water particles move in circular orbits.

Wave of translation   The turbulent advance of water created by breaking waves.

Wave period   The time interval between the passage of successive crests at a stationary point.

Weathering   The disintegration and decomposition of rock at or near the surface of Earth.

Welded tuff   A pyroclastic deposit composed of particles fused together by the combination of heat still contained in the deposit after it has come to rest and the weight of overlying material.

Well   An opening bored into the zone of saturation.

Wilson cycle   The complex cycle of ocean basin openings and closings named in honor of J. Tuzo Wilson, the Canadian geologist who proposed their existence.

Wind gap   An abandoned water gap. These gorges typically result from stream piracy.

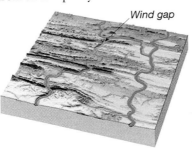

Wind gap

Xenolith   An inclusion of unmelted country rock in an igneous pluton.

Xerophyte   A plant highly tolerant of drought.

Yazoo tributary   A tributary that flows parallel to the main stream because a natural levee is present.

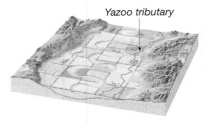

Yazoo tributary

Zone of accumulation   The part of a glacier characterized by snow accumulation and ice formation. The outer limit of this zone is the snowline.

Zone of aeration   The area above the water table where openings in soil, sediment, and rock are not saturated but filled mainly with air.

Zone of fracture   The upper portion of a glacier consisting of brittle ice.

Zone of saturation   The zone where all open spaces in sediment and rock are completely filled with water.

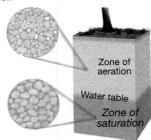

Zone of aeration
Water table
Zone of saturation

Index

Installation Instructions

Macintosh Information

Hardware and software requirements:

- Macintosh with 256 color graphics
- 13" or larger display (640×480 or higher)
- CD-ROM drive (double-speed)
- 4 MB of free RAM memory
- System 6.0.7 or newer

Refer to the Read Me file on the CD-ROM if you need information regarding how to run the program or how to install it to your hard disk.

Windows Information

Hardware and software requirements:

- IBM compatible 386 or better
- SVGA (256 colors in 640×480)
- CD-ROM drive (double-speed)
- 4 MB RAM (8 MB suggested)
- Windows 3.1 (enhanced mode) or Windows 95

Refer to the README.TXT file on the CD-ROM if you need information regarding how to run the program or how to install it to your hard disk.

Be certain that Windows is configured to operate in 256 colors; even with an SVGA card installed, Windows is often set up to use only 16 colors. You may find out what you are currently set to in the "Display" entry of the Windows Setup program in the Main group. Refer to your SVGA video card's manual regarding changing your Windows configuration to use 256 colors.

For technical support, please contact Tasa Graphic Arts, Inc. at 505-293-2727.

GEODe II

Geologic Explorations on Disk

YOU SHOULD CAREFULLY READ THE TERMS AND CONDITIONS BEFORE USING THE DISKETTE PACKAGE. USING THIS DISKETTE PACKAGE INDICATES YOUR ACCEPTANCE OF THESE TERMS AND CONDITIONS.

Prentice-Hall, Inc. provides this program and licenses its use. You assume responsibility for the selection of the program to achieve your intended results, and for the installation, use, and results obtained from the program. This license extends only to use of the program in the United States or countries in which the program is marketed by authorized distributors.

License Grant

You hereby accept a nonexclusive, nontransferable, permanent license to install and use the program ON A SINGLE COMPUTER at any given time. You may copy the program solely for backup or archival purposes in support of your use of the program on the single computer. You may not modify, translate, disassemble, decompile, or reverse engineer the program, in whole or in part.

Term

The License is effective until terminated. Prentice-Hall, Inc. reserves the right to terminate this License automatically if any provision of the License is violated. You may terminate the License at any time. To terminate this License, you must return the program, including documentation, along with a written warranty stating that all copies in your possession have been returned or destroyed.

Limited Warranty

THE PROGRAM IS PROVIDED "AS IS" WITHOUT WARRANTY OF ANY KIND, EITHER EXPRESSED OR IMPLIED, INCLUDING, BUT NOT LIMITED TO, THE IMPLIED WARRANTIES OR MERCHANTABILITY AND FITNESS FOR A PARTICULAR PURPOSE. THE ENTIRE RISK AS TO THE QUALITY AND PERFORMANCE OF THE PROGRAM IS WITH YOU. SHOULD THE PROGRAM PROVE DEFECTIVE, YOU (AND NOT PRENTICE-HALL, INC. OR ANY AUTHORIZED DEALER) ASSUME THE ENTIRE COST OF ALL NECESSARY SERVICING, REPAIR, OR CORRECTION. NO ORAL OR WRITTEN INFORMATION OR ADVICE GIVEN BY PRENTICE-HALL, INC., ITS DEALERS, DISTRIBUTORS, OR AGENTS SHALL CREATE A WARRANTY OR INCREASE THE SCOPE OF THIS WARRANTY.

SOME STATES DO NOT ALLOW THE EXCLUSION OF IMPLIED WARRANTIES, SO THE ABOVE EXCLUSION MAY NOT APPLY TO YOU. THIS WARRANTY GIVES YOU SPECIFIC LEGAL RIGHTS AND YOU MAY ALSO HAVE OTHER LEGAL RIGHTS THAT VARY FROM STATE TO STATE.

Prentice-Hall, Inc. does not warrant that the functions contained in the program will meet your requirements or that the operation of the program will be uninterrupted or error-free.

However, Prentice-Hall, Inc. warrants the diskette(s) on which the program is furnished to be free from defects in material and workmanship under normal use for a period of ninety (90) days from the date of delivery to you as evidenced by a copy of your receipt.

The program should not be relied on as the sole basis to solve a problem whose incorrect solution could result in injury to person or property. If the program is employed in such a manner, it is at the user's own risk and Prentice-Hall, Inc. explicitly disclaims all liability for such misuse.

Limitations of Remedies

Prentice-Hall, Inc.'s entire liability and your exclusive remedy shall be:

1. the replacement of any diskette not meeting Prentice-Hall, Inc.'s "LIMITED WARRANTY" and that is returned to Prentice-Hall, or
2. if Prentice-Hall is unable to deliver a replacement diskette that is free of defects in materials and workmanship, you may terminate this agreement by returning the program.

IN NO EVENT WILL PRENTICE-HALL, INC. BE LIABLE TO YOU FOR ANY DAMAGES, INCLUDING ANY LOST PROFITS, LOST SAVINGS, OR OTHER INCIDENTAL OR CONSEQUENTIAL DAMAGES ARISING OUT OF THE USE OR INABILITY TO USE SUCH PROGRAM EVEN IF PRENTICE-HALL, INC. OR AN AUTHORIZED DISTRIBUTOR HAS BEEN ADVISED OF THE POSSIBILITY OF SUCH DAMAGES, OR FOR ANY CLAIM BY ANY OTHER PARTY.

SOME STATES DO NOT ALLOW FOR THE LIMITATION OR EXCLUSION OF LIABILITY FOR INCIDENTAL OR CONSEQUENTIAL DAMAGES, SO THE ABOVE LIMITATION OR EXCLUSION MAY NOT APPLY TO YOU.

General

You may not sublicense, assign, or transfer the License of the program. Any attempt to sublicense, assign or transfer any of the rights, duties, or obligations hereunder is void.

This Agreement will be governed by the laws of the State of New York.

Should you have any questions concerning this Agreement, you may contact Prentice-Hall, Inc. by writing to:

Director of Multimedia Development
Higher Education Division
Prentice-Hall, Inc.
1 Lake Street
Upper Saddle River, NJ 07458

Should you have any questions concerning technical support, you may write to:

Tasa Graphic Arts, Inc.
11930 Menaul Blvd., NE
Suite 107
Albuquerque, NM 87112-2461

YOU ACKNOWLEDGE THAT YOU HAVE READ THIS AGREEMENT, UNDERSTAND IT, AND AGREE TO BE BOUND BY ITS TERMS AND CONDITIONS. YOU FURTHER AGREE THAT IT IS THE COMPLETE AND EXCLUSIVE STATEMENT OF THE AGREEMENT BETWEEN US THAT SUPERSEDES ANY PROPOSAL OR PRIOR AGREEMENT, ORAL OR WRITTEN, AND ANY OTHER COMMUNICATIONS BETWEEN US RELATING TO THE SUBJECT MATTER OF THIS AGREEMENT.